T0362271

Sodium Fast Reactors
with Closed Fuel Cycle

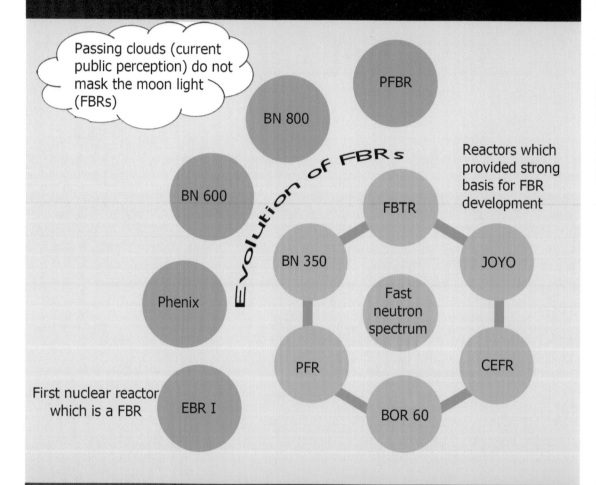

BALDEV RAJ | P. CHELLAPANDI | P.R. VASUDEVA RAO

Sodium Fast Reactors
with Closed Fuel Cycle

Sodium Fast Reactors
with Closed Fuel Cycle

BALDEV RAJ
P. CHELLAPANDI
P.R. VASUDEVA RAO

CRC Press
Taylor & Francis Group
Boca Raton London New York

CRC Press is an imprint of the
Taylor & Francis Group, an **informa** business

CRC Press
Taylor & Francis Group
6000 Broken Sound Parkway NW, Suite 300
Boca Raton, FL 33487-2742

First issued in paperback 2017

© 2015 by Taylor & Francis Group, LLC
CRC Press is an imprint of Taylor & Francis Group, an Informa business

No claim to original U.S. Government works

ISBN-13: 978-1-4665-8767-0 (hbk)
ISBN-13: 978-1-138-89304-7 (pbk)

This book contains information obtained from authentic and highly regarded sources. Reasonable efforts have been made to publish reliable data and information, but the author and publisher cannot assume responsibility for the valid_ ity of all materials or the consequences of their use. The authors and publishers have attempted to trace the copyright holders of all material reproduced in this publication and apologize to copyright holders if permission to publish in this form has not been obtained. If any copyright material has not been acknowledged please write and let us know so we may rectify in any future reprint.

Except as permitted under U.S. Copyright Law, no part of this book may be reprinted, reproduced, transmitted, or uti_ lized in any form by any electronic, mechanical, or other means, now known or hereafter invented, including photocopy_ ing, microfilming, and recording, or in any information storage or retrieval system, without written permission from the publishers.

For permission to photocopy or use material electronically from this work, please access www.copyright.com (http:// www.copyright.com/) or contact the Copyright Clearance Center, Inc. (CCC), 222 Rosewood Drive, Danvers, MA 01923, 978-750-8400. CCC is a not-for-profit organization that provides licenses and registration for a variety of users. For organizations that have been granted a photocopy license by the CCC, a separate system of payment has been arranged.

Trademark Notice: Product or corporate names may be trademarks or registered trademarks, and are used only for identification and explanation without intent to infringe.

Visit the Taylor & Francis Web site at
http://www.taylorandfrancis.com

and the CRC Press Web site at
http://www.crcpress.com

Contents

SECTION I Basis and Concepts

SECTION II   Design of Sodium Fast Reactors

SECTION III *Safety*

SECTION IV   Construction and Commissioning

SECTION V International SFR Experiences

SECTION VI   Fuel Cycle for SFRs

SECTION VII   Decommissioning Aspects

SECTION VIII   Domains of High Relevance to SFR:
Typical Examples

SECTION IX   Economics of SFRs with a Closed Fuel Cycle

Foreword

Nuclear power was harnessed for electrical power production for the first time in 1954. Six decades later, nuclear power contributes to nearly 14% of the global electricity generation, with almost all the power being generated using thermal reactors (water-cooled reactors) with uranium-235 as the fuel. It is known that fast reactors would be more efficient systems for the effective utilization of uranium resources to generate nuclear electricity, even though the accumulated experience with fast reactors is much less at only 400 reactor-years, compared to the 14,500 reactor-years experience accumulated with water-cooled reactors. Advanced countries, including the United States, United Kingdom, France, the Russian Federation, Japan, India, and China, have built and operated fast reactors. The advantages include their potential to breed fissile materials, burn long-lived minor actinides, and take the fuel to high levels of burnup. These features make them an important class of systems for providing sustainable nuclear energy in the coming decades.

The experience gained in a large number of countries and the vast knowledge base in associated technologies developed during the R&D in a number of countries have led to a high degree of confidence to design, construct, and operate fast reactors. International programs such as the Generation IV Forum (GIF or GEN IV) and the International Atomic Energy Agency (IAEA) supported International Project on Innovative Nuclear Reactors and Fuel Cycles (INPRO), which aim to foster sustainable nuclear energy operations in the world, have identified fast reactors as important systems for the future. It is generally known that sodium-cooled fast reactors have high potential for near-time commercial deployment within the next two decades. At the same time, it is also realized that fast reactors pose several technological challenges that demand concerted R&D efforts, particularly those related to addressing certain aspects of safety and economics. It is an accepted fact that fast reactors can be sustainable only with a closed fuel cycle. India made an early choice to follow this route, in order to maximize the utilization of the limited resources of uranium available in the country. The topic on nuclear fuel cycle has been a subject of intense R&D in a number of countries interested in fast spectrum reactors, including India.

Fast reactors provide an attractive system for countries like India that require very large and long-term contributions from nuclear electricity and the fact that sustainability has to be considered an important criterion in evolving reactor technologies. While information about the different aspects of fast reactors is available in many formats, including technical reports, papers in journals, and presentations in conferences, there are only a few books that deal comprehensively with the subject of fast reactors. A comprehensive source book providing information about various aspects of fast reactors and their associated fuel cycles would be of significant value for experts, as well as for younger-generation professionals aspiring to take up challenging R&D programs in designing and building fast reactor systems.

I am delighted that such a book on fast reactors has been made a reality by my colleagues, Dr. Baldev Raj, Dr. P. Chellapandi, and Dr. P.R. Vasudeva Rao. I congratulate them for their fine efforts in preparing this comprehensive treatise on this important subject. This book reflects the vast experience in several domains of sodium-cooled fast reactor technologies and associated fuel cycles across different countries. The topics covered include design aspects, codes and standards, technologies, allied related materials, and disciplines like manufacturing, control and instrumentation, and robotics. This book would, thus, provide not only the state-of-the-art knowledge on the subject but also highlight the challenges in the development of the fast reactor technology.

I am sure that the book would be a valuable source of information to scientists, engineers, and technologists working in the area of sodium-cooled fast reactor systems and also in the training and education of young scientists and engineers. This book is also bound to motivate young professionals in the nuclear field to aim at realizing the full potential of nuclear power through fast breeder reactors with closed fuel cycles. I hope the book will also engage the attention of other stakeholders, such as regulators and policy makers, apart from the academia.

I wish the readers a valuable outcome in terms of enhancing their knowledge and understanding on fast reactors.

R.K. Sinha
Secretary, Department of Atomic Energy (DAE)
Chairman, Atomic Energy Commission (AEC)
Mumbai, India

Preface

Among a few advanced nuclear reactor systems that could meet almost all the necessary requirements specified for a sustainable energy option, the sodium-cooled fast reactor depicts its competitiveness as well as readiness to be exploited within the next two decades. R&D continues to be pursued to realize enhanced economic competitiveness and safety. These kinds of reactors have already accumulated operating experience of about 400 reactor-years worldwide. This should be the reason for the growing interest among many countries to prioritize their efforts for the development of sodium-cooled fast reactors for realizing a sustainable and secure energy road map. It is well known that the design, construction, and commissioning of a sodium-cooled fast reactor pose several challenging aspects in science and technology. Currently, only a few countries in the world have developed a comprehensive knowledge in the relevant fields to the extent of applying them in demonstration and in commercial reactor systems. Hence, we felt a compelling need for a comprehensive book on this subject with state-of-the-art knowledge and up-to-date information and data, meeting expectations from the national as well as international community of nuclear scientists and technologists. This thought and purpose was our prime motivation for writing this book on the science and technology of sodium-cooled fast reactors with closed fuel cycles.

We have attempted to provide a comprehensive coverage on reactor physics, materials, design, safety analysis, validations, engineering, construction, and commissioning aspects. Apart from these, we have included a special chapter on allied sciences to highlight advanced reactor core materials, specialized manufacturing technologies, chemical sensors, in-service inspection, and simulators. Design aspects are presented systematically addressing all the essential ingredients with focus on reactor assembly including core and coolant circuits, fuel handling, instrumentation and control, energy conversion, and containment systems. Design codes and standards are presented with sufficient background for the readers to understand the underlying *mechanics* basis. Guidelines are provided for the selection of concepts, detailed design, analysis, and validation. Aspects covered in the chapters dealing with manufacturing and assembly provide sufficient basis to professionals from industries to enhance their capacity and capability for the construction of fast reactor power plants. Further, the science and technology aspects of fuel cycle technology aspects in Chapter 29. In general, this book can be considered as a source book and complementary to the books *Nuclear Reactor Engineering* written by Samuel Glasstone and A. Sesonske in 2004 and *Fast Spectrum Reactors* by Walter et al. in 2012. We sincerely hope that with the perspectives and knowledge gained through this book, a graduate engineer will be able to appreciate the design and development of sodium-cooled fast reactor systems and components. We are confident that the book will get serious attention from young, aspiring, and motivated minds in science and technology to pursue sodium fast reactors, thereby enabling them to become science and technology program leaders in the coming years, now that this technology is being harnessed for commercial energy production in many parts of the world.

The contributions provided by our colleagues, friends, and organizations from India and abroad are sincerely acknowledged. We record our appreciation to Dr. Gagandeep Singh and Ahringer Jennifer for their dedicated efforts to realize our dream of writing this book, which reflects our four-decade experience (individually and collectively) in working for the fast reactor science and technology with closed fuel cycle.

We sincerely welcome constructive criticisms from the readers, which will certainly help us revise this book for future editions.

Baldev Raj
P. Chellapandi
P.R. Vasudeva Rao

Authors

Dr. Baldev Raj, BE, PhD, served the Department of Atomic Energy, India, over a 42-year period until 2011. As distinguished scientist and director, Indira Gandhi Centre of Atomic Research, Kalpakkam, he has advanced several challenging technologies, especially those related to the fast breeder test reactor and the prototype fast breeder reactor. Dr. Raj pioneered the application of NDE for basic research using acoustic and electromagnetic techniques in a variety of materials and components. He is also responsible for realizing societal applications of NDE in areas related to cultural heritage and medical diagnosis. He is the author of more than 970 refereed publications, 70 books and special journal volumes, and more than 20 contributions to encyclopedias and handbooks, as well as the owner of 29 patents. He is immediate past president, International Institute of Welding and President, Indian National Academy of Engineering. He assumed responsibilities as the director of the National Institute of Advanced Studies, Bangalore in September 2014. He is a fellow of all science and engineering academies in India, member of the German Academy of Sciences, honorary member of the International Medical Sciences Academy, member of the International Nuclear Energy Academy, vice president, nondestructive testing (NDT), Academia International, and president-elect of the International Council of Academies of Engineering and Technological Sciences.

Dr. P. Chellapandi, BE (Hons.), MTech., PhD, is currently a distinguished scientist and director of the Reactor Design Group at the Indira Gandhi Centre for Atomic Research (IGCAR). He specializes in reactor design, thermal hydraulics, structural mechanics, safety analysis, and experimental simulations. He is one of the key persons involved in the design and development activities of the 500 MWe prototype fast breeder reactor (PFBR) since its inception. He has contributed significantly for the PFBR over a wide spectrum of design, analysis, qualification, and research activities, also involving academic and R&D institutions in the country. His current responsibilities include design of advanced oxide and metallic fast breeder reactors planned by the department. He is a senior professor at Homi Bhabha National Institute and has published about 130 journal papers. He is a fellow at the Indian National Academy of Engineering. He is the recipient of the Homi Bhabha Science and Technology Award, the Indian Nuclear Society Award, the Vasvik Award, the National Design Award in Mechanical Engineering from Institute of Electrical and Electronics Engineers (IEEE), the Agni Award for Excellence in Self-Reliance from Defence Research and Development Organization (DRDO), the Department of Atomic Energy (DAE) Group Achievement Award for the design, manufacture, and erection of PFBR reactor assembly components, and the Distinguished Alumnus Award from the Indian Institute of Technology (IIT), Chennai, India.

Dr. P.R. Vasudeva Rao, BSc., PhD, is the director of the Indira Gandhi Centre for Atomic Research (IGCAR) and the General Services Organization at Kalpakkam, India. After graduating from Vivekananda College, Chennai, India, he joined DAE in the 16th batch of Bhabha Atomic Research Centre (BARC) training school. He was instrumental in the setting up of the Radiochemistry Laboratory at IGCAR. He is an expert in the area of fast reactor fuel cycle, especially the back-end fuel cycle. He is a recipient of the Indian Nuclear Society Award (2007) for his contributions to the area of nuclear fuel cycle technologies. He was selected for the Material Research Society of India (MRSI) medal lecture in 1998 and MRSI-ICSC Superconductivity and Materials Science Senior Award in 2011. He was also selected for the award of the Silver Medal by the Chemical Research Society of India in 2011. He is a senior professor at the Homi Bhabha National Institute. He has nearly 250 publications in peer-reviewed international journals. His areas of interest include the development of technologies for fast reactors and associated fuel cycles, actinide separations, and education in the field of chemical sciences.

Section I

Basis and Concepts

1 Nuclear Fission and Breeding

1.1 INTRODUCTION

The basis of design and operation of nuclear reactors is governed by various interactions of neutrons with matter. Atomic and nuclear physics concepts are fundamental for understanding these nuclear reactions. This chapter presents the basic science behind the interaction of neutrons with atomic nuclei to understand the fast breeder reactor (FBR) concepts that would be dealt with in the subsequent chapters.

1.2 ABOUT THE NEUTRON

A heavy, dense, positively charged, extremely small region in the center of an atom is called the atomic nucleus. All the positive charges and more than 99% of the mass of an atom are concentrated in the nucleus, which was discovered in 1911 by Rutherford. At that time, only protons in the nucleus and electrons outside the nucleus were presumed to be the fundamental particles having opposite charges [1.1]. Rutherford conceived the existence of neutral particles along with protons in the nucleus. In 1932, English physicist James Chadwick repeated the high-energy radiation (highly penetrating radiation) experiments of Irene Joliot-Curie (daughter of Marie and Pierre Curie) and her husband Frederic Joliot-Curie using the polonium–beryllium (Po–Be) source, which concluded that the mysterious radiation consists of electrically neutral particles with almost the same mass as a proton. This provided the confirmation for the existence of the *neutral proton* predicted by Rutherford. Chadwick named this particle *neutron* (1932). Due to this discovery, it was clarified that the constituents of nuclei are protons and neutrons (nucleons). The total number of nucleons is termed *mass number* (*A*).

Neutron, thus, is a neutral subatomic constituent particle of every atomic nucleus except ordinary hydrogen. The mass of the neutron is slightly larger than that of the proton, which is $1.67492729 \times 10^{-27}$ kg [1.1]. Since neutrons are neutral particles, they are highly penetrating. Neutrons have a magnetic moment and also spin. They can be formed into polarized neutron beams. Neutrons are classified according to their energy as follows:

Ultracold	Below 2×10^{-7} eV
Very cold	2×10^{-7} eV $\leq$ E $< 5 \times 10^{-5}$ eV
Cold	5×10^{-5} eV $\leq$ E < 0.025 eV
Thermal	0.025 eV
Intermediate	Above thermal and up to 100 keV
Fast	Above 100 keV

One electron volt (eV) is the energy gained by an electron when it accelerates through 1 V of electric potential difference (1 eV = 1.602×10^{-19} J).

The neutron does not survive for long outside of the atomic nucleus. A bound neutron does not decay, whereas a free neutron decays by emitting an electron (β^-) and a neutrino (ν) with a half-life of about 10.3 min. The average lifetime of a free neutron is 14.9 min. The reaction involved in β^- decay is $n \rightarrow p + \beta^- + \bar{\nu} + 1.29$ MeV.

1.3   NUCLEUS STABILITY

Both the proton and the neutron are equally important in determining nuclear stability. In many stable nuclei of low mass number, up to 40, the number of neutrons and protons is equal or approximately so. In other words, the neutron/proton ratio is exactly or slightly larger than unity. With increasing atomic mass number, however, a nucleus is stable only if it contains more neutrons than protons. Thus, for the heaviest stable nuclei, with an atomic number of 80 or more, the neutron/proton ratio increases to about 1.5. It is also found that certain nuclei exhibit exceptional stability, which contain the so-called magic numbers of 2, 8, 20, 28, 50, and 82 protons or 2, 8, 20, 28, 50, 82, and 126 neutrons. Nuclei having magic numbers of both protons and neutrons are said to be *doubly magic*; examples are $_8O^{16}$, $_{20}Ca^{40}$, and $_{82}Pb^{208}$. In general, the magic nuclides are common in nature [1.2].

The nucleons are held by nuclear forces in the nucleus. Two types of nuclear forces exist between the nucleons: (1) attractive intranuclear forces between the nucleons within the short range of nucleus and (2) electrostatic or coulombic repulsive forces between the positively charged protons. These nuclear forces are very strong, short range, charge independent, and spin dependent. These forces dictate the energy that binds the nucleons together: protons and neutrons do not just form a simple collection in the nucleus, but they strongly combine with each other through a strong binding energy. The source of this binding energy is defined in the following.

In general, if two or more particles interact to combine together, then the total mass of the system would decrease to be less than the sum of the masses of the individual particles. The stronger the interaction becomes, the more the mass decreases. This decrease in the mass of the system is called the mass defect [1.3,1.4]. The mass defect of a nucleus of proton number Z and neutron number N is defined by

$$\Delta M = \left\{ \left(ZM_p + NM_n \right) - M(Z,N) \right\},$$

where
   M_p is the mass of the proton
   M_n is the mass of the neutron
   M is the mass of the nucleus of Z protons and N neutrons

The binding energy B(Z, N) can be obtained from the value of the mass defect by converting the mass to energy by employing Einstein's mass–energy relation:

$$B(Z,N) = \left\{ \left(ZM_p + NM_n \right) - M(Z,N) \right\} c^2,$$

where c is the speed of light. The mass defect and associated binding energy values per nucleon for various nuclides are shown in Figure 1.1. If $\Delta M < 0$, the nuclide is stable, and conversely, when $\Delta M > 0$, the nuclide is unstable. It is, therefore, shown in Figure 1.1 that very light elements (A < 20) and heavy elements (A > 180) are not stable and that maximum stability occurs around A ~ 50. Accordingly, the maximum B value is 8.79 MeV/nucleon at A = 56, which decreases to about 7.6 MeV for A = 238, as seen in the figure.

1.4   ENERGY FROM FISSION

The fission process is illustrated in Figure 1.2. When a neutron bombards a heavy nucleus, say uranium-235 (U-235), it absorbs the neutron, thus converting to uranium-236, which subsequently splits into two smaller nuclei (fragments) of almost equal mass, for example, barium-137

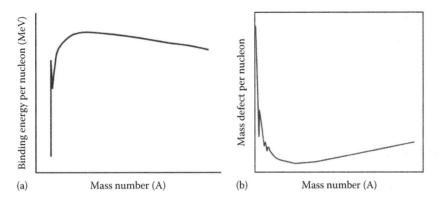

FIGURE 1.1 Schematic of (a) binding energy curves and (b) mass defect. (Trend curve is shown here; for more details, [1.3] can be referred.) (From Glasstone, S. and Sesonske, A., *Nuclear Reactor Engineering*, 4th edn, CBS Publishers, New Delhi, India, 2004.)

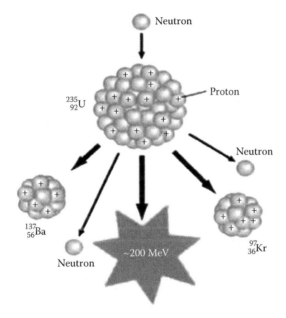

FIGURE 1.2 A typical fission process in U-235.

and krypton-97, while emitting some neutrons [1.4]. Apart from emitting fission fragments and neutrons, a huge amount of energy, larger than 200 MeV, is also released.

The fission process is possible with certain heavy nuclei; U-235 is one example. Fission can occur in other nuclei such as U-233 and Pu-239. Similarly, during the fission process, the nucleus does not always split into barium-137 and krypton-97. The number of emitted neutrons is also not always constant, but it ranges from one to many. Further, the emitted energy is not always constant but is almost 200 MeV.

The energy released can be estimated from the associated mass defect. The fragments are more stable than U-236. The new binding energy of the heavy nucleus when it splits is less than the binding energy before it splits. Excess energy is released in the process, which is nothing but a reduction in the net mass. The energy release can be quantified by employing Einstein's mass–energy relation. Assuming symmetric fission, the mass of each fragment is 118. From Figure 1.1, it can be seen that 0.001 amu is lost per nucleon in this event, and hence, the total mass loss = 236 × 0.001 amu = 236 × 0.001 × 1.66054 × 10⁻²⁷ kg = 0.392 × 10⁻²⁷ kg. By employing Einstein's mass–energy relation, E = 0.392 × 10⁻²⁷ × (3 × 10⁸)² = 3.53 × 10⁻¹¹ J = 3.53 × 10⁻¹¹/1.602 × 10⁻¹³ ≈ 220 MeV. Thus, a large amount of energy is released in fission processes of heavy nuclei. The major proportion, over 80% of the energy of fission, appears as kinetic energy of fission fragments and this immediately manifests as heat. The remaining 20% or so is liberated in the form of instantaneous gamma rays from the excited fission fragments and as kinetic energy of the fission neutrons [1.5]. The rest is released gradually in the form of energy carried by the beta particles and gamma rays emitted by the radioactive fission products as they decay over a period of time. This decay energy ultimately appears in the form of heat as the radiations interact with and are absorbed by matter. The energy released from complete fission of 1 g of uranium is equal to the energy from the complete burning of 3 tons of coal.

1.5   FISSION NEUTRONS AND ENERGY SPECTRUM

The neutrons released during fission can be divided into two categories: prompt neutrons and delayed neutrons. The former, which constitute more than 99% of the total fission neutrons, are released within 10⁻¹⁴ s of the instant of fission. The release of prompt neutrons probably takes place in the following manner. The excited compound nucleus formed by the capture of a neutron first breaks up into two nucleus fragments, each of which has too many neutrons for stability, as well as excess (excitation) energy of about 6–8 MeV required for the expulsion of a neutron. The excited unstable nuclear fragment frequently expels one or more neutrons—the prompt neutrons— within a very short time of its formation. The emission of prompt neutrons ceases immediately after fission occurs, but delayed neutrons are expelled continuously from the fission products over a period of a few hours, their intensity falling off with time. There are several groups of delayed neutrons, characterized by their respective half-life. For example, a group of delayed neutrons having a 56 s half-life is emitted from krypton-87, the radioactive decay of the fission product bromine-87. The parents of delayed neutrons (e.g., krypton-87) are called precursors having the same half-life (56 s for krypton-87).

The energy distribution of neutrons produced by fission is called fission spectrum. The fission spectra for prompt neurons and delayed neutrons are shown in Figure 1.3. Since the fission fragments having much higher excitation energy than the last neutron separation energy emit prompt neutrons, prompt neutrons have a higher energy spectrum, compared to the energy spectrum of delayed neutrons: the excitation energy of the delayed neutron emitter is usually much lower than

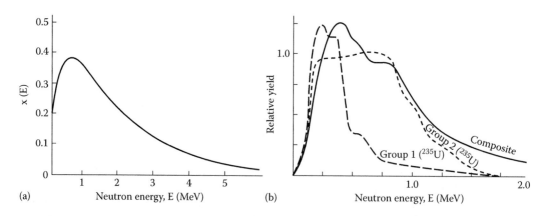

FIGURE 1.3   (a) Prompt fission spectrum. (b) Composite delayed neutron spectrum. (From Duderstadt, J.J. and Hamilton, L.J., *Nuclear Reactor Analysis*, John Wiley & Sons, 1976.)

that of the direct fission fragments, as seen in the figure. The average energy of prompt neutrons is 1000–2000 keV, compared to that for delayed neutrons, which lies in the range of 300–600 keV [1.6,1.7]; 99.8% of fission neutrons have energies less than 10 MeV as seen in the figure.

1.6 CHAIN REACTION

With a sufficient amount of fissile materials available, it is possible to continue nuclear fission reactions as a chain process. This happens when the number of neutrons produced per neutron lost in fissile material by absorption is greater than one. Such chain reactions are utilized in nuclear reactors to generate power. The principle of chain reaction is illustrated in Figure 1.4.

1.7 FISSILE AND FERTILE MATERIALS

Only three nuclides, which have sufficiently long half-life (a few hundreds of years), are fissionable by neutrons of all energies: uranium-233, uranium-235, and plutonium-239. Of these nuclides, uranium-235 is the only one that occurs in nature; the other two are produced artificially from thorium-232 and uranium-238, respectively, by neutron capture followed by two stages of radioactive decay, as illustrated in Figure 1.5. In the case of uranium-238, the most abundant naturally occurring isotope, the resulting nucleus after the capture of 1 neutron is uranium-239, which decays with the emission of negative beta particles (β^-). The resulting product Np^{239} is an isotope of an element of atomic number 93, called neptunium, which does not normally exist on earth to any detectable extent. Neptunium-239 is also beta-active and decays fairly rapidly to form the isotope Pu-239 of the element of atomic number 94, called plutonium. Plutonium occurs in nature in the merest traces only. A similar series of processes that occur in the case of thorium-232, another

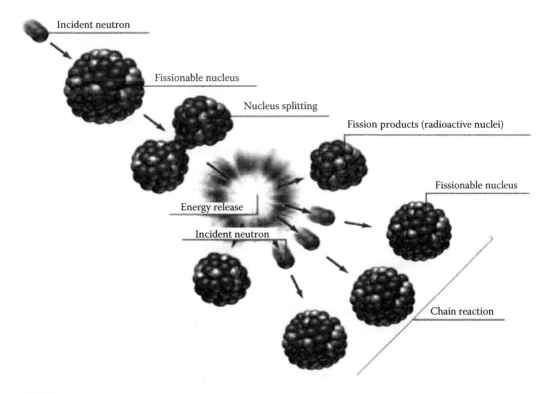

FIGURE 1.4 Principle of chain reaction.

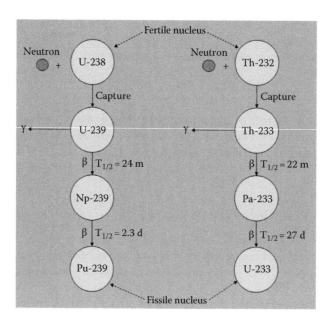

FIGURE 1.5   Generation of fissile nuclides from fertile nuclides.

abundant naturally occurring isotope, is illustrated in Figure 1.5. The product after the capture of a neutron is thorium-233, which undergoes two successive stages of beta decay: the first decay yields protactinium-233, and the second decay generates uranium-233, an isotope of uranium that is not found in any appreciable amount in nature.

Several other species are known to be capable of undergoing fission by neutrons of all energies, but they are highly radioactive and decay so rapidly that they have no practical value for the release of nuclear energy. In addition to nuclides, which are fissionable by neutrons of all energies, there are some that require fast neutrons to cause fission; among these, thorium-232 and uranium-238 are worthy of mention. For such nuclides, with neutrons below about 1 MeV energy, the main reaction is radioactive capture, but above this threshold value, fission also occurs to some extent. Since fission of thorium-232 and uranium-238 is possible with sufficiently fast neutrons, they are known as fissionable nuclides. On the other hand, uranium-233, uranium-235, and plutonium-239, which will undergo fission with neutrons of any energy, are referred to as fissile nuclides. Moreover, since thorium-232 and uranium-238 can be converted into fissile species, uranium-233 and plutonium-239, respectively, are also called fertile nuclides. Natural uranium contains mainly uranium-238 isotope (U-238) and also about 0.7% uranium-235 (U-235).

1.8   ABOUT BREEDING

Breeding is the process of converting a fertile nuclide to a fissile nuclide. Breeding ratio (BR) is the ratio of amount of fissile species produced to that of fissile species destroyed either through fission or through capture. Breeding is influenced by the following parameters:

- ν, the number of neutrons produced per fission
- η, the number of neutrons produced per neutron absorbed
- α, the ratio of capture/fission, expressed in terms of cross sections (σ_c/σ_f)

These parameters are related by $\eta = \nu/(1 + \alpha)$, which is the number of neutrons generated per neutron absorbed in a fissile atom [1.7,1.8]. A nuclear reactor can be a breeder in the broad neutron energy

TABLE 1.1

Neutron Yields "η" in Thermal and Fast Neutron Spectrum

Neutron Energy	Natural Uranium	Uranium-235	Uranium-233	Plutonium-239
Thermal	1.34	2.04	2.26	2.06
Fast	<1.00	2.20	2.35	2.75

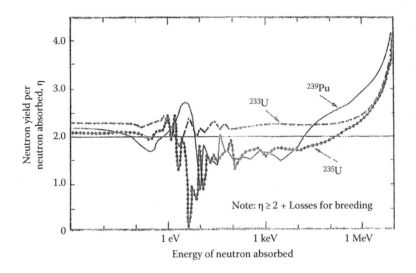

FIGURE 1.6 Neutron yields for various fissile atoms. (From Walter, A. and Reynolds, A., *Fast Breeder Reactors*, Pergamon Press, Inc., New York, 1981.)

spectrum, but adequate BRs can be achieved only by selecting the appropriate fertile and fissile isotopes for that spectrum. That is, the η value depends on the fuel materials and the neutron energy spectrum, as indicated in Table 1.1 as well as in Figure 1.6 [1.9].

The necessary condition for breeding is η > 2:

- One neutron for a new fission
- One neutron for a new conversion
- p for leakages or parasitic captures

Reserving one neutron to sustain the fission chain and also accounting for some unavoidable loss of neutron (p) for capture and leakage, the excess neutrons available for the fertile nuclei to generate fissile nuclei are [η − (1 + p)]. If [η − (1 + p)] = 1, we can produce another fissile atom to replace the one that got destroyed in the previous fission; if [η − (1 + p)] > 1, with the excess neutron [η − (1 + p) − 1], fissile atoms can be generated more than the atoms destroyed. This is the condition for breeding as depicted schematically in Figure 1.7 for Pu-239. The higher breeding in ^{239}Pu can be illustrated in the following example, a typical situation in an FBR. One hundred fissions of Pu-239 release nearly 300 neutrons, which will yield the following reactions:

- One hundred neutrons cause 100 new fissions, maintaining the chain reaction and consuming 100 fissile nuclei of Pu-239.
- Ninety neutrons are absorbed by capture in fertile material (U-238) in the core of the reactor, converting them into fissile nuclei (^{239}Pu).

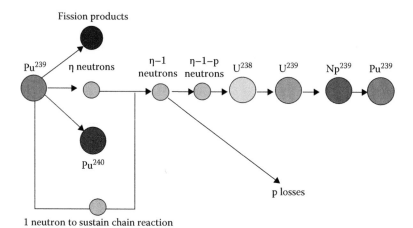

FIGURE 1.7 Condition for breeding. For breeding to be possible, the number of Pu-239 nuclei produced must exceed the number of Pu-239 consumed. In this figure, for one Pu-239 nuclei consumed, η–p–1 neutrons are produced. It is consequently necessary that η–p–1 > 1 and η > 2 + p > 2.

- Forty-five neutrons are absorbed by sterile capture, 25 of which are absorbed by fissile nuclei itself (parasitic capture).
- Sixty neutrons leak outside the core and most of them (~50 neutrons) are absorbed by capture in fertile material (^{238}U), converting them into fissile nuclei (Pu-239). Other neutrons (~10) are absorbed by sterile capture in blankets or in neutron shielding.

A large number of neutrons leaking out of the active core captured by the nuclei of U-238 in the blankets are the major contributors for the higher breeding in the FBR, which is demonstrated by calculating the conversion ratio (CR): the number of fissile nuclides generated from the fertile nuclides divided by the number of neutrons absorbed:

- CR in core = 100/(90 + 25) ≈ 1, that is, self-sustaining core without any external feed.
- CR total = (100 + 50)/(90 + 25) ≈ 1.3, that is, the breeding gain is 0.3 with blankets.

We may, therefore, conclude that it is the blanket portion of the core that makes the FBR breed higher. It is also worth to compare the situation in the case of a typical water-cooled reactor, say a pressurized water reactor (PWR). Even if the core of a PWR is surrounded by blankets, it does not significantly change its CR due to the low number of neutrons leaking out of the core, since they are slowed down in the core itself.

1.9 WORKING OF NUCLEAR REACTORS

The fission as well as the conversion of fertile materials into fissile elements takes place in a nuclear reactor, where the required number of fissions, proportional to the power to be generated, can be sustained by controlling the number of neutrons with the help of control rods. The control rods absorb the neutrons to maintain neutron balance at any given power of the nuclear reactor. While nuclear reactors employing thermal neutrons are called *thermal reactors*, the reactors employing fast neutrons are called *fast spectrum reactors* or simply *fast reactors*. In both kinds of reactors, the capture in fertile materials present in the reactor would result in conversion to fissile atoms. As seen in Table 1.1, the plutonium nuclides release more neutrons per neutron absorption upon fission with fast neutrons. Hence, the breeding potential of plutonium is much superior in the fast spectrum reactor.

In a nuclear reactor, the neutron population at any instant is a function of the rate of neutron production and the rate of neutron losses. Neutron production is essentially achieved through fission reactions. Neutron losses are due to nuclear absorption reactions, parasitic captures, and leakage from the reactor core. When a reactor's neutron population remains steady from one generation to the next (creating as many new neutrons as are lost), the fission chain reaction is self-sustaining and the reactor's condition is referred to as *critical*. When the reactor's neutron production exceeds losses, manifesting in increased power level, it is considered *supercritical*, and when losses dominate, it is considered *subcritical* and exhibits decreasing power [1.8,1.10].

1.10   REACTOR CONTROL AND SAFETY: REACTOR PHYSICS

Neutron multiplication in the core is governed by the following evolutionary equation:

$$\frac{dN}{dt} = \frac{\alpha N}{\tau},$$

where
   N is the number of free neutrons in a reactor core
   τ is the average lifetime of the neutron (before it either escapes from the core or is absorbed by a nucleus)
   α is a constant of proportionality, equal to the expected number of neutrons after one average neutron lifetime
   dN/dt is the rate of change of the neutron count in the core

The solution of this differential equation is given by

$$\frac{N}{No} = \text{Neutron multiplication factor} = \left(1+\alpha\right)^{1/\tau}.$$

If α is positive, then the core is supercritical and the rate of neutron production will grow exponentially until some other effect stops the growth. If α is negative, then the core is *subcritical* and the number of free neutrons in the core will shrink exponentially until it reaches an equilibrium at zero (or the background level from spontaneous fission). If α is exactly zero, then the reactor is critical and its output does not vary in time (dN/dt = 0 from earlier). Generally, α is as small as 0.01. The average neutron lifetime (τ) in a typical core is of the order of a millisecond. Then in 1 s, the reactor power will vary by a factor of $(1 + 0.01)^{1000}$, or more than 10,000. Such rapid variation would render it practically impossible to control the reaction rates in a nuclear reactor. Fortunately, the effective neutron lifetime is much longer than the average lifetime of a single neutron in the core due to the emission of delayed neutrons, which increase the effective average lifetime of neutrons in the core to nearly 0.1 s, so that a core with α of 0.01 would increase in 1 s by only a factor of $(1 + 0.01)^{10}$, or about 1.1, a 10% increase, which is a controllable rate of change. Most nuclear reactors are hence operated in a prompt subcritical, delayed critical condition: the prompt neutrons alone are not sufficient to sustain a chain reaction, but the delayed neutrons make up the small difference required to keep the reaction going. This has effects on how reactors are controlled: when a small amount of control rod is slid into or out of the reactor core, the power level changes at first very rapidly due to prompt subcritical multiplication and then more gradually following the exponential growth or decay curve of the delayed critical reaction. Furthermore, increases in the reactor power can be performed at any desired rate simply by pulling out a sufficient length of the control rod.

   The ratio of the number of neutrons in one generation to that of the previous one is called the effective multiplication factor (k). *Reactivity* (δk, which is also represented by ρ) is an expression of the departure from criticality: $\delta k = (k - 1)/k$ [1.3]. When the reactor is critical, $\delta k = 0$. When the reactor is subcritical, $\delta k < 0$. When the reactor is supercritical, $\delta k > 0$. Reactivity is commonly expressed in decimal or percentage or per cent mille (pcm) of $\Delta k/k$. However, it is often expressed in terms of \$. One dollar (1\$) reactivity is equivalent to the effective delayed neutron fraction β_{eff}. The value of k_{eff}, defined in neutron kinetics studies as 1\$, is of profound importance for core safety characteristics. When reactivity, defined as the deviation in k_{eff} from a critical state, exceeds +1\$, the core is in a supercritical state without the added reactivity of delayed neutrons and may undergo a rapid and potentially uncontrollable power excursion. Using a \$-based unit for reactivity coefficients, the safety characteristics of cores with different values of \$ (and the value of \$ can vary significantly) can be compared justly.

Reactivity coefficients are used to quantify the effect of variation in parameters on the reactivity of the core, which are defined as the amount of change in the reactivity for a given change in the parameter. While reactivity coefficients can be expressed as functions of a change in a reactor power or coolant flow, they are all fundamentally functionally dependent on temperature. The reactivity feedback from changes in temperature of the fuel, structural material, and coolant can, in a simplified manner, be summarized by five general feedback mechanisms that together control the core response to temperature transients. The reactivity feedback of a typical fast reactor is shown in Figure 1.8 [1.11]. The temperature change in each component will alter neutron absorption, leakage, and spectrum to some degree. Changes in each core component also affect all other components. Major effects are shown in Figure 1.8, but it does not represent all the effects occurring in the core. Some feedback effects are inherently positive (spectral hardening) or negative

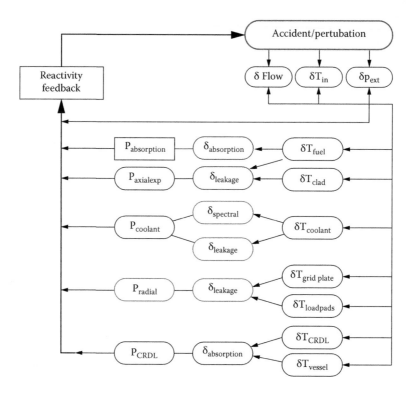

FIGURE 1.8 Reactivity feedback mechanisms. (Adapted from Qvist, S.A., Safety and core design of large liquid metal cooled fast breeder reactors, PhD dissertation, Nuclear Engineering, University of California, Berkeley, CA, 2013.)

(nonfissile absorption, leakage, thermal expansion of components) owing to its characteristics, while some feedback coefficients are design dependent (coolant and control rod driveline expansion). The major feedback effects are briefly described below:

- *Doppler*: The broadening of reaction cross sections by an increase in temperature is called the (nuclear) Doppler effect. Cross sections for all resonances (absorption, fission, and scattering) in all materials of the core are subject to the Doppler effect. While the total cross-sectional area of a resonance is always conserved, the shape of the resonance flattens with increasing temperature. This flattening is caused by an increase in the thermal oscillations of the nuclei. The range of the relative neutron/nucleus velocity widens because of the increased back-and-forth (Brownian) motion of the nucleus. This effect can be seen in Figure 1.9.
- *Fuel axial expansion*: As temperatures increase, the fuel rods or pellets thermally expand in all directions. The axial expansion effectively increases the height of the active core, and since fuel mass is conserved, it decreases the density of the fuel in the core, thus introducing negative reactivity.
- *Radial expansion of a grid plate and sub assembly (SA) bowing*: Due to the temperature increase, the grid plate expands, carrying all the assemblies loaded on it, and thus the core radially expands, introducing negative reactivity. The effect due to the bowing of the subassembly is dependent on the type of the core restraint system design, burnup level, core structural material swelling, etc. In a freestanding type of design, the reactivity feedback is normally negative. In a typical passive restraint core design, under the condition of high burnup, there could be a marginal positive reactivity feedback.
- *Control rod driveline expansion with respect to the reactor vessel*: This effect is dependent on the reactor design. However, in pool-type fast reactors, control rods are inserted into the core from the top structure and the core subassemblies are freely supported on the grid plate. When the temperature of the coolant increases, the control rods will thermally expand down into the core, which expands in the upward direction. Hence, the net thermal movements will cause a negative reactivity feedback.

The various reactivity effects are shown schematically in Figure 1.10.

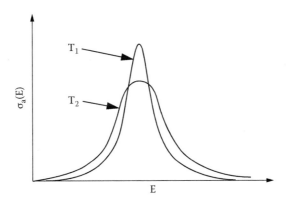

FIGURE 1.9 Doppler broadening of capture resonance cross section with $T_1 < T_2$. (From Lewis, E.E., *Fundamental of Nuclear Reactor Physics*, 1st edn, Academic Press, 2008.)

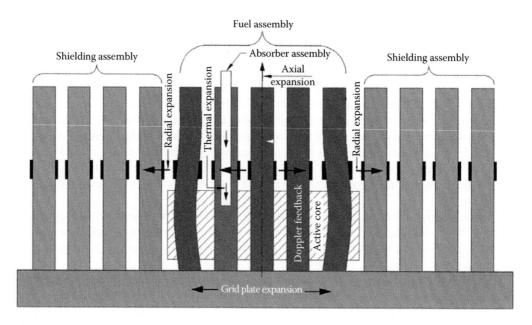

FIGURE 1.10 Schematic of fast reactor core reactivity feedbacks. (From Qvist, S.A., Safety and core design of large liquid metal cooled fast breeder reactors, PhD dissertation, Nuclear Engineering, University of California, Berkeley, CA, 2013.)

REFERENCES

1.1   Wong, S.S.M. (2004). *Introductory Nuclear Physics*, 2nd edn. Wiley-VCH Verlag Gmbh & Co KGaA, Weinheim, Germany.
1.2   Krane, K.S. (1988). *Introductory Nuclear Physics*. John Wiley & Sons, Wiley India Pvt. Ltd.
1.3   Glasstone, S., Sesonske, A. (2004). *Nuclear Reactor Engineering*, 4th edn. CBS Publishers, New Delhi, India.
1.4   Krappe, H.J., Pomorski, K. (2012). *Theory of Nuclear Fission*. Lecture Notes in Physics 838. Springer-Verlag, Berlin, Germany.
1.5   DOE Fundamentals Handbook. (1986). *Nuclear Physics and Reactor Theory*, Vol. 1 of 2, DOE-HDBK-1019/1–93, July 18, 1996. http://energy.gov/ehss/downloads/doe-hdbk-10191–93.
1.6   Lewis, E.E. (2008). *Fundamental of Nuclear Reactor Physics*, 1st edn., Academic Press, An imprint of Elsevier, London, U.K.
1.7   Duderstadt, J.J., Hamilton, L.J. (1976). *Nuclear Reactor Analysis*. John Wiley & Sons, New York, ISBN 0-471-22363-8.
1.8   Bell, G.I., Glasstone, S. (1970). *Nuclear Reactor Theory*. Van Nostrand Reinhold Company, New York.
1.9   Walter, A., Reynolds, A. (1981). *Fast Breeder Reactors*. Pergamon Press, Inc., New York.
1.10 Stacey, W.M. (2007). *Nuclear Reactor Physics*. Wiley-VCH Verlag Gmbh & Co. KGaA, Weinheim, Germany, ISBN 978-527-40679-1.
1.11 Qvist, S.A. (2013). Safety and core design of large liquid metal cooled fast breeder reactors. PhD dissertation. Nuclear Engineering, University of California, Berkeley, CA.

2 Fast Spectrum Reactor vis-à-vis Pressurized Water Reactors

2.1 INTRODUCTION

In the past and present energy scenarios, the role of pressurized water reactors (PWRs) is dominant compared to fast-spectrum reactors (FSRs). Hence, engineers, regulators, students, and others are very familiar with PWRs, which is not the case with FSRs. This chapter covers some basic reactor physics of FSR with reference to a typical PWR.

2.2 NEUTRONIC CHARACTERISTICS

Fission generates heat in a number of ways:

- The energy released by nuclear fission: The kinetic energy of fission products is converted to thermal energy when these nuclei collide with nearby atoms.
- The reactor materials absorb some of the gamma rays produced during fission and convert their energy into heat.
- Heat is produced by the radioactive decay of fission products and materials that have been activated by neutron absorption. This decay heat source will remain for some time even after the reactor is shut down.

Heat is extracted by the coolant for power generation. Heat generation and the removal process occur in the reactor core. Under thermal equilibrium condition, heat generation is equal to heat removal. Based on the process requirements, the core can be designed to sustain any power level, even nearly zero power by appropriately manipulating the number of neutrons and thereby the number of fissions in the core consisting of fissile, fertile, and absorber materials. These materials are surrounded by another class of materials to reflect/shield the neutrons leaking from the inner core. All these materials are placed in some specific configuration in the reactor core. In an FSR, the core is composed of the active core, blankets, absorber rods, neutron reflectors, and shields. While the blankets positioned at the top and bottom regions of the active core are called *axial blankets*, those surrounding the active core are called *radial blankets*. With this configuration, the neutrons leaking out in all directions are effectively captured by the fertile materials housed in blankets for breeding as well as fission reactions. Figure 2.1 depicts the schematic sketch of an FSR core configuration. The power distributions in the active core and blanket regions along the axial direction and across the core are depicted in Figure 2.2 at the beginning of life as well as at the end of life. Power generation in the blanket increases gradually and can attain a level close to the power generated in the active core. However, the power generated in the blankets is restricted due to safety considerations. The power generated in the blanket at the end of life is compared with the power distribution at the beginning of life of the core in Figure 2.2. In the case of thermal reactors, neutron leak is insignificant and no significant number of neutrons would be available for breeding. Hence, blankets are not introduced. However, in the case of fast breeder reactor (FBR) there will be more neutron leakage and fertile blankets are provided surrounding the core to profitably capture neutrons. In view of low cross sections, unlike thermal reactors, there is no large reactivity change of the core due to the accumulation of fission products.

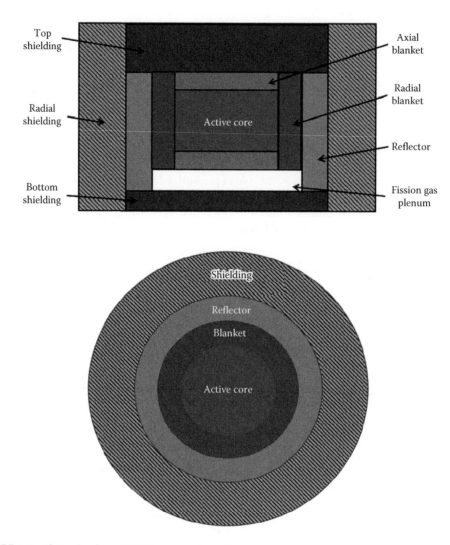

FIGURE 2.1 (**See color insert.**) FSR core configuration.

The neutron spectrum in FSR is complex in the sense that the associated energies are higher than those associated with thermal reactors. The average neutron energy in the case of FSR is 100 keV, which is four million times higher than the energy of the thermal spectrum. Hence, the core materials have low cross sections for all nuclear reactions, especially the fission cross section, that is, $\sigma_f \approx 2$ b for the fast spectrum compared to ≈ 500 b for the thermal spectrum [2.1]. In order to sustain the same number of fissions in both FSR and thermal reactors so as to generate the same thermal power, fissile material enrichment should be greater ($\approx 20\%$) for a typical FSR. FSR has a higher neutron flux (typically 10^{15} n/cm²/s, one order larger in magnitude than thermal reactors). Further to maintain the high energy spectrum of the neutron, the core should be as compact as possible, which implies that the FSR core volume is much smaller than that of the light water reactor (LWR), generating the same thermal power, which further result in high power density (specific power is 300–600 MW/m³ in FSR, about five times higher than that of LWR) and high linear power rating of fuel (400–500 W/cm of pin length). Hence, an efficient coolant is essential to extract the heat rapidly and liquid metal is the most appropriate choice. The important core neutronic characteristics of LWR and FSR are compared in Table 2.1 [2.2].

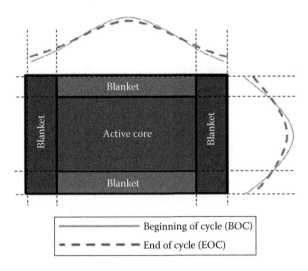

FIGURE 2.2 **(See color insert.)** Power distribution in the core including blankets.

TABLE 2.1
Comparison of LWR and FSR Core Characteristics

Parameter	Thermal Reactor	Fast Reactor
Fissile enrichment	0%–3% U^{235}	10%–30% Pu^{239}/U^{235}
Average neutron energy	~0.025 eV	~100 keV
Energy extracted: peak burnup, MWd/ton	~40,000	~100,000
Neutron flux, n/cm²s	10^{14}	5–10×10^{15}
Average core power density, W/cm³	~100	~300–400
Average fuel-specific power, kW/kg	~40	~100

Source: Waltar, A.E. et al., *Fast Spectrum Reactors*, Springer, 2012.

2.3 SAFETY CHARACTERISTICS

The FSR core is not in the most reactive configuration (MRC); that is, if there is any disturbance in the core configuration or if the core melts down, then there is a possibility of reactivity increase due to core compaction. However, in the case of LWR, only at a particular moderator to fuel ratio will keep the reactor in the MRC. If there is any change in the core configuration, there will be a decrease in reactivity. Over 99% of all neutrons, the so-called prompt fission neutrons, are generated directly during the time of fission, in the hard core. A residual part, <1% of delayed neutrons, is emitted by the decay of a few fission products. The prompt fission neutrons have an average energy of 2 MeV, corresponding to a velocity of 2000 km/s, which is almost independent of whether the fission is achieved by means of a fast or a thermal neutron, or whether a uranium and plutonium nucleus is split. In the fast reactor, the neutron energy loss, from their emergence to absorption in a fission, is relatively small, because they suffer only a few collisions. The effective lifetime, which is the mean time between two consecutive neutron generations, is equal to about 4.5×10^{-7} s (2.5×10^{-5} s in PWR), which is so short that it would not have been possible to control the chain reaction with a mechanically regulated apparatus, if the kinetics of the reactor had been defined by prompt neutrons. Hence, control in all the reactors is based on the time constants of delayed neutrons. The overall fraction of delayed neutrons is 0.35% for a typical FSR (0.6%–0.5% for PWR depending on the burnup).

Accordingly, the reactivity disturbances are related to the fraction of the delayed neutrons in reactor dynamic analyses. The delayed neutron fraction of 0.35% corresponds to an overproduction of neutrons by 0.35%, that is, K_{eff} = 1.0035, which corresponds to a reactivity of 1$.

In view of a shorter neutron lifetime, there is misconception regarding the requirement of very fast neutron absorber systems for the control of a chain reaction maintained by fast neutrons in FSR. The solution to the reactor kinetic differential equation with the consideration of delayed neutrons represented in Figure 2.3 provides better clarity [2.3]. The period shown on the ordinate is the time during which the power of the reactor increases by a factor e, if appropriate reactivity (>1) is inserted without considering any feedback effects. The reactor would attain the uncontrolled state. Under such uncontrolled situation, the FSR has a shorter period, compared to an LWR. However, under the controlled state of the reactor (reactivity introduction <1), the periods are practically the same for both FSR and LWR and are not distinguishable from each other from the reactor period point of view. The periods are so long that control is quite possible with technically simple mechanical regulating units. With reactivity higher than 1$, prompt neutrons determine the time behavior of the reactor (*superprompt criticality*). The periods are too short in both reactor types, in order to be able to interfere effectively with a fast shutdown. Thus, every reactor must be designed so that a superprompt critical condition cannot occur, and in all cases, this can be considered as a hypothetical event. If a superprompt critical excursion is arbitrarily postulated, it is shown that it is limited highly by feedback effects. In this case, the Doppler coefficient plays a prominent role, and the neutron absorption capability of U-238, available in the fuel material, increases as the fuel temperature increases. A basic study reported in 1963 shows that the excursion energy is small [2.4]. The theory based on this evidence has been verified in an extensive reactor experiment called SEFOR, United States [2.5].

Another misconception with regard to the lower delayed neutron fraction is that the safety margins available for the control are highly restricted in FSR. As opposed to this view, fast reactors are more stable even with a small number of delayed neuron fractions. First of all, these result from the benefits derived from fast neutrons. Variations in the operating parameters, for example, inlet temperature, coolant flow, and power, have a considerably less influence on the reactivity in the case of FSR. This may be obvious in the case of reactivity changes, which are produced due to a change in the reactor inlet temperature of around 1 K, which is ~0.0015$/K compared to 0.01–0.1$/K for LWR depending on the burnup. In spite of this, a short shutdown time (<1 s) is easily possible in FSR due to the short core height (one of the favorable consequences of high power density).

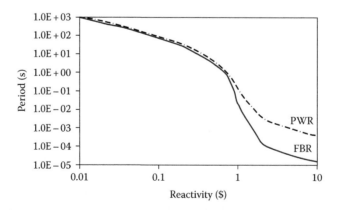

FIGURE 2.3 Period versus reactivity. (A trend curve is given here; for details, [2.3] can be referred.) (From Vossebrecker, V.H., *Special Safety Related Thermal and Neutron Physics Characteristics of Sodium Cooled Fast Breeder Reactors*, Warme Band 86, Heft 1, 1999.)

Lastly, high power density effects on the safety of FSR are addressed in this section. High power density signifies a low thermal capacity of the core. Hence, the consequences of the disturbances cause very fast temperature changes in the core, which in turn actuate the fast and continuous feedback by the Doppler and fuel expansion effects. Apart from this, high power density means small core dimensions, calling for a short distance travel of the control and safety rods and resulting in short shutdown times. Hence, the high power density of the FSR core is demonstrated as a safety-oriented advantage. A negative consequence of the high power density is that during the primary pump pipe rupture event without shutdown or with the total loss of coolant, high thermal gradients would result. However, any sudden stoppage of the coolant flow is physically impossible due to the high mass moment of inertia of the coolant and pumps, and further mismatch between the power and coolant flow would soon initiate a fast scram. Further, the simultaneous failure of pumps and scram is an extremely improbable accident, during which especially high heating rates are to be expected in the core. In spite of these, once initiation of sodium boiling occurs, it may lead to a rapid power increase and ultimately destruction of the core. Moreover, with the introduction of efficient temperature measurement systems at the core outlet with very small time delay, countermeasures are possible at the right time. This temperature monitoring is carried out in addition to neutron flux monitoring. In spite of these, a highly reliable shutdown system is considered to be an essential requirement in the FSR system to prevent any severe accident scenario.

2.4 GEOMETRIC FEATURES OF CORE

Geometrical features of the FSR core are characterized mainly by the type of coolant: liquid metal or gas. The geometrical arrangement of fuels, pellets in the case of LWR and liquid metal–cooled FSR and pebbles in the case of gas-cooled reactors, is shown in Figure 2.4. Pellets are housed within thin-walled tubes (*cladding*) and sealed at both ends which is called as *fuel pin*. A certain specific number of fuel pins are arranged within a *subassembly*. The coolant flows through each and every subassembly to extract the heat generated in it. There is no direct contact of the coolant with the fuel pellets with the cladding in between. Many such subassemblies form the reactor core. In the case of gas-cooled reactors, bubbles housed in a container form a "reactor core." There is direct contact between the gas and pebble surfaces. Adequate spacing is provided between the fuel pins to facilitate coolant flow. A long and thin wire is wound helically around the clad to achieve the required spacing between the pins after integration within the subassembly. Details of a typical fuel pin and fuel subassembly are shown in Figures 2.5 and 2.6, respectively. Each subassembly is mounted on the grid plate, which forms a common inlet plenum from which the coolant flows through the individual subassemblies. Figure 2.7 shows one subassembly mounted on the grid plate.

The subassemblies are generally freestanding on the grid plate in the case of a sodium coolant. However, in the case of heavy metals such as lead or lead–bismuth, there is a need to hold them

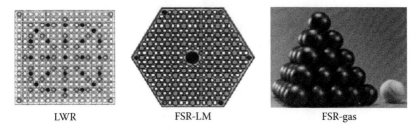

LWR   FSR-LM   FSR-gas

FIGURE 2.4 Geometrical arrangement of fuels.

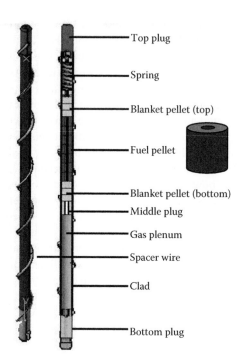

FIGURE 2.5   Typical FBR fuel pin.

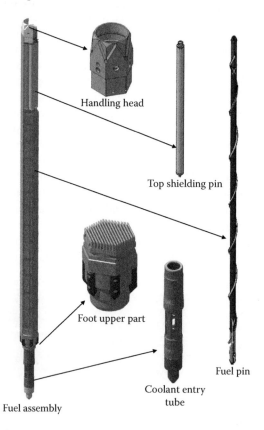

FIGURE 2.6   Typical FBR subassembly.

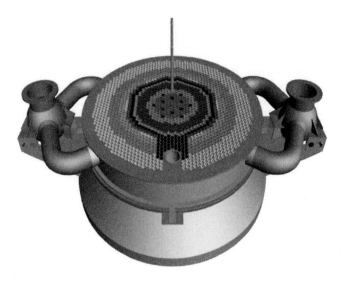

FIGURE 2.7   Typical view of a single subassembly located on the grid plate.

to resist gravity. Under the neutron and the temperature environment prevailing in the core, the subassemblies undergo gradual structural deformation due to void swelling and irradiation creep, which will lead to subassembly bowing and dilation. These deformations cause several problems affecting the core performance such as reactivity changes and enhanced extraction force for the subassembly. To overcome these problems, gaps are ensured between the subassembly faces by providing contact buttons. The gap needs to be optimized from allowable bowing, dilation, and neutronic considerations. The location of buttons, particularly elevation, is an important parameter in the design, which decides the bowing profile of the cluster of subassemblies as seen in Figure 2.8.

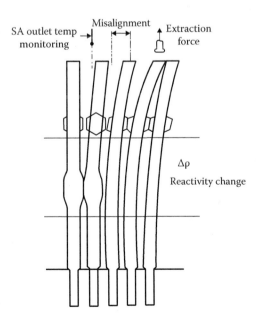

FIGURE 2.8   Schematic of the bowing profile of a cluster of subassemblies.

REFERENCES

2.1 Sesonske, G. (2004). *Nuclear Reactor Engineering and Reactor Design Basics*, Vol. 1. CBS Publishers and Distributors, New Delhi.

2.2 Waltar, A.E., Todd, D.R., Tsvetkov, P.V. (2012). *Fast Spectrum Reactors*. Springer, New York.

2.3 Vossebrecker, V.H. (1999). *Special Safety Related Thermal and Neutron Physics Characteristics of Sodium Cooled Fast Breeder Reactors*. Warme Band 86, Heft 1.

2.4 Tobita, Y. et al. (1999). Evaluation of CDA energetics in the prototype LMFBR with latest knowledge and tools. *Proceedings of ICONE-7*, Tokyo, Japan.

2.5 McKeehan, E.R. (1970). *Design and Testing of the SEFOR Fast Reactivity Excursion Device* (FRED), General Electric Co., Sunnyvale, CA. Breeder Reactor Development Operation, US Atomic Energy Commission (AEC), US. http://www.osti.gov/scitech/biblio/4050547/, doi:10.2172/4050547.

3 Description of a Fast Spectrum Reactor

3.1 INTRODUCTION

The reactor core with heat transport circuits, balance of plant (BoP), control systems, and fuel handling systems forms a nuclear power plant. The primary heat transport circuit facilitates continuous coolant flow through the core to maintain its temperature and transfer the heat removed from the core to the fluid in the BoP. BoP generates electricity from the heat derived from core. The pumps and heat exchangers in the primary circuit are the important components in the reactor. The primary coolant can be sodium or lead or lead–bismuth in the case of a liquid metal fast breeder reactor (LMFBR). Helium is normally used in gas-cooled reactors. Generally, water is used in BoP of LMFBR. In the case of a sodium-cooled fast reactor (SFR), gases such as nitrogen or carbon dioxide are being considered as an alternative coolant to water to circumvent the consequences of a large sodium water reaction. A typical fast spectrum reactors (FSR) is schematically shown in Figure 3.1. Its various general systems are described in the following text.

3.2 CORE AND REACTOR ASSEMBLY

The reactor core is the source of heat from nuclear fission. Schematically, the core consists of the central fuel region enveloped by the blanket region, which, in turn, is surrounded by neutron reflectors. Physically, the core is made up of different types of small assemblies that are replaceable. The fuel assemblies consists of fuel column in central region and to have better breeding, axial blankets are incorporated on either side of the fuel column in the fuel pin. The assemblies are arranged in a triangular pitch to minimize fissile inventory. Within the fuel region, absorber rods are located to control as well as shutdown the reactor.

In a pool-type LMFBR, the core is housed in a single vessel called main vessel (MV). It is closed by a top shield, which includes a roof slab, rotatable plugs, and control plug. The functions of the MV are as follows:

- Contains the bulk of the primary sodium
- Serves as the boundary for primary sodium and argon cover gas
- Supports the core support structure along with the loads coming on it
- Absorbs the energy and contains the radioactive products during an accident

The roof slab supports the main components like the MV, rotatable plugs, primary sodium pumps, intermediate heat exchangers, decay heat exchangers, and delayed neutron detectors. The control plug supports the core cover plate and houses the instrumentation and control equipment (core-monitoring thermocouples, absorber rod drive mechanisms, and failed fuel location modules).

The MV is surrounded by a safety vessel (SV), ensuring a safe level of sodium even in the case of an unlikely leak. The core subassemblies are supported on a grid plate, and their combined load is transferred to the MV through the core support structure. The assembly of the MV and its internals, SV, top shield, control plug, etc., is called the reactor assembly (RA).

RA is housed in a concrete vault lined with carbon steel. The SV is supported directly by the reactor vault independent of the support for the MV. On the outer surface of the SV, metallic

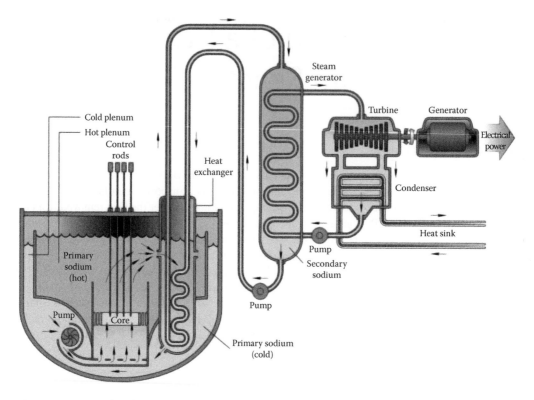

FIGURE 3.1   Typical FSR heat transport system.

insulation is provided to limit the heat transfer to the vault. A fuel transfer machine (FTM) is provided to transfer the fuel from in and out of the core. As part of a safety grade decay heat removal circuit, sodium–sodium heat exchangers are immersed in a hot pool to transfer the decay heat to the sodium–air heat exchangers. Generally, two groups of absorber rods are operated by separate drive mechanisms to have redundancy.

3.3   MAIN HEAT TRANSPORT SYSTEM

Figure 3.1 shows the flow sheet of the main heat transport system. Liquid metal is circulated through the core using primary sodium pumps. The hot primary sodium is radioactive and is not used directly to produce steam; instead, it transfers heat to the secondary sodium through four intermediate heat exchangers. The nonradioactive secondary sodium is circulated through independent secondary loops, each having a sodium pump, intermediate heat exchangers, and steam generators. In case of off-site power failure or nonavailability of the steam–water system, the decay heat is removed by a passive safety grade decay heat removal circuit.

3.4   COMPONENT HANDLING

Fuel handling is done after definite full power days with the reactor in shutdown condition and the coolant at lower temperature due to fuel depletion. Usually, certain portions of the core assemblies will be replaced based on the fuel cycle design. Usually, preheated fresh subassemblies are transferred to the core using special transfer machines. The spent fuel subassemblies are generally stored inside the MV for one campaign and then shifted to the ex-vessel cooling facility. Leak-tight shielded flasks are provided for the special handling of components like primary sodium pumps, intermediate heat exchangers, decay heat exchangers, and absorber rod drive mechanisms in case of any requirement. The components are decontaminated in a separate facility provided within the reactor containment building before they are taken for maintenance.

3.5 STEAM–WATER SYSTEM

The steam–water system uses a reheat, regenerative cycle using live steam for reheating. High-pressure superheated steam from the steam generators drives a turbo alternator, which is generally similar to that in conventional plants.

3.6 ELECTRICAL POWER SYSTEMS

The plant is provided with both off-site and on-site electrical power supply systems. A class III power supply system is provided with standby emergency diesel generators. Power supplies for instrumentation and control equipment are powered by class I and class II power supply systems in case of the nonavailability of class IV and class III systems.

3.7 INSTRUMENTATION AND CONTROL

The burnup compensation of reactivity is very small in FSRs. Hence, the reactor power is controlled manually in the case of FSR. Neutron detectors are provided above the core to monitor the neutron flux. Cover gas activity and delayed neutrons in the primary sodium are monitored for failed fuel detection. Thermocouples are provided to monitor the temperature of sodium at the outlet of each fuel subassembly. Flow delivered by sodium pumps is measured using a flowmeter, and safety action is taken on a power-to-flow ratio. Provisions are designed to ensure that there are at least two diverse safety parameters to shut down the reactor safely for each design basis event.

4 Unique Worthiness of SFR

4.1 INTRODUCTION

The nuclear fuel utilization factor is generally defined as the ratio of mass fissioned to the mass loaded in the core. In the fast breeder reactor (FBR) context, this can be defined as the ratio of the sum of the mass that has undergone fission yielding energy and the mass that has been converted into fissile material to the mass loaded in the core. Thermal reactors operate with a small quantity of fissile material, U^{235}, and dump a huge amount of the remaining material as *waste* without extracting its energy potential. However, the discharged fuel, containing some modest quantity of fissile material produced on fertile conversion, can be extracted through appropriate reprocessing technology. Once the fissile material and other materials such as fission products and minor actinides (MAs) are separated, the remaining material is called depleted uranium (DU), which can be used as fuel in FBRs. Thus, in the case of thermal reactors, the mass equivalent of energy released and the fissile material generated through the conversion process represent the utilization of fuel fed to the reactor. FBRs use DU, enriched with an adequate quantity of fissile material. Apart from this, DU is loaded in both the axial and radial blankets. Thus, the total mass fed to reactor is the sum of DU and fissile material fed into the active core and DU loaded into the blankets. As far as fuel utilization is concerned, it is the fissile material generated in the core and blankets as well as the mass equivalent of net energy released in the core. The mass of fission products represents the mass of fuel fissioned. With this understanding, the fuel utilization factor is defined for thermal and fast reactors. To get an idea on the values of various parameters involved in fuel utilization, let us consider PHWR, PWR, and FBR, each producing 1000 MWe power. Utilization in both open and closed cycle modes is considered. The various quantities referred are typical values (rounded off). The concepts of open and closed cycle modes are described in detail in Chapter 29.

4.2 URANIUM UTILIZATION IN THE OPEN FUEL CYCLE MODE

Let us consider a 1000 MWe PHWR operating with a thermodynamic efficiency of 28%. It should produce a thermal power of 3600 MWt. The initial fuel requirement is ~210 tons. For the typical fuel burnup of 7000 MWd/ton with a load factor of 95%, the annual natural uranium requirement works out to $3570 \times 365 \times 0.95/7000$, that is, ~177 tons. The irradiated fuel discharged contains ~98.91% DU, 0.7% fission products, 0.385% Pu, and 0.005% MA. Hence, the effective utilization of natural uranium is (0.7 + 0.385), that is, equal to ~1.1%. Similar estimates were made for PWR and FBR and the results are summarized in Table 4.1. The fuel for the PWR is uranium enriched with 4% U^{235} isotope. The average burnup that could be achieved in PWR is ~35,000 MWd/ton. For the FBR considered here, metallic fuel with 80% DU and 20% Pu and sodium coolant are

TABLE 4.1

Fuel Utilization in PHWR, PWR, and FBR in the Open Cycle Mode

Parameters	PHWR	PWR	FBR (BR = 1.4)	
Power (MWe)	1000	1000	1000	
Thermodynamic efficiency (%)	27.8	30	40	
Thermal power (MWt)	3600	3300	2500	
Fuel	Natural U	Enriched U (4%)	Core	Blanket
			U + Pu	DU
Initial fuel feed (tons)	210	70	16 + 4	20
Load factor (%)	95	80	85	
Average burnup (MWd/ton)	7,000	35,000	80,000	
Irradiation Period (efpd)	430	850	720	1500
Annual fuel feed (tons): AF	177	30	8 + 2 = 10	4.8
Annual spent fuel discharge				
Uranium (tons)	175.07	28.44	7.2	4.37
Plutonium (kg): PU	682	300	1800	400
Fission products (kg): FP	1240	1230	1000	30
Minor actinides (kg)	8.0	23	10	1
Fuel utilization factor (%): $\left[\dfrac{FP + PU}{AF}\right]$	1	5.0	21.8	

assumed. Both axial blankets housed within the core subassemblies and radial blankets kept surrounding the active core generally contain DU. Due to the use of a sodium coolant, which has high boiling point, a higher thermodynamic efficiency of the associated energy conversion system, that is, 40%, is considered compared to 28% for PHWR and 30% for PWR. It is clear from Table 4.1 as well as from Figure 4.1 that fuel utilization is very effective and waste production is minimum in FBRs, even if they operate in the open cycle mode. The specific annual feed of fuel (per unit energy production) is rather very low for FBRs compared to thermal reactors of the same electrical capacity. The core material flow is presented schematically for PHWR, PWR, and FBR in Figure 4.1.

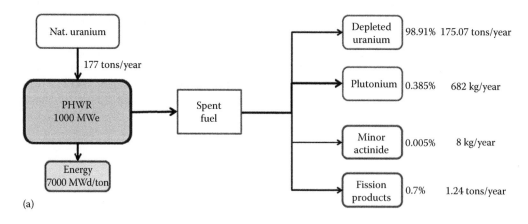

(a)

FIGURE 4.1 Fuel utilization and waste generation in (a) PHWR.   (*Continued*)

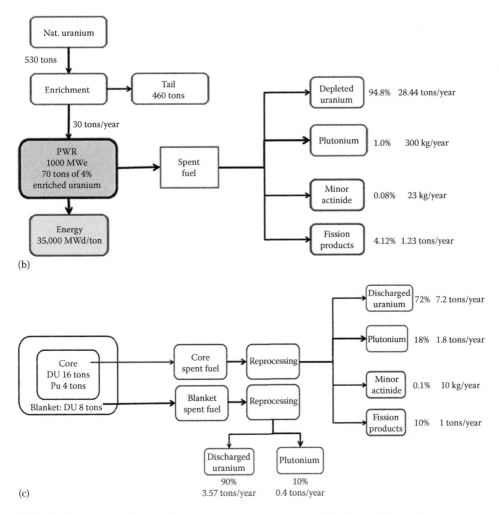

(b)

(c)

FIGURE 4.1 (*Continued*) Fuel utilization and waste generation in (b) PWR and (c) FBR.

4.3 URANIUM UTILIZATION IN THE CLOSED FUEL CYCLE MODE

Let us discuss the possibility of recycling the spent fuel in PHWR, PWR, and FBR. Recycling involves recovering U and Pu from the spent fuel, refabricating, and reloading back into the reactor. Table 4.2 gives the initial compositions of plutonium with respect to recycling in PHWR and PWR. Recycling in thermal reactors highly deteriorates the fissile content of plutonium in just one recycle. Further recycling would demand more plutonium mass due to increase in fertile contents. So recycling may not be economical in thermal reactors.

TABLE 4.2

Plutonium Vectors in Thermal Reactors

	PHWR				PWR			
Recycle	Pu^{239}	Pu^{240}	Pu^{241}	Pu^{242}	Pu^{239}	Pu^{240}	Pu^{241}	Pu^{242}
1	69	25	5	1	50	25	14	11
2	63	30	3	2	43	30	13	14

In the case of PWR, the discharged uranium contains about 0.8% U^{235}, which is slightly more than that of natural uranium, that is, more than that required for a PHWR. Hence, a sizeable quantity of spent fuel from PWR can be utilized in PHWR after removing fission products and higher actinides. In 40 years of reactor lifetime, a 1000 MWe PWR burns (5%) about 60 tons of fuel out of 1200 tons of fresh load, and it discharges about 1140 tons spent fuel. Recovering uranium from this discharge, another 10 tons can be reused as fuel in PHWR. However, while fueling a PWR for 40 years, an enrichment plant disposes about 7000 tons as tail (waste stream from the uranium enrichment process). Hence, the total amount of uranium dumped as waste, either in the form of enrichment tail or as spent fuel, is estimated to be about 8000 tons. Effective fuel utilization in a PWR, along with a credit of uranium recycling in PHWR, works to be about 0.9% of natural uranium—which is again not significant.

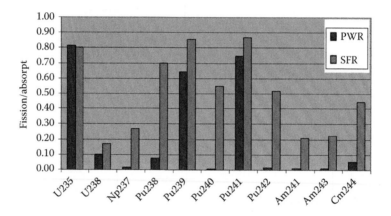

FIGURE 4.2 Comparison of the fissionability fraction in thermal and fast reactors. (From Pandikumar, G. et al., Scenario and preliminary assessment of minor actinide burning through SFR in the Indian context, Report CFBR/01119/DN/1004, Indira Gandhi Centre for Atomic Research, Kalpakkam, India, 2013.)

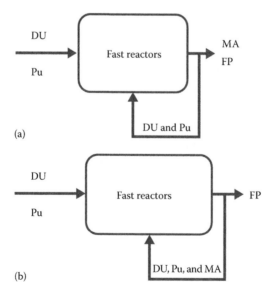

FIGURE 4.3 FBR closed fuel cycle schematic. (a) Depleted uranium and plutonium recycling and (b) fuel recycling with MA incineration.

Studies show that multiple recycling in FBR discharge fuel is feasible [4.1]. The variation of the plutonium vector with respect to recycling is not significant compared to thermal reactors. The only constraints are the reactivity worth, and the total mass must be maintained the same as that of fresh fuel loaded. Thus, recycling of spent fuel is very effective in FBRs. Figure 4.2 compares the fission fractions of the MAs in thermal reactors and fast reactors. Since most of the MAs are fertile nuclides, that is, susceptible for fission only with a high energy neutron, fast reactors are better for burning MAs [4.2,4.3]. The MAs enter as fuel in FBRs and contribute in the energy production, apart from transmuting to short-lived isotopes—an important benefit from FBRs. Figure 4.3 gives the schematic of Pu and MA recycling options in FBRs.

4.4   FUEL UTILIZATION IN THE FAST BREEDER REACTOR: A CASE STUDY

A scenario relevant to countries having a limited quantity of natural uranium is presented here. Let us assume that 100 kilotons of natural uranium is available as a nuclear resource. If this resource is fueled in PHWRs and PWRs, only 3 tons can be utilized at the maximum and about 97 tons would be discharged as waste. The waste is reprocessed to separate Pu, fission products along with MA, and the remaining material is DU. If FBRs start utilizing the DU enriched appropriately with plutonium extracted from reprocessing, they can produce enough fissile material for feeding another similar FBRs after every 10 years (a typical doubling time for metal-fueled reactors). Thus, FBRs would grow, each multiplying into two in every 10 years, until the entire DU is consumed. Figure 4.4 shows the flowchart of multiplication of FBRs. With all considerations, such as high breeding ratio, high burnup, shorter doubling time, shorter delays in reprocessing and fabrication, multiple fuel recycling (closed fuel cycle), loss of DU through nonfission reactions inside the reactor, and losses external to the reactors, it takes about 100 years to utilize the entire DU, if we start with one FBR and its multiplication. And it needs more than 2000 GWe FBRs (Figure 4.5) to burn the DU.

Now, a growth scenario is depicted by assuming that 10×1000 MWe FBRs are started simultaneously at some specified time scale. The estimation indicates that this would demand an initial inventory of about 40 tons of plutonium. In five reactor doubling period (50 years), the number of FBRs would multiply into 280 (considering 40 years of reactor lifetime). That is, within about 50 years, the power production would grow to about 280 GWe. Figure 4.5 shows the possible energy extraction using FBRs by utilizing the DU discharged from thermal reactors. To utilize the entire DU, it takes about 100 years and is achievable through 2000 FBRs, each with 1 GWe capacity. Figure 4.6 shows the DU resource utilization with respect to years. Thus, FBRs are essential for those countries having limited uranium and fissile materials to become aware of their significant nuclear contribution.

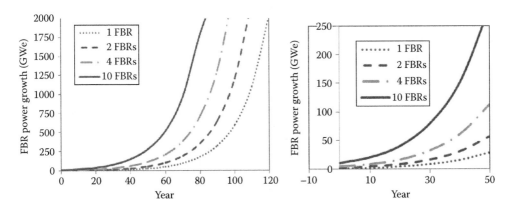

FIGURE 4.4   Energy extraction from DU utilization using FBRs.

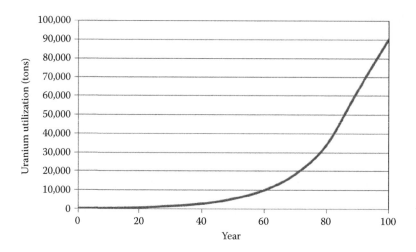

FIGURE 4.5 DU utilization in FBRs.

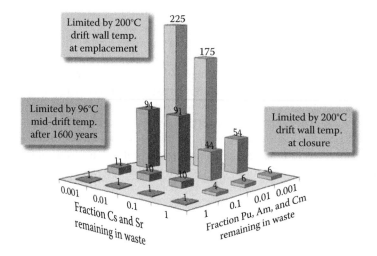

FIGURE 4.6 **(See color insert.)** Storage space required for the nuclear waste as a function of MA. (From Raj, B. et al., Assessment of compatibility of a system with fast reactors with sustainability requirements and paths to its development, IAEA-CN-176-05-11, FR09, Kyoto, Japan, 2009.)

4.5 HIGH-LEVEL RADIOACTIVE WASTE MANAGEMENT AND ENVIRONMENTAL ISSUES

In the previous sections, it has been demonstrated that the net fraction of long-lived isotopes is insignificant in the FBR. Figure 4.6 shows the assessment of the space required for the storage, published in [4.4], depending upon the MA contents. Since MAs can be recycled in FBRs, the space required for storing the waste from FBRs would be insignificant. Radiotoxicity from the wastes is estimated as a function of time and assessed with the radiotoxicity from natural uranium. The results are presented in Figure 4.7. It can be observed that radiotoxicity of DU is well below the radiotoxicity of the natural uranium ore, and the radiotoxicity from the fission products stored is also below that of the uranium ore before 300 years. However, the radiotoxicity of plutonium and MAs is of serious concern. If we allow plutonium and MAs for natural radioactive decay, it takes more than 10,000 years for the reduction of their radiotoxicity below the natural background toxicity. Moreover, storing the radioactive material for a long period under surveillance is impractical.

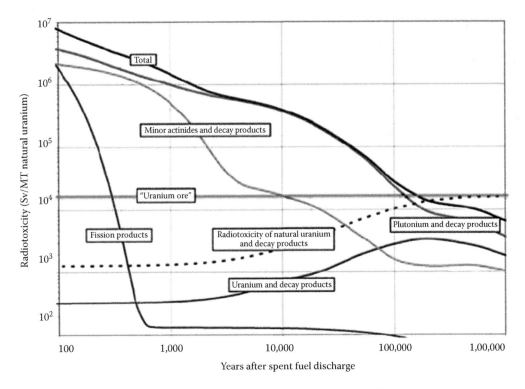

FIGURE 4.7 Radiotoxicity of spent fuel and natural uranium. (From http://www.hzdr.de/db/Cms?pOid=30396&pNid=2721.)

The fast reactors, with some minor modifications in the design, can also be utilized for burning the long-lived MAs. With the introduction of the MA burner, estimations show that the toxicity level of actinides could be reduced well before 400 years (Figure 4.8). The basic design aspects of MA burners are described in subsequent sections.

4.6 MINOR ACTINIDE BURNING DESIGN CONCEPTS

During reactor operation, high radiotoxic materials such as higher actinides and long-lived fission products are produced. The half-lives of the MAs are given in Table 4.3. The radiological hazards from long-lived MAs and fission products can be reduced by means of incineration and transmutation. Transmutation is a process of converting a long-lived radioactive nucleus, by neutron capture, to either a stable or short-lived nucleus. Transmutation is most appropriate for long-lived fission products. Incineration is the process of causing neutron-induced fission in actinides resulting in fission products. Compared to thermal reactors, FBRs have a higher potential for incinerating MAs (Figure 4.2), and hence, MA wastes could be effectively reduced by incineration in fast reactors.

The design of the fast reactor (hereinafter called as MA burner) can be modified for the deliberate inclusion of MAs for incineration. Despite the advantages, such modification demands higher fissile enrichment and greater burnup reactivity loss. The extent of MA addition is in general limited to about 5% of the total fuel due to the detrimental effect on some safety parameters. MA addition results in the reduction in the Doppler feedback, increase in coolant void coefficient/worth, and smaller effective delayed neutron fraction [4.5]. To keep the safety parameters within acceptable range and to maximize the MA burning capability, various design modifications including fuel types (oxide, metal, etc.) [4.6] and MA loading methods—homogeneous and heterogeneous mixing [4.7] (dedicated MA bearing fuel SA)—were investigated. It is observed that the metal FBR core with heterogeneous MA loading shows a higher MA transmutation rate.

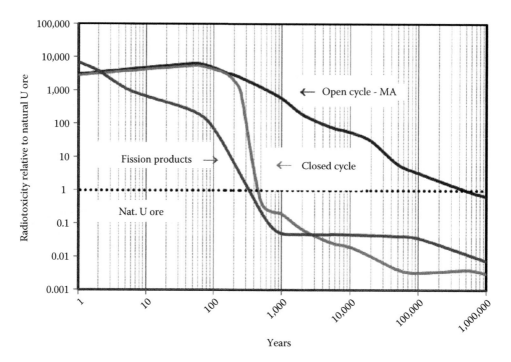

FIGURE 4.8 Radiotoxicity relative to the natural U ore with and without the MA burner. *Note*: MA production and depletion from 2052 onward considered. (From Pandikumar, G. et al., Scenario and preliminary assessment of minor actinide burning through SFR in the Indian context, Report CFBR/01119/DN/1004, Indira Gandhi Centre for Atomic Research, Kalpakkam, India, 2013.)

TABLE 4.3
Half-Life of Minor Actinides

Nuclide	Np237	Am241	Am42m	Am243	Cm242	Cm243	Cm244
Half-life	2.1×10^6 year	432 year	141 year	7380 year	163 day	29 year	18 year

TABLE 4.4
Half-Life of Certain Long-Lived Fission Products

Fission Products	Kr85	Sr90	Cs137	Tc99	Zr93	Cs135	Pd107	I129
Half-life year	10.7	28.9	30.2	2.10×10^5	1.50×10^6	2.30×10^6	6.50×10^6	15.7×10^6

Internationally, France, India, Russia, Japan, etc., have studied the feasibility of MA incineration in FBRs. It is reported that a metal-fueled FBR with an external feed of 5% MAs with respect to the total fuel could burn ~10% of the initially loaded MA. Based on the literature survey, it is found that a 1000 MWe metal-fueled FBR can burn 100 kg of MAs in a year. Although the reactor incinerates a sufficient amount of MAs, it must be remembered that it also produces 20 kg of MAs per year. Hence, the net burning rate of a 1000 MWe metal FBR would be 80 kg/GWey (~equivalent to the MA waste from 10 PHWRs). The remaining 90% of MA, left in the discharged fuel, must be recycled for effective MA burning. Thus, the MA burner can considerably reduce the amount of MA. Finally, the high-level waste would be left with a small amount of certain long-lived fission products. Table 4.4 gives the half-lives of long-lived fission products. Figure 4.9 schematically represents the multiple recycling of MAs in the fast reactor.

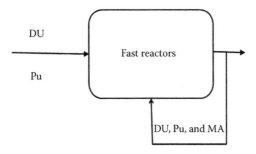

FIGURE 4.9   Fuel recycling with MA incineration.

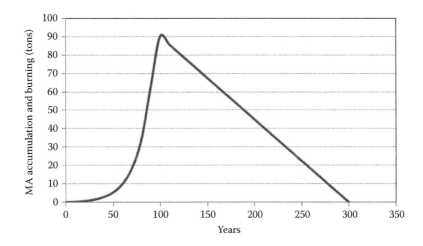

FIGURE 4.10   Accumulation of MA and incineration strategy.

4.7   TYPICAL MINOR ACTINIDE BURNING SCENARIO IN FAST SPECTRUM REACTORS (FSRs)

Energy extraction from the available 100 kilotons of natural uranium, can lead to the accumulation of about 90 tons of MA production with different combination of thermal and fast reactors. MA burning does not call for a new kind of design. FBRs have a significant potential for burning MAs. Another advantage of FBR is that addition of MA does not demand major modifications in the technical designs. That is, with the present FBR technology, a considerable amount of MAs can be reduced. Annually, a 1000 MWe FBR can effectively burn about 90 kg of MAs. Hence, in 40 years of lifetime, an FBR can burn about 3.6 tons of MA. With eight such FBRs, we can bring down the amount to a negligible value in 200 years through recycling. Figure 4.10 shows the accumulation of MAs over 100 years of reactor operation and the incineration strategy. Some amount of MAs that are left out could be burnt with those improved technology available in the future.

REFERENCES

4.1 Pandikumar, G., Gopalakrishnan, V., Mohanakrishnan, P. (2011). Improved analysis on multiple recycling of fuel in PFBR in a closed fuel cycle. *Pramana J. Phys.*, 77(2), 315–333.

4.2 IAEA. (2009). Advanced reactor technology options for utilization and transmutation of actinides in spent nuclear fuel, IAEA, Vienna publication, IAEA-TECDOC-1626. http://www-pub.iaea.org/mtcd/publications/tecdocs.asp.

4.3 Devan, K., Pandikumar, G., Harish, R., Gopalakrishnan, V., Mohanakrishnan, P., Srivenkatesan, R. (2009). Chapter 1 domain-I: Critical fast reactors with transmutation capability and with fertile fuels, IAEA, Vienna publication, IAEA-TECDOC-1626. http://www-pub.iaea.org/mtcd/publications/tecdocs.asp

4.4 Raj, B. et al. (2009). Assessment of compatibility of a system with fast reactors with sustainability requirements and paths to its development. IAEA-CN-176-05-11, FR09, Kyoto, Japan.

4.5 Minato, K. Transmutation in Fast Neutron Reactors. JNC, JEARI. http://www.neutron.kth.se/courses/GenIV/Chapter3.pdf

4.6 Ohki, S. (2000). Comparative study for minor actinide transmutation in various fast reactor core concepts. Oarai Engineering Centre, Japan Nuclear Cycle Development Institute, Madrid, Spain, December 11–13, 2000. https://www.oecd-nea.org/pt/docs/iem/madrid00/Proceedings/Paper31.pdf

4.7 Wakabayashi, T. (1998). Status of transmutation studies in a fast reactor at JNC. *Fifth OECD/NEA Information Exchange Meeting on Actinide and Fission Product Partitioning and Transmutation SCK-CEN*, Mol, Belgium.

4.8 Pandikumar, G., Devan, K., John Arul, A., Puthiyavinayagam P., Chellapandi, P. (2013). Scenario and preliminary assessment of minor actinide burning through SFR in the Indian context. Report CFBR/01119/DN/1004. Indira Gandhi Centre for Atomic Research, Kalpakkam, India.

4.9 OECD. (2006). Physics and Safety of transmutation Systems: A Status Report, Nuclear energy agency organisation for economic co-operation and development, NEA No. 6090. https://www.oecd-nea.org/science/docs/pubs/nea6090-transmutation.pdf

5 Design Objectives for the Efficient Use of Natural Uranium and Plutonium

5.1 INTRODUCTION

Fast reactors are a preferred option for effectively utilizing natural nuclear resources, in particular uranium and thorium. With the limited natural uranium available in certain countries, which are relying strongly on nuclear power, fast reactors are essential to meet their growing demands over a long period. This chapter provides the possibilities of a higher growth rate with fast spectrum reactors in terms of effective utilization of uranium and plutonium.

5.2 GROWTH

The current status of the nuclear power and projections indicated in Refs. 5.1–5.3 shows that there are now over 430 commercial nuclear power reactors operating in 31 countries, with over 370,000 mega watt (electrical) (MWe) of total capacity. About 70 more reactors are under construction. They provide over 11% of the world's electricity as continuous, reliable baseload power, without carbon dioxide emissions. Fifty-six countries operate about 240 civil research reactors, over one-third of these in developing countries. The total nuclear capacity is more than three times the total generating capacity of France or Germany from all sources. Sixteen countries depend on nuclear power for at least a quarter of their electricity. France gets around three-quarters of its power from nuclear energy, while Belgium, Czech Republic, Hungary, Slovakia, Sweden, Switzerland, Slovenia, and Ukraine get one-third or more. South Korea, Bulgaria, and Finland get more than 30% of their power from nuclear energy, while in the United States, United Kingdom, Spain, and Russia, it is almost one-fifth. Japan is used to relying on nuclear power for more than one-quarter of its electricity and is expected to return to that level. Among countries that do not host nuclear power plants, Italy and Denmark get almost 10% of their power from nuclear energy. About 70 further nuclear power reactors are under construction, equivalent to 20% of existing capacity, while more than 160 are firmly planned, equivalent to half of the present capacity.

The energy supply for the future has to be through cleaner means. Electricity demand is increasing at twice the rate of the overall energy use, signaling that electricity plants need higher deployment and attention. By 2035, it is likely to rise at 73%. While renewable energy sources may not be able to meet such large requirements, it would certainly suffice. The *World Energy Outlook 2012* reports that under the New Policies Scenario, coal demand increases 0.8% per year from 2010 to 2035, oil increases 0.5% pascal (pa), gas 1.6% pa, nuclear 1.9% pa, and renewables other than hydro and biomass 7.7% pa in Ref. 5.4. Nuclear power is most suitable given the urbanization witnessed worldwide leading to high-density electricity demand. The *International Atomic Energy Agency* (IAEA) in its "Annual Energy, Electricity and Nuclear Power Estimates for the Period up to 2050, September 2012 edition" [5.5] projects that nuclear capacity would increase from 370 to 456 GWe in 2030. The drivers for increased nuclear share are (1) increasing energy demand, (2) climate change, (3) security of supply, (4) economics, and (5) insurance against future fuel price rise.

Conscious of the environmental concerns, there will be nuclear renaissance in several countries. Moreover, it is expected that in the twenty-first century, the power industry will develop in the free market environment. The development of the entire power industry and nuclear power as its integral part will be influenced by the following main factors: economics, safety, radioactive waste management, nonproliferation of nuclear weapons, macroeconomic factors, restructuring of the electricity market, changing structure of power resources, and environmental protection. The growth of nuclear power will also depend on the status and maturity of the nuclear fuel cycle technologies. Parallel to the renaissance of nuclear energy, the concept of closed fuel cycle is also receiving close attention. It is now realized that reprocessing (or recycling) the spent fuel is essential for the effective utilization of resources and for reducing the environmental impact of nuclear power by reducing the requirement for waste repositories. Accordingly, several countries have renewed the emphasis on *closed fuel cycle* as a route to sustainability of the nuclear energy cycle. Many countries are expected to adopt a closed fuel cycle in the next century for effective utilization of uranium as well as partitioning and transmutation of transuranium elements rendering the fast reactor with a closed fuel cycle important.

Most of the present-day commercial power reactors are *thermal* reactors comprising about 13,000 reactor years of experience, and *fast reactors* have accumulated about 400 reactor years of experience. The world's first nuclear reactor was indeed a fast reactor, and further fast reactors had been developed more intensively during 1970–1980, and beyond 1980s, a steady decline is seen (refer to Section 1.4). This is mainly due to the low uranium price, high capital cost, and a few incidents mainly due to sodium (refer to Section 5.7). However, in recent years, fast reactors are being reexamined with renewed interest. Worldwide, initiatives are under way to develop the nuclear system, which would meet the energy requirements of the twenty-first century addressing the particular motivating factors of different regions of the world. For example, the *Generation IV International Forum (GIF)* and IAEA. International Project on Innovative Nuclear Reactors and Fuel Cycles (INPRO) are working on the energy system and its design aspects, which are suitable to meet aspirations. The factors that could be attributed for the motivation of the fast reactors and the associated objectives are

1. Effective utilization of natural uranium resources
2. Breeding, growth, and consequent energy security
3. Actinide burning arising from thermal reactors
4. Waste minimization and environmental consideration

The goals identified for Generation IV nuclear energy systems are reproduced in the following text [5.6].

5.2.1 SUSTAINABILITY

1. Generation IV nuclear energy systems will provide sustainable energy generation that meets clean air objectives and promotes long-term availability of systems and effective fuel utilization for worldwide energy production.
2. Generation IV nuclear energy systems will minimize and manage their nuclear waste and notably reduce the long-term stewardship burden in the future, thereby improving protection for the public health and the environment.

5.2.2 ECONOMICS

1. Generation IV nuclear energy systems will have a clear life-cycle cost advantage over other energy sources.
2. Generation IV nuclear energy systems will have a level of financial risk comparable to other energy projects.

5.2.3 Safety and Reliability

1. Generation IV nuclear energy systems operations will excel in safety and reliability.
2. Generation IV nuclear energy systems will have a very low likelihood and degree of reactor core damage.
3. Generation IV nuclear energy systems will eliminate the need for offsite emergency response.

5.2.4 Proliferation Resistance and Physical Protection

Generation IV nuclear energy systems will increase the assurance that they are a very unattractive and the least desirable route for diversion or theft of weapons-usable materials, and provide increased physical protection against acts of terrorism (www.gen-4.org/gif/jcms/c_9365/prpp).

Through the sustainability objective, resource conservation would be addressed, meeting the current generation requirements and without jeopardizing the future generations' needs. This would also address the allied aspects of sustainability, which are waste management and effective nuclear resource utilization. Apart from power generation, other applied areas would also be covered by these systems taking into account the overall primary energy use. Economical competitiveness is obviously dictated by the market forces. Safety is, as usual, the highest priority amidst the growing awareness on public safety. Proliferation resistance and physical protection aspects would address the concerns with respect to safe and intended usage. The major specific goals with respect to safety are (1) increased reliability, (2) low frequency of reactor core damage, and (3) elimination of the need for offsite emergency response. The fundamental criteria governing the nuclear energy system are pictorially depicted in Figure 5.1. Based on this comprehensive set of criteria, with the participation of industry, research institutions, and universities worldwide, proposals were called for and studied, yielding promising nuclear energy systems.

FIGURE 5.1 Gen IV fundamental criteria. (From Five GIF criteria Five International challenges, FR09 – Panel 2, December 2009. http://www-pub.iaea.org/mtcd/meetings/pdfplus/2009/cn176/cn176_presentations/panel_2/panel2-01.frigola.pdf.)

Enhanced sustainability is sought to be achieved mainly through adoption of a closed fuel cycle with reprocessing and recycling of plutonium, uranium, and minor actinides using fast reactors, providing significant reduction in waste generation and uranium resource requirements. The important parameters with respect to sustainability are (1) nuclear resource availability; (2) new nuclear resource exploration; (3) share of nuclear electricity; and (4) waste generation, radiotoxicity levels, and repository requirements. It is quite complex to work out a simple scenario covering all the aforementioned aspects. The parameters are linked with many other aspects, which include the reactor technology, fuel cycle strategy, and policy for waste. The once-through cycle is the most fundamental, which can be adopted as a reference for comparison. Existing uranium reserves would sustain for about 80–100 years. With new explorations, the nuclear capacity could be built up further but that would leave a huge repository of unspent fuel, leading to massive repository space requirements, which in turn proves to be expensive. The waste mass that in turn decides the repository space is shown in Figure 5.2a. The need for large uranium resources if the once-through cycle is adopted and the scenario of fast reactors with closed fuel cycle are depicted in Figure 5.2b. Adoption of closed fuel cycle would become essential due to the aforementioned counts. However, this would involve waste management strategies including transmutation and actinide separation and burning. This option of separation is sometimes viewed as a risk in the proliferation point of view. This leads to the consideration of appropriate closed fuel cycle technologies wherein the objectives of a closed fuel cycle would be achieved without resorting to separation. The aforementioned factors are kept in mind while devising the fuel cycle technologies. The Gen IV fuel cycle group defined four classes of fuel cycle in general: (1) the once-through fuel cycle, (2) a fuel cycle with partial plutonium recycle, (3) a fuel cycle with full plutonium recycle, and (4) a fuel cycle with full transuranic element recycle. Thus, the Gen IV fuel cycle concepts, which are under development, are in symbiosis with the existing light water reactor (LWR)

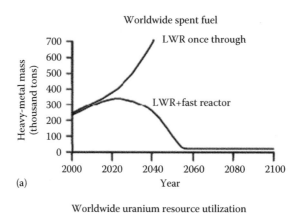

(a)

(b)

FIGURE 5.2 Fuel cycles with sustainability (a) waste mass and (b) U resource need. (From A technology roadmap for generation IV nuclear energy systems, GIF-002-00, December 2002.)

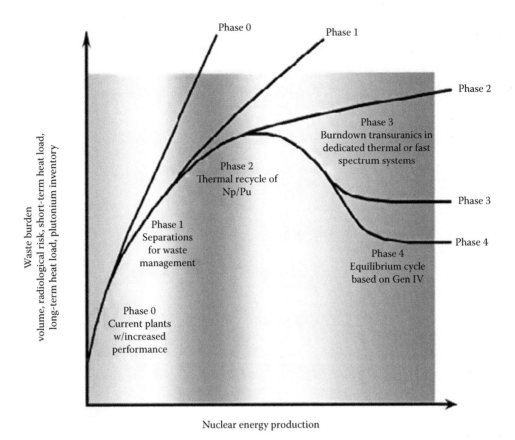

FIGURE 5.3 Waste burden reduction. (From Future Nuclear Energy Systems: Generation IV, 50th Annual Meeting of the Health Physics Society, Spokane, Washington, USA, July 2005.)

fleet in a flexible manner with economic competitiveness built in. The waste burden is sought to be reduced gradually in phases, as shown in Figure 5.3.

As an example, the approach adopted by French for the closed fuel cycle of fast reactors is shown in Figure 5.4 [5.7]. From the scenario of recycling only the U and Pu, homogeneous recycling is proposed for Gen IV systems against the heterogeneous recycle, which helps in achieving the major objectives of resource saving and waste minimization coupled with the nonproliferation objectives. With respect to economy, reduction of investment cost is proposed to be achieved by system simplification and compactness and optimization of in-service inspection (ISI), maintenance, and repair. In the safety domain, the risk of sodium water reaction is planned to be decreased or eliminated by optimizing the power conversion systems by opting for optimized steam generator and/or nitrogen/helium/supercritical CO_2 gas turbine. Large energy release due to severe accidents would be practically excluded by reduced sodium void effect, optimum core design, enhanced Doppler effect with a dense and high thermal conductivity fuel, and increased adoption of passive features in safety systems. The development of key technologies is under way currently.

There could be other possibilities of uranium resource usage and recycle: (1) use of spent pressurised water reactor (PWR) fuel in PHWRs, (2) use of enriched uranium fuel in fast reactors against the traditional mixed oxide (MOX), (3) use of depleted UO_2 in fast reactors as blankets, and (4) reprocessing of fast reactor blanket UO_2 in thermal reprocessing plants. The possibility is to be explored in line with the reactor fleet and the program pursued by individual countries.

While deciding the future nuclear system design, though the safety goals have been defined prior to the Fukushima accident, a reassessment and strengthening of the systems based on the

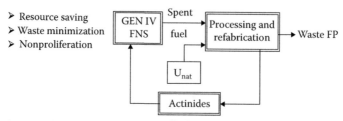

➤ Resource saving
➤ Waste minimization
➤ Nonproliferation

➤ Develop international nonproliferation standards to allow for diverse fuel cycle processes
➤ Keep all options open as they could be deployed in sequence

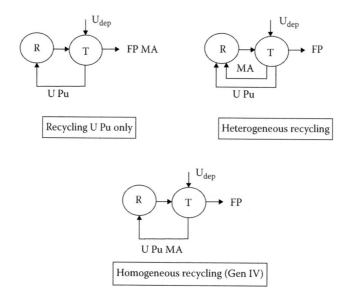

FIGURE 5.4 Strategy in France with respect to closed fuel cycle of fast reactors. (From Renault, C., ENEN, 4th generation nuclear reactor systems for the future, saclay, *The Generation IV International Forum and Fourth Generation Nuclear Systems*, Nuclear Energy Division, CEA, Paris, France, 2007. xa.yimg.com/kq/groups/16979795/889597805/name/01_renault.pdf.)

lessons learned are planned by various countries. In the wake of the 2011 accident, several utilities conducted stress tests of the plants and demonstrated the safety margin in the design of existing and upcoming plants. Some system designs have been revisited due to safety concerns. The major areas of safety that are relooked and are being strengthened are with respect to margins against beyond-design-basis earthquakes, flooding of plant systems, loss of heat sink, loss of offsite electrical power especially for a prolonged period, and severe accident management. In this regard, future fast reactor designs, which are currently under development, incorporate adequate design measures to demonstrate safety against extreme natural events. Current sodium-cooled fast reactors have adequate cooling capability within the reactor vessel owing to its inherent potential. The coolability aspect of the spent nuclear fuel has been given the highest attention. Barring these differences arising from the specific characteristics of sodium fast reactors, other measures have been implemented in thermal reactors such as extended offsite emergency power provision; guards against external flooding are also thought of in fast reactors. Besides these, the sodium chemical accident potential in sodium fast reactors is also reassessed based on which the design measures are being worked out. Alternative coolant materials such as lead are also actively considered, and currently, few lead-cooled nuclear reactor systems are under development such as ALFRED, MYRRHA (Belgium), BREST-OD-300 (Russia), ELFR (Europe),

ELECTRA (Sweden), and G4M (United States) [5.9]. Another measure to prevent severe accident is through achieving either a negative or a small positive sodium void coefficient by evolutionary design of the reactor core.

5.3 PERFORMANCE AND FUEL CONSUMPTION ASPECTS

Higher and effective utilization of natural uranium resource up to 60%–70% is possible only in the fast breeder reactor (FBR) system with multiple recycling, as explained in the following text in comparison with PHWR using natural uranium. In PHWR, a maximum of 0.7% of the uranium resource is used. In a closed fuel cycle through FBRs, the fuel can be recycled any number of times. But considering the fissile material quality, reprocessing loss, and fertile feed for every recycle, around 10 recycles are realizable. In every cycle, about 7 at% of heavy elements are burnt corresponding to a reasonable average burnup of 70,000 mega watt day per ton (MWd/ton) (~7 at%). Assuming 7 recycles are possible out of 10, nearly 49% of fissile atoms can be burnt in FBR, which gives a ratio of about 70 times utilization factor. This means that 1 kg of natural uranium would generate about 3,645,600 kWh in FBR, compared to only about 52,080 kWh possible in PHWR, assuming an efficiency of 31%. It is worth mentioning that with advanced fuel with high burnup (peak burnup 200,000 MWd/ton) and fuel cycle losses of 1%, it is possible to realize an even higher utilization factor. Further, with higher in-core breeding, the fuel residence time (cycle length) can be increased, and hence, high burnup is possible. High burnup results in better fuel utilization due to the reduction of fissile material loss during reprocessing.

World uranium reserves are classified into two broad categories, namely, known conventional and identified resources and undiscovered conventional resources. The known reserves category consists of the reasonably assured reserves (RARs) and inferred reserves (IRs). The undiscovered category consists of the estimated additional resources (EAR) category and speculative resources (SR). The total identified resources (reasonably assured and inferred) as of January 1, 2009, is 5.4 million tons falling under the USD 130/kgU price tag. The total undiscovered resources are reported to be more than 10.4 million tons, thus totaling about 16 million tons [5.9–5.12]. The current usage is about 68,000 tonsU/year. At this rate of consumption, the current identified resources themselves would supply for about 80 years in open cycle mode for the thermal reactors and the total resources would last for about 240 years. If the resources are deployed in close cycle mode and in fast reactors, the period becomes multifold. Looking at the energy growth projected, there is likely to be a severe constraint on fossil fuels, which would eventually lead to large-scale deployment of nuclear energy on a long-term basis. To cater to such demands, adoption of water reactors with the open cycle mode will not be the right solution and fertile isotopes have to be exploited, for which fast reactors are ideal. With fast neutron reactors, the only reserves, U-238 currently stored as a tail-end product from enrichment plants, allow for increased energy reserves up to a factor of about 50 in comparison to the current water reactor technology. Therefore, fast reactors are the ideal vehicles for efficient use of uranium resources. A more detailed treatment about fuel utilization and fuel consumption with respect to different types of reactors is given in Chapter 3.

REFERENCES

5.1 World Nuclear Organisation. Information Libriary. http://www.world-nuclear.org/info/Current-and-Future-Generation/World-Energy-Needs-and-Nuclear-Power/, (accessed September 2013).

5.2 World Nuclear Organisation. Information Libriary. http://www.world-nuclear.org/info/Current-and-Future-Generation/The-Nuclear-Renaissance\ , (accessed August 2011).

5.3 World Nuclear Organisation. Nuclear Power in the World Today. http://www.world-nuclear.org/info/Current-and-Future-Generation/Nuclear-Power-in-the-World-Today/, (accessed April 2014).

5.4 OECD/IEA, (2012). *World Energy Outlook 2012*. International Energy Agency, Paris, France. http://www.iea.org/publications/freepublications/publication/WEO2012_free.pdf

5.5   IAEA. (2013). Energy, electricity and nuclear power estimates for the period up to 2050. Reference data series. IAEA, Vienna, Austria.

5.6   U.S. DOE Nuclear Energy Research Advisory Committee and the Generation IV International Forum. (2002). A technology roadmap for generation IV nuclear energy systems. GIF-002-00, December 2002. http://www.gen-4.org/PDFs/GIF_Overview.pdf

5.7   Renault, C. (2007). ENEN, 4th generation nuclear reactor systems for the future, saclay. *The Generation IV International Forum and Fourth Generation Nuclear Systems*, Nuclear Energy Division, CEA, Paris, France.

5.8   Department of Nuclear Energy. (2013). Status of innovative fast reactor designs and concepts. A supplement to the IAEA Advanced Reactors Information System (ARIS). *Nuclear Power Technology Development Section Division of Nuclear Power*, IAEA, Vienna, Austria.

5.9   Red Book. Uranium: Resources, production and demand, RAF3007, *Workshop on Uranium Data Collection & Reporting*, July 2010, Ghana, http://www.iaea.org/OurWork/ST/NE/NEFW/documents/RawMaterials/RTC-Ghana-2010/5.RedBook.pdf.

5.10 Price, R. and Blaise, J.R. (2002). Nuclear fuel resources: Enough to last? NEA updates, NEA News 2002 No. 20.2, http://www.oecd-nea.org/nea-news/2002/20-2-Nuclear_fuel_resources.pdf.

5.11 World Nuclear Organisation. Supply of uranium, http://www.world-nuclear.org/info/Nuclear-Fuel-Cycle/Uranium-Resources/Supply-of-Uranium/, (accessed October 2014).

5.12 OECD. (2010). A Joint Report by the OECD Nuclear Energy Agency and the International Atomic Energy Agency. Uranium 2009 resources, production and demand. Organisation for economic Co-Operation and Development, NEA No. 6891.

5.13 Kevan D.W. (2005). Future Nuclear Energy Systems: Generation IV, 50th Annual Meeting of the Health Physics Society, Spokane, Washington, USA, July 11, 2005.

5.14 Frigola, P. (2009). Five GIF criteria Five International challenges, FR09 – Panel 2. http://www-pub.iaea.org/mtcd/meetings/pdfplus/2009/cn176/cn176_presentations/panel_2/panel2-01.frigola.pdf, December 2009.

6 Prospect of Various Types of FSRs

6.1 INTRODUCTION

Many reactor types were considered initially; however, the list was downsized to focus on the most promising technologies and those that could most likely meet the goals of the Gen IV initiative. The following are the typical fast reactors:

- Sodium-cooled fast reactor (SFR)
- Lead-cooled fast reactor (LFR)
- Molten salt reactor (MSR) (epithermal)
- Gas-cooled fast reactor (GFR)

Relative to the current nuclear power plant technology, the claimed benefits of these reactors include

- Nuclear waste that remains radioactive for a few centuries instead of millennia
- 100–300 times more energy yield from the same amount of nuclear fuel
- The ability to consume existing nuclear waste during the production of electricity
- Improved safety upon operation

6.2 SODIUM-COOLED FAST REACTORS

The SFR uses liquid sodium as the reactor coolant, allowing high power density with low coolant volume. SFR is built on some 390 reactor-years experience with sodium-cooled fast neutron reactors over five decades and in eight countries. Most plants so far have had a core and a blanket configuration, but new designs are likely to have all the neutron action in the core. Other research and development (R&D) is focused on safety in loss of coolant scenarios and improved fuel handling. The SFR utilizes depleted uranium (DU) as the fuel matrix and has a coolant temperature of 500°C–550°C, enabling electricity generation via a secondary sodium circuit, the primary one being at near-atmospheric pressure. Figure 6.1 shows the schematic sketch of an SFR.

Favorable Points	Unfavorable Points
High thermal inertia	Positive void coefficient
Good temperature margin to boiling	Criticality issues during core meltdown
Unpressurized reactor vessel	Violent Na reactions with water and air
Long research and operation experience	Reaction of liquid sodium with mixed oxide (MOX) fuel

6.3 LEAD-COOLED FAST REACTORS

The LFR is a flexible fast neutron reactor that can use DU or thorium fuel matrices and burn actinides from light water reactor (LWR) fuel. Liquid metal (Pb or Pb–Bi eutectic) cooling is at low pressure by natural convection (at least for decay heat removal [DHR]). Fuel is metal or nitride, with full actinide recycle from regional or central reprocessing plants. A wide range of unit sizes

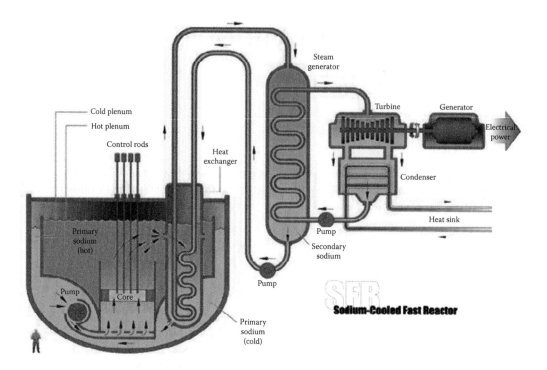

FIGURE 6.1 Sodium-cooled fast reactor. (From IAEA, Status of innovative fast reactor designs and concepts, a supplement to the IAEA advanced reactors information system [ARIS], Nuclear Power Technology Development Section, Division of Nuclear Power, Department of Nuclear Energy, [TAREF] formed under NEA, Vienna, 2013.)

is envisaged, from factory-built *battery* with 15–20-year battery life for small grids or in developing countries to modular 300–400 mega watt (electrical) (MWe) units and large single plants of 1400 MWe. An operating temperature of 550°C is readily achievable, but 800°C is envisaged with advanced materials to provide lead corrosion resistance at high temperatures, enabling thermo-chemical hydrogen production. Figure 6.2 shows the schematic sketch of the LFR.

Favorable Points	Unfavorable Points
High thermal inertia	High melting temperature
High boiling temperature	Problematic cleaning and decontamination
Unpressurized reactor vessel	Corrosive and toxic
Good passive safety	Activation of coolant
Compatible with water and air	

6.4 MOLTEN SALT REACTORS

In an MSR, the uranium fuel is dissolved in the sodium fluoride salt coolant, which circulates through graphite core channels to achieve some moderation and an epithermal neutron spectrum. The reference plant is up to 1000 MWe. Fission products are removed continuously and the actinides are fully recycled, while plutonium and other actinides can be added along with U-238, without the need for fuel fabrication. The feed and removal of fuel are accomplished through a collection and injection system in such a way that the critical mass is maintained in the reactor core. The coolant temperature is 700°C at very low pressure, with 800°C envisaged. A secondary coolant system is used for electricity generation, and thermochemical hydrogen production is also feasible.

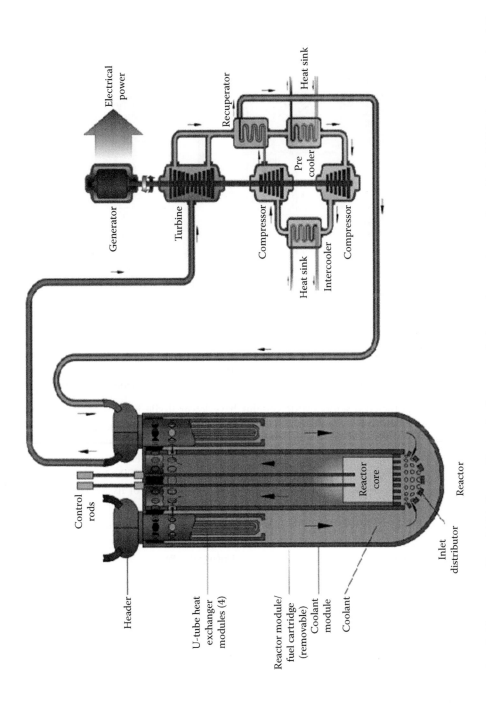

FIGURE 6.2 Lead-cooled fast reactor. (From IAEA, Status of innovative fast reactor designs and concepts, a supplement to the IAEA advanced reactors information system [ARIS], Nuclear Power Technology Development Section, Division of Nuclear Power, Department of Nuclear Energy, [TAREF] formed under NEA, Vienna, 2013.)

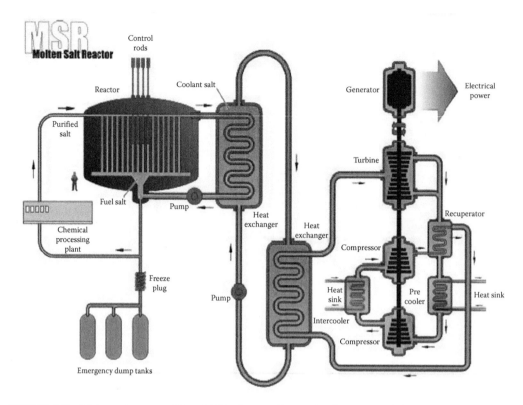

FIGURE 6.3 Molten salt reactor. (From IAEA, Status of innovative fast reactor designs and concepts, a supplement to the IAEA advanced reactors information system [ARIS], Nuclear Power Technology Development Section, Division of Nuclear Power, Department of Nuclear Energy, [TAREF] formed under NEA, Vienna, 2013.)

Compared with solid-fueled reactors, MSR systems have lower fissile inventories, no radiation damage constraint on fuel burnup, no spent nuclear fuel, no requirement to fabricate and handle solid fuel, and a homogeneous isotopic composition of fuel in the reactor. These and other characteristics may enable MSRs to have unique capabilities and competitive economics for actinide burning and extending fuel resources. Figure 6.3 shows the schematic sketch of the MSR. Attractive features of the MSR fuel cycle include the following: high-level waste comprising of fission products only, hence shorter-lived radioactivity; small inventory of weapons-grade fissile material (Pu-240 being the dominant Pu isotope); low fuel consumption; and safety due to passive cooling up to any size.

Favorable Points	Unfavorable Points
No risk of core melting	Complicated neutronics: fuel in entire primary circuit.
Reactor control fuel flow	Irradiation of heat exchanger.
Online removal of fission products	Salts are corrosive.
Thorium usage	High salt melting temperature.
Limited waste production	Complicated reactor start-up.
Simplified reprocessing	No experience with reactors.

6.5 GAS-COOLED FAST REACTORS

Like other helium-cooled reactors that have operated or are under development, GFRs will be high-temperature units—850°C. They employ a similar reactor technology to the very high temperature reactor (VHTR), which is suitable for power generation, thermochemical hydrogen production, or other process heat. To generate electricity, helium will directly drive a gas turbine (Brayton cycle). It has a self-generating (breeding) core with fast neutron spectrum and no fertile blanket. Robust nitride or carbide fuels would include depleted uranuim (DU) and any other fissile or fertile materials as ceramic pins or plates, with a plutonium content of 15%–20%. As with the SFR, used fuel would be reprocessed on site and all the actinides recycled repeatedly to minimize production of long-lived radioactive wastes. An alternative GFR design has lower-temperature (600°C–650°C) helium cooling in a primary circuit and supercritical CO_2 at 550°C and 20 mega pascals (MPa) in a secondary system for power generation, reducing the metallurgical and fuel challenges associated with very high temperatures. Figure 6.4 shows the schematic sketch of the GFR.

Favorable Points	Unfavorable Points
Thermal negative feedback	Low thermal inertia
Resistant fuel barrier (ceramic fuel)	High-pressure system
Inert coolant	Complicated fuel and cladding
	High coolant flow (vibrations)
	No operating experience
	DHR complications

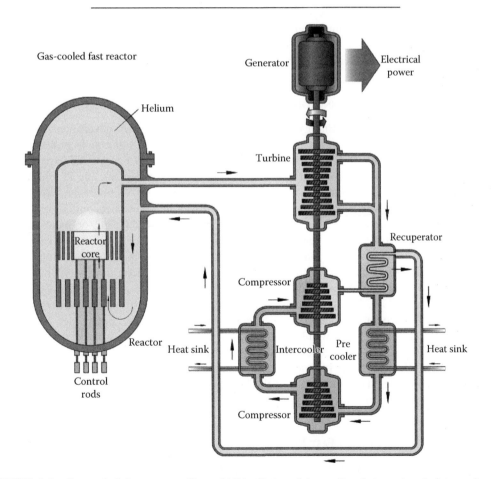

FIGURE 6.4 Gas-cooled fast reactor. (From IAEA, Status of innovative fast reactor designs and concepts, a supplement to the IAEA advanced reactors information system [ARIS], Nuclear Power Technology Development Section, Division of Nuclear Power, Department of Nuclear Energy, [TAREF] formed under NEA, Vienna, 2013.)

6.6 COMPARISON OF ADVANCED FAST REACTORS WITH SFRs

Reactor Type	Coolant	Temperature (°C)	Pressure	Uses
SFR	Sodium	550	Low	Electricity
LFR	Lead or Pb–Bi	480–800	Low	Electricity and hydrogen
MSR	Fluoride salts	700–800	Low	Electricity and hydrogen
GFR	Helium	850	High	Electricity and hydrogen

6.7 FAST REACTORS THAT EVOLVED POST-FUKUSHIMA

In the wake of the 2011 earthquake and tsunami that devastated coastal regions of northern Japan and the Fukushima Daiichi nuclear power plant, future reactors have the following additional features [6.1]:

- Passive safety features for shutdown and DHR systems
- Long-term coolability of core (diversified systems)
- Pessimistic load combinations
- Inclusion of low-probability severe accidents
- Robust severe accident management strategy

With these concepts, the following reactor designs have been developed.

6.7.1 ADVANCED SODIUM TECHNOLOGICAL REACTOR FOR INDUSTRIAL DEMONSTRATION

In mid-2006, the French Atomic Energy Commission was appointed by the government to develop two types of fast neutron reactors that are essentially Generation IV designs: an improved version of the sodium-cooled type (SFR), which already has 45 reactor-years operational experience in France, and an innovative gas-cooled type. Both would have fuel recycling, and in mid-2009, it was recommended that the sodium-cooled model, Advanced Sodium Technological Reactor for Industrial Demonstration (ASTRID), should be a high priority in R&D on account of its actinide burning potential (Figure 6.5) [6.2]. ASTRID is envisaged as a 600 MWe prototype of the commercial series of 1500 MWe SFR reactors, which is likely to be deployed from about 2050 to utilize the half million tons of DU that France will have by then and burn the plutonium in used MOX fuel. ASTRID will have high fuel burnup, including minor actinides (MAs) in the fuel elements, and while the MOX fuel will be broadly similar to that in pressurised water reactor (PWRs), it will have 25%–35% plutonium. It will use an intermediate sodium coolant loop, but whether the tertiary coolant is water/steam or gas is an open question. Four independent heat exchanger loops are likely, and it will be designed to reduce the probability and consequences of severe accidents to an extent that is not now done with FSRs. ASTRID is called a *self-generating* fast reactor rather than a breeder in order to demonstrate low net plutonium production. ASTRID is designed to meet the stringent criteria of the Gen IV International Forum in terms of safety, economy, and proliferation resistance.

6.7.2 JAPAN SODIUM-COOLED FAST REACTOR (FIGURE 6.6) [6.3]

The *Japan Atomic Energy Agency* is now implementing a conceptual design study for the Japan sodium-cooled fast reactor (JSFR) and R&D on innovative technologies in the *Fast Reactor Cycle Technology Development* (*FaCT*) project. The plant design aims to achieve economic competitiveness, enhanced safety, and reliability, which are crucial targets for the commercialization of

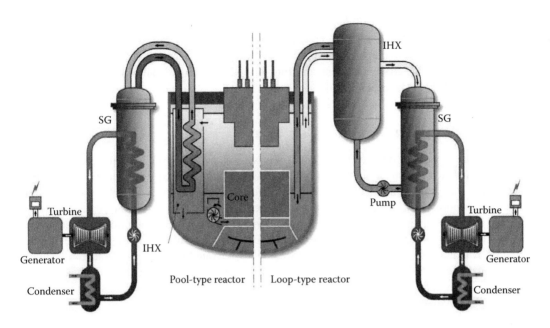

FIGURE 6.5 ASTRID reactor (600 MWe—French reactor). (From IAEA, Status of innovative fast reactor designs and concepts, a supplement to the IAEA advanced reactors information system [ARIS], Nuclear Power Technology Development Section, Division of Nuclear Power, Department of Nuclear Energy, [TAREF] formed under NEA, Vienna, 2013.)

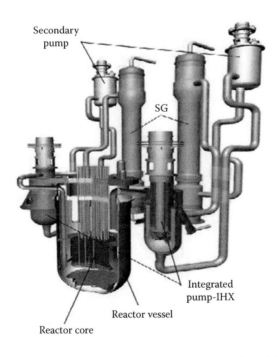

FIGURE 6.6 Japan sodium-cooled fast reactor (1500 MWe—Japan). (From IAEA, Status of innovative fast reactor designs and concepts, a supplement to the IAEA advanced reactors information system [ARIS], Nuclear Power Technology Development Section, Division of Nuclear Power, Department of Nuclear Energy, [TAREF] formed under NEA, Vienna, 2013.)

fast reactors. In the FaCT project, commercialization is aimed at around 2050 for an SFR with a power output of 1500 MWe accompanied by innovative technologies. With this process, the demonstration reactor with a power output assumed to range from 500 to 750 MWe is planned to start operation at around 2025. The conceptual design results for both the commercialized and demonstration reactors are scheduled to be presented in 2015.

6.7.3 ADVANCED FAST REACTOR (FIGURE 6.7) [6.4]

The advanced fast reactor (AFR-100) core concept was developed targeting a small electrical grid to be transportable to the plant site and operable for a long time without frequent refueling. The reactor power rating was strategically decided to be 100 MWe, and the core barrel diameter was limited to 3.0 m for transportability. The design parameters were determined by relaxing the peak fast fluence limit and bulk coolant outlet temperature to beyond irradiation experience assuming that advanced cladding and structural materials developed under U.S. Department of Energy (DOE) programs would be available when the AFR-100 is deployed. With a derated power density and U–Zr binary metallic fuel, the AFR-100 can maintain criticality for 30 years without refueling. Evaluated reactivity coefficients provide sufficient negative feedbacks, and the reactivity control systems provide sufficient shutdown margins.

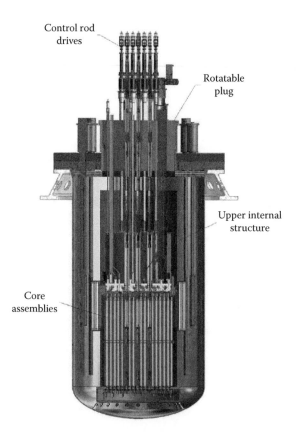

FIGURE 6.7 Advanced sodium-cooled fast reactor (100 MWe—United States). (From IAEA, Status of innovative fast reactor designs and concepts, a supplement to the IAEA advanced reactors information system [ARIS], Nuclear Power Technology Development Section, Division of Nuclear Power, Department of Nuclear Energy, [TAREF] formed under NEA, Vienna, 2013.)

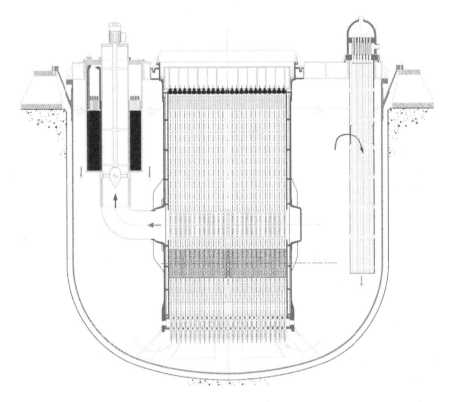

FIGURE 6.8 European lead-cooled fast reactor (ELFR) (600 MWe—Europe). (From IAEA, Status of innovative fast reactor designs and concepts, a supplement to the IAEA advanced reactors information system [ARIS], Nuclear Power Technology Development Section, Division of Nuclear Power, Department of Nuclear Energy, [TAREF] formed under NEA, Vienna, 2013.)

6.7.4 European Lead-Cooled Fast Reactor (Figure 6.8) [6.5]

The European lead-cooled system (ELSY) reference design is a 600 MWe pool-type reactor cooled by pure lead. The ELSY project demonstrates the possibility of designing a competitive and safe fast critical reactor using simple engineered technical features while fully complying with the Generation IV goal of sustainability and MA burning capability. Sustainability was a leading criterion for option selection of core design, focusing on the demonstration of the potential to be self-sustaining in plutonium and to burn its own generated MAs. Safety has been one of the major focuses all over the ELSY development. In addition to the inherent safety advantages of the lead coolant (high boiling point and no exothermic reactions with air or water), a high safety grade of the overall system has been reached. In fact, the overall primary system has been conceived in order to minimize pressure drops and, as a consequence, to allow DHR by natural circulation. Moreover, two redundant, diverse, and passive operated DHR systems have been developed and adopted.

REFERENCES

6.1 Takeda, T., Shimazu, Y., Foad, B., Yamaguchi, K. (2012). Review of safety improvement on sodium-cooled fast reactors after Fukushima accident. *Nat. Sci.*, 4, 929–935. doi:10.4236/ns.2012.431121.

6.2 Beils, S., Carluec, B., Devictor, N., Luigi Fiorini, G., François Sauvage, J. (2011). Safety approach and R&D program for future French sodium-cooled fast reactors. *J. Nucl. Sci. Technol.*, 48(4), 510–515.

6.3 Ohiki, S. (2012). Conceptual core design study for Japan sodium-cooled fast reactor: Review of sodium void reactivity worth evaluation. *Technical Meeting on Innovative Fast Reactor Designs with Enhanced Negative Reactivity Feedback Features, IAEA TWG-FR*, Vienna, Austria.

6.4 Kim, T.K., Grandy, C., Hill, R.N. (2012). A 100 MWe advanced sodium-cooled fast reactor core concept. *PHYSOR 2012: Conference on Advances in Reactor Physics—Linking Research, Industry and Education*, Knoxville, TN.

6.5 Alemberti, A. (2012). ELFR: The European lead fast reactor design, safety approach and safety characteristics. *Technical Meeting on Impact of Fukushima Event on Current and Future Fast Reactor Designs*, HZDR, Dresden, Germany.

6.6 IAEA. (2013). Status of innovative fast reactor designs and concepts: A supplement to the IAEA advanced reactors information system (ARIS). Nuclear Power Technology Development Section, Division of Nuclear Power, Department of Nuclear Energy, Vienna, Austria.

Section II

Design of Sodium Fast Reactors

7 Choice of Materials and Their Performance

7.1 INTRODUCTION

Fast breeder reactor (FBR) technology involves a multidisciplinary approach to solve the various challenges in the areas of fuel and materials development. Fuels for FBRs have significantly higher concentration of fissile material than in thermal reactors, with a matching increase in burn-up. The design of the fuel is an important aspect that has to be optimized for efficient, economic, and safe production of power. FBR components operate under hostile and demanding environment of high neutron flux, liquid sodium coolant, and elevated temperatures. Resistance to void swelling, irradiation creep, and irradiation embrittlement are therefore major considerations in the choice of materials for the core components. Structural and steam generator materials should have good resistance to creep, low-cycle fatigue, creep–fatigue interaction, and sodium corrosion.

7.2 FUEL

The reactor core houses the fuel. Fast reactor fuels, in particular, have the capability of performing to high burn-ups, that is, 100–200 GWd/ton. This is three times higher than that of typical LWR. This high burn-up leads to higher fission product generation and in turn larger fuel swelling and larger fission gas release. Toward achieving high compactness, the fuel pin diameters are chosen to be smaller compared with LWR fuels, and hence, the fuel must be capable of sustaining higher temperature gradient as well as specific powers (factor of 4 higher than LWR). The following are the desirable features of an ideal fuel for fast reactors:

- High specific power: Thermal conductivity and melting point of fuel should be higher.
- High burn-up: High resistance to radiation damage.
- High fuel atom density: To reduce the core dimensions and fuel inventory.
- Good compatibility with cladding and coolant.
- Higher negative prompt Doppler coefficient: To provide inherent safety feature for the core.
- No metallurgical phase change below melting point that could exhibit significant changes in thermo-physical-chemical properties.
- High breeding ratio and uranium utilization: Fuel should have higher neutron yield.
- Ease of fabrication.

Some features may contradict each other. For example, the metals have high conductivity but low melting point; high fuel atom density but harder neutron energy spectrum; and hence, less negative Doppler coefficient.

7.2.1 OXIDE FUEL

Oxide fuels gained importance in FBRs because of their higher burn-up potential and vast irradiation experience as driver fuels in thermal reactors. Oxide fuels have demonstrated very satisfactory dimensional and radiation stability, as well as chemical compatibility with cladding and coolant materials in LWR applications. However, the environment in an FBR, that is, higher temperatures and larger exposure, could alter the satisfactory performance. For example, the oxygen-to-metal ratio (O/M) is maintained at marginally above 2.0 for LWR oxide fuel; but in FBRs in order to avoid oxidation of clad, the O/M ratio is maintained hypo-stoichiometric. The atom density of oxide fuels is low, which dictates the breeding ratio to be less when compared with metal fuels. Also, the oxide fuels have very low thermal conductivity because of which high temperature gradient exists in the fuel. The positive point with oxide fuel is that its melting point is very high, which enables it to operate at higher linear power in spite of its low thermal conductivity.

The low thermal conductivity gives rise to large radial thermal gradients in fuel pellets, often of the order of 2000–4000°C/cm. This gradient causes the fuel microstructure to change quite rapidly after the reactor power has been raised to its operating level. The thermal conductivity of fuel is a function of temperature, density, O/M ratio, and Pu fraction in the fuel and the local morphology of the fuel matrix.

7.2.1.1 Physical Properties

The thermal conductivity values for MOX fuel with typical composition of 20% Pu with O/M ratio of 1.97 and density of 90% as a function of temperature are given in Table 7.1. Similarly, values for specific heat and thermal expansion are also given in Table 7.1. The solidus temperature of MOX fuel is dependent on Pu fraction, and with burn-up the solidus temperature comes down. The melting point correlation is as follows:

$$T_{m, nominal} = 3120 - 655.3 \times P + 336.4 \times P^2 - 99.9 \times P^3 \text{ (K)}, \qquad (7.1)$$

where P is the mole fraction of PuO_2 in mixed oxide.

With burn-up,

$$T_m = T_{m, nominal} - 0.5 \times \text{burn-up (MWd/kgHM)}.$$

7.2.1.2 Swelling

Fuel swelling happens because of solid and gaseous fission products generated in the fission process. The solid fission products get dispersed in the fuel and cause the fuel to swell, which is in the range of 0.15%–0.45% per at% burn-up. Greater fuel swelling results from the gaseous fission products that are insoluble in the fuel and form bubbles that migrate and coalesce, giving rise to fuel swelling, and they migrate from the bulk of the fuel to the central void and the fuel-clad gap and the plenum. Thus, the net fuel swelling is derived from the balance between fission gas retention versus release that depends

TABLE 7.1

Physical Properties of MOX Fuel as a Function of Temperature

Temperature (K)	Thermal Conductivity (W/m K)	Specific Heat (J/kg K)	Thermal Expansion Coefficient (10^{-6} m/m K)
373	5.283	283	10.904
673	3.879	313.2	10.99
973	3.065	333.9	11.507
1273	2.543	366.3	11.701
1573	2.231	431.2	12.134
1873	2.13	550.9	12.803
2173	2.24	749.1	13.706
2473	2.531	1050.9	14.84
2698	2.835	1361.3	15.84

Source: Carbajo, J.J. et al., *Nucl. Mater.*, 299, 181, 2001.

on grain structure, porosity distribution, temperature, and temperature gradient. To accommodate this swelling, the fuel pellets are fabricated with built-in porosities with which the pellet density will be around 90%. The radial gap between fuel and clad also accommodates swelling. Oxide fuel releases most of the gaseous fission products because of greater mobility at high operating temperature, and hence the swelling rate is low (1.2%–1.7% volume increase per at% burn-up). Hence, to achieve high burn-up of 10 at%, 15%–20% void volume is required in the fuel pin. Thus, the smear density of oxide fuels is in the range of 80%–85% by which larger burn-up can be achieved.

7.2.1.3 Fission Gas Release

Fission gas is produced during the fission process either directly from fission or from decay of the fission products. For a fast reactor, about 0.27 is fission yield in the form of stable fission gas, that is, for every fission process that generates fission products out of which 27% yield as stable fission gases. The fission gas generation as a function of burn-up is shown in Figure 7.1. Xenon constitutes the maximum portion of the fission gas followed by krypton. The process of nucleation of fission gas bubble, its growth, and diffusion is very complex. The fission gas thus generated is either retained in the fuel matrix or is released to the plenum through cracks. The portion retained in the matrix causes swelling of the fuel, and the escaped fission gas pressurizes the plenum.

Swelling and release are complementary phenomena, and their rates are closely related to local temperatures. In the hot center region of the fuel pin, most of the fission gases are released once they are generated, and the swelling is small. Near the cold periphery region, swelling is higher because the mobility of gas atoms is very low and the released ratio is low. It also needs to be mentioned that the steep temperature gradient in a fuel rod provides a powerful driving force for the gas bubble migration and eventually changes the fuel structure. Generally, up to 1300 K, there is no significant fission gas release thereafter, it increases, and especially above 1900 K, significant portions are released to fission gas plenum. So, most of the fission gas produced in the periphery of the pellet is retained due to its low temperature, which is a concern in the case of fuel transients that may lead to sudden release of the fission gas. The fission gas release in Phenix fuel pin as a function of burn-up is shown in Figure 7.2.

7.2.1.4 Fuel Clad Compatibility

The occurrence of fuel clad chemical interaction (FCCI) depends on the oxygen potential of the fuel and cladding inner surface temperature. Oxygen potential of the MOX fuel is a function of Pu content, O/M ratio, and burn-up. As burn-up increases, O/M ratio and oxygen potential increase. It has been reported that the higher the initial O/M ratio at a given burn-up, the lower is the temperature at which FCCI occurred. O/M ratio can also affect migration of actinides.

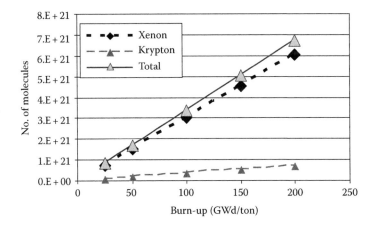

FIGURE 7.1 Fission gas generated as a function of burn-up.

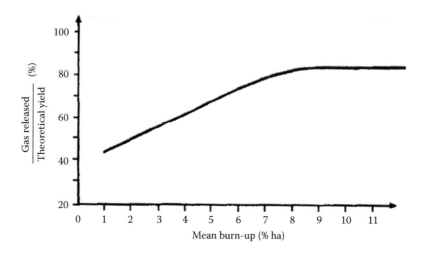

FIGURE 7.2 Dependence on burn-up of the fission gas release rate in unstrained pins.

7.2.1.5 Fuel-Coolant Compatibility

In the case of clad breach, the reaction of sodium with fuel is a concern. Sodium is likely to react with MOX fuel forming compounds of the type sodium urano-plutonates that have low density. This reaction is more probable with hyperstoichiometric fuels.

7.2.1.6 Performance at Higher Burn-Up

The major issues pertaining to the use of oxide fuels for fast reactors are the increase in O/M ratio with increasing burn-up and oxidation of the clad and corrosion due to the fission products. Lower initial O/M ratio, pin designs with adequate plenum to accommodate higher fission gas release, annular pellets for reduced fuel temperatures, and suitable cladding thickness to compensate the corrosion are the critical features required for achieving a high burn-up in oxide fuels.

Behavior of fast reactor mixed oxide fuels can be categorized into three stages. During the beginning of life (3–4 at% burn-up), owing to the poor fuel-cladding heat transfer characteristics and low thermal conductivity, high temperatures and high radial thermal gradients are observed. The fuel microstructure and geometry evolve significantly with an increase in burn-up. These include pore migration toward the center of the pellet, redistribution of oxygen in the fuel, radial redistribution of plutonium by an evaporation–condensation mechanism, closure of the fabrication gap, creation of a central hole for the initially solid pellet, etc. At moderate burn-ups (7–8 at%), the residual gap of a few micrometers is filled gradually with fission gases that pollute the bond gas helium, decreasing its thermal conductivity, leading to an increase in the central temperature. When fuel reaches a burn-up of 7–8 at%, the gas present in this gap is replaced by a mixture of compounds formed from fission products such as molybdenum and cesium, resulting in initiation of FCCI. At a higher burn-up value (12 at%), large volumes of fission gases are generated and oxide fuels release most of the gases to the plenum generating stress on the cladding. Oxidation and internal cladding corrosion due to fission products can reach significant depths of 100–150 μm. Mechanisms relating to the corrosion of clad are created by the combination of cesium and tellurium fission products.

7.2.2 Metal Fuel

The first FBR constructed used metal fuel composed of uranium, plutonium, or their alloys. The principal advantages with the metal fuels are high thermal conductivity, high density, and good breeding ratios. The main disadvantage with the metal fuel is its low melting point and high swelling that limit the peak burn-up. By decreasing the smear density, higher burn-ups can be achieved;

but the advantage of high density of the fuel is sacrificed. High radial gaps limit the peak linear heat rating of the fuel due to high-temperature drop in the gap. To decrease the gap drop, generally the gap between fuel and clad is filled with sodium.

7.2.2.1 Physical Properties

The physical properties of metal fuel as a function of temperature are given in Table 7.2.

7.2.2.2 Fission Gas Release

Krypton and xenon are the gaseous fission atoms formed during fission reaction. They are completely insoluble in the fuel matrix and precipitate as bubbles. The fission products, which are gaseous atoms, generated are released from the fuel when they reach any space that is connected to free volume. The amount of gases produced depends on the burn-up. In FBRs, due to high burn-ups, large fission gas plenums are provided in the pins.

Due to fission, gas atoms are uniformly generated in the fuel matrix. Due to diffusion of the gas atoms, gas bubble nucleation takes place, which is proportional to gas concentration in the matrix. The nucleated bubble slowly grows as a closed bubble due to diffusion of the gas atoms from the matrix. The small closed bubbles grow and form bigger closed bubble. At threshold burn-up, a portion of closed bubbles interconnect and form an open bubble through which the gas is released to the plenum. Subsequent to the open bubble formation, gas release takes place in two modes. First, through the diffusion of the gas atoms to the open bubbles; second, through further conversion of the closed bubbles to the open bubbles. Whereas the former process directly releases the gas without creation of additional bubbles, the later creates more open bubbles. The swelling of the matrix is proportional to the volume occupied by the closed bubbles and open bubbles in the fuel matrix. The distribution of various bubbles in the fuel matrix is shown schematically in Figure 7.3. The percentage fission gas release as a function of burn-up for metal fuel is shown in Figure 7.4. Details can be seen in Ref. 7.2.

TABLE 7.2
Physical Properties of Metal Fuel as a Function of Temperature

Temperature (°C)	K (W/m K)	Cp (J/kg K)	E (Pa)	Poisson's Ratio	Coefficient of Linear Expansion (K⁻¹)
350	17.8	169.4	8.44E+10	0.28	1.76E−05
500	21.9	185.6	6.70E+10	0.31	1.76E−05
600	24.6	236.9	5.54E+10	0.34	1.92E−05
700	27.3	191.1	3.06E+10	0.36	2.24E−05
800	30.0	191.4	2.98E+10	0.38	2.69E−05
900	32.7	191.4	2.90E+10	0.40	3.15E−05
1000	35.4	191.4	2.82E+10	0.43	3.60E−05
1100	38.1	191.4	2.74E+10	0.45	4.06E−05

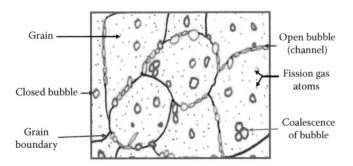

FIGURE 7.3 Fission gas bubbles in fuel matrix.

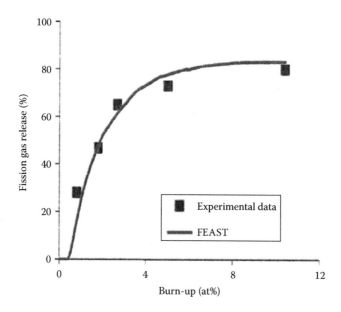

FIGURE 7.4 Fission gas release measured and predicted by Feast code. (From Karahan, A., Modelling of thermo-mechanical and irradiation behaviour of metallic and oxide fuels for sodium fast reactor, PhD report, MIT, Cambridge, MA, 2009.)

7.2.2.3 Swelling

Due to accumulation of gas bubbles in the fuel matrix, the volume of the slug increases. As already discussed, bubbles consist of two types: closed bubble and open bubble. After crossing some threshold burn-up, some portion of the closed bubbles will convert to open bubbles, that is, they are connected to the plenum by interlinking each other and thus forming a channel. The increase in the matrix volume takes place because of both closed bubble and open bubble. Apart from swelling due to bubble, solid fission product accumulated during the fission process also increases the volume of the matrix. The increase in the volume of the slug at the middle and top sections of the slug is shown in Figure 7.5.

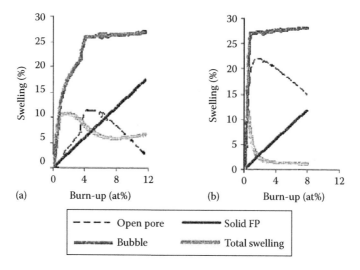

FIGURE 7.5 Swelling at middle and top of the fuel slug. (a) Middle of fissile column and (b) top of fissile column. (From Karahan, A., Modelling of thermo-mechanical and irradiation behaviour of metallic and oxide fuels for sodium fast reactor, PhD report, MIT, Cambridge, MA, 2009.)

7.2.2.4 Fuel-Cladding Chemical Interaction

FCCI is a complex multicomponent diffusion problem. It involves the inter-diffusion of fuel and cladding constituents at operating temperatures. Specifically, inter-diffusion has been characterized by the diffusion of Fe and Ni when available as cladding constituents into the fuel with a corresponding diffusion of lanthanide fission products into the cladding. The potential problem of inter-diffusion of fuel and cladding components is essentially twofold: the weakening of cladding mechanical properties and the formation of relatively low-melting-point compositions in the fuel. This inter-diffusion can be controlled by the presence of zirconium between fuel and clad. Zirconium can be added to the fuel matrix, which with irradiation migrates toward the periphery and forms a barrier to FCCI. The thickness, uniformity, and rate of formation of the Zr-rich layer appear to be the controlling factors. It seems that a thick and uniform Zr retards the formation of molten phases.

Another way is to have a Zr layer co-extruded along with clad for a thickness of ~150 μm, which retards the inter-diffusion. The advantages of having Zr as a liner instead of Zr in fuel matrix are as follows:

1. Neutron economy is improved in view of lower volume fraction of Zr in the fuel pin, and hence higher breeding ratio.
2. Fuel thermal conductivity is not affected.

At the same time, the presence of Zr liner without cracking under fuel clad mechanical interaction (FCMI) pressure with irradiation has to be confirmed by irradiation experiments.

7.2.3 Carbide Fuels

Carbide fuels have a very good thermal conductivity and also a high melting temperature that make them attractive for use in power reactors. Carbide fuels are manufactured through the sintering route to achieve the required porosity. Due to high thermal conductivity, the thermal gradient in the fuel is very less, because of which pore migration and restructuring phenomena are absent in the carbide fuels. Also, due to small thermal gradient, cracking of fuel is not predominant as observed in the oxide fuels.

7.2.3.1 Physical Properties

Thermal conductivity of carbide fuel:

The correlation for thermal conductivity as a function for 90% dense fuel is given next, which is based on the measurements:

$$K = (-4.55617998 \times 10^{-14} \times T^4) + (1.15593 \times 10^{-10} \times T^3) + (1.87943 \times 10^{-8} \times T^2)$$
$$- (2.8604 \times 10^{-5} \times T) + (0.06253) \text{ and } T \text{ in K.}$$

7.2.3.2 Fission Gas Release

Due to the lower operating temperatures and stronger fuel, the fission gas release in the carbide fuels is less compared with oxide and metal fuels. Typical release as a function of burn-up is shown in Figure 7.6. From the existing experimental results, it can be observed that low- and medium-density fuels release more fission gas than high-density fuels. Unstoichiometric fuel contains more voids than stoichiometric fuel and hence can contain more gas. Impurities hinder gas diffusion and therefore bubble formation and also bubble motion.

The important factors that affect the fission gas release rate from the carbide fuel are temperature, burn-up, porosity, and grain size. Remaining parameters such as stoichiometry, pore-size distribution, and cracking also affect the fission gas release to some extent.

7.2.3.3 Swelling

With increasing burn-up, the gap between the fuel and clad decreases due to the swelling of the fuel. Initially, the increase in the clad diameter is very small due to threshold swelling limit. After radial

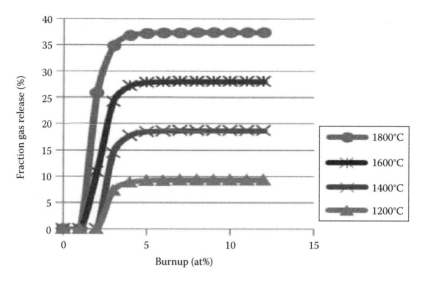

FIGURE 7.6   Fission gas release in carbide fuels as a function of burn-up.

gap closure, the clad expansion takes place due to FCMI. Carbide fuels are very hard, and creep effect is negligible in releasing the mechanical stresses developed in the fuel. Hence, FCMI induces high interface pressure on the clad in the case of carbide fuels. Also, carbide fuels exhibit higher swelling strains due to their poor release of fission gas. The predominant factors influencing the swelling are temperature and burn-up. Factors such as stoichiometry, porosity, and grain size also influence to some extent. The swelling rate correlation as a function of centerline temperature of the fuel is [7.3]

$$\dot{S} = \frac{1}{0.7653 + 2.11 \times 10^{-4}\,T - 2.75 \times 10^{-7}\,T^2} \quad 300 \le T < 1650,$$

$$= \frac{0.26}{1 - 8.3814 \times 10^{-4}\,T + 1.755 \times 10^{-7}\,T^2} \quad 1650 \le T \le 2250,$$

$$\dot{S} = Swelling\ rate\ \frac{\%\Delta V}{V}\ per\ atom\,\%\ burn\text{-}up,$$

$$T = Temperature\ K.$$

7.2.4   Nitride Fuel

Although mixed carbide has been the most studied fuel throughout the world, from the reprocessing point of view, nitride is preferred due to its better compatibility with the PUREX process. Swelling of nitride fuel is higher than oxide fuel, and hence, less smear density has to be chosen. Assuming a swelling rate of roughly 1.1%–1.6% per at% (instead of 0.6 per at% for oxide), it is necessary to limit this smear density to 75%–78% in order to reach an irradiation of roughly 150 GWd/ton. This can be obtained by adjusting either the density of the fuel or the value of the fuel-cladding gap. High thermal conductivity and melting temperature allow improved pellet thermal behavior, despite the choice of a low smear density, that is, high linear power can be achieved. With helium as a bonding in pins, linear power of ~700 W/cm is achievable. This can be increased to ~900 W/cm if sodium is used as the bonding medium (sodium is compatible with nitride fuel). By providing a safety margin, even the helium-bonded pins can be operated safely at 450–500 W/cm linear power. Finally, since its density is higher than that of oxide fuel, higher breeding ratios are achieved in nitride fuels.

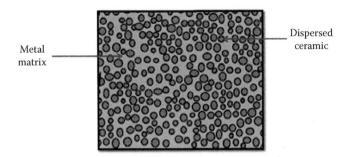

FIGURE 7.7 Cermet fuel.

7.2.5 Cermet Fuel

Cermet fuel consists of particles of ceramic fuel uniformly distributed in a metal matrix. The advantages of oxide fuels are easy fabrication, good reactor performance, and well-established reprocessing technology. The major disadvantages are low thermal conductivity, low density, and hence low breeding ratio and longer doubling times. To overcome the disadvantages of oxide fuels, the solution may lie in the cermet fuel like (U-PuO_2), where PuO_2 is dispersed in uranium metal matrix, combining the favorable features of oxide fuel and metal fuels. A typical cermet fuel is shown in Figure 7.7. Some of the advantages of cermet fuel are as follows:

1. High thermal conductivity compared to ceramic fuel because of the metallic matrix.
2. Only localized damages inside the matrix (fissile component in particulate form), resulting in lesser global damage.
3. Higher coefficient of thermal expansion compared to ceramic fuel, which is good from safety considerations.
4. The fuel can be operated at a higher temperature compared to binary U-Pu fuel because of U matrix.
5. Uranium and T91 clad eutectic formation temperature above 700°C (compared to 650°C with austenitics) gives added flexibility in safe operation of fuel.
6. The manufacture by powder metallurgy route would be easier. Yield is expected to be higher than metallic fuel fabricated by melting and casting route (to be confirmed with experience).

7.2.6 General Intercomparison

The general inter-comparison of properties of all types of uranium fuel is shown in Table 7.3.

7.3 CORE STRUCTURAL MATERIALS

The core of a typical fast reactor consists of several thin-walled small diameter fuel pins, which are bundled and encased in a wrapper tube. The fuel pin is a structural container of the nuclear fuel stacked in the form of pellets and thus provides the first barrier of safety for the fissile material and the radioactive fission products. Hence, its structural integrity is the foremost requirement in the design. Fuel pins are separated from one another to ensure adequate coolant flow by helically wound spacer wires. Typically, the fast reactor fuel pins are smaller in diameter compared to thermal reactors mainly from the considerations of fissile inventory and higher surface area facilitating high heat transfer. Fuel pins are arranged as a cluster in triangular lattice pitch forming a closely packed hexagonal configuration. The wrapper tube, apart from providing a means of structural support for the pin bundle, facilitates appropriate coolant flow past the fuel pins to remove the nuclear heat.

TABLE 7.3

Inter-Comparison of Properties of Uranium Fuel Types

Parameter	Oxide	Carbide	Metal	Nitride
Density (g/cm³)	Low (10.97)	Medium (13.6)	High (19.1)	Medium (14.3)
Melting temperature (K)	High (3138)	Medium (2780)	Low (1405)	High (3035)
Thermal conductivity (W/[m K])	Low (3)	Medium (21.6)	High (35)	Medium (20)
Thermal expansion coefficient (K⁻¹)	Low (10E–6)	Low (12E–6)	High (19E–6)	Low (9E–6)
Specific heat (J/[kg K])	High (271.5)	Medium (200.7)	Low (108.5)	Medium (204.8)

The wrapper tube is also alternatively known as duct or hexcan. The assembly of pin bundle and duct along with bottom foot, top head, and several miscellaneous parts within the duct is called sub-assembly (SA). In a power plant scale fast reactor, there are hundreds of SAs in the core, including fuel, absorber, blanket, reflector, and shielding SAs. The materials for the principal components of fuel SA, choice of materials, advanced materials development, design requirements, design safety limits (DSL), and structural design criteria form the subject of this section.

7.3.1 Environment

The principal components of an SA are subjected to intense and severe working conditions in terms of nuclear, thermal, mechanical, and chemical environment during their life in the core. The fuel elements are subjected to strenuous conditions of high temperature, high pressure, conditions of contact with chemically noncompatible fission products, and the coolant in case of a pin failure and above all subjected to high irradiation by medium-to-hard spectrum neutrons. In terms of severity, the three basic and important components of the core, viz. spacer wire, wrapper, and clad tube, face the conditions in the increasing order. In addition to individually getting affected, the SA deformation in whole due to the environmental conditions or in combination such as clad-spacer wire, pin cluster-duct influences the design to a greater extent. The conditions of temperature, pressure, and medium in contact vary slightly for the clad, spacer wire, and duct. But, in terms of neutron dose, these three components have more or less the same neutron dose. With respect to irradiation effects, these components are treated commonly. They only differ in the manner in which the effect of irradiation manifests such as bowing, dilation, bundle–duct interaction, pin–wire interaction, and pin ovalization. These aspects are covered in Chapter 8, which deals with core design.

The maximum rated fuel clad tubes experience temperatures in the range of 350°C–700°C along their length under steady-state operating conditions. Under plant transient conditions following pump trips, power failure, etc., the temperatures can rise up to even 1000°C or more for a short duration range from seconds to minutes. The duct is subjected to relatively lower temperature range of 350°C–650°C. The fuel clad is subjected to internal pressure due to fission gas generation, and hence, the pressure varies with respect to its residence time in the reactor, that is, as a function of its burn-up. Typically, the initial fill gas pressure will be around 0.1–0.15 MPa, and at the end of life, it increases up to 5–6 MPa depending on the burn-up. Duct is subjected to a coolant internal pressure of 0.2–0.7 MPa. Both the clad and duct are subjected to the effects of flowing sodium. The coolant velocity is dependent on the type of SA design in which many parameters are involved. Typically, the coolant velocity inside the bundle ranges from 5 to 9 m/s. This coupled with the impurities in sodium causes corrosion, although it is very small due to strict control on sodium chemistry.

The clad in addition faces the corrosive action of radioactive fission products on inside surface, which is crucial in the structural design of the clad. Depending on the type of steel and fuel used, the effect of corrosion will vary. For a typical oxide fuel with austenitic stainless steel clad, reduction in the clad thickness would be in the range of 50–80 μm.

The core materials are subjected to the effects of high neutron flux. The clad sees the highest flux. The mean neutron energies in a typical oxide-fueled sodium-cooled fast reactor would be in the range of 100–200 keV for power size reactors and 300–500 keV for test reactors. The radiation-induced microstructural changes in the clad lead to changes in physical properties as well as significant degradation in mechanical properties. This environment dictates the choice of materials for cladding and wrapper.

7.3.2 EFFECTS OF IRRADIATION

Accumulated neutron exposure (ϕ) over the period (t) is called "fluence (ϕt)." ϕ is a function of neutron energy level (E). Effects of $\phi(E)$.t at the energy level E on the metal matrix are inducing atomic displacements from original lattice position, which in turn depends upon the associated cross section of constituents of the metal. A typical displacement cross section for Fe is illustrated in Figure 7.8 [7.4]. As the neutron flux levels in fast reactors are about one or two orders of magnitude higher (~10^{15} n/cm² s⁻¹) than that in thermal reactors, the core materials are subjected to a high fast neutron fluence coupled with high temperatures, resulting in high levels of atomic displacements as discussed earlier. Thus, the neutron irradiation damage is quantified by the unit displacements per atom (dpa), that is, the number of times an atom is displaced from its lattice site on an average. A schematic of atomic displacements is shown in Figure 7.9. Displacements of atoms from their lattice sites are the primary cause, which is also aided by the nuclear transmutation to both solid and gaseous products. Since the rate at which the displacements occur is a function of neutron flux and neutron energy, dpa depends on the type of fuel, core size, temperature, etc. FBR core structural materials mainly austenitic stainless steels have demonstrated their capacity to withstand over 100–120 dpa for the whole core. In select experimental irradiations, even higher dpa (up to 160 dpa) has been sustained [7.5]. Development of materials to withstand higher and higher dpa is continuously aimed.

7.3.3 IRRADIATION DAMAGE

The four major effects that are important from core design considerations with respect to austenitic stainless steels are phase stability, void swelling, irradiation creep, and changes in mechanical properties (especially yield and ultimate tensile strength and ductility). These phenomena are

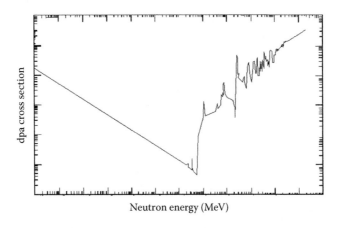

FIGURE 7.8 Displacement cross section for iron (ENDF/B-VI).

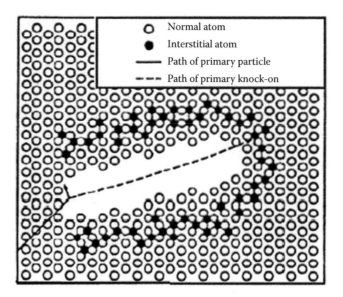

FIGURE 7.9 Schematic of atomic displacement.

intimately linked, and it has been shown that void swelling depends sensitively on the evolution of phases in austenitic stainless steels and is the dominant determinant of both irradiation creep behavior and mechanical properties. Alterations in chemical composition and microstructure that influence void swelling also exert a strong effect on irradiation creep. These effects are discussed subsequently.

7.3.3.1 Phase Stability

During irradiation, three general classes of phases can form in type 316 SS and its variants. The first class is radiation-enhanced or radiation-retarded thermal phase. In this group, precipitate phases form due to thermal aging, which includes M_6C, $M_{23}C_6$, and MC carbides, σ and χ intermetallics. The second class is radiation-modified phase, which is different from the composition found during thermal aging. These are primarily M_6C, Laves, and FeTiP. The third class is radiation-induced phase, which is uniquely produced by reactor irradiation, and the phases in this category include Ni_3Si (γ') and G-phase ($M_6Ni_{16}Si_7$) silicides and needle-shaped MP, M_3P, or M_2P phosphides. Radiation-induced segregation occurs due to the presence of a large supersaturation of vacancies and interstitials. The mechanisms of radiation induced segregation are solute drag effect and inverse Kirkendall effect [7.6]. The various precipitates formed due to segregation involving elements like Ni and Si are intimately linked to the onset of void swelling.

7.3.3.2 Void Swelling

Due to neutron irradiation, a system of vacancies and interstitial atoms are produced within the metal matrix. The vacancies are mobile in nature. The mobility increases with temperature. Once the vacancies start collecting and nucleating at selected sites, the embryo void forms and grows in size as more and more vacancies start accumulating. Once the vacancies attain the level of super saturation, bigger voids are formed. Void formation occurs predominantly in austenitic stainless steels in the temperature range of 673–973 K under neutron irradiation in FBRs. Helium produced from (n,α) reactions in Ni helps as a catalyst and stabilizes the voids. The swelling does not occur below a certain temperature because vacancies are not adequately mobile. Beyond certain temperature, also swelling will be less as vacancies and interstitials recombine together due to their very high mobility. The formation and growth of voids are sensitive to nearly all metallurgical variables

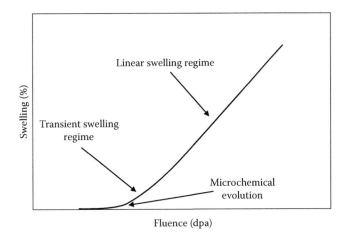

FIGURE 7.10 Typical material swelling behavior.

like chemical composition and thermomechanical history and irradiation parameters like fluence, dose rate, and irradiation temperature [7.7]. Typical swelling behavior of austenitic stainless steel is shown in Figure 7.10, schematically depicting the various regimes. The fluence dependence of swelling can be described as a low swelling transient period followed by an acceleration to a regime of near linear swelling. As can be seen, below threshold fluence, swelling is nil due to insufficient quantity of vacancies. The fluence corresponding to the start of linear swelling regime is characterized as the incubation fluence. Metallurgical parameters are found to be sensitive only in the duration of incubation and transient regimes. The steady-state swelling regime is relatively insensitive to these variables with the swelling rate of most austenitic alloys being ≈1% per dpa over a wide range of irradiation temperatures.

7.3.3.3 Irradiation Creep

Irradiation creep is caused due to the combined effects of an external nonhydrostatic stresses and the presence of both interstitial atoms and vacancies at very large supersaturation levels. The main characteristic of irradiation creep is that it occurs at relatively low temperatures where the thermal creep is negligibly small. Stress-induced preferential absorption (SIPA) of interstitials at dislocations is considered as the dominant mechanism of irradiation creep. The process of SIPA leads to development of creep deformation, which is proportional to the applied stress and the irradiation dose. Thermal creep can also be altered due to microstructural changes as a result of irradiation.

There is ample evidence to show that in most of the austenitic stainless steels there is a direct coupling between the creep and swelling rates after swelling begins [7.8]. An important consequence of this coupling is that swelling-resistant materials are also resistant to irradiation creep. In fact, solutes such as phosphorus, which reduce swelling, also reduce creep strain in austenitic stainless steels. Nickel content in austenitic stainless steels influences both void swelling and the creep. Creep affects both clad and the wrapper. Clad diameter increases due to swelling and creep. The diametral deformation on the clad as a function of the stress in the clad is shown in Figure 7.11 for 316 SS to highlight the increase in the deformation as a function of stress, which is gradually increasing with burn-up [7.9].

The life limit on the clad is dictated by the creep rupture. The creep rupture behavior of materials is predicted by the Larsen–Miller parameter. The creep rupture life depends on stress, temperature, and material. A schematic representation of a typical Larsen–Miller curve is shown in Figure 7.12. This curve is essential and useful for the design studies as it can be used for any combination of the parameters involved, viz. temperature and stress.

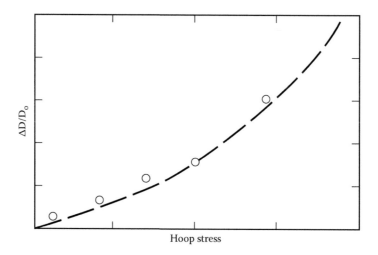

FIGURE 7.11 Clad deformation due to creep in 20% CW 316SS. (For details [7.9] can be referred.)

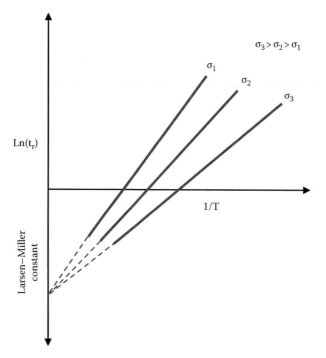

FIGURE 7.12 Schematic of Larsen–Miller curve.

7.3.3.4 Mechanical Properties

The mechanical properties, such as yield and ultimate strength and the ductility, are affected by irradiation. The effect happens in two ways: (1) due to helium embrittlement and (2) radiation hardening. Both effects occur in the matrix in a combined way.

Upon high irradiation, the material tends to become brittle due to the production of helium by (n,α) reaction with Ni, which settle in the intergranular boundaries. This is obviously dependent on temperature of irradiation, prior microstructural condition, and neutron fluence (dpa). Temperature is a very important variable as it determines the stability, distribution, and morphology of the defect structures such as dislocation loops, network dislocations, voids and helium bubbles, and in turn embrittlement. In particular, at high-temperature regime, the mobility of helium in the metal matrix is sufficiently high enough to cause intergranular-related failures.

Upon irradiation, more defects are produced in the materials as mentioned, which hardens the materials. This leads to an increase in both yield and ultimate strength, which in turn reduce ductility. An increase in yield strength observed in 316 SS in EBR-II is shown in Figure 7.13 as a typical example [7.10]. In the low-temperature regime, for annealed austenitic stainless steels, there is an increase in yield strength and ultimate tensile strength resulting in ductility reduction. The increase in yield strength is reported to be much more pronounced compared to the increase in ultimate tensile strength. The radiation hardening and the ductility reduction are shown in Figure 7.14. The reduced ductility at high temperatures is associated with an increased tendency for fracture to occur along the grain boundaries. This has been attributed to the phenomenon of helium embrittlement [7.11]. At temperatures and doses where void swelling is significant, it exerts a strong influence on mechanical properties [7.12]. This correlation between ductility and swelling could be attributed to either the direct effect of voidage in promoting brittle fracture or to the indirect effect of radiation-induced segregation due to the presence of voids. Radiation-induced segregation could lead to instability of austenitic matrix due to nickel segregation at void surfaces facilitating transformation to martensite, which would induce embrittlement. Thus, the measures to improve swelling resistance of austenitic stainless steels would also enhance resistance to embrittlement.

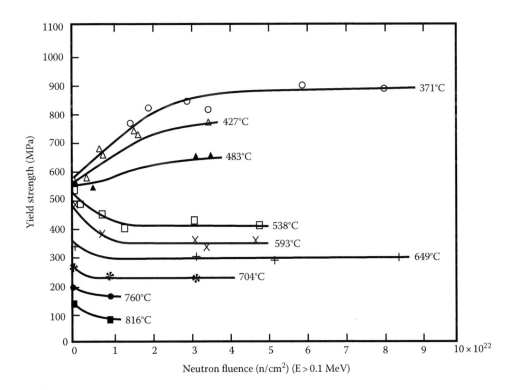

FIGURE 7.13 Yield strength variation in AISI 316SS.

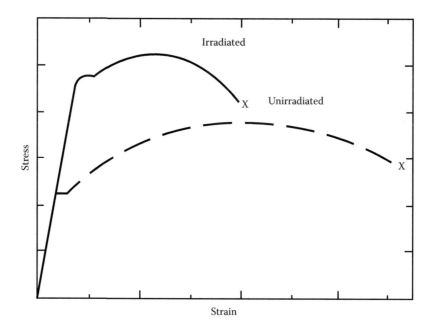

FIGURE 7.14   Radiation hardening and ductility reduction.

7.3.4   CHOICE OF MATERIALS

The primary design requirements on the clad and wrapper materials are (1) good resistance to irradiation effects, (2) adequate mechanical properties during the course of operation and also at the end of life, and (3) better compatibility with coolant and fuel material. The major requirements arise mainly from the considerations of fuel pin integrity, coolability, and structural safety margin and easy handling of subassemblies. Accordingly, the selection criteria for materials will be applied. The major selection criteria for the basic fuel pin components are indicated in the following table [7.13]. Basically three classes of materials have been employed in the fast reactors so far: (1) austenitics, (2) ferritics, and (3) high nickel alloys.

S No.	Component	Major Requirements
1.	Clad	Swelling resistance, creep strength, high mechanical strength, adequate ductility, compatibility with fuel, fission products and coolant
		Easy fabrication and weldability
2.	Wrapper	Swelling resistance, creep strength, moderate mechanical strength, adequate ductility, compatibility with coolant
		Easy fabrication and weldability
3.	Spacer wire	Swelling resistance, moderate mechanical strength, compatibility with coolant
		Easy fabrication

Table 7.4 lists the clad and wrapper materials used in major FBRs in various countries. Table 7.5 gives the detailed chemical composition of the various cladding materials used in the fast reactors worldwide [7.14].

The material development on the austenitic stainless steel series has been mostly based on varying the alloy composition to derive the intended requirements such as reduced void swelling. The presence of nickel in solution is suggested to enhance effective vacancy diffusion coefficient and therefore leads to reduced void nucleation rate. Increasing nickel and reducing chromium would

TABLE 7.4

Materials Selected for Cladding in Major FBRs

Reactor	Country	Clad Material	Wrapper Material
CEFR	China	ChS-68CW	EP-450
EFR	Europe	ALM1 or PE16	EM10 or Euralloy
Rapsodie	France	316	
PHÉNIX	France	15-15 Ti	EM10
SUPERPHÉNIX	France	15-15 Ti	EM10
KNK-II	Germany	1.4970	1.4981
FBTR	India	316 (CW)	316 L (CW)
PFBR	India	20% CW D9	D91
JOYO	Japan	316 (20% CW)	316(20% CW)
MONJU	Japan	PNC 316 (20% CW)	PNC 316 (20% CW)
JSFR	Japan	ODS	PNC-FMS
BN-350	Kazakhstan	EI-847	16Cr-11Ni-3Mo
		ChS-68CW (from 1987)	EP-450 (from 1987)
KALIMER	Republic of Korea	HT9	HT9
BR-10	Russian Federation	EI-847	18Cr–9Ni–Ti
BOR-60	Russian Federation	ChS-68CW	EP-450
BN-600	Russian Federation	ChS-68CW (from 1987)	EP-450 (from 1987)
BN-800	Russian Federation	ChS-68CW-I stage	EP-450
		EK-181-II-stage	
BN-1200	Russian Federation	EK-164CW-I stage	EP-450
		EK-181-II stage	
		ODS-III stage	
DFR	United Kingdom	Niobium	—
PFR	United Kingdom	STA Nimonic PE 16	PE16/FV448
EBR-II	United States	316	—
Fermi	United States	Zr	—
FFTF	United States	316 (20% CW), HT9	316 (20% CW), HT9

increase stability of the γ phase against precipitation of undesirable phases that also remove certain solutes responsible for conferring void swelling resistance to these steels. Figure 7.15 compares the performances of a few clad materials as observed in Phenix [7.15]. There is an increase in the swelling resistance on moving from unstabilized 316 SS to 316 Ti, 15-15 Ti (D9). The major difference in these alloys is the increase in the incubation dose for swelling. Cold-worked 15-15 Ti has reached a record dose of 140 dpa without excessive deformation. The presence of fine carbides of elements like Ti and Nb provides recombination sites for point defects and hence reduces their super saturation and consequently void swelling. Ti/C ratio is known to play an important role in determining the irradiation behavior of austenitic steels. Maximum swelling resistance in cold-worked 15-15 Ti of a high carbon grade (C 0.08–0.12 wt%) is obtained when the Ti/C ratio is below the stoichiometric composition, that is, when the material is understabilized (i.e., Ti content of less than four times the carbon content in weight percentage) [7.16]. The reason for this behavior is the synergistic interrelation between freely migrating carbon and the formation of finely dispersed TiC particles. The fine TiC particles are redissolved by recoil dissolution and then continuously contribute to trapping mechanism as long as the steel is understabilized.

Solute elements like titanium, silicon, phosphorous, niobium, boron, and carbon play a dominant role in determining void swelling resistance by extending the duration of the transient regime. This observation has led to the development of advanced core structural materials such as D9 and D9I. Silicon improves void swelling resistance of austenitic stainless steels through its role as a fast

TABLE 7.5

Typical Chemical Composition of Cladding Materials

Austenitic Stainless Steels														
304 SS	0.05	18	10	0.3	0.4	1.5	—	—	—	—	—	—	—	—
316 SS	0.05	17	13	2	0.6	1.8	—	—	—	—	—	—	0.002	—
Japan														
ENC 316	0.055	16.0	14.0	2.50	0.80	1.80	—	0.08	0.10	0.028	—	—	—	—
EMC 1520	0.06	15.0	20.0	2.50	0.80	1.90	—	0.11	0.25	0.025	—	—	—	—
France														
316Ti	0.05	16	14	2.5	0.6	1.7	—	—	0.4	0.03	—	—	0.005	—
15-15Ti	0.1	15	15	1.2	0.6	1.5	—	—	0.4	0.007	—	—	0.005	—
15-15Ti$_{opt}$	0.1	15	15	1.2	0.8	1.5	—	—	0.4	0.03	—	—	0.005	—
15-15Ti mod	0.085	14.9	14.8	1.46	0.95	1.50	—	—	0.50	0.007	—	—	0.004	—
United States														
D9	0.052	13.8	15.2	1.50	0.92	1.74	—	—	0.23	0.003	—	—	—	—
D9I	[a]	13.5	15.5	1.8–2.22	0.8	2.0	—	—	0.25	0.025–0.04	0.005	—	—	—
ASTM A771	0.03–0.05	12.5–14.5	14.5–16.5	1.5–2.5	0.5–1.0	1.65–2.35	—	0.05 max.	0.1–0.4[b]	0.04 max.	0.005–0.01	0.01 max.	0.004–0.006	—
India														
D9(PFBR)	0.035–0.050	13.5–14.5	14.5–155	2.0–2.5	0.50–0.75	1.65–2.35	—	0.05 max.	5.0C–7.5C	0.02 max.	—	0.01 max.	—	—
D91	0.04–0.05[c]	13.5–14.5	14.5–15.5	2.0–2.5	0.7–0.9	1.65–2.35	—	0.05 max.	0.25	0.025–0.04	—	—	0.004–0.006	—
Germany														
1.4970	0.1	15	15	1.2	0.4	15	—	—	0.5	—	—	—	0.005	—
United Kingdom														
FV548	0.09	16.5	11.5	1.4	0.3	1	—	0.7	—	—	—	—	—	—
Russian Federation														
EI-847	0.04–0.06	15–16	15–16	2.7–3.2	<0.4	0.4–0.8	0.1–0.3	<0.9	0.2–0.5	<0.02	—	—	—	—
ChS-68	0.05–0.08	15.5–17	140–15.5	19–2.5	0.3–0.6	1.3–2	—	—	0.2–0.5	<0.02	—	—	0.002–0005	—
EK-164	0.05–0.09	15–16.5	18–19.5	2–2.5	0.3–0.6	1.5–2	0.15	0.1–0.4	0.25–0.45	0.01–0.03	—	—	0.001–0.005	0.15Ce
Nickel-Base Alloys														
PE16	0.13	16.5	43.5	3.3	0.2	0.1	—	—	1.3	—	—	—	—	1.3 (A1)
INC706	0.01	16	40	0.02	0.09	0.4	—	3	1.5	—	—	—	—	—
12RN72HV	0.1.	19	25	1.4	0.4	1.8	—	—	0.5	—	—	—	0.0065	—

(Continued)

TABLE 7.5 (Continued)

Typical Chemical Composition of Cladding Materials

Ferritic–Martensitic Alloys

United Kingdom														
F1	0.15	13.0	0.47	—	0.30	0.45	—	—	—	—	—	—	—	—
FV607	0.13	11.1	0.59	0.93	0.53	0.80	0.27	—	—	—	—	—	—	—
CRM-12	0.19	11.8	0.42	0.96	0.45	0.54	0.30	—	—	—	—	—	—	—
FV448	0.10	10.7	0.64	0.64	0.38	0.86	0.16	0.30	—	—	—	—	—	—
France														
F17	0.05	17.0	0.10	—	0.30	0.40	—	—	—	≤0.008	≤0.008	0.020	—	—
EM10	0.10	9.0	0.20	1.0	0.30	0.50	—	—	—	≤0.008	≤0.008	—	—	—
EM12	0.10	9.0	0.30	2.0	0.40	1.00	0.40	0.50	—	≤0.008	≤0.008	—	—	—
T91	0.10	9.0	<0.40	0.95	0.35	0.45	0.22	0.08	—	≤0.008	≤0.008	0050	—	—
Germany														
1.4923	0.21	11.2	0.42	0.83	0.37	0.50	0.21	—	—	—	—	—	—	—
1.4914	0.14	11.3	0.70	0.50	0.45	0.35	0.30	0.25	—	—	—	0.029	0.007	—
1.4914 mod	0.16–0.18	10.2–10.7	0.75–0.95	0.45–0.65	0.25–0.35	0.60–0.80	0.20–0.3.0	0.10–0.25	—	—	—	0.010 max.	0.0015 max.	—
United States														
HT9	0.20	11.9	0.62	0.91	0.38	0.59	0.30	—	—	—	—	—	—	0.52 (W)
403	0.12	12.0	0.15	—	0.35	0.48	—	—	—	—	—	—	—	—
Japan														
PNC-FMS	0.2	11	0.4	0.5	—	—	0.2	0.05	—	—	—	0.05	—	—
Russian Federation														
EP-450	0.1–015	12–14	<0.3	1.2–1.8	<0.6	<0.6	0.1–0.3	0.25–0.55	—	—	—	—	0.004	—
EK-181	0.1–0.2	10–12	<0.1	<0.01	0.3–0.5	0.5–0.8	0.2–1	<0.01	0003–0.3	—	—	—	0.003–0.006	(1–2)W (0.05–0.3)Ta
ChS139	0.18–0.2	11–12.5	0.5–0.8	0.4–0.6	0.2–0.3	0.5–0.8	0.2–0.3	0.2–0.3	0.003–0.3	—	—	—	0.003–0.006	(1–1.5)W

a   Fe balance adjusted such that Ti/(C + N) = 5.

b   Fe balance target Ti = 0.25%.

c   Fe balance adjusted such that Ti/C = 4–5.

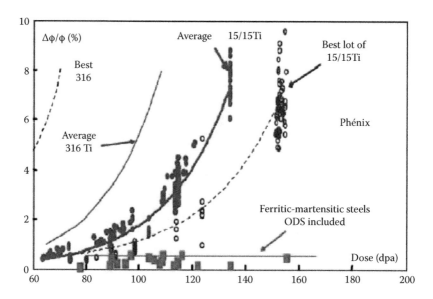

FIGURE 7.15 Phénix clad swelling behavior.

diffusing species, which reduces supersaturation of vacancies by their relatively fast migration to sinks. Similarly, the improvement in swelling resistance by phosphorous additions arises due to a decrease in void density caused by the uniform distribution of needle-shaped phosphides.

Another means of swelling reduction is by mechanical cold working. The degree of cold work that influences dislocation density is an important parameter determining the void swelling behavior of austenitic stainless steels. Since dislocations act as sinks for point defects, their density influences void swelling. Cold work also influences swelling behavior by providing sites for nucleation of precipitates as well as for trapping of helium bubbles. The cold work is also specified with a view to enhancing creep strength. Normally, optimum cold work of 20% is specified for the austenitic stainless steels.

Another class of alloys that has been studied extensively for their irradiation behavior is nickel-base superalloys (PE16, IN706, etc.). Most of the studies have shown that swelling is low in these alloys. But nickel-base alloys suffer from serious irradiation embrittlement due to helium.

Ferritic steels have been envisaged in the long term to support high performance from the fuel pins. The ferritic steels display excellent swelling-resistance characteristics, although their high-temperature creep strength is poor, especially at temperatures above 650°C. Many varieties of ferritic steels such as EM10, EM12 HT9, and Gr91 have been employed in the fast reactors. Due to their poor creep strength, these classes of materials were never considered for clad. However, they have been used as wrapper as its design requirements are met by these steels. Even this use was found to be limited due to the DBTT issue due to irradiation. This particular fact prompted several R&D initiatives revolving around change in chemical composition and appropriate heat treatments. It has been realized through various means such as ensuring fully martensitic structure and also by the optimization of austenitizing temperature. More details on this can be had from Refs. 7.17–7.20.

7.3.5 Advanced Materials and Improvements

For economic viability, the target burn-up required for FBRs is more than 20 at% of heavy metal (200,000 MWd/ton), and this can be achieved only by the availability of materials resistant to void swelling, irradiation creep, and irradiation embrittlement, as well as satisfying the high-temperature mechanical properties. The residence time of the fuel SA, and hence the achievable fuel burn-up, is limited by either the void swelling of the hexagonal wrapper material or creep strength of clad. Since fuel cycle cost is strongly linked with burn-up, selection of materials resistant to void swelling and irradiation creep is very important. The objective of materials development is to increase residence time of fuel elements in the core with a view to increasing the burn-up.

7.3.5.1 Advanced Austenitic Stainless Steels

Since the deformations due to swelling and irradiation creep are dependent on chemical composition and fabrication history of austenitic stainless steels as mentioned earlier, it is important to fix specifications with narrow limits on chemical composition to avoid differences in deformations arising out of heat to heat variations for different core components. Gradients in swelling and creep lead to interaction of core components that can affect the reactivity of the core, the movement of control rods, the flow of coolant, and the ability to withdraw and replace components as needed. It has been noticed that irradiation creep and embrittlement are related to swelling. Reduction of swelling would solve most of the engineering problems arising out of irradiation effects. Hence, there have been concerted efforts to develop swelling-resistant austenitic stainless steels with optimization of their chemical composition [7.21].

The basic approach to develop materials to withstand high burn-up of the fuel is as follows: optimize chemistry, cold work to introduce defects like dislocations or coherent, stable precipitates to delay onset of swelling. The same microstructural features improve high-temperature creep behavior also. Very often, the development of such engineering steels required the development of characterization methods, suitable for detection of features at the atomic level, the expertise in which needs to be developed, like the high-resolution microscopy, convergent beam electron diffraction, and positron lifetime spectroscopy. Optimization of Ti/C ratio based on creep property is shown in Figure 7.16, and high-resolution micrograph of TiC in austenite matrix is shown in Figure 7.17 [7.22].

7.3.5.2 Ferritic Steels

For doses above 120 dpa, austenitic stainless steels could not be employed as void swelling is found to be detrimental. It is also well known that ferritic/martensitic steels such as modified 9Cr-1Mo and Sandvik HT9 exhibit high void swelling resistance than conventionally used austenitic stainless steels.

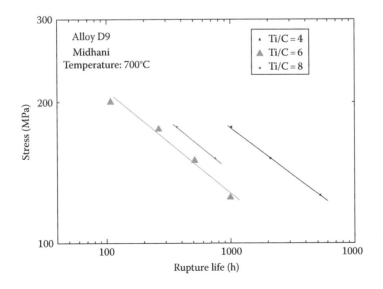

FIGURE 7.16 Optimization of Ti/C ratio.

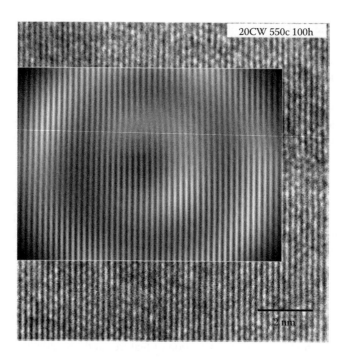

FIGURE 7.17   High-resolution micrograph.

There is a great incentive in developing ferritic–martensitic steels for cladding applications at higher temperatures ~923 K. Toward this goal, extensive efforts are underway to improve the creep strength of 9%–12% Cr steels. Alloys capable of operation up to 873–893 K have already been developed, and it is likely that their temperature capability could be extended for temperatures ~923 K with further modifications. This, coupled with the fact that for metallic fuels clad temperatures would be lower, 9%–12% Cr steels with improved creep strength appear attractive.

Ferritics, with their radiation resistance superior to austenitics, has a major challenge to overcome: temper embrittlement in ferritics. Having understood the mechanism of embrittlement, two methods are possible to overcome temper embrittlement: reduce the amount of tramp elements or "grain boundary engineering" methods. Temper embrittlement in ferritic steels, associated with segregation of impurities to grain boundaries and consequent weakening, poses a serious limitation in its successful employment at low temperatures. Certain special grain boundaries, especially the coincident site lattice (CSL) type, have low energy and are known to be resistant to segregation of impurities and precipitation of embrittling phases at/near the boundaries. An effective network of connected CSL grain boundaries can be visualized to block the propagation of an intergranular crack. Theoretical studies to identify the fractional amount of CSL boundaries required to prevent such a crack network based on the percolation approach have been attempted in India. Using a simulated 3D Poisson-Voronoi polyhedral grain structure to represent an equiaxed structure in real materials, a Monte Carlo simulation has been carried out to arrive at the statistical probability of a percolating crack cluster for different fractional amounts of CSL grain boundaries. The percolation threshold has been calculated for the first time for such a simulated microstructure. By comparing the experimentally measured CSL boundary fraction and the percolation threshold value, the propensity of the material to intergranular fracture has been assessed. Presently, an attempt is being made in India to increase the fraction of CSL boundaries in the 9Cr-1Mo ferritic–martensitic steel through suitable thermomechanical treatments.

7.3.5.3 ODS Steels

Oxide dispersion–strengthened steels are other alternative with the potential of having the advantage of ferritic–martensitic steel with respect to high swelling resistance but being able to push operating temperatures to 923 K and beyond from creep strength point of view. Essentially, in the ODS steels, the distribution of stable Y_2O_3 oxide particles can be controlled on a nanoscale that serves as a strong block for mobile dislocations and as a sink for radiation defects at the particle–matrix interfaces. However, ODS steels are credited with bamboo-like grain structure and strong deformation texture, which gives rise to anisotropic mechanical properties, especially an inferior biaxial creep rupture strength. There are indications that the controlling of grain morphology by means of recrystallization processing and development of equiaxed structure through austenite–martensite phase transformation successfully provide improved strength and ductility in the hoop direction of ODS ferritic–martensitic steels. There are other problems with the ODS steels besides anisotropy. At present, the literature is devoid of information on the production of thick-walled parts or large-diameter tubing. Fabrication processes for these materials for heavy sections still need to be established, and this means addressing the problem of joining the materials. ODS alloys are both expensive and difficult to form into tubes and other complex shapes. Therefore, much research and development will still be necessary on forming of these materials before these alloys will be ready for structural applications on a larger scale.

The development of ODS steels is oriented in such a way that the creep resistance is sought to be improved while retaining the excellent swelling resistance and the high thermal conductivity property. Presently, yttria (Y_2O_3) is considered the potential dispersoid. The high performance is envisaged to be achieved by finely dispersing the particles that would cause hindrance to dislocation mobility and also act as the trapping sites for the point defects produced by irradiation. The particle size employed is 2–3 nm, which also contributes to the improved creep rupture strength. Currently, 9Cr-ODS steel with composition of $9Cr–0.12C–2W–0.2Ti–0.37Y_2O_3$ has received international consideration. Clad tubes (with 8.5 mm OD and 0.5 mm wall thickness) of this composition have crossed test period of about 24,900 h without failure under creep test conditions of 105 MPa hoop stress at 973 K.

In a study on 9Cr-ODS (base composition Fe–0.12C–9Cr–2W) with 0.12%–0.2% Ti and 0.30%–0.37% yttria, martensitic 9Cr-ODS steels are also found to contain Y–Ti–O complex oxides similar to those in ferritic 12Cr-ODS steels. Maximum in creep strength has been observed at 0.20% Ti and 0.37% yttria (Figure 7.18) [7.23]. This has been again ascribed to the refinement of oxide particles and decrease in inter-particle spacing with increase in Ti and yttria contents.

Commercial production route involves mechanical alloying of rapidly solidified pre-alloyed powder and oxide powder particles (of nanoscale), followed by consolidation with hot extrusion, or hot-rolling or hot-isostatic pressing. Final heat treatments are required to remove anisotropy so as to produce clad tubes with good biaxial creep strength and ductility (in hoop direction). Current issues to be resolved are production of thin clad tubes at room temperature by cold-pilgering and elimination of complete microstructural anisotropy, particularly in 12-Cr ferritic ODS steels. Modification of final grain morphology in final clad tubes from 5 to 10 μm in 12Cr-ODS steels and about 1 μm in martensitic 9Cr-ODS steels is to be carried out to achieve further increase in creep strength, joining techniques and long-term stability of nanoscale oxide particles under neutron irradiation. R&D efforts should focus on these aspects. Further, DBTT is of main concern below 473 K. As of now, DBTT data are not available for the third-generation ODS steels. But it has been reported in 1DS (second-generation ODS steel) ODS steel that, in unirradiated condition, DBTT is around 230 K and moreover associated fracture surfaces (even in lower self region) revealed ductile (fibrous) features. Irradiations at and below 775 K caused no appreciable change in total absorbed energy for 1DS. Though irradiations above 793 K resulted in considerable decrease in USE and LSE, embrittlement is not significant and LSE is found to be quite high.

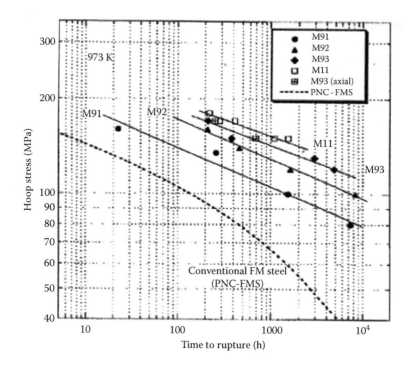

FIGURE 7.18   Creep strength of ODS steels.

7.3.5.4   Accelerator as a Means of Material Development

Study of radiation damage up to high doses of about 150 dpa would require 3–4 years in a nuclear reactor (Figure 7.19) [7.24]. In contrast, a few hours in an accelerator can effectively simulate some crucial aspects of the neutron damage. The ion-irradiation damage rate produced thus is characterized by many techniques like precision profilometry, position lifetime spectroscopy, and high-resolution electron microscopy (HREM), besides theoretical modeling, which enables a meaningful

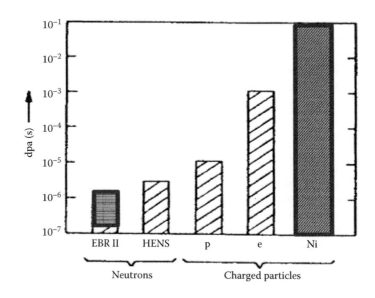

FIGURE 7.19   Damage efficiency.

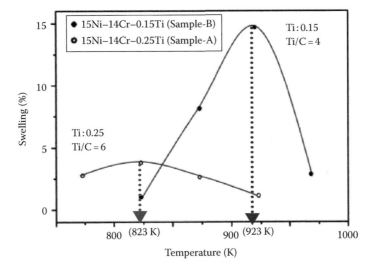

FIGURE 7.20 Void swelling.

inter-comparison of results from accelerator and nuclear reactors. Accelerator experiments can play a useful role in short-listing the most potential swelling-resistant candidate alloys for further extensive research. In this regard, the case of alloy D9 titanium-modified high-chromium stainless steel carried out in India is briefly mentioned.

Alloy D9 has been investigated quite extensively to optimize its chemistry and structure. A 1.7 MV Tandetron accelerator is employed for inducing ion-irradiation damage. The details of accelerators based on ion beam simulation are dealt in Section 31.4. Figure 7.20 shows void swelling at 100 dpa as a function of irradiation temperature for 20% cold-worked D9 alloy containing 0.25 Ti (Ti/C = 6) and 0.15 Ti (Ti/C = 4) [7.25]. The void swelling was measured by heavy ion irradiation of 30 appm helium preimplanted samples using a damage rate of 7×10^{-3} dpa/s. In addition to shift in peak swelling temperature, the swelling at peak is also increased to 15% for the alloy containing 0.15% of Ti. The presence of nanometer-sized coherent TiC precipitates embedded in the austenite matrix serves as sink to radiation-generated point defects and thus leading to an enhancement in swelling resistance. The higher volume fraction of TiC precipitates in D9 containing higher Ti makes it more resistant to swelling. Theoretical models and numerical solutions using multiscale multiphysics modeling concepts are also another means of understanding the material behavior under irradiation that will be useful for preliminary filtering of candidate alloy variations.

7.3.6 Safety Limits With Respect To Materials

For a typical power fast reactor, the DSLs that are adopted for clad and wrapper are described later. The criteria could vary between the countries depending on the material and the target performance parameters.

Design of components with sufficient margin is the first step in defense-in-depth principle adopted to ensure reactor safety. This requires consideration of all the loads in the components during normal operation as well as during off normal events. All the design basis events occurring during the lifetime of the components concerned are typically classified into four categories based on the frequency of occurrences, although some variations in the categorization are practiced by different countries. It is essential to define the DSL for the four categories of events in order to assess that the design provisions of the system/components are adequate to ensure safety. The limits imposed are generally in terms of temperatures, radiation doses, and structural design parameters. Limits are imposed on component/systems that could be broadly classified into three major groups. The first group comprises the core components whose performance conditions are vital for reactor safety, and they form the primary barrier for the radioactive fission products. The second group comprises hot pool and cold pool systems and components with limits defined in terms of temperatures and structural parameters. This is the second barrier for the fission products in case of failure of first barrier, that is, clad. The third group comprises radiation limits to personals with respect to plant personnel, public at site boundary, and public at neighborhood. This section mainly discusses and summarizes the philosophy adopted and typical DSLs used for fuel, clad, and coolant.

For temperature limits concerning fuel, the criteria criterion is to avoid fuel melting or limit the melting to as small a volume as possible. Hence, the limits for fuel are normally derived based on the extent of melting. Derivation of these limits in countries, especially France and the United States, is based on fuel transient experiments undertaken. For temperature limits concerning the clad, the criterion is to retain its integrity. The clad material is subjected to stresses and operating at high temperatures and hence undergoes damage. The main damage is from creep considerations. Hence, for defining clad temperature limits, one suitable approach is to use cumulative damage fraction (CDF) concept, for which structural design criteria for highly irradiated components and data on material behavior under transients are very important. Since the limits are on the basis of damage, time duration also is to be defined for transients. For limits concerning the coolant, the philosophy is to avoid bulk coolant boiling or to limit the coolant boiling to the local spots. Hence, the limits for the coolant are normally derived based on coolant boiling. The limit to coolant boiling in local spots should be defined in such a way that it does not lead to systematic clad melting leading to material relocation.

The DSLs adopted for the cladding in various fast reactors for normal operation and for design basis events (upset/emergency and faulted condition) are given next [7.26–7.28]. For the transient events falling in between the normal operation and faulted condition DBE, in addition to temperature limits, time duration also will be normally specified.

Operating Condition (°C)	FFTF	Monju	Phenix	PFBR	EFR
Normal condition	700	—	—	700	—
Design basis (upset/emergency)	815–870	830	800	800/900	740–780
Design basis (faulted)	980	—	—	1200	Case by case

7.3.7 STRUCTURAL DESIGN CRITERIA

The structural design criteria for the design of clad are important in view of its stringent duty and also due to the fact that they undergo high levels of irradiation. ASME Pressure Vessel Code Section III, Division I, Subsection NH4 and RCC-MR code (French code for fast reactors) that govern the design of out-of-core structural components for FBRs cannot be applied for the design of fuel subassemblies. Therefore, design has to be based on the rules applicable for highly irradiated components. Development of such rules requires extensive data on both out-pile and in-pile material data. A typical approach adopted for the structural design of clad along with the design rules based on American RTD standards is described later [7.29–7.31].

For the clad, the applicable failure mechanisms are (1) creep rupture, (2) tensile instability due to membrane loading, (3) collapse in bending/possible cracking in outer fibers in bending, (4) ratcheting due to secondary stress cycling/possible cracking due to large secondary stress, (5) cracking at strain concentrations due to short-term loads (static notch weakening), (6) unstable crack propagation, and (7) excessive deformation limited by functional requirements.

Two types of structural analysis are usually done for the clad to verify the design conformance to the structural design criteria: elastic and inelastic analysis (IEA). In the case of core components, because of neutron irradiation, elastic analysis may include swelling and irradiation-induced creep that are not structurally damaging but cause only deformation (similar to thermal strain, but irreversible). This type of analysis is called elastic/in-pile creep/swelling analysis (EICSA). The limits followed for this type of analysis are the same as that for elastic analysis. The second type of analysis is the detailed IEA. While in EICSA the limits are on stresses, in the IEA the limits are on strains. A typical set of design rules showing the relationship between protection against failure mechanisms and corresponding design rules for EICSA and for IEA is summarized in Tables 7.6 and 7.7.

TABLE 7.6
Failure Mechanisms and Associated Design Rules for EICSA

Failure Mode	Governing Parameter	Critical Value	Factor of Safety	Design Rule
Tensile instability	P_m	S_y or S_u	1.18 and 1.7 for category 1 and 2	$P_m < 0.85\ S_y$ or $< 0.6\ S_u$ for category 1 and 2 loads
			0.9 and 1.25 for category 3	$P_m < 1.1\ S_y$ or $< 0.8\ S_u$ for category 3 loads
Failure in bending	$P_m + P_b$	$K\ S_y$ or S_u	Same as above	Same as above
Ratcheting	$P_m + P_b + Q$	Shake down limits or S_u	1.7 for category 1 and 2 and 1.25 for category 3 on S_u	$P_m + P_b + Q < 0.6\ S_u$ for category 1 and 2 $P_m + P_b + Q < 0.8\ S_u$ for category 3 loads
Localized rupture	$P_m + P_b + Q + F$	S_u	1.25 for category 1 and 2 and 1.1 for category 3	$P_m + P_b + Q + F < 0.8\ S_u$ for category 1 and 2 loads $P_m + P_b + Q + F, < 0.9\ S_u$ for category 3
Damage due to creep and fatigue	Time duration, Δt and no. of cycles, n	Rupture life t_d and fatigue limit N	4 on CDF for category 1, 2 for category 1 and 2 and 1.3 for category 1, 2, and 3	$\Sigma \Delta t/t_d + \Sigma n/N < 0.25$ for category 1 and <0.5 for category 1 and 2 and <0.75 for category 1, 2, and 3
Brittle fracture	K_I	K_{IC}	—	$K_I < K_{IC}$

TABLE 7.7
Failure Mechanisms and Associated Design Rules for IEA

Failure Mode	Governing Parameter	Critical Value	Factor of Safety	Design Rule
Tensile instability and Ratcheting	ε_m^p	$\varepsilon_u/2$	3 for category 1 and 2 loads 1.5 for category 1, 2, and 3	$\Sigma \varepsilon_m^p/(\varepsilon_u/2) < 0.33$ for category 1 and 2 $\Sigma \varepsilon_m^p/(\varepsilon_u/2) < 0.66$ for category 1, 2, and 3
Localized rupture	ε_t^p	ε_f/TF	2 for category 1 and 2 and 1.25 for category 1, 2, and 3	$\Sigma \varepsilon_t^p/(\varepsilon_f/TF) < 0.5$ for category 1 and 2 $\Sigma \varepsilon_t^p/(\varepsilon_f/TF) < 0.8$ for category 1, 2, and 3
Damage due to creep and fatigue	Time duration, Δt and no. of cycles, n	Rupture life t_d and fatigue limit N	4 on CDF for category 1, 2 for category 1 and 2 and 1.3 for category 1, 2, and 3	$\Sigma \Delta t/t_d + \Sigma n/N < 0.25$ for category 1 and <0.5 for category 1 and 2 and <0.75 for category 1, 2, and 3
Brittle fracture	J-integral	J_c	—	$J < J_c$

In deciding the stress limits, especially for elastic analysis (the IEA is linked to strain limits), the ductility reduction under neutron dose requires careful considerations. In materials in which extensive strain hardening takes place (austenitic stainless steels that are used in core components), exceeding yield stress in a few special cases does not pose a problem because of the capacity of large ductility to absorb moderate strains. In fact, the rules for elevated temperature service allow for exceeding yield point in bending at outer fibers and also in case of secondary stresses in the shale-down regime. However, when the ductility becomes very low (~1%), exceeding yield stress even at local points can be detrimental. In IEA, such limits can be explicitly defined. But in elastic analysis, such limits can be imposed only through stress limits and these are sought to be prescribed by linking with ductility of the material that is found to be well correlated with the ratio S_y/S_u of the irradiated material. The irradiated data in the case of ductility are shown in Figure 7.21 [7.32]. This underlines the importance of such data for evolving the design.

Typical approach adopted is briefly explained. The stress limits to be satisfied for EICSA are given in Table 7.5. Two S_y/S_u thresholds have been considered in these limits: $S_y/S_u = 0.7$, beyond which the ductility of the irradiated material reduced below 5%, and $S_y/S_u = 0.85$, beyond which the ductility of the irradiated material reduced below 1%. These ductility transitions are significant while considering limits in bending and for secondary stresses. When the ductility is greater than 5% ($S_y/S_u < 0.7$), the membrane stress limits, bending stress limits, and secondary stress limits are all based on S_y. When the ductility is low, then these limits will be based on S_u. When ductility is very low, limits for bending loads (for $\varepsilon_u < 5\%$) and for secondary stresses (for $\varepsilon_u < 1\%$) are similar to those for membrane stresses. Higher stresses may be permitted for austenitic stainless steel when ductility is high, based on limit load approach and shake-down mechanisms, respectively. The strain limits to be satisfied for IEA are given in Table 7.6. To protect against plastic tensile instability due to loads that occur over the life of component, a summation rule is followed as the ductility of the material varies depending on the neutron dose. The limit on ε_m^p can be suitably assumed as equal to $\varepsilon_u / 2$.

Another alternative approach is based on CDF, which uses a life fraction rule to find out the damage of the clad that includes damage occurring in steady-state operation and transient state. Traditionally, CDF would include the creep and fatigue damage. But, in the case of fuel pin, the fatigue fraction will be insignificant, and hence, only creep is normally considered. Damage fraction at any particular interval of time is the ratio between the interval time and maximum time if

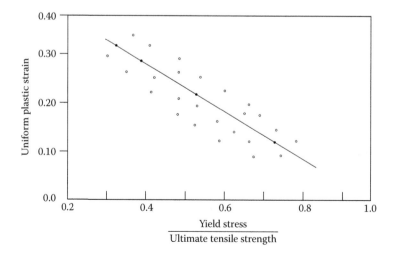

FIGURE 7.21 UE versus ratio of YS/UTS.

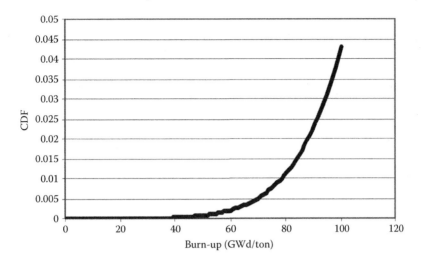

FIGURE 7.22 CDF versus burn-up (typical fuel pin).

subjected to total damage that would happen. If the damage fraction is integrated for the whole life of the clad, it gives the total damage of the clad. Cladding breach or rupture is implied by CDF of unity:

$$CDF = \int_0^t \frac{dt}{t_r(\sigma, T, \phi t)},$$

where

t_r is the time to rupture that is maximum time if subjected to the given stress, temperature, and fluence clad will operate without failure in hours

σ is stress in MPa

T is temperature in K

ϕt is fluence in n/cm^2

To find out the t_r, correlations are available in the literature for steady-state as well as transient conditions. A typical correlation that can be used for 20% CW D9 material to find out the t_r in the case of steady-state operation is given next [7.33]:

$$\log(t_r) = \frac{(5.4042 - \log(\sigma))}{2.244 \times 10^{-4} T} - 13.5.$$

During reactor operation, different categories of events are envisaged. So, CDF is also apportioned for different category events as a design approach. The approach could vary between the countries based on the accepted criteria by the regulators. Typically, for the Indian PFBR, the damage fraction under normal operation is limited to 0.25 and 0.5 is allotted for transient events, covering category 2 and 3 events. The remaining 0.25 is available for spent fuel storage and handling [7.34]. Figure 7.22 shows the CDF for the PFBR fuel pin as a function of burn-up as an example.

7.4 REACTOR STRUCTURES

In an FBR, apart from core, there are structures to support the core and contain the coolant as well as to facilitate circulation of sodium to remove the nuclear heat from the core. Further, circuits are required to facilitate the transport of heat from sodium to fluid (water/gas) to energy conversion

TABLE 7.8

Principal Selection Criteria for FBR Structural Materials

General Criterion	Specific Criteria
Mechanical properties	Tensile strength
	Creep
	Low cycle fatigue
	Creep–fatigue interaction
	High cycle fatigue
Design	Inclusion in RCC-MR/ASME design codes
Structural integrity	
Weldability	
Workability	
International experience	
Easy availability	
Lower cost	

system to generate electricity. The materials of construction of energy conversion system are of conventional type and hence not discussed. The focus in this section is to discuss the materials employed for the systems/components involved in nuclear steam supply system (NSSS). These components have direct contact with the coolant transferring the heat. They are vessels, tanks, pumps, heat exchangers, piping, pipe fittings, etc. Unlike the core structural materials, these structures should be designed for the entire plant life (40–60 year). Some structures see mild neutron dose (<5 dpa) and high temperature (~600°C) in addition to system pressures (<1 MPa except steam generators). These structural materials constitute high percentage of the total mass of the NSSS and accordingly the material cost. The principal selection criteria for FBR structural materials are listed in Table 7.8. Since most of the structures perform core supporting and core cooling functions, they are categorized under higher nuclear class categories (safety classes 1 and 2). Figure 7.23 depicts some typical materials and their safety classes adopted for a pool-type SFR in India. Hence, the choice of materials is very crucial in the FBR, especially for achieving the economy essential for the commercial exploitation.

7.4.1 Materials for the Reactor Systems Including Piping

Low alloy steels are not considered suitable for structural components of the primary heat transport system since they do not possess adequate high-temperature mechanical properties. Among stainless steels, ferritic stainless steels are not suitable because of (1) inadequate high-temperature mechanical properties, (2) sigma phase embrittlement at high temperatures, and (3) difficulty in welding due to grain coarsening. Martensitic stainless steels are prone to notch sensitivity, low ductility, and susceptibility to embrittlement between 693 and 823 K, and are therefore not considered. Austenitic stainless steels are chosen as the major structural materials in view of their adequate high-temperature mechanical properties, compatibility with liquid sodium coolant, good weldability, availability of design data, good irradiation resistance, and above all the fairly vast and satisfactory experience in the use of these steels in sodium-cooled reactors. Designers of FBRs all over the world prefer monometallic construction for liquid sodium systems because of the concern of interstitial element transfer (carbon in particular) through liquid sodium due to the differences in thermodynamic activity in bimetallic system. Hence, austenitic stainless steels are employed in the entire liquid sodium system even if the temperatures of some components are low enough to use less-expensive ferritic steels. Table 7.9 lists the structural materials selected for major components such as reactor vessel, intermediate heat exchanger (IHX), and piping in

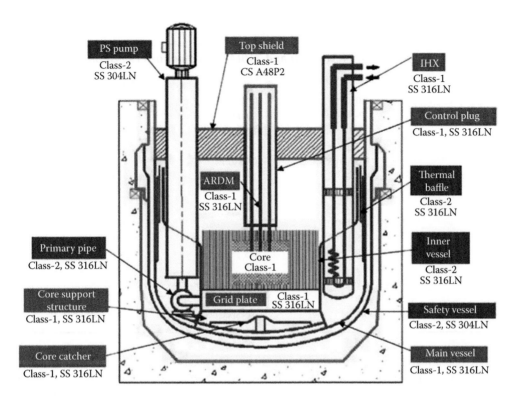

FIGURE 7.23 **(See color insert.)** Safety classifications and materials of SFR reactor assembly components.

TABLE 7.9
Materials Selected in SFR for Major Components

Reactor	Country	Reactor Vessel	IHX	Primary Piping Hot Leg/Cold Leg	Secondary Piping Hot Leg/Cold Leg
Rapsodie	France	316 SS	316 SS	316 SS/316 SS	316 SS/316 SS
Phenix[a]	France	316LSS	316 SS	—/316 SS	321 SS/304 SS
PFR[a]	United Kingdom	321 SS	316 SS	—/321 SS	321 SS/321 SS
JOYO	Japan	304 SS	304 SS	304 SS/304 SS	2.25Cr–1 Mo 2.25Cr–1 Mo
FBTR	India	316 SS	316 SS	316 SS/316 SS	316 SS/316 SS
BN-600[a]	Russia	304 SS	304 SS	—/304 SS	304 SS/304 SS
SPX-1[a]	France	316LN	316LN	—/304LN	316LN/304 SS
FFTF	United States	304 SS	304 SS	316 SS/316 SS	316 SS/304 SS
MONJU	Japan	304 SS	304 SS	304 SS/304 SS	304 SS/304 SS
SNR-300	Germany	304 SS	304 SS	304 SS/304 SS	304 SS/304 SS
BN-800[a]	Russia	304 SS	304 SS	—/304 SS	304 SS/304 SS
CRBRP	United States	304 SS	304/316	316 SS/304 SS	316H /304H
DFBR	Japan	316FR	316 FR	316FR/304 SS	304 SS/304 SS
EFR[a]	Europe	316LN	316LN	—/316SS	316LN/304 SS

[a] For Pool type reactor, there is no hot leg piping.

TABLE 7.10

Chemical Composition of 316 L(N) and 316FR Specified for EFR, DFBR, and SPX

Element	316L(N) SS (EFR)	316FR (DFBR)	316L(N) SS (Superphenix)
C	0.03	0.02	0.03
Cr	17–18	16–18	17–18
Ni	12–12.5	10–14	11.5–12.5
Mo	2.3–2.7	2–3	2.3–2.7
N	0.06–0.08	0.06–0.12	0.06–0.08
Mn	1.6–2.0	2.0	1.6–2.0
Si	0.5	1.0	0.5
P	0.025	0.015–0.04	0.035
S	0.005–0.01	0.03	0.025
Ti	NS	NS	0.05
Nb	NS	NS	0.05
Cu	0.3	NS	1.0
Co	0.25	0.25	0.25
B	0.002	0.001	0.0015–0.0035
Nb + Ta + Ti	0.15	Not specified	Not specified

currently operating or designed FBRs all over the world [7.35]. The grades selected include 304, 304L, 316, 316L, 321, 347, and their equivalents. Stabilized austenitic stainless steels 321 and 347 are less popular since their welds are prone to cracking during welding, during reheating, and also in service. These steels have also poor creep ductility. Table 7.10 gives the chemical composition of structural materials selected for European Fast Reactor (EFR), Demonstration Fast Breeder Reactor (DFBR, Japan), and Superphenix (France).

In high-temperature sodium, austenitic stainless steels have good resistance to general corrosion and localized corrosion. Localized corrosion is absent since the surface of steel is clean in sodium (no passive film) and electrochemical reaction is not possible in nonaqueous medium. However, mass transfer of metallic elements in SS can take place under the influence of nonmetallic impurities in liquid sodium such as oxygen and carbon. Oxygen leads to the formation of sodium chromite and carbon results in carburization or decarburization, which in turn influence the mechanical properties. Impurities other than carbon and oxygen such as chloride, calcium, and potassium are known to influence corrosion. Fast reactors and sodium loops have been successfully run for many years testifying to the long-term compatibility of austenitic stainless steels with sodium at elevated temperatures.

316L(N) SS is used for components experiencing relatively higher temperatures (above 770 K) while 304L(N) SS is selected for the rest of the structural components since the cost of 304L(N) SS is less by 20% in terms of material cost (about 15% if extra thickness required is also taken into account). During prolonged operation at elevated temperatures, stainless steels undergo microstructural changes such as precipitation of carbides and brittle intermetallic phases. Embrittlement will not be a problem for components operating below 700 K since precipitation is extremely sluggish at these temperatures. Table 7.11 gives the chemical compositions of types 304L(N) and 316L(N) SS specified for typical FBR along with the ASTM and RCC-MR specifications, which indicates that these specifications are more stringent than the ASTM specifications. The chemical composition ranges have been narrowed down to reduce the scatter in mechanical properties. The composition limits have been revised to meet specific property requirements. Chromium, molybdenum, nickel, and carbon contents have been specified taking into account intergranular corrosion resistance criteria developed from operating experience with nuclear reactors (both light water and fast reactors). Lower limits have been specified for carbon and nitrogen to ensure that the mechanical properties

TABLE 7.11

FBR Specification for 304L(N) and 316L(N) SS with ASTM A240 and RCC-MR

Element	304L(N)		316L(N)		
	ASTM	FBR	ASTM	FBR	RCCMR
C	0.03	0.024–0.03	0.03	0.024–0.03	0.03
Cr	18–20	18.5–20	16–18	17–18	17–18
Ni	8–12	8–10	10–14	12–12.5	12–12.5
Mo	NS	0.5	2–3	2.3–2.7	2.3–2.7
N	0.1–0.16	0.06–0.08	0.1–0.16	0.06–0.08	0.06–0.08
Mn	2.0	1.6–2.0	2.0	1.6–2.0	1.6–2.0
Si	1.0	0.5	1.0	0.5	0.5
P	0.045	0.03	0.045	0.03	0.035
S	0.03	0.01	0.03	0.01	0.025
Ii	NS	0.05	NS	0.05	—
Nb	NS	0.05	NS	0.05	—
Cu	NS	1.0	NS	1.0	1.0
Co	NS	0.25	NS	0.25	0.25
B	NS	0.002	NS	0.002	0.002

Inclusion Contents (max.)

Type	Thin	Thick
Type A (sulfide)	1	0.5
Type B (alumina)	2	1.5
Type C (silicate)	2	1.5
Type D (globular oxide)	3	2.0
A + B + C + D	6	4.0

NS, not specified.

match those of 304 and 316 SS grades for which design curves are available in RCC-MR/ASME codes. An upper limit has been specified for carbon to ensure freedom from sensitization during welding. The upper limit for nitrogen is lowered to 0.08 wt% compared to 0.16 wt% in ASTM and ASME specifications mainly on consideration of improving weldability and minimizing scatter in mechanical properties. Phosphorus, sulfur, and silicon are treated as impurities as they have adverse effects on weldability. Therefore, acceptable maximum limits are reduced to values that can be achieved in steel-making practice.

Considering the adverse effects of titanium, niobium, copper, and boron on weldability, maximum permissible limits have been imposed, although no such limits exist in ASTM specifications. A minimum level has been specified for manganese to improve weldability. Upper limit has been specified for cobalt to reduce Co^{60} activity induced by neutron irradiation so as to facilitate ease of eventual maintenance of the components of primary sodium system. In addition to more stringent composition limits, a specification for inclusions has been added keeping in view that sulfide inclusions are most detrimental especially from welding considerations and globular oxides are least harmful. A grain size finer than ASTM No. 2 is specified so as to achieve optimum high-temperature mechanical properties.

For grid plate, though temperatures are not in the creep region, 316L(N) SS is preferred over 304L(N) SS in view of better ductility after irradiation. Studies carried out at IGCAR (up to 10,000 h), which are consistent with the international experience, have shown that creep rupture strength of 316L(N) grade is superior to 316 SS; it has generally lower creep rates than type 316 SS. The improvement in properties is attributed to solid solution strengthening by nitrogen and precipitation

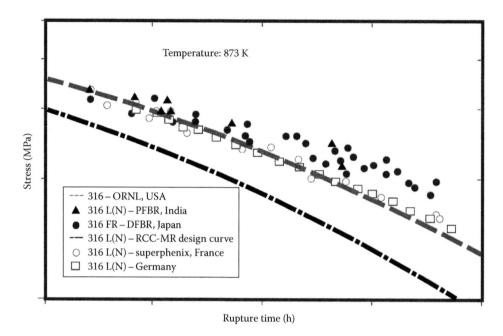

FIGURE 7.24 Creep rupture strengths of SS 316 and 316L(N). (From Schirra, M.C., *Fracture of Engineering Materials*, Parker, J.D. (ed.), The Institute of Materials, London, UK, p. 612; Brinkman, C.R., ORNL/CP-101053, National Technical Information Service, Alexandria, VA, 1999.)

strengthening by fine carbides. Figure 7.24 shows a comparison of the creep rupture strength of 316L(N) with 316 SS. The 316L(N) data relate to long-term creep tests (up to 60,000 h) from breeder programs in Germany [7.36], Japan [7.37], France [7.37], and India. The reference 316 SS data are taken from the long-term creep program of ORNL, USA [7.37]. RCC-MR design curve for expected minimum stress to rupture is superimposed in Figure 7.24. Creep rupture strength of 316 L(N) SS is better especially for longer rupture times.

Reasonable amount of data are available in the open literature on the LCF and creep–fatigue interaction behavior of nitrogen-alloyed 316 SS intended for fast reactor applications. Comparative evaluation of strain-controlled LCF behavior of nitrogen-alloyed 316FR (which is the Japanese fast reactor grade steel) and 316 SS shows that continuous cycling LCF life is not significantly influenced by minor variations in carbon and nitrogen contents (Figure 7.25) [7.37]. However, the minor changes in carbon and nitrogen have been found to have substantial effects on creep–fatigue interaction behavior. Hold-time tests conducted at tensile peak strain at elevated temperatures clearly indicated that while increasing dwell time reduced fatigue lives in both 316 SS and its nitrogen-alloyed version, the extent of such reduction is smaller for nitrogen-alloyed 316 SS (Figure 7.26) [7.37]; the creep–fatigue interaction tests on 316FR SS had been performed concurrently in Japan and the United States. Results of studies carried out at IGCAR are consistent with the behavior reported by other countries.

7.4.2 Welding Consumables and Properties of Welds

Welding is extensively employed in the fabrication of FBR components. Weld metal cracking and heat affected zone (HAZ) cracking are major areas of concern in welding austenitic stainless steels. Weld metal cracking can be controlled by optimizing the chemical composition of the welding consumables. The optimized chemical composition for PFBR for 316(N) SS welding electrodes is given in Table 7.12 along with the ASME specifications for E-316 SS. Carbon in the range of 0.045–0.055 wt% and nitrogen in the range of 0.06–0.1 wt% are specified to provide weld joints with improved creep

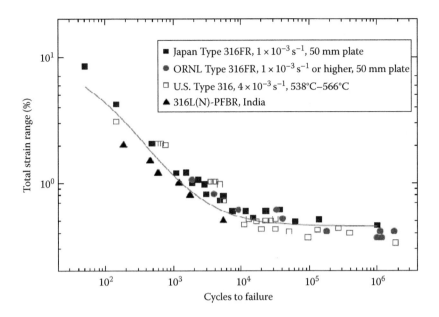

FIGURE 7.25 Fatigue behavior of 316 and 316FR SS. (From Brinkman, C.R., ORNL/CP-101053, National Technical Information Service, Alexandria, VA, 1999.)

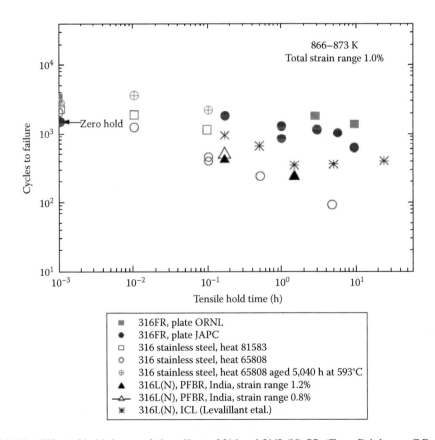

FIGURE 7.26 Effect of hold time on fatigue lives of 316 and 316L(N) SS. (From Brinkman, C.R., ORNL/CP-101053, National Technical Information Service, Alexandria, VA, 1999.)

TABLE 7.12

Typical Specifications for Modified 316 Electrodes and Permissible Limits for Delta-Ferrite as per WRC-92 FN Diagram (Single Values Specified Are Maximum Permissible)

Element	ASME SFA 5.4 E316	FBR 316(N)
C	0.08	0.045–0.055
Cr	17–20	18–19
Ni	11–14	11–12
Mo	2–3	1.9–2.2
N	Not specified	0.06–0.10
Mn	0.5–2.5	1.2–1.8
Si	0.9	0.4–0.7
P	0.04	0.025
S	0.03	0.02
Ti + Nb + Ta	NS	0.1
Cu	0.75	0.5
Co	Not specified	0.2
B	Not specified	20 ppm
δ-Ferrite	3–10 FN	3–7 FN

strength and freedom from sensitization in the as-welded state. In addition, ferrite in the weld metal is specified to be between 3 and 7 ferrite number (FN) to promote ferritic solidification mode. A minimum of 3 FN is specified to ensure freedom from hot cracking in the weld metal. Because delta-ferrite undergoes phase changes to carbides and brittle intermetallic phases at high temperatures, an upper limit of 7 FN has been specified. Nitrogen in the specified range has no detrimental effect on weldability of 316 L(N) stainless steel. HAZ cracking is avoided by specifying lower permissible limits for P, S, and Si and also by specifying limits on B, Ti, and Nb, which are not specified in the ASTM standards for the base metal. 316(N) SS electrodes would be utilized for welding of both 316L(N) SS and 304L(N) SS base materials. This would avoid any mix-up of electrodes in welding if a different electrode is selected for 304L(N) SS. 16-8-2 Filler wire will be used for TIG welding since this composition has better microstructural stability, creep strength, and toughness. 16-8-2 Electrodes for MMA welding are not easily available.

Evaluation of creep properties of 316 (N) weld metal has shown that its creep rupture strength and ductility are better than those of 316 SS weld metal. Figure 7.27 shows a comparison of creep rupture strengths of 316 and 316(N) weld metals at 923 K; about 30% increase in rupture strength was observed by alloying with nitrogen [7.38]. There is no information available in the open literature on the LCF behavior of 316(N) welds and their joints. Detailed investigations conducted at IGCAR at 873 K revealed the LCF life in the order base metal > weld metal > weld joint. The rank order of lives exhibited by 316 (N) and its weld joint is very much similar to those shown by stainless steels 304, 316, and their weld joints. The poor strain-controlled fatigue resistance of weldments is attributed to the presence of coarse grains in the HAZ, which act as a metallurgical notch leading to shortening of the crack initiation phase.

7.4.3 Materials for Steam Generator

The very high reactivity of sodium with water makes the steam generator a key component in determining the efficient running of the plant and demand high integrity of steam generator components. The high integrity of SG components can be achieved by choosing a proper material followed by an optimized design and fabrication. It is decided to manufacture the steam generator in mono-metallic

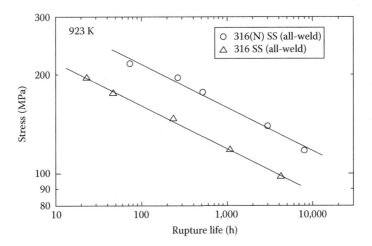

FIGURE 7.27 Comparison of creep rupture strengths of 316 and 316(N) SS weld metals. (From Sasikala, G. et al., 31A, 1175, 2000.)

material (tube, shell, and thick-section tube sheet/plate) since employing a single structural material enhances the reliability of the critical tube to tube sheet welds. Modified 9Cr-1Mo ferritic steel has been selected for all the steam generator components of PFBR. The selection of modified 9Cr–1Mo steel is based on several important considerations, and these are described in the following paragraphs.

The principal selection criteria of materials for steam generator are given in Table 7.13. These include the general criteria as well as the criteria directly related to the use of materials in sodium-heated steam generator. Materials selected for LMFBR steam generator application should meet requirements of high-temperature service such as high-temperature mechanical properties including creep and low-cycle fatigue, resistance to loss of carbon to liquid sodium, which leads to reduction in strength, resistance to wastage in case of small leaks leading to sodium–water reaction, and resistance to stress corrosion cracking (SSC) in sodium and water media.

TABLE 7.13
Principal Selection Criteria for SFR Steam Generator Material

General Criteria	Criteria Related to Use in Sodium
Mechanical properties	Mechanical properties in sodium
Tensile strength	Susceptibility to decarburization
Creep strength	
Low cycle fatigue and high cycle fatigue	
Creep–fatigue	
Ductility	
Ageing effects	
Inclusion in pressure vessel codes or	Corrosion under normal sodium chemistry condition,
availability of adequate data	fretting, and wear
Corrosion resistance under storage (pitting)	Corrosion resistance In the case of sodium water
normal and off-normal chemistry conditions	reaction (stress corrosion cracking, self-enlargement
	of leak and Impingement wastage)
Workability	
Weldability	
Availability	
Cost	

TABLE 7.14

Materials Selected for Steam Generator in Major FBRs

Reactor	Sodium Inlet (K)	Steam Outlet (K)	Tubing Material	
			Evaporator	Superheater
Phenix	823	785	2.25Cr–1 Mo	321 SS
			2.25Cr–1 Mo stabilized	
PFR	813	786	2.25Cr–1 Mo stabilized	316 SS
			Replacement unit: 2.25Cr–1Mo	Replacement unit in 9Cr–1Mo
FBTR	783	753	2.25Cr–1Mo stabilized	
BN-600	793	778	2.25Cr–1Mo	304 SS
SPX-1	798	763	Alloy 800 (once through integrated)	
MONJU	778	760	2.25Cr–1Mo	304 SS
SNR-300	793	773	2.25Cr–1Mo stabilized	2.25Cr–1Mo
				Stabilized
BN-800	778	763	2.25Cr–1Mo	2.25Cr–1Mo
CRBR	767	755	2.25Cr–1Mo	2.25Cr–1Mo
DFBR	793	768	Modified 9Cr–1Mo (once through integrated)	
EFR	798	763	Modified 9Cr–1Mo (once through integrated)	

The choice of materials for FBR steam generators in operation or under design worldwide is presented in Table 7.14. The sodium inlet and steam outlet temperatures for these steam generators are also included in this table. In the case of PFBR steam generator, the sodium inlet temperature is 798 K, whereas the steam outlet temperature is 766 K. It can be seen from this table that 2.25Cr–1Mo steel, either plain or stabilized grade, has been used in evaporators, whereas in superheaters austenitic stainless steels have also been used. However, the recent trend favors the use of modified 9Cr–1Mo steel for LMFBR steam generator applications.

For PFBR steam generator, a range of materials starting from ferritic steels (2.25Cr–1Mo, Nb stabilized 2.25Cr–1Mo, 9Cr–1Mo [grade 9], modified 9Cr–1Mo [grade 91]), austenitic stainless steels (AISI 304/316/321), and alloy 800 were examined. In view of the poor resistance to aqueous SCC, austenitic stainless steels of 300 series were not considered for the steam generator. Alloy 800 shows better resistance to SCC than austenitic steels, but it is not immune to SSC in chloride and caustic environments. Therefore, ferritic steels are the most preferred for steam generator applications. Among the ferritic steels, 2.25Cr–1Mo and 9Cr–1Mo steels and their variants were considered for the steam generator. The basis for choosing modified 9Cr–1Mo (grade 91) and its relative properties with other candidate materials are described in the following paragraphs.

Before describing the various properties of grade 91 steel, which led to its selection for PFBR steam generator, it is important to mention the material specification for steam generator, which is close to that specified in ASTM. The chemical composition (Table 7.15) is controlled within close limits to avoid scatter in the mechanical properties. Lower limits are specified for residual elements, like sulfur, phosphorous, and silicon, to improve weldability and reduce the inclusion content to ensure a high degree of cleanliness.

The strength of grade 91 steel is found to be higher than those of 2.25Cr–1Mo and plain 9Cr–1Mo steels. The creep strength of modified 9Cr–1Mo steel is significantly higher than those of the conventional 2.25Cr–1Mo and plain 9Cr–1Mo steels over a wide temperature range and is greater or equal to that of AISI type 304 austenitic stainless steel up to 873 K (Figure 7.28) [7.39]. Comparison of creep strengths of various grades of ferritic steels widely used for high-temperature applications suggests that modified 9Cr–1Mo steel exhibits higher creep strength than most of the other materials (Figure 7.29) [7.40]. In Figure 7.29, open circle is used to represent nine different grades of ferritic steels that converge at long test durations. Though for shorter creep lives, 12Cr–1Mo–1W–0.3V

TABLE 7.15

PFBR Specifications for Modified 9Cr–1Mo Steel Tubes and Welding Consumables

Element	Base Metal (Tube)	Filler Wire (Weld Deposit)	Electrode
C	0.08–0.12	0.08–0.12	0.08–0.12
Cr	8.00–9.00	8–9.5	8.0–9.5
Mo	0.85–1.05	0.85–1.05	0.85–1.05
Si	0.20–0.50	0.2–0.4	0.2–0.3
Mn	0.30–0.50	0.5–1.2	0.5–1.2
V	0.18–0.25	0.15–0.22	0.15–0.22
Nb	0.06–0.10	0.04–0.07	0.04–0.07
N	0.03–0.07	0.03–0.07	0.03–0.07
S	0.01 max.	0.01 max.	0.01 max.
P	0.02 max.	0.015 max.	0.01 max.
Cu	0.10 max.		0.25 max.
Ni	0.20 max.	0.6–1.0*	0.6–1.0*
Al	0.04 max.		0.04 max.
Sn	0.02 max.	—	—
Sb	0.01 max.	—	—
Ti	0.01 max.	—	—

Note: Ni + Mn less than or equal to 1.5.

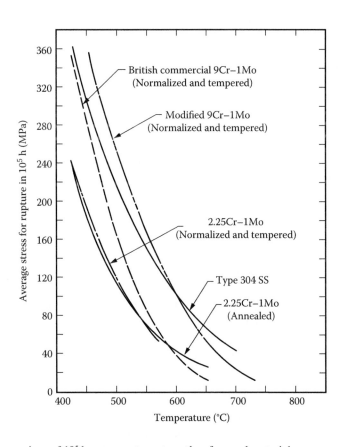

FIGURE 7.28 Comparison of 10^5 h creep rupture strengths of several materials.

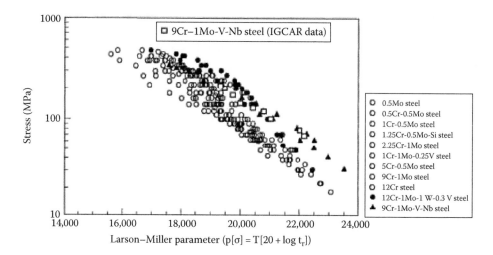

FIGURE 7.29 Creep-rupture strength of 11 types of ferritic heat-resistant steels. (From Kimura, K. et al., in *Advances in Turbine Materials, Design and Manufacturing*, Strang, A. et al. (eds.), The Institute of Materials, London, UK, 1997, pp. 257–269.)

exhibits higher creep strengths, for larger creep lives, its creep strength approaches that of low-alloy ferritic steels because of microstructural degradation. Modified 9Cr–1Mo steel does not exhibit such a drastic reduction in creep strength at longer durations due to the stability of the microstructure; its creep strength remains higher than that of several ferritic steels at longer test durations. This is the most important aspect favoring the selection of modified 9Cr–1Mo steel for steam generator. Further, the higher creep strength of modified 9Cr–1Mo steel allows the use of comparatively thinner tubes. The thinner tube and higher thermal conductivity of this material reduce the heat transfer area requirements of the steam generator.

7.4.4 Welding Consumables for Modified 9Cr–1Mo Steel

For the welding of modified 9Cr–1Mo steel, consumables having composition closely matching with that of the base metal are normally employed. However, achieving required toughness in the weld metal after PWHT has been a problem in this steel, especially in the case of shielded metal arc (SMA) welds. Therefore, in the AWS/ASME specification for the consumables, minor modifications have been made for Ni, Mn, Nb, V, and N. In addition, the specification calls for the determination of RTNDT and the use of only alloyed core wire for making SMA welding electrodes. The specified composition is given in Table 7.15.

7.4.4.1 Use of Modified 9Cr–1Mo Steel for Thick Sections

Apart from superior elevated temperature mechanical properties of 9Cr–1Mo grades of steels as compared to 2.25Cr–1Mo steel, the ability and tolerance for providing nearly a constant microstructure over a large section size is another aspect in the selection of modified 9Cr–1Mo steel for steam generator. This in turn ensures only a small variation in mechanical properties with increasing thickness or between the surface and the centre of thick section products.

7.4.5 Tri-Metallic Transition Joint

As the main structural and piping material is austenitic 316LN SS and the steam generator material is modified 9Cr–1Mo steel, a dissimilar weld involving these two materials is inevitable in the fabrication of steam generator. A large number of premature failures of the direct joint involving austenitic stainless steel and ferritic steel operating at high temperature have been reported in the past, mainly from fossil power plants. The failures are mainly attributed to (1) large differences in the thermal expansion coefficients of these two steels, which lead to generation of thermal stresses during start-up and shutdown; (2) difference in the creep strength of these materials; and (3) carbon migration from ferritic steels to austenitic steels leading to the formation of soft zone near the interface. It has been shown that the introduction of an intermediate piece of material, having thermal expansion coefficient value between austenitic SS and ferritic steel, can significantly reduce the thermal stresses generated. (Coefficient of thermal expansion for 316 L(N) SS and 9Cr–1Mo steel are 18.5 and 12.6 μm/m/K, respectively.) Accordingly, a tri-metallic joint configuration in which an alloy 800 spool piece welded to 316L(N) SS pipe on the one side, and modified 9Cr–1Mo steel pipe on the other side, is chosen for this dissimilar joint. Although alloy 600 is another material being used for transition joints, alloy 800 has been preferred over alloy 600 since the material is included in ASME code and also was the choice of transition joint (2.25Cr–1Mo–SS) for CRBRP steam generator. For welding alloy 800 to modified 9Cr–1Mo, Inconel 82/182, welding consumable is recommended. For welding alloy 800 to 316 L(N) SS, 16-8-2 filler wire is selected.

7.5   COOLANT

The need to preserve the hardness of the neutron spectrum in a fast reactor precludes the use of moderating materials such as water as coolant. Since the fast reactors have a high specific power dictated by economic consideration, only materials having high heat transfer coefficient such as liquid metals could be used as coolants. In addition to sodium, there are other liquid metallic coolants that were considered as prospective but not used due to their disadvantages. Liquid sodium has many attractive features as a fast reactor coolant, some of which are listed below:

- Liquid sodium has a very high thermal conductivity and a very high heat transfer coefficient. It has a high temperature coefficient density. Thus, even with the normally available fuel surface area, the liquid sodium can remove the large heat from the core even under conditions of natural convection that sets up easily under incidents of sodium during pump failure.
- Sodium has a high boiling point (1156 K). It is possible to keep the pressure of the coolant system near atmospheric pressure. Designing such low-pressure systems is easy and inexpensive. Moreover, there are no safety issues involved with loss of system pressure since the stored energy in the system is less. It can be operated at high outlet temperatures with good margin to boiling making greater conversion efficiencies possible.
- Pure sodium is highly compatible with austenitic stainless steel, which is the major structural material in fast reactors. It has very low corrosion rate on SS at the operating temperatures (~1 μm/year), which is quite tolerable.
- It has a low melting point (371 K), which means that keeping it as a liquid in the coolant circuits is easy and requires less preheating as compared to lead. Moreover, being a solid at room temperature, it is easy to freeze the coolant pipeline for cutting during repair.
- Sodium is very easy to purify just by passing it through a low-temperature section called cold trap. Since the solubility of various impurities in sodium is a strong function of temperature, the impurities precipitate at the cold section easily.
- Sodium has no long-lived activation product. ^{24}Na (n, γ product) and ^{23}Ne (n, γ product) are all short-lived. ^{22}Na (n, 2n product) does not build up to high levels as it has a high neutron capture cross section for thermal neutrons.
- The values of viscosity and density of liquid sodium are very close to that of water. This makes it easy to test sodium system components such as pumps using water as test fluid.
- Sodium has good radiation and thermal stability.

- It has a good heat capacity and low density leading to low pumping power requirements.
- Sodium is not a biologically toxic material.
- Sodium is available in plenty in nature. It is cheap and easy to produce.

The only problem with sodium is its high chemical reactivity and for medium- and large-size reactors its reactivity worth is positive. It reacts readily with oxygen in air to form oxides and reacts violently with water. This high chemical reactivity and the high exo-thermicity of sodium fire in the event of a sodium leak apparently pose a great danger. However, the operating experience for 20 fast reactors the world over, which is equivalent to around 380 reactor years, testifies to the possibility of coping with the problem without much difficulty through design measures. The leakage possibilities are minimized by taking into account, while designing sodium systems, the peculiar conditions such as high and fluctuating secondary stresses, high cycle fatigue, and prolonged exposure to high temperatures. By adopting proper design codes for thin-walled vessels and pipe work and by employing nondestructive in-service inspection and leak detection methods, it is possible to prevent sodium leaks and ensure system leak tightness. This also helps in minimizing radioactivity release to the environment.

7.5.1 Physical Properties

Salient physical properties of sodium are given in Table 7.17, and corresponding properties of water are also given for comparison. Melting point of sodium is approximately 2 K below the boiling point of water, and the melting is accompanied by a volume increase of 2.17%. Molten sodium appears like mercury but is actually lighter than water. Because of its high boiling point, sodium has a large liquidus range (1057 K), which is next only to that of lithium among other alkali metals. Its thermal conductivity (0.84 J/cm/s/K) is approximately 2 orders higher than that of water, and its specific heat (1383 kJ/kg/K) is 2.5 times less than that of water. As seen from Table 7.17, viscosities of sodium and water are not very different. Hence, operations such as pouring, mechanical pumping, stirring, and other general laboratory manipulations of molten sodium are very similar to those of water. Electrical resistivity of sodium is very low (in fact, among alkali metals, it has the lowest electrical resistivity), and this enables it to be efficiently pumped by electromagnetic principles.

7.5.2 Nuclear Properties

7.5.2.1 Radiological

Liquid sodium has a very low absorption cross section for fast neutrons. The absorption cross section of ^{23}Na is 0.87 mb for fast neutrons. The (n,γ) activation product is ^{24}Na, which is a strong beta-gamma emitter. Fortunately, it has a short half-life of 15 h. It is possible to approach the radioactive sodium coolant circuit for maintenance and repair after about 10 half-life periods from reactor shutdown. However, during reactor operation, the activity level is in mCi/cm³. The other activation product of ^{23}Na is ^{22}Na by (n, 2n) reaction. This has a rather long enough half-life of 2.7 years. However, it has a high absorption cross section for thermal neutron, which prevents it from building up above a few mCi/cm³ levels. The (n, p) reaction product of ^{23}Na is ^{23}Ne, which has a very short half-life of 38 s. This is quantitatively released to the cover gas space.

7.5.2.2 Reactivity Coefficients of Sodium Coolant and Reactor Safety

Sodium is sufficiently nonmoderating and can preserve the hardness of the neutron spectrum. It has a moderating ratio of only 0.89 compared to 62 for water. One nuclear property of the coolant that is important for reactor safety is its reactivity coefficient for both sodium temperature and void.

7.5.2.3 Temperature Reactivity Coefficient and Sodium Void Coefficient

The temperature and power coefficients of reactivity of any reactor should be negative so that temperature and power transients are self-limiting. Further, the time constant associated with them should be small so that the reactivity feedbacks are available promptly and the system is stable. Upon temperature rise or loss of sodium, the neutron spectrum hardens, resulting in the following effects: larger leakage of neutrons leading to negative reactivity changes, increased fast fission from fertile materials leading to positive reactivity changes, and increase in "η" (neutrons emitted per neutron absorbed) leading to positive reactivity changes.

Loss in density of sodium leads to reduction in absorption of neutrons and hence the positive reactivity effect. However, this effect is negligibly small. Calculations that have been confirmed experimentally indicate that for small-size reactors like FBTR, the net coolant reactivity coefficient is negative, while for medium- and large-sized reactors like Superphenix, it is positive. The contribution of coolant temperature reactivity coefficient is only 20% of the total. In power coefficient, only 4% contribution comes from positive reactivity coefficient of sodium and the rest is from the fuel. Thus, any transient is undercooling or reactivity-initiated incidents are mainly arrested by fuel reactivity coefficient not by coolant reactivity coefficient. Also, the time constant associated with the fuel and coolant reactivity coefficients is small and so poses no stability problem.

7.5.3 CHEMICAL PROPERTIES

Being an alkali metal, sodium metal is highly reactive. On exposure to dry air, solid sodium first reacts with oxygen to form voluminous sodium oxide layer, which retards the rate of further reaction. Sodium does not react with nitrogen and does not form any stable nitride. It reacts with humid air very fast, and the major reaction product is sodium hydroxide, which subsequently reacts with carbon dioxide in air to yield sodium carbonate. Since these reaction products are highly hygroscopic, uncontrolled exposure of sodium to moist air can lead the metal coming into contact with water (or water-rich phases), resulting in explosive sodium–water reaction. Liquid sodium reacts instantaneously with gaseous oxygen. It burns in air to yield sodium oxide. Under oxygen-deficient conditions, the product formed is sodium monoxide, Na_2O. When oxygen supply is unlimited and temperature is low, the product formed would be sodium peroxide, Na_2O. Liquid sodium reacts with gaseous hydrogen and forms its hydride, NaH. Reaction rate with gaseous hydrogen is slow at temperatures below 523 K and is very fast at temperatures above 673 K. Na_2O and NaH dissolve in liquid sodium before precipitating as a separate phase. Solubility of these compounds in sodium increases with increase in temperature.

Liquid sodium is generally contained in austenitic and ferritic steel circuits. Sodium is quite compatible with them under pure conditions. Solubility of the steel constituents—iron, chromium, nickel, manganese, molybdenum, etc.—in sodium is low and is in ppm ranges only.

However, higher oxygen contents in sodium (>5 ppm) would lead to enhanced corrosion and mass transport phenomenon. Carbon is an important interstitial alloying element in structural steels, and sodium can act as a medium for its transfer from one section of structural components to another. This can arise due to the presence of temperature gradient in a sodium circuit with structural components made of single alloy and/or the presence of multialloy components in that circuit. Carbon transport is kinetically slow at temperatures below 673 K and can become very significant above 873 K. Reaction of sodium with water/steam is of special importance. Heat from secondary sodium is transferred to water across a ferritic steel tube wall in the steam generator of a fast reactor. In case of a steam leak into sodium circuit, sodium reacts instantaneously with steam/water forming sodium hydroxide and hydrogen. The exothermic nature of the sodium–water reaction and the caustic nature of the product can lead to tube material wastage and cause the steam leak to expand rapidly. If these leaks are not detected in time and remedial action not taken, the growing flame can cut the adjoining steam carrying tubes and lead to sodium–water explosions. Such a blow out in PER resulted in the destruction of 40 steam tubes and the one at BN-350 resulted in a long shutdown. However, these incidents did not lead to any catastrophe or nuclear release and the events were completely manageable. Sodium hydroxide is stable in sodium only above 693 K. Sodium hydroxide in excess sodium decomposes to yield Na_2O and NaH below this temperature. Hydrogen formed in the sodium–water reaction can dissolve further in sodium, the slow kinetics of which has been mentioned earlier.

7.5.3.1 Specifications and Chemical Quality Control

Pure sodium is quite compatible with austenitic stainless steel. However, the presence of impurities even in ppm levels can lead to enhanced corrosion and mass transfer. Carbon in sodium leads to carburization of components. Trace metals can get activated and add to activity burden. On account of these, the sodium is purified to very high-purity levels before charging it into coolant circuits.

REFERENCES

7.1 Carbajo, J.J., Yoder, G.L., Popov, S.G., Ivanov, V.K.J. (2001). A Review of the Thermophysical Properties of MOX and UO2 Fuels. *J. Nucl. Mater.*, 299, 181.

7.2 Karahan, A. (2009). Modelling of thermo-mechanical and irradiation behaviour of metallic and oxide fuels for sodium fast reactor. PhD report, MIT, Cambridge, MA.

7.3 Mikailoff, H. (1974). L'element Combustible Carbure a Joint Helium Pour la Fillere a Neutrons Rapides: Problems Poses par le gonflement du combustible. BIST 196.

7.4 IAEA. (2012). Summary report of the technical meeting on primary radiation damage: From nuclear reaction to point defects. Nuclear Data Section, INDC(NDS)-0624, Vienna, Austria. October 1–4, 2012.

7.5 IAEA. (1999). Status of liquid meal cooled fast reactor technology IAEA, Vienna. IAEA-TECDOC-1983. April 1999.

7.6 Garner, F.A. (1993). Irradiation performance of cladding and structural steels in liquid metal reactors. In *Nuclear Materials, Part 1, Materials Science and Technology*, B.R.T. Frost (ed.). VCH Publishers, Weinheim, Germany, pp. 420–543.

7.7 Mansur, L.K. (1986). Swelling in irradiated metals and alloys. In *Encyclopedia of Materials Science and Engineering*. Pergamon Press, Oxford, UK, Chapter 6, p. 4834.

7.8 Garner, F.A. (1984). Recent insights on the swelling and creep of irradiated austenitic alloys. *J. Nucl. Mater.*, 122–123, 459.

7.9 Puigh, R.J. et al. (1982). An in-reactor creep correlation for 20% cold worked AISI 316 stainless steel. *Effects of Irradiation on Materials: Eleventh Conference*, ASTM, Philadelphia, PA, ASTM STP 762, pp. 108–121.

7.10 Garner, F.A. (1994). Irradiation performance of cladding and structural steels in liquid metal reactors. In *Material Science and Technology*, Cahn, R.W. et al. (eds.). VCH, Weinheim, Germany, Volume No. 10A, pp. 419–543.

7.11 Ulmaier, H., Trinkaus, H. (1992). Helium in metals—Effects on mechanical properties. *Mater. Sci. Forum*, 97–99, 451–472.

7.12 Yilmaz, F., Hassan, Y.A., Porter, D.L., Romanenko, O. (2003). Swelling and mechanical property changes in Russian and American austenitic steels in EBR-II and BN350. *Nucl. Technol.*, 144(3), 369–378.

7.13 Mannan, S.L., Chetal, S.C., Raj, B., Bhoje, S.B. (2003). Selection of materials for prototype fast breeder reactor, Indira Gandhi Centre for Atomic Research, *Transactions of the Indian Institute of Metals*, Kalpakkam, India, 56(2), 155–178. http://www.igcar.gov.in/igc2004/PFBR.pdf.

7.14 IAEA. (2012). Structural material for liquid metal cooled fast reactor fuel assemblies—Operational behavior. IAEA Nuclear Energy Series No. NF-T-4.3.

7.15 Le Flem, M. et al., CEA. (2011). Advanced materials for fuel cladding in sodium fast reactors: from metals to ceramics. *Presentation to 3rd MATGEN Summer School*, Lerici, Italy, September 19–23, 2011.

7.16 Herschbach, K., Schneider, W., Ehrlich, K. (1993). Effects of minor alloying elements upon swelling and in-pile creep in model plain Fe-15Cr-15Ni stainless steels and in commercial DIN 1.4970 alloys. *J. Nucl. Mater.*, 203, 233–248.

7.17 Waltar, A.E., Todd, D.R., Tsvetkov, P.V. (2012). *Fast Spectrum Reactors*, New York, USA, Springer Press.

7.18 Dubuisson, P., Gilbon, D., Seran, J.L. (1993). Microstructural evolution of ferritic-martensitic steels irradiated in the fast breeder reactor Phénix. *J. Nucl. Mater.*, 205, 178.

7.19 Klueh, R.L., Harries, D.R. (Eds.) (2001). High chromium ferritic and martensitic steels for nuclear applications. ASTM, Philadeplphia, PA, p. 135.

7.20 Raj, B., Mannan, S.L., Vasudeva Rao, P.R., Mathew, M.D. (2002). Development of fuels and structural materials for fast breeder reactors. *Sadhana*, 27(Part 5), 527–558.

7.21 Jayakumar, T., Laha, K., Mathew, M.D., Saroja, S., Kartik, V. (2013). Development of advanced fuel cladding material for Indian sodium cooled fast reactors. ICAPP, Korea, Japan, 1(2), 826–832, April 2013.

7.22 Raj, B. (2007). A perspective on R&D in austenitic stainless steels for fast breeder reactor technology at Kalpakkam. *International Symposium on Advances in Stainless Steels*, ISAS 2007, April 9, 2007, Chennai.

7.23 Ukai, S., Mizuta, S., Fujiwara, M., Okuda, T., Kobayashi, T. (2002). Development of 9Cr-ODS martensitic steel claddings for fuel pins by means of ferrite to austenite phase transformation. *J. Nucl. Sci. Technol.*, 39(7), 778–788, July 2002.

7.24 Ullmaier, H., Schilling, W. (1980). *Physics of Modern Materials*. IAEA, Vienna, Austria, p. 301.

7.25 David, C., Panigrahi, B.K., Rajaraman, R., Balaji, S., Balamurugan, A.K., Nair, K.G.M., Amarendra, G., Sundar, C.S., Raj, B. (2007). Effect of titanium on the void swelling behavior in (15Ni-14Cr)-Ti modified austenitic steels studied by ion beam simulation. *J. Nucl. Mater*, 392(3), 578.

7.26 Gyr, W., Friedel, G., Friedrich, H.J., Pamme, H., Firth, G., Lauret, P. (1990). EFR decay heat removal system design and safety studies. *International Fast Reactor Safety Meeting*, Snowbird, UT, pp. 543–552.

7.27 Simpson, D.E., Little, W., Peterson, E. (1967). Selected safety considerations in the design of the fast flux test facility. *Proceedings of the International Conference on the Safety of Fast Reactors*, Aix-en-Provence, France, September 1967, pp. 19–22.

7.28 Del Beccaro, R., Mitchell, C.H., Heusener, G. (1992). The EFR safety approach. *International Conference on Design and Safety of Advanced Nuclear Power Plants*, Tokyo, Japan, October 1992, pp. 29.1/1–29.1/8.

7.29 Nelson, D.V., Abo-El-Ata, M.M., Stephen, J.D., Sim, R.G. (1978). Development of design criteria for highly irradiated core components. ASME-78-PVP-78, ASME Pressure Vessels and Piping Conference, Montreal, Canada, June 1978.

7.30 Wei, B.C., Nelson, D.V. (1982). Structural design criteria for highly irradiated core components. In *Pressure Vessel & Piping Design Technology—A Decade of Progress*. ASME Publication, New York.

7.31 Desprez, D. et al. (1987). Design criteria for FBR core components—An overview of the methodology developed in France. *Conference on FBR Systems: Experience Gained and Path to Economic Over Generation*, Washington, DC.

7.32 Shibahara, I., Omori, T., Sato, Y., Onose, S., Nomura, S. (1993). Mechanical property degradation of fast reactor fuel cladding during thermal transients, ASTM STP 1175, ASTM, Philadelphia, PA.

7.33 Puthiyavinayagam, P., Govindarajan, S., Chetal, S.C. (1994). Design data for 20% CW D9 material. Internal Document. IGCAR, Kalpakkam, India.

7.34 Puthiyavinayagam, P., Roychowdhury, D.G., Govindarajan, S., Chellapandi, P., Singh, O.P., Chetal, S.C. (2002). Design safety limits in prototype fast breeder reactor. *First National Conference on Nuclear Reactor Safety*, Mumbai, India, November 2002.

7.35 Fast Reactor Database. (2006). IAEA, IAEA-TECDOC-1531, Vienna. ISSN 1011–4289. http://www-frdb. iaea.org.

7.36 Schirra, M.C. In *Fracture of Engineering Materials*, Parker, J.D. (ed.). The Institute of Materials, London, UK, p. 612.

7.37 Brinkman, C.R. (1999). ORNL/CP-101053. National Technical Information Service, Alexandria, VA.

7.38 Sasikala, G., Mathew, M.D., Bhanu Sankara Rao, K., Mannan, S.L., Trans, A., (2000). Creep deformation and fracture behavior of types 316 and 316L(N) stainless steels and their weld metals. *Trans, A.*, 31A, 1175.

7.39 Sikka, V.K. (1984). Development of Modified 9Cr-1Mo Steel for Elevated temperature Service. *Proc. of Topical Conf. on Ferritic Alloys for Use in Nuclear Energy Technologies*, Davis, J.W., Michel, D.J. (eds.). TMS-AIME, Warrendale, Pennsylvania, USA, pp.317–324.

7.40 Kimura, K., Kushima, H., Abe, F., Yagi, K., Irie, H. (1997). Advances in turbine materials, design and manufacturing. In *4th International Charles Parsons Turbone Conference*, Strang, A. et al. (eds.). The Institute of Materials, London, UK, pp. 257–269.

8 System and Components

8.1 INTRODUCTION

The sodium fast reactor consists of numerous systems and components that have to work in synchronization with each other to produce the desired reactor power. The major systems include the reactor core system, nuclear steam supply system (NSSS), instrumentation and control (I&C) system, and energy conversion system (ECS). In this chapter, the concepts involved in reactor core design, which includes the pin and subassembly (SA) design, functions, and descriptions of various components that are involved in the primary and secondary heat transport system followed by components in the NSSSs, I&C system, etc., are discussed.

8.2 REACTOR CORE

In the core, several nuclear reactions, predominantly fission, capture, breeding, reflection, and absorption, take place. The core is amidst a cloud of fast neutrons: the specific characteristic of the fast reactor core. The heat generated in the core is removed by the passage of appropriate coolant, and the neutrons emanating from the core are reflected partially and the neutrons leaking finally from the core are being captured in the shielding materials positioned around the core. The core is made up of a group of SAs consisting of fuel in the inner most ring surrounded axially as well as radially by blankets, followed by reflector and shielding zones, beyond which a few storage locations are provided in some design concepts. Figure 8.1 shows an arrangement of an SA, that is, a typical core layout adopted from Kim et al. [8.1]. The SA consists of fuel, coolant, structural, and shielding materials. The important design requirement is to maximize the fuel volume fraction in order to have a high-energy neutron spectrum, optimum coolant volume fraction, and minimum structural materials such that the core is made compact. These requirements led to the choice of a hexagonal shape for the SA to house a bundle of thin-walled tubes called cladding; each tube is filled with a certain quantity of fuel or blanket or shielding materials in the form of slugs or pellets. Cladding filled with fuel slugs/pellets is called a fuel pin. In order to facilitate smooth coolant flow, an optimum space is provided between the pins by introducing either wires on the outer surface of each of the pins or perforated single or multiple plates spaced uniformly along the axial direction of each of the SA. The pins are arranged in a triangular pattern with possibly minimum pitch to further achieve compactness. Pressurized coolant is fed to each of the SA from the common inlet plenum based on the heat it generates, for which a specific design feature is introduced: introduction of a set of orifice plates is a widely adopted option. The foot of the SA is inserted into the respective sleeve provided in the grid plate, which forms an inlet coolant plenum.

In the section, design constraints including design safety limits, fuel requirement, coolant flow rate, and core design aspects of fuel pins and ducts are elaborated. Further, reactor physics, thermal hydraulics, and structural mechanics aspects are covered. For illustration purposes, a typical pool-type 500 MWe capacity sodium-cooled fast reactor component (SFR-500) is considered without losing the generic aspects.

8.2.1 DESIGN CONSTRAINTS

As for as core design is concerned, the reactor power and the mixed mean temperatures of the sodium coolant at the SA inlet and SA outlet form the basic inputs. Thermal power (P_t) to be generated in the core is computed from the electric power (P_W) by considering the thermodynamic

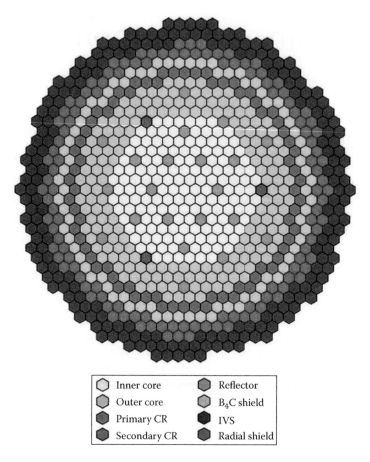

	Inner core		Reflector
	Outer core		B$_4$C shield
	Primary CR		IVS
	Secondary CR		Radial shield

FIGURE 8.1 Core layout.

efficiency (η) as $P_t = P_w/\eta$. Further, for the chosen fuel, coolant, and structural materials, certain temperature limits are specified as the design safety limits, which need to be respected. The number of pins and the number of fuel subassemblies (FSAs) are derived respecting the design safety limits and giving due considerations to economy. The following are the essential design safety limits.

Under 115% of nominal power with an overpower margin, the highest temperature at the central line of the fuel pellet (fuel hot spot temperature) should be less than the minimum melting point of the material. The highest clad midpoint temperature (clad hot spot temperature) should be less than a specified fraction of the melting point of the cladding material. The highest coolant temperature (coolant hot spot temperature) should be less than the boiling point of the coolant at the respective pressure.

The central line temperature limits the heat that can be allowed to generate in the fuel pellet. To quantify this aspect, the concept of linear heat rate (χ), which is the heat generated per unit length of the fuel, is defined. The expression for χ is derived from the volumetric heat generation rate, $\dot{Q}$, and the quantity of energy generated per unit volume per unit time, which is proportional to the neutron flux and fissile content of the fuel. Then χ is defined as

$$\chi = \dot{Q}\pi r^2$$

where r is the radius of the solid pellet. The linear power can also be derived in terms of the central line temperature (T_c), outer surface (T_s), and conductivity of the fuel (K_T) by solving the following 1D heat transfer equation in cylindrical coordinates:

$$\frac{1}{r}\frac{\partial}{\partial r}\left(K_T r \frac{\partial T_r}{\partial r}\right) = -\dot{Q}$$

where

    r is the fuel radius
    K_T is the thermal conductivity of the fuel expressed as a function of temperature
    T_r is the temperature at any radius r
    $\dot{Q}$ is the heat generation per unit volume

For an annular pellet, the K integral is given by,

$$\int_{T_s}^{T_c} K_t dT = \frac{\chi}{4\pi}\left[1+\left(\frac{2r_i^2}{\left(r_f^2 - r_i^2\right)}\right)\ln\left(\frac{r_i}{r_f}\right)\right]$$

where

    r_f is the pellet radius
    r_i is the central hole radius of the pellet

For a solid pellet (r_i tending to zero), the linear heat rate χ turns out to be independent of the radii. Accordingly,

$$\chi = \int_{T_s}^{T_c} K_t dT$$

While the central line temperature could be a fraction of the melting point of the fuel material, the surface temperature depends on the heat dissipation at the surface, which in turn depends upon the gap conductance between the fuel pellet surface and the inner surface and the cladding and heat transfer characteristics of the clad and coolant. The gap conductance would be high when the pellet surface touches the cladding (e.g., mechanically bonded fuel pins in the case of metal fuel) and/or the gap is filled with a highly conducting fluid (e.g., sodium-bonded fuel pins in the case of metal fuel). In the case of oxide fuels, due to the high corrosion of sodium with the fuel material, sodium bonding is avoided. Alternatively, the space between the fuel pellet and cladding is filled with helium gas. With regard to heat removal, liquid metal coolants are the best choice, for example, sodium. Thus depending upon the choice of fuel material, coolant, and fuel–cladding design options, "χ" could be limited. Here, "χ" would also be restricted while meeting the design specification of the mixed mean temperature of the coolant at the core outlet plenum. With the main design objective of maximizing the fuel volume fraction with respect to the previously mentioned constraints, the design of fuel pins, ducts, and core configurations is worked out. Finalizing the design involves several iterations.

With specific reference to SFR-500, the nominal electric power is 500 MWe and η is 40%, and hence, the thermal power to be generated is 500/0.4 = 1250 MWt, which is to be mainly delivered by the FSAs. The remaining SAs, particularly blanket SAs, contribute some small fraction of power. The core inlet temperature is 670 K, which is the temperature of the cold plenum. Though the design is targeted to achieve the near-equal temperatures of sodium coming out of each SA by an appropriate manipulation of the coolant flow, termed as flow zoning, it is practically impossible to achieve the same SA outlet temperature from each of the SAs. The SA outlet temperatures of sodium vary location to location by a narrow margin ($\pm 10°C$). However, the mixed mean temperature at the plenum just above the core takes into consideration the design specification. This value is 820 K in the case of SFR-500. The maximum linear heat rate that could be achieved with respect to the associated constraints is 450 W/cm.

8.2.2 MAJOR STEPS IN THE CORE DESIGN

The reactor thermal power, core power, peak linear power, and allowable neutron displacement per atom (dpa) for the clad and wrapper are the basic input data for the core design. Further, the active core height is selected based on several optimization studies of the reactor design, particularly the reactor assembly diameter. The choice of lower core height results in more number of pins and larger core diameter and vice versa. Smaller height is also preferable for safety considerations, which will be discussed later. Further, the maximum allowed fissile content (e.g., % plutonium in the fuel) is specified taking into account material availability and other technical considerations, such as chemistry with structural materials and coolant. The breeding ratio is to be specified giving due considerations on the potential of the fuel material: the oxide core gives the lowest value and the metallic core yields the highest value. Further, the breeding requirement depends upon the country's strategy. The numbers of fuel pins that will be accommodated in an SA (217 or 271) are also selected in the beginning of the core design, with respect to the constraints on the SA size. Another important core design parameter is the cycle length, that is, refueling interval.

The core design starts with knowing the parameters stated earlier. First, guess is made on the fuel pin diameter and spacer wire diameter. The axial and radial form factors, which define the average power w.r.t. axial and radial power distributions, are assumed with an understanding on the distributions (radial is Gaussian and axial is cosine distributions in a broad sense). From the peak linear heat power and knowing the axial and radial form factor, the average linear power is computed. The number of pins for the chosen active core height is then computed to yield the specified core thermal power, and subsequently the number of SAs is computed with respect to the maximum number of pins planned per SA. In a similar way, assuming pin dimensions, the axial blanket column and the number of radial blanket SAs are estimated. Once the number of fuel and blanket SAs is known, the core layout indicating the active fuel zone and radial and axial blanket zones is arrived at. Subsequently, the active core volume and volume fractions of the fuel, steel, and coolant are computed.

Reactor physics computations are made for the core layout for the various volume fractions computed as previously presented. The computations provide actual radial and axial form factors, fissile fuel enrichment required, control rod worth, and optimum locations and resulting breeding ratio. The computations also indicate the time duration during which the specified fuel enrichment can sustain the criticality condition. After this cycle time, there is a need to refuel by replacing a few numbers of FSAs. In case the fuel enrichment and cycle time are not acceptable, fuel pin diameters need to be changed and the whole analysis is repeated with the updated axial and radial form factors. Control rod worth and its locations facilitate update of the core layout and various volume fractions. The breeding ratio can also be improved to get the desired value by changing the fuel and blanket pin diameters and blanket quantity.

The finalized core design at the end of the iterations provides peak burnup, that is, the maximum energy that can be extracted from the unit quantity of the fuel in the active core and associated neutron dose (dpa) to the core structural materials. If the peak burnup is not acceptable, the targeted cycle length should be modified. Else, to enhance the burnup limit, an alternate material or advanced structural material may have to be used. The main ingredients of the core design are presented in Figure 8.2.

After finalizing the core design, detailed flow and temperature distributions along as well as across the SA, the pressure drop, and in turn the pumping power required are computed through thermal hydraulics analysis. Apart from this, the structural integrity of the fuel and blanket pins

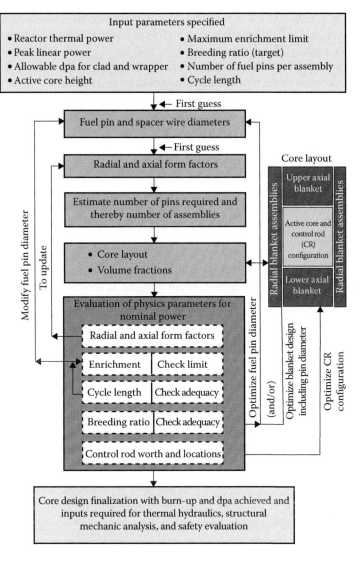

FIGURE 8.2 Core design flowchart.

and SA, irradiation-induced structural deformations, SA extraction forces, and cumulative damage fractions are all computed through detailed structural mechanics investigation. Apart from nominal steady loadings, the core is analyzed for partial power generation and transient conditions including accidental loading situations. These aspects are presented in detail in Chapters 9 and 12.

8.2.3 SELECTION OF FUEL PIN DIAMETER

The core is comprised of several fuel rods along with passages for the coolant to flow to remove the heat generated in the rods. The fuel rods are not exposed directly to the coolant to preclude any possibility of the fissile material being carried away by the coolant and also to contain the fission gases generated during the fission. Hence, the rods in the pellet/slug are inserted into a thin tube called "cladding." Claddings containing fuel pellets are called fuel pins. Each of the fuel pin containing the pellets is welded at both ends with end plugs. The fuel type can be ceramic or metal, based on which the gap between the fuel and clad can be filled by a "heat-conducting medium," either gas or liquid metal, specifically helium or sodium. The thickness of the clad tube should be minimum so that a higher fuel volume fraction is achieved, but at the same time, the clad thickness should be sufficient enough to withstand the generated fission gas pressure at high temperatures up to the design target burnup.

In any fast reactor core design, the selection of the fuel pin diameter is the starting point. One of the important parameters that influence the selection of the fuel pin diameter is the specific fissile inventory, that is, the mass of the fissile material required to generate the unit power, M_o/P, and this should be minimized to reduce the doubling time. The fissile specific inventory can be expressed as

$$\frac{M_o}{P} = \frac{e\pi R_f^2 \rho_f}{\chi}$$

where
> e is fissile enrichment in the fuel
> R_f is the radius of the fuel pellet
> ρ_f is the fuel density
> χ is the linear power

Having selected "χ," the influence of other parameters is examined in Table 8.1. The trend of the fissile-specific inventory variation with the fuel pin diameter is shown in Figure 8.3.

TABLE 8.1
Influence of Various Parameters on the Core Performance

Parameter	Effects
e	Enrichment cannot be altered as this influences the criticality. Also increasing "e" reduces the breeding ratio and M_o/P increases.
ρ_f	ρ_f *decrease* increases porosity and so more leakage and "e" increases. Also the fuel conductivity "K" reduces, which brings down the linear power χ. So M_o/P increases. [Ps: $K_P = K(1-\alpha P)/(1+\beta P)$, where $P = (1 - \rho/\rho th)$ is porosity]. α and β are correction factors for porosity that are experimentally determined. ρ_f *increase* brings down neutron leakage and so "e" can be reduced, also increasing "K" and making linear power χ increase. Though M_o/P reduces more than the relative increase in ρ_f, this is unfavorable, since porosities are intentionally fabricated into the fuel matrix to accommodate swelling.
R_f	R_f *decrease* results in increase in "e"; however, the combined value eR_f^2 decreases more than the increase in "e." This is the only favorable choice to reduce M_o/P to reduce doubling time and the fissile inventory.

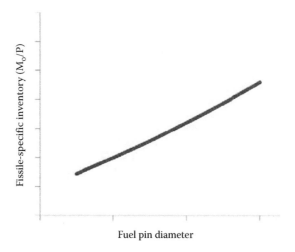

FIGURE 8.3 Fissile-specific inventory variation with fuel pin diameter.

When selecting the pin diameter, the cost should also be considered. Beyond a certain size, small pins become increasingly expensive to fabricate, posing difficulty in accommodating a spacer wire and also the coolant flow area. In the case of a sodium coolant, the coolant with a very high heat transfer coefficient (h), the fuel pin diameter is a weak function w.r.t. the coolant flow. Even for a low flow, the heat transfer is effective due to molecular conduction.

8.2.4 Fuel Pin Constitutions

Fuel pins are cylindrical tubes filled with fuel having a blanket material on either side. The fuel and the blanket are made of pellets for a ceramic fuel or blanket material and of slugs for metallic fuels/blanket. The bottom end of the fuel/blanket is either supported on a middle plug crimped to the tube or just stand on another tube or rests on the bottom end plug. To prevent motion of the pellets during transportation and handling of the SA, the top end of the blanket is held against a spring supported by a spring support that is seated on the top axial blanket on one side and against a top plug on the other side. Highly pure helium is used as the bond gas between the pellets and the clad inner wall to improve heat transfer. The pin is sealed at both ends by welds.

A plenum to accommodate the released fission gases from the fuel with burnup is provided at the top or bottom of the fuel column or at both ends. The length of the fuel fission gas plenum is based on the mechanical design of fuel pin, which will be discussed later. The fission gas plenum is about the same height as the core. In U.S. design (fast flux test facility [FFTF] and clinch river breeder reactor project [CRBRP]), the plenum is above the core, whereas in French design (Phénix, superphénix [SPX]), the small upper and large lower plenums are provided. Provision of the lower plenum for fission gases helps in reducing the fuel pin length. The top plenum excludes the possibility of fission gases passing through the core in case of any fuel pin failure. The constituents of a typical fuel pin are shown in Figure 8.4.

8.2.5 Spacing of Pins

After selecting the fuel pin diameter, the fuel pins have to be spaced apart for the coolant to pass through. The spacing between the pins should provide a sufficient flow area for the coolant and also accommodate the swelling of the clad tubes that resulted from design target burnup. Pin spacers are provided for this purpose. Two concepts are considered in fast breeder reactor (FBR) fuel pins: wire-wrapped or grid spacers. In wire-wrapped spacers, the fuel pin is wrapped around by a spacer wire along the length with a certain helical pitch and is welded/locked at the top and bottom of the pin in the end plug region. In grid spacers, the grids are arranged in the shape of a honeycomb [8.2]. A series of spacers are stacked at an appropriate interval. Multiple grid spacers are kept at two or three alternate axial levels and are normally anchored to the inner surface of the SA duct. The schematic of both spacer wire and grid spacer concepts is shown in Figure 8.5. The pros and cons of the both concepts are discussed later.

Grid spacers have some specific advantages. The spacers positioned only a few elevations consume less steel volume across the core region, and hence, a higher fuel volume fraction is achievable. The bundle pressure drop will be lower since the coolant passes through a straight rather than swirling movement. The design has flexibility to locate the spacer axially away from the position of the peak cladding temperature. In the case of fuel melting, the grid spacers will act as a mechanical obstruction to some extent for the molten fuel and hence possible accumulation of molten fuel near the peak active core region. Thus, this feature has a great disadvantage: it enhances the chance of flow blockage. Further, fabrication cost is also high, since carefully aligned physical indentures must be made in the assembly of the duct wall to axially affix the grids. On the contrary, in the case of wire-type spacers, the molten fuel or any small mechanical obstructions entering the coolant channel will be hydraulically removed from the active core region and hence no possibility of flow blockage.

FIGURE 8.4   Constituents of a typical fuel pin.

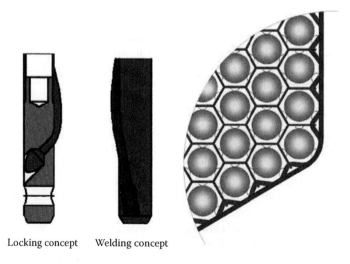

Locking concept   Welding concept

FIGURE 8.5   Helical spacer wire and honeycomb grid spacer concept.

Mixing of the coolant is better because of the swirling motion caused by the helically wound spacer wire, increasing the SA average outlet temperature. Mechanical vibrations of the pins are less as it has more support points, that is, contact points between wires. However, the associated pressure drop is high. Further, fabrication is easy and hence involves lesser manufacturing cost. However, the fuel volume fraction obtained is less. In summary, wire-type spacers are adopted universally, mainly from the safety considerations with respect to flow blockage. Hence, the wire spaces accumulated vast experience at high-burnup conditions.

The spacer wire diameter has to be selected judiciously: a higher pin pitch provides more coolant flow area and can accommodate swelling but at the cost of reduced fuel volume fraction. The reduction in fuel volume fraction will impose an increase in the fuel inventory. The variation of the coolant flow area and fuel volume fraction with the spacer wire diameter is shown in Figure 8.6. Also, the probable trend variation of the fuel inventory with the fuel volume fraction is shown in Figure 8.7 for constant fuel enrichment. As the coolant flow area increases, the pressure drop in the bundle comes down, but the fuel inventory will increase and hence optimum dimensions are to be selected.

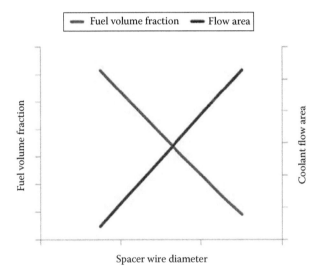

FIGURE 8.6   Fuel volume fraction and coolant flow area with spacer wire diameter.

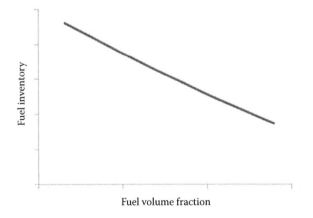

FIGURE 8.7   Fuel inventory variation with fuel volume fraction.

8.2.6 CONCEPT OF SUBASSEMBLY

A certain number of fuel pins are grouped and housed with a duct or hexagonal sheath to facilitate effective cooling and handling. The duct or hexagonal sheath housing certain numbers of fuel pins is called SA. In order to achieve a high fuel volume fraction, pins are to be arranged closely, for which a triangular pitch arrangement is a preferred choice. This in turn results in a hexagonal shape. The duct has the following functions:

- Facilitates the coolant flow past the fuel pins
- Allows for placing the orifices in the individual SAs to achieve the required flow zoning
- Provides structural support for the pin bundle
- Provides a mechanical means to load the pins into the core
- Provides a barrier to limit the propagation of a rupture of a few pins to the rest of the core

The number of pins per SA is decided based on the following factors [8.3,8.4]:

- The reactivity worth of an SA is to be limited to avoid a large reactivity swing during refueling.
- The decay heat level is to be optimum. Too large decay heat would result in high temperature rise posing difficulty in handling.
- The assembly weight should be minimum for handling considerations.
- *Mechanical performance*: A large sheath results in more stress and deformation due to coolant pressure and also larger swelling and bowing.
- Cost reduces by reducing the number of SAs with the increase in the number of pins per SA.
- *Criticality during shipment*: The SA should remain subcritical even when flooded with water.
- *Burnup*: The average burnup goes down for a given peak burnup due to more power gradient in the larger SA.
- *Refueling time*: Larger assemblies require a few steps, hence less time.

Usually 217 or 271 pins are chosen for large reactors, that is, 9 or 10 rows of fuel pins.

8.2.7 SA COMPOSITION

SA is composed of the foot, body, and head. The foot is a cylindrical tube with slots for coolant entry at the bottom. The body is basically a hexagonal sheath housing a cluster of fuel pins (217 or 271) arranged in 9 or 10 rows with a triangular pitch. The body in turn is connected to the foot through a support block. The head of the SA is connected to the end of the body. The head houses the axial top shield in the form of a cluster of pins to prevent neutron dose on the components above the SA—the control plug (CP), roof slab, heat exchangers, and primary pumps—from the neutrons streaming out of the core. Since the shielding pin bundle is positioned above the SA above the fuel pin bundle, it also facilitates mixing of the coolant emerging out of the fuel pin bundle before leaving the SA. Thus, the coolant entering the SA through the radial slots provided in the foot passes through the space between the fuel pins (removing the fission heat) and gets thoroughly mixed, and further flow passes the shielding bundle before leaving the SA through the head exit. The foot also has a set of flow-regulating devices—orifice plates—above the coolant entry slots. By appropriately designing the orifices, the flow required to get the desired temperature at the SA outlet can be achieved. The schematic of the FSA is shown in Figure 8.8.

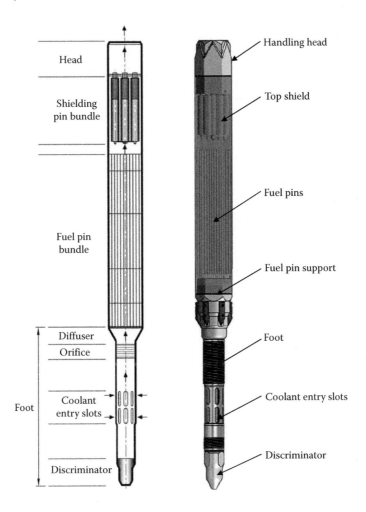

FIGURE 8.8 Typical FSA.

8.2.8 Flow Distribution to the Core

Coolant flow allocation is based on power generation. Power generation in the core varies due to variation in the flux. As the power distribution in the reactor is not uniform, the flow through each SA has to be allocated such that at the outlet of the SA, the temperature is nearly uniform. There are two methods generally followed: either to send flow proportional to the power or to change the power of the SA by keeping a constant flow. For peripheral SAs, the power decreases because of the drooping characteristics of the flux. To make the SA power uniform, enrichment needs to be increased in the outer SA zone.

However, it is very complicated either to allocate different flows or to manage different enrichments for each SA individually. Indeed, an optimized design is preferred in which some SAs having similar powers are grouped together and assigned the same flow, and also few different enrichment zones are provided to get a more uniform power.

A typical example with two enrichment zones and few flow zones is shown in Figure 8.9. The variations in the power profile in radial and axial directions are shown in red color. The power generated is zero above and below the active core because of the absence of fission reactions. Power is maximum at the center of the core. Toward the periphery, initially, power falls and increases at the

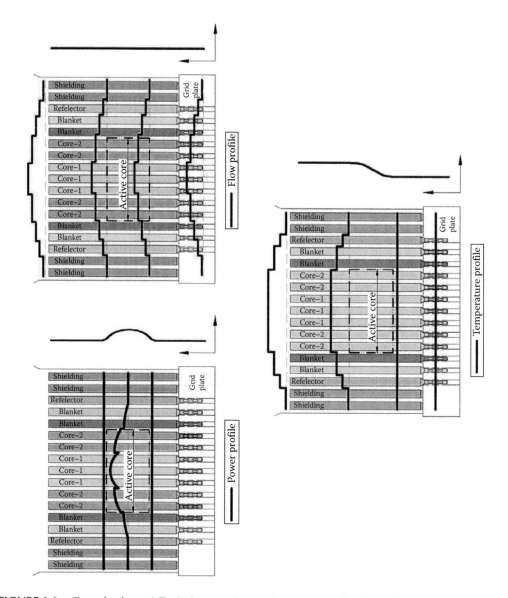

FIGURE 8.9   (**See color insert.**) Typical power, flow, and temperature distribution in the core.

second enrichment zone and decreases further. The flow profiles in different SAs are shown in blue color. Central subassemblies (CSAs) receive more flow because of higher power. The flow requirement decreases for peripheral SAs, which is controlled by the orifice. The flow remains constant along the length of the SA.

The temperature profile is shown in green color. At the inlet and up to the core bottom, the temperature is constant. At the active core region, the coolant temperature increases and is constant thereafter. Thus, at the exit of the SAs, a more or less uniform temperature is achieved for the FSAs. The blanket SA's power changes with the cycle length due to breeding, and hence, its outlet temperature varies with the cycle length. For the rest of the SAs, the temperature increase is insignificant due to negligible power production.

Flow enters the foot of the SA through the grid plate sleeve holes. Multiple holes can be provided at the sleeve to avoid gross flow blockage. Flow entry into the foot can be radial or axial. Radial entry provides hydraulic locking of the SA that will not allow the SA to lift under any flow

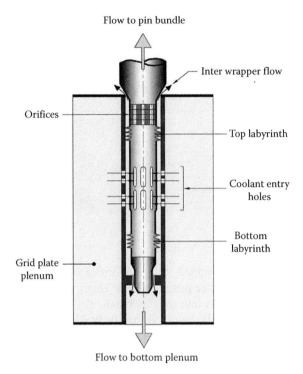

FIGURE 8.10   Flow distributions at the SA foot.

conditions. In the case of axial entry, additional springs are provided at the foot of the SA to ensure locking. Typical flow distribution at the foot of the SA with radial entry is shown in Figure 8.10. To minimize flow leakage between the sleeve and foot, labyrinths are provided. In the foot of the SA, flow-regulating devices are kept according to the SA power. Orifice plates are widely used as flow-regulating devices. Orifice plates can be of machined hole design or honeycomb type. The selection of the hole size, number of plates, and its orientation dictates the coolant flow in the SA. A typical honeycomb orifice used in FBRs is shown in Figure 8.11. The height of the orifice stack depends on

FIGURE 8.11   Typical single honeycomb orifice.

FIGURE 8.12 Orifice stack.

the extent to which pressure is to be dropped. A typical orifice stack for peripheral SA is shown in Figure 8.12. After passing through orifice plates, the flow enters into a divergent foot region where the flow gets developed and then enters into the fuel bundle region. The flow mixes in the bundle region because of pin spacers and finally comes out of the bundle region and then passes through the top shielding bundle and leaves axially at the head of the SA.

8.3 NUCLEAR STEAM SUPPLY SYSTEM

The heat generated in the core has to be transported to the steam generators (SGs) to produce steam. The system of components that facilitates such heat transport phenomenon is called NSSS. The main components of the NSSS are primary coolant pumps to develop a sufficient pressure head to drive the coolant through the core and heat exchangers to transport the heat ultimately to water to convert it to steam. Thus, the pumps, heat exchangers, and associated pipelines form primary and secondary circuits. Particularly in the case of SFRs, an intermediate loop is normally introduced to eliminate the possibility of entry of any reaction products to the core subsequent to a large sodium–water reaction in the SG. Such products have high moderating characteristics, which introduce high positive reactivity in the core having high fissile materials. Apart from this, the sodium–water reaction generates high transient pressure around the reaction zone within the SG that could propagate along the sodium in the pipeline: the concern is that they can induce vibrations on core SAs and consequent reactivity oscillations. With the presence of an intermediate loop, such scenarios can be ruled out with analytical justifications. The major components in the secondary sodium circuit are intermediate heat exchangers (IHXs), secondary sodium pumps (SSPs), SGs, and tanks. The primary and secondary circuits along with the core constitute the NSSS. Figure 8.13 shows the schematic sketch of NSSS. The reactor assembly also called reactor block is one of the important systems of the primary circuit. The main functions of the reactor assembly are to support the core, facilitate core cooling, and reactor control and maintain the safe state of the reactor. The layout of components in the primary sodium circuit depends upon the pool-type or loop-type concepts. In this section, design features of pool- and loop-type concepts as well as important reactor assembly components, pumps, and SGs are described [8.5,8.6].

8.3.1 Pool versus Loop Concepts

The pool concept is featured by nearly all the primary sodium coolant contained inside a single reactor vessel [8.7]. Therefore, this vessel encloses the primary sodium pumps (PSPs) and IHXs, in addition to the internal structures surrounding the core, and devoted either to its feeding and

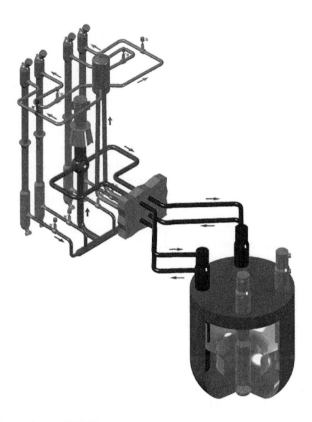

FIGURE 8.13 Schematic sketch of NSSS.

supporting or separation of the various hydraulic plenums. As opposed to the pool primary system, the loop concept is characterized by some major primary components kept outside the reactor vessel. Several variants of the loop concept exist. In the earlier existing loop-type SFRs, the primary pumps and the IHXs, without grouping, are located adjacent to the reactor vessel and connected together by inlet and outlet pipes through nozzles. The major features of the pool-type concept are as follows:

- The nozzles, especially at the bottom part of the reactor vessel, are eliminated. The top-entry concept is typically a kind of solution with penetrations of the primary pipes through the reactor roof (either U-shaped or L-shaped pipes).
- Shortened primary pipes reducing the number of bends.
- Primary pumps and IHXs can be located into dedicated component vessels or gathered in a single PSP–IHX component.

In addition, recently innovative conceptual designs are evolved combining the features of the pool and loop concepts in order to sum their own benefits and to remove their respective drawbacks. These variants can be considered as hybrid designs, but strictly speaking they are variants of the loop concept. A top-entry concept conceived in demonstration fast breeder reactor (DFBR)—a Japanese reactor—can be referred in this context. A detailed analysis of either of pool or loop concepts shows that the reactor power level can influence the choice of design variants. The largest possible factory-made components were considered as having relevant interest by some designers. They later promoted the modular plant concept characterized by several reactor power units coupled to a power block (i.e., single balance of a plant), complying with commercial-size unit requirements. On the contrary, if the power units of the plant site are separated, as for the large-size nuclear plant units, then only the possible size effects are expected, but without common equipments for a given power block. Generally speaking, the modular concept can be applied to any component; for example, the modular SG can be implemented on a large-size power unit. The meaning of the modular concept in the present case is particular to the reactor primary system. For either the pool or loop primary circuit concept, an intermediate circuit is usually added between the primary circuit and the ECS. For the loop concept, the number of intermediate circuits has to be compatible with the number of primary loops, whereas for the pool concept, the number of intermediate circuits is quite an open design option, at least based on reactor size concerns. The recent interest in eliminating the inter-mediate system for economic reasons gives rise to several alternative conversion systems (e.g., ECS without the water–steam cycle) to exclude concerns of the sodium/water interaction risks. However, the feasibility of SFR without the intermediate system is not yet demonstrated, but proposals are made with the loop concept, for which it is reported to be more adoptable than the pool concept.

The advantages and drawbacks of each concept depend in practice on the design variants. However, one can notice some general trends allocated to each main concept but to be considered carefully for the previous reason. Motivation and challenge for a pool-type concept can be perceived as opposed to the loop concept as follows:

- There is no accident scenario of loss of the primary coolant. The primary sodium inven-tory is managed by safety provisions (e.g., guard vessel).
- The large thermal inertia of the reactor block contributes to slow down any transient loss of heat sink.
- There is no risk to break the hydraulic loop from the core outlet toward the core inlet.
- A very efficient natural circulation of the primary circuit is expected, as a flow backup at the reactor shutdown state, in case of loss of the forced flow mode (e.g., pumps trip).
- There exists a cold sodium plenum at the pump's suction upstream, which acts as buffer against either thermal shocks or gas entrainment toward the core.
- In practice, there is no risk of radioactive sodium fire, except for limiting events leading to a hypothetical core disruptive accident (HCDA).
- Good mechanical behavior of the primary containment against energetic HCDA.
- Ease of radiation protection in normal operation.

On the other hand, matters on competitiveness and on flexible operational conditions remain as a challenge:

- Limited access for inspection and repair of the under sodium internal equipments
- Seismic behavior of sodium-free level and large structures
- Reactor-block compactness limitation due to integrated large components

Nevertheless, slight differences are expected between pool and loop concepts, based on existing technology, as regards the construction cost.

In the same way, motivation for the loop concept and challenges as opposed to the pool concept can be perceived as follows:

- Ease of access for maintenance and repair of those of the primary components located outside the reactor block (e.g., IHX).
- Compactness of the reactor block (e.g., vessel diameter) and reduction of the primary loop number.
- Potential for further construction cost reduction and for innovative change of primary and intermediate equipment (e.g., pump design, integrated components, intermediate circuit change).
- Any rotating pump shaft is away from the core vicinity.

In return, the designer has to solve the following issues:

- Prevention of loss of the primary coolant (e.g., pipe integrity) and provisions to maintain the hydraulic loop through the core whatever the abnormal operating conditions are
- Potential consequences of a loss of coolant accident (LOCA), inside (e.g., gas entrainment) and outside the primary circuit (e.g., active sodium fire)
- Provision against asymmetric operating faults (e.g., trip of one of the pumps leading to reverse flow)
- Suitable implementation of several decay heat removal (DHR) systems in order to cope with different accidental configurations of the primary circuit
- Additional provisions regarding radiological protection

8.3.2 CONCEPTUAL DESIGN OF NSSS COMPONENTS

The choice of design option is crucial at the start of the conceptual design stage in order to meet the requirements to achieve safety and economy concurrently [8.5,8.6]. Hence, the concepts selected should be simple, mature, and robust and should be based on operating experiences, availability of relevant design criteria, analysis capability, international trends, availability of materials, and constructability including transportation. Newer materials and concepts should be introduced carefully after detailed analysis and thorough validations to meet both functional and safety requirements. Standard and proven concepts should be adopted wherever possible. The functions of each component should be defined comprehensively. The operating conditions, environments, safety classifications, design requirements, and interface constraints should be identified clearly. While selecting the SFR components, whose design is largely governed by neutronics, complex pool hydraulics, structural mechanics, seismic design requirements, shielding requirements, etc., it is preferable to introduce a judicious mix of flexibility and rigidity. From the point of long and reliable operation of the plant, needless redundancies and rigid geometrical constraints should be avoided. The geometrical features should be selected as far as possible to have complete access for inspection and to achieve reliable technology. The synergy and interplay of various governing performances need to be understood in a scientific manner with regard to constraints in realization of technology.

The primary circuit is constituted by the core and coolant circuit. This circuit is housed within the reactor vessel also called the "main vessel." The main vessel has a top cover called "top shield." The main vessel, apart from the core and coolant, houses the grid plate and core support structure (CSS) and accommodates the PSP, IHX, and control plug. The grid plate supports the core and facilitates circulating the sodium to the core SAs to remove the nuclear heat. The PSP provides the required pressure head to the cold sodium to flow through the core SAs. The top shield supports the PSP, IHX, control plug, and in-vessel fuel handling system (called transfer arm). The control plug, in turn, accommodates absorber rod drive mechanisms (ARDMs). In the pool-type concept, there is a need of a vessel called inner vessel, which separates the hot and cold sodium in the main vessel. Thus, the inner vessel ensures desired flow through the IHX and targeted design temperatures in the hot and cold pools. The absence of the inner vessel or a large leak in the inner vessel will result in higher stabilized overall temperatures of the pool (which will exceed the structural temperature limits) in the process of heat removal by IHX from the mixed pool. The IHX, SG, and SSP are the main components in the secondary circuit. The heat transport circuit components of typical pool- and loop-type reactors are shown in Figure 8.14.

Apart from the previously mentioned circuits, the DHR circuit, consisting of dedicated heat exchangers immersed in a hot pool (in the case of the pool concept), and air heat exchangers (AHXs), which transport the heat to the atmosphere (ultimate heat sink) during DHR operations, are important systems from the point of view of safety. Depending upon the system, this circuit consists of decay heat exchangers (DHXs) and pumps. Compared to thermal reactors, the SFR system operates with a high operating temperature. In view of higher steam temperatures, the thermodynamic

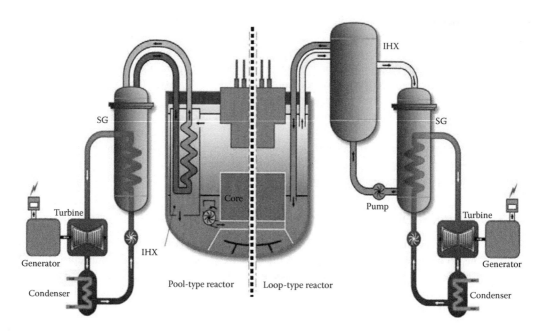

FIGURE 8.14   Heat transport circuits of pool- and loop-type reactors.

efficiency is around 40%, which implies less thermal pollution and less loss of nuclear heat. This is one of the main advantages of SFR power plants, compared to water reactors.

In the following paragraphs, the design features of main NSSS components are described. The various design options are covered in detail in Ref. 8.5. The material choice is dealt in Chapter 7.

8.3.2.1   Reactor Assembly Components in the Pool-Type Concept

The main vessel along with its internals and the top shield along with the components supported on it constitute the reactor assembly [8.8]. The main vessel is the most important component in the reactor assembly system. The main vessel is filled with a sufficient quantity of sodium with the cover gas space to accommodate the net volume change of sodium at various operating temperatures: typically, shutdown at 470 K, isothermal conditions at 670 K, normal operation at 800 K (average), and peak thermal transient at 850 K. The core is positioned approximately at the center of the main vessel. The core SAs are generally freestanding on the grid plate: the foot of each SA is inserted into its respective sleeve embedded in the grid plate. The grid plate, apart from accommodating all the sleeves, acts as a cold sodium plenum to distribute the coolant to each of the SAs. The sodium enters the individual SAs through the holes in the sleeves and subsequently to the SA through the slots in its foot. The hot sodium emerging out from the SAs after absorbing the nuclear heat is collected in the hot plenum within the inner vessel. Thus, the inner vessel separates the hot pool from the cold sodium pool within the main vessel. The inner vessel is basically a bottom-supported, self-standing thin vessel bolted or welded to the outer periphery of the grid plate. The inner vessel also provides passages (also called "standpipes") for the IHXs such that the inlet windows are directly exposed to the hot pool and exit widows are directly exposed to the cold pool. With these features, the hot pool sodium can enter the IHX through the inlet windows and leave the IHX through the bottom exit windows to mix with the cold sodium pool. The sodium is pressurized by the PSP and fed to the pump header, which is generally spherical in shape. Subsequently, the pressurized sodium flows to the grid plate plenum through primary pipes. In order to take care of random failure of a pipe (a safety case in the plant dynamic study), more than one pipe is often chosen to connect the pump header to the grid plate. This arrangement closes the primary sodium circuit. The pump impellers are immersed in the cold pool within the diffuser. Sodium from the cold pool flows to the impeller first and gets pressurized within the diffuser before getting collected in the pump header through a nozzle called "receptacle." The impeller is driven by the motor supported on the top shield, the closure for the main vessel. The shaft connecting the impellers and drive system passes through the pump standpipes in the inner vessel. It is worth mentioning that the hot sodium flows from the hot pool to the cold pool through the IHX windows. To facilitate this arrangement, the elevation of the IHX standpipe is kept below the respective inlet window. In order to avoid sodium flow bypass from the hot pool to the cold pool, there is a seal arrangement (the details can be seen in the section on IHX). The flow bypass is avoided through pump passages by extending the pump standpipes sufficiently above the hot pool level. The PSPs and IHXs are supported on the top shield.

Another important component in the reactor assembly is the control plug (also called "above core structure"). The essential functions of the structure are as follows:

- To maintain benign pool hydraulics: With the absence of this structure, the sodium jets emerging from the core could produce high turbulence in the hot pool, free-level fluctuations, gas entrainment, etc.
- To support and provide passages for ARDMs.
- To accommodate thermal sleeves for the core-monitoring thermocouples (TCs).
- To accommodate neutron-monitoring detectors and failed fuel–locating modules.

The control plug is also supported on the top shield. Another important feature in the design is an internal arrangement for maintaining the entire main vessel at cold pool temperatures. The cooling of the vessel is carried out to minimize the formation of undesirable carbides and sigma phases in the austenitic stainless steel material, thereby enhancing the structural reliability.

Finally, the vessel is closed at the top by the top shield. The top shield provides a leak-tight barrier between the cover gas and reactor containment building (RCB). It also provides biological and thermal shielding in the top axial direction of the reactor. It consists of a stationary part (roof slab) and rotatable plugs. The roof slab supports the components, such as IHXs and PSPs. Depending upon the design concepts, there can be single or multiple rotatable plugs, which house the mechanisms for the absorber rod drives and fuel handling machines. A typical reactor assembly, integrated with components and sodium, described in this chapter is shown schematically in Figure 8.15.

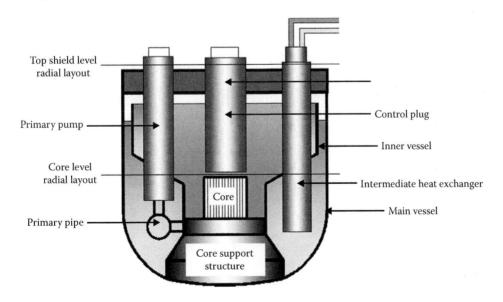

FIGURE 8.15 Reactor assembly of a typical pool-type SFR.

It is worth mentioning that the reactor assembly houses an in-vessel fuel handling system and also a transfer pot for loading fresh FSAs after removing the spent FSAs. The details of the entire fuel handling scheme and associated components are dealt in a separate section (Section 8.4). A core catcher to accommodate molten coolant debris generated consequent to a severe accident is also placed within the space below the CSS. The details of a core catcher are covered in Chapter 5.

8.3.2.2 Reactor Block in the Loop-Type Concept

The reactor block houses the core; the grid plate, which is usually integrated with the reactor inlet plenum; and above-core support structure [8.9]. The hot primary sodium emerging from the core outlet is sent to the IHXs through piping welded to the nozzle at the upper portion of the reactor vessel. The primary sodium after transporting the heat to the secondary sodium in the IHX is pumped back to the reactor through inlet nozzles by the PSPs. Thus, the primary sodium circuit components, PSPs and IHXs, are arranged outside the reactor vessel. The rotatable plugs that enable the fuel handling scheme form the upper part (top shield) of the reactor block. In view of a relatively small diameter of the vessel and in turn the grid plate, there is no need for any strong back such as the CSS for the grid plate. Since the hot sodium comes out from the reactor vessel, there is no need of the inner vessel. These features make the reactor block very compact (Figure 8.16).

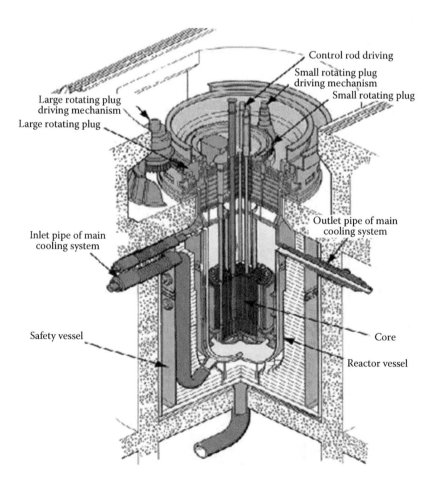

FIGURE 8.16 Reactor block of a typical loop-type SFR. (From Sylvia, J.L. et al., *Nucl. Eng. Des.*, 258, 266, 2013.)

8.3.2.3   Reactor Assembly Support Arrangements

The other important decision is the option of supporting the reactor assembly either at the top or bottom (Figure 8.17). There should not be any significant constraints for thermal expansions that would take place under various operating temperature conditions, which is essential for support of the system and components. Most of the SFRs, except Russian reactors where the bottom-supported-vessel concept is adopted [8.10], adopt the top-supported-vessel concept. With this, the vessel is supported at the periphery of the top shield through a cylindrical/conical skirt assembly, allowing the vessel to move freely in the downward direction to accommodate various thermal expansions. Both top support and bottom support arrangements have relative merits and demerits that require careful consideration. The top-supported vessel is free to expand in both radial (except at the support location) and axial directions leading to lower stress levels of thermal origin. On the other hand, the loss of this advantage, together with the requirement of a large-diameter bellow to accommodate the axial thermal expansion between the support and the top cover, might weigh against the option of the bottom-supported vessel especially for larger power (>500–800 MWe) stations. However, the advantages of a bottom-supported vessel for seismic-induced loading are obvious in view of the possibility of bringing support elevation in line with the mass center of the reactor assembly, and the overall weight of the top shield could be reduced by transferring loads of the pumps and heat exchangers in the vicinity of the bottom support. In the bottom-supported-vessel design, the diameter of the top cover can be reduced (e.g., Beloyarsk Nuclear Power Station-600 [BN-600]) by supporting the primary sodium equipment over the extended cantilever portion of the reactor vault (RV), thereby relieving the reactor vessel of such a function. In the loop-type design built so far, the top-supported-vessel designs have been predominantly favored except in BN-350, wherein a support closer to the core level has been chosen. Generally, for both loop and pool designs, the bottom-supported option has been largely favored in Russian designs.

8.3.2.4   Design Features of Reactor Assembly Components

8.3.2.4.1   Main Vessel

The main vessel is an important component in the reactor assembly as it forms a boundary for the radioactive primary sodium and argon cover gas [8.11]. It supports the core SAs through the CSS and grid plate. Thus, it also forms a major part of the core support path. In the pool type, the vessel is generally made of cylindrical shell with bottom dished head and supported at its top periphery to

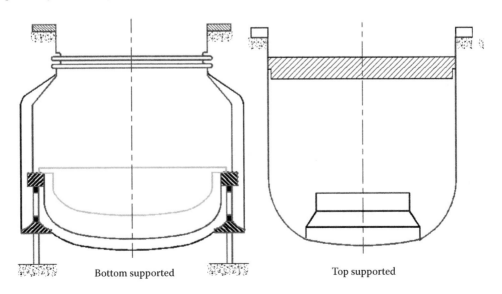

FIGURE 8.17   Reactor assembly support options.

facilitate free thermal expansions under various operating temperatures (shutdown, normal operation, DHR condition, and thermal transients). In BN-type reactors in Russia, the main vessel is supported at the bottom, and hence, the vessel moves upward due to thermal expansions, which are absorbed in the bellows incorporated near the top shield. There are design variants w.r.t. the shape of the bottom dished head and top shield arrangements. The main vessel diameter and height are important parameters that have a high impact on safety and economy especially in the case of the pool-type concept. Owing to its large capacity requirements to accommodate the entire primary circuit and relatively low mechanical loads that decide the structural wall thickness, the optimization of the vessel geometry results in a large-diameter thin-shell structure. The vessel diameter is governed by the core diameter, and thermal baffles associated with the main vessel cooling system form tolerances achievable during manufacture of the main vessel, inner vessel, and thermal baffles, positioned coaxially. Experiences indicate that it is easier to achieve tight tolerances with smaller diameter and large tolerances need to be specified relatively for the large-dimensioned vessel. This calls for larger radial gaps between the shells to meet the functional requirements, which in turn results in a larger diameter of the main vessel, due to the cascading effects. An alternative could be a machined main vessel, by which the vessel can have a perfect geometrical shape with a possibly minimum diameter, one of the options proposed for the European fast reactor (EFR).

Another interesting aspect of the design is the main vessel cooling circuit. The temperature of the main vessel is kept low, possibly close to the cold pool temperature to enhance its reliability. This also helps mitigate temperature gradients during all operating conditions, thereby causing stresses/strains and creep/fatigue damages. To maintain a lower temperature, the vessel is cooled by diverting a fraction of sodium from the cold pool, the concept employed in PFBR, BN-600, Phénix, and SPX-1. Figure 8.18 shows the cooling arrangements adopted in a few international SFRs. The kind of cooling circuit concept adopted in these designs incorporates a weir shell over which cold sodium flows to cool the adjacent main vessel. The circuit generally consists of a feeding collector (plenum adjacent to the main vessel) and restitution collector (plenum adjacent to the feeding collector). The sodium after passing through these two collectors joins back to the cold pool. Apart from this, it also eliminates the possibility of sodium level and temperature fluctuations (responsible for thermal ratcheting and fatigue) on the surface of the main vessel. Hence, practically the vessel is subjected to insignificant creep/fatigue damage and ratcheting (more details are presented in Chapter 10). In spite of these advantages from the structural integrity point of view, elimination of the weir cooling system has several advantages for economic considerations, such as lesser system, reduced vessel diameter, and elimination of structural joints. Hence, alternative design concepts are

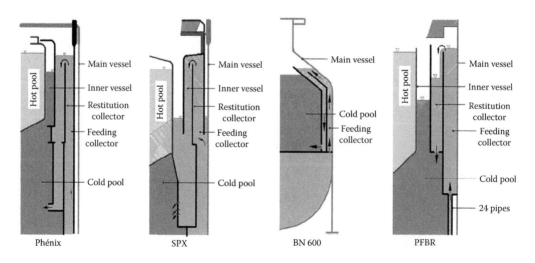

FIGURE 8.18 Main vessel cooling arrangements.

studied internationally, eliminating the cooling system or use of advanced structural materials to have enhanced high-temperature strength (e.g., high-nitrogen steel).

8.3.2.4.2 Inner Vessel

The main objective of an inner vessel incorporated into pool-type reactors is to separate the hot sodium pool from the cold pool. In a loop-type concept where both hot and cold pools do not coexist, there is no need of an inner vessel. Geometrically, the inner vessel is composed of a lower cylindrical shell, upper cylindrical shell, and conical shell connecting the upper and lower cylindrical shells, called "redan." The inner vessel is designed with penetrations in the redan portion to allow the passages of pumps and heat exchangers to enter into the cold pool. The inner vessel is generally a self-standing structure, supported at its base on the grid plate. Figure 8.19 depicts a few geometrical options of the inner vessel conceived in international SFRs. Either single-wall or double-wall concepts are being adopted in international SFRs. The single-wall concept is generally a self-standing vessel, which carries both mechanical and thermal loads in an optimum way. On the contrary, in the double-wall concept, mechanical loads on one vessel and thermal loads on another vessel are shared as in SPX. The inner vessel, which is in contact with the hot pool, absorbs thermal gradients in view of its flexibility, and a relatively rigid outer vessel, which is in contact with the cold pool, supports the mechanical load due to the sodium pressure and self-weight including standpipes.

An important aspect to be studied in the design is the appropriate choice of a sealing system to be provided to minimize the leakage of hot sodium to cold sodium. The pumps can penetrate the vessel through long passage tubes (pump standpipes) covering the entire hot pool isolating the hot sodium completely. Such arrangement (isolation of hot pool) is not possible for IHX penetrations, since the hot sodium should enter into the IHX to transport its heat to the secondary sodium, without any significant bypass to the cold sodium pool. This has been achieved in SPX in a novel way using an argon pocket (Figure 8.20a). Though it has many advantages in terms of good leak tightness, construction simplicity, and ease of disassembly and assembly of IHX for maintenance, gradual loss of argon due to diffusion into the hot liquid sodium and consequent fear of argon entry into the core has led to not use this option for future reactors. Alternatively, mechanical seals (piston ring type) are used, which can be designed to minimize sodium leakage to an insignificant level. Figure 8.20b shows argon and mechanical seal arrangements. Mechanical seals are employed in BN-600 and PFBR. This kind of option calls for higher IHX standpipe diameter, hence the diameter of the main vessel.

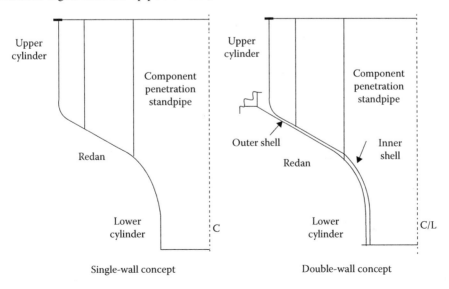

FIGURE 8.19 Geometrical options of inner vessel.

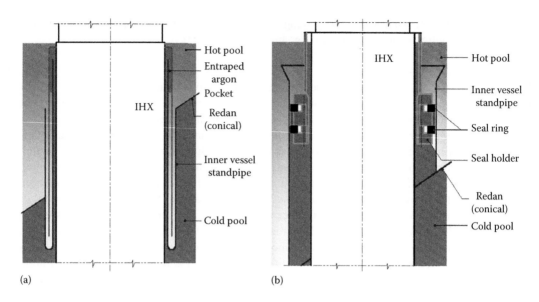

(a)   (b)

FIGURE 8.20   Inner vessel sealing options: (a) argon seal and (b) mechanical seal.

The overall shape of the inner vessel is largely influenced by thermal hydraulics and structural mechanics considerations. Minimization of thermal stratification in the hot sodium pool and through-wall temperature gradients are the main thermal hydraulics issues to be addressed in the shape optimization and elevation of the torus portion (called "redan") of the vessel. From the structural mechanics point of view, buckling of redan under mechanical loads combined with seismic-induced dynamic pressure and thermal gradients is the critical issue to confirm the shape of the vessel and the thickness [8.12]. Details of thermal hydraulics and structural analyses that led to shape optimization and elevation are presented in Chapter 10.

8.3.2.4.3   Grid Plate

The grid plate supports the entire core SAs and acts as an inlet plenum for the uniform distribution of cold sodium to the foot of each of the core SAs. Coolant entry into the grid plate is through primary pipes. The salient features of the grid plate are the box-type structures with vertical sleeves connecting the top and bottom plates; sleeves designed, apart from facilitating the required flow to the various SAs, to provide sufficient bending rigidity to the grid plate by structurally connecting the top and bottom plates over the entire surfaces; and supplying an optimum quantity of sodium to the bottom plenum and thereby to the main vessel cooling system. The grid plate should be very rigid so that the horizontality of the plate and in turn the verticality of SAs will be ensured under all static and dynamic loading conditions including seismic events. The grid plate is supported by the CSS at the outer skirt and is also resting on the CSS at several intermediate loading pads (Figure 8.21). The outer shell connecting the grid plate and CSS provides the sealing boundary for the cold sodium pool at the periphery.

One of the evolving designs is to choose the plenum with a possibly minimum diameter. This is achieved by restricting the number of sleeves; those having perforations to facilitate radial entry of the coolant to the core SAs (fuel, absorber, blanket, and storage) are integrated with the top and bottom plates. The reflector and shielding SAs, which do not call for cooling, are directly supported on the top plate through spike arrangement (Figure 8.22). This concept, apart from facilitating a smaller diameter for the plenum as well as for the CSS, brings significant manufacturing simplicity and material saving, thereby significant economy. This is being considered for future FBRs being designed in India [8.13].

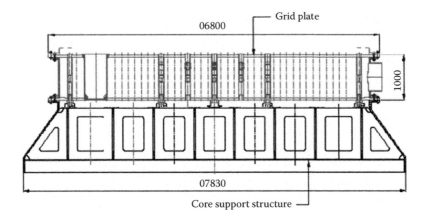

FIGURE 8.21 Grid plate on the CSS.

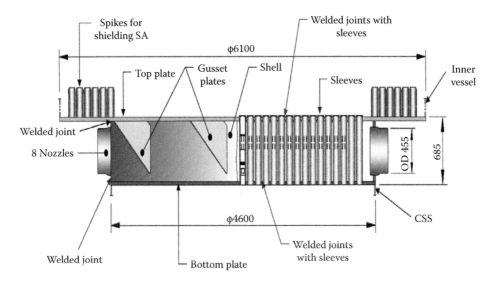

FIGURE 8.22 Welded grid plate with smaller cold plenum.

Another feature that requires careful evaluation in the grid plate design is the type of construction of the grid plate, that is, bolted or welded. There are basically four junctions in the grid plate: (1) top plate with inner vessel, (2) top plate with intermediate shell, (3) intermediate shell with bottom plate, and (4) bottom plate with CSS. Junctions 2 and 3 with the top and bottom plates with an intermediate shell are to be welded to maintain the high pressure in the plenum. Bolted or welded options are considered for junctions 1 and 4. The welded option is preferred for compactness and thermal shock considerations. Further, the welded option is favored to totally avoid coolant leakage from the pressure plenum and some marginal economy. However, the decision of choosing this option largely depends on the manufacturing capability, since the bolted option is favorable due to its relatively ease of manufacture.

8.3.2.4.4 Core Support Structure

There is an essential requirement to maintain the horizontality of the grid plate so as to maintain the verticality of the core SAs under all operating conditions. This calls for a strong backup with high bending rigidity for the large grid plate, adopted especially in pool-type reactors. The strong backup structure is also called as CSS. It is preferable to design such structure with minimum weight, which

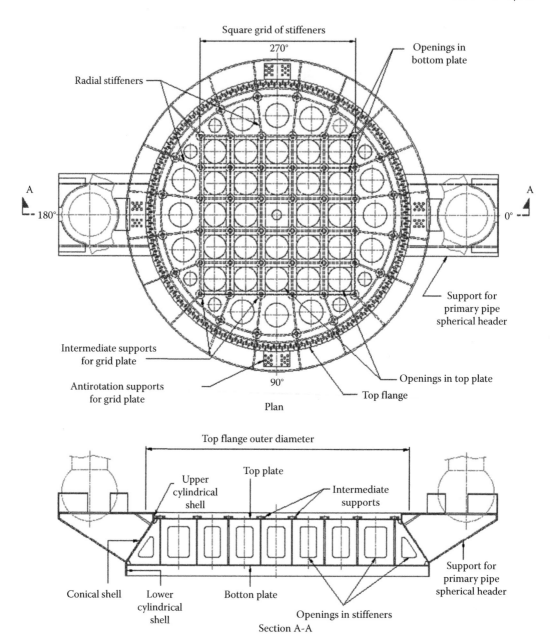

FIGURE 8.23 Stiffener pattern on CSS.

implies that a stiffened box-type structure is the better choice. The selection of stiffener configuration is an important design choice that should be arrived at through comprehensive structural optimization studies. For EFR and PFBR, the stiffener layout consists of a square grid pattern in the central zone with the radial stiffener at the periphery (Figure 8.23). Several spacer pads or loading pads are embedded on the top plate of the CSS, which facilitates transmission of the grid plate loads uniformly to the vertical stiffeners, then through the outer skirt, and finally to the main vessel bottom dished head. This sort of arrangement provides a nearly uniform stiffness for the plate over the FSA zone, which minimizes "compaction" and "flowering" modes of core displacements, particularly during seismic events. The CSS concepts adopted in various SFRs are described in Ref. 8.11.

8.3.2.4.5   Top Shield: Roof Slab and Rotatable Plugs

The top shield is nothing but the top cover of the main vessel. It is a leak-tight barrier between the cover gas and the reactor containment building, providing biological and thermal shielding in the top axial direction of the reactor. It consists of a stationary part, called roof slab, and rotatable plugs. In the pool-type design, the stationary part supports the components, such as IHXs and PSPs. In the loop type, it forms the interface between the rotatable portion and the reactor vessel. Depending on the design concepts, there can be single or multiple rotatable plugs, which house the mechanisms for the absorber rod drives and fuel handling machines. The components hanging from the top shield are generally very long, amplifying a small thermal/mechanical bowing of the top shield parts, resulting in large lateral displacements at the bottom locations, which should be restricted so that the absorber rods, pumps, and IHX can be moved up and down without any significant mechanical interferences. This is possible with the choice of rigid design concepts (e.g., box-type design). The components mounted on the top shield penetrate to enter the reactor pool. Freezing of sodium in the annular spaces between the component penetrating the roof slab and adjoining shells embedded in the roof slab causes major problems during fuel handling and removal/replacement of the components for maintenance. To avoid freezing of sodium aerosols in the narrow annular gaps, it is preferred that the temperature of the top shield is maintained above the sodium freezing point. To facilitate the plug rotations during fuel handling operations without any unacceptable leaks of radioactive cover gas to reactor containment building, proper seals should be provided. Thus, designing for high rigidity, managing the sodium freezing problems, and sealing are the major challenges in the design and construction of the top shield.

Regarding loadings, self-weight along with the weights of components it supports is the main load under normal operation. Further inertial forces generated during seismic events, weights of shielded flasks used for removal of components for maintenance, and transient forces due to sodium slug impact under core disruptive accident (CDA) add additional major loads to the component and decide basic thickness requirements. It is preferable to keep the top shield temperatures low so that personnel access over it is possible for various maintenance operations. However, sodium freezing is not avoided, and hence, a minimum temperature of about 100°C is always maintained. This is called the warm roof concept. However, it is possible to maintain the structure at a low temperature by providing proper insulation, which is called the "cold roof concept." Thus, there are three types: cold roof (T = 40°C–60°C), warm roof (T = 100°C–120°C), and hot roof (T = 200°C). U.S. reactors have adopted the hot roof concept, while French reactors initially adopted the cold roof and later switched over to the warm roof. In the cold roof, special insulation (as in SPX-1) is provided to keep the bottom and penetration inner surfaces hot to avoid sodium deposition. However, this is a very costly and not very reliable solution. Among these three, the warm roof concept is preferred in view of concerns of sodium freezing at narrow annular penetrations. Further, irrespective of the concepts chosen, the temperature gradients across the height/thickness should be limited so that the tilting of components hanging from it would be within acceptable values.

There are various options in the conceptual design of the roof slab that may be categorized based on (1) the temperature of the operation, (2) type of construction, (3) type of support, and (4) location of rotatable plugs. The roof slab can be constructed in different ways: box type with radial/circumferential stiffeners and shielding material within the box structure or using a thick plate. The box-type construction is a welded construction using plates, and the required structural strength is achieved by choosing proper combinations of height and number of radial and circumferential stiffeners. Concrete is generally filled as the shielding material within the box structure. Alternatively, the top shield can be constructed by welding of thick plates [8.14], and shielding is achieved by choosing the appropriate plate thickness. Thus, the chosen plate thickness should meet both shielding and structural requirements. The cooling circuit can be embedded within the box structure to enable cooling of the structure, and thus, the absolute temperature as well as the gradient can be controlled. In the case of thick plate construction, adequate insulation should be provided at the bottom surface to ensure minimum heat transfer to the slab. The roof slab is connected to the reactor vessel by using a stepped construction at the periphery. Generally, carbon steel is chosen as the structural material for economic reasons, leading to a dissimilar weld between the reactor vessel and roof slab. The roof slab along with all the components is to be supported over the RV. It can rest through either rollers, screw jack, or welded skirt. The support should accommodate the thermal expansion of the roof slab. The load distribution in the vault as well as the roof slab shall be as uniform as possible. It should provide leak tightness between the concrete vault and the RCB atmosphere. The rotatable plugs facilitate the fuel handling operation, and the location of the same can be concentric or eccentric with the core based on the fuel handling option selected. The layout options of the components over the top shield are dealt in Section 8.4. A few design concepts used in other international SFRs are highlighted in Ref. 8.11.

8.3.2.4.6 Control Plug

In order to monitor the temperatures of sodium jets emerging from the core SAs as well as monitor the neutrons, several sensors/systems are positioned above the core. These sensors/systems are supported on a structure called "control plug." This is designed to be a part of the top shield of the reactor assembly and functions like the top shield: a leak-tight barrier between the reactor containment building and reactor internals and biological and thermal shield. To perform its function, it accommodates the components/equipments such as ARDMs, core-monitoring TCs, neutron detectors, and failed fuel location systems. The lower portion of the control plug is partially immersed in the hot sodium pool. Since it is positioned just above the core, it influences the thermal hydraulic behavior of the hot pool. It promotes mixing of sodium exiting from the FSAs and reduces perturbation of the sodium-free level so as to prevent gas entrainment. It absorbs significant mechanical energy release under CDA.

The overall geometry consists of several vertical tubes to provide passages for accommodating ARDMs, TC sleeves, sodium sampling tubes, etc., which are positioned properly within a skirt assembly. The skirt assembly consists of a perforated lattice plate at the bottom most elevation to maintain the structural stability of the TC, the core cover plate, and the stay plates at intermediate elevations. The diameter of the control plug is determined by the number of core (fuel/blanket) SAs that are under temperature monitoring and would be roughly equal to the core diameter. Figure 8.24 shows a 3D view of typical control plug parts. The height is decided by the elevation difference between the top shield and core top. The geometrical shape of the core cover plate, the gap between the core cover plate and the top end of core SAs, perforations in the skirt shell, and perforations in the shroud tubes influence the hot pool hydraulics significantly. Thermal stratification in the hot pool, gas entrainment in the vicinity of the sodium-free level, and thermal loadings on the inner vessel are influenced by the geometry of the control plug. The flow through the control plug is a critical parameter. While higher flow is preferred to minimize thermal gradients in various parts of the control plug under thermal transients, lower flow is better from flow-induced vibration and gas entrainment points of view. Hence, the flow into control plug should be optimized through detailed thermal hydraulics analysis supported by experimental validations to ensure that thermal transient effects, gas entrainment, and flow-induced vibration risks are minimum and acceptable. Further, an optimum gap has to be provided between the thermowells of core-monitoring TCs and the head of SAs so that free rotation of the plugs is possible during fuel handling operation.

Many features are added to enhance the reliability of the plug. Stay plates provide structural rigidity to the shell and support the shroud tubes for ARDMs. Thermal shields at the bottom of the plug protect the core cover plate from thermal striping. The thicknesses of the shell and the plates are optimized by detailed thermomechanical analysis during normal operation and seismic condition including fluid–structure interaction effects. Welding is avoided in highly stressed locations and in the shell near the sodium-free level. The middle assembly acts as a thermal and biological shield. The purpose of shielding is to limit the radiation level over the control plug. Concrete is not considered a shielding material, since its operating temperature is high. The possible alternatives are graphite, steel balls, and steel plates. It may be a single thick plate or multiple plates of the same

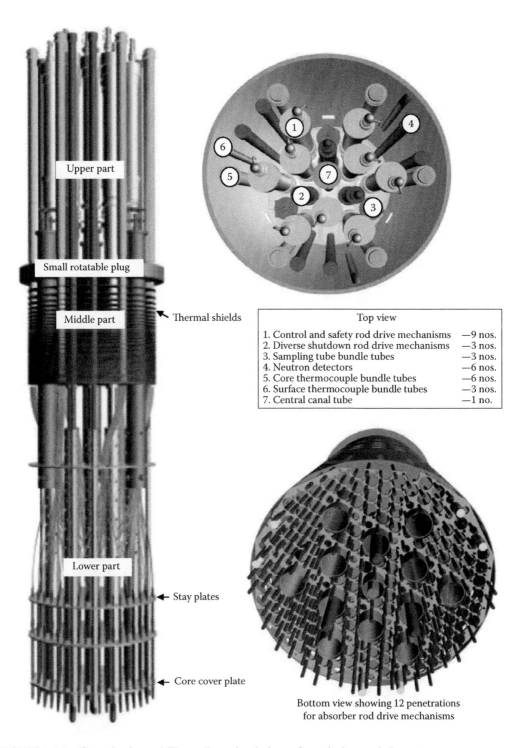

Upper part

Small rotatable plug

Middle part — Thermal shields

Lower part

Stay plates

Core cover plate

Top view

1. Control and safety rod drive mechanisms —9 nos.
2. Diverse shutdown rod drive mechanisms —3 nos.
3. Sampling tube bundle tubes —3 nos.
4. Neutron detectors —6 nos.
5. Core thermocouple bundle tubes —6 nos.
6. Surface thermocouple bundle tubes —3 nos.
7. Central canal tube —1 no.

Bottom view showing 12 penetrations
for absorber rod drive mechanisms

FIGURE 8.24 **(See color insert.)** Three-dimensional views of a typical control plug.

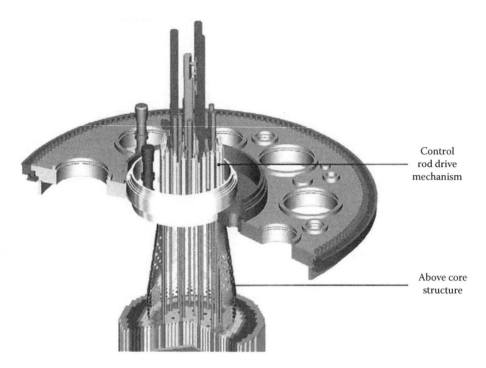

FIGURE 8.25   Control plug integral to small rotating plug.

total thickness distributed over the plug height. The cellular convection in the annular gap between the middle assembly region of the shell and the housing shell produces circumferential temperature dissymmetry and may lead to shifting of core-monitoring TCs due to tilting of the control plug. The mechanism box assembly covers all the components, which are supported on the control plug.

A few typical control plug concepts adopted in international SFRs are presented in Ref. 8.11. In the evolutionary design of SFRs, the control plug integral to small rotating plug has been conceived (Figure 8.25) for EFRs. With the integral control plug adopted in EFR, the main vessel diameter can be reduced significantly. The option of a spilt control plug as shown in Figure 8.26 is being studied for Japanese SFR (JSFR) to have a compact reactor assembly by having the provision to lift

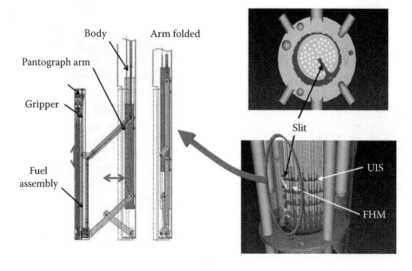

FIGURE 8.26   Option of spilt control plug.

the control plug above the top shield during fuel handling operations to facilitate free access to the fuel transfer machine to handle the core SAs below the control plug region.

8.3.2.4.7 Primary Sodium Pump

PSP performs an important safety function of circulating the coolant across the core to remove the nuclear heat under all operating conditions of the reactor [8.11]. Design and selection of materials and manufacturing technology for sodium pumps differ to a large extent from conventional pumps because these pumps operate relatively at high temperatures and have high reliability. Basically two options are available, that is, centrifugal and electromagnetic (EM) pumps. Centrifugal pumps are widely used for applications for varying the flow. Extensive R&D results and large field data are available for the performance and testing of this kind of pumps. EM pumps are generally used for low-flow (<1000 kg/s) and low head applications. EM pumps do not have any moving parts, which are preferred for trouble-free operations. But they are less efficient compared to centrifugal and positive displacement pumps. Also, the design and manufacturing of these pumps require more R&D. The technology of making the duct of EM pumps for large flow has still not fully matured. However, for smaller flows required for auxiliary circuits, EM pumps are employed in fast reactors. Hence for sodium-cooled reactors, centrifugal pumps are widely used for the main circulating pumps and EM pumps for some of the auxiliary circuits. Further, any option on the pump should be studied with caution whether it is technologically viable to design and manufacture. Availability of castings of high quality, manufacture of long and slender precision items, quality sealing in hostile environment, etc., are some of the technological constraints to be given due consideration. In terms of economy and safety, the number of pumps has to be chosen. Depending on the type of reactor such as pool or loop, the number of pumps will have a different effect. For a required flow condition, if the number of pumps is high, the size of the pump will be smaller, benefiting the reactor vessel in the case of pool-type reactors. If the number of pumps is low, the size of the pumps will be larger. The number of pumps and the flow rates in various fast reactors are given in Table 8.2.

It is seen from Table 8.2 that three pumps are generally used for medium- and large-size pool-type reactors. In order to account for many small leak paths within the reactor block (Figure 8.27),

TABLE 8.2
Details of Primary Sodium Pumps of International SFRs

Reactors	No. of PSPs	Power (MWt)	Core, ΔT (K)	Specified Flow/Pump, Q_a (m³/min)	Theoretical Flow/Pump, Q_t (m³/min)	Q_a/Q_t
Rapsodie	2	40	106	10.2	12.25	0.83
Kompakte Natriumgekühlte Kernreaktoranlage (KNK-II)	2	58	165	10	11.41	0.88
EBR-II	2	62.5	102	34.1	19.9	1.71
FFTF	3	400	143	56	60.5	0.93
Monju	3	714	132	100	117	0.85
PFR	3	650	161	84	87.4	0.96
Phénix	3	563	165	63	73.9	0.85
PFBR	2	1250	150	247.8	270.6	0.92
BN-600	3	1470	170	161.7	187	0.86
BN-800	3	2100	193	205	235.5	0.87
SPX-1	4	2990	150	290	314	0.92
CDFR	4	3800	171	310	360.8	0.86
EFR	3	3600	150	450	519	0.87

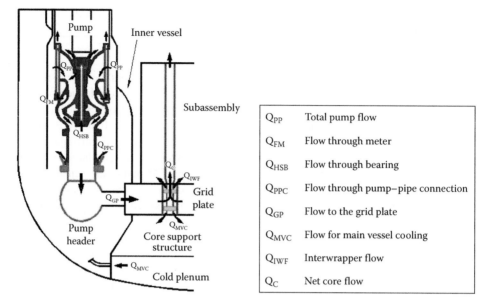

Q_{PP}	Total pump flow
Q_{FM}	Flow through meter
Q_{HSB}	Flow through bearing
Q_{PPC}	Flow through pump–pipe connection
Q_{GP}	Flow to the grid plate
Q_{MVC}	Flow for main vessel cooling
Q_{IWF}	Interwrapper flow
Q_C	Net core flow

FIGURE 8.27 Apportionment of flow from the primary pump.

the PSP is required to supply additional flow, apart from the core flow. As for flow-induced vibration and lifting of core SAs, the margin on the maximum flow required should be kept minimum, for which a realistic assessment of leakage flows is essential. This has been ensured for almost all the reactors as seen in Table 8.2, where it is found that the ratio of the specified flow rate to the theoretical flow rate is close to unity. The salient features of PSPs are described in the following text.

In case the reactor operation with one of the pumps under maintenance is foreseen toward achieving a high capacity factor, termed as (n − 1) operation, there is a need of a nonreturn valve (NRV) to prevent the reverse flow in the affected pump, thereby eliminating the possibility of bypass of the core flow. However, the provision will increase the axial length of the pump and may cause safety concerns in case of stuck-up. Hence, the (n − 1) operation is not proposed particularly in the case of a small number of pumps (a minimum of two pumps could suffice), thereby eliminating the need of an NRV in the pump. In view of excellent experience, this does not generally affect the plant capacity/availability factor.

The primary pump should circulate the coolant even if the normal power supply is not available in removing the decay heat from the core. To accomplish this, a few features are provided on the pumps: provision of class III power supply to the pump, incorporation of flywheel with adequate mass inertia, and provision of a pony motor. The flow halving time of the flywheel is to be chosen such that for a certain period of time, the pump shaft will be running with the energy provided from the flywheel and will ensure the required flow through the core. The pump can be operated with the pony motor, even when no class IV and class III power supply is available, using the battery supply for a limited period of time.

The higher the speed, the lesser will be the size of the pump. But there is limitation on selecting higher speed. The selection speed can be decided only after getting the suction conditions such as the net positive suction head available (NPSHA), which is a plant factor depending on the operating pressure (i.e., cover gas pressure in the case of sodium pumps). The NPSHA for fast reactors are normally small compared to other reactors because fast reactors are low-pressure systems. Another factor that is important in deciding the speed of the pump is the net positive suction head requirement (NPSHR), which is a factor based on the pump design. This NPSHR is a function of speed. The NPSHA should always be more than the NPSHR with some margin. Hence, if we choose a higher speed for the pump, the NPSHR will be high and the margin on the NPSHA will be lower.

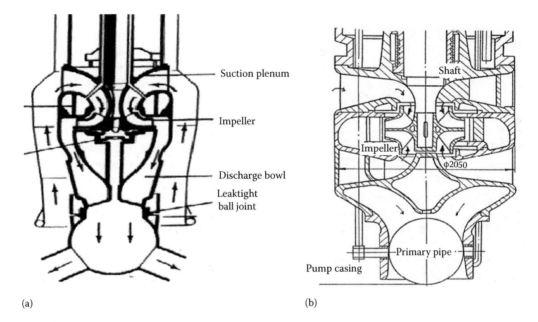

FIGURE 8.28   (a) Single- and (b) double-suction pumps.

The NPSHR can be calculated based on the flow and speed conditions (also measured experimentally). Another factor that contributes to the selection of the speed is the design of the shaft. The natural frequency of the shaft should be at least 25% away from the maximum operating speed. Hence, a very high speed will complicate the shaft design. Lower speed is generally preferred. In SFRs, valves are not permitted to adjust the flow due to consequences of possible stuck-ups. Rather, the speed of the pump is varied to meet the different reactor power levels. There are different options available to control the speed of the pump. One of these options is the Ward Leonard drive system, having a dc motor driven by a dedicated dc generator, whose output voltage can be varied to get the desired speed. The main disadvantage is that the whole system is bulky. Another option is the variable speed drive system. This system can control the speed within ±1 rpm. This fine control is used to minimize flow fluctuations and subsequent temperature fluctuations. There are designs with pumps that can run only at a few discrete speeds (typically two speeds).

Most of the sodium pump impellers are of single-suction type. The choice of single or double suction (Figure 8.28) as well as bottom or top suction will influence the maximum diameter of the pump. A double-suction pump permits operation at higher speeds than a single-suction pump and is therefore more compact. However, this advantage is offset by the complicated layout of hydraulics and is therefore generally not preferred. A double-suction impeller has more entry losses associated with the top suction impeller and an involved discharge layout associated with the bottom suction, resulting in difficulties to get high-quality castings. A single-stage top suction impeller simplifies the hydraulic layout considerably compared to the bottom suction concept (in particular for a pool-type reactor). However, this results in more entry losses, as the sodium flow has to turn at 180° before entering the impeller eye, which may reduce the NPSHA.

The seals and bearings normally use a lubricant to remove the heat generated. Oil has a potential of a major fire hazard, and hence, it is required to minimize the usage of oil and a good design feature to prevent oil spillage. Leakage of oil in the reactor is to be prevented by the proper design of oil collection and its recovery. Oil-free bearings, such as magnetic bearings, are being considered in later designs (e.g., EFR).

The top and bottom portions of the pump encounter different temperatures and create differential thermal movements. To accommodate these movements, various options are available. These options are (1) tilting sleeves at the bottom/top or combinations of these, (2) spherical support, and (3) sliding supports at the top or bottom or a combination of these options. To accommodate axial thermal expansion, flexibility is provided at the pump to pipe connection location.

The shaft transmits the torque from the drive motor to the impeller. The shaft of vertical sodium pumps is generally lengthy especially for pool-type reactors because of the submergence of the impeller, traverse of the reactor cover gas space, and radiation shielding requirements. To meet the simultaneous requirement of the torque and critical speed, the shaft is made of composite construction with a hollow middle portion welded on either side to solid ends. A shaft based on the critical speed alone would be uniformly hollow. However, in order to provide shielding against radiation, the shaft is made solid at the top and bottom from hydraulic considerations. The diameters at various sections are fixed based on torque and critical speed considerations. An important feature of shaft fabrication is the stress-relieving heat treatment to be carried out after the welding of the hollow and solid ends in a vertical furnace (to avoid the sagging of the long shaft) to remove the residual stress. If residual stresses are not removed, the shaft may distort during operation. The hollow shaft is evacuated after heat treatment to eliminate convection currents inside the shaft. Precision manufacture and balancing of the pump assembly, especially the rotor part, should be done to have vibration characteristic to the desired level as problems of shaft vibration have been reported in BN-600.

In order to avoid overhang in long shafts, the bottom bearing is generally located underneath the sodium-free level closer to the impeller. Therefore, it becomes necessary that the bottom bearing be lubricated by sodium. Because of the rather low kinematic viscosity of the sodium at the operating temperature and the fact that maximum rotational speed is limited, the bearing used for this purpose is of hydrostatic type. These bearings get the sodium feed from the pump discharge itself. There is some flexibility in selecting the clearance for a hydrostatic bearing (hydrostatic bearing can permit more clearance; higher clearance is required for sodium-operated pumps to avoid any blocking due to impurities) as compared to a hydrodynamic bearing, the only limitation being the permissible leakage flow. The top bearing is of conventional type as it is not immersed in sodium.

Mechanical seals are used for sealing between the rotating shaft and the stationary parts of the pump to prevent escape of radioactive cover gas on the one hand and entry of air into the main vessel on the other hand. Stuffing box-type seals are not employed because of its leak characteristic. The material selected should have (1) good compatibility with liquid sodium, (2) good weldability, (3) good formability, and (4) high temperature strength. Based on these requirements, SS 304 LN is widely used as a principal material for sodium pumps. Wherever metal to metal contact is inevitable, the surfaces are hardfaced. The materials used for the major parts of the pumps operating at cold temperatures are listed in Table 8.3.

A generic geometrical feature of a typical pump is schematically shown in Figure 8.29. The PSP is generally a vertical centrifugal pump with a top suction impeller having a sodium-free level. The bottom portion of the standpipe fixed to the inner vessel permits suction from the cold pool. Subsequently, the suction bell directs the sodium to flow downward to the impeller. The impeller delivers sodium to an axial diffuser from where it is led to the discharge nozzle. From the nozzle, sodium is fed to the spherical header that supplies sodium to the grid plate through primary pipes.

TABLE 8.3
PSP Materials (Typical)

Component	Material (Typical)
Hydraulic parts (impeller, diffuser, suction, and discharge casing)	SS 304 L
Shaft, flanges, journal	SS 304 LN forging
Bolts for sodium service	SA-453 grade 660 class B
Bolts for nonsodium service	A-193 grade B7
Flywheel	SA-508 grade 2 (carbon steel forging)
Hardfacing	Colmonoy (nickel-based alloy)

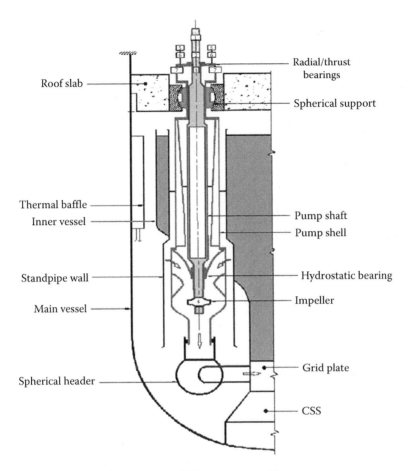

FIGURE 8.29 Schematic of a primary pump within reactor assembly.

The pump assembly consists of the pump shell and shaft that are located inside a standpipe penetrating into the cold pool. The impeller is mounted on the shaft, and the suction bell is attached with the pump shell, at the bottom. The shaft and pump shell are connected at the top by an assembly of thrust and radial bearing. In order to provide guide to the shaft at the bottom, there is a hydrostatic bearing between the shaft and the suction bell, just above the impeller level. Thus, the shaft rotates freely within the shell. The shell, in turn, is supported on the spherical bearing assembly at the top so that the shell can tilt to accommodate the cumulative radial expansion of the header, primary pipes, and grid plate assembly. The discharge pipe, which is attached with the pump shell, is inserted into the nozzle of the spherical header with close tolerance so that the pump shell as well as the shaft can expand freely in the axial direction. The flange of the spherical support is bolted to the roof slab to ultimately transmit the entire loads to the roof slab. The pump shaft is coupled to the drive motor shaft by means of a flexible coupling. The drive is an ac induction motor driven by variable speed. Typically, the speed is varied between 15% and 100% of the nominal value.

8.3.2.4.8 Safety Vessel

Sodium leakage in the main vessel can cause a great concern on safety due to the loss of coolant to remove the decay heat [8.15]. Hence, to avoid any unacceptable loss of coolant from the main vessel, a safety vessel is provided surrounding the main vessel with an optimum annular nominal gap so that the sodium level in the main vessel does not fall below the inlet windows of DHXs. Further, with the minimum gap between the main vessel and the safety vessel selected, it should be possible to have full access for the periodic in-service inspection (ISI) of the main vessel outer surface

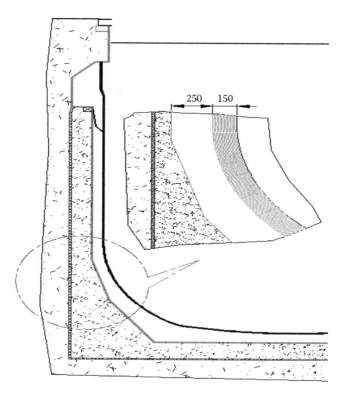

FIGURE 8.30   Safety vessel supported on the RV.

(Figure 8.30), after accounting for the possible deformations of the vessel due to thermal ratcheting in particular. The gap is filled generally with nitrogen and is monitored for any sodium leak from the main vessel. The weight of the reactor internals including the core does not act on the safety vessel, even under an unlikely event of main vessel leak; only the hydrostatic pressure of sodium would act on the safety vessel. Hence, the thickness of the vessel can be selected lower than that of the main vessel. The minimum nominal radial gap selected is 300 mm for PFBR and 700 mm for SPX-1. The choice of material can either be austenitic stainless steel 304 LN or even carbon steel from economic considerations. Figure 8.31 shows the sodium level variations when the main vessel leaks in case of a typical SFR.

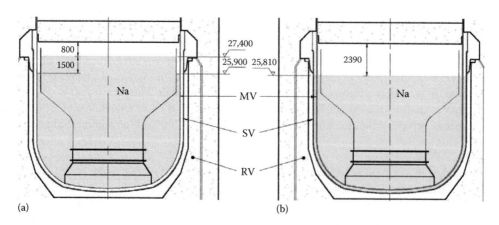

FIGURE 8.31   Sodium-free levels when main vessel leaks. (a) Normal sodium free level and (b) sodium free level with leakage.

There is a need to minimize the heat loss from the hot pool and thereby reduce the heat load for the RV cooling circuit. To achieve this, it is not preferred to mount thermal insulation on the main vessel surface, to avoid any difficulty of inspection. Alternatively, the outer surface of the safety vessel can be used to support thermal insulation panels. Thermal insulation panels should function reliably over the entire plant life of the reactor. Based on thorough analysis, thermal insulation panels made up of specified numbers of thin polished stainless steel sheets arranged in parallel are found to be suitable for this important function.

8.3.2.5 Secondary Sodium Circuit

In a typical secondary sodium circuit, IHX to derive heat from the primary circuit, SGs to transport heat to the feedwater to generate steam, and sodium pumps to drive the sodium are the main components. Apart from these, the sodium circuit in general should have expansion tanks to accommodate the sodium expansions during various operating conditions. In order to minimize the consequences of a large sodium–water reaction in SGs—high pressure and spread of corrosive reaction products within the system—two provisions are introduced in the circuit: (1) rupture disk assembly and (2) fast dumping system. To collect the sodium in case of such an event, dump tanks are also incorporated at the lowest elevation. The entire sodium circuit is double walled with guard piping arrangement in order to contain any inadvertent sodium leak in the system. The interspace is maintained with inert gas and is compartmentalized to limit the quantity of coolant leak. The sodium leak detection system is an important ingredient of the secondary sodium circuit. Leak detection is usually carried out by provision of spark plug detectors, wire-type leak detectors, sodium aerosol detectors (SADs), etc. Generally, more than one leak detection system is incorporated to increase the reliability of leak detection. Volumetric expansion of sodium and thermal expansion of components/piping are the critical issues in the design of a sodium piping system. The expansion tank is provided with adequate cover gas space to accommodate the volumetric expansions of sodium under various operating/transient conditions. The overall piping layout is finalized with adequate flexibility to minimize thermal expansion effects with appropriate supporting arrangements to withstand sodium–water reaction pressures as well as seismic forces and moments. Figure 8.32 shows a typical secondary sodium circuit in a pool-type SFR along with major components.

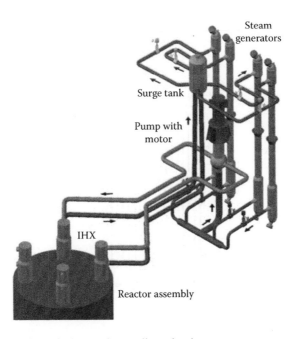

FIGURE 8.32 Schematic of a typical secondary sodium circuit.

The hot leg of the piping, from IHXs to SGs, is generally constructed of SS 316 LN grade material and the cold leg is of SS 304 LN grade. The austenitic stainless steel grades such as SS 321 and SS 347 are avoided due to problems associated with reheat cracking as experienced in PFR. The material of SGs is generally made of appropriate ferritic steel, and austenitic stainless steels are not used due to stress corrosion cracking. Some of the unique features of sodium piping are as follows:

- Thin-walled piping in view of low design pressure to mitigate thermal stresses and economy considerations.
- Joints without flanges to avoid unequal thickness thereby mitigating transient thermal stresses. Bolted joints are not allowed to prevent the risk of sodium leaks. Hence, all welded joints without flanges are adopted.
- Bellow seal or frozen seal valves to prevent sodium leaks.
- Lined with heaters for preheating purpose prior to sodium filling in the circuit.

8.3.2.5.1 Intermediate Heat Exchangers

The major functions of IHXs are to transfer heat from primary sodium to secondary sodium, to take part in the DHR, to provide a leak-tight barrier between primary sodium and secondary sodium, and to prevent the effects of sodium–water reaction in SG (high-pressure, hydrogenous materials, and reaction products) from reaching the core [8.16,8.17]. IHX is a vertical shell and tube-type countercurrent sodium to sodium heat exchanger. The schematic arrangement of IHX is shown in Figure 8.33. Generally, the primary sodium flows on the shell side and the secondary sodium flows on the tube side. In a typical pool-type reactor, the shell side is designed such that the pressure drop is less and limited to the sodium-level difference between the hot pool and cold pool during operation. The tube length of IHX is limited by the size of the reactor assembly for a pool-type FBR.

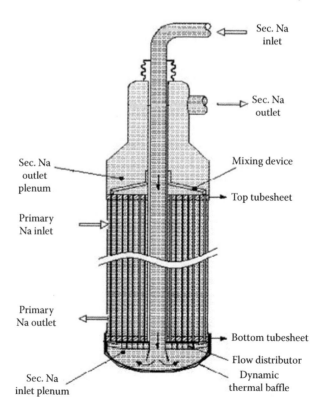

FIGURE 8.33   Schematic sketch of IHX.

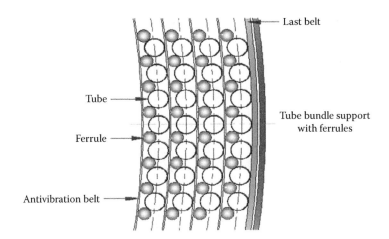

FIGURE 8.34 Arrangement of antivibration belts with ferrules in IHX.

The tube size is optimized between the pressure drop and the overall size of IHX. Hence, a shorter length is preferable. In a pool-type reactor, the whole IHX has to be removed for any maintenance, but for a loop-type reactor, the design allows for removal of the tube bundle alone instead of removing the whole IHX. The tubes are supported at regular intervals by antivibration belts to minimize flow-induced vibration. The ferrule type of support (Figure 8.34) is used for supporting tubes in the antivibration belt, which offers the least flow resistance as compared with the other support arrangements (e.g., baffle plates). The design is made conducive to natural circulation even when the SSP is not operating to enable decay heat removal.

Either the bend tube concept or straight tube concept is used in SFRs. The bend tube concept is adopted in IHX of Fermi (United States), Rapsodie (France), KNK (Germany), and PFR (United Kingdom), whereas the straight tube concept is adopted in Phénix (France), SNR (Germany), EBR-II (United States), and BN-600 (Russia). The straight tube concept is preferable for manufacturability as compared to the bend tube concept. While there is no problem of accommodating thermal expansion in the case of the bend tube concept, the straight tube design can accommodate very limited differential expansions between the bundle to shell and tube to tube in a bundle. Hence, in the case of the straight tube design, the temperature difference across the tube bundle should be minimum. There is a difficulty in achieving this. The hot primary sodium flows to the IHX through the top window and gets bypassed, that is to say that more flow in the periphery and lesser flow deep into the tube bundle due to increase in resistance toward the inner row. Hence, sodium flow uniformity across the tube bundle is not practically possible. To cope up with the situation, a flow distributor (Figure 8.35a) is provided on the tube side fluid for flow partitioning with higher flow on the outer row of tubes as compared to the inner row of tubes to reduce the temperature difference across the tubes. Along with this, a flow mixing device (Figure 8.35b) is also provided at the secondary sodium outlet over the top tube sheet to reduce the exit temperature difference across the tube bundle during steady-state and transient situations. Thus, by incorporating these two devices, the problem of differential thermal expansion in the straight tubes is mitigated significantly. The primary side is designed for maximum pressure due to a severe accident loading condition, and the secondary side is designed for maximum pressure generated during a large sodium–water reaction in SGs.

8.3.2.5.2 Steam Generator

In the SGs incorporated in the secondary sodium circuit, the heat from the hot secondary sodium is transported to the feedwater to convert it into steam [8.18–8.20]. In the case of once-through SGs, the steam coming out is superheated, which is sent straight into the turbogenerator to produce electricity. On the other hand, there are designs where there are three modules for each SG unit: evaporator, superheater, and reheater. Accordingly, there is a need of moisture separators.

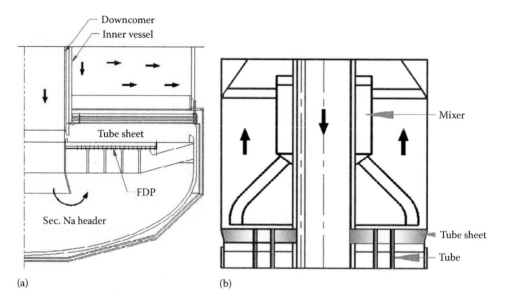

FIGURE 8.35 Flow distributor and flow mixing device at IHX: (a) flow distributor at bottom reader and (b) flow mixing device at exit.

Generally, in SFR, SGs are once-through type, taking the benefits of high-temperature sodium entering the SG (>500°C) and reduction in the number of accessories like isolation valves and leak detection systems.

The SG of a typical SFR is a vertical, once-through, countercurrent, shell and tube–type heat exchanger with sodium on the shell side and water/steam on the tube side. The schematic sketch of a typical SG is shown in Figure 8.36. The option of sending sodium on the tube side was studied by

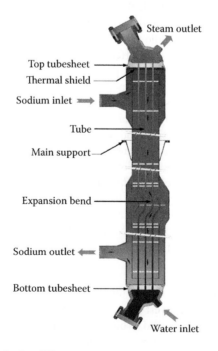

FIGURE 8.36 Schematic sketch of an SG.

Russians due to the benefit of least propagation of damage in case of a tube leak, and it is discouraged due to very large material consumption that affects the economy for a commercial-scale reactor. Orifices are to be provided in each tube at the feedwater inlet for flow instability considerations. Tubes are to be supported at various locations with appropriate intervals to minimize flow-induced vibration. For a straight tube design, the tubes are to be provided with thermal expansion bends to accommodate differential thermal expansion between the shell and tubes and among the tubes. Suitable detectors are to be provided at the sodium outlet of an SG for detection of water/steam leak. Tube to tube sheet joint is by internal bore butt welding with raised spigot type, which provides access for 100% radiography of the weld joint. The top and bottom tube sheets are to be protected by thermal shields from thermal shocks during plant transients. The SG shell is usually designed for the design basis sodium–water reaction pressure.

Longer lengths of seamless tubes are usually preferred in order to reduce the number of tube to tube sheet weld joints, which improves the plant availability. The tube length is usually limited by the manufacturability and transportability of the equipment. Temperature distribution along the tube length for the straight tube SG is depicted in Figure 8.37 for nominal power condition. The heat transfer in SG is governed by the tube side heat transfer coefficient. The typical heat transfer coefficient on the shell side (sodium side) is around 23,000 W/m² K throughout the tube length for full power condition. For the tube side, the heat transfer coefficient varies for various flow regimes (preheating regime, nucleate boiling regime, film boiling regime, and superheating regime). The water side heat transfer coefficient reduces to a minimum value at the steam exit end. This reduction is expected because with the increase in steam temperature, the density of steam decreases and the velocity and dynamic viscosity increase, which finally reduce the Nusselt number and hence reduce the heat transfer coefficient. The outer shell of the SG is designed for transient pressure generated within the tube bundle during a large sodium–water reaction. The tubes are designed for pressure and temperature of the steam produced. Ferritic steels, 2(1/4) Cr–1Mo or 9Cr–1Mo (for higher temperatures), are the structural materials generally used for SGs to avoid stress corrosion cracking problems. Various SG concepts employed in international SFRs are presented in Ref. 8.11.

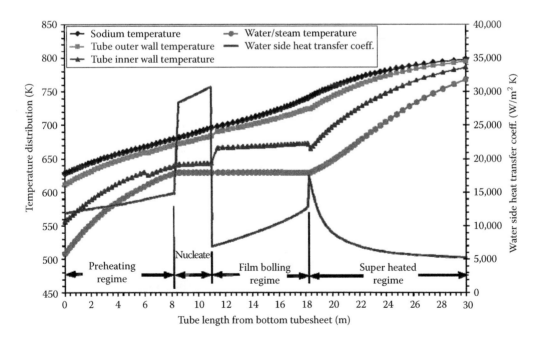

FIGURE 8.37 Temperature distributions along the SG tube length.

8.3.2.5.3 Secondary Sodium Pump

SSP is very similar to the primary pump, except for the height of the pump shaft, which is relatively small. It is a mechanical centrifugal, vertical shaft–type, single-stage bottom suction pump and is housed in a fixed shell called pump tank having sodium-free level. The space above the sodium is filled with an inert cover gas as a protective barrier against air ingress or sodium leak. The shaft is guided by a hydrostatic bearing at the bottom and a thrust bearing at the top. To prevent the leakage of argon cover gas, mechanical seals, cooled by oil, are provided at the top. The pump is capable of operating using variable speed drive as in the case of PSPs. A flywheel is provided in the shaft to have some specific flow halving time (typically ~4 s) taking into account the thermal transients in secondary circuit components and the smooth onset of natural convection in the primary sodium circuit. Figure 8.38 shows a schematic sketch of the SSP for a typical pool-type SFR.

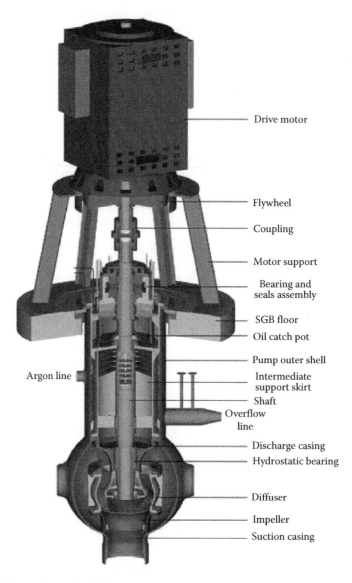

FIGURE 8.38 Schematic sketch of SSP.

8.3.2.5.4 Electromagnetic Pump

The mechanical sodium pump imparts pumping energy to sodium by means of mechanical movement of the impeller [8.21]. There is a relative motion between the stationary parts and the moving impeller of the pump; hence, complete sealing of sodium is difficult. The presence of physical moving parts in mechanical pump makes it more prone to failure, which demands more maintenance. One of the alternatives to mechanical pumps is EM pump. The good electrical conductivity of sodium is exploited in EM pumps. EM pumps have no moving part, which makes it maintenance free. Sodium is hermetically sealed in EM pumps, which eliminates the problem of sodium leakage. The working principle of EM is as follows. An electric current is forced (either conduction or induction) to flow through the liquid metal. When the current carrying the liquid metal is placed in the magnetic field, it experiences an EM body force (Lorentz force). The direction of the force is decided by Fleming's left-hand rule: Force (F) = BIL, where B is the flux density, I is the current, and L is the length of the current-carrying portion of the liquid metal. All the parameters mentioned previously are 90° to each other. The previously mentioned principle is exploited in all types of EM pumps in some form.

EM pumps are classified into two broad categories: conduction and induction pumps. Conduction pumps are the simplest type in construction where the current through the liquid metal is fed directly from the source. Conduction pumps are further classified as dc conduction pump (DCCP) and ac conduction pump (ACCP). The induction pumps are those where the current is induced in the liquid metal by transformer action. Induction pumps can be broadly classified as flat linear induction pump (FLIP) and annular linear induction pump (ALIP) and are detailed in the following text.

8.3.2.5.5 DC Conduction Pump

Figure 8.39 shows a simple construction of DCCP [8.22]. In DCCP, the current is forced through the sodium kept in the pump duct placed in the magnetic field. Hence, the force is developed in the sodium, which pumps the sodium. The magnetic field is produced by the electromagnet or permanent magnet placed such that the direction is mutually perpendicular to the current and direction

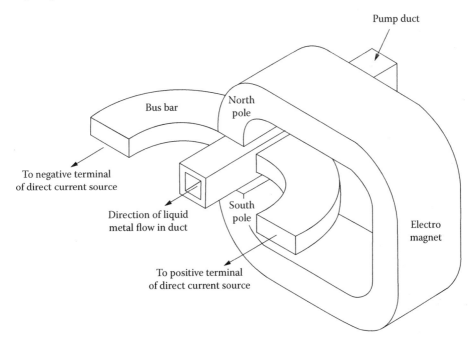

FIGURE 8.39 Schematic sketch of dc conduction EM pump.

of the desired flow. The conducting nature of the duct material and liquid metal contributes to the leakage current parallel to the duct current. The fringing flux at the end also contributes to the breaking action due to the eddy current. This, in addition to other ohmic loss, gives rise to very low efficiency. These types of pumps find application where the ambient temperature is very high. The operating current of this type of pump will be of the order of a few kA, and the operating voltage will be of the order of a few volts.

8.3.2.5.6   AC Conduction Pump

The working principle of ACCP is similar to DCCP except for the current and the flux that will be of alternating nature. In ACCP, there are two windings placed in a laminated silicon steel core: primary multiturn coil and secondary single-turn coil. The sodium placed in the duct, which obtains its high conduction current, is a part of a single-turn secondary coil. The primary winding produces a pulsating field placed in the same core. This field reacts with the alternating current resulting in a unidirectional pumping force. The pressure developed in the pump pulsates at twice the supply frequency as explained in the following:

$$F = BIL$$

Substituting

$$B = B_{max}\sin\omega t \quad \text{and} \quad I = I_{max}\sin(\omega t \pm \theta), \quad \omega = 2\pi ft$$

where f is the supply frequency. Then the force equation yields

$$F = (B_{max}\sin\omega t) \times I_{max}\sin(\omega t \pm \theta) \times L$$

By using trigonometric identities, we can write

$$F = B_{max} \times I_{max} \times \left[\frac{(1-\cos 2\omega t)\cos\theta \mp (\sin 2\omega t \sin\theta)}{2} \right]$$

The term $2\omega t$ gives rise to double-frequency pulsation in developed pressure, which can create noise and vibration.

The ACCP finds its application in various experimental liquid metal loops having a low head and flow rate. Conduction pumps are not normally used except in certain stringent process requirements due to its lower efficiency.

8.3.2.5.7   Flat Linear Induction Pump

The FLIP is the simplest form of a linear induction pump. Figure 8.40 indicates a schematic of a FLIP. The FLIP consists of a duct, carrying sodium, and a three-phase stator winding placed in slots perpendicular to the direction of flow that generates a traveling magnetic field. A copper bar is brazed at both sides of the duct along the length that plays the role of an end ring like normal cylindrical induction machines. When the stator is excited from a three-phase supply, it creates a traveling magnetic field along the pump duct, inducing electric current in the conducting liquid sodium. The interaction between the traveling magnetic field of the stator and induced current in the liquid metal produces an EM body force, pumping the liquid sodium through the duct.

8.3.2.5.8   Annular Linear Induction Pump

Figure 8.41 shows a general assembly of an ALIP [8.23]. In ALIP, the annular duct contains the liquid sodium. The stator consists of a three-phase circular distributed winding over the duct. The coils are placed in the slots of laminated stator stacks. The flux return path is through the laminated

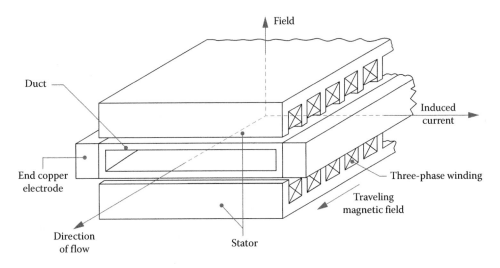

FIGURE 8.40 Schematic sketch of FLIP.

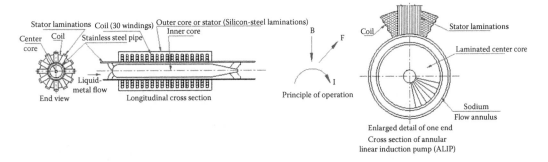

FIGURE 8.41 Schematic sketch of ALIP.

center core. The working principle is similar to that of FLIP. The stator produces a linear travel-ing magnetic field along the length of the annulus, which induces current in the sodium filled in the annulus. Interaction of this current with the stator magnetic field produces a pumping force. The ALIPs are divided into two categories from the construction point of view: the flow-through type and reflux type. In the flow-through ALIP, sodium enters into one end of the annulus and leaves from the other end. In the reflux type, the entry and exit of sodium are on one side only. Replacement of winding can be done only by cutting the sodium pipeline in a once-through-type ALIP. In the reflux type, replacement of winding can be done without cutting the duct/pipeline since both the sodium inlet and outlet are at one end of the pump.

Figure 8.42 indicates a typical head versus flow characteristic of EM pumps. It can be seen that the flow through the pump can be varied by varying the voltage applied to the pump. In FLIP and ALIP, the sodium duct (made of stainless steel) works as short-circuited secondary; hence, when-ever voltage is applied to the pump, short-circuiting current will be flowing in the duct causing heating. Because of this, this type of EM pump can be applied with a certain voltage without the establishment of sodium flow and without damage to the pump duct. If a higher voltage is to be applied to the pump, then establishment of sodium flow that removes heat generated in the duct is a must. The region of this operation is called the forbidden zone of the pump. Figure 8.43 depicts photographs of typical DCCP and ALIP.

The efficiency of the induction-type EM pumps is of the order of 5%–20%. Higher efficiency is achieved at higher flow rates. However, it has lower efficiency as compared to mechanical pumps.

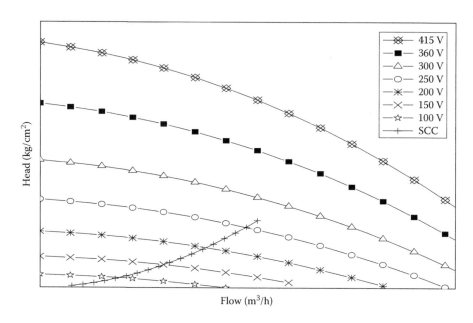

FIGURE 8.42 Typical head versus flow characteristic of the EM pump.

FIGURE 8.43 Photographs of EM pumps.

Due to this, EM pumps are used for comparatively low-capacity requirements such as experimental sodium loops and purification/other auxiliary sodium circuits of the reactor. EM pumps were used in EBR-II. Mechanical pumps are used in primary and secondary circuits of sodium-cooled fast reactors due to their higher efficiency. A flywheel is attached in PSPs that facilitates pumping of sodium in the core for a short time (few seconds) due to the stored kinetic energy even after a sudden electrical power failure in the pump. This is very critical aspect of reactor safety from the immediate DHR point of view, which gives an upper hand to mechanical pumps over EM pumps in primary circuit since EM pumps stop pumping immediately after the power failure. Pumping of sodium even after power failure may be obtained by providing a large-capacity battery bank to the EM pump, but this methodology is yet to be proven and yet to be accepted from the safety aspect. Mechanical pumps should be used carefully in case the sodium has impurities such as oxides or hydrides that could plug narrow zones in the pump (e.g., hydrostatic bearing location) and cause seizure of the pump. This problem does not exist in EM pumps because the gap in the duct of EM pumps where sodium flows is of the order of a few millimeters, which cannot be plugged by impurities. Hence, EM and mechanical pumps complement each other. Today, EM pumps are used in purification circuit and mechanical pumps are used in primary circuit by virtue of their better efficiency and safety.

8.3.2.6 Decay Heat Removal Circuit

Most of the fission products generated in the core are not stable, and they continue to undergo further nuclear processes to achieve stable states [8.11,8.24]. These processes lead to the release of alpha, beta, or gamma radiation, which is converted into the thermal movement of atoms of the material with which they interact. Thus, heat continues to be generated in a nuclear reactor core even after shutdown, which is referred to as decay heat. As time progresses, the quantity of unstable fission produced keeps reducing and hence the decay heat also reduces with time. There is substantial heat generation by radioactive decay, especially immediately following shutdown. The major source of heat production immediately after reactor shutdown is due to the beta decay of radioactive elements produced recently in the fission process. In addition to the fission product decay, some decay heat is produced by beta decay of U^{239} and Np^{239} plus smaller amounts from decay of activation products (e.g., steel, sodium) and higher-order actinides such as Cm^{242}. The decay heat is a function of the core composition and burnup, besides the operating power history. Quantitatively, immediately after reactor shutdown, the decay heat can be as high as 5% of the core power at which the reactor was operating. About 1 h after shutdown, the decay heat reduces to about 1.7% of the previous core power. After a day, the decay heat falls to 0.7%, and after a week it will only be 0.4%. A 0.1% reactor power continues to be produced in the core even after 120 days. The nuclear decay power as a function of time after reactor shutdown is given in Figure 8.44. There will be substantial decay heat produced in spent fuel elements even after several months. Therefore, special considerations are required for its safe removal during storage of spent fuel. An approximation for the decay heat curve valid from 10 s to 100 days after shutdown is given as

$$\frac{P}{P_0} = 0.066\left(\left(\tau - \tau_s\right)^{-0.2} - \tau^{-0.2}\right)$$

where

P is the decay power
P_0 is the reactor power before shutdown
τ is the time since the reactor starts
τ_s is the time of reactor shutdown measured from the time of start-up (in seconds)

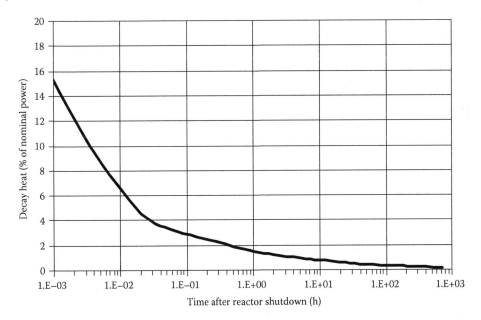

FIGURE 8.44 Nuclear decay power as a function of time.

Continuous cooling of sodium pools should be ensured for the removal of decay heat from the reactor core. Otherwise, the reactor core and reactor assembly components could reach unsafe temperatures over a period of time, resulting in major structural damage. Therefore, DHR is one of the most important safety functions and must be accomplished with high reliability. Safety criteria followed for plant design require the nonavailability of the DHR function to be $<10^{-7}$/reactor-year. Apart from this, the following safety criteria are generally postulated:

- The DHR system should have suitable redundant loops to fulfill the requirements assuming a single failure criterion.
- The DHR must be ensured after safe shutdown earthquake (SSE) and severe accident conditions, particularly after CDA for long-term coolability.
- The number of active elements in the heat removal path may be reduced to a minimum, implying the need of introduction of more passive systems.
- Various systems should be sufficiently separated physically and functionally different, to prevent common cause and cross-linked failures.
- Apart from DHR circuit components and piping, all the systems and components supporting the DHR functions, such as the main vessel and top shield, should be designed with high reliability.

The DHR can be fulfilled in various ways as shown in Figure 8.45. When offsite (grid) power is available and all the heat transport circuits are intact, the preferred path to remove decay heat is through the normal heat transport path consisting of primary sodium, secondary sodium, and water/steam circuits. The steam generated is directly sent to the main condenser bypassing the turbine. This path is utilized in almost all the SFRs around the world. When the flow rate reduces to ~20% of the nominal flow rate, the main condenser needs to be bypassed with an auxiliary condenser. In case of loss of the condenser cooling water circuit, SG cooling can still be achieved by sending low-pressure water to the storage tank and releasing the resulting steam to the atmosphere. In order to realize this option, sufficient feedwater inventory (for up to 8–12 h of station blackout [SBO] situation) in the storage tank should be available. This option is utilized

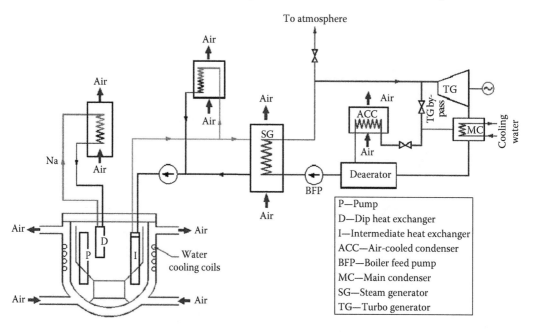

FIGURE 8.45 DHR options.

in Phénix with 12 h of water storage capacity. If the steam/water system is not available, air cooling of the SG outer surfaces by natural or forced circulation can be resorted to. A variant to this would be an air cooler in the secondary sodium circuit. All the previously discussed options require availability of power supply, condenser cooling water circuit, steam–water circuit, and secondary sodium circuit. Also, all these options consist of a large number of active components (pumps, valves, etc.) in the heat removal paths. Hence, the reliability of DHR functions is limited and less. Moving closer to the reactor, cooling of the RV by air/water coils can be considered. This concept is called RV auxiliary cooling system, and in Phénix and Superphénix (France), it is accomplished with water as coolant. But the typical heat removal rate for the system is limited (~4 MWt) for a typical 500 MWe plant. However, this concept shall be used in smaller and loop-type sodium-cooled fast reactors.

The next option is to remove heat directly from the hot pool of the reactor by providing a dip heat exchanger in the primary sodium system. This option will be available even if all the secondary sodium circuits are unavailable. Sodium heated in the dip heat exchanger tubes rejects the heat to air in the Na–air heat exchanger. The primary sodium flow in the dip heat exchanger, sodium flow in the intermediate sodium circuit, and airflow in the Na–air heat exchanger may be of forced convection or natural convection. The natural convection option has no active components, and hence, the reliability of the DHR function is high. Brief descriptions of various options are given in the following.

8.3.2.6.1 Decay Heat Removal through the Normal Heat Transport System

During normal operating conditions of the reactor, heat produced in the core is removed through the heat transport system consisting of sodium systems and steam–water system. It is the SG in which nuclear heat is used to convert water into steam to expand in the turbogenerator. The unused heat is finally rejected to the environment in the condenser. It is always desirable to use the normal heat transport system for decay removal after safety control rod accelerated movement (SCRAM). In such case, the heat removal function can be performed only by operating most of the systems of the normal heat removal system. Moreover, the usage of systems designed for full power management, for removal of low power of the order of 1%, leads to serious operational problems and economic penalty. Therefore, an auxiliary heat removal system of smaller capacity overriding in the normal heat removal system is generally envisaged. This kind of system known as operating grade decay heat removal system (OGDHRS) is used in the PFBR design [8.25]. This system as shown in Figure 8.46 uses the normal SG for heat removal from the sodium system, and a smaller condenser, desirably air cooled, will remove the heat out to the environment. This requires forced circulation of water through the SG.

8.3.2.6.2 Decay Heat Removal through the Steam Generator Outer Surface

In one of the alternative systems for DHR, heat is directly rejected to the environment from the SG itself. Decay heat is transferred through secondary coolant loops to the SGs. Here, SGs are housed in special casings and airflow established through the casing either by forced means or natural can remove the decay heat as adopted in FBTR and PHÉNIX reactors. This system has the benefit of eliminating the steam–water system for the DHR function. However, both the primary and secondary sodium systems are required to be operational.

8.3.2.6.3 Decay Heat Removal through Heat Exchangers in the Secondary Sodium System

Decay heat can also be removed through air-cooled heat exchangers mounted in the secondary sodium circuit. This type of system is used in Superphénix. Decay heat is removed by sodium to air exchangers mounted on a secondary loop bypass line and designed to operate even in the event of an air-supply failure. Sodium circulates by forced convection. The cooling airflow is normally supplied by a motor-driven fan. However, in the event of unavailability of the fan, natural air circulation can

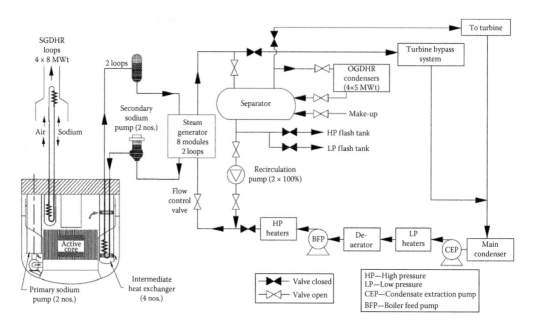

FIGURE 8.46 Typical operating grade DHR system.

be obtained by means of the draft created in a stack placed over the exchanger. A similar system has been used in BN-600 [8.26] and Monju [8.27], which requires the operation of a secondary sodium circuit for DHR.

8.3.2.6.4 Direct Reactor Auxiliary Cooling System

In this system, decay heat is removed through immersed DHXs in the primary sodium pool. The immersion-type DHR system requires the use of an intermediate sodium loop as water or air cannot be admitted as the cold fluid in immersion heat exchangers. Thus, this kind of system involves an immersion cooler dipped in a sodium pool, an intermediate sodium loop, and an intermediate sodium to air heat exchanger. The intermediate sodium flow can be either by natural convection or by forced convection using an EM pump. The airflow through the AHX can also be either by natural convection using a chimney of sufficient height or by forced convection using a blower or fan. PFBR adopts this concept (Figure 8.47) for the alternative DHR, which is known as the safety grade decay heat removal (SGDHR) system [8.28]. In PFBR, natural circulation of sodium is used in the intermediate loop. A tall chimney is used for natural convection of air. Thus, the DHR system is totally passive for PFBR. A similar system is proposed for SNR-2, KALIMER, and BN-1600 [8.29]. Another option is to design the DHR system with forced convection. The system will have a lower heat removal capacity when it is used under natural convection. This concept is adopted in PFR [8.30], SPX-1, SPX-2, DFBR, and EFR [8.31].

8.3.2.7 Reactor Vessel Auxiliary Cooling System

This system removes heat passively from the reactor containment vessel and dissipates it to the ambient air by the inherent processes of natural convection in fluids, heat conduction in solids, and thermal radiation. Heat is removed from the core and transported to the reactor vessel wall by natural convection of primary sodium. This kind of system can be economically accommodated in medium-rated reactor systems up to 500 MWe. In reactor systems of larger capacity, the vessel size, which is determined based on accommodating various components of the primary system, becomes inadequate to achieve the desired DHR capacity. This type of system has been used by S-PRISM [8.32], Phénix, SPX-1, and KALIMER.

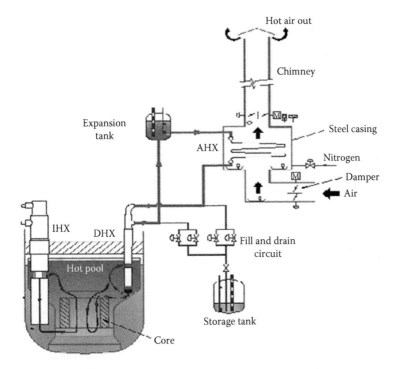

FIGURE 8.47 Typical DRACS.

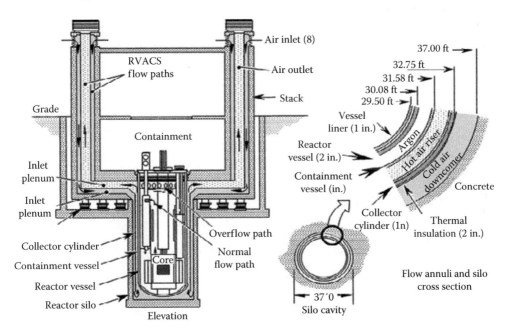

FIGURE 8.48 Reactor vessel auxiliary cooling system.

The reactor vessel auxiliary cooling system (RVACS) of a prism reactor is shown in Figure 8.48. This system can dissipate all of the reactor's decay heat through the reactor vessel and containment vessel walls by radiation and convection to naturally circulating air outside the containment vessel without exceeding structural temperature limits. This system will be in operation even during the normal operation of the reactor, removing a small amount of heat to the environment continuously.

But it functions at intended high heat removal capacity only when the vessel temperature has significantly increased. The primary sodium flow through the reactor core is maintained by natural circulation. The decay heat generated in the core is removed by the primary sodium and transferred to the reactor vessel. The heat is transferred to the containment vessel by thermal radiation and natural convection. The heat is then transferred to the air between the containment vessel and the collector cylinder by convection. The natural convection flow of air established in the system ultimately removes the heat to atmospheric air.

8.3.2.7.1 Decay Heat Removal System in Various Reactor Designs

In the CRBRP design [8.33], there are three backup systems to remove decay heat in case of non-availability of the normal heat sink. The first one is the protected air-cooled condenser (PACC) system that cools the steam drum directly. A second heat sink can be made available by opening the safety relief valve in the steam line, thereby venting steam to the atmosphere. The third system is a completely separate overflow heat removal system (OHRS) provided to extract heat directly from the in-vessel primary loop. The heat sink for this system is provided by an air-cooled heat exchanger. This system provides DHR in situations where the SG is not available. The DHR system of a loop-type JSFR consists of a combination of one loop of a direct reactor auxiliary cooling system (DRACS) and two loops of a primary reactor auxiliary cooling system (PRACS) as shown in Figure 8.49. The heat exchangers of DRACS are located in the reactor vessel's upper plenum. Each heat exchanger of PRACS is located in the IHX upper plenum. These systems operate fully by natural convection and are activated by the opening of dc-power-operated dampers of air coolers. In the innovative design of the DHR system for JSFR, the primary system and the cold pool are thermally coupled by the PRACS, which is composed of heat exchangers, fluidic diodes, and connecting pipes as shown in Figure 8.50. The fluidic diode reduces leakage flows under primary loop forced circulation.

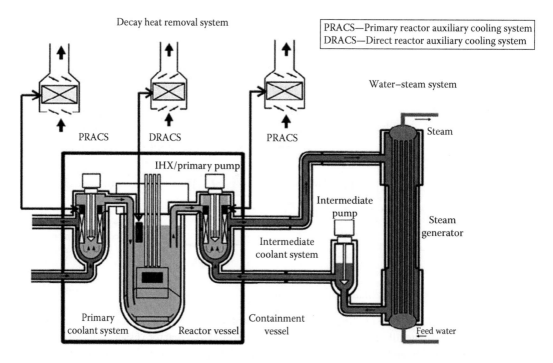

FIGURE 8.49 Primary reactor auxiliary cooling system.

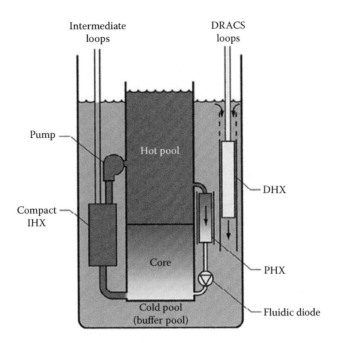

FIGURE 8.50 Innovative design of the DHR system.

8.3.2.7.2 Decay Heat Removal under Power Failure Conditions

Nuclear reactor systems are provided with four grades of power supply. They are classified as classes I–IV. Class IV is the normal power supply available from the grid that can be expected to be unavailable in the plant for several hours. Class III supply is obtained from a more reliable power source whose unavailability is limited to a few hours. This is normally derived from diesel generators in the plant. Class II supply is obtained from a further reliable source, essentially battery banks, and cater to the ac loads in the plant. Class I is the most reliable power supply provided for instrumentation in the plant. Even in case of complete failure in all the power sources (SBO), the DHR function should not be affected. The reactor heat transport system is arranged in such a way that decay heat can be removed when forced cooling is unavailable. This is achieved through natural convection. Most of the reactor systems adopt this concept for the alternative DHR system. The backup DHR systems of SNR-300 and Superphénix were designed in such a way that adequate heat removal is possible even with passive operation of the active systems, that is, with natural convection in the sodium to air heat exchangers as well as the sodium loops. It has been demonstrated under actual SFR operating conditions that natural convection can, in fact, be established rapidly enough to prevent fuel damage. Successful tests carried out at EBR-II, FFTF, PFR, and Phénix reactor systems, in which pumps were completely cut off at partial power, demonstrated that natural convection can be readily established after loss of pumping capability [8.34–8.36].

8.3.2.7.3 Decay Heat Removal Following a Severe Accident

Consequent to a severe accident involved, melting of nearly the whole molten core debris moves downward and ultimately settles on the core catcher provided at the bottom of the main vessel. The core catcher is a structural member that collects, supports, and maintains the core debris in a subcritical configuration. The decay heat generated by the debris bed settled on the core catcher is transferred to the surrounding sodium. This would result in significant rise in sodium pool temperatures if heat is not removed continuously from the sodium. Therefore, any of the DHR systems needs to be operational even after severe accident conditions in the plant. In a pool-type reactor

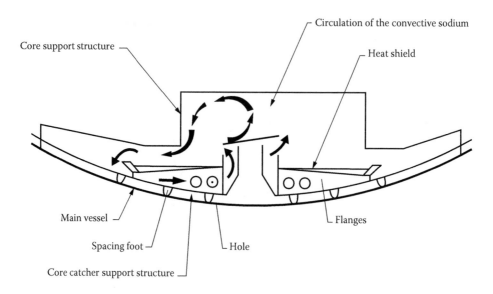

FIGURE 8.51 DHR paths from the core catcher.

system employing the DRACS-type DHR system, the success of the heat removal function depends on the establishment of a natural convection path between the heat source and the heat exchangers of the DRACS. This type of system is advantageous over other types due to the availability of the heat sink in close proximity in the primary pool. The perforation that is generated in the grid plate and CSS due to melt through of the molten fuel could facilitate a favorable path for establishing natural convection flow between the core debris and safety grade heat exchangers immersed in the hot pool as shown in Figure 8.51. The decay heat is designed to withstand transient pressures generated under a severe accident.

8.3.2.8 Sodium Purification Circuit

The impurity type and amount vary between the primary sodium circuit and secondary sodium circuit. In the primary sodium system, the impurity ingress is due to oxygen and moisture in the argon cover gas and oxygen and water vapor adhere to the fresh surfaces introduced into the sodium system (FSAs during fuel handling, pump/IHX due to maintenance, etc.). In the secondary sodium system, apart from impurity ingress through cover gas and maintenance operation, hydrogen diffuses from the water side to the secondary sodium side through SG tubes. In the SG, due to the water side corrosion of the ferritic steel tubes used, hydrogen is derived as per the following equation:

$$3Fe + 4H_2O \rightarrow Fe_3O_4 + 4H_2 \uparrow$$

In the previously mentioned chemical reaction, hydrogen diffuses to the sodium side through the tube wall, due to the hydrogen concentration difference between the water side and sodium side. This hydrogen load to the secondary sodium is to be removed continuously by cold trap; otherwise, it will mask the water/steam leak detection in the SG. In case of tube failure, water will leak into the sodium side and cause a sodium–water reaction, generating hydrogen. Hence, water leak detection in the SG is monitored by the hydrogen concentration level in sodium by hydrogen meters.

In a pool-type reactor, the purification of primary radioactive sodium shall be placed in the reactor pool (in-vessel) or outside the reactor pool (ex-vessel). In the ex-vessel concept, the radioactive primary sodium is taken out of the reactor pool and is purified in the cold trap and is returned to the reactor pool. The cold trap, economizer, EM pump, flowmeter, plugging indicator, and associated piping are kept in a shielded cell outside the reactor pool but inside the reactor containment building

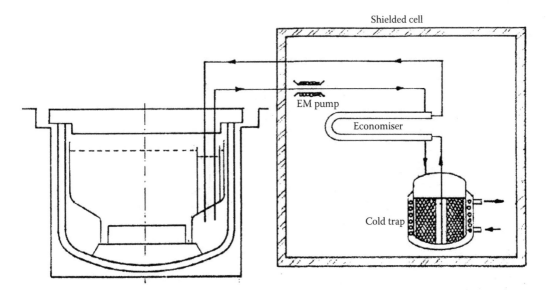

FIGURE 8.52 Ex-vessel purification circuit.

(Figure 8.52). In the ex-vessel purification circuit, the sodium could be taken either from the hot pool or cold pool. The cold pool location would be economical, as the piping and components shall be designed for cold pool sodium temperature. It needs an enormous amount of concrete for shielding. The cold trap needs to be kept in a lead-shielded box inside a concrete cell. Design measures like double envelope or inerted steel cabin have to be provided to take care of radioactive sodium leaks and siphoning of radioactive sodium from the reactor pool. In the in-vessel concept, the cold trap, economizer, EM pump, and flowmeter are housed in a tank. This assembly is located in the reactor pool and is supported from the roof slab. The EM pump takes sodium from the reactor pool and passes through the economizer and cold trap wire mesh. After purification, the sodium returns to the pool through the economizer. Sodium can be taken from the hot pool or cold pool, though the present design for in-vessel purification adopts sodium taken from the hot pool to avoid additional penetration in the inner vessel. In this concept, no radioactive primary sodium comes out of the reactor pool, an important design feature with regard to safety. A study shows that the reactor containment size shall be reduced if in-vessel purification is adopted.

In the ex-vessel concept, roof slab penetration is required only for small primary sodium pipelines that exit the reactor vessel. The in-vessel purification concept could influence the main vessel size. However, this does not seem to be applicable for the designs that have incorporated the in-vessel purification concept (SPX-1). The in-vessel purification system shall be accommodated in the available space in the reactor vessel without influencing the main vessel dimension. In the ex-vessel concept, the purification components are placed outside the reactor pool, in the shielded cells inside the reactor containment building, influencing the reactor containment building size. A preliminary study shows that the reactor containment size shall be reduced if in-vessel purification is adopted. In the ex-vessel concept, design features (double envelope/inerted steel cabin) have to be provided for radioactive sodium leaks and siphoning of radioactive sodium from the reactor pool (in case of leak in the purification system that goes undetected). In the in-vessel concept, no radioactive primary sodium is transported out of the reactor pool; therefore, the possibility of radioactive sodium leak in the reactor containment building is precluded. The siphoning of sodium from the reactor vessel is avoided. In the ex-vessel concept for the maintenance of purification components, reactor shutdown is not required. In the in-vessel concept for the maintenance of the components in the purification system and sodium purity monitoring systems, reactor shutdown is required because the shield plug in the roof slab has to be removed. However, the primary sodium

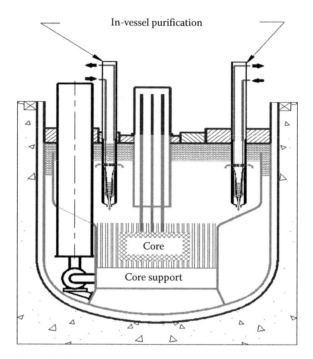

FIGURE 8.53　In-vessel purification circuit.

purification circuit need not run all the time and the maintenance/replacement shall be carried out at the time of the next shutdown for fuel handling/maintenance. The in-vessel concept is preferred due to increased safety and improved economics. The schematic of the cold trap for the in-vessel purification concept is shown in Figure 8.53.

8.3.2.8.1　Cold Trap

Sodium purification is carried out using the cold trap, which works on the principle of decrease in solubility of impurities in sodium with temperature. The saturation solubility curves for oxygen and hydrogen in sodium are shown in Figure 8.54. The schematic arrangement of the cold trap is

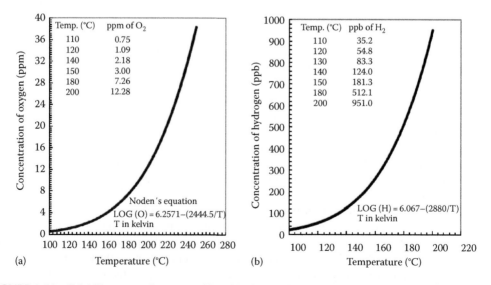

FIGURE 8.54　Solubility curves for oxygen (a) and hydrogen (b) in sodium w.r.t. temperature.

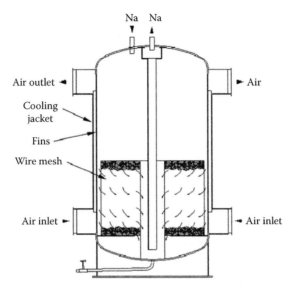

FIGURE 8.55 Schematic of the cold trap.

shown in Figure 8.55. In the cold trap, sodium is cooled to the cold point temperature (temperature below the saturation solubility of impurities present in sodium) to form a compound and passed through a wire mesh packing where the compound precipitates and is retained. The coolant used to cool sodium may be air, nitrogen, or thermofluid. NaK may be provided in the jacket (Figure 8.56) between the sodium in the cold trap and the thermofluid coolant tube, which acts as a thermal

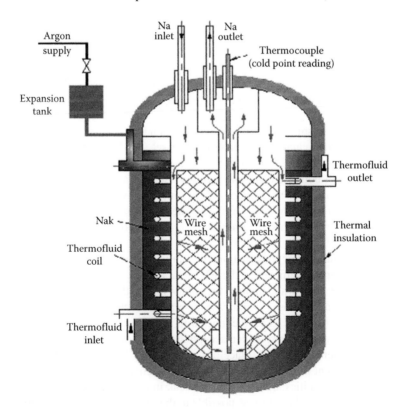

FIGURE 8.56 NaK jacket in the cold trap.

bond between the sodium and the thermofluid. An NaK jacket is provided to avoid thermofluid oil entry into the sodium system in case of a leak in the cooling coil tube. If air/nitrogen is used as a coolant, the coolant flows over the fins provided over the shell of the cold trap. The cold trap is cooled from the outside. As the sodium flows from top to bottom, it gets cooled to the cold point temperature; the impurities get precipitated and get trapped in the wire mesh. The wire mesh in the cold trap acts as a nucleation site for crystallization of the sodium oxide or sodium hydride compound and gets trapped. The purified sodium passes through the central riser pipe and joins the main system. The trapping efficiency in the cold trap is a function of the sodium residence time in the wire mesh region of the cold trap. A minimum residence time of 6 min is preferred. Effective utilization of the wire mesh volume in the cold trap is taken as 20% of the total mesh volume provided before a major increase in the pressure drop across the wire mesh packing. The required wire mesh volume is achieved from the impurity load, residence time, and 20% utilization. The density of the wire mesh packing used is ~400 kg/m^3. The sodium flow rate through the cold trap is based on the time required to purify the sodium to an acceptable limit after refueling/maintenance operation and residence time of sodium in the cold trap.

The impurities that precipitated in the cold trap can be removed by the regeneration method for reuse of the cold trap for economic considerations. The impurities settling in the cold trap for the primary sodium circuit are highly radioactive in nature, no regeneration of the cold trap in the primary sodium purification circuit is anticipated, and hence the cold trap is designed and sized for the life of the reactor. The impurities that have settled in the cold trap of a secondary sodium circuit are nonradioactive and mainly sodium hydride, and the secondary sodium cold trap can be regenerated. The secondary cold trap shall be sized for a specified period. The secondary cold trap shall be regenerated and reused periodically. At about 450°C, the sodium hydrides deposited in the wire mesh decompose into sodium and hydrogen. The hydrogen gas formed can be removed from the cold trap sodium-free surface by argon purging or vacuum. The vacuum regeneration method is preferred over the argon purging method as regeneration is faster and there is no need for a large quantity of argon.

8.3.2.9 Reactor Vault

The reactor assembly or the reactor block rests and transfers the load to a reinforced concrete cylindrical structure known as an "RV." The RV provides support for the reactor assembly, safety vessel, and other components of the reactor assembly (e.g., inclined fuel transfer machine), provides biological and thermal shielding in the radial direction beyond the safety vessel, and provides containment for leaks both the main vessel and safety vessel. The RV is designed to serve the previously mentioned functions and is also carefully shaped to limit the fall in the level of sodium (in the unlikely event of a leak in both the main and safety vessels). Further, the RV is lined on its inner surface with steel plates called liner. The liner prevents moisture/water from coming in contact with the safety vessel. It acts as a leak-tight barrier for the inert gas within the annulus between the vault and the safety vessel. Apart from this, the liner also serves as a left-in shuttering during the concreting process of vault construction.

With concrete as the material of construction, the vault requires cooling to maintain its structural functions. In order to improve efficiency, the RV is fabricated into two portions: the inner wall and outer wall. While the inner wall supports the safety vessel, the outer wall supports the reactor assembly. In view of the vault's close proximity to the reactor assembly, the temperature of the inner wall is higher than the outer wall. Hence, the inner wall is embedded with a cooling arrangement to maintain the temperature within the specified acceptable values under all operating conditions. In a typical cooling arrangement, the outer surface of the vault liner has embedded cooling pipes close to the liner. The arrangement for cooling is provided with adequate isolation and redundancy provisions in anticipation of difficulties arising out of the leak in an embedded cooling pipe. Provisions for venting are also incorporated as close as possible to the liner to avoid gas accumulation near the liner panels and to vent the gases/water vapor generated during reactor operation. This arrangement

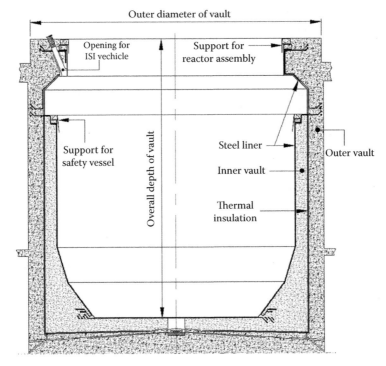

FIGURE 8.57 Schematic of the reactor vault for pool-type SFR.

avoids inward bulging of the liner panels. Since the outer wall is a major load-carrying structure, it is preferable to keep its temperature as low as possible. A gap filled with a thermal insulation material (e.g., thick expanded polystyrene insulation) separates the outer wall from the inner wall. Hence, the temperature of the inner wall is higher than the outer wall. Hence, the temperature of the inner wall is higher than the outer wall. Accordingly, design provisions should be made to accommodate the resulting thermal expansion without any appreciable straining in the concrete. Further, the coaxiality between the inner and outer walls should be maintained, particularly under seismic loading, which induces horizontal forces. Any relative radial displacement that can alter the radial gap between the main vessel and safety vessel may pose problems for the free access of the ISI equipment that monitors the healthiness of main vessel surfaces. The structural design of the vault is carried out in accordance with civil design structural codes with high levels of safety to meet various operating/incident and accident conditions along with appropriate seismic loading. Figure 8.57 shows the schematic of the RV for a typical pool-type SFR.

8.3.2.10 Operating Temperatures and Design Life

With regard to economy, a longer design life with higher temperatures is always preferred. Since the boiling point of sodium is high, ~1160 K, it is possible to choose a higher mixed mean temperature at the core outlet while maintaining a higher margin against coolant boiling. This facilitates generating steam at higher temperatures and higher pressures in SGs. Besides higher operating temperatures, longer design life is also preferred from economic considerations. Hence, choosing a higher operating temperature along with longer plant life is essential, which is very much possible in the case of SFRs. However, there are a few technical constraints resulting from material degradation, high-temperature failure modes and structural mechanics, and availability of design codes and standards. Material degradation is mainly due to the effects of sodium, thermal aging, and neutron irradiation dose. Sodium is practically noncorrosive once oxygen and carbon impurity is controlled. Exposure to flowing sodium produces changes in material properties due to carburization and decarburization. Prolonged exposure to high temperature causes precipitation of carbides and sigma phase formation in stainless steels. This in turn causes reduction of ductility and fracture toughness. The major high-temperature failure modes, which are specific to SFRs, are (1) creep/fatigue damage, (2) high cycle strain controlled fatigue damage, (3) thermal ratcheting, and (4) loss of tube wall thickness in IHX and SG tubes. More details of these failure modes and analysis methods are presented in Chapter 10. The various aspects involved in choosing optimum parameters and permissible operating temperatures along with the longer plant life for the SFR are investigated in Ref. 8.37. The highlights are only presented below.

The components operating at high temperatures are fuel and blanket SAs, inner vessel, control plug, IHX, ARDMs, secondary sodium hot leg piping, SGs, steam pipes, and turbines. Of these, the steam pipes and turbines are designed for high-temperature and high-pressure steam conditions (810 K and 20 MPa, respectively) in modern fossil power plants, and hence they are not the limitations. The design of fuel and blanket SAs, particularly cladding, under high temperature for longer life in the core is very important for the development of high burn-up fuel, which in turn reduces the fuel cycle cost. This is governed by selection of advanced materials, rather than the design optimization and analysis methodology, etc. Furthermore, the core SAs are replaceable in the course of reactor operation. Then the overall layout of the piping is designed to minimize the thermal expansion stress in the secondary circuit, though these are nevertheless significant. Thermal transients lead to the need for relatively thin-walled piping to minimize thermal transient stresses but are adequate to withstand the internal pressure (<1 MPa). However, the thermal expansion stresses concentrate at bends and lead to potential problems of creep buckling. It is worth mentioning here the primary Na outlet/inlet temperatures for various FBRs in the world: 833/658 K for Phénix, 833/673 K for PFR, 818/668 K for SPX-I, 823/653 K for BN-600, 802/670 K for Monju, and 818/668 K for EFR.

As far as plant life is concerned, the nominal design life of most of the nuclear reactors in the world is from 25 to 40 years. From engineering assessments of many plants, it has been established that they can operate for much longer. Many reactors have been granted license renewals for operating life extension from the original value in the United States, France, United Kingdom, and Russia. In Japan, plant lifetimes up to 70 years are envisaged. In the 1950s during the developing phase of the nuclear power plants (NPPs), reactors were very conservatively engineered with a useful lifetime of only 20–25 years, but finally they were authorized to operate for 50 years. Consistent plant life extension policies, well over design lifetimes, are being implemented all over the word, in order to take the maximum advantage of existing plants and increasing their performances. The possibility of a longer life is due to the improved mechanical and structural behaviors of the component and excellent operating experience.

Loss of tube wall thickness in SG and IHX at different locations—tube, weld, and belt support location—due to corrosion and fretting wear is a critical factor in choosing longer design life. With the available data on fretting wear rate and the possibility of ISI of heat exchanger tubes by which the loss of tube wall thickness at support location can be monitored regularly, heat exchanger tubes (SG and IHX) can have a design life of 60 years.

ASME Section III, Division 1, Subsection NH, French Code RCC-MR, and American and French design codes provided criteria and rules for the design of high-temperature components for the design life of 40 years. More details about these codes are provided in Chapter 10. Based on these considerations, it is possible to choose confidently the primary outlet temperature around 820 K and design life of 40 years. Further, R&D is required for a 60-year plant life.

8.4   REACTOR MECHANISMS

Critical mechanisms in nuclear reactors are those associated with shutdown and fuel handling systems. The design of mechanisms operating in liquid metals in particular involves many challenges, in view of their opaqueness, self-welding (galling) characteristics with metallic parts, and corrosion effects. Operating experiences of liquid metal–cooled fast reactors (LMFBRs) indicate that many fuel handling incidents affected the reactor availability factor significantly. In this section, the operating principles of shutdown and fuel handling systems are brought out with design variance and challenges. Design provisions to prevent galling and certain passive design features to improve the reliability of the shutdown systems (SDSs) are also discussed.

8.4.1 Shutdown Systems

The SDS controls the fission reaction by absorbing the neutrons in the reactor core. The system includes (1) absorber rods, (2) drive mechanisms to drive or drop the absorber rods into the active core region, and (3) plant protection system (PPS) consisting of instrumentation to monitor plant parameters, signal processing, power supply, and safety logic to trigger the shutdown action. When control rods are placed at the lowest elevation in the core, the reactor is in shutdown state. To make the reactor critical, the rods should be raised to a certain elevation. To achieve a desired power level, the rods are to be raised further. Rising of the control rods by their drive mechanisms removes the neutron absorbing material from the active core region, resulting in controlled multiplication of the neutronic population, thereby increasing the reactor power to the desired level. In case there is a demand from the PPS to rapidly shut down the reactor, the rods should be dropped suddenly. This is termed as SCRAM. Thus, the SDS enables power raising/lowering/shutdown of the reactor on demand during operation and rapid shutdown (SCRAM) to ensure safety during design basis events. Such system also facilitates maintenance of the desired power level with increasing fuel burnup.

8.4.1.1 Absorber Materials

The shutting down mode is usually effected by the introduction of a neutron poison, a material that absorbs the neutrons and reduces the effective neutron multiplication taking place in the reactor core. The choice of a particular neutron poison material depends on its absorption cross section, which indicates its effectiveness in absorbing neutrons. The absorption cross section of a material varies with the neutron energy. Materials having a high absorption cross section in the neutron range of operation of the reactor are preferred. Popular neutron poison materials include boron, lithium, cadmium, tantalum, europium, and gadolinium stainless steel. Neutron absorbers in the form of solid rods, called control rods, are the ones used commonly in several reactor systems. In view of the relatively low absorption cross section with a high-energy neutron spectrum, highly enriched absorber materials are used in fast spectrum reactors. For example, the use of 60%–90% enriched boron carbide (B_4C) is quite common in sodium-cooled fast reactors.

With regard to the shape and size, control rods are generally in the form of long rods, connected to drive mechanisms, either individually or as a bank (group). Alternate forms of control rods include articulated rods (smaller segments linked together), liquid poisons, and powder or granules dispersed in gas/liquid. The number and worth of control rods are fixed based on reactor physics calculations taking into account the size of the core, the distribution of various materials present in the core, reactivity worth to compensate consumption of fissile material due to reactor operation on power (burnup), and margin to ensure safe shutdown of the reactor (shutdown margin).

8.4.1.2 Description of a Typical Shutdown System

Figure 8.58 shows schematically a typical arrangement of a control rod and its drive mechanism. The rod is connected to a mobile part, which is connected to an upper part through an electromagnet. The upper part houses the drives for controlled movement of the rod. For rapid shutdown, de-energization of the electromagnet results in decoupling of the mobile part along the control rod, which then falls by gravity and gets inserted into the core. Toward the end of the fall, the rod is decelerated through a dashpot provided in the mobile assembly, avoiding any damage to the rod and the core due to possible impact, when the rod is brought to rest at the end of the travel. The control rod drive mechanism is also provided with instrumentation to sense the presence of the rod (load cell), travel or position of the rod during normal raising/lowering (potentiometers/synchro/encoder), and friction during movement and possible overload (load cells).

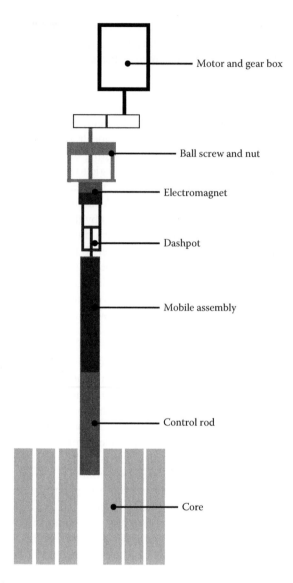

Motor and gear box

Ball screw and nut

Electromagnet

Dashpot

Mobile assembly

Control rod

Core

FIGURE 8.58   Typical arrangement of a control rod and its drive mechanism.

8.4.1.3   Diversity of Shutdown Systems

Generally, two independent and diverse SDSs are provided to ensure that the reactor is definitely shut down on demand. The objective is to have overall reliability of the SDS such that the combined non-availability of both SDSs shall be <10^{-6}/reactor-year, which is verified by reliability calculations. The diversity of the design is effected at all levels including feedback signals for shutdown, logic system, type of control rod (solid/articulated rod), drive arrangements (ball screw and nut/rack and pinion), location of dashpot (inside/outside reactor), location of electromagnet (outside reactor/in sodium), or type of system itself (liquid poison dispersed in coolant or articulated rods instead of solid rods). Under modified or deformed core geometry that can happen during a seismic event, the articulated rods (Figure 8.59) are much easier to insert into the core as compared to a solid control rod.

   The rods and their mechanisms are categorized broadly into primary and secondary systems (Figure 8.60). In some reactors, the primary rods are also termed as control and safety rods (CSRs), and associated drive mechanisms are called control and safety rod drive mechanism (CSRDM). These rods execute the functions of power raising/lowering/shutdown of the reactor on demand

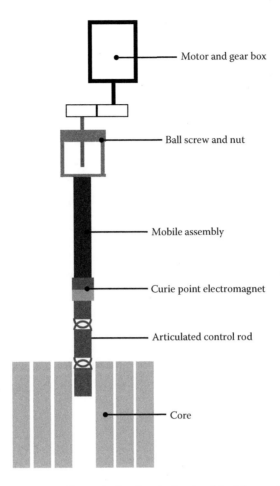

FIGURE 8.59 Typical arrangement of an articulated control rod and its drive mechanism.

during operation and also SCRAM. The secondary rods are termed as diverse safety rods (DSRs), and their actuating mechanisms are called diverse safety rod drive mechanism (DSRDM). These rods only shut down the reactor. Table 8.4 summarizes the diverse features of these two SDSs.

The PPS calls for reactor shutdown once certain identified plant parameters exceed the threshold values. Typical plant parameters that actuate the PPS include power, period, reactivity (all based on the neutron flux [ϕ]), power to flow ratio (P/Q), reactor inlet temperature (θ_{RI}), delayed neutron detector (DND), mean temperature rise across the core ($\Delta\theta_M$), and deviation in the SA outlet temperature ($\delta\theta_I$) and CSA outlet temperature (θ_{CSA}). They are grouped into two, and the parameters under each group actuate the PPS independently as depicted in Figure 8.61. Diversity is also provided in both hardware and software aspects.

8.4.1.4 Essential Safety Criteria

The total reactivity worth of the SDS is such that in the shutdown state, with both primary and secondary absorber rods in the active core, the reactor is subcritical with a high margin of K_{eff}. Further, even under a postulated event that adds significant reactivity to the core, for example, replacement of the most reactive control rod by most reactive FSAs, termed as fuel handling error, K_{eff} is restricted to less than unity (typically $K_{eff} = 0.95$). Let n_p be the number of primary rods. Even under the assumption that one rod is not functional and also all the secondary rods are not dropped from the highest elevation in the core (while producing nominal power), the reactor can be shut down positively with a drop of, that is, $(n_p - 1)$ rods only. Similarly, if n_s is the number of secondary rods, even

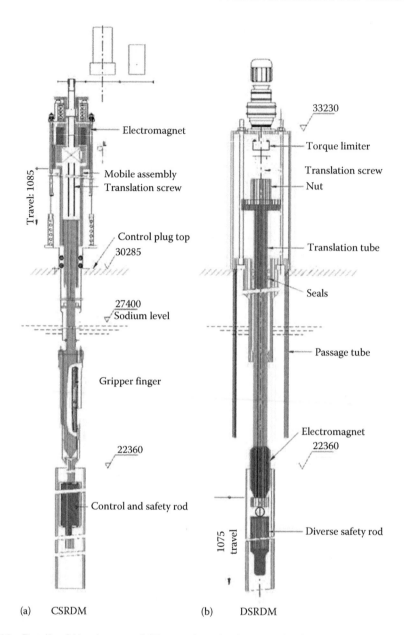

FIGURE 8.60   Details of (a) primary and (b) secondary shutdown mechanisms.

under the assumption that one of the secondary rods is not functional and all the primary rods are not dropped, the reactor can be shut down positively with a drop of $(n_s - 1)$ rods only. Thus, each kind of system (primary/secondary) can independently shut down the reactor even with one of its rods is not available. With primary rods present inside the core, withdrawal of all the secondary rods does not lead to reactor criticality. The primary worth thus takes care of reactivity loss due to temperature rise, power rise, burnup compensation, and uncertainty margin.

8.4.1.5   Passive Design Features

Failure analysis of SDSs indicates that failures of the PPS and failure to insert the rod into the core due to core deformation are the two major causes for nonavailability of the SDS on demand.

TABLE 8.4

Comparison of Diverse Features of the Shutdown System

S. No.	Parameter	CSRDM	DSRDM
1	Function	Control and safety	Safety
2	Position of absorber rod during normal operation	Partly inside the core	Above the core
3	Clearance between rod sheath and wrapper	Minimum required	Relatively more to ensure free fall
4	Location of SCRAM release electromagnet	In argon space	In hot sodium
5	Part released on SCRAM	Mobile assembly with control safety rod	Diverse safety rod only
6	Deceleration device for control rod	Oil dashpot in air	Sodium dashpot within DSR sheath
7	Safety logic	Pulse coded logic	Conventional solid-state logic with fine impulse test (FIT)

The nonavailability of the PPS is due to internal and external factors for which adequate redundancy and diversity are provided. To facilitate rod insertion into the deformed core, articulated rods or liquid poison/granule injection into the core is being considered.

Even though a combined failure probability of the SDS is $<10^{-6}$/reactor-year, there are certain design basis events that when combined with the failure of the SDS lead to severe consequences and are categorized under beyond-design-basis events. To mitigate this, passive devices in the SDS are being implemented so that the event is terminated without the need for external intervention. The popular passive devices under development are listed in Table 8.5 [8.38]. The important ones are as follows:

- *Curie point electromagnet-based system*: It makes use of the natural principle of demagnetization of electromagnets at Curie point temperature, so that when the reactor coolant temperature exceeds the Curie point temperature, the control rods are released and fall under gravity into the core (Figure 8.62a).
- *Injection of liquid poison/granules into the core*: When the coolant temperature exceeds the specified safety limit, the injection can be realized by incorporating a thermal fuse or rupture disk assembly (Figure 8.62b).
- *Hydraulic suspended absorber rod system*: In the case of pump seizure, the sudden reduction of coolant pressure supporting the rods will allow an automatic drop of rods into the active core region (Figure 8.62c).

8.4.2 Fuel Handling System

The fuel handling system includes mechanisms/machines provided for refueling the reactor core. The refueling scheme of a reactor envisages replacement of a certain amount of fuel at defined intervals. Accordingly, it has mechanisms/machines to remove the spent fuel from the reactor and replace them with fresh ones once the available fissile material is not adequate to sustain the power level due to burnup. The fuel handling interval is linked to the physics of core management. A batch of SAs is replaced at the end of each operational campaign fixed in terms of effective full power days (EFPD). The batch size may vary depending on the number of replacement cycles of a full core. Additionally, a proportion of blanket SAs and control rods are also replaced depending on their cycle life.

8.4.2.1 Fuel Handling Operations

The fuel handling operations involved are (1) operations within the reactor vessel and (2) outside the reactor vessel. The starting point or ending point for the fuel handling operation is the transfer of

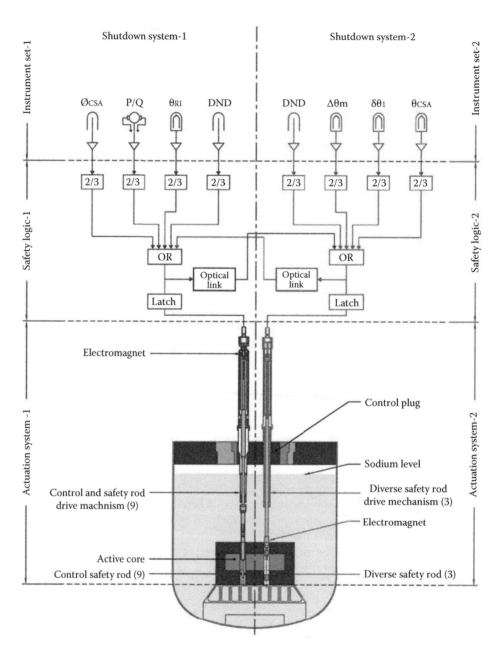

FIGURE 8.61 Diversity in SDSs.

SAs in a cask housing the SAs on a truck. The cask housing fresh fuel/blanket/absorber rod/special SAs is taken from the truck coming from the fuel fabrication plant. Similarly, after irradiation and adequate cooling, the spent SAs are placed in a cask and loaded in the truck, which takes it to the reprocessing plant.

In FBRs, the refueling campaign is carried in the reactor shutdown condition, when all the control rods are fully inserted into the core and the core has sufficient negative reactivity. In a typical pool-type FBR, the temperature of sodium is reduced to 180°C–200°C during fuel handling in order to reduce the influence of sodium aerosols on operation of fuel handling mechanisms. In this temperature range, there is less formation of sodium aerosols, leading to less deposition problems

TABLE 8.5
Passive Shutdown System in Fast Reactors

S. No.	System	Principle
1	Lithium expansion module (LEM)	The thermal expansion of lithium under coolant outlet temperature causes injection of lithium liquid column into the active core portion.
2	Lithium injection module (LIM)	Melting of a thermal fuse seal causes injection of liquid lithium into the active core.
3	Curie point electromagnet	Demagnetization of electromagnet at Curie temperature leads to fall of the control rod into the core under gravity.
4	Gas expansion module (GEM)	Increase in neutron leakage under loss of flow due to the absence of sodium with resultant negative reactivity addition.
5	Hydraulic suspended rod	Loss of flow causes automatic insertion of rod into the core since normally the control rod is kept under suspension due to coolant flow.
6	Enhanced thermal effects	Thermal expansion due to rise of coolant outlet temperatures leads to enhancement of the thermal expansion mechanism (Japan), shutdown by thermal expansion of sodium (Germany), and PS subassembly with thermal principle–based actuation devices (Russia).

Source: Reproduced from Burgazzi, L., *Nucl. Eng. Des.*, 260, 47, 2013. With permission.

during rotation of rotatable plugs and during movement of fuel handling machines. The temperature selected is also sufficiently high to avoid freezing of sodium at stagnant locations within the sodium pool. During fuel handling, the cover gas pressure is also reduced to a few mbar with respect to the RCB pressure in order to reduce leakage through the dynamic seals of rotatable plugs. Typically, the cover gas pressure during fuel handling is 0.2–0.5 kPa (2–5 mbar) with respect to the RCB pressure.

Online refueling, that is, fuel handling when the reactor is in operation, is possible only in natural uranium–fueled reactors where the inherent worth of a fuel bundle is very less so that refueling during the reactor operation does not result in large reactivity swings. On the other hand, reactor cores utilizing higher fissile enrichment (>2%) have to shut down for refueling, which is not practically possible for FBR conditions. The amount of reactivity associated with a single SA is considerable, and hence, a large reactivity swing could occur during SA loading while the reactor is in power, creating problems in reactor control. In addition, generally FBRs have a vertical reactor configuration with the control rods along with their drive mechanisms housed in a control plug located above the core. Since it is difficult to provide direct sealing for the high-temperature and radioactive sodium present within the reactor vessel, an argon cover gas is used above the sodium level. Maintaining the leak tightness during refueling with the reactor in operation is also difficult to achieve. Apart from this, the configuration of the reactor is such that to gain access to the core SAs directly below the control plug, it becomes necessary to move the control plug away from the centerline of the core during fuel handling. This requires the control rods to be detached from their drive mechanisms resulting in the need for the reactor to be shut down. Once the time taken for fuel handling is kept close to the time required for other planned maintenance activities of the reactor, offline fuel handling does not significantly influence the availability of the reactor. The overall scheme of core SA handling is schematically shown in Figure 8.63.

8.4.2.1.1 *Fresh Fuel Handling*

The fresh fuel has radioactivity due to gamma activity from residual fission products, low-energy gamma rays from plutonium and americium, and neutron emissions largely from the (α, n) reactions from the fuel. Hence, the fresh FSA requires moderate shielding. From the cask, the SA is transported to an inspection bench for checking of the SA identification number, enrichment level of the SA using gamma scanning, dimensional variation check, visual checking, SA flow zone identification by measurement of dimensions of predetermined steps provided in the SA foot, and a gross flow blockage test by measuring the resistance to flow of an instrument air circulated through the SA. After storage in the fresh SA storage bay, during the refueling campaign, they are preheated

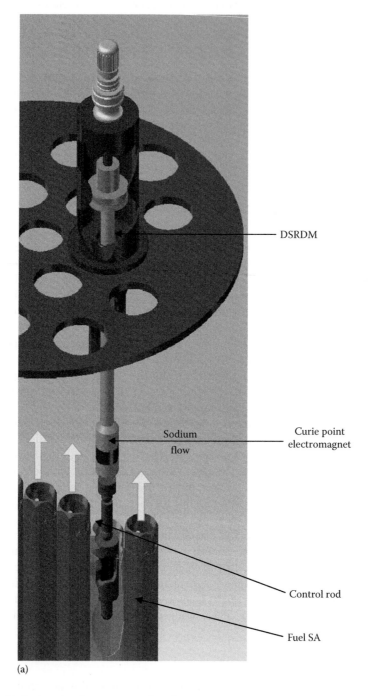

(a)

FIGURE 8.62 (a) Curie point electromagnet–based system. *(Continued)*

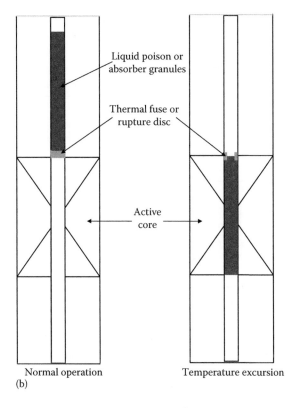

Normal operation
(b)

Temperature excursion

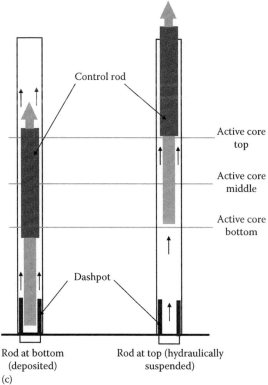

Rod at bottom
(deposited)
(c)

Rod at top (hydraulically
suspended)

FIGURE 8.62 (*Continued*) (b) Liquid/granule injection into the core. (c) Hydraulically suspended absorber rod.

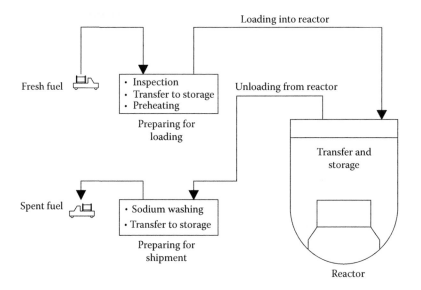

FIGURE 8.63 Core SA handling scheme.

to ~150°C to drive the moisture sticking to the SAs, reducing the load on the cold trap equipment provided as part of the purification circuit to remove impurities from the primary sodium. The SAs are then loaded into the reactor. In the case of fresh fuel, the following are the various operations to be accomplished by the fuel handling system:

- Transfer of the fresh fuel/blanket SA from the truck to outside the reactor vessel and further to a specific location in the core called in-vessel transfer port (IVTP)
- Transfer of the spent FSA from the core to the IVTP earmarked within the core layout, and after adequate cooling, transfer the same to outside the reactor
- Transfer of fresh fuel/blanket SA from the IVTP to any specified location in the core

8.4.2.1.2 Spent Fuel Handling

The major challenge during handling of a spent SA is its high decay heat shortly after reactor shut and its high radioactivity. The decay heat is strongly dependent on the specific power (kW/kg of U + Pu) and cooling time (days after discharge from the reactor) and is less dependent on burnup (MWd/ton). This is illustrated in Table 8.6 from published data [8.39]. The high decay heat leads to rise in temperature of the fuel clad, and if an adequate heat removal mechanism is not provided during handling, it may lead to loss of integrity of the clad with the resultant release of radioactivity

TABLE 8.6
Decay Heat in kW for 100 kg Fuel (U + Pu)

S. No.	Fuel Burnup (MWd/ton)	Specific Power (kW/kg)	Days after Discharge from Reactor		
			10	40	100
1	100,000	400	77	41	21
2	50,000	400	67	31	15
3	100,000	100	23	12	6
4	50,000	100	21	11	5.5

Source: Takahashi, N. et al., *Nucl. Technol.*, 86, 7, 1989.

into the working environment. With specific reference to SFRs, handling has to be done under sodium and argon environments. While handling in sodium, the decay heat up to ~90 kW can be dissipated without loss of integrity of the clad [8.40]. However, the spent SA has to be handled in an inert gas environment prior to its washing to remove the sodium sticking to the SA. This calls for storage of the spent SA either within the reactor vessel (in-vessel storage) or outside the reactor vessel (ex-vessel storage) in order to reduce its decay heat to a level compatible with reprocessing plant requirements.

There are two options to store the spent FSA immediately after removing from the active core region: in-vessel storage and ex-vessel storage. In the case of in-vessel storage, the spent SAs with high decay heat are stored in internal storage locations located on the periphery of the core. The SAs are stored for one or more campaigns till the decay heat is reduced to a level suitable for subsequent handling. Initially, dummy SAs are provided in these storage locations, which are progressively replaced with spent SA during successive fuel handling campaigns. In-vessel storage is provided in Phénix, PFBR, Superphénix 2, and EFR. In the case of ex-vessel storage, the spent SA is transferred for storage outside the reactor vessel. This can be in the form of a vessel filled with sodium with provisions to store the SA vertically. The vessel has provision for cooling to remove the decay heat transferred from the SA and a dedicated purification system to maintain purity of the sodium. Transfer of SA between the reactor and the ex-vessel storage takes place in a sodium-filled pot. With this method, SAs having a large decay heat (~25–35 kW) can be transferred out of the reactor vessel. An adequate number of storage locations are provided in the storage vessel to store fresh assemblies also prior to a fuel handling campaign. During a handling campaign, the spent SAs are exchanged with a fresh SA between the reactor vessel and the storage vessel. The advantages of ex-vessel storage are shorter fuel handling time (subsequent transfers from the storage vessel to the outside are done in parallel with reactor operation), capability to remove SAs with large decay heats; and the shortest time required for full core unloading, whenever the need arises. The disadvantages are that the storage vessel is a replica of a reactor vessel with its own associated cover gas/cooling and purification systems, large sodium inventory, and additional machines to transfer the fuel within the storage vessel resulting in high overall cost. Ex-vessel storage in a large sodium vessel is adopted for Phénix and Superphénix 1. In Superphénix 1, there was a sodium leak due to failure in the welds of the storage vessel, and ultimately, the sodium storage option was deleted [8.41]. An economic alternative to ex-vessel sodium storage is water pool storage. The SAs are stored vertically in a steel-lined concrete tank filled with water. The water serves to provide both shielding and cooling purposes. It has a dedicated purification system to maintain the required water chemistry to avoid corrosion of the fuel clad of spent fuel stored in the pool. Special ventilation systems are also provided to control humidity and activity near the pool area.

Before water pool storage or after sodium storage, the SAs are washed in washing pits to remove the sodium sticking to the SA. This is done to avoid the load of SA cleaning on the reprocessing plant.

For very-long-term external storage (without reprocessing), dry storage in inerted casks is also adopted. In this method, the spent fuel after sodium cleaning is canned in leak-tight inert casks made of steel/concrete and the casks are in turn stored for long term in open storage. The storage is provided with adequate physical security to ensure safety during storage.

8.4.2.2 Fuel Handling System Components

In an equilibrium cycle, a certain number of spent fuel and blanket SAs are replaced with fresh ones. The absorber assemblies are also replaced after a certain interval once the required enrichment is decreased. Before removal of the assemblies from the core one by one, they are first placed in a temporary location. For this purpose, on the periphery of the core, an IVTP is incorporated. The handling diameter is defined as the diameter of a circle enveloping the centers of the IVTP on one side and farthest in-vessel storage on the other side. The in-vessel handling scheme covers the SA located within the handling diameter. The SA located outside the handling diameter is generally shielding SAs, which are permanent, not required to be replaced normally. If they are to be replaced, a separate provision has to be made to handle them. Subsequent operation called "ex-vessel transfer" refers to the transfer of SAs from the IVTP to a location outside the reactor from where the SA is transferred by other machines for subsequent sodium cleaning and storage. The components associated with both in-vessel and ex-vessel transfers are described in the following.

8.4.2.2.1 Components for In-Vessel Handling Operations

The transfer of fuel within the reactor vessel is carried out using combined rotation of one or more rotatable plugs and using one or more in-vessel handling machines (IVHMs). The principle of in-vessel handling using two rotatable plugs—small rotatable plug (SRP) and large rotatable plug (LRP)—is illustrated in Figure 8.64. The rotatable plugs are nested in construction, that is, they are located and also move one within the other. Hence, the SRP is located within the LRP and is supported over the LRP with a suitable provision to enable rotation (bearing/roller support). Similarly, the LRP is supported over the fixed roof provided for the reactor vessel. The plugs require a suitable support arrangement incorporating bearing/roller support and dynamic seals for cover gas sealing. The centerline of the IVHM is indicated by point A. During SRP rotation by an angle β, point A moves to point B and it again moves to point C during LRP rotation by an angle α. Rotation of the SRP moves point A to the center of the core (point O) thus enabling positioning of the IVHM anywhere on the line traced between points A and O. Rotation of LRP after achieving the required radius of positioning of the IVHM using SRP rotation gives the required angle of rotation to position the handling machine over the desired core location. Thus, to cover the full core, the SRP needs to be rotated by a maximum of 180° while the LRP needs to be rotated by 0°–360°.

There are three types of IVHMs:

1. Straight pull–type machine
2. Offset arm–type machine
3. Pantograph-type machine

Generally, the IVHM is also provided with a 0°–360° rotation. Provision of an offset arm–type or pantograph-type machine gives an additional radius of rotation, which is supplementary to

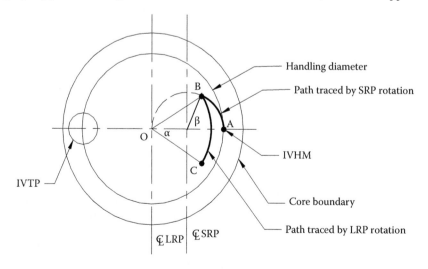

FIGURE 8.64   Principle of in-vessel handling. *Note:* CL, centerline.

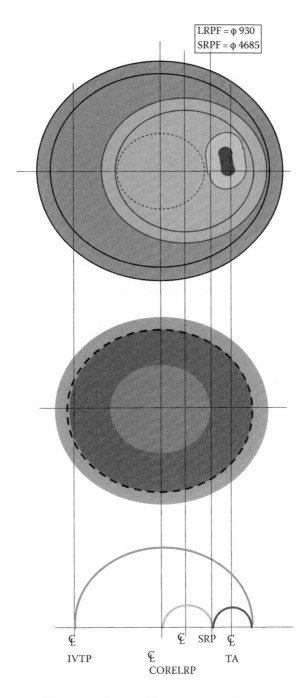

FIGURE 8.65 In-vessel handling using offset arm. *Note:* CL, centerline.

SRP rotation. This is illustrated in Figure 8.65, where use of both the offset arm (transfer arm) and SRP rotations covers the full range of required radius of rotation. With the use of a straight pull–type machine, a minimum of two rotatable plugs is required (Figure 8.64). With offset arm–type and pantograph-type machines, the use of a single rotatable plug is possible. The handling schemes for offset arm/pantograph types of IVHMs with a single rotatable plug are illustrated in Figure 8.66.

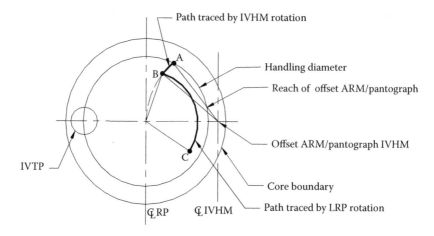

FIGURE 8.66 Handling scheme with offset arm/pantograph-type machine. *Note:* CL, centerline.

The four basic functional requirements of the IVHM are:

1. Opening/closing of the gripper finger to hold/release the SA
2. Hoisting of the gripper up and down to raise/lower the SA out/into the core
3. Rotation of the machine to supplement SRP rotation in order to get the required radius of rotation
4. Rotation of the gripper to orient the SA (due to its hexagonal shape, it is necessary to orient the SA with the shape of the vacancy in the core) prior to lowering of the SA into the core

The three types of IVHMs differ in the location of gripper finger axis with respect to the machine hoisting axis. For the straight pull type, both the axes are coincident. For the offset arm type, the hoisting axis is offset by a fixed distance from the gripper axis. In the pantograph type, the hoisting axis is offset but variable from the gripper axis. This is illustrated in Figure 8.67.

FBTR, Superphénix 1, EFR, Joyo, DFBR, BN-350, BN-600, FFTF, and CRBRP have adopted the straight pull type; Phénix, PFBR, SPX-2, and Monju have adopted the offset arm type; and PFR, DFBR, and JSFR have adopted the pantograph type. The straight pull type is the simplest in design and operation. The offset arm and pantograph types have relatively complex loading due to the eccentric application of load on the gripper with respect to the hoisting function of the machine. However, all the three options have been designed and used in various reactors as indicated earlier and hence are technically feasible. The pantograph machine is the most complex among the three types but gives great economical advantage of use of single rotatable plug and the smallest LRP flange diameter as seen in recent innovative designs in Japanese reactors (JSFR).

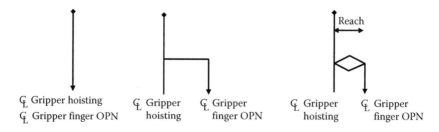

FIGURE 8.67 Different types of IVHMs. *Note:* CL, centerline.

8.4.2.2.2 Components for Ex-Vessel Handling Operations

The scheme has two options: flask transfer and cell transfer. In flask transfer, a leak-tight argon-filled flask is connected to a canal in the roof above the IVTP and the SA is gripped and hoisted up into the flask. Leak-tight valves, one on the roof and another on the flask, are provided. The valves are kept open for SA movement and are closed during transfer of SA to the washing pit. Before opening the flask and canal valves, the interspace between the valves is flushed with argon. The advantage of flask transfer is that it employs a simple straight pull–type gripper for handling operations, and with additional cooling provisions, there is a possibility of handling SA with a higher decay power. The flask is handled by the building crane or moved on dedicated rails. The drawbacks of flask transfer are that it is a time-consuming process due to repeated coupling/decoupling and argon flushing and also leads to longer fuel handling time. The possibility of air entry to the primary system is high. The method of flask transfer is illustrated in Figure 8.68. In cell transfer, a dedicated cell is provided linking the canal in the reactor roof and a transfer location outside the reactor. The cell is filled with argon. The SA is transferred through the inert gas–filled cell using a crane or a transfer machine. Figure 8.69 shows the method of cell transfer. In another variant, a rotating transfer lock is provided between the reactor and the outside portion. The lock has two inclined ramps through which the SA is hoisted up. The SA after it is hoisted up is either transferred to the other side of the ramp either by rotation or by swinging. This concept is adopted in PFBR, Phénix, Superphénix 1, and EFR. The cell transfer method is preferred due to reduction in fuel handling time and better leak tightness. However, the fueling machine employed is complex in construction and there is a limitation to transfer SA with higher decay power. Also, additional special provisions are required to take care of stuck conditions of SA during transfer.

8.4.2.3 Major Safety Requirements during Fuel Handling

8.4.2.3.1 Prevention of Criticality in Storage

The array of spent FSAs could become a neutronically critical system in view of a relatively large quantity of fissile material and water storage, compared to the thermal reactor core of similar size. This can be avoided by proper spacing between the adjacent SAs so as to maintain the geometrically safe configurations under all normal and accident conditions including flooding and earthquake. Generally, under such conditions, the calculated K_{eff} shall not exceed a conservative value, say 0.95.

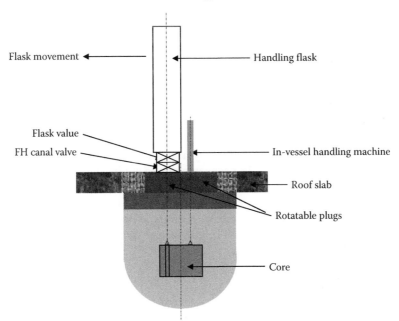

FIGURE 8.68 Flask transfer concept.

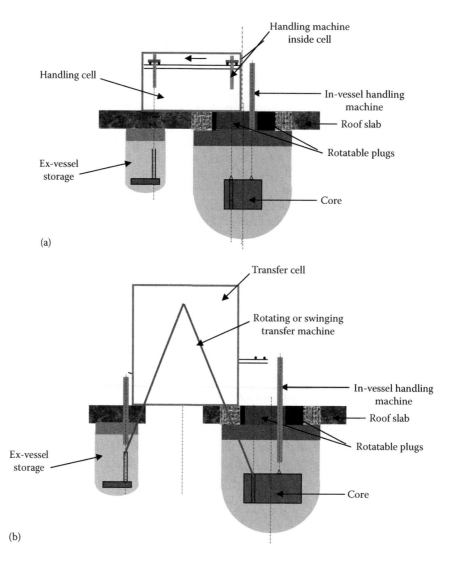

FIGURE 8.69 **(See color insert.)** Cell transfer concept: (a) using fixed cell and (b) using rotating or swinging transfer lock inside transfer cell.

8.4.2.3.2 Cooling of SA in External Storage and during Handling

The SA pins would get heated up due to the decay heat generated mainly by the fission products. It is essential to remove the heat so that the peak temperature of pins does not exceed the applicable design safety limit. While the SA is immersed in sodium, heat removal is not of great challenge. However, while it is in the inert gas space during handling, the temperature may rise if adequate cooling is unavailable. Cooling is a great challenge in the case of SA stuck-up during handling. Hence, a reliable cooling arrangement should be provided in storage and during handling to maintain integrity of the SA, particularly the clad.

8.4.2.3.3 Radiation Protection

All fuel handling machines and storage facilities should be provided with adequate shielding to limit radiation exposure of the operation and maintenance personnel.

8.4.2.3.4   Failed SA Handling and Storage

In the case of an SA with failed fuel pins, provisions should be made in the design for safe handling and storage of failed SAs considering the release of radioactivity, fuel–coolant compatibility, DHR, contamination of operation area, etc. Generally, such SA will be canned in a leak-tight secondary container.

8.4.2.3.5   Specific Design Aspects

The speed of insertion of fresh fuel, which contains large fissile inventory, should be such that reactivity insertion rates are within limits. It should be ensured through proper design provisions that the handling machine distinguishes between the control SA and FSA at the time of latching to prevent inadvertent substitution of one for the other. Further, provisions should exist to ensure that during the handling of an SA, the rest of the core is not disturbed. Design provisions should be made on machines to prevent dropping of the SA during handling. Interlocks shall be provided on the handling machines to prevent wrong operations. In addition to the previous text, IVHMs should have provision for clear identification of the type of SA being handled. IVHMs should also have provision to detect wrong loading of an SA on the grid plate.

8.4.2.3.6   Operational Errors

Provisions should be made in the design to check operational errors (like wrong positioning of rotating plugs, incorrect operation of machine) by diverse means.

8.4.2.3.7   Transport Casks

Provisions should exist in the plant to receive fresh SA and dispatch irradiated SA in transport casks in a safe manner meeting all the statutory requirements.

8.4.2.4   Fuel Handling Experiences and R&D Requirements

The major challenge in fast reactor fuel handling has been the remote operation of fuel handling machines inside the reactor. The opacity of sodium makes visualization of fuel handling operations impossible. Operation of fuel handling machines in fast reactors has been trouble free, except for two major incidents in FBTR and Joyo, where rotatable plugs were rotated with the fueling machine coupled to the reactor.

In FBTR, during transfer of the FSA from the third ring to the storage location in 1987, the foot of the assembly got bent as it was projecting below the guide tube. This resulted in curtailed movement of the gripper tube and finally jamming of the FSA along with the gripper tube inside the guide tube. A complex mechanical interaction occurred within the reactor vessel causing damage to the fuel handling machine gripper, the FSA held by the gripper, guide tube, and several reflector SAs [8.42]. The bent SA was extracted through the damaged guide tube with extra force. The guide tube was cut and removed in two pieces using specially designed tools (Figure 8.70). Modifications were implemented: mechanical stopper for fuel handling gripper and redundant interlocks for plug rotation authorization. It took 2 years to recover from this incident, and the reactor was restarted in May 1989. After incorporating these modifications, there were no further incidents and the operation of the system has been trouble-free.

In the Joyo reactor, the top of the irradiation test SA "MARICO-2" was protruding above the core and got bent during plug rotation (Figure 8.71). The projecting SA also caused damage to the upper core structure. Efforts are under progress to rectify the damage [8.43].

To overcome the limitations of opacity of sodium and to detect the absence of any coupling of the core with the above core mechanisms, an under sodium ultrasonic scanner (USUSS) is deployed [8.44]. The device sends ultrasonic waves in the radial direction and looks for reflections back from protruding objects. Depending on the time taken for receipt of the reflecting echo, the possibility of

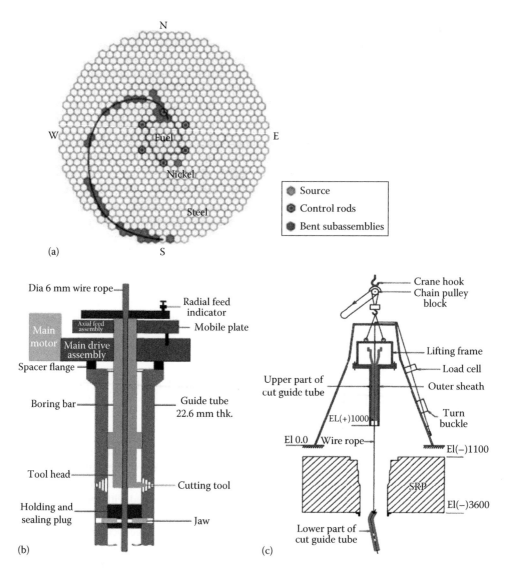

FIGURE 8.70   Fuel handling incident in FBTR: (a) path of damaged subassemblies, (b) remote cutting of guide tube, and (c) retrieval of guide tube.

protrusions and its distance from the origin of the ultrasonic wave is arrived at. Depending on the signal and after suitable signal processing, the shape of the object is also built up. The same object is used to view downward from the top of the core to arrive at the shift of the centerline of the SAs due to bowing. Though USUSS is fairly developed, much remains to be done regarding improving sensitivity and improving signal processing to arrive at detecting all types of protrusions.

Galling (self-welding) of metal surfaces is another major challenge governing the smooth operation of reactor mechanisms. The surfaces in contact under sodium at high temperatures develop a tendency to fuse or self-weld together especially under high contact stress. To avoid galling of the surfaces, hardfacing of the mating surfaces is carried out. Hardfacing involves providing a coating on the metal surface that resists self-welding. Popular hardfacing techniques include hardchrome plating, stellite, and colmonoy deposition. Hardchrome plating involves depositing a chromium-rich layer using the electrochemical process. The deposit is hard (~60 HRC) and wear resistant. Stellite coating involves depositing a cobalt-rich layer of hardness ~40 HRC. However, under irradiation, the cobalt becomes radioactive and poses problems of dose to maintenance personnel/during

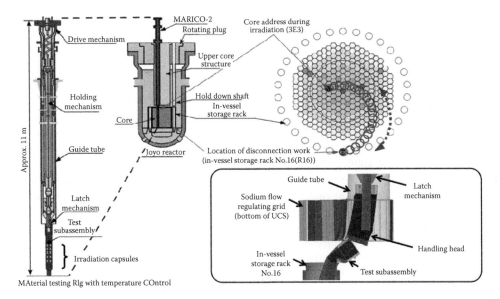

FIGURE 8.71 MARICO-2 test SA bending in Joyo.

decommissioning. Instead, a nickel-rich deposit of colmonoy has been found to have a hardness of ~50 HRC. Colmonoy develops very less induced radioactivity as compared to stellite. Proper selection of material pairs and coating for material pairs is important for surfaces in contact under high-temperature sodium to avoid galling effects especially for parts under sliding/rolling contact like guiding surfaces, rollers, pins, and journal or rolling bearings.

8.5 INSTRUMENTATION AND CONTROL SYSTEM

The purposes of the I&C systems in any power plant are to

- Assist the operator in controlling the plant in a manner consistent with specified power and safety targets
- Monitor the plant and warn of deviations from normal
- Provide independent safety and control actions (and shutdown actions if required)
- Prevent further undesirable consequences of an accident without operator intervention and then provide appropriate facilities for whatever action is necessary

In general, the main functions of the I&C system in any plant are to monitor, control, and protect the plant.

8.5.1 BASIC FUNCTIONS OF I&C

In any conventional power plant, instruments like TCs; RTDs; and pressure, level, and flow instruments are used. In addition, in NPPs, special instruments are used to monitor neutron flux and radiation levels, etc. Further, I&C plays critical roles for the post shutdown monitoring, isolation of reactor containment building during accident conditions, and also post accident monitoring (PAM) purposes.

A special feature of breeder reactors is sodium instrumentation, which requires special sensors for measurements in highly chemically reactive, hot, and radioactive sodium. Neutron flux is monitored using high-temperature fission chambers (HTFCs) from shutdown to 200% of nominal power (Pn). A failed fuel detection system is provided to monitor the presence of fission products in sodium and argon cover gas. All potentially radioactive areas are monitored for any release of radioactivity to the environment. In fast reactors, fast response for reactivity, core temperature measurement (CTM), and failed fuel detection are very crucial. The instrumentation provided for measurement of sodium level, flow, leak detection from capacities and pipelines, and detection of water/steam leak in the SGs is very specific to fast reactors.

Instrumentation for reactor components and core monitoring of fast reactors poses further challenge as all sensors are to be routed out through the limited openings provided at the top of the reactor vessel. Since the signals are large in number, routing of cables on the top of the reactor is difficult due to shortage of space. Additionally, the cables routed from the reactor to the periphery of the reactor are to be disconnected to facilitate rotation of rotating plugs for fuel handling operations. Trailing cable system is required to make some important signal available even during rotation of plugs.

Signals acquired from sensors are conditioned, processed, and presented to the operator for smooth running during normal operation and to protect the plant during design basis events. I&C is designed to provide automatic action to protect both the plant and the environment. In summary, the basic functions of the I&C system are measurement, indication, recording, control, and protection.

The I&C system is a hybrid consisting of hardwired and computer-based systems. In general, processing of all the parameters initiating safety actions is hardwired. The exception is the core temperature monitoring system, wherein computers are used in view of the requirement for dynamic processing. Computer-based systems are generally preferred for the I&C of other systems to take advantage of the advancement in the field of computer technology.

8.5.2 General Design Features

Many design features are adopted along with good engineering practices to meet the safety criteria of safety-critical and safety-related instrumentation systems.

Range of measurement: All the instruments are provided with measurement ranges suitable for monitoring normal operation and anticipated operational occurrences. The ranges of instruments are selected such that the practice of using the middle two-thirds of ranges is followed for all instruments, except in special cases.

Redundancy: For the process variables that lead to plant shutdown on crossing a safe limit, redundant instrumentation is provided.

In situ testing: To the extent possible, all the safety-critical instruments are provided with built-in test signals to carry out periodical surveillance checks. Healthiness of the instrumentation for the safety-critical systems is monitored, and intercomparison (triplicate instruments) of signals is carried out through a computer as a part of an online surveillance check.

Fail-safe design: To the extent possible, the systems are designed such that any malfunctioning or failure of the system component does not lead to unsafe condition of the plant. The plant is brought to a safe shutdown on failure of important equipment/instrument.

Single failure criteria: The system is designed such that failure of a single equipment does not endanger plant safety. Most of the critical parameters are monitored through triplicate instrument channels.

Common mode failure: Triplicate channels of a safety-critical parameter are powered through different divisions of power supplies, the cables are routed through separate cable trays in independent routes, and the instrument channels are located in different cabinets located in separate rooms as far as possible.

Seismic qualification: I&C systems are classified into seismic category 1, 2, or 3 (uncategorized) depending on its importance in ensuring safety during the seismic incident. Seismic category 1 systems are designed to withstand SSE, and category 2 systems are designed to withstand operation basis earthquake (OBE). Instrument support system structures, which form part of the main equipment (e.g., thermowell), are designed as per the classification of the equipment. To meet the seismic requirements, an assembled equipment (such as instrument channels, relays) is subjected to rigorous seismic tests as per the site-specific seismic spectrum to establish their suitability for use in the system. I&C systems are qualified primarily by testing. Equipment for which testing is not feasible are qualified by seismic analysis.

Safety classification: Instruments are classified as safety-critical, safety-related, and nonnuclear safety (NNS) systems depending on its role in preventing radioactivity release to the environment. The selection of instruments, its signal processing, and qualification requirements are influenced by the safety class of the instrumentation system.

8.5.3 TYPE OF INSTRUMENTATION

Instruments used in a fast reactor can be divided as nuclear, sodium, and conventional instruments. The sensors used are single, redundant, triplicate, or diverse based on the type of application. Where safety is of prime concern, each measurement is triplicate and two-thirds voting logic employed for initiating safety actions. Diverse instruments are provided for protecting the reactor from any single abnormal event, wherever required redundant sensors are provided to improve plant availability. All other parameters are measured using single sensors.

8.5.3.1 Nuclear Instruments

Nuclear instrumentation mainly consists of the following:

* Neutron flux monitoring system
* Failed fuel detection and location system (FFDLS)
* Radiation monitoring system (RMS)

8.5.3.1.1 Neutron Flux Monitoring System

The core status of a reactor is monitored in all states—shutdown, fuel handling, start-up, intermediate, and power ranges through the neutron flux monitoring system. Neutron detectors are placed in the control plug and below the safety vessel. HTFCs with a sensitivity of 0.2 cps/nv are provided in the control plug. These detectors are installed in the housing tubes provided in the control plug and are replaceable. These are suitable for operation up to 570°C and are used in the low power range only due to the influence of gamma during high-power operation of the reactor. Standard fission chambers with a sensitivity of 0.75 cps/nv are installed in tubes provided below the safety vessel and are replaceable. These are used in the power ranges from 5% to 100%. The output signals from these detectors are processed by start-up, intermediate, and power range channels. Three of the detectors in the control plug and three of detectors below the safety vessel are used in tandem for safety actions during all states of reactor operation. The processed signals (such as linear power, period, and reactivity) from these channels initiate automatic reactor SCRAM during operational occurrences.

The neutron flux at the core center varies from 5×10^7 nv at shutdown to 8×10^{15} nv at nominal power (1250 MWt). The neutron flux at detector location below the safety vessel is 1.34×10^5 nv at nominal power. The neutron signal is calibrated against the thermal power after every fuel handling operation at 90% of target power. The location of neutron detectors in a typical FBR is shown in Figure 8.72.

8.5.3.1.2 Failed Fuel Detection and Location System

The FFDLS monitors fuel pin integrity and provides an early warning alarm in case of increase in cover gas activity and initiates reactor trip in case of abnormal increase in fission product activity in the coolant sodium to prevent failure propagation in case of fuel pin clad failure. Fuel pins are

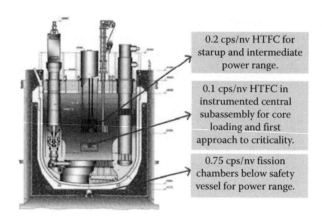

0.2 cps/nv HTFC for starup and intermediate power range.

0.1 cps/nv HTFC in instrumented central subassembly for core loading and first approach to criticality.

0.75 cps/nv fission chambers below safety vessel for power range.

FIGURE 8.72   Location of neutron detectors in a typical FBR.

provided with a leak-tight metal cladding for containment of the fission products. FFDLS consists of gaseous fission product detection (GFPD), delayed neutron monitoring in the sodium pool, and system for location of failed fuel.

For GFPD, the cover gas is sampled through gas flow ion chambers with associated instrument channels that normally provide an early warning. A high-purity germanium detector with multi-channel analyzer (offline), which is a high-resolution gamma spectrometer, is used for the analysis of active gaseous fission products data that along with the history of fuel residence time in the core facilitate identification of the failed fuel. Generally, it is a slow process lasting for days to months for a small fuel failure releasing fission gas to escalate to the size of releasing DN precursors in sodium. Hence, GFPD is not connected to initiate SCRAM.

Failed fuel detection based on delayed neutron monitoring consists of eight identical blocks placed at the inlet of each of four IHXs. Each block consists of three HTFCs (surrounded by graphite to thermalize) for detection of the delayed neutrons. B_4C shielding is provided to reduce the effect of the streaming core neutrons. Output of the detectors is processed to provide trip on failed fuel detection within 60 s to prevent failure propagation.

For failed fuel location, three modules are provided in the control plug. Each module consists of a selector valve that samples sodium from the designated SA outlets. An optical encoder attached to the shaft of the selector valve indicates the respective SA outlet sampled by the system. The failed fuel location system is operated whenever fuel failure is indicated by the delayed neutron detection (DND) system. The system scans the sodium sample from all FSA in ~8 h.

8.5.3.1.3 Radiation Monitoring System

The RMS checks the radiation levels in all potentially radioactive areas and initiate alarm/inter-lock actions as applicable. It includes monitors for gamma, gas activity, particulate activity, stack gas activity, effluent discharge, contamination during normal operation, and postaccident periods. They act as an early warning system to keep the release of activity to the environment and limit the exposure of personnel well below the permissible limits. The area gamma monitors in RCB and the activity monitors on the downstream of RCB filters initiate RCB isolation in case of activity release above the permissible limits.

8.5.3.2 Sodium Instruments

Sodium is solid at room temperature (M.P. 98°C). The high electrical conductivity property of sodium is used for leak detection, level, and flow measurements. Most of the sodium sensors are noncontact type and rated to work in high temperatures (up to 750°C) and radioactive environments (neutron flux, 10^9 n/cm^2/s and gamma flux, 10^3 Sv/h).

8.5.3.2.1 Temperature Measurement

K-type (chromel–alumel) TCs (4, 2, and 1 mm diameter) are used in different applications. These TCs are selected as they have very good radiation resistance and an almost linear characteristic over a required range of temperature.

CTM is provided for detection of abnormal power/flow in FSA, core design validation (physics and engineering), and burnup management. The basic function of CTM is to measure coolant temperature changes and initiate a safety action whenever there is partial plugging in FSAs (inadequate core cooling), error in core loading or fuel enrichment, or orifice error. Mineral insulated, SS sheathed, and ungrounded junction chromel–alumel TCs of 1 mm diameter are used. Considering safety and availability, two TCs as a probe are provided in one thermowell at the outlet of each of the FSA. The response time of the sensor with thermowell is 6 ± 2 s. Real-time computers (RTCs) scan each of the FSA outlet sodium temperature once in a second. The mean core outlet temperature (θ_M), mean temperature rise across the core ($\Delta\theta_M$), and the deviation in individual SA sodium outlet temperature ($\delta\theta_I$) above the expected value are calculated online. SCRAM is initiated when the respective set points are crossed. Refer to Figure 8.73 for the signal processing scheme.

By using a probe with three TCs for core temperature monitoring, the instrumentation system gets simplified. Future reactors will have this design of TC probe, which will provide independence in the signal processing system also.

In the CSA, six TCs are provided without a thermowell. These TCs are in direct contact with sodium (response time 150 ms). Six TCs are provided at the suction side of the two PSPs to measure the core inlet temperature.

TCs are used to monitor temperatures of the inner vessel, thermal shield, outer surface of the main vessel, safety vessel, top shield, roof slab, SRP, LRP, and control plug. This helps in monitoring the temperature gradient across them and also for design validations of temperature distributions on various components.

The sodium temperature in tanks and other equipment is measured by TCs in thermowell. Single-walled sodium pipes are provided with line heaters and monitored using surface TCs. A similar arrangement is made for external surfaces of all sodium capacities. The temperature measurement constitutes the largest number of measuring instruments for sodium.

8.5.3.2.2 Leak Detection

Leakage of sodium is hazardous because of its reaction with oxygen and moisture in the surrounding atmosphere. Leakage from primary systems can cause radioactive contamination. Leak detectors are provided on pipelines, tanks, and other capacities. Sodium leak detection systems are designed to meet the requirements of ASME Section XI, Division 3, which specifies that sodium leak at the rate of 100 g/h is to be detected in 20 h for air-filled vaults and 250 h for inert vaults to avoid long-term corrosion effects.

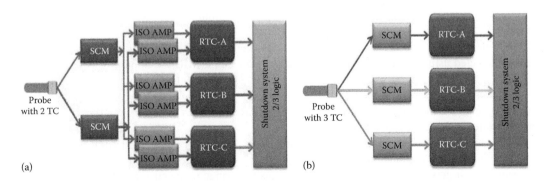

FIGURE 8.73   Signal processing scheme: (a) scheme in PFBR and (b) scheme in FBR-1&2.

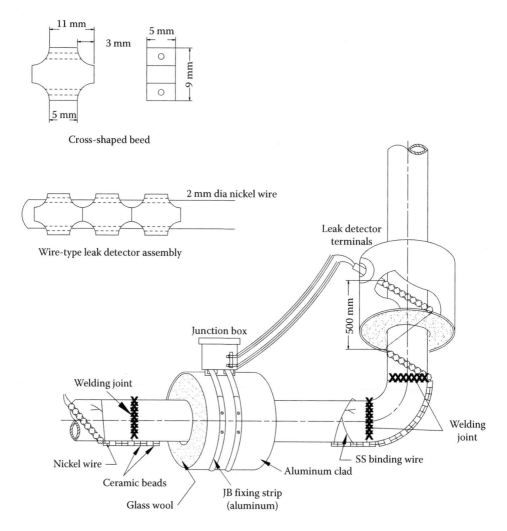

FIGURE 8.74   Leak detector assembly.

Leaks from single-walled equipment and pipes containing sodium are detected using beaded nickel wire held along the outer wall of the equipment or pipes. Refer to Figure 8.74 for leak detector assembly. When sodium leaks, it bridges the gap between the wall and the nickel wire and the electronic circuit detects its presence. Another method of leak detection based on this principle is the spark plug leak detector (SPLD) used in guard pipes, sodium service valves and leak collection trays, etc.

Sodium pipelines within RCB have guard pipes filled with nitrogen. For such pipes, sodium leak is detected using mutual inductance–based leak detectors (MILDs). Sodium gets collected in a pocket. The probe consists of a primary winding excited with a high-frequency current. The presence of sodium changes the secondary output, which signals sodium leak. This method is also used to detect sodium leak in intermediate spaces, for example, the interspace between the main vessel and the safety vessel.

SADs are employed as an area monitor for sodium leaks. Air is sampled from areas where sodium piping and equipment are located and passed through an SAD. Sodium aerosol present in the sampled air gets ionized within the detector causing change in the current, initiating an alarm. The sensitivity of this detector is better compared to other detectors.

Conventional smoke detectors are also provided as area monitors that can detect the smoke generated from a sodium fire. In addition to these detectors, a major sodium leak can be confirmed from the fall in sodium level in various sodium tanks through the mutual inductance–type level probes.

8.5.3.2.3 Level Measurement

The sodium level is monitored in all sodium capacities including the main vessel. Leakage of sodium from tanks and capacities causes a decrease in sodium level. Mutual inductance–type probes for continuous and discontinuous level measurement are used. Continuous level probes are used for level indication, and discontinuous level probes (switches) are used for safety interlocks and alarms.

These probes are provided in pockets welded to the tanks. There are two windings: one primary and one secondary wound on the former. The primary is excited with a constant ac current of the order of 100 mA at 3 kHz. Due to the transformer action, an electro motive force (emf) is induced in the secondary. When sodium is not outside the pocket, the induced emf in the secondary is high. When the sodium level rises and covers the probe, it acts as a short-circuited secondary and an emf is induced in it. This induced emf sets up the eddy current in sodium that produces a magnetic field in the opposite direction to the main field. So the flux linked with the secondary is reduced, and hence, the voltage induced in it also decreases. This decrease in the secondary voltage is proportional to the increase in the level.

Probes of similar design are used as level switches. By providing three or four sets of winding on the same probe at different levels, it can detect corresponding discrete levels. As these probes are installed in pockets, they are not in contact with sodium. Maintenance is easy. In the mutual inductance (MI) type, a number of discrete levels can be detected by installing a single probe. Refer to Figure 8.75 for a schematic of level measurement. However, MI level probes with its long length pose manufacturing and handling difficulties.

RADAR level transmitter is the alternate instrument for monitoring the sodium level. The instrument consists of a small cone antenna located inside the tank and the associated electronics

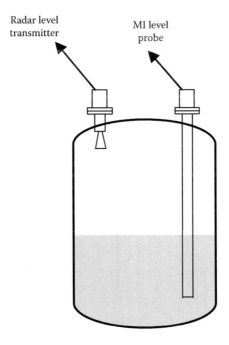

FIGURE 8.75 Schematic of level measurement.

located outside. This is easy to handle at the site. There is no effect on the performance of the instrument due to sodium vapor deposition on the antenna. The feasibility of continued use in sodium is established.

8.5.3.2.4 Flow Measurement

Conventional flowmeters like orifice plates, venturi tube, and rotameter cannot be used for sodium application as they need welded, screwed, or flanged connection and/or pressure tapping. Permanent magnet flowmeters (PMFMs) are used for sodium flow measurement in sodium pipelines. By using permanent magnet, the requirement of power supply is eliminated for the normal working of the flowmeter.

8.5.3.2.5 Permanent Magnet Flowmeters

PMFM works on the principle of a direct current (dc) generator following the well-known Faraday's law of induction, that is, "If a conductor moves in a magnetic field a potential is developed across the conductor which is proportional to its velocity and strength of the magnetic field."

$$\text{Induced emf (E)} = BLV \text{ (V)}$$

where
  B is the magnetic flux density
  L is the length of the conductor, and in case of this flowmeter, L is the diameter of pipe D
  V is the velocity of the conductor movement

Figure 8.76 explains the working principle of PMFM.

Some correction factors are to be applied to the basic formula to take care of various uncertainties. They are

1. K_1, to account for the short circuiting of the pipeline and the liquid metal
2. K_2, to take care of the end effect of the magnetic field
3. K_3, to take care of the temperature effect on the magnetic field

Hence, the formula becomes $E = K_1 K_2 K_3 BDV$ volts.

Advantages of PM flowmeters are as follows: no pressure drop in the circuit, no moving part (hence less maintenance), no penetration in the pipe, a passive device, and simple signal processing.

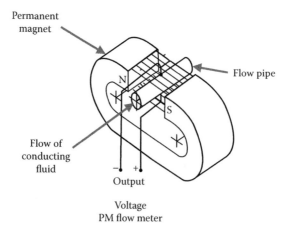

FIGURE 8.76 Permanent magnet flowmeter.

8.5.3.2.6 Eddy Current Flowmeter

Eddy current flowmeters are used for sodium flow measurements in the primary pump discharge. This is an indirect measure of core flow. It is made in the form of a probe and inserted in the roof slab. It consists of three windings: one primary winding at the middle and two secondary windings on both sides of the primary. The primary is excited with a constant ac current of the order of 100 mA. The flux set up by primary links with the secondary and emf are induced in them. When sodium is static outside the probe, equal voltages are induced in the secondary windings. The secondary windings are connected in reverse, and hence, the output voltage is zero. The movement of sodium causes eddy current. Due to this effect, the secondary output is altered. The difference in the secondary output is proportional to the velocity of sodium.

The temperature of sodium affects the operation of the eddy current flowmeter, so temperature compensation is provided in the electronic circuit. The temperature effect at high frequencies is minimal, so a high-frequency excitation current is selected to excite the primary winding. Refer to Figure 8.77 for the schematic of an eddy current flowmeter.

8.5.3.2.7 Speed Measurement

The speed of PSPs is monitored using an EM pickup coil. A toothed wheel is mounted on the shaft of the pump (impeller side of the coupling), with three pickup coils located close to the wheel. The rotating wheel induces pulses in the coil, creating a signal, which is counted using a counting circuit. Counts are a measure of pump speed. Such signal is part of the reactor protection system and is used to detect failure of the pump due to power supply failure, pump drive failure, and pump seizure.

8.5.3.2.8 Sodium Purity Monitors

Sodium is continuously purified to reduce corrosion of structural materials, for plugging of narrow flow passages in colder regions, and for mass transfer of radioactive materials. In the case of secondary sodium, purification helps keep hydrogen background levels in sodium low to improve the sensitivity of hydrogen-in-sodium leak detection system. The H_2 and O_2 impurity levels are maintained <10 ppm. Sodium purification is accomplished by cold trapping, and sodium purity is monitored

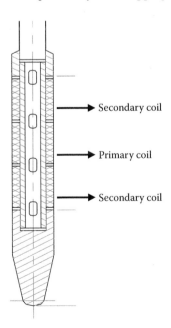

FIGURE 8.77 Eddy current flowmeter.

using plugging indicators. The principle of measurement is based on the property of decreased solubility of impurities such as sodium oxide and sodium hydride in sodium as the temperature of sodium is lowered. The plugging indicator comprises of an annular pipe with an orifice at one end and cooled by air externally. As the temperature of flowing sodium decreases, plugging of orifices takes place with reduction in flow. Temperature (referred as the plugging temperature) is measured when flow reduces to 80% of the unplugged flow.

8.5.3.2.9   Steam Generator Leak Detection System

The function of the SG leak detection system is to detect water/steam leak into sodium as early as possible. In the event of an escalating leak, it prevents further water/steam addition by depressurizing and draining the SG and isolating and inerting with nitrogen. Sodium-heated SG has a strong impact on plant availability as a small tube leak risks the damage to neighboring tubes due to the exothermic reaction and formation of corrosive sodium hydroxide. In the event of a leak of water/steam into sodium, hydrogen is evolved. Elaborate methods are employed to measure hydrogen concentration both in argon cover gas and in sodium, while the leak is still in the initial stages.

SG tube leaks are classified as small, medium, and large depending on the leak rates. Hydrogen-in-sodium detection (HSD) systems based on electrochemical hydrogen meters are provided at the outlet of each SG unit and at the common sodium line to the pump tank to detect small water/steam leaks into sodium. Two hydrogen-in-argon detection (HAD) systems based on thermal conductivity detectors are provided in the cover gas space of the surge tank. This is to detect small water/steam leaks (up to 10 g/s) during start-up and shutdown as the HSD is not effective for sodium temperatures below 623 K as hydrogen formed does not get dissolved in sodium and remains as bubbles that escape into the cover gas. Pressure switches are provided in the pump tank for medium leaks (10 g/s–2 kg/s). Spark plug–type detectors are provided downstream of rupture disks to detect large leaks (>2 kg/s).

8.5.3.3   Conventional Instruments

These represent instrumentation provided for steam, water, turbine, and plant auxiliary systems. Besides resistance temperature detector (RTD), TC, pressure, flow, level, and analytical instruments, turbovisory instruments are provided to measure vibration of bearing pedestal, speed, eccentricity, differential expansion, overall expansion of turbine, and position of governor valves, isolation valves, speeder gear, and load limiting gear.

8.5.4   Signal Processing

Data acquired from the sensors can be processed by either analog or digital technology. While analog systems offer proven reliability with several operating years of experience, the digital computer–based monitoring and display systems offer operator-friendly human–machine interfaces (HMIs) and reduction in cabling and are cost competitive.

8.5.4.1   Safety-Critical Systems

Signal processing is totally hardwired except where dynamic processing is required like core temperature monitoring for which triplicate computers are used. Safety-critical I&C systems are instruments for core temperature, flow, and neutron flux measurement, failed fuel detection, protection logics, safety grade DHR damper logic, and RCB isolation damper logic. Triplicate instrument channels are provided to monitor each parameter. Signal processing and cable routing are physically separated. Safety actions are taken on two out of three logics. Power supply to the triplicate instrument channels is also from independent divisions of I&C power supply.

8.5.5 Safety-Related Systems

The safety-related system plays a complementary role to the safety-critical system in achievement and maintenance of safety. Dual redundant systems are used for processing of signals. In case of failure of the first system, the second system takes over automatically with the help of the switchover logic. Reactor start-up checks, control rod drive mechanisms, primary sodium systems, etc., are examples of this class.

8.5.6 Nonnuclear Safety

NNS systems are not directly involved in tripping the plant but aid in alerting or warning the operators of abnormal process behavior. For this class, commercial off-the-shelf products such as programmable logic controllers (PLCs) or remote data loggers configured as distributed digital control system (DDCS) can be used. This offers the advantage of multiplexing by placing the PLC close to the measuring points in the field. But for the purpose of standardization, most of the process systems use dual redundant systems similar to safety-related systems.

8.5.7 Control Architecture

A three-level hierarchical architecture consisting of the field, local control centers (LCCs), and control room (CR) is adopted. The I&C system sensors and controllers are located in the field. The functions of the components in this level are data acquisition, logic functions, and closed-loop control. Communication between the components in the field and the LCC (the next level) is through the digital communication bus and/or over direct cables. When environmental conditions do not permit installation of electronic units in the field, it is located at the LCCs. The LCC level, which is the second level of the hierarchy, accommodates system cabinets of hardwired systems and computer-based control systems to process the field signals. The LCC also houses network cabinets to establish a data highway to link the components in the field with the CR.

8.5.8 Control Room

The CR complex includes the CR, handling control room (HCR), computer room, and shift charge engineer room. The backup control room (BCR) is located away from the main CR to ensure that both CR and BCR will not become inhabitable at a time. The CR is the highest level in the control system hierarchy. The CR is located in a convenient location to provide easy access to operators under all plant conditions. Plant operations are carried out from the CR. The commands and control set points fed by the operator in the CR are transferred to the field, and actions taken by the field components are reflected in the HMI provided in the CR. Abnormal conditions for all the safety-critical parameters and important parameters of safety-related and NNS systems are annunciated in window alarms in CR panels.

8.5.9 Process Computer

The process computers located in the CR complex receive the online plant data from the RTC systems located in various LCCs of different buildings over a dual fiber-optic data highway. The plant computer updates the database; stores the data on their hard disks; archives the data on the tape library; prints the alarm messages on the printer; does plant-level calculations like thermal balance, reactivity balance, and derived parameters; and in turn sends the data to display stations mounted on the console and panel over a dual Ethernet interface. They also receive control commands like set point change for proportional integral derivative controller (PID) loops, open/close command for valves, and on/off commands for equipments from display stations/keyboards, validate them, and in turn send it to the RTCs located in the LCCs for control action. The entire application software running on the process computer is developed in line with applicable guides and subjected to internal and external verification and validation.

8.5.10 REACTOR PROTECTION SYSTEM

The reactor protection system is designed to protect the plant from various postulated initiating events. Protection logic receives signals from various systems such as neutronic instrumentation, failed fuel detection, and temperature and flow measurement. SCRAM is effected by two sets of absorber rods with two drive mechanisms. It is effected by de-energizing electromagnets and dropping the absorber rods under gravity for both systems. The first set of rods called the CSRs is used for start-up, controls the reactor power, controls shutdown, and shuts down the reactor. The second set of rods called the DSRs is used only for shutdown of the reactor. SCRAM is followed with an automatic drive down of the CSR.

Shutdown is effected by two independent protection logics that process all the events requiring shutdown and initiate fast insertion of the respective absorber rods into the core. Each of the SDS is capable of independently bringing the reactor to a safe shutdown state.

In order to achieve fast response, the protection logic for the two SDSs is built with a solid-state logic on two different principles. The SDS1 uses conventional solid-state logic built with programmable logic devices (PLDs), with online fine impulse test (FIT). The SDS2 employs pulse coded logic (PCL) technology, where the logic state 1 is encoded as a sequence of pulses rather than a voltage level. In PCL, the presence of a pulse train at the logic output state keeps the electromagnet energized, and if logic is stuck up at 0 or 1 anywhere in the chain, it results in a trip state of the chain. This technique has a self-diagnostic feature, and hence, no online testing is provided. The systems are designed with an instrument response time of ~100 ms including logic circuits. This along with the response time of ~100 ms of the electromagnet ensures actuation of the SDS within 200 ms on the occurrence of an event. The absorber rod drop time is designed to be within 1 s.

In order to further improve the reliability, optical cross-link between the two SDSs at the output stage is provided, that is, the output of one system is fed to the other SDS and vice versa through an optical switch that provides galvanic isolation between the logics. In order to avoid spurious trips, all trip parameters are passed through a latching circuit of 50 ms time delay in each protection logic. The latching time selected (50 ms) is less than the response time of ~100 ms of the electromagnets of CSRDM and DSRDM and forms a part of it.

The SDSs are designed with an overall reliability of 1×10^{-6}/reactor-year. I&C systems are subjected to environmental qualification, electro magnetic interference/electro magnetic compatibility (EMI/EMC) tests, and seismic testing to improve the overall reliability.

8.5.11 POSTACCIDENT MONITORING

The instrumentation for post accident monitoring is meant for monitoring the status of the core and RCB in the case of a CDA. Normal process upsets and DBEs such as transient overpower are not considered as design basis events for the design of PAM. In addition, common mode events such as fire and flooding are not treated as PAM events.

The function of the system is to monitor the following parameters, during and after an accident: core neutron flux, main vessel temperature, RCB temperature, RCB gamma activity, RCB sodium aerosol activity, and RCB pressure.

Instruments for PAM have ranges to cover the maximum values that can attain under the worst-case conditions (i.e., under CDA, which is a beyond-design-basis event). The instruments are also qualified to survive the conditions associated with the CDA, such as temperature, pressure, and radiation level. Signal processing electronics for these instruments are located outside the reactor building to protect it from high radiation during CDA.

8.5.12 MAIN CONTROL LOOPS

There are three major control loops for the control of a reactor power, primary and secondary sodium flow and turbine inlet pressure and temperature. The inlet conditions of the turbine are to be maintained constant at all reactor power levels for high steam cycle efficiency. This demands that sodium temperatures be maintained at specific values at the outlet of the SG and the core for each selected power level. The previously mentioned requirements are achieved by effecting the speed control of the PSP and SSP.

8.5.12.1 Reactor Power Control

The plant is designed for baseload operation. The reactor power is controlled manually and does not follow any measured variable automatically. The reactor is set at a desired power level by manually adjusting the CSRs one at a time. Manual control is safe and adequate as the reactivity changes are small. Additionally, the pool provides a large thermal inertia and the flow variations are negligible as the variable speed drives are not affected by grid frequency variations between 47.5 and 51.5 Hz, justifying the adequacy of manual control.

8.5.12.2 Primary and Secondary Coolant Flow Controls

Primary and secondary coolant flow controls are used for maintaining the sodium temperature difference across the reactor core and SG, respectively. Depending on the target power level, the speeds of the primary and secondary pumps are set manually from the CR from 20% to 100%. However, maintaining the speed at the set value is automatic. The speed of the PSPs is run down automatically following the reactor trip to reduce thermal shocks.

8.5.13 GROUNDING

A well-designed grounding system is essential for effective functioning of I&C systems. The plant is provided with three grounding systems: safety ground (G1), shield ground (G2), and signal ground (G3). This is implemented as per IEEE 1050 "Guide for Instrumentation and Control Equipment Grounding in Generating Stations," 2004.

8.5.14 I&C POWER SUPPLY

Electric power supply is classified in the following:

Class IV power supply: ac power supply to the auxiliaries that can tolerate prolonged power interruption without affecting the safety of the plant.

Class III power supply: Supply to the auxiliaries that can tolerate short (3 min) power interruption. Under normal conditions, this power supply is drawn from the class IV power supply, and in case of loss of class IV power supply, emergency diesel generators provide the backup.

Class II power supply: No-break ac power supply to auxiliaries. This is derived from the class III power supply buses through a rectifier/charger and inverter. Battery backup is provided at the input of the inverter to provide the no-break ac power supply during the interruption of class III power supply.

Class I power supply: No-break dc power supply to the auxiliaries. This is derived from the class III power supply buses through a rectifier/charger. Battery backup is provided at the output of the rectifier/charger to provide no-break dc power supply during the interruption of class III power supply.

Class I and class II power supplies are required for I&C loads. These power supply systems are required for an extended duration to take care of the requirements of SBO conditions in the plant.

8.5.15 Seismic Instrumentation

Seismic instrumentation is provided in the plant for the following activities:

- To record the ground motion data during a seismic event and evaluate the need for post-earthquake inspection of important components
- To collect data on the behavior of structures and components during earthquakes to verify the adequacy of the design analysis
- To trigger alarms in the CR to assess the degree of severity of the seismic event and to decide whether the plant can continue to operate or should be shut down

Strong motion seismic instrumentation locations as per U.S. nuclear regulatory commission (NRC) "Regulatory Guide 1.12 for NPP Instrumentation for Earthquakes" are

- Free field
- Containment foundation
- Two elevations on a structure inside the containment
- An independent seismic category 1 structure foundation where the response is different from that of the containment structure
- An elevation on the independent seismic category 1 structure selected in the fourth item

8.5.16 Future Directions

Continuous updating of I&C systems is required with advancement in technologies. Also, it is required to address obsolescence. The I&C system of fast reactors provides a lot of opportunities for further improvement. An abundant scope exists for the development of innovative instrumentation. Some of the areas are indicated in the following that provide excellent opportunities for researchers and students:

- Development of a signal processing system with high reliability for locating an area in the roof slab where ambient temperature is high and for being a safety-critical system are challenging.
- Development of a reliable wireless instrumentation system for reactors.
- Development and engineering of radio detection and ranging (RADAR)-based level probes for sodium applications.
- Development of an integrated core temperature and eddy current flowmeter probe to enable continuous monitoring of flow through individual FSAs.
- Development of simplified nuclear instrumentation electronics using techniques like field programmable gate array (FPGA) and other I^{2}C bus.
- Development and qualification of a large quantity of instruments/hardware imported at present.
- Development of under sodium viewing systems.
- Development of signal processing systems for acoustic SG leak detection.

8.6 ENERGY CONVERSION SYSTEMS

Power conversion systems (PCSs) are coupled to nuclear reactors through SGs that use the heat generated in the core to generate steam that in turn drives turbine generators to produce electricity. SGs are heat exchangers used to convert water into steam from heat produced in a nuclear reactor core, located after the primary reactor coolant loop, thus forming the link between the NSSS and balance of the plant. The primary coolant is heavy water in pressurized heavy water reactors and liquid sodium in fast reactors. There are reactors that do not incorporate any PCS and are used purely for research purposes. The world's first nuclear plant of any type that generated electricity was the experimental fast breeder reactor EBR-1, designed by Argonne National Laboratory. Table 8.7 gives a list of experimental, prototype, and demonstration reactors with and without PCSs attached to the nuclear systems.

Today, all fast breeder reactors in operation employ the Rankine cycle with reheat and regeneration in their power conversion cycle. The PCSs of SFRs are superior to those of other nuclear reactors as they have higher thermomechanical efficiency and consequently reduce the thermal pollution on the environment. PCSs operate on cycles followed in conventional thermal power

TABLE 8.7

Experimental, Prototype, and Demonstration Fast Reactors with and without PCS

S. No.	Reactor	Country	Criticality Date	Thermal Rating (MW)	Electrical Rating (MW)	PCS Incorporated
1	Clementine	United States	1946	0.025	—	No
2	EBR-1	United States	1951	1.2	0.2	Yes
3	BR-1/2	Union of Soviet Socialist Republics (USSR)	1956	0.1	—	No
4	BR-5/10	USSR	1958	5/10	—	No
5	Dounreay (DFR)	United Kingdom	1959	60	15	Yes
6	LAMPRE	United States	1961	1	—	No
7	Fermi (EFFBR)	United States	1963	200	65	Yes
8	EBR-II	United States	1963	62	20	Yes
9	Rapsodie	France	1967	40	—	No
10	SEFOR	United States	1969	20	—	No
11	BOR-60	USSR	1969	60	12	Yes
12	KNK-II	Germany	1977	58	21	Yes
13	Joyo	Japan	1977	100	—	No
14	FFTF	United States	1980	400	—	No
15	FBTR	India	~1983	50	15	Yes
16	PEC	Italy	~1985	118	—	No
17	BN-350	USSR	1972	150	1000	Yes
18	Phénix	France	1973	250	568	Yes
19	PFR	United Kingdom	1974	250	600	Yes
20	BN-600	USSR	1980	600	1470	Yes
21	Superphénix 1	France	1983	1200	3000	Yes
22	SNR-300	Germany	1984	327	770	Yes
23	Monju	Japan	1987	300	714	Yes
24	CRBRP	United States	~1988	375	975	Yes
25	Superphénix 2	France	~1990	1500	3700	Yes
26	CDFR	United Kingdom	~1990	1320	3230	Yes
27	SNR-2	Germany	~1990	1300	3420	Yes
28	BN-1600	USSR	~1990	1600	4200	Yes
29	DFBR	Japan	~1990	1000	2400	Yes

Source: Walter, A.E. and Reynolds, A.B., *Fast Breeder Reactors*, Pergamon Press, New York, 1981.

plants, but with additional systems and controls that increase reliability and minimize the effect of transients arising in the PCS on the nuclear system. One of the biggest problems in the commercial deployment of SFRs is the sodium–water reaction, due to the choice of sodium as the reactor coolant and water–steam cycle as the power cycle. SGs are subjected to a high pressure difference, high temperature, and large heat transfer area, which increase the potential hazard for leakage and hence the reaction. Design features that ensure integrity of the SG, component material selection, and stringent quality assurance during manufacturing and erection are essential to prevent the possibility of a sodium–water reaction. In addition, extra safety systems have to be provided to detect, mitigate, and rapidly terminate an SG leak to protect against potential investment loss and to preserve the integrity of the coolant boundary. Systems that are added to assure that incipient sodium leaks in the SG are detected include advanced integrated leak detection systems, fast-acting steam and water side isolation and blow down systems, and a sodium–water reaction pressure relief system.

In order to preclude the previously mentioned risk, studies have been initiated within the SFR community on the possibility of adopting gas-based PCSs, involving several gases and their mixtures as potential media. Inert gases like helium and neon are highly utilized in the gas cycle; however, these have high diffusibility and are costly. Nitrogen also provides an inert atmosphere and is quite inexpensive. Marginally higher cycle efficiencies than nitrogen (η = 38.5%) are reported typically at turbine inlet conditions of 515°C, 180 bar, and reactor sodium inlet temperature of 395°C for He–Ni (η = 38.9%), He–Ar (η = 38.6%), and Ne (η = 39.2%) [8.45]. A carbon dioxide–based cycle has better efficiency than helium and nitrogen; however, it is not completely benign with sodium [8.46]. The choice of fluid depends on many parameters like operating temperature, cycle pressure, thermomechanical efficiency, resulting compactness of PCS, and cost.

8.6.1 POWER CYCLE

8.6.1.1 Rankine Cycle with Regeneration and Reheat

In line with the conventional fossil fuel power plants, the Rankine cycle is the most common choice for the PCS. The process of reheating the steam between turbine stages and heating of feedwater before it enters the SG (regeneration) increases the cycle efficiency. The functions of the steam–water system are to supply feedwater to the SG at a required temperature, pressure, and chemistry, to utilize reactor thermal energy from the SG and produce steam at rated parameters and quality to drive turbogenerator to produce rated power, and to reject the unutilized thermal energy of steam into the ultimate heat sink through a condenser. For a coastal site, maintaining the discharged seawater temperature within the environment (MoEF) stipulations is an important consideration. The schematic of a typical Rankine cycle with reheat and regeneration is shown in Figure 8.78 along with a representation on the T–S plot. The PCS based on the Rankine cycle coupled with the nuclear reactor is shown in Figure 8.79.

Fast reactors have the advantage of high operating temperatures enabling superheated steam cycles similar to conventional fossil fuel plants having higher efficiency. The option of steam–steam or sodium–steam reheat and regenerative feedwater heating is also mostly employed to further improve cycle efficiency. Further as a preventive safety measure, the increased sodium–water interfaces associated with sodium reheat, along with the increased number of tube to tube sheet joints and the additional piping, valves, and leak detection provision, are avoided, and integral SG units are normally employed. In the steam reheat option, live steam is used to reheat the partially expanded steam before admission to intermediate pressure (IP) and low pressure (LP) turbine stages, even though the temperatures achieved in this method are lower than those achievable with sodium reheat. Feedwater heaters preheat the feedwater entering the SG using steam bled from the turbine to improve the thermodynamic efficiency by increasing the average temperature at which heat is added to the cycle. The cycle is optimized for the number of heaters w.r.t. gain in

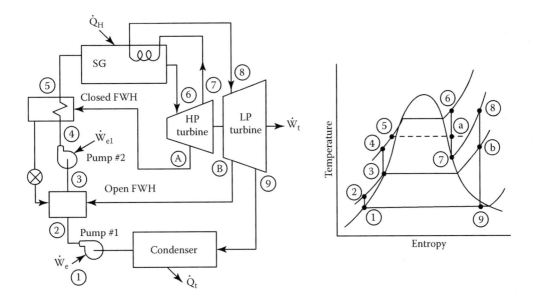

FIGURE 8.78 Rankine cycle with regeneration and reheat: (a) Rankine cycle schematic and (b) temperature-entropy diagram. 1–2, pumping from the condenser; 2–3, regeneration using steam (B); 3–4, pumping from the deaerator; 4–5, regeneration using steam (A); 5–6, heat exchange in steam generator; 6–7, expansion in HP turbine; 7–8, reheating of steam; 8–9, expansion in LP turbine; and 9–1, steam condensation.

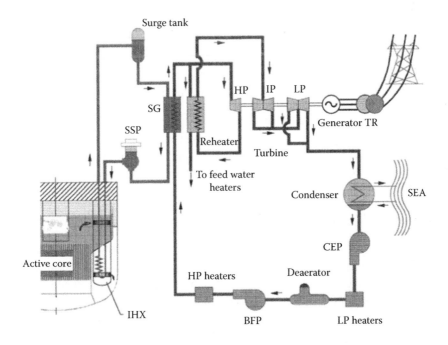

FIGURE 8.79 PCS based on the Rankine cycle with the reactor system.

efficiency versus increase in capital cost. The salient features of PCS employing a Rankine cycle are as follows:

1. The steam cycle shall incorporate reheat and regeneration to improve efficiency and also reduce the moisture at the last stage blades to <12%.
2. A minimum flow through the once-through SG shall be maintained and the operating pressure chosen to avoid instability-related effects.
3. An offline steam–water separator suitably sized as percentage of full flow may be used during start-up and shutdown to eliminate erosion of bypass valves during two-phase steaming.
4. The minimum temperature of feedwater admitted to the SG should be at least 150°C to avoid sodium freezing in the SG.
5. For adequate redundancies, 3% × 50% boiler feed pumps, two turbo driven and one motor driven, are preferred configurations for a plant of capacities >500 MWe. The boiler feed pump should be capable of dry run for a minimum duration, say at least 10 min, to save the pump in case of any exigency. This is achieved by selection of an advanced class of boiler feed pumps with lesser shaft diameter, smaller bearing span, larger wear ring clearance, and larger diameter of the balancing drum.
6. The feedwater system shall be designed to have cold end ΔT value, say <150°C, during normal operation to reduce the thermal stresses at the SG inlet. This can be achieved by using auxiliary steam line for heating in the HP heater during start-up or by keeping the deaerator at higher pressure. The deaerator pressure during the normal operation needs to be considered while deciding on this aspect.
7. Feedwater chemistry is based on all-volatile treatment (AVT) and is stringent for once-through SGs.
8. The total relieving capacity of the steam system in the case of the turbine trip shall be accommodated by the turbine bypass system and instrumented atmospheric steam discharge valves with a suitable margin. The mechanical safety valves of the SG should not participate in normal-pressure relief operations during such transient operations and should be designed to act for SG protection during any overpressurization event.
9. For cycles with once-through SGs, the DM water makeup will be in the order of 1%. The condensate polishing unit should be full flow with 3% × 50% to cater to online polishing to maintain strict water chemistry.
10. Chlorination shall be done in seawater cooling systems to reduce the effects of biofouling and the free chlorine limit at discharge shall be 0.5 ppm.

8.6.1.2 Closed Brayton Cycle with Intercooling

The Brayton cycle is being considered as a favorable candidate for PCSs of next-generation SFRs. The Brayton cycle has much smaller-sized components and a simpler layout compared to the currently used Rankine cycle. The compactness of turbomachineries and heat exchangers contributes to the reduction in capital cost due to space reduction. The efficiency is either similar or superior to those of the Rankine cycle [8.47]. The schematic of a typical Brayton cycle with intercooling is shown in Figure 8.80 along with a representation on the T–S plot. The Brayton cycle coupled to the reactor system is shown in Figure 8.81.

The major advantage of the Brayton cycle is replacement of the once-through SG of the Rankine cycle with sodium to gas heat exchangers. The heat exchangers used are compact printed circuit heat exchangers (PCHEs), which are effectively monolithic blocks of stainless steel containing embedded flow channels. The concerns about tube failures of shell and tube heat exchangers are thus eliminated [8.48].

For Brayton cycles, heavy-duty axial flow gas turbines can be chosen. The overall pressure ratio of these units varies from 5:1 to 35:1. Turbine inlet temperatures can be as high as 1350°C. However, the temperatures in nuclear PCSs are limited due to the limitations on temperature in the reactor

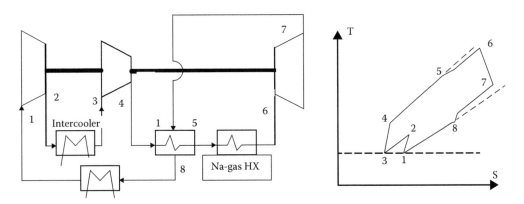

FIGURE 8.80 Brayton cycle with intercooling: 1–2, compression no. 1; 2–3, intercooling; 3–4, compression no. 2; 4–5, heater from GT exhaust; 5–6, heat exchange in Na–gas heat exchanger; 6–7, expansion in turbine; 7–8, cooling of GT exhaust with regeneration to compressed air; 8–1, after cooling of GT exhaust before entering the compressor.

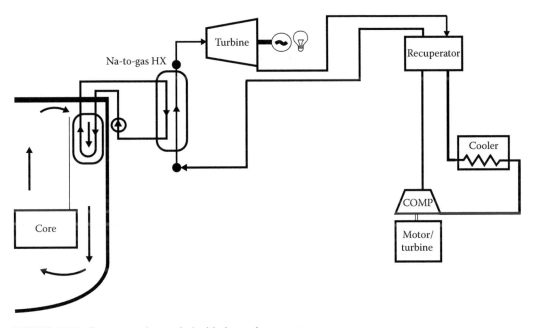

FIGURE 8.81 Brayton cycle coupled with the nuclear reactor.

primary/secondary coolant system. Both single-flow and split-flow gas turbines with high isentropic efficiencies can be employed.

The compressors used are mainly of two types—centrifugal and axial—providing continuous flows at high pressure ratios. The pressure ratio for centrifugal compressors can be up to 1.9 per stage. The adiabatic efficiency of up to 85% can be achieved. Compressor efficiency is very important in the overall performance of the gas turbine as it consumes 50%–60% of the power generated by the gas turbine. Use of intercoolers reduces the compressor work significantly by increasing the density of the working media. Axial compressors provide large flow rates at high pressure ratios; however, their stable operating range is narrow. To avoid this problem, multiple centrifugal compressors with intercooling are employed.

The use of regenerators in conjunction with industrial gas turbines substantially increases cycle efficiency. The working pressure is kept high to keep the size of the heat exchanger small. Of all the heat exchangers, PCHEs may prove to be the best choice because of their high effectiveness and compact size as compared to shell and tube heat exchangers of the same capacity.

The addition of a heat exchanger merely causes a slight reduction in the specific work output due to the additional pressure losses. The intercooling and regeneration heat exchangers substantially increase the overall efficiency and reduce the stage optimum pressure ratios.

8.6.1.3 Choice of Power Cycle and Working Fluid

The Brayton cycle configuration, design of sodium–gas heat exchanger, and other components of the PCS need to be further studied in entirety for technoeconomic viability. At present, there are no established operating power plants employing the Brayton cycle. Further, the indigenous technological base for manufacturing large-sized gas turbines and associated power cycle equipment is limited.

The French have initiated the ASTRID 1500 MWt/600 MWe program with the nitrogen-powered Brayton cycle. ABTR is planned as a 250 MWt/95 MWe reactor with supercritical carbon dioxide–powered Brayton cycle with the Rankine cycle as a backup. However, it is reported that there is no increase in cycle efficiency on adopting the Brayton cycle over the Rankine cycle at turbine inlet temperatures below 525°C. Therefore, the well-proven steam–water cycle–based PCS is currently retained as a practical solution for fast reactors planned in the immediate future.

8.6.2 Special Features of Power Conversion Systems for SFRs

8.6.2.1 Steam Generator and Associated Systems

The SG with sodium on the shell side and water on the tube side is in itself a special component as it calls for safety systems to monitor, detect, prevent, and protect the system from a potential sodium–water reaction. The SG tube side depressurization system forms a very important part of the steam–water system. On leak detection, protective action is initiated by cutting out the feedwater supply, isolating the main steam header and depressurization of the SG through steam–water dump valves at the feedwater inlet end of the SG. A line diagram of a depressurization circuit is shown in Figure 8.82. All the valves involved in these actions are quick-acting pneumatic gate valves. Isolation valves V1 and V2 are normally open, and dump valves V3 and V4 are normally closed. The steam–water mixture dumped during depressurization is diverted to the dump tank where steam and water are separated and the steam vented to the atmosphere. In order to have a high reliability for the dumping on a demand, the dump valve is duplicated.

After quick depressurization of the SG on the steam/water side, the nitrogen circuit provided is used to pressurize the SG tubes with nitrogen at a pressure slightly above the sodium side pressure under a full load operating condition, to prevent secondary sodium from the shell side entering the tubes.

8.6.2.2 Steam–Water Separator

Achieving steam quality admissible to the turbine in a fast reactor takes a very long time (~14–18 h) due to controlled power raising operation. During start-up, hot water is first discharged from the SG outlet followed by saturated steam and finally superheated steam. In order to prevent moist steam from eroding the turbine bypass valves, a steam–water separator is provided to separate the steam and water from the mixture.

The steam separator is a vessel in which the steam–water mixture is made to pass through a series of turboseparators arranged in a row by a separating chamber. The steam separated from the separator is sent to the turbine bypass, and the separated condensate is collected through the downcomer nozzle into a storage tank that drains the condensate to the condenser through a flash tank. The separator is taken offline once superheated steam evolves out of the SG.

8.6.2.3 Decay Heat Removal System through the Steam–Water System

The steam–water system is the ultimate heat sink path in the power plant. When the normal heat transport path is available, heat can be removed using the turbine bypass system. The quality deteriorates

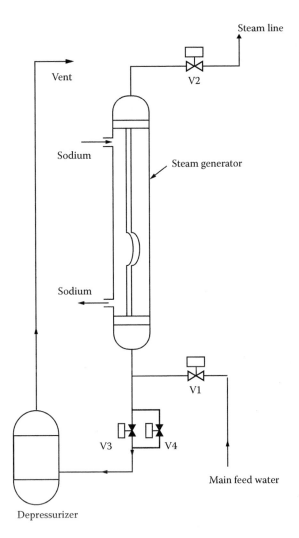

FIGURE 8.82 SG tube side depressurization circuit.

progressively after reactor shutdown and the steam reaches the saturation state. The steam–water separator is valved in and the system now operates in a smaller loop to remove the decay heat of the reactor, attain the cold shutdown condition, and maintain the reactor in isothermal condition. The smaller closed loop that rides on the steam–water system is shown in Figure 8.83. This smaller circuit system is sized for a fraction (typically 20%) of the total rated flow. The water held up in the separator as a result of separation of two-phase steam from the SG is circulated back to the SG by means of recirculation pumps. The boiler feed pumps are taken off the circuit at this juncture. The steam separated from the two-phase mixture in the separator is sent to the auxiliary condensers that may be either air or water cooled. The steam condenses in the auxiliary condensers and returns back to the separator. The temperature of the system is reduced in a controlled manner by varying the speed of the fans/pump (or throttling valve) that causes forced circulation of air/water across the auxiliary condenser. However, this system does not serve or support any safety function and hence is not a safety-related system.

In order to meet the stringent safety function of DHR, a nuclear safety class 1 DHR system, directly removing heat from the hot pool of the reactor, is used. This is a passive circuit that comes into play when either of the steam–water system, offsite power, or secondary sodium system becomes unavailable.

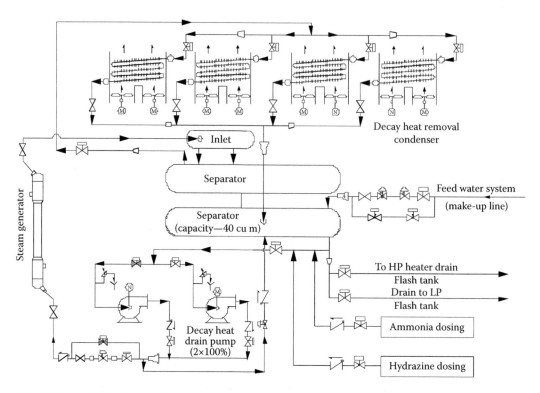

FIGURE 8.83   DHR circuit riding on the steam–water system.

8.6.3   System Classification and Categorization

The PCS is classified as NNS class and nonseismic category.

REFERENCES

8.1   Kim, Y.-i., Lee, Y.B., Lee, C.B., Chang, J., Choi, C. (2013). Design concept of advanced sodium-cooled fast reactor and related R&D in Korea. *Sci. Technol. Nucl. Install.*, 18 pp., Article ID 290362. http://dx.doi.org/10.1155/2013/290362.

8.2   De Paz, J.F. (1975). Pressure drop and volume fraction of grid and wire spaced subassemblies, 24 pp., ANL-AFP-13 United States, October 1975.

8.3   Walter, A.E., Reynolds, A.B. (1981). *Fast Breeder Reactors*. Pergamon Press, New York.

8.4   Olander, D.R. (1976). *Fundamental Aspects of the Nuclear Reactor Fuel Elements, TID-26711*. ERDA, Springfield, VA.

8.5   Chetal, S.C., Bhoje, S.B., Kale, R.D., Rao, A.S.L.K., Mitra, T.K., Selvaraj, A., Sethi, V.K., Sundaramoorthy, T.R., Balasubramaniyan, V., Vaidyanathan, G. (1995). Conceptual design of heat transport systems and components of PFBR-NSSS. *Conference Technical Committee Meeting on Conceptual Designs of Advanced Fast Reactors*, Kalpakkam, India, October 1995.

8.6   Chetal, S.C., Balasubramaniyan, V., Chellapandi, P., Mohanakrishnan, P., Puthiyavinayagam, P., Pillai, C.P., Raghupathy, S., Shanmugham, T.K., Sivathanu Pillai, C. (2006). The design of the prototype fast breeder reactor. *Nucl. Eng. Des.*, 236(7–8), 852–860.

8.7   Chikazawa, Y., Kotake, S., Sawada, S. (2011). Comparison of advanced fast reactor pool and loop configurations from the viewpoint of construction cost. *Nucl. Eng. Des.*, 241(1), 378–385.

8.8   Aithal, S., Sritharan, R., Rajan Babu, V., Balasubramaniyan, V., Puthiyavinayagam, P., Chellapandi, P., Chetal, S.C. (2009). Design and manufacture of reactor assembly components of 500 MWe PFBR. *Conference Peaceful Uses of Atomic Energy (PUAE 2009)*, New Delhi, India, September–October 2009.

8.9   Srinivasan, G., Kumar, K.V.S., Rajendran, B., Ramalingam, P.V. (2006). The fast breeder test reactor-design and operating experiences. *Nucl. Eng. Des.*, 236(7–8), 796–811.

8.10 Nevskii, V.P., Malyshev, V.M., Kupnyi, V.I. (1981). Experience with the design, construction, and commissioning of the BN-600 reactor unit at the Beloyarsk Nuclear Power Station. *Sov. Atom. Energy*, 51(5), 691–696.

8.11 IAEA-TECDOC-1531. Fast reactor database 2006 update. International Atomic Energy Agency (IAEA), Vienna, Austria, December.

8.12 Chellapandi, P., Chetal, S.C., Raj, B. (2008). Investigation on buckling of FBR vessels under seismic loadings with fluid structure interactions. *Nucl. Eng. Des. (Elsevier)*, 238(12), 3208–3217, December 2008.

8.13 Chellapandi, P., Puthiyavinayagam, P., Balasubramaniyan, V., Ragupathy, S., Rajan Babu, V., Chetal, S.C., Raj, B. (2010). Design concepts for reactor assembly components of 500 MWe future SFRs. *Nucl. Eng. Des.*, 240(10), 2948–2956, October 2010.

8.14 Mitra, A., Rajan Babu, V., Puthiyavinayagam, P., Vijayan Varier, N., Ghosh, M., Desai, H., Chellapandi, P., Chetal, S.C. (2012). Design and development of thick plate concept for rotatable plugs and technology development for future Indian FBR. *Nucl. Eng. Des.*, 246, 245–255, May 2012.

8.15 Chellapandi, P., Chetal, S.C., Raj, B. (2012). Numerical simulation of fluid-structure interaction dynamics under seismic loadings between main and safety vessels in a sodium fast reactor. *Nucl. Eng. Des.*, 253, 125–141, December 2012.

8.16 Gajapathy, R., Velusamy, K., Selvaraj, P., Chellapandi, P., Chetal, S.C., Sundararajan, T. (2008). Thermal hydraulic investigations of intermediate heat exchanger in a pool-type fast breeder reactor. *Nucl. Eng. Des.*, 238(7), 1577–1591, July 2008.

8.17 Athmalingam, S. (2011). Intermediate heat exchanger for pfbr and future fbr, International Atomic Energy Agency, Technical Working Group on Fast Reactors, Vienna, Austria, 18 pp; IAEA Technical Meeting on Innovative Heat Exchanger and Steam Generator Designs for Fast Reactors, December 21-22, 2011. http://www.iaea.org/NuclearPower/Downloadable/Meetings/2011/2011-12-21-12-22-TM-NPTD/6_India-IHX-for-PFBR-and-future-FBRs.pdf

8.18 Chetal, S.C., Vaidyanathan, G. (1997). Evolution of design of steam generator for sodium cooled reactors. *HEB 97*, Alexandria, Egypt, April 1997.

8.19 Srinivasan, R., Chellapandi, P., Jebaraj, C. (2010). Structural design approach of steam generator made of modified 9Cr-1Mo for high temperature operation. *Trans. Indian Inst. Metals*, 63(2–3), 629–634, April 2008.

8.20 Muller, R.A. et al. (1975). Evolution of heat exchanger design for sodium cooled reactors. *Atom. Energy Rev.*, 13, 215.

8.21 Baker, R.S. (1987). *Handbook of Electromagnetic Pump Technology*. Elsevier, New York.

8.22 Nashine, B.K., Dash, S.K., Gurumurthy, K., Rajan, M., Vaidyanathan, G. (2006). Design and testing of D.C. conduction pump for sodium cooled fast reactor. *14th International Conference on Nuclear Engineering, ICONE 14*, Miami, FL.

8.23 Sharma, P., Sivakumar, L.S., Rajendra Prasad, R., Saxena, D.K., Suresh Kumar, V.A., Nashine, B.K., Noushad, I.B., Rajan, K.K., Kalyanasundaram P. (2011). Design, development and testing of a large capacity annular linear induction pump. *Energy Procedia*, 7, 622–629.

8.24 Dixit, A.S., Bhoje, S.B., Chetal, S.C., Selvaraj, P. (1986). Decay heat removal for PFBR. *International Conference on Science and Technology of Fast Reactor Safety*, Guernsey, U.K., May 1986.

8.25 Satish Kumar, L., Natesan, K., John Arul, A., Balasubramaniyan, V., Chetal, S.C. (2011). Design and evaluation of operation grade decay heat removal system of PFBR. *Nucl. Eng. Des.*, 241(12), 4953–4959, December 2011.

8.26 Kochetkov, L.A. et al. (1991). Operating experience on fast breeder reactors in the USSR. *International Conference on Fast Reactors and Related Fuel Cycles*, Kyoto, Japan, October–November 1991.

8.27 Ninokata, H., Izumi, A. (1990). Decay heat removal system of the Monju reactor plant and studies related to the passive actuation and performances (invited paper). *Proceedings of the International Fast Reactor Safety Meeting*, Snowbird, UT, August 1990, Vol. II, pp. 319–330.

8.28 Athmalingam, A., John Arul, A., Parthasarathy, U., Kasinathan, N., Sundaramoorthy, T.T., Selvaraj, A., Chetal, S.C. (2002). Decay heat removal in prototype fast breeder reactor. *First National Conference on Nuclear Reactor Safety*, Mumbai, India, November 2002.

8.29 Mitenkov, F.M., Samoilov, O.B. (1991). Advanced enhanced safety PWR of new generation. *IAEA Technical Committee Meeting*, Vienna, Austria, November 1991.

8.30 Gregory, C.V., Bell, R.T., Brown, G.A., Dawson, C.W., Hampshire, R.G., Henderson, J.D.C. (1979). Natural circulation studies in support of the Dounreay PFR. *Proceedings of the IEEE Region 6 Conference*, Sacramento, CA, pp. 1599–1606.

8.31 Gyr, W. et al. (1990). EFR decay heat removal system design and safety studies. *International Fast Reactor Safety Meeting*, Snowbird, UT, 1990, Vol. 3.

8.32 Boardman, C.E., Dubberley, A.E., Carroll, D.G., Hui, M., Fanning, A.W., Kwant, W.A. (2000). Description of the S-PRISM plant. *ICONE 8*, Baltimore, MD, April 2000.

8.33 Graham, J. (1975). Nuclear safety design of the Clinch River breeder reactor plant. *Nucl. Safety*, 16, 5, September–October 1975.

8.34 Sackett, J.I. (1997). Operating and test experience with EBR-II, the IFR prototype. *Prog. Nucl. Energy*, 31(1/2), 111–129.

8.35 Agarwal, A.K., Guppy, J.G. (1991). *Decay Heat Removal and Natural Convection in Fast Breeder Reactors*. Hemisphere Publishing Corporation, Washington, DC.

8.36 Beaver, T.R. et al. (1982). Transient testing of the FFTF for decay heat removal by natural convection. *Proceedings of the LMFBR Safety Tropical Meeting*, European Nuclear Society, Lyon, France, July 1982, Vol. II, pp. 525–534.

8.37 Bhoje, S.B., Chellapandi, P. (1995). Operating temperatures for an FBR. *Nucl. Eng. Des.*, 158(1), 61–80, September 1, 1995.

8.38 Burgazzi, L. (2013). Analysis of solutions for passively activated safety shutdown devices for SFR. *Nucl. Eng. Des.*, 260, 47–53.

8.39 Takahashi, N. et al. (1989). Study of an advanced fuel handling system. *Nucl. Technol.*, 86, 7–16, July 1989.

8.40 Blanks, D.M. (1985). Fuel handling options for commercial fast breeder reactors. *International Conference on Engineering Developments in Reactor Refueling*, Newcastle-upon-Tyne, U.K., May 1985.

8.41 IAEA document. Ref. for SPX-1 storage drum deletion.

8.42 Suresh Kumar, K.V. et al. (2011). Twenty five years of operating experience with the fast breeder test reactor. *Energy Procedia*. 7, 323–332.

8.43 Takamatsu, M., Ashida, T., Kobayashi, T., Kawahara, H., Ito, H., Nagai, A. (2013). Restoration work for obstacle and upper core structure in reactor vessel of experimental fast reactor "Joyo". IAEA-CN-199/103, Japan Atomic Energy Agency (JAEA). http://www.iaea.org/NuclearPower/Downloadable/Meetings/2013/2013-03-04-03-07-CF-NPTD/T9.2/T9.2.takamatsu.pdf.

8.44 Sylvia, J.L. et al. (2013). Ultrasonic imaging of projected components of PFBR. *Nucl. Eng. Des.*, 258, 266–274.

8.45 Latge, C. (2013). Energy conversion systems for SFR. *Presentation at IGCAR*, India, January 2013.

8.46 Zhang, H. et al. (2009). Investigation of alternate layouts for supercritical carbon dioxide Brayton cycle for a sodium-cooled fast reactor. *ICAPP*, Tokyo, Japan.

8.47 Yoon, H.J. et al. (2012). Potential advantages of coupling supercritical CO_2 Brayton cycle to water cooled small and medium size reactor. *Nucl. Eng. Des.*, 245, 223–232, January 2012.

8.48 Chang, Y.I., Finck, P.J., Grandy, C. (2006). Advanced burner test reactor. Preconceptual design report. Nuclear Engineering Department. ANL-ABR-1 (ANL-AFCI-173), Argonne National Laboratory, U.S.

9 Design Basis

9.1 INTRODUCTION

Sodium coolant remains in liquid state up to about 1160 K at ambient pressure, while the normal operating temperature in sodium-cooled fast reactor (SFR) does not generally exceed ~820 K. Hence, there is no need of pressurization to maintain the sodium in liquid state. This is why the design pressure of SFR system components is low. However, there exists a large temperature difference across the core (~150°C) between inlet and outlet sodium. This large temperature also reflects in the hot and cold legs of the sodium piping. In the case of pool-type reactors, this temperature difference coexists in the hot and cold sodium pools housed in the reactor vessel. During thermal transients such as power failure and pump trip event, the hot pool components and hot leg piping could face colder shock and the cold pool components and cold leg piping could face hot shock. In view of high heat transfer characteristics of sodium, the rapid changes occurring in the coolant are transmitted completely to the structure with insignificant film drop. Apart from these, austenitic stainless steels, the structural materials generally chosen, have low thermal conductivity, low thermal diffusivity, and large thermal expansion coefficients. These are factors responsible for causing certain structural mechanics failure modes of specialized nature.

In view of low operating pressure, thin-walled vessels and piping are generally chosen. This is also preferable to mitigate thermal stresses, apart from economic considerations. Further, the vessel dimensions are large, particularly for a pool-type SFR, as the reactor vessel houses the entire primary sodium circuit. To be specific, the main vessel is a large-diameter thin-walled vessel. Apart from the main vessel, other internal vessels housed inside the main vessel such as inner vessel and thermal baffles also come under the category of large-diameter thin-walled vessels. As a result of high inertial force and pressure generated within the large sodium mass housed in the main vessel during seismic events, thin-walled vessels are prone to buckle. In order to have adequate factor of safety against buckling risks, the vessel may call for additional thickness. The buckling behavior of thin shells subject to complex dynamic pressure distributions generated due to fluid structure interactions in association with high temperatures and sharp temperature gradients leads to a unique failure mode; the investigation of such phenomenon calls for special formulations and solution methodologies. Thus, the seismic-induced stresses decide the minimum wall thickness for various components and piping.

Thermomechanical loadings decide the life of the components in SFR. The temperature fluctuations lying in the frequency range of 1–10 Hz due to either level fluctuations in the vicinity of sodium free level or due to thermal stratifications or thermal striping (these are explained subsequently) induce high-cycle thermal fatigue that could enhance the damage caused by low-cycle fatigue (LCF). Such complex interactions are unique to SFR. The special failure modes of specialized nature have been consolidated systematically in Figure 9.1, based on extensive literature review, in-depth discussions among experts, and the operating experiences accumulated over 400 reactor years worldwide. This chapter provides more details of these failure modes and design and analysis methodologies to prevent them.

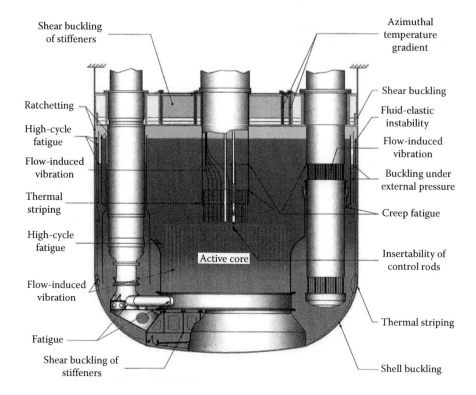

FIGURE 9.1 (**See color insert.**) Failure modes considered in design.

9.2 FAILURE MODES

9.2.1 THERMAL STRIPING

Sodium exhibits very efficient heat transfer to structures. In the zones where the hot and cold sodium happen to meet, mixing would not be complete. The temperatures in the nonmixing zone fluctuate randomly in spite of large turbulent diffusion. This phenomenon is called *thermal striping*. The surrounding structures have to sustain these temperature fluctuations on their surface. A typical situation is depicted in Figure 9.2, which occurs on the above core structures facing the sodium jets emerging from the core subassembly (SA) outlets. While the sodium jets emerging out from the fuel

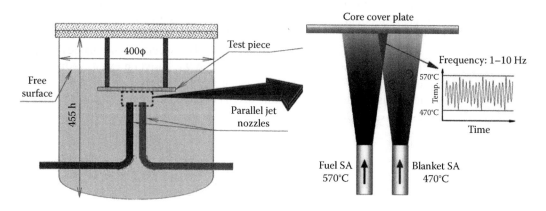

FIGURE 9.2 (**See color insert.**) Thermal striping phenomenon near core cover plate.

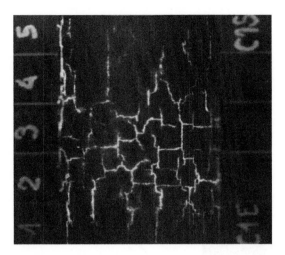

FIGURE 9.3 Thermal striping cracks.

SAs are hotter, the sodium jets from the blanket or control SAs are relatively lower due to very low heat generated within them. The thermal striping causes high-cycle fatigue (HCF) damage. The appearance of a network of resulting cracks at the surface is depicted in Figure 9.3.

9.2.2 THERMAL STRATIFICATIONS

Since both hot and cold sodium pools coexist in the pool-type concept, sodium layers with high ΔT (150 K maximum) exist in narrow transition regions of hot and cold pools during operating as well as transient conditions. This phenomenon is termed *thermal stratification*. Further, such stratification is also caused by different temperatures of sodium jets arising from core SAs. In the compact arrangement of fast reactor core, the fuel, breeder, and internally stored spent fuel SAs are located in the same layout. In spite of the elaborate flow zoning exercises, the temperatures of sodium exiting from the fuel SA (850 K typical), breeder SA (750 K typical), and spent SA (680 K typical) are vastly different. The transition zone between the hot and cold sodium pools is influenced by the presence of cold flow due to leakage from the feet of the SAs. The temperature difference between the hot and cold pools induces thermal flux through the inner vessel structures. With these thermal conditions prevailing in the pool, coupled with the large thermal expansion coefficient of sodium (2.8×10^{-4}/K) and the large size of the reactor pool, the Richardson number in the pool is of the order of unity, indicating that the inertial and buoyancy forces are of similar magnitudes. As a result, a stratification interface is developed, as shown in Figure 9.4. Thermal stratification can occur in the transition area between the hot and cold zones of the reactor. This region is annular and bounded laterally by the core periphery and the redan portion of the inner vessel. The thermal behavior of this zone is influenced by two conflicting phenomena: (1) cold boundary layer due to the heat flux across the redan shell and cold sodium flow from peripheral SAs of the core that generates a stable cold layer and (2) the main sodium in the hot pool that causes penetration of hot sodium into this area. This hot main sodium flow is the consequence of the recirculation created by the core outlet flow. The equilibrium between these two effects causes mixed convection and a stratified flow in this region, as shown in Figure 9.4.

Thermal stratification is also observed in pools during transient conditions. The limiting transient states for a pool-type reactor are safety control rod accelerated movement (SCRAM). During such transients, cold sodium jets will be injected into the hot sodium pool due to slow flow reduction. The sodium temperature at the core outlet falls rapidly (reduction in the temperature of the order of 130 K at 15–20 K/s). During these transient conditions, depending on the operating strategy

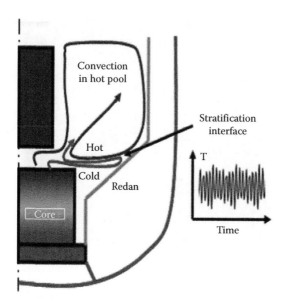

FIGURE 9.4 Thermal stratification in hot pool.

adopted for the reactor, the primary sodium flow rate could also vary. Hence, under these conditions, coupled with the temperature difference effects, the flow reduction effects will augment the net buoyancy forces imposed on the main sodium flow in the hot pool. This can, in turn, generate considerable changes in the flow pattern in the hot plenum and can even lead to the formation of a stratified hot layer above the inlet windows of the intermediate heat exchangers (IHX).

Thermal stratification effects may also be observed in the cold pool during transient conditions. The changes in the flow pattern may be brought about by the influence of buoyancy forces caused by the temperature difference between sodium streams at the outlet of IHX coupled with flow changes under transient conditions. These effects can produce, depending on operating conditions, considerable changes of flow in the cold plenum and may even lead to the formation of a hot stratified layer above the lower part of the pump assembly. Apart from the primary sodium pool, thermal stratification can also develop in sodium pipelines if the conditions are conducive, especially during low flow transients.

9.2.3 FREE-LEVEL FLUCTUATIONS

SFR designs generally include the presence of a free level of sodium in some components. An inert gas (usually argon) at low pressure is filled in the space over the free level of sodium. These provisions allow for thermal expansion of sodium during temperature transients and guaranty inert chemical conditions above sodium. The different vessels and tanks that are concerned whatever the type of concept (loop or pool) are:

- The main vessel itself
- The inner vessel as well as stand pipes (in pool type)
- Different baffle shells (for instance, in the case of design with dedicated circuit for cooling down the main vessel at the temperature of core inlet)
- The main primary pipes (hot, cold) in the loop type that may cross the free level in the main vessel, etc.
- The dedicated expansion tanks and/or the main steam generator (SG) vessels in the secondary circuits that transfer the heat to the energy conversion system

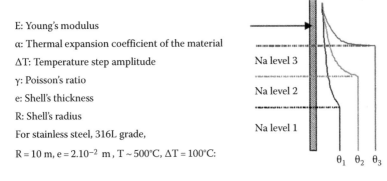

E: Young's modulus

α: Thermal expansion coefficient of the material

ΔT: Temperature step amplitude

γ: Poisson's ratio

e: Shell's thickness

R: Shell's radius

For stainless steel, 316L grade,

$R = 10$ m, $e = 2.10^{-2}$ m, $T \sim 500°C$, $\Delta T = 100°C$:

$\pi/\beta \sim 1$ m, $\sigma_b \sim 95$ MPa, $\sigma_h \sim 160$ MPa

FIGURE 9.5 Free-level fluctuations: parameters.

Furthermore, one can expect that the temperature gradient can vary and move along the shell (at free level: due to the thermal expansion of sodium associated with the transient and/or variation of the pump speed depending on the trip considered, or in the hypothesis of stratification, by evolution of the level) as depicted in Figure 9.5.

9.2.4 CELLULAR CONVECTION

Natural convection of gases/liquids in vertical enclosures is a normal phenomenon, where the fluid absorbs heat from hot wall during its upward travel and deposits the heat on the cold wall during its downward travel. During these upward and downward movements, boundary layers develop over the hot and cold vertical walls. When the width of the annular space (e) reduces or the height of the enclosure (H) increases (i.e., as the aspect ratio, e/H reduces), the hot wall and cold wall boundary layers approach each other. At some specific value of aspect ratio that depends upon the temperature for the given geometry, both the boundary layers would interact. This interaction is the condition for onset for the breakdown of symmetry in the flow structure leading to asymmetric convection. This phenomenon is termed cellular convection [9.1]. For example, the top shield penetrations face this phenomenon, which is depicted in Figure 9.6. In this figure, the hot shell represents the outer shell of IHX and the cold shell represents the roof slab penetration shell, cooled by air. The annular space is filled with argon cover gas. In such conditions, the type of convection pattern formed in the annular space is unsymmetric in nature with multiple cells/loops. The number of convection loops formed in the annulus depends on many factors: temperature difference between the source and sink, width of the annulus, ratio of length to height of the annulus, and cooling conditions on the various boundary surfaces of the annulus. Based on experimental studies performed, the geometric parameters of the cylindrical annuli where the convective flow turns out to be unsymmetric have been arrived at. When the aspect ratio of the annulus is <0.21 (i.e., e/H < 0.21), unsymmetric flow pattern has been observed [9.2]. A more general relationship has been arrived at relating boundary layer thickness (δ) to the gap width as e/δ < 3.

The number of loops of convective flow formed depends on the ratio of circumference (πd) and height of the annulus [9.3]. When this ratio is of the order of two, the number of loops formed is one, that is, one hot leg and one cold leg. When this ratio is of the order of four, the number of loops formed is two. However, a clear dependency relation has not yet been established. The vertical penetration depth of cellular convection depends strongly on e/H and cooling conditions. The convection is weak if e/H is small and top shield cooling is strong.

Due to the asymmetric nature of the cellular convection, the hot argon enters the annulus at one circumferential segment of the penetration. It gets cooled along its vertical and circumferential

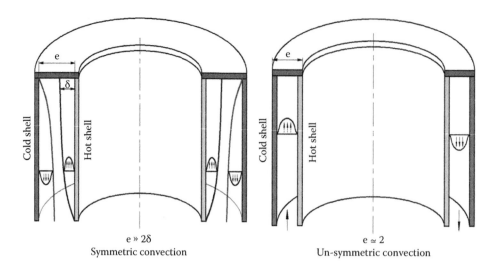

Cold shell   Hot shell   Cold shell   Hot shell

$e \gg 2\delta$ $e \simeq 2$
Symmetric convection   Un-symmetric convection

FIGURE 9.6 Condition for onset of cellular convection.

travel in the annulus and then leaves the annulus in some other annular segment. As a result, the component wall becomes hot in one sector and cold in another. This leads to differential thermal expansion of the structures. As the components of SFR are thin and long structures (~10 m long), it amplifies the tilt and adverse interaction of the components with their mating parts inside the pool, as well as stress. This tilt is very critical in components like control drive mechanisms.

When the number of loops is one, the maximum and minimum temperatures occur along the shell exactly at the opposite meridian lines, under which the component tilts without change of circularity (insignificant stresses). In the design, maximum allowable tilting is specified for smooth functioning of mechanical seals in IHX and pump–pipe connections in the spherical header. When the number of loops is more, there would be multiple maxima and minima along the circumference, resulting in changes of circular shape (ovality), without causing significant titling. However, the resulting ovality causes high hoop stresses, which are to be limited from thermal fatigue considerations. For very narrow annulus, possible in the case of machined penetrations, flow through annulus is restricted and hence a stagnant region would be developed at the upper part of the annulus. Hence, the convective flow pattern would be restricted at the lower part of the annulus, which does not have any significance. Hence, machined penetrations are the preferred solution.

9.2.5 Locations of Cellular Convection

The top shield forms a cover for the primary sodium pool with argon gas between its bottom and free level of sodium. Outside of the top shield is in the ambient environment of the reactor containment building. The top shield has many penetrations for components IHX, DHX, sodium pimps, control plug, rotating plugs, etc. These penetrations form narrow vertical cylindrical annuli that open at the bottom. The temperature of sodium pool is ~820 K during nominal condition, and the top shield is cooled to maintain its temperature at ~120°C (in the reactor adopting hot roof concept) and at ~50°C (in reactors adopting cold roof concept). The gap width of the annular penetrations has to be as narrow as possible to have compact top shield of small diameter and to reduce the heat load on the top shield. The height of the annular penetrations is typically 1.5–2.0 m. This large height is required from the considerations of biological shielding offered by the concrete that fills the box-type top shield. As a result, the bulk temperature of argon cover gas attains a value in between that of sodium and top shield. The gap width of the annular penetrations is very small (~10–25 mm) compared to the height (~2 m); hence, the natural convection of argon taking place in the annuli is asymmetric, as shown in Figure 9.7, because the hot argon enters the annulus at one circumferential

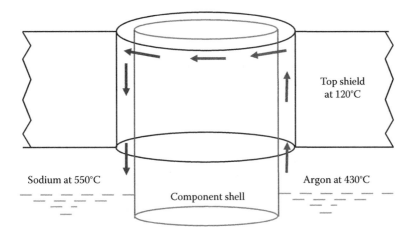

FIGURE 9.7 Direction of argon flow during asymmetric convection.

location and leaves the gap at the opposite side. It is worth mentioning that if the gap would have been wide, both upward and downward flows occur within the annular gap, resulting in symmetric flow pattern over the entire circumference.

9.2.6 FLUID-ELASTIC INSTABILITY OF THIN SHELLS

The main vessel that contains a large quantity of radioactive primary sodium and supports the reactor core is a critical structure. In order to enhance its structural reliability, apart from choosing high ductile and high strength material (typically SS 316 LN), the temperature is kept below creep regime, that is, <700 K during normal operation by means of weir cooling circuit (Figure 9.8). A specified fraction of cold sodium, leaking from the core inlet plenum, namely, grid plate,

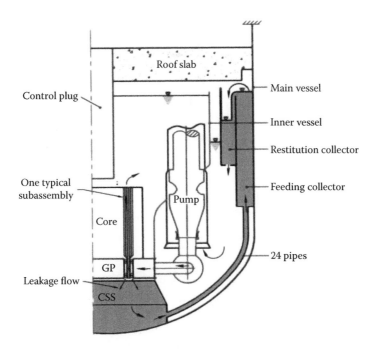

FIGURE 9.8 Main vessel weir cooling system.

flows through the annulus between the main vessel and weir shell to remove the heat transmitted from the hot pool and thereby maintain the main vessel temperature below 700 K. Under some critical combinations of overflow rate and fall height, the weir shell undergoes fluid-elastic instability. Such vibrations were noticed during hot commissioning of SPX, the French fast breeder reactor (FBR) with 1200 MWe capacity. The mechanism of fluid-elastic instability phenomenon is explained later.

Among many vibration modes of FBR shells, the vibration mode that has the lowest natural frequency corresponds to a circumferential wave number (n) lying in the range of 4–10 Hz. Accordingly, if the weir shell is disturbed from its equilibrium position, it naturally vibrates with a particular wave number n. For the stable flow configuration, the vibration decays exponentially to zero. On the contrary, the shell vibrates with exponentially increasing amplitudes for the unstable system. If the dynamic fluid forces causing vibration are developed from the shell displacements, then the resulting unstable vibration is termed as fluid-elastic instability. The fluid-elastic instability that affects the weir shell is mainly due to the sloshing of the liquid free levels that are associated with the feeding and restitution collectors. The mechanism is illustrated schematically in Figure 9.9. Let us assume that the weir shell is perturbed to vibrate with one of the natural frequencies of wave number n, that is, the shell moves radially inward and outward alternatively over the circumference. On those sectors corresponding to feeding collector, where the shell moves outward, the liquid is accelerated in the upward direction due to the dynamic pressure developed on the outer surface of the weir shell. Hence, the free level rises above the mean level, resulting in an increased overflow rate. At the same time, on the alternative sectors, the liquid level falls below the mean level, resulting in either decreased or no flow condition. If we look at the corresponding situations on the restitution collector, while the weir shell moves outward, the free level falls below its mean level and the level raises when the weir moves inward. At any circumferential location, the level difference between the free surfaces of feeding and restitution collectors is the instantaneous fall height. The delay time depends upon the overflow rate and associated fall height. Thus, the flow rate, fall height, and delay time have Azimuthal

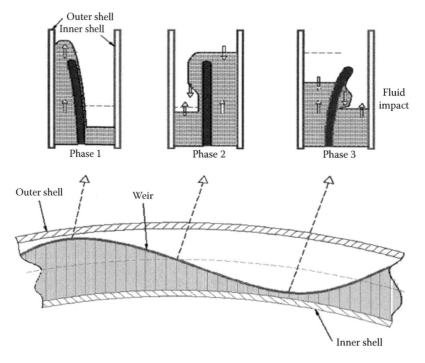

FIGURE 9.9 Scheme of weir instability phenomenon.

and temporal variations. This is one important aspect responsible for the instability. If the kinetic energy, imparted by the liquid falling from the weir crest on the free surface of the restitution collector, is greater than the energy dissipated due to structural and fluid damping, then any perturbation introduced on the shell will develop to cause instability. For the given weir flow configuration, this situation arises under some critical combinations of flow rate and fall height. The critical combinations of overflow rate and fall height that initiate the weir shell instability are plotted in the form of stability charts.

9.2.7 BUCKLING OF THIN SHELLS

It is well known that large-diameter thin-walled shell structures are prone to buckle under compressive stresses. In SFRs, pool-type concept has such vessels—main vessel, inner vessel, thermal baffles, and safety vessel—which have to be checked against this failure mode (Figure 9.8). Further, box-type structures such as top shield, grid plate, and core support structure constituted by plates have buckling risks (Figure 9.8). For these vessels/plates, generally lower wall thickness is selected due to (1) low design pressure at the operating temperatures (sodium coolant can remain in liquid state up to about 1160 K), (2) requirement of adequate flexibility under various operating conditions (820 K for hot pool and 670 K for cold pool under steady state, isothermal temperature of 500 K during shutdown and thermal transient of about 15–20 K/s during reactor SCRAM), and (3) economic considerations since the capital cost is directly linked to the steel consumption. The diameter to thickness ratio (D/h ratio) of the main vessel of a few international reactors is presented in Table 9.1. The table indicates that the D/h ratios lie in the range of 430–875; thus, they are considered as slender. Loads such as self-weight, sodium pressure head, temperature, and thermal gradients are generally not critical under normal operating conditions. However, the forces generated under seismic events impose important loadings to the structures to cause buckling. The reasons for this are elaborated in the following text.

The vessels—main vessel, inner vessel, thermal baffles, and safety vessel—are generally overhanging with *cantilever-type* support arrangements as shown in Figure 9.10. For example, let us consider a top-supported main vessel, carrying heavy liquid mass and dead load due to core and other internal structural mass (1150 tons of sodium with 700 tons of other load acting at the vessel bottom for prototype fast breeder reactor [PFBR]). This vessel would have natural frequency lying in the range of 5–10 Hz where seismic force amplification is generally the maximum. Further, there exists a relatively thin annulus of liquid between (1) inner vessel to inner baffle, (2) inner baffle to outer baffle, (3) outer baffle to main vessel, and (4) main vessel to safety vessel under the event of main vessel leak. The radial annulus gap to diameter ratio lies around ~1/100, and the sodium confined in this space enhances added mass to the adjoining shells to reduce the natural frequencies as well as generates high dynamic pressure during seismic events. The existence of large free fluid surfaces is the source of sloshing, which generates significant convective masses and forces under seismic events. Apart from seismic forces, stationary as well as cyclic loads do cause buckling that are of different kinds in these shells. The nature of buckling of thin shells of SFR can be broadly classified as follows:

TABLE 9.1
Dimensions of Main Vessel and Primary Sodium Masses of International SFRs

Parameter	Phenix	PFR	BN-600	SPX-1	PFBR	EFR
Diameter (m)	11.85	12.25	12.92	21	12.9	17.2
Thickness (mm)	15	25	30	24	25	35
Diameter/thickness	790	490	431	875	516	491
Primary Na mass (tons)	800	850	770	3200	1150	2200

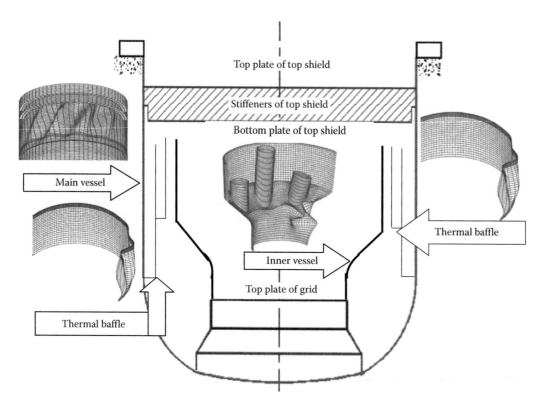

FIGURE 9.10 Thin shells/plates in SFR, having risk of buckling.

- *Shear buckling*: Caused by shear stresses in plates and shells
- *Shell mode buckling*: Constituted by circumferential waves caused by compressive membrane stress due to axial and bending stress in shells
- *Localized shell mode buckling*: Caused by local axial membrane compressive stress caused by axial thermal stress gradients, in shells near sodium free level
- *Progressive buckling*: Caused by combined steady axial stress of mechanical origin and cyclic variation of axial thermal stresses in shells
- *Creep buckling*: Caused by creep deformation of shells operating at high temperature
- *Bowing mode*: Created on plates subjected to compressive membrane stress

The structural portions and associated buckling modes are presented in Table 9.2.

9.3 CODES AND STANDARDS

Design codes provide rules/criteria/empirical correlations to protect the components against certain generic failure modes. Codes and standards (C&S) are necessary for regulating the design, construction, and operation of nuclear power plants in general. The C&S are essential to ensure normal operation, avert accidents, and limit the effects of an accident. The availability of these C&S is also a basic condition to institute clear and satisfactory relationships between regulators, designers, manufacturers, and operators. The codes described in this chapter are restricted to the mechanical design of nuclear steam supply system (NSSS) components. Accordingly, applicable failure modes are addressed: gross yielding, tensile rupture, fatigue damage and buckling (time independent and temperature dependent) and creep strain, creep rupture, creep buckling, and creep and fatigue damage (time and temperature dependent). Since the nuclear components are subjected to severe thermomechanical loadings, the design code addresses the possible failure modes due

TABLE 9.2
Important Buckling Modes for Various Thin Shells

Governing Loads	Buckling Mode
Main vessel—cylindrical portion	
Shear forces generated under seismic excitations	Shear mode
Cyclic axial temperature distribution in the vicinity of free level due to variation of levels	Progressive buckling
Main vessel—bottom dished head	
Combined effects of hydrostatic pressure of sodium and axial force acting at triple point under normal and seismic loading conditions	Shell mode
Inner vessel—torus/redan portion	
Combined mechanical and axial and through-wall temperature gradients	Creep buckling with s hell mode
Inner vessel—upper cylindrical portion	
Cyclic axial temperature distribution in the vicinity of free level due to variation of levels	Progressive buckling
Thermal baffles—top cylindrical portion	
Cyclic axial temperature distribution in the vicinity of free level due to variation of levels	Progressive buckling
Thermal baffles—bottom portion above the support	
Combined effects of hydrostatic pressure of sodium under normal and seismic loading conditions	Shell mode
Grid plate—top plate	
Compressive stress developed under hot shock	Bowing
Roof slab—top plate	
Compressive stress developed due to self-weight, weight of components mounted on it, plus inertial forces generated under seismic events	Bowing
Roof slab—stiffener plate	
Shear force transmitted by stiffeners under normal and seismic loading conditions	Shear buckling
Roof slab—bottom plate	
Sodium slug impact pressure plus compressive stress developed under core disruptive accident	Bowing (concave upward)

to static as well as cyclic temperature gradients. Accordingly, the *design by analysis* approach is adopted. The analysis calls for the determination of detailed stresses due to mechanical loadings to derive primary stresses and thermal gradients to derive secondary stresses and strains. The code provides appropriate permissible stresses separately for the primary and secondary stresses. While primary stress limits ensure adequate wall thickness, the secondary stress limits protect against deformation-related failure modes accumulated over the plant life.

9.3.1 CIVIL STRUCTURES

Civil structures include reactor containment, SG, turbine, fuel, electrical, diesel generator, control and service buildings, and other related steel structures. These are classified into the following categories as per the American Society of Civil Engineers (ASCE) (1998):

- DC1   Pressurized concrete reactor vessels (not applicable to PFBR)
- DC2   Containment structures; referred further as class 2
- DC3   Internal structures of reactor building, auxiliary and safety-related balance of plant (BOP) building and structures; referred further as class 3
- DC4   Nonsafety class structures; referred further as nonnuclear service (NNS)

The following are a few important design and construction codes:

- IS-456   *Code of practice for plain and reinforced concrete for general building construction* for NNS structures
- IS-1893   *Criteria for earthquake-resistant design of structures*
- ANSI/AISC-690   *Nuclear facilities—safety-related structures for design, fabrication, and erection of structural steel for buildings* for steel structures of class 2 and 3 categories
- IS-800   *Code of practice for use of structural steel in general building construction* for NNS steel structures
- IS-875   Code of practice for design loads (other than seismic) for buildings and structures (part 1), imposed loads (part 2), wind loads (part 3), and special loads (part 5)
- AERB/SS/CSE-1   Design of concrete structures important to safety
- AERB/SS/CSE-2   Design, fabrication, and erection of steel structures important to safety
- AERB/SS/CSE   Civil structures important to safety
- ASCE: 4-98   Seismic analysis of nuclear safety–related structures

9.3.2 Mechanical Components: NSSS

Mechanical NSSS components basically constitute of vessels, tanks, roof slab, rotatable plugs, pumps, piping, heat exchangers, valves, vessel supports, etc. Depending upon the safety classification of components, appropriate codes, standards and guides (CSG) are defined in this section. Several C&S that have been developed for nuclear reactor design in various countries are used for the design of their FBR components. ASME Boiler and Pressure Vessel Code, Section III Division 1 for the design of pressure vessel in the temperature range <700 K for austenitic stainless steel and 650 K for carbon and ferritic steels, is the basic code. For the design of high-temperature components, Section III code rules are not directly applicable because it does not consider the failure modes due to creep. For the high-temperature applications, ASME design and construction are given in ASME Code Section III, Subsection NH (1998) [9.4]. In addition, French code RCC-MR is available exclusively for FBR applications [9.5]. For PFBR, the choice is between RCC-MR and ASME Code Subsection NH. RCC-MR (1993) has been selected as the design and construction codes for the following reasons:

- RCC-MR is based on the French experiences of design, construction, and operation of Rapsodie, Phenix, SPX, and design of European fast reactor (EFR).
- RCC-MR is completely relevant to the FBR components. It takes into account the special features of the components in geometric and/or loading conditions, that is, thin-walled structures, low pressure, and high thermal loading.
- Rules for ratcheting, buckling, and creep effects are more appropriate for FBR situations.
- Design rules are given for absorber rod drive mechanisms and fuel handling systems, high-temperature design procedures for flanges and heat exchangers, and fatigue damage assessment for the components having crack-like defects.
- RCC-MR provides material specifications, design data, and manufacturing requirement, for principal materials selected for PFBR. In addition, material characteristics are fully given in Appendix A3, including cyclic stress strain curves, weld strength reduction factors, and special technical appendices for the simplified methods.

However, provisions of ASME codes would be employed for limited situations for which the rules are not yet defined in RCC-MR. Nevertheless, it is admissible to use the corresponding ASME code for class 2 and 3 nonsodium components. For the aspects not covered in both the codes, interim rules followed in other reactors are adopted, which are described later.

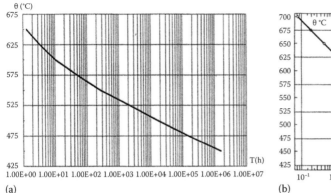
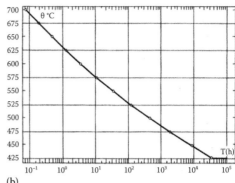

(a)   (b)

FIGURE 9.11 Creep crossover curves for (a) SS 316 LN and (b) SS 304 LN. (From RCC-MR, Design and construction rules for mechanical components for FBR nuclear islands, Section I, Subsection B, Class 1 Components, AFCEN, 1993.)

It is interesting to note that although ASME and RCC-MR approaches are nearly similar in several aspects, they differ very much specifically in the following two main aspects.

9.3.2.1 Definition of Significant Creep and Negligible Creep Regimes

In ASME, the time-dependent failure modes should be considered once temperature exceeds a specified value for the particular materials: 700 K for austenitic stainless steels, for example. In RCC-MR, the significant creep zones are indicated as a function of time and temperatures. Such chart is called *creep crossover curve*. The creep crossover curve for the austenitic stainless SS 316 LN as well as for SS 304 LN is shown in Figure 9.11.

9.3.2.2 Effects of Cyclic Secondary Stresses in Association with Primary Stresses

It is well known that depending upon the magnitudes of primary and secondary stress values, the structures exhibit several complex deformation modes reflecting elastic cycling, shakedown, plastic cycling, and ratcheting phenomena. In ASME, the combinations of primary stress and cyclic secondary stresses and the associated deformation modes are analytically derived and presented in the form of charts: Bree's diagram at noncreep regimes and *O'Donnel & Porowski diagram* at significant creep regimes. In RCC-MR, the effects of cyclic secondary stresses on primary stresses are consolidated in a diagram called *efficiency index diagram*. While the ASME introduces the *core stress concept* through the O'Donnel & Porowski diagram, RCC-MR brings the concept of effective primary stress. Details can be seen in Ref. 9.5.

Provisions of ASME codes are employed for some aspects for which the rules are not yet defined in RCC-MR. Nevertheless, it is admissible to use the corresponding ASME code for class 2 and 3 nonsodium components. For the aspects not covered both in RCC-MR and in ASME, certain ad hoc rules given in the literature are employed. The summary of a few important rules is presented next.

9.4 DESIGN CRITERIA FOR THE ASPECTS NOT COVERED IN RCC-MR/ASME

For the aspects not covered either in RCC-MR or in ASME—effects of sodium, radiation damage, thermal ageing, corrosion, weld strength reduction factors for modified 9Cr–1Mo, design rules for thermal striping and core disruptive accident, and leak before break procedure for sodium boundary components—certain ad hoc rules given in the literature are used. The summary of these rules is as follows.

9.4.1 SEISMIC DESIGN CRITERIA

In addition to satisfying the various codal limits on the stresses/strains, for the earthquake loading, the components should perform satisfactorily during and/or after the earthquake. To assure that, some additional requirements based on their functions are imposed on certain components whenever required. The general functional requirements after operating basis earthquake (S1) are that no damage is tolerated in parts that cannot be inspected or easily repaired. Similarly, the general functional requirement after safe shutdown earthquake (S2) is (1) there should not be any major leaks in active circuits and (2) there should not be any large sodium fire. In addition, there are specific requirements on seismic response parameters as given next:

- Relative vertical displacement between the control rod gripper and fuel SAs top is limited to 12 mm corresponding to 0.5$ reactivity addition.
- Relative horizontal displacement between the core top and control and safety rod drive mechanism (CSRDM) gripper is limited to 25 mm in order to ensure the insertability of CSRDM into the core after accounting for the possible deformations due to the effects of temperature, ratcheting, ageing, radiation, etc.
- Vertical seismic acceleration at the grid plate level has to be limited to 0.9 g to avoid the lifting of the fuel SAs.
- Clearance in the hydrostatic bearing has to be maintained in order to maintain the fluid film in the hydrostatic bearing and hence to avoid pump bearing seizure.
- Relative displacement between the stand pipe top and IHX/pump is limited to 50 mm to avoid mechanical interactions and possible structural damages.

9.4.2 EFFECT OF SODIUM

The effects of reactor quality sodium on the mechanical properties of SS 316 LN, SS 304 LN, and modified 9Cr–1Mo (T91) steel base metal are negligible. Hence, the material data generated in air could be safely applied for the design of components with wall thickness larger than 2 mm. The influence of corrosion, carburization, decarburization, formation of ferritic substructure layers, and sensitization is considered for thin-walled components (t < 2 mm). For the component parts having thickness <2 mm, for example, IHX tubes having thickness of 0.8 mm, the loss of thickness due to corrosion may be significant. Hence, extra thickness is provided to consider the effects of corrosion. The loss of thickness due to sodium corrosion is quantified for PFBR in Ref. 9.6. Similar exercise to consider the loss of thickness due to sodium corrosion is carried out for SG tubes (nominal wall thickness is ~2.3 mm) in the thickness selection [9.7].

9.4.3 EFFECT OF LOW-DOSE IRRADIATION

The major effect of neutron irradiation is loss of ductility at lower temperatures. Literature compiling views of experts indicates that the neutron dose level <1 dpa is negligible for the cold pool components [9.8]. Specifically, the grid plate has to be confirmed for the effects of irradiation. If this dose is exceeded, the axial shielding may have to be increased in core SAs. Recent literature indicates that the limit of 1 dpa is highly pessimistic and even up to 4 dpa can be acceptable. However, more irradiation data are required to employ this limit in the design.

At higher temperatures, neutron dose affects indirectly. For the high-temperature components specifically, above core structural parts, the thermal neutron irradiation produces helium, mainly by the reaction with boron present in the steel at high temperature. Helium affects creep rupture strength and ductility of austenitic stainless in the creep regime. The interim recommendation for design activities is to apply a stress reduction factor that varies as a function of helium content as indicated in Figure 9.12 [9.8].

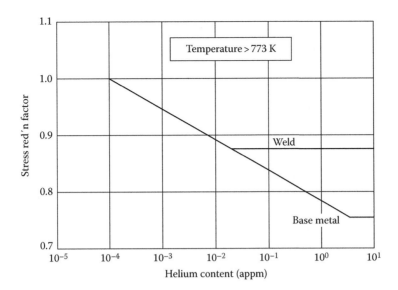

FIGURE 9.12 Effect of helium on creep rupture strength for SS 316 LN and SS 304 LN. (From Escaravage, C. et al., EFR-DRC proposal to introduce in design work in low dose neutron irradiation effects, *Proceedings of the IAEA Specialists Meeting on Influence of Low Dose Irradiation on the Design Criteria of Fixed Internals in Fast Reactor*, IAEA-TECDOC-817, 1995.)

9.4.4 EFFECT OF HIGH-DOSE IRRADIATION

The core components that are subjected to high dose of neutron irradiation (>5 dpa), a set of design rules with associated set of mechanical properties has been formulated based on limited information available on the practice followed in other countries and documented [9.9]. The salient features of recommendations are given next.

The reduction of ductility of the material due to neutron irradiation has the maximum impact on the design rules. Irradiation-induced creep and swelling of material should also be included in the analysis for deformation of components. Two types of analyses can be followed: elastic/in-pile creep/swelling analysis (EICSA) or inelastic analysis (IEA). For EICSA, limits are stipulated on stresses while for IEA the limits are on strains. The failure mechanisms and associated design rules for EICSA and IEA are given in Tables 9.3 and 9.4, respectively [9.10].

9.5 THERMAL HYDRAULIC DESIGN CRITERIA

As already mentioned, for the structural design, various design codes are adopted. These codes specify the allowable stress limits. However, such limits are not explicitly available for thermal hydraulic parameters. But the thermal hydraulic parameters form input for estimation of stress values in the structures. The following are important thermal hydraulic limits established for the 500 MWe pool-type PFBR. Since some of these limits are design specific, such limits for other reactor designs need to be worked out, though the limits presented later can be taken as a basic guideline.

9.5.1 DESIGN RULES FOR THERMAL STRIPING

The thermal cycles are caused by a special phenomenon called *thermal striping* that occurs mainly at the core cover plate of control plug and also at mixing *Tee* junctions in the sodium pipelines. The frequency of fluctuation ranges from 0.01 to 10 Hz, which yields cumulative cycles in the order of 1.26×10^9 for a typical plant, designed for 4 years with 75% load factor (LF). These temperature

TABLE 9.3

Failure Mechanisms and Associated Design Rules for EICSA

Failure Mode	Governing Parameter	Critical Value	Factor of Safety	Design Rule
Tensile instability	P_m	S_y or S_u	1.18 and 1.7 for categories 1 and 2; 0.9 and 1.25 for category 3	$P_m < 0.85 S_y$ or $< 0.6 S_u$ for categories 1 and 2; $P_m < 1.1 S_y$ or $< 0.8 S_u$ for category 3
Failure in bending	$P_m + P_b$	$K S_y$ or S_u	1.18 and 1.7 for categories 1 and 2; 0.9 and 1.25 for category 3	$P_m < 0.85 S_y$ or $< 0.6 S_u$ for categories 1 and 2; $P_m < 1.1 S_y$ or $< 0.8 S_u$ for category 3
Ratcheting	$P_m + P_b + Q$	Shake down limits or S_u	1.7 on S_u for categories 1 and 2; 1.25 on S_u for category 3	$P_m + P_b + Q < 0.6 S_u$ for categories 1 and 2; $P_m + P_b + Q < 0.8 S_u$ for category 3 loads
Localized rupture	$P_m + P_b + Q + F$	S_u	1.25 for categories 1 and 2; 1.1 for category 3	$P_m + P_b + Q + F < 0.8 S_u$ for categories 1 and 2 loads; $P_m + P_b + Q + F < 0.9 S_u$ for category 3
Damage due to creep and fatigue	Time duration, Δt and no. of cycles, n	Rupture life t_d and fatigue limit N	4 on CDF for categories 1 and 2; 1.3 for categories 1, 2, and 3	$\Sigma \Delta t/t_d + \Sigma n/N < 0.25$ for category 1; < 0.5 for categories 1 and 2; < 0.75 for categories 1, 2, and 3
Brittle fracture	K_I	K_{IC}	—	$K_I < K_{IC}$

Note: Factors of safety given for damage due to creep and fatigue are on cumulative damage fraction (CDF).

TABLE 9.4

Failure Mechanisms and Associated Design Rules for IEA

Failure Mode	Governing Parameter	Critical Value	Factor of Safety	Design Rule
Tensile instability and ratcheting	ε_m^p	$\varepsilon_u/2$	3 for categories 1 and 2; 1.5 for categories 1, 2, and 3	$\sum \varepsilon_m^p/(\varepsilon_u/2) < 0.33$ for categories 1 and 2; $\sum \varepsilon_m^p/(\varepsilon_u/2) < 0.66$ for categories 1, 2, and 3
Localized rupture	ε_t^p	ε_f/TF	2 for categories 1 and 2; 1.25 for categories 1, 2, and 3	$\sum \varepsilon_t^p/(\varepsilon_f/TF) < 0.5$ for categories 1 and 2; $\sum \varepsilon_t^p/(\varepsilon_f/TF) < 0.8$ for categories 1, 2, and 3
Damage due to creep and fatigue	Time duration, Δt and no. of cycles, n	Rupture life t_d and fatigue limit N	4 on CDF for category 1; 2 for categories 1 and 2; 1.3 for categories 1, 2, and 3	$\Sigma \Delta t/t_d + \Sigma n/N < 0.25$ for category 1; < 0.5 for categories 1 and 2; < 0.75 for categories 1, 2, and 3
Brittle fracture	J-integral	J_c	—	$J < J_c$

fluctuations cause HCF on the component wall. There are a few critical locations particularly in the hot pool where both LCF and HCF cycles are simultaneously imposed on the metal wall for which matured design rules are not available to ensure the structural integrity. The damage becomes very severe for the welded structures particularly with defects even though the defects are within acceptable limits of the applicable inspection codes.

With an objective of establishing thermal striping limits for PFBR, a fracture mechanics-based method has been derived taking into account the frequency-dependent through-wall stress decay. The UK fatigue design procedure is used to obtain the crack size from the cumulative creep–fatigue damage. Based on the method, thermal striping limits have been derived as a function of accumulated creep–fatigue damage and frequency. Thermal striping limits are more stringent for lower frequencies and larger creep–fatigue damages. Since the frequency spectrum is not available, the limits are derived conservatively based on 0.01 Hz, the lowermost frequency possible. Furthermore, the limits are found to be independent of structural wall thickness for thickness >5 mm (5 mm case has been interpreted from the analysis for 30 and 10 mm thickness).

Based on investigations of thermal striping failures that occurred in the operating reactors—Phenix, SPX, and BN-600—as well as the simulated test results from international test facilities, SUPERSOMITE and FAENA, a factor of 1.2 has been applied [9.10]. Finally, a design chart is recommended that specifies the thermal striping limits for PFBR as a function of creep–fatigue damage [9.11]. The following are some important limits extracted from the design chart:

- For the core cover plate of control plug wherein the LCF damage is negligible (~0.0), the acceptable thermal striping limit is 60 K. For the lower portion wherein the damage is ~0.2, the acceptable limit is 50 K.
- For the locations on the inner vessel or main vessel wherein the accumulated fatigue damage may be moderate (0.5), the acceptable thermal striping limit is 40 K.
- Acceptable amplitude of oscillation of stratified sodium layers on the inner vessel redan surface is ±270 mm.
- Acceptable amplitude of free-level oscillation on upper shell of inner vessel is ±55 mm.

9.5.2 Temperature Asymmetry in Cold Pool

During plant transients associated with one secondary loop, temperature in the cold pool associated with the affected loop becomes hotter or colder than that of the unaffected loop. This leads to circumferential temperature difference in cold pool structures, namely, main vessel and grid plate. Flow coast-down characteristics of secondary sodium pump and cold pool capacity need to be arrived at such that the circumferential temperature difference (temperature asymmetry) is <30 K in the structures. Similarly, sodium flow in the main vessel cooling system is to be distributed to respect this temperature asymmetry limit.

9.5.3 Free-Level Fluctuation

Due to the large surface area (>100 m²) interfacing with argon, hot pool free surface oscillates. The height of inner vessel above the mean hot pool sodium level should be adequate to avoid overflow of hot pool sodium to cold pool. The amplitude of fluctuation has to be minimum so that the height of inner vessel can be minimum. Nominal temperature of hot pool is 547°C and that of argon cover gas is 430°C. When the level fluctuates, structures partly immersed in hot pool and partly exposed to cover gas see alternating temperatures. From HCF considerations, the amplitude of fluctuations has to be <50 mm.

9.5.4 Free Surface Velocity in Pool

To avoid stratification risk in the pool, sodium velocity in the bottom part of the pool has to be high. At the same time, velocity at the free surface has to be limited to 0.5 m/s to avoid gas entrainment. Entrainment of gas has to be limited to the extent that it does not cause any reactivity changes in the core. Continuous limited flow of argon through the core is not any serious concern. But segregation of minute bubbles in grid plate to agglomerate into a larger bubble and its sudden movement into the core are to be avoided.

9.5.5 High-Cycle Temperature Fluctuation

The extent of stratification in hot pool is a function of ratio of buoyancy forces to inertial forces. The stronger the inertial forces, the lower the axial temperature gradient will be. The stratification interface normally oscillates. Based on detailed structural mechanics calculations, it has been established that the gradient has to be limited to <300 K/m. From thermal striping considerations, the peak-to-peak temperature fluctuations on structures have to be limited to 60 K for control plug and 40 K for main vessel and inner vessel.

9.5.6 Heat Loss to Top Shield

The heat loss to top shield has to be minimum to minimize the heat load on the top shield cooling circuit. By the provision of thermal shield, direct radiation heat load and argon convection heat load are reduced by ~50%. But this, in turn, increases the bulk temperature of argon cover gas. Increase in bulk temperature will enhance cellular convection and the associated temperature asymmetry. The bulk cover gas temperature will also affect axial temperature gradient in the structures. Cellular convection of argon in narrow gap penetrations of top shield has to be managed to have temperature asymmetry <30 K. Annular gap size and cooling conditions of top shield are to be optimized to respect this limit.

REFERENCES

9.1 Goldstein, S. et al. (1979). Thermal analysis of the penetrations of a LMFBR. *Fifth International Conference on Structural Mechanics in Reactor Technology*, Berlin, Germany.

9.2 Timo, D.P. (1954). *Free Convection in Narrow Vertical Sodium Annuli*. Knolls Atomic Power Laboratory Report: 1082, Schenectady, NY.

9.3 Mejane, H., Durin, M. (1982). Natural convection in an open annular slot. *Proceedings of the Seventh International Heat Transfer Conference*, Meichen, Germany.

9.4 ASME. (1998). Class 1 Components in elevated temperature service. Section III, Div 1, Subsection NH, ASME Press.

9.5 RCC-MR. (1993). Design and construction rules for mechanical components for FBR nuclear islands. Section I, Subsection B, Class 1 Components, AFCEN.

9.6 Rajendran Pillai, S. et al. (1999). High temperature corrosion of clad and structural materials in sodium. IGCAR internal Report: PFBR/MCG/CSTD/Dec.

9.7 Rajendran Pillai, S. et al. (December 1999). Corrosion loss of SG tubes of modified 9Cr–1Mo steel. IGCAR internal Report: PFBR/MCG/CSTD/R-2.

9.8 Escaravage, C. et al. (1995). EFR-DRC proposal to introduce in design work in low dose neutron irradiation effects. *Proceedings of the IAEA Specialists Meeting on Influence of Low Dose Irradiation on the Design Criteria of Fixed Internals in Fast Reactor*, IAEA-TECDOC-817, Gif-sur-Yvette, France.

9.9 Govindarajan, S. (2000). Structural design criteria for high dose neutron irradiation. IGCAR internal Report: PFBR/31100/DN/1007.

9.10 Gelineau, O. et al. (1994). Thermal fluctuation problems encountered in LMFBRs. *Specialists Meeting on Correlation between Material Properties and Liquid Metal Cooled Fast Reactors*, AIX-en-Provence, France.

9.11 Chellapandi, P., Chetal, S.C., Raj, B. (2009). Thermal striping limits for components of sodium cooled fast spectrum reactors. *Int. J. Nucl. Eng. Des.*, 239, 2754–2765, August 2009.

10 Design Validations

10.1 INTRODUCTION

In view of the high demand for long and reliable operations with due considerations on economy, several innovative concepts and evolutionary features are incorporated in the current sodium-cooled fast reactor (SFR) designs. Design of SFR components adopts the philosophy of *design by analysis* to comply with the design codes and standards, which themselves are not sufficiently matured, that is, still evolving. Analysis through computer codes calls for several specialized and advanced concepts in the domains of structural mechanics and thermal hydraulics (refer to Chapter 9). Hence, the SFR design has to be thoroughly validated by adequate experimental programs. The computer codes employed in the design are validated normally by solving common benchmark problems and comparing the code to code results. Further, to assess their overall performance, scaled-down or full-scale experiments are conducted. Finally, the components that are designed and manufactured are subjected to various stages of qualification tests before integration with the reactor.

In this section, a list of commonly as well as specially developed computer codes in the domain of structural mechanics and thermal hydraulics are addressed highlighting the validation procedure adopted. A few international benchmark exercises are also presented. Further, a few basic experiments in the field of mechanics that are specific to SFR and large-scale experiments conducted for the qualification of critical components are highlighted. Adequate references are cited for the benefit of advanced reading. Validation of computer codes and safety experiments are addressed in the chapters under safety.

10.2 STRUCTURAL ANALYSIS CODES AND STRUCTURAL DESIGN METHODOLOGY

Since the SFR components are operating at high temperatures and designed for a long reliable operation, the components are designed by analysis. For conventional analysis such as structural design optimization, buckling, and vibration, several commercial codes now available are employed. However, for the investigation of specific failure modes dealt in Chapter 9, dedicated in-house computer codes are developed and validated adopting *code-to-code comparison*, experimental benchmarks, component tests, and evaluation exercises. For illustrating these aspects, a few of the validation exercises undertaken for the structural design of 500 MWe prototype fast breeder reactor (PFBR) are presented in this section. The specific issues addressed are material constitutive models based on viscoplastic theory, thermal ratcheting in thin vessels subjected to moving temperature gradients, creep–fatigue life prediction by RCC-MR design code rules, fluid elastic instability of thin shells, nonlinear vibration of core subassemblies (SAs), parametric instability of thin shells subjected to liquid sloshing under seismic events, and buckling of thin shells representing SFR geometries, namely, main vessel and inner vessel.

10.2.1 Material Constitutive Models Based on Viscoplastic Theory

Under SFR environment, depending upon the values of primary and secondary stress ranges, the material point in the components exhibits several complex behaviors: elastic cycling, shake down, and ratcheting. Hence, inelastic analysis is essential for predicting accurately stress and strain history in the SFR components subjected to complicated mechanical and thermal loadings and thereby

TABLE 10.1

Guidelines to Choose Appropriate Material Models

Collapse Mode → Behavior Model ↓	Excessive Deformation, Plastic Instability	Progressive Deformation	Creep–Fatigue
Perfect plastic + creep rule	Suitable (1)	Avoid	Avoid
Isotropic strain hardening + creep rule	Suitable (2)	Avoid	Avoid
Linear kinematic hardening + creep rule	Avoid	Use with care (3)	Use with care (6)
Combined hardening (Chaboche viscoplastic, etc.)	Suitable (2)	Use with care (4), (5)	Suitable

Notes: (1) Model used mostly for limit analysis; (2) identification with minimum monotonic tensile curves for the material; (3) results may not be conservative; (4) satisfactory results although often too conservative; (5) identification with reduced cyclic curves except where the strain amplitude is small, in which case, the mean monotonic curves should be used; and (6) results may not be conservative if the hold times are on residual stress states.

for the accurate estimation of ratcheting and creep–fatigue damages. This calls for use of realistic material constitutive models to be employed in the computer codes. Such constitutive models should be able to simulate accurately the complex material behavior. The model to be considered would depend on the physical phenomenon and on the collapse mode to be analyzed. Table 10.1 provides some broad guidelines to choose appropriate models.

For the structural analysis of SFR components made of austenitic stainless steels (particularly SS 316 LN) operating at high temperature, a *23-parameter viscoplastic model* proposed by Chaboche and Nouailhas [10.1] is employed. For the ferritic steels (particularly for modified 9Cr–1Mo steel, the structural material for steam generators [SGs]), a *20-parameter model* proposed by Chellapandi and Ramesh [10.2] is employed. The governing equations for these two models are given in Appendix 10.A, along with the values of material parameters identified based on several uniaxial tests. In this main section, the prediction capabilities of these two viscoplastic models are presented.

10.2.1.1 Prediction Capability of Viscoplastic Models

In order to demonstrate the prediction capability of viscoplastic models, five sample problems over a wide range of rate sensitivity domains under complex monotonic and cyclic loadings are solved, of which two are analytic solutions and the other three are practical problems. The geometries of the problems are modeled with eight-noded isoparametric elements with 2 × 2 Gaussian points.

Solutions of the first two academic problems obtained by an in-house computer code, *CONE*, are compared with analytical solutions, and the remaining three are compared with results from EVPCYC [10.3], ABAQUS [10.4], and SYSTUS [10.5] computer codes.

10.2.1.2 Problem No. 1: Biaxial Stress and Cycling

A single element subjected to homogeneous biaxial loading cycle is displayed in Figure 10.1. For the finite element analysis, one eight-noded isoparametric plane stress element is selected. The evolution of viscoplastic strains when subjected to stress cycling (case 1) and strain cycling (case 2) is shown in Figures 10.2 and 10.3, respectively. Numerical simulations are compared with analytical solutions in these figures, which show that the predictions are quite good. The problem has been taken from Ref. 10.6.

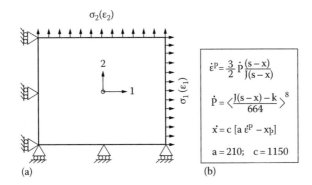

The material equations shown are:

$$\dot{\varepsilon}^P = \frac{3}{2} \dot{p} \frac{(s-x)}{J(s-x)}$$

$$\dot{p} = \left\langle \frac{J(s-x)-k}{664} \right\rangle^8$$

$$\dot{x} = c \, [a \, \dot{\varepsilon}^P - x\dot{p}]$$

$$a = 210; \quad c = 1150$$

FIGURE 10.1 (a) Geometrical and (b) material details of benchmark problem for biaxial tests.

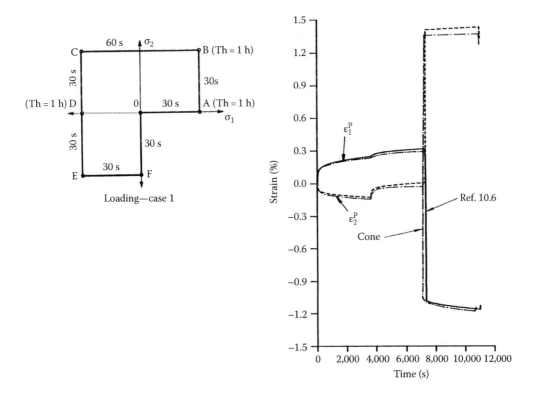

FIGURE 10.2 Evolution of viscoplastic strain for stress cycling (case 1 of problem 1).

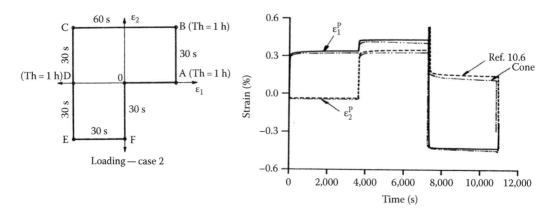

FIGURE 10.3   Evolution of viscoplastic strain for strain cycling (case 2 of problem 1).

10.2.1.3   Problem No. 2: Viscoplastic Behavior of Pressure Vessel

A long thick cylinder under internal pressure is analyzed for long-term viscoplastic behavior (Figure 10.4). The geometry is modeled with 10 isoparametric elements with deformation constraint in the axial directions. This problem is also taken from Ref. 10.6. Predictions of radial and tangential viscoplastic strains extracted at the Gaussian point nearest to the inner surface are compared with the reference solutions depicted in Figure 10.5. Here, again the predictions are excellent.

10.2.1.4   Problem No. 3: Complex Uniaxial Behavior

Prediction of creep deformation under various stress level (case 1) and cyclic stress–strain behavior under a strain-controlled test with various hold times (case 2) for the material stainless steel type 316 LN. The problem is modeled with one eight-noded isoparametric element with appropriate

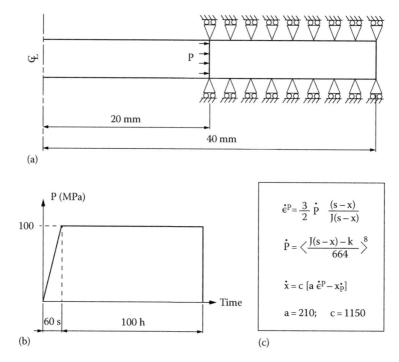

FIGURE 10.4   Details of benchmark problem no. 2 (pressure vessel): (a) geometry, (b) loading, and (c) material model.

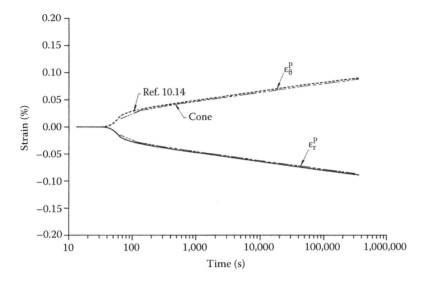

FIGURE 10.5 Evolution of viscoplastic strain for problem no. 2.

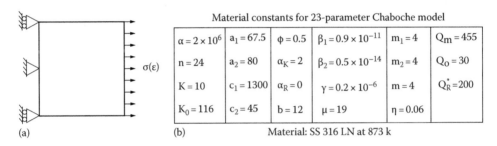

	Material constants for 23-parameter Chaboche model					
$\alpha = 2 \times 10^6$	$a_1 = 67.5$	$\phi = 0.5$	$\beta_1 = 0.9 \times 10^{-11}$	$m_1 = 4$	$Q_m = 455$
$n = 24$	$a_2 = 80$	$\alpha_K = 2$	$\beta_2 = 0.5 \times 10^{-14}$	$m_2 = 4$	$Q_0 = 30$
$K = 10$	$c_1 = 1300$	$\alpha_R = 0$	$\gamma = 0.2 \times 10^{-6}$	$m = 4$	$Q_R^* = 200$
$K_0 = 116$	$c_2 = 45$	$b = 12$	$\mu = 19$	$\eta = 0.06$	

(a) (b) Material: SS 316 LN at 873 k

FIGURE 10.6 (a) Geometrical and (b) material details of benchmark problem no. 3.

boundary conditions imposed to simulate the uniaxial behavior. The material data have been taken from Ref. 10.7. The geometrical and material data are defined in Figure 10.6. The comparisons of creep strains predicted for case 1 are shown in Figure 10.7. In Figure 10.8, the evolution of maximum stress is illustrated as a function of the number of cycles in the strain-controlled test with various hold times. The cyclic stress–strain loops for the first 10 load cycles are depicted in Figure 10.9. In all the cases, the prediction capability of the code is satisfactory.

10.2.1.5 Problem No. 4: Viscoplastic Behavior of Circular Notch

Viscoplastic analysis of a round bar with a circular notch is subjected to axial tension with small deformation assumption. The problem is taken from Ref. 10.8. The details of geometry and material data and loadings are illustrated in Figure 10.10. Results by viscoplastic analysis are compared with those obtained from analysis with the ABAQUS code. For the current analysis, 180 eight-noded axisymmetric isoparametric elements are used to model the one-fourth symmetric portion of the section with appropriate boundary conditions at the lines of symmetricity. The original and deformed geometries are shown in Figure 10.11. The distribution of radial, axial, and hoop strains and stresses at the minimum section of the notched specimens are plotted in Figures 10.12 and 10.13, respectively, along with the respective prediction by the ABAQUS code.

10.2.1.6 Simulation of Thermal Ratcheting in Thin Shells

The benchmark experiment called *VINIL* conducted at CEA, Cadarache [10.9], is analyzed for demonstrating the adequacy of the *23-parameter Chaboche viscoplastic model*. VINIL deals

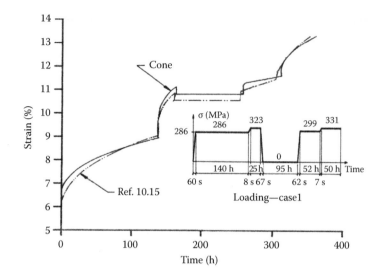

FIGURE 10.7 Evolution of viscoplastic strain for problem no. 3 (case 1).

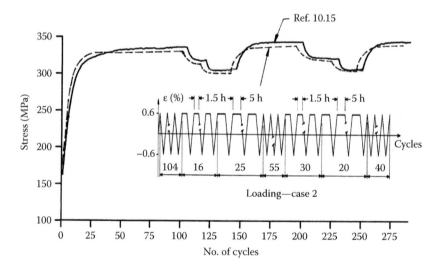

FIGURE 10.8 Evolution of peak stress for problem no. 3 (case 2).

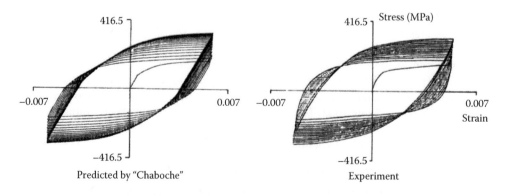

FIGURE 10.9 Numerical simulation of cyclic stress–strain behavior of SS 316 LN.

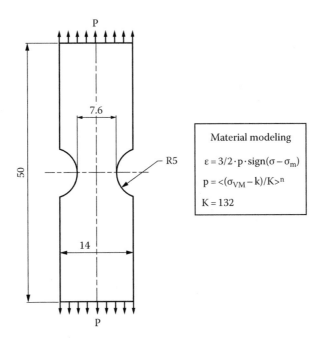

FIGURE 10.10   Details of benchmark problem no. 4 (circular notch).

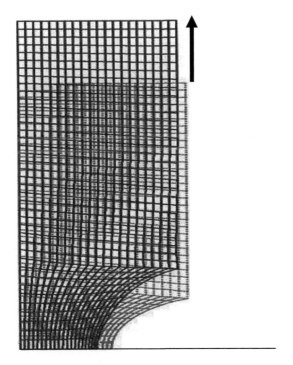

FIGURE 10.11   Original and deformed finite element mesh.

with the simulation of ratcheting in a cylindrical shell of 800 mm diameter, 1.2 mm thick and 400 mm length, made of SS 316 LN, as illustrated in Figure 10.14. The sodium level and associated temperature variations are shown in Figure 10.15. It is seen that the initial sodium free level is at mid-elevation and the temperature of sodium is 200°C and the argon space above is also at 200°C. Sodium temperature is raised to 620°C within 40 min without changing the elevation of

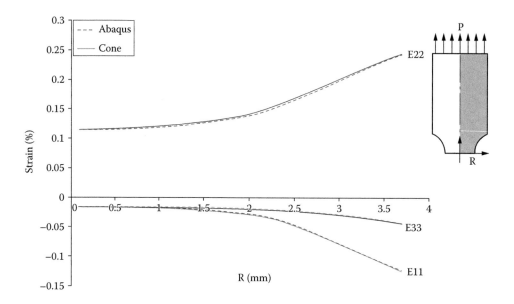

FIGURE 10.12 Strain distribution along the central section.

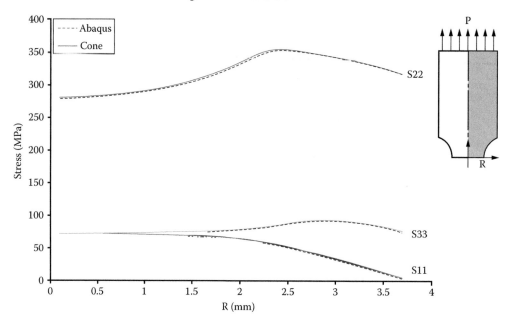

FIGURE 10.13 Stress distribution along the central section.

sodium free level. Subsequently, the vessel is lowered by 35 mm to simulate the raising of free level within 15 s and kept for 50 min at the same elevation where creep occurs. Subsequently, the vessel is raised to its original elevation within 15 s to simulate the lowering of free level. Finally, the sodium is cooled slowly to 200°C to bring back to its initial condition. The argon is kept at 200°C during the entire sequences of loading. The time variations of sodium free level and its temperature are shown in Figure 10.15, which are the important input data for the analysis. The vessel of 400 mm length is divided into 544 eight-noded isoparametric axisymmetric elements with 2181 nodes. Two elements are used through thickness direction. The minimum size of the element is 0.2 mm near the vicinity of sodium free level to accurately predict the sharp variation

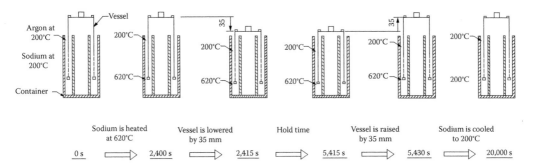

FIGURE 10.14 VINIL: simulation of ratcheting in thin shell.

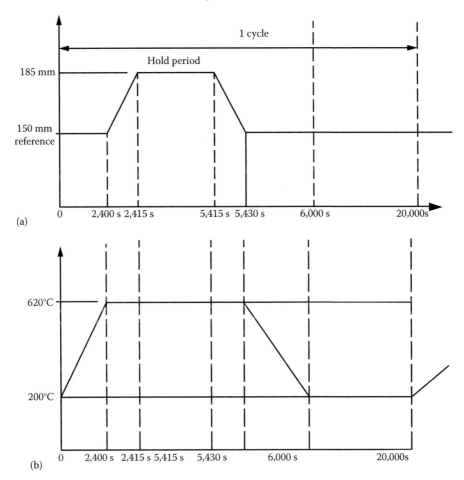

FIGURE 10.15 Details of level and temperature variation for VINIL benchmark: (a) sodium free level variation and (b) sodium temperature variation.

of discontinuity stresses. All the degrees of freedom except the radial displacements are arrested at the top edge. At the bottom edge, only rotations are arrested.

First, a transient heat conduction analysis is performed to determine the temperature distribution across the metal wall over the length at various time intervals. The heat transfer coefficient of sodium in the stagnant condition is 1000 W/m² K (corresponding to natural convection), and under motion, it is 2000 W/m² K (forced convection). For argon, a constant value of 10 W/m² K is considered. Sufficiently small time step is selected for a stable solution respecting the Fourier number

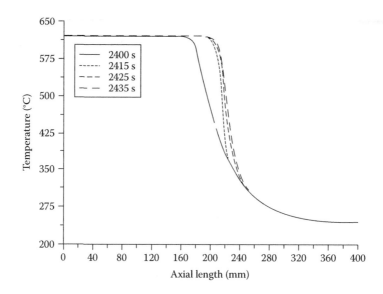

FIGURE 10.16 Evolution of axial temperature gradient.

(thermal diffusivity × time step/square of minimum element size) to be <0.1. Figure 10.16 depicts the evolution of temperature distribution during and after the vessel is lowered. It is worth noting that the axial temperature gradient is sharp immediately after the free level reaching the highest elevation; subsequently, the gradient reduces marginally due to heat conduction along the axial direction.

Knowing the temperature profile at every solution time step, corresponding stress analysis is carried out. For realistically determining stresses and strains, viscoplastic analysis is done using the *23-parameter Chaboche model*. The material constants used in the analysis are given in Table 10.2. The results are compared with international benchmark results in Figure 10.17. With the view of possible variations in the material and temperature data, the exact reproduction of deformation is not possible. In spite of this, it is noted that the comparison is satisfactory.

10.2.1.7 Prediction Capability of 20-Parameter Viscoplastic Model

Salient features of modified 9Cr–1Mo that motivate development of a new model are as follows:

- The material exhibits marked strain-rate sensitivity at elevated temperature. The strain-rate sensitivity increases with increasing temperature.
- The material displays strain softening at larger strains in elevated temperature, during monotonic tests. However, under lower-temperature monotonic loading, the material primarily strain hardens. The temperature over which monotonic hardening/softening behavior changes abruptly lies in the range of 650–750 K.
- The material cyclically softens in strain-controlled cyclic tests. The amount of cyclic softening appears to be independent of temperature.
- Under constant stress loading, the material exhibits an apparently very low primary creep stage in the 750–900 K temperature range. Though the amount of primary creep is insignificant, it increases with increasing temperature. Secondary creep rates also increase with increasing temperature and stress.

In order to represent these features, the 23-parameter Chaboche viscoplastic model described in the previous section, essentially developed for SS 316 LN, has been modified to model the mechanical behavior of grade 91 steel at high temperature. The major modifications are elimination of (1) an exponential term, which is to simulate the strain insensitivity in the intermediate temperature range, (2) coupling

TABLE 10.2
Material Parameters of Chaboche Viscoplastic Model

Parameters	Temperature (K)			
	293	473	773	873
E (MPa)	192,000	178,000	161,000	145,000
N	0.3	0.3	0.3	0.3
k (MPa)	200	135	95	35
a_R	1	1	1	1
b	20	12	12	12
Q_o	30	40	70	40
Q_{max}	390	460	495	460
M	19	19	19	19
H	0.04	0.04	0.04	0.04
g (MPa^{-mr} s^{-1})	0	0	0	2×10^{-7}
Q_R^* (MPa)	200	200	200	200
M	2	2	2	2
c_1 (MPa)	65,000	65,000	65,000	65,000
a_1	1,300	1,300	1,300	1,300
c_2 (MPa)	1,950	1,950	1,950	1,950
a_2	50	50	50	50
FY	0.5	0.5	0.5	0.5
b_1 (MPa^{-m1} s^{-1})	0	0	0	10^{-12}
m_1	4	4	4	4
b_2 (MPa^{-m1} s^{-1})	0	0	0	2×10^{-13}
m_2	4	4	4	4
K_o (MPa/s)	10	10	10	70
a_k	0	0	0	1
n	24	24	24	24
a	0	0	0	0

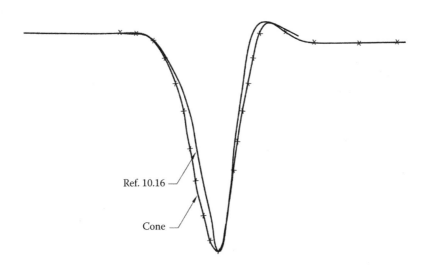

Ref. 10.16

Cone

FIGURE 10.17 Prediction of ratcheting.

TABLE 10.3
Values of Material Parameters of Chaboche Model for Grade 91 Steel

	Temperature (K)				
Constants	298	673	773	823	873
n	0.0	0.0	0.0	0.0	0.0
k	41.7	41.7	10.5	6.2	4.5
K_0	369.9	306.0	514.0	783.4	1076.6
α	1.3	1.3	2.5	4.7	8.5
a_1	150.0	150.0	146.5	141.1	105.0
c_1	7000.0	7000.0	7000.0	7000.0	7000.0
β_1	41.7	41.7	10.5	6.2	4.5
m_1	0.0	0.0	0.44E–25	0.61E–15	0.76206E–11
a_2	117.5	117.5	64.8	48.9	27.9
c_2	500.0	500.0	500.0	500.0	500.0
β_2	41.7	41.7	10.5	6.2	4.5
m_2	0.0	0.0	0.92E–25	0.1E–14	0.96319E–11
a_3	266.6	173.4	120.7	82.4	83.6
c_3	37.5	37.5	37.5	37.5	37.5
β_3	41.7	41.7	10.5	6.2	4.5
m_3	0.0	0.0	0.96E–24	0.95E–14	0.16254E–09
b_1	30.0	30.0	30.0	30.0	30.0
Q_1	−65.0	−65.0	−65.0	−65.0	−65.0
b_2	0.3	0.3	0.3	0.3	0.3
Q_2	−15.0	−15.0	−15.0	−15.0	−15.0

between kinematic and isotropic hardening, and (3) plastic strain memorization effects, and inclusion of (1) third term in the kinematic hardening variable to account for the large strain range effects, (2) two terms in the isotropic softening variable, and (3) effect of isotropic hardening in the viscous stress. With these, the model for grade 91 steel now involves only 3 kinematic hardening tensorial variables (X_1, X_2, and X_3) and 2 isotropic softening scalar variables (R_1, R_2). In total, 20 material parameters are used to define the material behavior. The constitutive equations are stated in Appendix 10.B along with the material parameters over a temperature range of our interest (Table 10.3).

In order to demonstrate the prediction ability of this model, several uniaxial test data are predicted. Figure 10.18 shows the monotonic stress–strain curves in the temperature range of 723–873 K

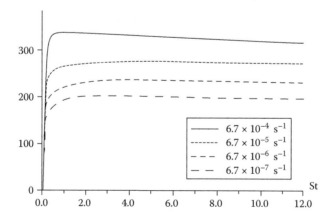

FIGURE 10.18   Simulation of tensile curves of modified 9Cr–1Mo.

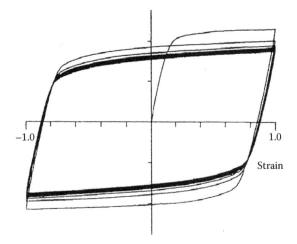

FIGURE 10.19   Simulation of cyclic stress–strain loops for modified 9Cr–1Mo.

up to 12% strain. It is seen in this figure that the monotonic hardening decreases with increasing temperature. Above 823 K, a slight monotonic softening can be noted. The strain rate sensitivity of this material can be seen in Figure 10.18. The cyclic stress–strain hysteresis loops are plotted in Figure 10.19 for various temperatures (298–873 K) under strain-controlled cycling of ±1%, at a constant strain rate of 6.7×10^{-4} s^{-1}. Cyclic consolidation curves shown in Figure 10.20 depict clearly the cyclic softening behavior of the T91 material. The softening behavior is more or less independent of temperature. Some creep curves of SFR interest are shown in Figure 10.21. All these simulations match satisfactorily with the uniaxial data published [10.10]. More details of the model development, validation, and applications can be seen in Ref. 10.10.

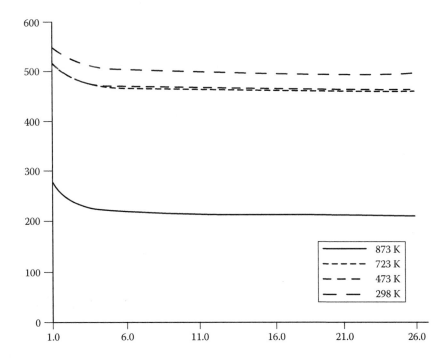

FIGURE 10.20   Simulation of cyclic softening of modified 9Cr–1Mo steel.

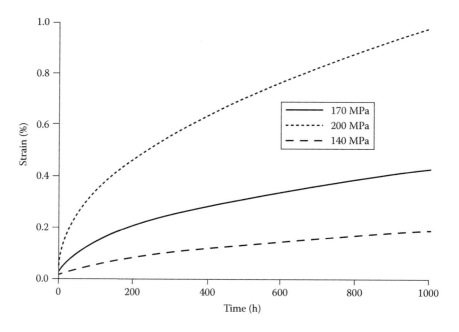

FIGURE 10.21   Simulation of creep curves for modified 9Cr–1Mo steel.

10.2.2   Creep–Fatigue Damage Assessment Procedure of RCC-MR

This benchmark test series deals with experimental validation of the procedure by application of load cycles on circular plates with thick hub welded at the center. The outer periphery of the plate is held by a circular ring of balls on either side to simulate perfect simply supported condition. Displacements are applied at the center. One typical load cycle consists of pulling the rigid central hub upward to the maximum 3 mm displacement, holding the hub at that position for 2 h, pushing downward to 6 mm so that the plate is subjected to 3 mm displacement in the reverse direction, holding it as such for 2 h, and finally pulling back to its neutral position. Tests are conducted at 873 K (600°C) isothermal condition. Thus, the applied load cycles develop high stress–strain concentrations on the fillets at the plate–hub junctions and in turn high creep–fatigue damage. The test setup and typical load cycles are described in Figure 10.22. The facility consists of a 10-ton loading frame, a split furnace, and an actuator operated through the motor along with the computer and programmable logic controller (PLC). There are load cell and linear variable differential transformer (LVDT) sensors to measure the load developed in the plate and displacements, respectively.

The complex aspect of the problem is the relaxation characteristic of multiaxial stress (specifically biaxial state in the present problem) at the weldments. CAST3M issued by CEA is employed for deriving peak stresses and strains [10.11]. Subsequently, RCC-MR: 2002 and 2007 versions [10.12] are adopted for determining creep and fatigue damage values based on *simplified analysis* route. The validation is carried out with the experimental data on load–deflection curves, peak strains, and number of cycles to initiate cracks at the peak location. The specimens are investigated for the crack at critical location at regular intervals by interrupting the test. The number of cycles, when the shape of hysteresis loop changes is taken as the number of cycles to cause failure. Further, visual inspection was carried through viewing window online. Tests are conducted on several such plates. A typical load–deflection curve of one of the plate is shown in Figure 10.23. Among the tests on various plates, an average number of cycles required to cause the failure is noted to be 86 cycles (172 h). In most of the tests, the location of crack is at the fillet, at 20.6 mm from the center of the plate and on both sides of the plate as shown in Figure 10.24. The metallurgical investigation depicts intergranular creep cracks with fatigue crack, which specifies the crack propagation as a mixed mode.

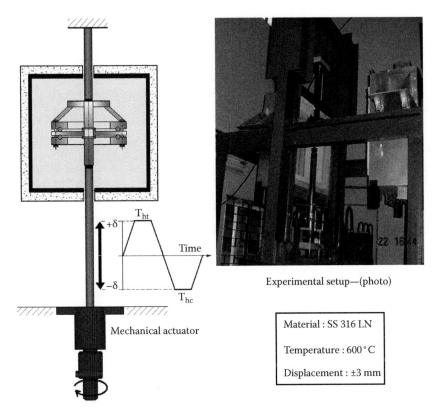

FIGURE 10.22 Creep–fatigue damage simulation on a circular plate: a test setup.

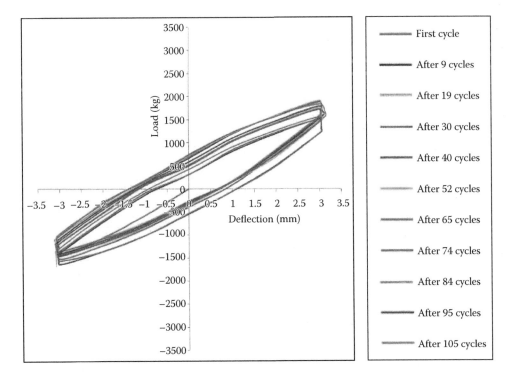

FIGURE 10.23 **(See color insert.)** Load versus deflection (with 1 h hold time) plot of the plate at 873 K.

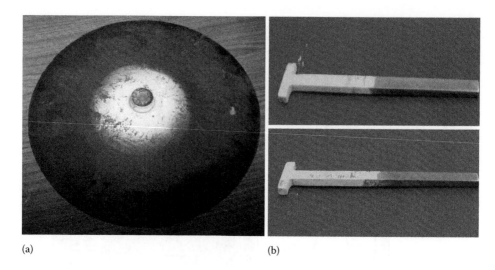

(a)　　　　　　　　　　　　　　　　(b)

FIGURE 10.24　Cracks on both sides of the plate: (a) full plate and (b) cut sections.

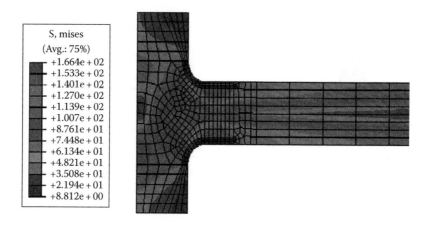

S, mises
(Avg.: 75%)
+1.664e + 02
+1.533e + 02
+1.401e + 02
+1.270e + 02
+1.139e + 02
+1.007e + 02
+8.761e + 01
+7.448e + 01
+6.134e + 01
+4.821e + 01
+3.508e + 01
+2.194e + 01
+8.812e + 00

FIGURE 10.25　**(See color insert.)** Stress distribution after creep relaxation (primary + secondary stress).

The required stresses and strain values are numerically computed using CONE with axisymmetric quadrilateral finite elements. The finite element model (FEM) and stress distribution in the same are shown in Figure 10.25. The peak stress location from finite element analysis of the plate is at 20.60 mm from the center, and RCC-MR procedure for creep–fatigue damage evaluation is followed. As per RCC-MR, elastic plastic strain is calculated from elastic stress result. Creep strain per cycle is obtained by considering a hold time of 2 h/cycle. The effect of stress relaxation on the creep–fatigue damage evaluation was studied by using the design codes RCC-MR: 2002 and RCC-MR: 2007. Effect of stress relaxation on the creep damage is found to be significant as seen in the following.

Creep damage without considering relaxation is 1.352 and 0.017 by considering relaxation as per the procedure recommended in RCC-MR: 2002. The corresponding values are estimated as 1.247 and 0.015 as per RCC-MR: 2007. Fatigue damage per cycle is calculated from the allowable number of cycles as per the design code. Fatigue damage as per RCC-MR: 2002 is 0.00233 and 0.00202 per cycle as per RCC-MR: 2007. The number of permissible cycles taking into account the creep–fatigue damage interaction predicts the failure number of cycles as 45 or 90 h as per RCC-MR: 2002 and 51 cycles or 102 h as per RCC-MR: 2007. As indicated earlier, these values are close to the number of failure cycles observed as 86 (an averaged value). RCC-MR: 2007 predictions are found to be closer to the test data.

Let us discuss the relaxation behavior of peak stresses developed due to displacement controlled loading. The stress relaxes with the passage of time because of small-scale permanent deformation due to creep strain. Norton's law (power law), the best for secondary creep phenomenon, is used to model the creep deformation. The stress relaxation rate is given by

$$\dot{s}_r = -\frac{E}{C_r}\dot{\varepsilon}_{fl}(\sigma_k)$$

In secondary creep region, strain rate is less compared to primary regime; hence, Norton's law predicts less stress relaxation if loading condition leads to creep deformation mainly due to primary creep. It is expected that viscoplastic analysis could provide realistic relaxation behavior, and thereby, the life prediction could be closer to the test data.

10.2.3   LIFE PREDICTION OF COMPONENTS WITH CRACK-LIKE DEFECTS

The design codes specify stringent inspection requirements to ensure high quality of structural materials and manufacturing standards. In view of the fact that welds are weak links in a structure, the codes do not permit welds without adequate and reliable inspection methodologies. However, there are a few difficult locations for preservice inspection where crack-like defects, termed *geometrical singular points*, are unavoidable. Figure 10.26 shows a few such locations in the reactor assembly of a typical SFR. These geometrical singular points persist in the fuel pin-end plug welding, rolled and welded joint of the intermediate heat exchanger (IHX), and plate shell weld junctions in the control plug. Considering the practical difficulties in completely excluding such singular points, RCC-MR edition provides special design rules called the σ_d *approach* to ensure that singular points can be adequately addressed in class 1 components. With the σ_d approach, the state of stress and strain at the characteristic distance d ahead of a crack tip is used to assess the basis of acceptance. For the present study, the recommendation of RCC-MR: 2002 is addressed. Since, for a component operating at high temperature, creep damage at the critical weld locations could be a

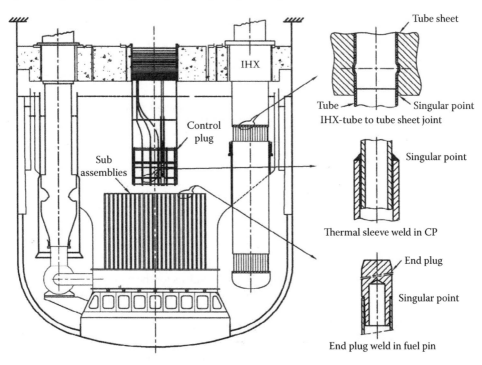

FIGURE 10.26   Welds with singular points in SFR assembly.

life-limiting factor, creep damage is alone considered in the benchmark exercise. Accordingly, the design procedures for the estimation of creep damage at welds with crack-like defects addressed in Appendix A16 of RCC-MR (A16) based on the σ_d approach are presented [10.12].

The benchmark exercise deals with experimentally generated creep crack growth data from precracked standard compact tension (CT) specimens machined from welded SS 316 LN plates [10.13], to assess the validity of the A16 approach. As per the procedure, it is required to determine the equivalent stress, called σ_d based on the Rankine theory, that is, the maximum principal stress at the characteristic distance d ahead the crack tip. The crack initiation life is the minimum time to rupture corresponding to σ_d at the specified temperature T. The recommended value for the characteristic distance (d) is equal to 50 μm for the austenitic stainless steels in A16. However, the development of a visible crack (0.1–0.2 mm size) is generally considered as crack initiation. Using the distance of 50 μm for calculating σ_d may be considered a conservative value for the assessment. For the present investigation on experimental data, 0.2 mm is considered for defining crack initiation life. The value of σ_d should be computed taking into account the effect of the singularity and plasticity. Using the simplified method recommended in A16, σ_d is derived from elastically computed stresses by applying Neuber's rule to account for plasticity. The elastic stresses in turn are derived from associated stress intensity factors (K) using Creager's formula. K can be computed either using equations given in A16 for a few standard geometries or by numerical methods for complex geometries/loadings.

10.2.3.1 Experimental Details

The plane CT specimens are manufactured from a single heat of welded plates made of type 316 LN steel plates using a matching manual metal arc (MMA) weld combination. The interface between the parent metal and the MMA weld metal is located parallel to the central line of the specimens. The specimens are precracked at room temperature to generate initial crack lengths of 17.58 and 17.41 mm. Except for the initial crack size, the geometrical and loading conditions are the same for both specimens. The details for one specimen are shown in Figure 10.28. The creep crack propagation tests were conducted in air at 823 K by applying a constant axial load of 20 kN for the duration of the test. The times taken for creep crack growth of 0.2 mm were found to be 300 and 400 h, respectively, which are considered as creep crack initiation times for the two specimens.

10.2.3.2 Numerical Prediction Steps

10.2.3.2.1 Step 1: Linear Elastic Stress Intensity Factor

For the CT specimen, the stress intensity factor (K_I) is computed following section A16.8221.2, which recommends the following equation:

$$K_I = F_b \cdot \sigma \cdot \sqrt{\pi a} \tag{10.1}$$

where
$$F_b = (2 + a/w) \cdot F_1 / (1 - a/w)^{3/2} \cdot \sqrt{\pi a / w}$$
$$F_1 = (0.886 + 4.64 \cdot (a/w) - 13.32 \cdot (a/w)^2 + 14.72 \cdot (a/w)^3 - 5.6 \cdot (a/w)^4)$$
$$\sigma = N_1 / (w \cdot B)$$

The symbols are defined in Figure 10.27 itself. The width (w) is equal to 38 mm, thickness (B) is equal to 19 mm, and axial force (N_1) is equal to 20 kN. K_I values for the two specimens whose initial crack lengths equal to 17.58 and 17.41 mm are 46.67 and 46.07 MPa√m, respectively.

10.2.3.2.2 Step 2: Elastically Computed Characteristic Stress (σ_{de})

The value of σ_{de} is equal to the maximum principal stress = $K_I / \sqrt{2\pi d}$. The value of σ_{de} for specimen 1, whose K_I is 46.67 MPa√m, is 2633 MPa at d = 50 μm. The corresponding value for specimen 2, whose K_I is 46.07 MPa√m, is 2600 MPa.

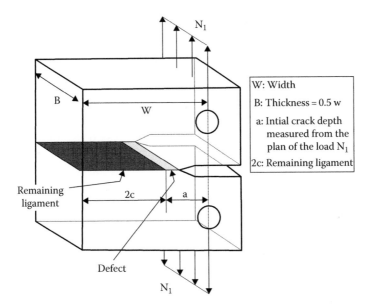

FIGURE 10.27   Details of benchmark problem.

10.2.3.2.3   Step 3: Elastoplastic Characteristic Stress (σ_d)

σ_{VM} is determined using *Neuber's rule*. Based on detailed elastoplastic analysis, the procedure used to compute elastoplastic stress and strains as per RCC-MR code has been modified. The modified procedure is illustrated in Figure 10.28. As per the modified Neuber's rule, von Mises stress and strain corresponding to elastically computed peak von Mises stress are computed by the following equation:

$$\sigma_{VM} \cdot \varepsilon_{VM} = \sigma_{VMe} \cdot \left[\frac{\sigma_{VMe}}{E} + B(\sigma_{ref})^{1/\beta} \right] \tag{10.2}$$

where

E and B are elastic and plastic moduli of the material, respectively

β is the strain hardening index in a Ramberg–Osgood equation

The values of β and B are taken from Ref. 10.13 as 0.138 and $(1/289.2)^{1/\beta}$ MPa, respectively. The empirical correlation recommended in Table A16.8221.3 of A16 is used for determining σ_{ref} for a

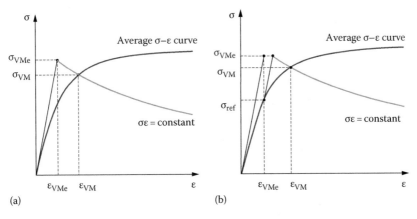

FIGURE 10.28   Determination of elastoplastic stress by Neuber's rule. (a) RCC-MR: Original and (b) RCC-MR: Modified.

CT specimen under the assumption of von Mises plane strain conditions. The elastically computed σ_{He} can be corrected to obtain σ_H using the following equation:

$$\sigma_H = \sigma_{He} \cdot \left[\frac{\sigma_{VM}}{\sigma_{VMe}}\right] \qquad (10.3)$$

10.2.3.2.4   Step 4: Multiaxial Creep Damage Criteria

The governing stress for creep damage estimation (σ_{eq}) is expressed as a function of both the von Mises stress (σ_{VM}) and the hydrostatic stress component (σ_H), as given in Equations 10.3 and 10.4, respectively.

10.2.3.2.5   Step 5: Relaxation of Stresses

Secondary stresses generally undergo creep relaxation during sustained conditions at high temperature. For the present case, relaxation is considered for the von Mises stress. Since the value of the von Mises stress at the crack tip is relatively small, the relaxation is limited and there is no concern of nonconservatism due to relaxation below the value required for equilibrium. The relaxation of the starting stress σ_{VM} is computed using the following equation recommended in RCC-MR 3262.1:

$$\frac{d\sigma_{VM}}{dt} = -\frac{E\dot{\varepsilon}_c'}{3} \quad \text{and} \quad \dot{\varepsilon}_c' = K\sigma_{VM}^n \qquad (10.4)$$

where K and n are constants associated with Norton's law, which are provided in RCC-MR: Appendix Z as $K = 9.722 \times 10^{-30}$ and $n = 10.4707$ at 823 K. The relaxation of σ_H is assumed to vary in a manner similar to σ_{VM}.

10.2.3.2.6   Step 6: Prediction of Creep Crack Initiation Life

The triaxial stress components at the characteristic distance determined using Creager's equations given in A16 are as follows:

- *For specimen 1 at d = 0.2 mm*: $\sigma_x = 1313$ MPa, $\sigma_y = 1313$ MPa, and $\sigma_z = 788$ MPa. Application of Neuber's rule yields $\sigma_{VM} = 174$ MPa, $\sigma_H = 376$ MPa, and $\sigma_{eq} = 301$ MPa. Time variation of σ_{eq} and accumulation of creep damage are depicted in Figures 10.29 and 10.30, respectively. From Figure 10.30, the time at which creep damage is equal to 1 is 435 h (T_{rd}).
- *For specimen 2 at d = 0.2 mm*: $\sigma_x = 1297$ MPa, $\sigma_y = 1297$ MPa, and $\sigma_z = 778$ MPa, which yields $\sigma_{VM} = 173$ MPa, $\sigma_H = 374$ MPa, $\sigma_{eq} = 300$ MPa, and $T_{rd} = 451$ h. Compared to the experimental data for these cases (300 and 400 h), the improved procedure predictions (435 and 451 h, respectively) show satisfactory experimental validation of the adopted procedure.

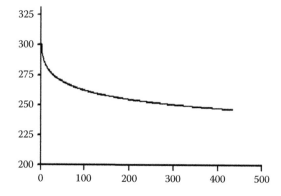

FIGURE 10.29   Relaxation of σ_{eq}.

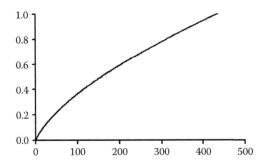

FIGURE 10.30 Cumulative creep damage.

10.2.4 Fluid Elastic Instability of Thin Shells

In this validation study, the fluid elastic instability of a 1/15 scale weir-coolant system that simulates the thermal baffle of the Japanese demonstration fast breeder reactor (DFBR) is analyzed to compare the test data presented by Fujita et al. [10.14]. Water is used in place of sodium. The geometrical details are given in Figure 10.31. The weir shell is made of polyvinyl chloride (PVC) whose thickness is 6 mm. The weir shell is fixed at the bottom. The material properties are as follows: for PVC material, $\rho = 1700$ kg/m^3, $E = 2.5$ kPa, and $\nu = 0.45$ and for water, $\rho = 1000$ kg/m^3, $\mu = 0.001$ N s/m^2, and $c = 1435$ m/s.

Free vibration analysis is the first step. For this, the weir shell along with rigid inner and outer shells and water filled in the annuli are considered for the analysis. The free level difference between

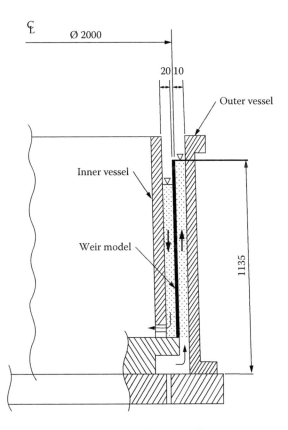

FIGURE 10.31 Japanese benchmark problem. *Note*: CL, center line.

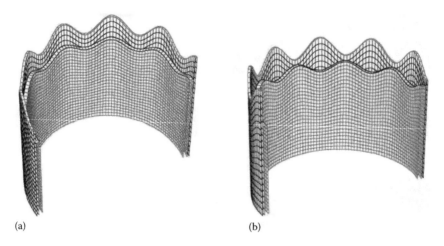

(a)   (b)

FIGURE 10.32 Sloshing mode shapes for Japanese benchmark problem: (a) in-phase mode and (b) out-of-phase mode.

feeding and restitution collectors is a parameter in the analysis. The free analysis is performed using *Fourier option* for the selected wave number n. Since the interest is on the sloshing mode shapes, all the in-phase and out-of-phase sloshing modes are extracted. The mode shapes corresponding to n equal to 10 are depicted in Figure 10.32. Further, dominant shell vibration mode is also extracted. The natural frequencies corresponding to these three modes are plotted as a function of wave number (n) in Figure 10.33. The shell has the lowest natural frequency (~1.7 Hz) corresponding to n equal to 5. The effect of fluid structure interaction is measured by the difference between the natural frequencies of in-plane and out-of-plane sloshing modes. Accordingly, the results shown

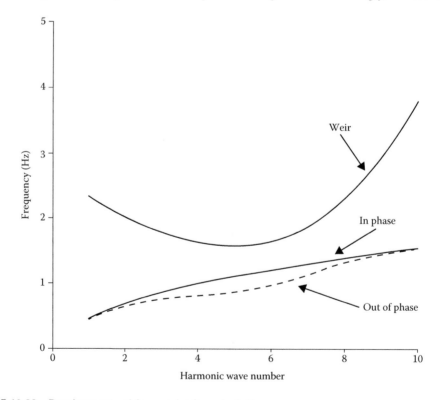

FIGURE 10.33 Dominant natural frequencies for weir shell.

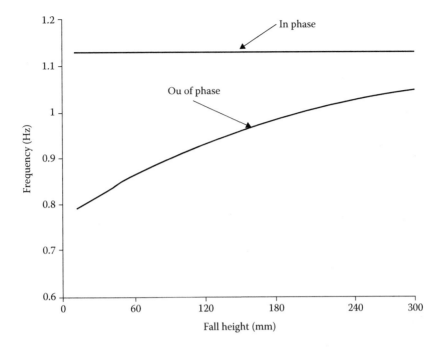

FIGURE 10.34 Sloshing frequency versus fall height.

in Figure 10.33 indicate that the interaction is stronger for the n equal to ~5. The instability modes observed in the experiment [10.14] correspond to the wave number ~5. The sloshing frequencies are computed corresponding to wave number 5 for the fall heights ranging from 10 to 300 mm and are presented in Figure 10.34. While the in-phase sloshing frequency is independent of fall height, the out-of-phase frequency is increased with increasing fall height. The difference between these frequencies is higher at lower fall height, which implies that instability due to sloshing occurs at lower fall heights.

The next step is dynamic response analysis. The important input data for the analysis are v_f and τ, which are computed for the range of overflow rates and fall heights. The results are presented in Figures 10.35 and 10.36, respectively. The results are presented in Figures 10.35 and 10.36,

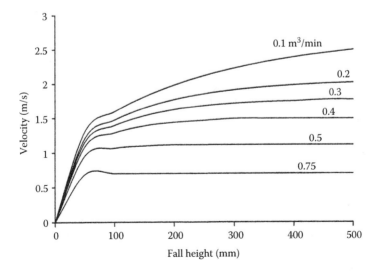

FIGURE 10.35 Impact velocity versus fall height.

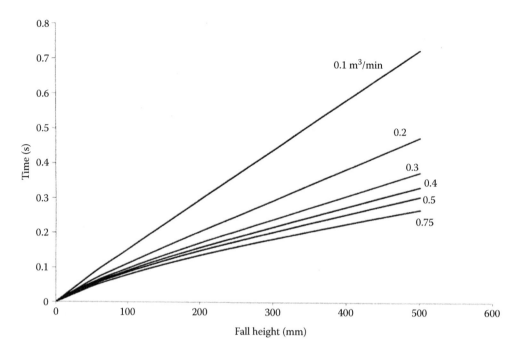

FIGURE 10.36   Delay time as a function of fall height and flow rate.

respectively. As seen in Figure 10.35, the impact velocity gets stabilized after flowing over a certain fall height; under this condition, the friction force is just equal to gravity force. Hence, the delay time, also called as fall time, increases monotonically with a linear variation along the stabilized regime (Figure 10.36). To start the calculation, the shell is disturbed from its equilibrium position by imposing a slight velocity perturbation corresponding to one of the mode shapes. Subsequently, the dynamic displacements of weir shell and heights of free levels are predicted. The global modal damping coefficient ξ is assumed as 1%–3%. The evolution of wave heights of feeding and restitution collectors as well as the weir crest displacements for the ξ values, 1% and 3%, are depicted in Figure 10.37. The damping delays the evolution of amplitudes and reduces the stabilized vibration amplitudes.

In order to compare the results with test data, some specific data sets are compared. With the absence of many important input data such as actual material properties, damping values, surface roughness, and weir crest geometry, point-by-point comparison is difficult. Hence, comparison is done based on overall predictions of fall time and displacements. Figure 10.38 shows the comparison of calculated fall time with test data. The prediction is very good for higher flow rate. For the lower flow rate, the fall time has been underestimated by theory. The actual shell surface roughness that decides the friction factor is needed to improve the prediction. However, the predictions are satisfactory. The dynamic displacements predicted by the nonlinear analysis are compared with the test data. Since damping is not known, analysis is carried for two damping values, that is, 1% and 3%. In the test, the instability zones were observed over the fall height of 0–600 mm. The maximum displacements vary from 1 to 3 mm for the flow rates varying from 0.1 to 1 m³/min, respectively. At 100 mm fall height, vibration is minimum. The nonlinear analysis indicates that the instability zones lie over the fall height of 50–350 mm. Vibration amplitude has a minimum value at the fall height of about 150 mm. The maximum amplitude is 1.2 and 7.5 mm for ξ = 1% and 3%, respectively. With the absence of many important measured input data, these theoretical predictions seem to be very good. The analysis carried out for a typical SFR indicates that the instability is absent when the damping value is equal to or more than 1.0% as seen from Figure 10.39, which shows the dynamic response of weir shell [10.15].

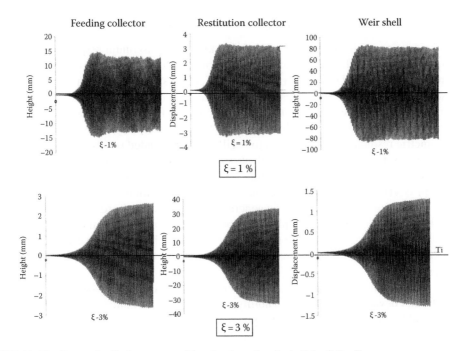

FIGURE 10.37 Dynamic displacements of free levels and weir shell for 0.6 m³/s.

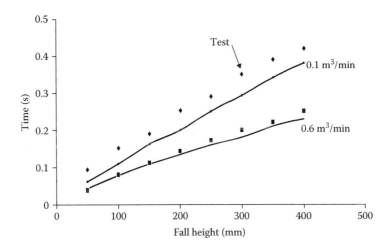

FIGURE 10.38 Prediction of fall time.

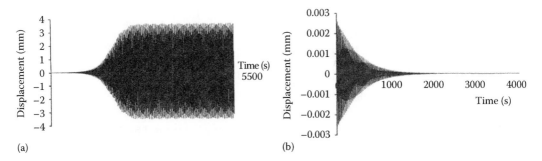

FIGURE 10.39 Displacement of PFBR weir shell during fuel handling: (a) ξ = 0.5% and (b) ξ = 1%.

10.2.5 PARAMETRIC INSTABILITY OF THIN SHELLS

In many engineering applications, it is essential to know the maximum response of the system and the nature of the oscillations under external excitations during design in order to meet the fundamental requirements. Under external excitations, the system may show maximum response under resonance and/or parametric resonance. Resonance corresponds to the tendency of the system to oscillate with greater amplitude when the external excitation frequency is equal to the natural frequency of the system. In resonance, the rate of increase in system amplitude is linear. Parametric resonance [10.16] refers to oscillatory unbounded out-of-plane displacements in a mechanical system due to time-dependent variation of the system parameters under cyclic in-plane force, lower than the critical buckling force corresponding to peak value. The response of the system is orthogonal to the direction of external excitation. In parametric resonance, the rate of increase in system response is generally exponential and grows without limit. This exponential unlimited increase in amplitude is potentially dangerous to the system. Although parametric resonance is secondary, the system may undergo failure near the critical frequencies of parametric resonance. Parametric resonance is also referred to as parametric instability or dynamic instability.

Flexible components and structures, such as core SAs and thin-walled shell structure in reactor assembly, can develop parametric instability under seismic earthquake. In reactor assembly, the thin thermal baffle under random hydrodynamic pressure of sodium under horizontal seismic excitation, the sloshing displacements of sodium free surface, and the displacements of core SAs under vertical seismic excitations can undergo parametric instability (Figure 10.40). The governing dynamic finite element equation for parametric instability is expressed in the following matrix form:

$$M\ddot{X} + C\dot{X} + K_E X + K_G X = 0 \tag{10.5}$$

where
  M, C, K_E, and K_G are mass, damping, elastic stiffness, and geometric stiffness matrices, respectively
  X is the displacement vector and the dot represents differentiation w.r.t. time

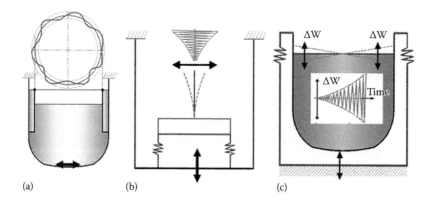

(a)   (b)   (c)

FIGURE 10.40 Parametric instabilities in reactor assembly system: (a) thermal baffle, (b) core subassemblies, and (c) sodium free level.

Equation 10.5 is called Hill's equation. The stability analysis of Equation 10.5 is investigated using Hsu's criteria [10.17]. To apply Hsu's criteria, Equation 10.5 is transformed to the following equation using mode superposition technique:

$$\ddot{\eta}_n + 2\xi\omega_{mn}\dot{\eta}_n + \omega_{mn}^2\eta_n + \sum_{m=1}^{M}\left(\sum_{s=0}^{S} d_{nms}\cos\omega_s t + \sum_{s=0}^{S} e_{nms}\sin\omega_s t\right)\eta = 0 \qquad (10.6)$$

where

η_n is the projected displacement
ω_{mn} is the natural frequency of the system
ξ is the modal damping factor
ω_s is the external excitation frequency
d_{nms} and e_{nms} are the components of projected geometric stiffness matrix

To understand the phenomenon, a benchmark experiment was carried out. A shake table test to understand the dynamic stability of sloshing in plane free surface of liquid in rectangular tanks is conducted in Structural Mechanics Laboratory, IGCAR to see the slosh response for various amplitudes and frequencies of the external excitation. The motivation behind this benchmark exercise is to study the instability in liquid free surface under base excitation, which is of special interest to pool-type SFRs, where the possibility of setting up unstable sloshing of the liquid free surface of the sodium pool in the large diameter vessel cannot be ruled out under seismic excitations. Figure 10.41 shows the shake table experimental setup and the stability chart investigated for stability analysis of the sloshing using Equation 10.5.

A container was excited vertically with different frequencies and amplitudes lying in stable and unstable regions in the stability chart (Figure 10.42). Figure 10.42 shows the slosh response of free surface for excitation parameters lying in stable and unstable regions. The behavior of the free surface was very much according to the stability chart plotted. To further validate the studies, a numerical simulation using arbitrary Eulerian–Lagrangian finite element scheme was carried out, and the numerical results are in good match with the experimental results. Further, the parametric instability studies in slender structures (beam, plate, shells), free surface of fluid, and fluid-filled thin shells were carried out numerically [10.18].

Tank filled with water on shake table

(a)

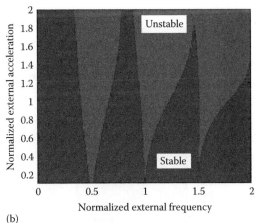

(b)

FIGURE 10.41 (a) Shake table test for sloshing in a rectangular tank and (b) stability chart.

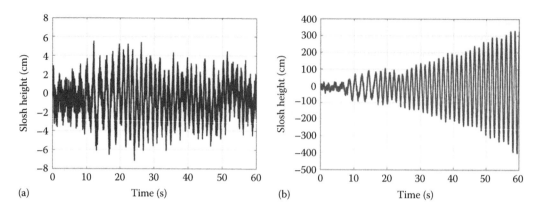

(a)   Time (s)   (b)   Time (s)

FIGURE 10.42   Sloshing responses under (a) stable and (b) unstable situations.

10.2.6 VALIDATION OF SEISMIC ANALYSIS THROUGH MOCK-UP STUDIES

A few international computer codes and mock-ups are presented in this section. The SFR structures are broadly divided into two categories: core SAs and out-of-core structures, which are mainly vessels. The important aspects that need to be validated in the numerical simulations are linear and nonlinear effects of fluid structure interactions. Accordingly, the following four problems have been chosen here.

10.2.6.1 Seismic Response of SFR Core Subassembly Cluster

SFR cores are composed of several SAs of various types. These SAs are separated by small gaps and immersed in a fluid. During a seismic event, impacts may occur between SAs inducing a nonlinear behavior of the core. The two phenomena that mainly influence the response of the core are the fluid structure interaction and the impacts between SAs. To study the core response, methods have been developed and validated through tests on the RAPSODIE core mock-up [10.19]. The aim of this section is to present the mock-up test results under seismic excitations and compare them with the corresponding results obtained from the analysis considering the global fluid structure interaction effects.

The RAPSODIE core mock-up details are shown in Figure 10.43. The mock-up is composed of 91 fuel SAs located at the central portion (1 central assembly and 5 rows around) and 180 neutronic

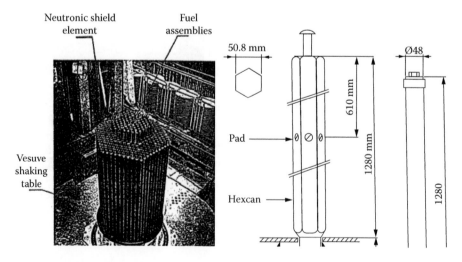

FIGURE 10.43   RAPSODIE mock-up and details of core and neutronic shield SAs.

shield elements (4 rows) surrounding the fuel assemblies. In order to perform tests in water, the mock-up is surrounded by a rigid vessel. The fuel assembly is composed of a cylindrical spike and a hexagonal can containing the fuel pins. The total length of the assembly is 1.5 m. There are two localized contacts between the spike and the diagrid. The weight of the assembly (20 kg) is supported at the lower contact point, and at the upper one, there is a small technological clearance necessary for the introduction of the spike into the diagrid. At about 0.6 m from the top of the assembly, the pads are located. The distance between two adjacent hexcans is 1 mm except at the pad level where it is reduced at 0.1 mm. The neutronic shield elements are composed of steel cylinders clamped at the upper level of the diagrid. A detailed description of the mock-up can be found in Ref. 10.20.

In the present study, the seismic response of core SAs subjected to a constant level of excitation, that is, 100% operating basis earthquake (OBE), is studied. The test results are compared with the nonlinear numerical model, incorporated in the computer code CORALIE (developed by CEA). A detailed presentation of the COALE can be found in [10.21]. In this code, each assembly is represented by its first eigenmodes. The impacts phenomena are represented by a nonlinear spring system acting only during the impact. This stiffness must take into account two aspects: the local deformation of the assembly at the impact point and the truncation effect of the modal basis. A dissipation of energy due to the impact is represented by an additional shock damper placed in parallel with the shock stiffness. The fluid coupling is only taken into account between two adjacent assemblies by using coupling masses deduced from experiences on a small bundle [10.22,10.23]. The fluid couples the motion of the core with the motion of the vessel surrounding the core: as this vessel is very stiff, its motion is supposed to be null. It may be noticed that the added and coupling masses are equal for all assemblies of the same type, independently of their position in the core and uniformly distributed on the length of the assemblies. Such model supplies acceptable agreement with test results, for maximal displacements on the central row (Figure 10.44) and for the time history of the central fuel assembly (Figure 10.45). The first observation from Figure 10.45 is that the level of displacements decreases from the central row to the external one. This can be related with the fact that the sum of the gaps for the external row is smaller than for the central one. Apart from this, the correlation becomes bad (especially for the time history) for external assemblies for which impacts are strong.

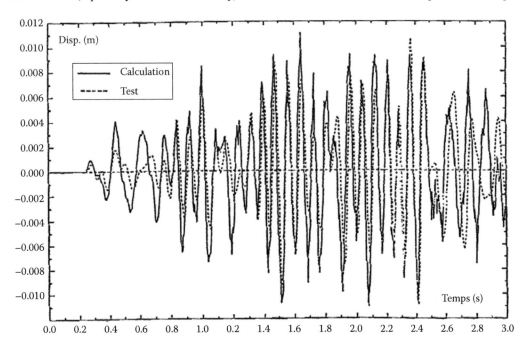

FIGURE 10.44 Maximal displacement in central row of RAPSODIE mock-up.

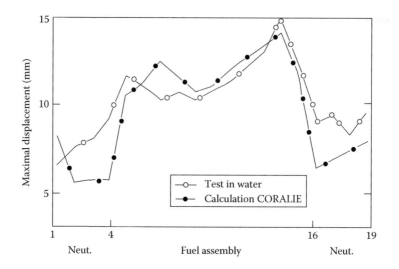

FIGURE 10.45   Time history of central fuel subassembly.

10.2.6.2   Investigation of Seismic Response Uplift of SFR Core under Seismic Events

SFR core SAs are individually fitted into circular tubes held by the core supporting plate. They are freestanding and are not restrained mechanically in the vertical direction. As a result, core components may suffer uplift when they are excited by a large vertical earthquake motion. Therefore, it is an important subject in the seismic design to study the dynamic uplift behavior of core components under seismic excitation. Toward this, an analytical model is used to estimate the vertical motion of core components during an earthquake. To simplify the theoretical treatment, a core component is replaced with a lumped mass in consideration of the collision between the core component and the core supporting plate. The dynamic characteristics of the collision are represented by linear spring and damper. Further, the buoyancy of the fluid, the frictional force, and the fluid resistance force acting on the core component are also taken into consideration. Details of the analytical formulation can be seen in Ref. 10.24.

To examine the dynamic uplift phenomena, the vertical vibration test was carried out by using a small-scale model. The objective of the test is to demonstrate the applicability of the analytical model for the vertical vibration of the core SAs. The scheme of the experimental model is shown in Figure 10.46. A core component model is freestanding on the support block, and its horizontal motion is restricted by load cells with a narrow gap. In the simulation analysis for this testing, the measured horizontal load applied to the core component model is included as a vertical frictional force. The upward force acting on the core component under actual pressurized condition is simulated by using a coil spring. Spring extending length is long enough for the deformation due to the uplift to be neglected. Water level is determined in consideration of the actual ratio of buoyancy to the core component weight. A cylindrical vessel housing this test equipment is mounted on the vibration table.

The comparison of time histories for response displacement of the core SA is shown in Figure 10.47 for the random as well as sinusoidal excitations, and it is thereby confirmed that the simple structural modeling for the core SA is reasonable.

Similar tests were conducted for experimentally quantifying the maximum acceleration level that can initiate core lift-off for PFBR [10.25]. The specific feature of this experiment is the application of vertical excitations and constant fluid pressure of about 1 MPa simulating the flowing sodium effects (Figure 10.48). The vertical excitations are generated as random signals to represent the floor response time history at the grid plate level (Figure 10.49). It is absorbed from the vertical displacement histories of the SA depicted in Figure 10.49 that the lift-off could occur if peak acceleration value is greater than about 1.5 g.

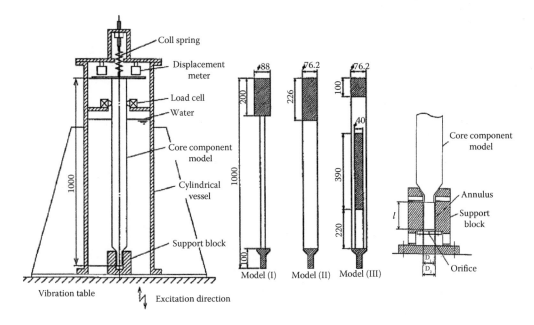

FIGURE 10.46 Details of test setup and SA model used in uplift studies.

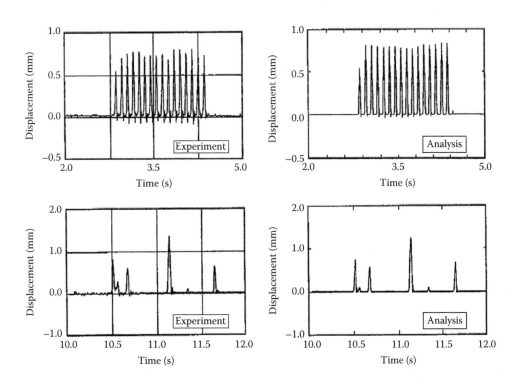

FIGURE 10.47 SA uplift (measured and predicted) in terms of displacements.

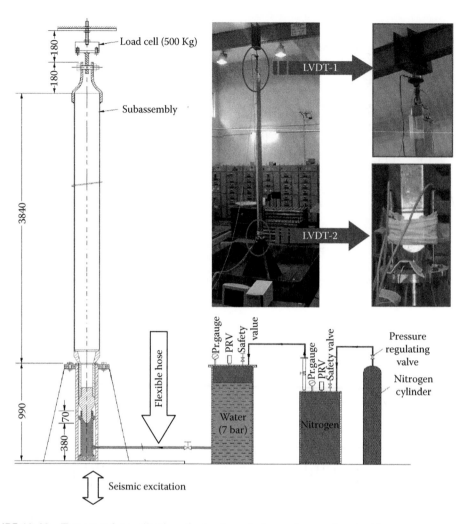

FIGURE 10.48   Test setup for application of seismic excitation and upward coolant force.

10.2.6.3   Seismic Response of Flexible Tanks with Fluid Structure Interaction Effects

A 12 × 6 ft tank, shown on the shaking table in Figure 10.50, is welded from sheets of aluminum
3 ft high and 0.080 and 0.050 in. thick. The tank is attached to the shaking table by four bolts
located at a 3 ft radius about the center of the bottom plate. In addition, anchoring devices are provided, which could clamp the edge of the base plate rigidly to the table. Thus, two different tank
systems are studied: (1) with the tank wall free to uplift and (2) with the base of the tank wall fully
clamped. Four pressure transducers are located in the south tank wall on the excitation axis at 2,
18, 35, and 53 in. above the base. The Fourier component representation of the tank deformations
provides an effective basis for plotting the distorted shapes at any selected instant of time. Distortion
sequences during intervals of intense response are shown in Figure 10.51 for the tank with base free.
Figure 10.51 clearly shows the three-lobe deformation pattern that develops with the occurrence of
base uplift, the top rim of the tank deflecting inward as the base uplifts. Generally, current thin shell
liquid storage tank designs are based on hydrodynamic pressures predicted by Houser's approximate analysis, which assumes that the liquid in the tank may be divided into one part that moves
directly with the tank and a second part that moves relative to the tank at the first sloshing mode frequency. Pressures induced by these two masses of water are designated impulsive and convective,
respectively, and because the tank is presumed rigid, the impulsive pleasure is in direct response to

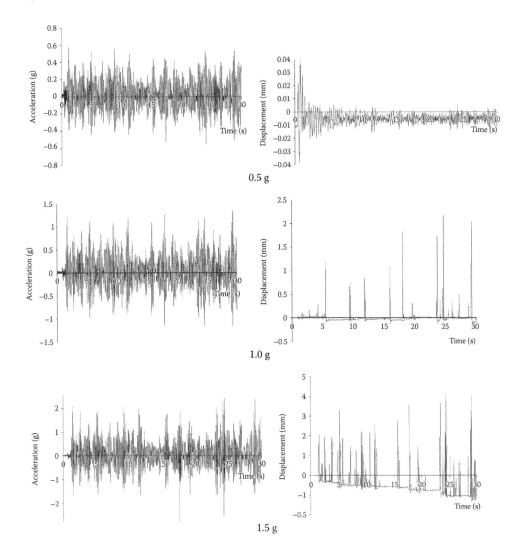

FIGURE 10.49 Vertical accelerations at grid plate and displacement of SA w.r.t. grid plate.

the base acceleration. The sloshing liquid, on the other hand, responds to the base acceleration as a single degree-of-freedom system, amplified or attenuated as indicated by the response spectrum of the input motions. Correlations of the observed dynamic pressures with analytical results calculated with the Housner model [10.26] are shown in Figure 10.52. The correlation obtained near the base in the fixed base case, shown in Figure 10.52c, is seen to be quite satisfactory, which demonstrates that the rigid tank assumption is valid near the clamped base. Pressures observed at mid-height for the fixed base case, shown in Figure 10.52d, are similar to the analytical prediction but are amplified by a factor of nearly two due to the flexibility of the tank wall. The results for the free base case, shown in Figure 10.52a and b, demonstrate major deviations of the actual tank behavior from the assumed response mechanisms due to the influence of uplifting.

In summary, the test results show that the fluid–structure interaction produces significant cross-section distortion of the tank, even though the primary seismic loads tend to induce only translation of the circular section. The source of nut-of-round distortion is clear for the free base case because uplift obviously destroys the axial symmetry by reducing the effective stiffness in the uplifted region. In the fixed base case, the cause of section distortions is less evident, but probably it is associated with initial imperfections in the tank geometry. For example, the four-lobe Fourier

FIGURE 10.50   Circular flexible tank filled with water mounted on shake table.

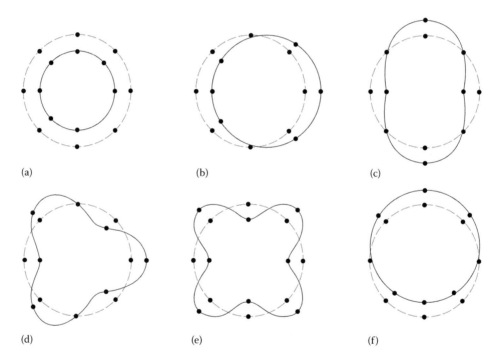

(a)   (b)   (c)

(d)   (e)   (f)

FIGURE 10.51   Fourier decomposition of top rim deformation patterns: (a) constant radial; (b) cos(0), radial; (c) cos(80), radial; (d) cos(30), radial; (e) cos(40), radial; and (f) sin(0), radial.

component, which is strongly represented in the dynamic response for this case, may be associated with the vertical weld lines in the tank that are located at 90° intervals. Linear analysis of the influence of such imperfections seems feasible, but it has not been done here; moreover, it will be difficult to measure the imperfections in the fluid-filled tank. Hydrodynamic pressures predicted by elementary rigid tank theory show good agreement with experiment only where the tank wall displacements are small. Some modification of the theory is needed to account for tank deformations.

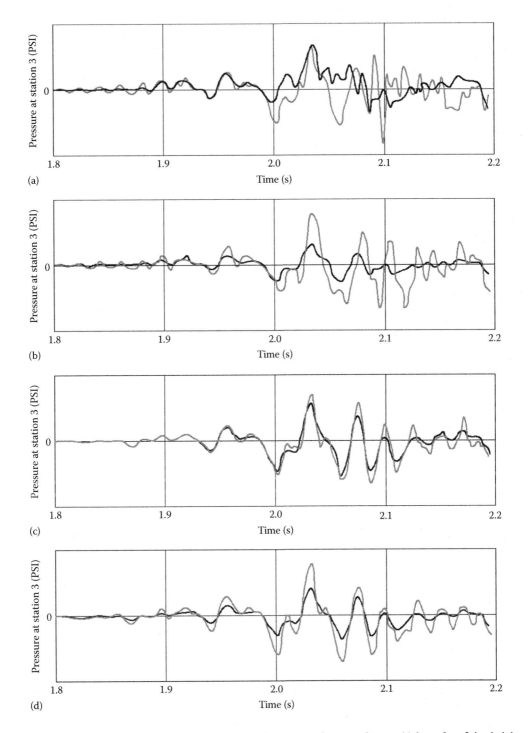

FIGURE 10.52 Hydrodynamic pressure on the tank: measured versus theory: (a) base free 2 in. height, (b) base free 35 in. height, (c) base fixed 2 in. height, and (d) base fixed 35 in. height. (From Housner, G.W., Finite element analysis of fluids in containers subjected to acceleration. Nuclear Reactors and Earthquakes, TID 70211, U.S. Atomic Energy Commission, Washington, DC, 1969.)

Perhaps cantilever column deflections excluding section distortion, as proposed by Veletsos and Yang [10.27], provide some improvement, but it appears from these results that section distortion is a significant factor both in the hydrodynamic pressures and in the stress response. Thus, it may be concluded that further study of the response mechanisms associated with out-of-round distortions for both fixed and free base conditions is essential in order to make improvements in the seismic design procedures of liquid storage tanks.

10.2.6.4   Seismic Study on Large Pool-Type SFR

A pool reactor concept that has been designed during a 3-year period of feasibility study is provided in Ref. 10.28. Major features of the concept are as follows:

1. Core is suspended from the roof slab inner edge with the core support cylinder to reduce relative displacement between the cure and the control cods under vertical seismic excitation. And the load applied to the main vessel is significantly reduced.
2. A main vessel lateral support that consists of shear keys is attached to reduce the response acceleration by increasing the natural frequency of the main vessel vibration over the resonance in the peak floor response spectrums (~12 Hz).
3. A lateral seismic support of the cure support structure using fluid-coupled vibration effects is introduced. To evaluate seismic characteristics of the pool reactor having the aforementioned features, the following items are important: (a) fluid inertial effect, (b) fluid–structure-coupled vibration, and (c) load transmission characteristics of the main vessel lateral support shear keys.
4. Overall vibration characteristics of the pool reactor structure.
5. Verification of validity and applicability of analytical models, which explain the earlier phenomena.

Four separate vibration tests, indicated in Figure 10.53, were conducted to investigate the previously discussed items:

Test No. 1: Vibration test of the main vessel containing water.
Test No. 2: Vibration test of the assembled model without pumps and IHXs containing water.
Test No. 3: Vibration test of the full assembled model containing water.
Test No. 4: Vibration test of the full assembled model containing water with the main vessel lateral support shear key structure.

For the numerical predictions, a simple *beam model* and a detailed *3D shell element* are employed. In the beam model, structures were modeled as beams with equivalent stiffness, and masses of structure were distributed on each node points. Added mass distributions for the top suspended fluid-filled vessel were determined based on Housner's theory [10.29], which derives the fluid impulsive pressures developed on a rigid cylinder container subjected to horizontal acceleration. The main vessel lateral support was also modeled as spring elements having equivalent stiffness derived by FEM analysis. For the shell model shown in Figure 10.54, the NASTRAN finite element program was employed to compute natural frequencies of the test models containing fluid. The fluid effects were taken into account by using the virtual mass method in NASTRAN. Three-dimensional shell elements (QDAD4 and TRIA3) were used in the analyses. The main vessel lateral support was modeled as spring elements and distributed circumferentially on the vessel grid.

Vibration characteristics (resonance frequencies, vibration mode, etc.) were examined both in the experimental results and analytical results. A comparison of experimental and analyses results of resonance frequencies of the main vessel and upper inner structure (UIS) is shown

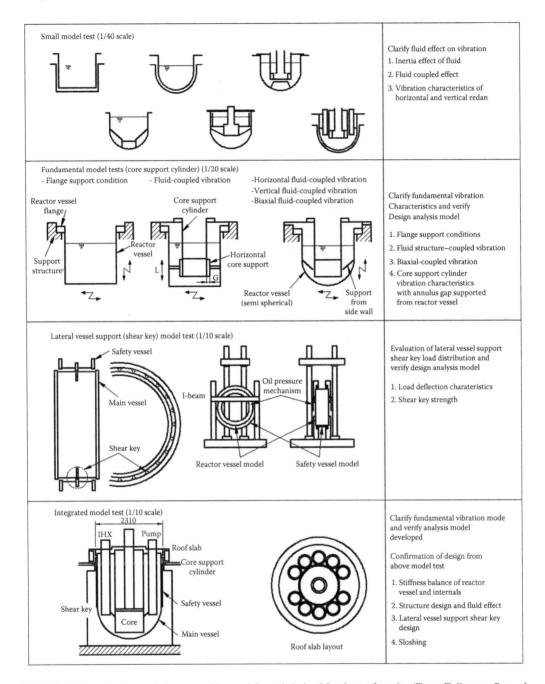

FIGURE 10.53 Various reactor assembly models and their objectives of study. (From Fujimoto, S. et al., *Experimental and analytical study on seismic design of a large pool type LMFBR in Japan, Transactions of Eighth International Conference on Structural Mechanics in Reactor Technology (SMiRT-8),* Brussels, Belgium, 1985, Vol. EK1, pp. 315–320.)

in Table 10.4. As a result, it was concluded that two analytical models mentioned earlier were able to simulate the fluid inertial effect and the fluid coupled vibration of the structure, and also the applicability for the seismic design analysis was confirmed. As for the natural frequency of vibration of the sloshing, theoretical solutions, which are derived by analyses using simplified two concentric cylindrical models of the hot plenum, were compared with the test data (Table 10.5).

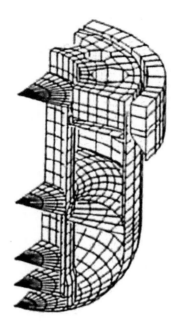

FIGURE 10.54 Finite element model. (From Fujimoto, S. et al., Experimental and analytical study on seismic design of a large pool type LMFBR in Japan, *Transactions of Eighth International Conference on Structural Mechanics in Reactor Technology* (*SMiRT-8*), Brussels, Belgium, 1985, Vol. EK1, pp. 315–320.)

TABLE 10.4

Natural Frequency Values: Comparison of Test Data with Numerical Predictions

Test No.	Main Vessel—First Mode			UIS—First Mode		
	Test	Beam Model	3D Model	Test	Beam Model	3D Model
1	21.8	23.1	22.9	—	—	—
2	19.3	19.1	19.2	24.5	23.1	26.2
3	18.6	18.3	18.8	24.7	23.1	26.2
4	24.3	25.4	25.4	24.8	22.9	26.5

TABLE 10.5

Sloshing Characteristics: Comparison of Test Data with Numerical Predictions

Test No.	Inner Radius	Outer Radius	Test	Analysis
1	0	1.10	0.609	0.60
2	0.502	1.10	0.501	0.49
3	0.502	1.10	0.501	0.45

It was concluded that the analytical results gave a good agreement with the test data for natural frequency of vibration of the sloshing, and the internal structure gave a less influence for the natural frequency of vibration. The analytical model predictions on the seismic response characteristics (natural frequencies and participation factors) are shown in Figure 10.55. In the analysis, a fluid dynamic effect of the fluid coupled seismic support of the core support structure was included in the model.

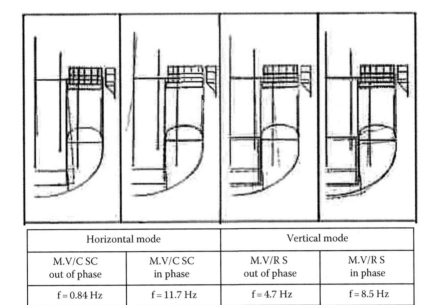

Horizontal mode		Vertical mode	
M.V/C SC out of phase	M.V/C SC in phase	M.V/R S out of phase	M.V/R S in phase
f = 0.84 Hz	f = 11.7 Hz	f = 4.7 Hz	f = 8.5 Hz
β = 1.34	β = 11.05	β = 7.29	β = 15.29

FIGURE 10.55 Mode shapes, natural frequencies, and participation factors β. (From Fujimoto, S. et al., *Experimental and analytical study on seismic design of a large pool type LMFBR in Japan, Transactions of Eighth International Conference on Structural Mechanics in Reactor Technology (SMiRT-8)*, Brussels, Belgium, 1985, Vol. EK1, pp. 315–320.)

10.2.6.5 Elastic Instability of Thin Shells under Seismic-Induced Forces

Elastic instability causes buckling. In the reactor assembly of SFR, the main vessel, inner vessel, and thermal baffles are thin-walled shell structures prone to buckle. The buckling of thin shells under randomly varying seismic-induced forces calls for sophisticated analysis techniques in time domain analysis. The convectional seismic responses analysis provides time-dependent stress and pressure distributions, which is performed in the first phase. Subsequently, elastic instability is investigated by *bifurcation analysis*. Figure 10.56 shows a 3D finite element mesh generated for the seismic analysis, which includes thin vessels, and fluid and fluid-free surfaces. The dynamic pressures developed on the main vessel, inner vessel, and inner and outer thermal baffles, derived from the analysis at a critical instant, are depicted in Figure 10.57. Under the kind of pressure fields generated on the surfaces shown in Figure 10.56, the upper cylindrical portion of the main vessel is subjected to shear buckling (Figure 10.58a). The inner vessel under the combined action of static pressure of 4 m of sodium head and dynamic pressure developed under vertical seismic excitations causes buckling at the toroidal portion (Figure 10.58b). The inner and outer thermal baffles adjacent to the main vessel, which are subjected to dynamic pressure distribution developed under horizontal and vertical excitations, undergo asymmetrical buckling as shown in Figure 10.58c and d. More details can be seen in Ref. 10.30.

The elastic instability has been simulated on 1/13 scaled down models of the main vessel cylindrical portion and inner vessel by imposing the respective peak forces. The prediction of shear buckling mode and critical buckling load for the main vessel straight portion is shown in Figure 10.59. More details can be seen in Ref. 10.31. Elastic instability of the torus shell portion of the inner vessel subjected to pressure and axial forces acting along the stand pipes is simulated experimentally and numerically. A typical comparison is shown in Figure 10.60. More details can be seen in Ref. 10.32. The critical buckling load mode shapes are predicted by using CAST3M computer code. It is found that the CAST3M code used for the shear buckling analysis of thin shells overpredicts the critical shear buckling load by 20% for some cases indicated in Figure 10.60 [10.31].

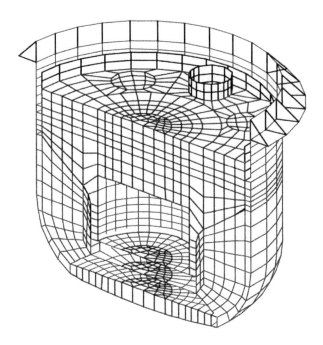

FIGURE 10.56 Finite element mesh of reactor assembly.

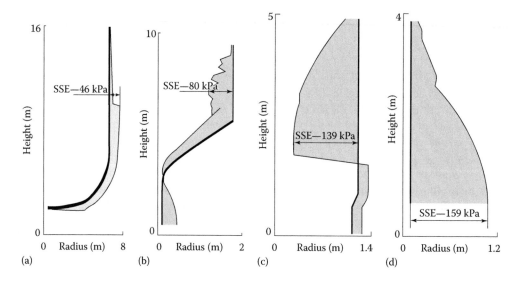

FIGURE 10.57 Peak dynamic pressure distribution at a critical time step w.r.t. buckling: (a) main vessel, (b) inner vessel, (c) inner baffle, and (d) outer baffle.

10.3 THERMAL HYDRAULIC CODES

In this section, a few international benchmark solutions specific to SFR situations are presented.

10.3.1 PREDICTION OF FREE SURFACE VELOCITY LIMIT FOR ONSET OF GAS ENTRAINMENT

Sporadic entrainment of argon gas in primary sodium can lead to reactivity perturbations in the core. One of the free surface parameters that affects argon gas entrainment is free surface velocity. It is essential to control the free surface velocity below certain limits to avoid the risk of gas entrainment.

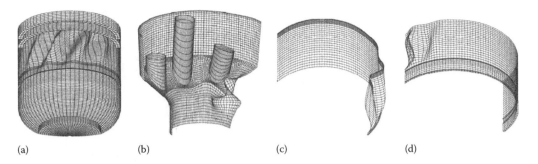

(a) (b) (c) (d)

FIGURE 10.58 Buckling mode shapes of thin vessels of reactor assembly: (a) main vessel, (b) inner vessel, (c) inner baffle, and (d) outer baffle.

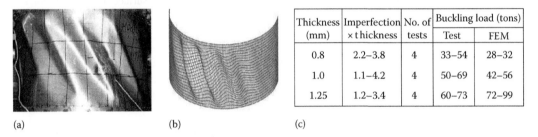

Thickness (mm)	Imperfection ×thickness	No. of tests	Buckling load (tons)	
			Test	FEM
0.8	2.2–3.8	4	33–54	28–32
1.0	1.1–4.2	4	50–69	42–56
1.25	1.2–3.4	4	60–73	72–99

(a) (b) (c)

FIGURE 10.59 Experimental and numerical simulation of shear buckling of thin shells: (a) test mode shape, (b) predicted mode, and (c) comparison of critical buckling load.

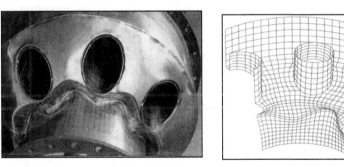

Loading, P	Geometrical imperfection (mm)				P_{exp}	P_{FEM}	$\dfrac{P_{exp}}{P_{FEM}}$
	Lower shell	Torus	Cone	Upper shell			
Pressure (MPa)	−1.2 to 1.4	−1.4 to 1.4	−2.7 to 1.5	−1.3 to 1.7	0.9	0.95	0.95
Axial force, t	−2.5 to 1.3	−1.2 to 1.1	−2.9 to 1.7	−1.8 to 2.5	1.61	1.40	1.15
Pressure+ axial force	−1.2 to 1.1	−1.7 to 1.4	−1.1 to 1.3	−1.8 to 0.9	0.175 1.180	0.151 1.180	0.16

FIGURE 10.60 Simulation of buckling mode under pressure and axial force for inner vessel.

Realistic modeling of gas entrainment by computational fluid dynamics (CFD) simulation requires the use of two-phase flow models that can capture the interface between the argon gas bubbles and liquid sodium. However, performing CFD studies for large-size reactor pools is very demanding in terms of computer memory and time. To circumvent this difficulty, small-size models having geometric similarity with reactor hot pool have been studied by CFD simulations employing volume of fluid (VOF) method. Based on these parametric studies, it is found that gas entrainment risk can be

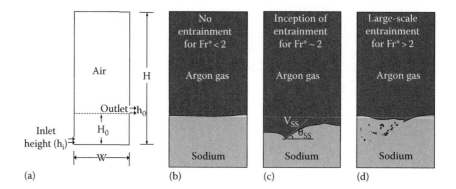

FIGURE 10.61 CFD study of gas entrainment in ideal models to arrive at a free surface velocity criterion for onset of argon entrainment in sodium. (a) CFD model sketch, (b) no entrainment, (c) inception of entrainment, and (d) large scale entrainment.

avoided if the modified Froude number (Fr^*) at the free surface is <2.0 [10.33]. According to this criterion, the limiting free surface velocity to avoid gas entrainment in SFR hot pool is found to be 0.5 m/s. The idealized geometry and free surface profiles as a function of Fr^* are depicted in Figure 10.61.

10.3.2 Development of Porous Body Model for Internal Blockage in Subassembly

A fast reactor fuel SA consists of a large number of pins. The fuel pins are very thin and are densely arranged in a triangular pitch. The hydraulic diameter for the coolant flow in the SAs is very small (~3 mm). Because of the miniature gap, there is a finite chance for the formation of local blockage in a coolant subchannel. In the case of subchannels with local blockage, there is risk of local temperature shoot-up and sodium boiling with consequent changes in the reactivity. Since a partial blockage in an SA would threaten the safety of the SA, a detailed understanding of local temperature distribution in the SA and the capability of core temperature monitoring system to detect the partial blockage are essential. CFD simulation of fuel SA with helical wire wrap itself demands a huge computational time. Hence, it is difficult to have a detailed model of the debris/particles that form the local blockage. It is practical to adopt a porous body model for the debris. However, such models have to be validated against suitable benchmark experiments. Toward this, a porous body model for partial internal flow blockages [10.34] was analyzed in CFD to predict the sodium temperatures in the blocked subassembly. The predicted sodium temperature within the porous blockage has been compared against the benchmark SCARLET-II experiments of Olive and Jolas [10.35]. SCARLET-II experiment was conducted with 19-pin wire-wrap bundles including a central blockage comprising 6 subchannels of 60 mm long along the flow direction, as shown in Figure 10.62. The cross-sectional view of the CFD model geometry considered is also highlighted in the same figure. Pin diameter is 8.5 mm, and the pins are arranged in triangular pitch of 9.79 mm with an electrical heating of 45 kW/pin. Porosity of the blockage particles is 0.32. The objective of the experiment was to provide maximum temperature within blockage. Sodium temperature predicted in the blockage region is compared with the measured data in Figure 10.62, demonstrating the reliability of the model.

10.3.3 Validation of Natural Convection in a Series of Horizontal Cylindrical Fins

Sodium valves are used in secondary sodium systems to isolate SG units that are faulty. These are frozen seal butterfly valves. This valve has a circular disk, perfectly fitting in the valve seat provided in sodium pipes. The disk is attached to a cylindrical stem and the stem is covered by a sheath, which itself is attached to the pipeline, as depicted in Figure 10.63. There is positive clearance (forming an annular path) between the stem and the sheath. Due to the pressure in the sodium line,

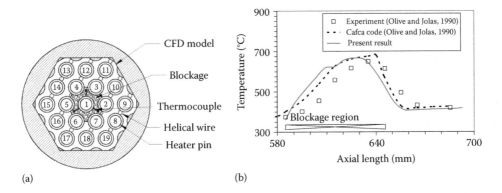

(a)   (b)

FIGURE 10.62 Validation of present calculation with SCARLET-II experiment: (a) SCARLET-II test section and (b) comparison of temperature in the blocked region. (From Olive, J. and Jolas, P., *Nucl. Energy*, 29, 287, 1990.)

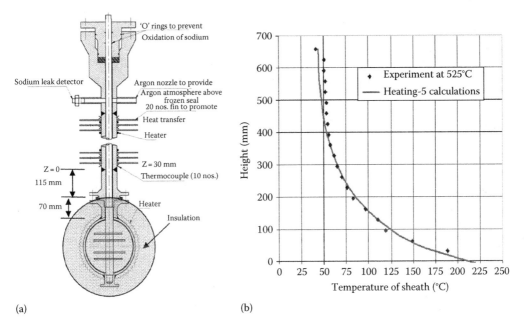

(a)   (b)

FIGURE 10.63 Details of (a) frozen seal sodium valve tested at SILVERINA facility and (b) comparison of predicted and measured sheath temperatures for sodium at 525°C.

sodium rises in this annular path. To avoid sodium leaking through this path, the sodium is forced to freeze in this path. To promote freezing of sodium, heat transfer from the valve sheath has to be enhanced. This is achieved by providing circular fins attached to outer surface of the sheath. It is essential to ensure that freezing is achieved solely by natural convection of air. Natural convective cooling of fins is passive and hence is highly reliable. Too large a number of fins are uneconomical while too less a number would not lead to sodium freezing. Hence, the number of fins required and the fin spacing are to be determined by an integrated conjugate thermal analysis of the valve assembly. One of the important inputs for this analysis is knowledge of natural convective heat transfer coefficient over fins. As the fins are stacked one over the other, conventional correlations for Nusselt number are not applicable. The adjacent horizontal fins of sodium valve form open-ended cavities (see Figure 10.63). Natural convection of air takes place through these constricted cavities. For these configurations, heat transfer coefficient correlation has been developed based on parametric studies by CFD calculations [10.36]. Using these correlations, conjugate heat transfer analysis of the entire valve assembly including all the fins is carried out using the thermal analysis code HEATING5 [10.37], which considers surface-to-surface radiation interaction among the fins. Based on these studies, acceptable values for the number of fins and length and gap between them are arrived at. This is followed by experimental verification of the selected valve assembly in SILVERINA sodium loop at IGCAR. The stem temperature measured in the facility and that predicted by the computer code are compared in Figure 10.63, demonstrating validity of the heat transfer correlations developed by CFD studies.

10.3.4 COVER GAS THERMAL HYDRAULICS

The roof slab, a major portion of the top shield in a typical pool-type SFR, supports many components such as IHX, DHX, primary sodium pumps, control plug, and rotating plugs. The penetrations provide annular spaces between the roof slab and the components. These gaps are filled with argon that communicates with the argon cover gas above the free surface of sodium. These annular gaps are subjected to axial temperature gradient that induces cellular convection. Therefore, heat and mass transfers between the free sodium surface and roof slab structures must be known to correctly design the cooling circuit (thermal loadings and sodium deposits). Sodium aerosols are formed in the cover gas space that can affect the total heat transfer through the annulus. Because of mass transfer, sodium deposits may exist, especially in the cooler parts. Basic understanding of the cover gas heat and mass transfer phenomena (for SPX-1 reactor) was obtained through a large mock-up test [10.32], GULLIVER facility (Figure 10.64) which represents the upper closing of the reactor. It has been noted that aerosol formation depends mainly on sodium temperature and the temperature gradient between the sodium and the roof slab. Engineering-size models of the roof slab are found essential to understand the thermal performances of the roof slab. Development of theoretical models and their verification on specific experimental models remain necessary, particularly for sodium aerosol mass concentration, particle size distribution, and thermal radiation in sodium vapor–filled areas. Two-dimensional CFD-based thermal hydraulic codes have been developed to predict cellular convection of argon gas and temperature distribution in various shells that form the annuli. These codes have been

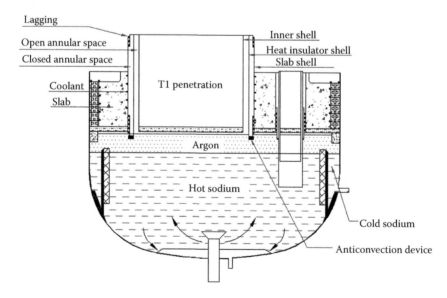

FIGURE 10.64 GULLIVER experimental setup for cellular convection in roof slab. (From Wakamatsu, M. et al., *Nucl. Sci. Technol.*, 32[8], 752, 1995.)

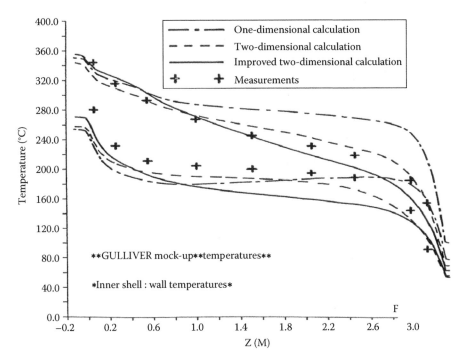

FIGURE 10.65 Inner shell temperatures along hot and cold legs in GULLIVER.

validated against the GULLIVER experimental loop. Figure 10.65 depicts inner shell temperature along the hot and cold legs of the cellular convection loop predicted by 1D and 2D codes along with the measured data.

10.3.5 THERMAL STRIPING

Thermal striping is a random temperature fluctuation generated by the mixing of flow streams at different temperatures. In an SFR, it occurs in the upper plenum as a result of the mixing of sodium coolant jets from fuel, blanket, and control SAs. It is necessary to consider the temperature cycling in the context of fatigue on the structures positioned close to the mixing region. More details of this phenomenon are presented in Chapter 9. To understand thermal striping, many fundamental experiments have been carried out using different fluids. One of the important benchmark experiments is that of Wakamatsu et al. [10.38], wherein experiments simulating the upper core structure and impinging jets from fuel SAs and control rod channels have been performed (Figure 10.66). Sodium and water were used as simulated fluids to examine the differences in their physical properties. Based on the temperature fluctuation data in fluid and solid surface, a simplified model has been developed for quantification of temperature attenuation in the boundary layer. The ratio of surface temperature fluctuation to fluid temperature fluctuation is 30%–80% in water tests and it is 20%–50% in sodium tests. These data are used to develop CFD-based computational models for predicting temperature fluctuations in the reactor. Prediction of thermal striping requires advanced turbulence models, like direct numerical simulation (DNS) and large eddy simulation (LES). The peak-to-peak temperature fluctuations in sodium (2 mm away from the test piece) (Figure 10.67) and in the test piece predicted by the CFD code employing DNS are depicted in Figure 10.67a. In this figure, thermocouple location 11 is the interface between the 2 jets. The maximum peak-to-peak temperature fluctuation occurs at the interface. Also, there is a large attenuation of temperature fluctuation within the boundary layer. Figure 10.67b depicts the measured temperature fluctuation in the fluid and in the solid.

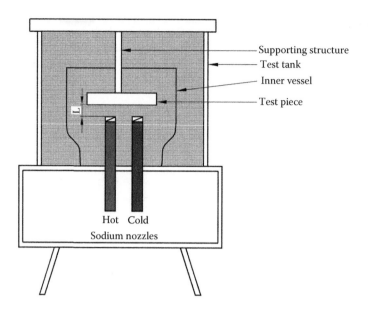

FIGURE 10.66   Thermal striping benchmark experimental facility (dimensions in mm).

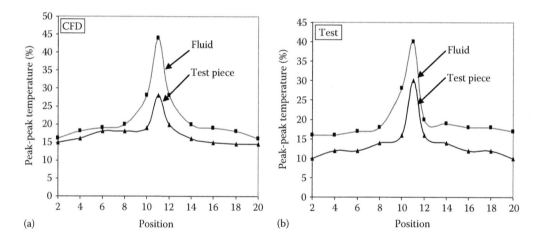

FIGURE 10.67   Amplitude of temperature fluctuations in the fluid and test piece. (a) CFD analysis and (b) actual measurement.

10.4   LARGE-SCALE EXPERIMENTAL VALIDATIONS

10.4.1   Validation of Phénix End-of-Life Natural Convection Test

Pool-type SFR that was in operation in France was shut down permanently in the year 2009. The French nuclear design group carried out several tests in the reactor before its permanent shutdown to generate valuable data and knowledge based on various aspects of reactor design and operation. Under this activity, natural convection tests in the primary circuit of the reactor were carried out. This exercise was open to international collaboration through an International Atomic Energy Agency (IAEA)-sponsored collaborative research project. Several countries including India participated in this exercise [10.39]. The main objective of this activity was to improve the participants' analytical capabilities in various fields of research and design of SFR. Due to the favorable thermophysical properties of sodium coolant, natural convection phenomenon is adopted for the safe and reliable means of decay heat removal. Elimination of external pumping power in this system ensures high level of reliability. Decay heat removal system is one of the most important safety functions that is required to be accomplished with high reliability; thus, the system design should also be carried out using high-performance tools and models. This can be achieved through validation and improvement of mathematical tools and models adopted in the design against benchmark data. The international benchmark on the natural convection test was organized with *blind* calculations as a first step, and then *posttest* calculations and sensitivity studies were compared with reactor measurements. Eight organizations from seven member states took part in the benchmark: ANL (United States), CEA (France), IGCAR (India), IPPE (Russian Federation), IRSN (France), KAERI (Korea), PSI (Switzerland), and University of Fukui (Japan). Each organization performed computations and contributed to the analysis and global recommendations.

The natural convection test in Phénix reactor was performed on June 22–23, 2009. Phénix is a pool-type SFR composed of a main primary reactor vessel, three sodium secondary circuits, three tertiary water/steam circuits, and one turbine. Phénix nominal power was 560 MWt/250 MWe. However, during the ultimate tests in 2009, one secondary circuit was out of operation and the reactor was operating at a reduced power of 120 MWt before the natural convection test. A schematic of the reactor assembly is shown in Figure 10.68.

Before the beginning of the test, the reactor was operating at full power (350 MWt) for one day, and then the reactor power was manually decreased from 350 to 120 MWt and the reactor was stabilized. The usual nominal core flow rate is 3000 kg/s at full power, but it was reduced to 1280 kg/s for the test. Only four IHXs were in operation during the ultimate tests, the two other ones being replaced by inactive components called DOTE. The standard instrumentation available in the reactor and a special instrumentation for the final tests were used simultaneously.

The standard instrumentation is used to measure

- Primary and secondary pumps speed, which can be used to estimate the flow rate via the pumps characteristics, when the pumps are operating
- Secondary mass flow rate on each secondary loop

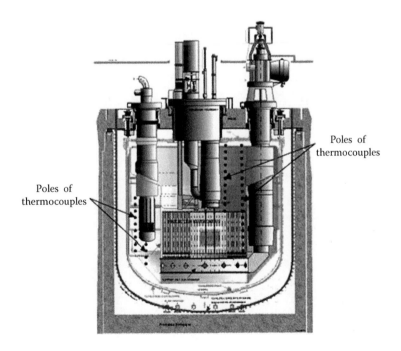

FIGURE 10.68 Schematic of reactor assembly of Phénix.

- Inlet temperature to each primary pump
- Fuel SAs outlet temperature measured a few centimeters above each fuel SA
- IHX inlet and outlet temperatures on both the primary and secondary sides

The test was carried out from this power level. The natural convection test in the primary circuit is divided into two phases: (1) with no significant heat sink in the secondary circuits, except the heat losses along the piping and through the casing of the SG, and (2) with significant heat sink in the secondary circuits, by opening the casing of SG at the bottom and at the top, which involves efficient air natural circulation in the SG casing. The main objective of the computational studies is to qualify the prediction capability of the onset of sodium natural circulation in the primary circuit. Computational studies have been carried out using system codes developed based on classical 1D approach, as well as using coupled 1D/3D code systems.

The transient phenomenon in the 3D geometry of the primary circuit of Phénix reactor is analyzed by solving the governing differential equations of flow physics by numerical means on a computational mesh. The tool selected has the capability of solving steady, transient, laminar, turbulent, compressible flow phenomena along with heat transfer (convection and conduction) even in porous medium. One-dimensional models for the secondary circuit are added to the code through user-defined program modules. The representative computational mesh adopted for the analysis is shown in Figure 10.69. The classical analysis approach using system dynamics codes adopts models for reactor core, primary sodium circuit, secondary circuit, sodium pumps, heat exchangers, and sodium pools. Thermal models are based on heat balance between various sections exchanging the heat such as fuel and sodium through the clad in SA, primary sodium and secondary sodium through the tube wall in IHX, sodium and ambient air through the pipe wall and insulation in sodium piping, and secondary sodium and water through the tube wall in SGs. Hydraulic model is based on momentum balance between various flow segments in the primary and secondary sodium circuits. Torque balance is adopted for the modeling of pump with the characteristics derived from

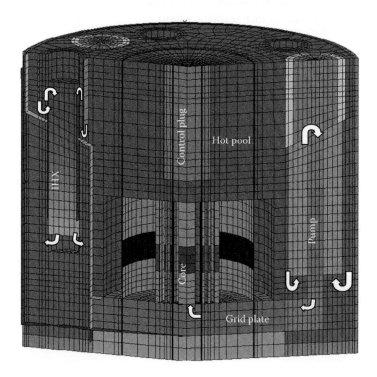

FIGURE 10.69 Computational mesh for 3D analysis.

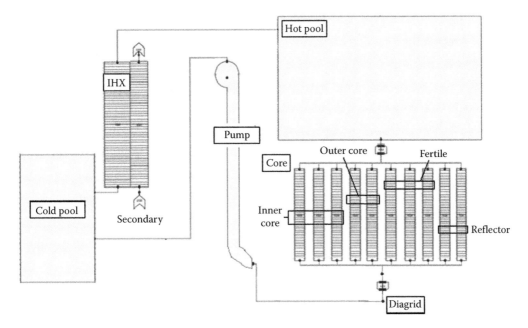

FIGURE 10.70 Nodalization scheme in the system dynamics model.

generalized homologous characteristics. Fluid levels in the tanks are modeled through dynamic mass balance. Neutronic model for the core is based on point kinetics approximation. Transient solution is obtained by prompt jump approximation. Detailed models are also incorporated for the calculation of various reactivity feedback effects due to radial expansion of grid plate where SA are supported, control rod expansion, volumetric expansion of sodium inside the fuel and blanket portion of core, axial clad steel expansion, axial fuel expansion, and Doppler effect due to changes in fuel temperature. The representative nodalization scheme adopted for modeling primary sodium circuit of Phénix reactor is shown in Figure 10.70.

The codes used by the benchmark participants are listed as follows:

- ANL with SAS4A/SASSYS-1 1D code
- CEA with CATHARE_2 1D code
- IGCAR with commercial CFD code for blind and posttest exercises, and DYANA-P 1D code only for posttest exercise
- IPPE with GRIF 3D code
- IRSN with CATHARE_2 1D code
- KAERI with MARS-LMR 1D code
- PSI with TRACE 1D code
- University of Fukui with NETFLOW++ 1D code, only for posttest exercise

Natural convection flow evolution in the core predicted by different participants is shown in Figure 10.71. Large deviation can be observed between the predictions made by different partici- pants. Unfortunately, there is no measurement of natural convection flow rate in the reactor to be compared with computational results. The predicted values of core inlet temperature are shown in Figure 10.72. The predicted values of primary coolant temperature at the inlet and outlet of heat exchanger in the primary sodium circuit are shown in Figures 10.73 and 10.74, respectively. The inlet temperature is not well predicted by the computational codes. This is due to the strong influ- ence of thermal stratification effects in the hot pool, which is difficult to predict accurately with the system codes. It is also influenced by the relative location of its measurement in the experiment and

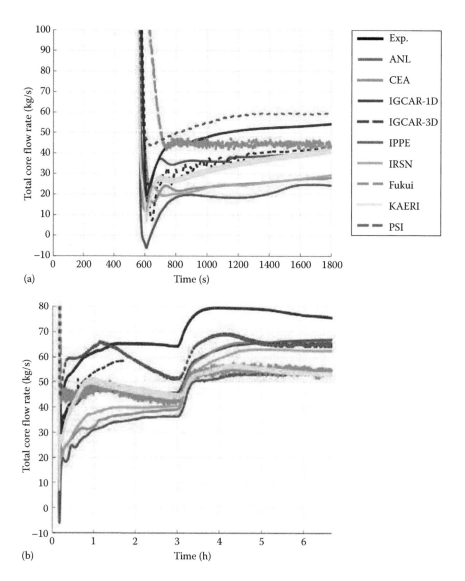

FIGURE 10.71   (See color insert.) Prediction of natural convection evolution in the core by different participants: (a) represents the zoomed view at the initial time period of figure and (b) for more understanding.

monitoring location in the numerical study. Other plant parameters predicted by different codes match reasonably well with experimental data.

The main lessons learned from the Phénix test and the coordinated research project (CRP) benchmark are as follows.

The effectiveness of intrinsic safety features of a fast reactor such as (1) core power reduction before SCRAM due to negative reactivity feedback effects, (2) rapid onset of natural convection through the core after the forced cooling is stopped, (3) significant contribution of heat absorption by the thermal inertia of sodium pools and heat removal through various heat loss paths, and (4) efficient heat removal through SG natural convection air cooling was well demonstrated during the test.

Discrepancies were observed in the code-to-code comparisons of the core flow rate predictions in the natural convection regime, but there was overall coherence between computed and measured temperature evolutions. Limitations were observed in the capability of 1D system codes to deal with

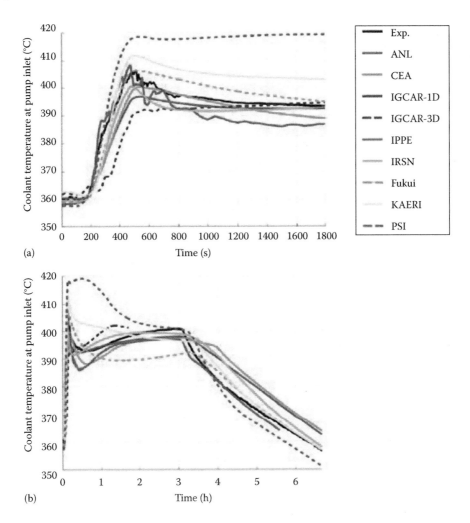

FIGURE 10.72 **(See color insert.)** Evolution of core inlet temperature: (a) represents the zoomed view at the initial time period of figure and (b) for more understanding.

buoyancy effects such as thermal stratification in plena or local recirculation. Use of 3D codes or coupling 1D system with 3D codes is required for a better estimation of such complicated flow conditions.

The CRP has pointed out the need for carrying out a complete validation process for the numerical codes used for prediction natural convection decay heat removal in the areas of coupled 1D/3D model, meshing, pressure drop, and heat transfer correlations. Sensitivity analysis is useful to detect the most important parameters that have an influence on the main issues as the cladding temperature and the core outlet temperature. Analytical tests will give information on correlations to be used in natural convection. Reduced scale water models can be developed to qualify the code on more global representative geometries. Further in-reactor experimental tests are needed to verify the adequacy of the codes to compute real reactor conditions. The natural convection test in Phénix reactor has been used as one common support in this process.

10.4.2 Pool Hydraulic Experiments for French Reactors

Thermal hydraulics of the upper plenum is one of the noblest parts of the fast reactor thermal hydraulic activity as many challenges are concentrated in this region. Starting at the core outlet level, an important safety requirement is the validation of the core outlet temperature measurements.

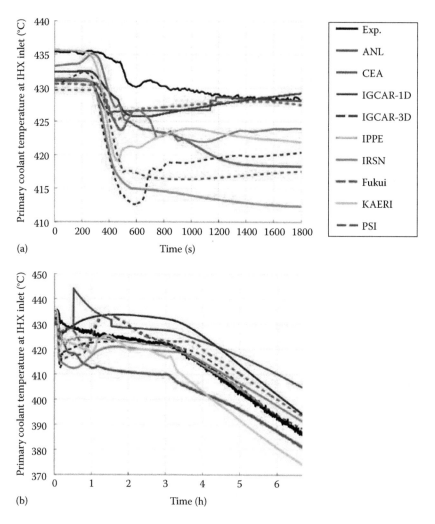

FIGURE 10.73 **(See color insert.)** Primary coolant temperature predicted at heat exchanger inlet: (a) represents the zoomed view at the initial time period of figure and (b) for more understanding.

Thermocouples located above each fuel SA can play a key role in the prevention of reactor accident due to a flow blockage. It is quite important to ensure that the thermocouples measure the effective SA outlet temperature for all operating conditions. This question is raised because the thermocouples are generally located a few centimeters above the core outlet level, as they are mounted on special devices supported by the control plug. The SA jets at high velocity and the radial flow below the control plug may induce vibration. A detailed velocity and pressure field with information on the fluctuating values is needed to evaluate the risk of flow-induced vibration. The porosity of the upper core structure is a parameter that will influence the thermal hydraulic behavior in the whole hot pool. It will also have a significant effect on the recirculating flow in the interwrapper region.

At the sodium free surface, level oscillations could occur as a consequence of the hydraulic behavior in the upper plenum, with the associated thermal fatigue on the inner vessel and other internal structures due to the temperature gradient at the free surface. The region of the upper plenum located along the inner vessel that separates the upper plenum and the lower plenum may be concerned by a significant temperature gradient and possible thermal stratification. The clad failure detection requires a reliable system to avoid a large quantity of fission products to be dispersed in the primary circuit. The thermal hydraulic behavior of the upper plenum including the control

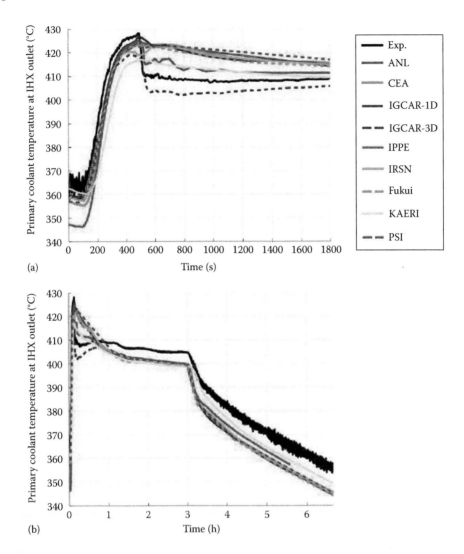

(a)

(b)

FIGURE 10.74 (See color insert.) Primary coolant temperature predicted at heat exchanger outlet: (a) represents the zoomed view at the initial time period of figure and (b) for more understanding.

plug is an important parameter to determine the transfer function between the SA and the detector. The transit time and the signal attenuation are directly dependent on the velocity field and the turbulence characteristics in the upper plenum.

The use of experimental facilities remains very useful for qualification process and the rapid test of design adjustments. The thermal hydraulics of core outlet region has been simulated by reduced scale mock-ups using water. In the 1980s, CEA has built a water facility known as PERIGEE for studying the thermal hydraulic behavior of the core outlet region and the whole upper plenum of Superphénix reactor [10.40]. This facility was a 90° sector of hot pool at a reduced scale of 1/5, without any simulation of the interwrapper region. In the 1990s, a new water facility known as JESSICA was built for studying thermal hydraulics of core outlet region of the European fast reactor (EFR). This facility (Figure 10.75) was also a 1/3 scale 90° sector with a representation of the interwrapper region to estimate the recirculating flow and its influence on the core outlet behavior.

During the Superphénix project and the EFR project, the experimental approach was the unique way to estimate the thermal hydraulic behavior of the control plug. The highly complicated internal geometry of this component including the control rods and numerous instrumentations led to a situation

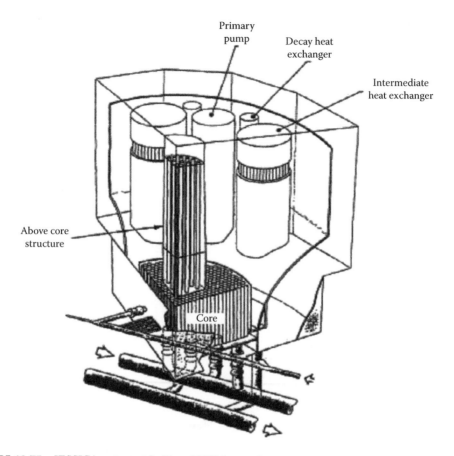

FIGURE 10.75 JESSICA water test facility of EFR hot pool.

FIGURE 10.76 COLCHIX water model of EFR hot pool.

FIGURE 10.77 COCO water model of EFR cold pool.

that use of numerical approach is nearly impossible. So, water test facilities as for the control plug were used, at a reduced scale between 1/8 as COLCHIX experiment (Figure 10.76) and 1/3 as JESSICA experiment. These facilities were designed to enable the measurement of the flow distribution in the control plug structure in steady-state condition. They also allowed the simulation of transient regimes with imposed core flow rate and temperature evolutions. Hence, the temperature evolution in the upper core structure was measured simultaneously with plenum temperature evolution. These simultaneous measurements were used to estimate the thermal stresses on the control plug shell.

For the prediction of transient conditions in cold pool, numerical simulation is largely used to estimate the evolution of the flow and temperature fields. From an experimental point of view, water mock-up has been used for qualification. Richardson similarity is required as buoyancy influence is important in the cold pool. Figure 10.77 shows the COCO water test facility used for the EFR project. It was a 1/10 reduced scale model where various transient regimes were simulated. The heat exchanger tube bundle was partly represented to provide a reasonable velocity profile at the heat exchanger outlet.

10.4.3 THERMAL STRATIFICATION OF SODIUM IN POOLS

The experimental studies devised for studying thermal stratification should take into account turbulence and low Prandtl number of liquid metals specifically. Sodium experiments such as CORMORAN [10.41] in France have been performed to assess a generic validation on thermal stratification and heat transfer correlation for various flow regimes. Tests on CORMORAN model and supporting computational work were conducted to investigate the thermohydraulic behavior in a rectangular cavity and to improve the understanding of the thermal hydraulics of the Corps-Mort region of the EFR. Because of the geometrical distortion between CORMORAN and the reactor, the purpose of the tests is not to directly simulate the characteristics of the stratification but to study the main phenomena involved, that is, (1) the position of the stratification, (2) the values and the instabilities of vertical temperature gradient in the vessel, and (3) the temperature fluctuations at the stratification boundary. These experimental results are used to validate the computational

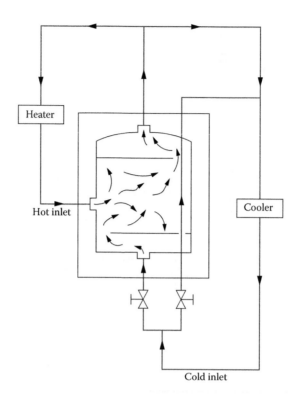

FIGURE 10.78   CORMORAN setup.

approach used in the assessment of stratification problems in the reactor. The CORMORAN model (Figure 10.78) is a rectangular cavity fed with hot sodium. At the bottom of the model, there is a structure with horizontal and vertical plates representing the Corps-Mort structure. The right-side wall of the cavity simulates the redan, which can be cooled by forced convection on its external side. Two configurations were used to cool the test section: (1) heat transfer through the redan wall with cold sodium flowing in the side duct and (2) injection of a small cold flow into the larger of the two chambers directly under the Corps-Mort horizontal baffle. Fluctuations near the stratification interface were evaluated from sequential measurement and from signal processing.

Two series of numerical computations were conducted with the finite volume version of the TTIO_U Code [10.42]. Both approaches start with the solution of unsteady filtered Navier–Stokes equations. The difference between the two series of tests is principally the turbulence model: k-ε model and LESs. The purpose of 2D computations using the standard k-ε turbulence model is to calculate the time-averaged temperature gradient and stratification interface. At the centerline, the comparison between computed temperature and experimental measurements is good. Near and on the cold wall, the vertical gradient agrees well but the interface position is slightly underestimated. The purpose of 3D computations using LES is to evaluate temperature instabilities and fluctuations, in complement of time-averaged results. This provides simulations of turbulent fluctuations and global instabilities, which helps in the better understanding of the physics. Figure 10.79 shows the comparison of a TRIO_U calculation and the CORMORAN cavity experiment performed at CEA with a thermal stratification configuration.

10.4.4   CFD Prediction of Thermal Stratification in MONJU Reactor

In a fast reactor with breeder SAs and internally stored spent fuel SAs, the temperatures of sodium exiting from the fuel are vastly different. Due to this temperature difference coupled with large thermal expansion coefficient of sodium and huge size of the reactor pool, the Richardson number

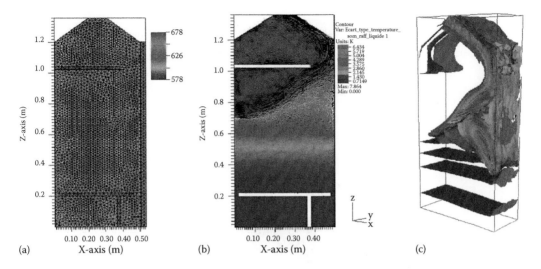

FIGURE 10.79 **(See color insert.)** TRIO_U calculation of CORMORAN sodium experiment: (a) mesh, (b) velocity and temperature, and (c) temperature fluctuation.

in the pool is of the order of unity. This leads to formation of thermal stratification. Turbulence model is a critical issue in CFD analysis, especially for stratified flow conditions. Most of the turbulence models have been developed mainly for forced convective flows. Identification of suitable turbulence model for buoyancy-dominated flows, assessment of predictive capabilities of standard turbulence models for liquid-metal applications with stratification, and establishing the adequacy of mesh and approximation made in the geometry for small-scale structures in the pool are all ongoing research activities in the domain of CFD for fast reactor applications. Experimental data measured in large size sodium loops or data measured in operating reactors are ideal for CFD model validation. Toward this, IAEA has proposed a CRP-based transient hot-pool temperature distributions, measured in MONJU reactor during the event of a simulated reactor trip followed by *loss of condenser vacuum*. The CRP is participated by France, Russia, the United States, Korea, China, and Japan.

MONJU is a loop-type sodium-cooled Japanese fast reactor. Measurements carried out in the hot pool of MONJU during steady-state and transient conditions indicate strong stratification of sodium. As a part of this collaborative project, 3D CFD model of MONJU hot pool has been developed, and steady and transient thermal hydraulic analyses have been performed. The 3D CFD model of MONJU upper plenum consists of upper core structure and control rod guide tube represented explicitly. Direction-dependent pressure loss coefficients are used to account for the resistance offered by other fine-scale structures. The flow holes in the inner barrel are also modeled explicitly (Figure 10.80a). High Reynolds number standard k-ε model and Reynolds stress turbulence (RST) model have been used to simulate the effects of turbulence. Two profiles for the holes have been considered in the CFD model: sharp-edge holes (Figure 10.80b) and round-edge hole (Figure 10.80c). The predicted transient temperature distribution in the pool at a particular instant of 10 min as predicted by the k-ε model is depicted in Figure 10.81. Formation of stratified front in the pool due to the injection of cold sodium from core, upward movement of stratification interface, and flow bypass to outlet nozzle through holes in the inner barrel are very well predicted by the various CFD codes adopted by various countries. Both the turbulence models predicted identical rate of interface movement and nearly equal magnitudes of the axial temperature gradient, suggesting that the standard high Reynolds number k-ε model is good for mixed convective flows of sodium.

Vertical temperature distribution predicted at the thermocouple rack location as predicted by the CFD model is compared against the plant data in Figure 10.82. The interface moves upward as time progresses. When the perforations are modeled as sharp-edge holes, it is clear that the CFD

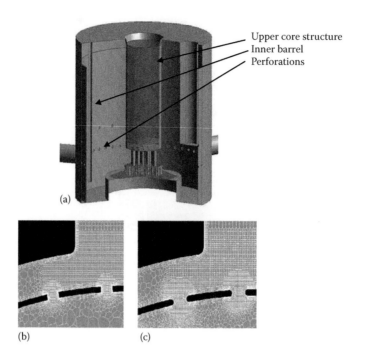

FIGURE 10.80 MONJU hot pool: (a) schematic, (b) sharp-edge holes, and (c) round-edge holes.

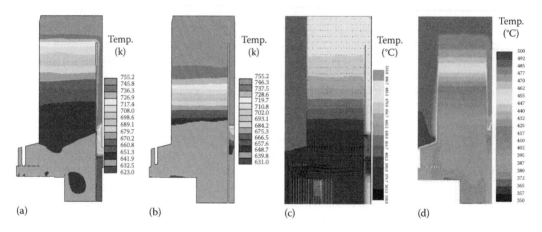

FIGURE 10.81 **(See color insert.)** Transient temperature field in MONJU hot pool at 10 min: (a) India (sharp edge), (b) India (round edge), (c) Russia (round edge), and (d) China (sharp edge).

simulation overpredicts that the upward movement of the stratification and temperature gradient at the interface is sharper in numerical simulations than in the measurement. However, when the perforations are modeled as round-edge holes, the CFD predictions agree closely with the test data. This suggests that finer details of the geometry need to be properly modeled for accurate prediction of thermal stratification in SFR pools.

10.4.5 THERMAL STRIPING

Several studies have been carried out internationally to understand the effects of thermal striping. Water and sodium experiments [10.43,10.44] were carried out in Japan on a configuration of triple parallel jets. Detailed temperature and velocity fields were measured by a movable thermocouple

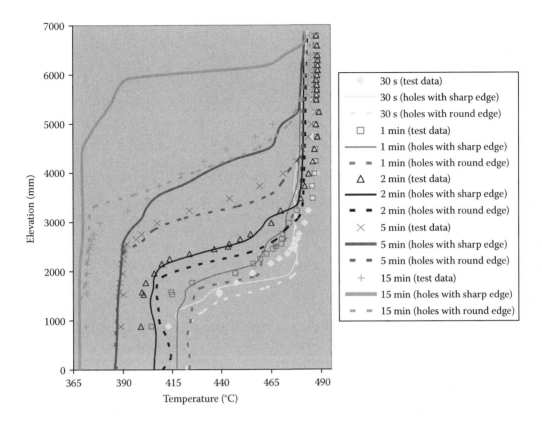

FIGURE 10.82 **(See color insert.)** Vertical temperature distribution at thermocouple rack position.

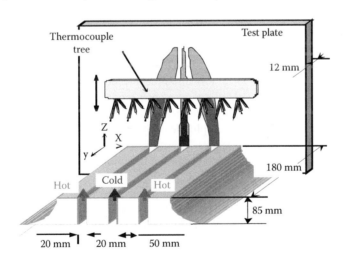

FIGURE 10.83 Schematic of parallel triple jet test section in sodium.

tree and particle image velocimetry (only in the water experiment). A schematic of the test section of the sodium experiment is shown in Figure 10.83. This flow geometry is a simplified model of an outlet of control rod channel surrounded by core fuel SAs. In the triple jet experiment, a vertical wall stands in parallel to the three jets. Detailed flow visualization results were obtained in water experiments. Figure 10.84 depicts a 1/15 of a second time sequence of the triple jet under homogeneous temperature condition. The images have been taken with laser-sheet (argon laser) illumination from

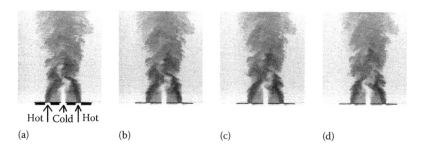

Hot | Cold | Hot

(a)   (b)   (c)   (d)

FIGURE 10.84   Visualization of triple parallel jet flows: (a) 0, (b) 1/15, (c) 2/15, and (d) 3/15 s.

the side and uranine dye added to the water. The side-to-side swaying in unison of the three jets was observed. Overall temperature distributions in the sodium test are similar to those in the water test. The temperature distribution in the downstream region, near the wall, is steeper than at midplane. The profiles in the sodium experiment are quite similar to those in the water except at the downstream midplane. The fluctuation intensity in the water decays earlier in the downstream midplane compared to that in the case of sodium. Theoretical analyses were carried out by using a numerical code based on finite difference method. Temperature profile and fluctuation characteristics were found to be well predicted [10.45].

Korea Atomic Energy Research Institute (KAERI) has provided detailed experimental data for the thermal striping phenomenon [10.46], mainly to test the turbulence models and LES approach. The test sections used are planar double-jet and planar triple-jet arrangements, and the working fluid is air. Several experiments are performed by varying the inlet temperatures and velocities. Ushijima et al. [10.47] carried out numerical calculations for a coaxial jet with different temperatures using the high-Reynolds-number differential stress and flux model. The time average of the velocity and temperature generally agreed with the measured data; however, some discrepancies were found in the turbulence quantities, such as the turbulent heat fluxes. Nishimura et al. [10.48] performed a numerical calculation for the triple-jet experiment and found that the low-Reynolds-number differential stress and flux model is able to simulate appropriately the experimental results most importantly with respect to the oscillatory motion and consequently the mean profiles of the flow pattern, while the k-ε model consistently under predicted the extent of mixing.

Another study has been carried out to test three turbulence models (Figure 10.85) such as the shear stress transport (SST) model [10.49], the two-layer model [10.50], and the elliptic relaxation (V2-f) model [10.51]. It was seen that only the elliptic relaxation model is capable of predicting the oscillatory behavior of temperature. The amplitude of the temperature fluctuation predicted by this model is smaller than the experimental values. Thermal striping in SFR has been studied for a long time in France with reference to the crack observed in the pump vessel of the Phénix reactor. The LES model implemented in TRIO_U is well adapted to determine the amplitude and frequency of the fluctuations. Air, water, and sodium tests were performed on core outlet mixing regions, mixing tees, and other mixing regions for the validation of the computer code. Figure 10.86 shows the comparison of a TRIO_U calculation and the WATLON water mixing tee experiment at JAEA [10.52]. Figure 10.87 shows the comparison of a TRIO_U calculation and the sodium PLAJEST mixing experiment performed at JAEA [10.53].

10.4.6   SUBASSEMBLY THERMAL HYDRAULICS

In the case of fast reactors, the fuel pins are arranged in a triangular pitch and the pin bundle is housed inside a hexagonal sheath. Helically wound spacer wires are provided over each fuel pin to avoid pin-to-pin contact and to guard the pin bundle against flow-induced vibration. The helical spacer wires also promote mixing of sodium among various subchannels. Direct contact between spacer wire and fuel pin leads to large-scale circumferential variation in heat transfer coefficient

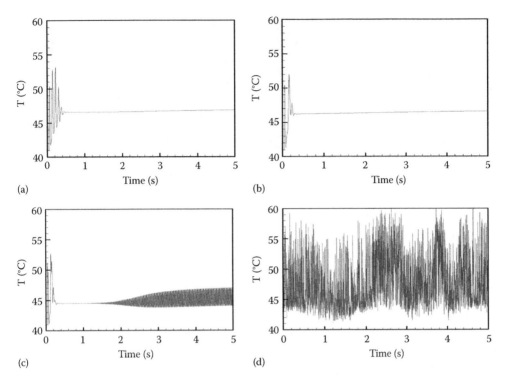

FIGURE 10.85 Temperature evolution during the first 5 s: (a) SST model, (b) two-layer model, (c) V2f model, and (d) experiment.

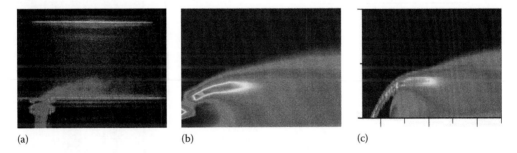

FIGURE 10.86 TRIO_U calculation of WATLON experiment at JAEA [10.53]: (a) experimental visualization, (b) experimental prediction of temperature fluctuation, and (c) TRIO_U prediction of temperature fluctuation. (Copyright October 2008, the American Nuclear Society, La Grange Park, Illinois, FL.)

around a fuel pin with the associated hotspots in the fuel clad. Spacer wires lead to generation of circumferential flow in the bundle, which exhibits periodic variations along the flow direction. In the early periods of SFR design, it was felt almost impossible to carry out a 3D CFD simulation of flow and temperature fields inside the fuel pin bundle of reactor size. To overcome this difficulty, a subchannel analysis approach was proposed. In this approach, the temperature, pressure, and velocity in a subchannel are averaged, and a representative thermal hydraulic condition specifies the state of a subchannel. Flow and temperature distributions in the core are obtained by modeling and solving the conservation equations in a subchannel. Therefore, it is essential to model the intersubchannel heat transfer between the adjacent subchannels as accurately as possible. In such models, at any axial plane, the SA is divided into two control volumes. The unknowns are two axial velocity components, one circumferential velocity, two temperatures, and one pressure. The subchannel mass, momentum, and energy equations are solved to find these unknown variables. Subchannels

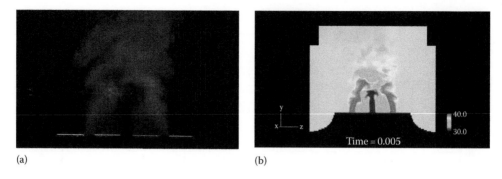

(a)   (b)

FIGURE 10.87 TRIO_U calculation of PLAJEST experiment at JAEA: (a) experimental visualization and (b) TRIO_U prediction of temperature field.

exchange momentum and energy, but the effect of mass exchange is accounted by eddy viscosity (ε). The magnitudes of ε and circumferential velocity are taken from experiments. They are in general functions of Reynolds number, pitch to diameter ratio, and helical pitch of the wire wrap.

Lorenz et al. [10.54] carried out hydraulic tests in a 91-pin bundle with the objective of providing database for verifying thermal hydraulic codes and supplement mixing data. Dye was injected axially into the centroid of a prescribed subchannel, and the dye front and location of maximum dye concentration at each elevation were observed. It was found that the dye front moved faster than the wire-wrap angle for some axial distance, followed the wire wrap for some distance, and moved slower than the wire wrap for some distance (Figure 10.88). This pattern repeats itself for

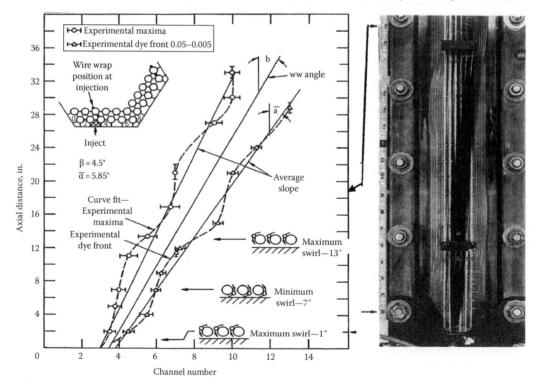

FIGURE 10.88 Swirl flow visualized by dye injection in a 91-pin bundle using water [10.55]. (From Lorenz, J.J. et al., Experimental mixing studies and velocity measurements with a simulated 91 element LMFBR fuel assembly, Technical Memorandum, ANL-CT-74-09, Karlsruhe, Germany, 1974; Copyright November 1973, the American Nuclear Society, La Grange Park, IL.)

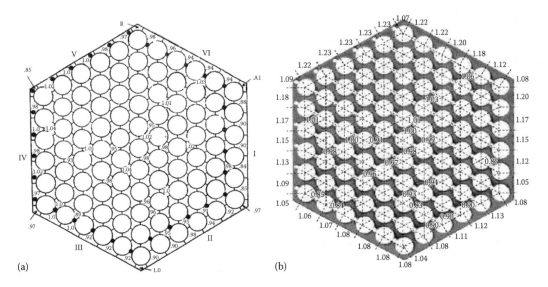

FIGURE 10.89 Subchannel velocities in a 91-pin bundle: (a) measured versus (b) CFD simulation.

every axial pitch of wire wrap. The effective swirl velocity is 1.3 times that dictated by the wire angle. The mean value of subchannel axial velocity measured in the 91-pin bundle is depicted in Figure 10.89 [10.55]. The same computation using a high-fidelity 3D parallel CFD calculation is also depicted in Figure 10.86 [10.56]. It can be seen that there is a fair degree of match indicating that the CFD codes are capable of predicting intricate flow physics in fuel pin bundles.

10.4.7 QUALIFICATION OF JET COOLING SYSTEM FOR TOP SHIELD

The top shield that forms the top cover for the reactor and provides support for all the components that enter into the reactor vessel receives heat from the hot sodium pool through multimodal heat transfer involving conduction through immersed components, convection, radiation, as well as heat deposition through sodium aerosols. The heat from the top shield is removed through its internal cooling system as well as loss to surrounding structures like safety vessel and reactor vault through complex heat exchange involving the surfaces of main vessel top portion, safety vessel, safety vessel insulation, and reactor vault liner and from its top surface to the reactor containment building (RCB) air. There are conflicting requirements of temperature limits for the materials such as steel, concrete, neutron detectors, cables, and elastomer seals. Further, to avoid freezing of sodium aerosols in the narrow annular gaps, a minimum temperature of 100°C needs to be ensured for top shield. Restricting the temperature difference to ~20°C across the top shield is important to limit its thermal bowing and thereby to limit the slope at primary sodium pumps and IHX locations.

Toward achieving and maintaining the temperature limits specified, an air cooling system is integrated within the box structure of the top shield (Figure 10.90). The inlet cooling air coming out of the jet plate impounds on the bottom plate as air jets and removes the heat from it. While rising through the narrow annular gaps surrounding penetrations, the cooling air also removes heat deposited to the penetrations through cellular convection of argon cover gas. On reaching the collection plenum below the top plate, the hot air rejects some heat to the top plate so that the top plate is heated and the temperature gradient across the top and bottom plates of the top shield is maintained within the limits specified. Considering the complexity involved, a detailed experimental investigation was carried out to validate the thermal design adopted for the top shield. The experimental facility depicts the reactor's upper structure consisting of roof slab, rotatable plug, control plug, and surrounding vault region with reduced diameter while maintaining the material thicknesses, height, and annular gaps as in the reactor (Figures 10.91 and 10.92). From thermal considerations, facility simulates the reactor system

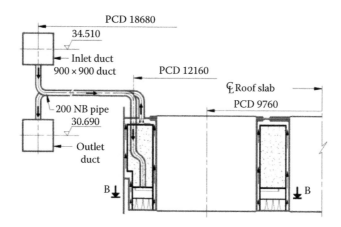

FIGURE 10.90   Schematic of top shield cooling system. *Note:* CL, center line.

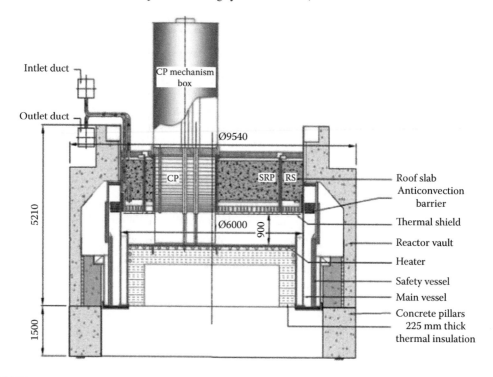

FIGURE 10.91   Cross-sectional view of integrated top shield facility.

in 1:1 scale. To simulate the hot sodium pool of the reactor, a heater plate embedded with electrical heaters is provided below the top shield. Further, a polished sheet has been fixed over the heater plate to simulate the sodium emissivity as close as possible. Based on the heat flux estimated for different heater plate temperatures, it is observed that at a temperature of 450°C, the heat flux to top shield is simulated as in the reactor. At this temperature, the steady-state temperature readings from ~500 thermocouples, embedded at various locations within the facility, are extracted and analyzed.

Figure 10.93 shows the plot of roof slab bottom plate temperature and differential temperature across roof slab for different cooling flows at a heater plate temperature of 450°C. From the plot, it is clear that with the rated cooling flow, the bottom plate temperature can be maintained at the design requirement of 120°C whereas the differential temperature across the roof slab can be limited to <20°C, as specified. Temperature distribution along the inner shell of the roof slab is very important

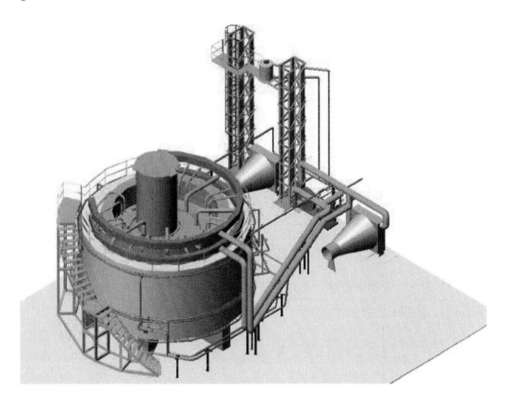

FIGURE 10.92 Three-dimensional view of integrated top shield test facility.

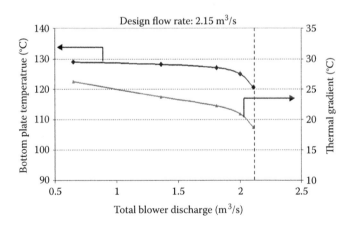

FIGURE 10.93 Roof slab bottom plate temperature and thermal gradient.

to understand the cellular convection phenomenon expected to occur in the narrow annular gaps, the amount of heat transferred to the component, and the structural implications of nonuniform distribution. From the plot (Figure 10.94), it is observed that for an annulus diameter of 4.1 m with gap varying from 16 to 30 mm, two loops of cellular convection are occurring with a maximum temperature gradient of ~25°C. Using the experimental data from the facility, CFD analysis was carried out and numerical tools were validated. Figure 10.95 shows the comparison of experimentally measured circumferential temperature distribution in the roof slab inner shell (at a heater plate temperature of 350°C) with predicted temperature distribution. Comparison is shown at elevation of 28,600 mm. In this figure, both the experimental data and CFD prediction show the formation of three convective loops.

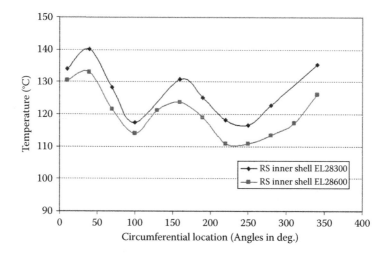

FIGURE 10.94 Circumferential temperature distribution in roof slab (small rotatable plug).

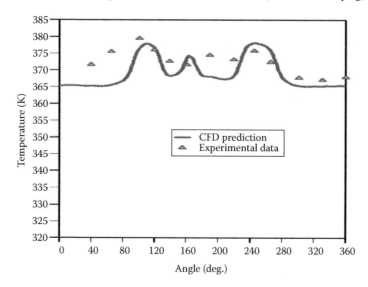

FIGURE 10.95 Temperature distribution along the inner shell of roof slab.

10.4.8 CELLULAR CONVECTION OF ARGON IN TOP SHIELD PENETRATIONS

Toward understanding the various features of cellular convection—the conditions for onset, the number of convection cells, their dependence on geometric parameters of annulus, and temperature conditions—many experimental investigations have been carried out by the SFR community. A detailed survey of literature can be obtained from Paliwal et al. [10.57]. Hemnath et al. [10.58] studied cellular convection of argon in the COBA experimental test facility. The COBA test vessel has two parts: the lower vessel and the upper vessel. Liquid sodium is contained in the lower vessel and an annulus of 20 mm gap width is formed in the upper vessel as shown in Figure 10.96. The upper vessel height is 1000 mm, and upper vessel can be dismantled into inner shell and outer shell. The upper vessel simulates a typical component penetration in the SFR top shield. To achieve a different axial temperature difference in the annulus wall, a surface heater is put on the top of the outer shell. Stainless steel–sheathed chromel–alumel thermocouples are used to measure the outer shell circumferential temperature at the top, middle, and bottom. Eight thermocouples are distributed

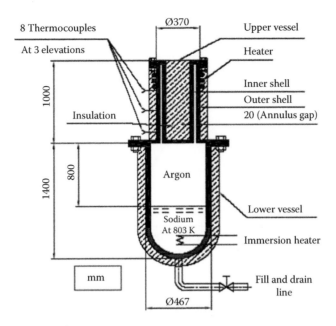

FIGURE 10.96 Schematic of COBA experimental facility.

circumferentially at each height. An immersion heater in the lower vessel maintains sodium at 803 K. The sodium level in the lower vessel is maintained such that cover gas height is 800 mm above the sodium free level. The argon cover gas pressure inside the test vessel is maintained at 100 millibar above the atmosphere. Fiber glass wool insulation is provided over the circumference of the upper and lower vessels to minimize the heat loss.

In order to validate the computational method, the experiment has been simulated using the commercial CFD code STAR-CD. For modeling turbulence, the low Reynolds number k-ε turbulence model has been used. The sodium pool is assumed to be at a constant temperature. All the outer walls are subjected to convective cooling boundary condition with heat transfer coefficient as 4 W/m² K and ambient temperature as 40 K. The inner surface of the inner shell is assumed to be adiabatic as it is filled with insulating material. Figure 10.97a depicts the predicted velocity distribution of argon

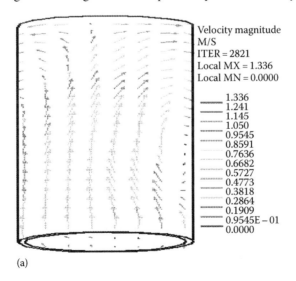

(a)

FIGURE 10.97 (a) Argon velocity (m/s) distribution in the COBA annulus.   (*Continued*)

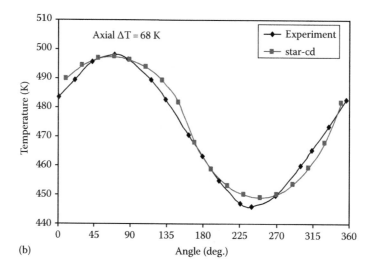

(b)

FIGURE 10.97 (*Continued*)   (b) Outer shell temperature at the bottom of the annulus.

in the experimental annulus. Argon enters the annulus at one circumferential location, flows in the upward direction, takes a turn in circumferential direction, and descends from some other circumferential location. From the predicted temperature of argon gas in the annulus, the circumferential temperature difference in gas was found to decrease with height and maximum circumferential temperature difference occurred at the bottom of the annulus. The measured outer shell temperature at the bottom of the annulus along with the predicted data is shown in Figure 10.97b. It can be seen that the predicted shell temperature compares well with the measured data. The code predicts a double-cell flow pattern (i.e., one hot leg and one cold leg), which is in agreement with the observed data. This study confirms that cellular convection can be predicted to a reasonable accuracy by CFD codes by 3D simulations with low Reynolds number k-ε turbulence model.

10.4.9   Qualification of Control Rod Drive Mechanisms for PFBR

For the 500 MWe PFBR, there are nine control and safety rod drive mechanisms (CSRDMs). Full-scale prototype CSRDMs along with control and safety rod (CSR) have been tested for their performance and endurance in air and sodium at IGCAR. The qualification procedure adopted for one typical component is illustrated in this section (Figure 10.98).

The number of endurance test cycles of scram operations to be carried out for qualifying the performance of CSRDM and CSR is worked out based on the guidelines given in ASME, Section III, Division I, Appendices, II-1500. The required number of cycles of testing depends on the test temperature. For example, it is worked out to be 3460 at 845 K (572°C), 4351 at 823 K (550°C), or 5062 at 803 K (530°C). This is in agreement with the number of tests carried out in other reactors. Based on the earlier guidelines, the actual number of test cycles is decided depending upon the test temperature. Initially, extensive testing was carried out on prototype CSRDM and CSR to check and ensure all the intended functions in air and then in hot argon at 473 K.

As part of the seismic qualification, CSRDM along with CSR has been successfully tested for OBE and safe shutdown earthquake (SSE) in water. Full insertion of CSR within the stipulated time and healthy functioning of CSRDM during and after testing have been demonstrated. A dedicated facility for conducting the seismic testing of tall and slim absorber rod drive mechanisms (ARDMs) has been constructed (Figure 10.99). This facility is mainly a concrete structure (reaction wall) having a height of 14 m, 6 m below the ground level and 8 m above the ground level. Three rigid support structures have been designed, fabricated, and erected. These act as a rigid interface between reaction wall and CSRDM/CSR SA. The mechanism (~12 m length) and grid

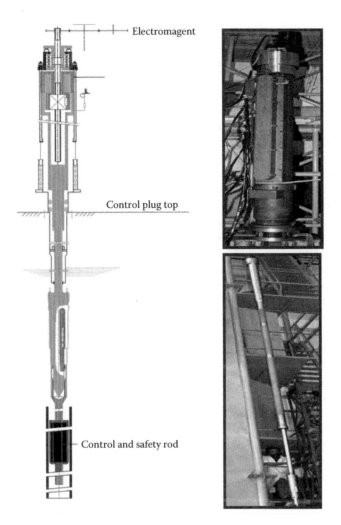

FIGURE 10.98 CSRDM and its qualification stage.

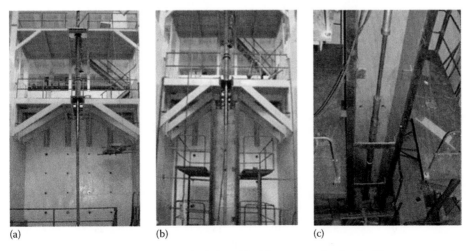

(a)   (b)   (c)

FIGURE 10.99 CSRDM test setup in pseudo shake table: (a) upper portion in air, (b) upper portion in water, and (c) lower portion in water.

plate sleeve were supported on uniaxial sliding facilities, which are in turn mounted on rigid support structures. Three actuators with maximum displacement of ±50 mm and maximum capacity of 5 tons were used to shake the mechanism at three locations: control plug top level, grid plate level, and button level of hexagonal SA. Experiments have been conducted to measure the natural frequency of individual components and the coupled system. The results matched well with the theoretical models.

In order to measure the drop time during the seismic events, a robust procedure was well established. Upon receiving the scram signal, the electromagnet (EM) is de-energized and the mobile assembly of CSRDM along with CSR is released to fall under gravity. At the end of free fall travel of 835 mm, the mobile assembly is decelerated by an oil dashpot for the remaining 250 mm travel. The free fall time is the most important parameter as the negative reactivity added during this travel of CSR ensures shutdown of the reactor with sufficient margin. After testing in air, the testing was done with the portion of CSRDM up to sodium level immersed in water to simulate the added mass effect. A vertical tank with dismountable cover on one side was fabricated and erected on the box reaction wall such that it encloses the CSRDM and CSR SA. The tank enclosed the system from grid plate support level to above free sodium level and is stationary. A dismountable cover is provided in the front side of the tank in order to enable dismantling and erection of CSRDM without disturbing the tank. Rubber gaskets were used to achieve leak tightness of tank. Leak tight and flexible connections between the stationary tank and the actuator at two locations—at button level and grid plate level—were achieved using rubber bellows. The rubber bellow at button level accommodates the axial movement of the actuator. The one at the grid plate level accommodates lateral movement between the SA and tank. The bellows were independently tested before coupling with the test setup for twice the value of displacement and pressure expected during testing.

CSRDM along with CSR and CSR SA was tested for 69 OBE in water with different delay times between the start of seismic excitation and scram (τ). The maximum drop time was recorded when EM was de-energized at 4 and 9 s after the seismic excitation starts. There was an increase of 255 ms when compared to the normal drop in water. The consolidated result of testing is shown in Figure 10.100. Comparison of time versus displacement during normal drop and drop with OBE in water is shown in Figure 10.101. Normal functioning of the system was checked and ensured during every session of the testing. Further, tests were conducted at half SSE to find out the time delay τ at which maximum drop time occurred. With 13 such drops, it was found that maximum

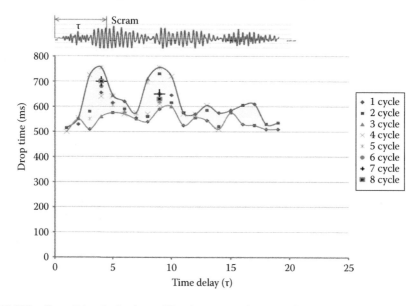

FIGURE 10.100 Consolidated seismic qualification test results under OBE.

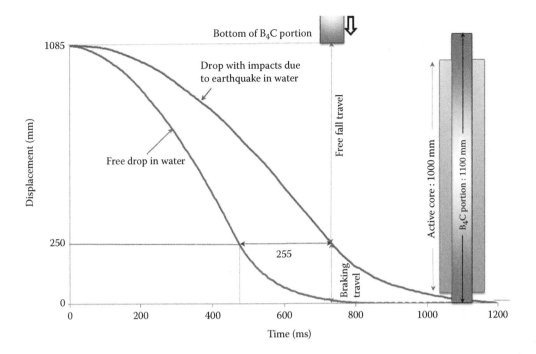

FIGURE 10.101 Time versus displacement during normal drop and drop with OBE.

drop time occurs when EM was de-energized at 4 s after the seismic excitation starts. Full SSE at $\tau = 4$ s was done for three times. One more SSE was done at $\tau = 9$ s. The maximum free fall measured is 180 ms. After every SSE, CSRDM was tested for its normal functionality. The results confirm the healthy functioning of CSRDM even after SSE.

10.5 EXPERIMENTAL FACILITIES FOR QUALIFICATION OF SFR COMPONENTS IN INDIA

Many experimental facilities have been designed and operated at IGCAR, to cater to the testing and qualification requirements of FBTR and PFBR. The details with respect to large sodium test facilities and water test facilities that are currently in operation at IGCAR are mentioned as follows:

1. Large component test rig (LCTR)
2. Sodium water reaction test facility (SOWART)
3. Steam generator test facility (SGTF)
4. *S*afety gr*A*de *D*ecay *H*eat remov*A*l in sodium (*NA*) experimental facility (SADHANA)
5. *S*c*A*led *M*odel *R*eactor *A*ssembly *T*hermal hydraulics (SAMRAT) water test facility

10.5.1 LARGE COMPONENT TEST RIG

LCTR facility (Figure 10.102) was constructed to carry out full-scale testing of large-sized critical components of PFBR in sodium under simulated reactor operating conditions. LCTR is housed in the 43 m tall high bay of Engineering Hall-III. Two large storage tanks with a total sodium hold up to 100 tons are positioned inside a 6 m deep steel-lined dump pit. It consists of four test vessels of different capacities in which independent test conditions can be maintained, heater vessel with 200 kW immersion heaters, 150 kW sodium to air heat exchanger, 20 m³/h capacity flat linear induction pump for sodium circulation, and permanent magnet flowmeters for flow measurement, air-cooled cold trap, plugging indicator, etc. Sodium sampling is done using nickel tube samplers.

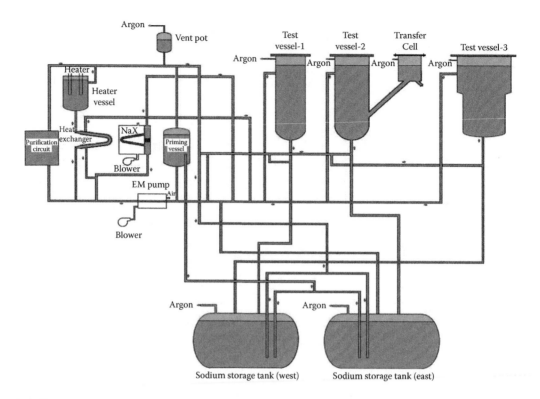

FIGURE 10.102 Flow sheet of LCTR.

Argon cover gas system and compressed air system for pneumatic actuated valves form the auxiliary systems. The pipelines and components are provided with surface heaters, thermocouples, and wire-type leak detectors. Both continuous and discontinuous type level sensors are used in the test vessels. Sodium aerosol detectors are used for detection of sodium fire in the loop area. Commercial fire alarm system based on smoke detectors is also installed. A PLC-based data acquisition and control system is used for monitoring all the parameters in the control room. Leak collection trays are provided below the sodium components and piping.

This facility was commissioned in 1994, and the cumulative operating hours of the test facility till date is 75,000 h. The maximum operating temperature of the facility is 600°C, and the maximum sodium flow rate is 20 m³/h. The material of construction of this test facility is austenitic stainless steel, grade 316. Qualification tests for fuel handling systems carried out in LCTR are described in the following.

10.5.1.1 Qualification of Inclined Fuel Transfer Machine for PFBR

In PFBR, the primary side of inclined fuel transfer machine (IFTM, Figure 10.103) consists of primary tilting mechanism (PTM), which is bolted to the grid plate, primary ramp (PR), shield plug, interconnecting piece, bellows, and primary gate valve. Performance testing of PR and PTM was carried out in various stages: in air at room temperature, in hot air, and in sodium at 200°C. The testing was limited to ~10% of the total number of cycles expected in the reactor life of 40 years as the same components are to be installed in the reactor after testing. The performance of the system was monitored continuously by measuring the current drawn by motor (measure of motor torque) and monitoring the system for noise and vibration. The effect of the reactor operation at rated temperature of 547°C on the aerosol deposition in cold regions of gate valve was also simulated by means of two dwell period tests, each of 150 h duration at 547°C. The dwell periods were interspersed between the cyclic tests. The system was tested for a total of 510 cycles in sodium. The performance of the system

(a)   (b)

FIGURE 10.103   IFTM and its components: (a) PTM inside TV-2 with TP and (b) PR with liner.

was along expected lines. On completion of sodium testing, the components were cleaned in situ in the test vessel using water vapor–carbon dioxide process, removed from the vessel, thoroughly water washed, chemical cleaned, transported, and erected on pile in PFBR.

10.5.1.2   Qualification of In-Vessel Fuel Handling Machine for PFBR

Transfer arm (TA) is the in-vessel fuel handling machine of PFBR (Figure 10.104). The testing consisted of lifting the SA, rotating the machine, lowering the SA into the transfer pot, and then reversing these steps to lower the SA back into the mock core. One cycle of operation involved gripping the SA from the central location in grid plate, raising the SA by 4.6 m, rotating TA by ~87° in order to align the SA over the transfer pot, and lowering by 4.6 m in order to rest the SA inside the transfer pot. The testing was done at a gripper hoist speed of 30 mm/s. The test vessel is provided with a grid plate and small core consisting of six SAs (each of 4.5 m in height and 260 kg weight) and a central SA. The vessel is provided with a top flange with oblong opening for insertion of TA. Before commencing testing in sodium, 20 cycles of testing were done in the empty vessel at room temperature followed by 36 cycles at elevated temperatures ranging from 60°C to 170°C. This was followed by four cycles of tests in argon environment at temperatures from 60°C to 170°C just prior to sodium filling. The vessel was filled with sodium, and 300 cycles of testing were successfully completed in sodium at the fuel handling temperature of 200°C after which the temperature was raised to 550°C and the machine parked at this temperature for 100 h. Testing was resumed in sodium at the fuel handling temperature of 200°C, and 300 cycles of testing were once again completed. On completion of sodium testing, the components were cleaned in situ in the test vessel using water vapor–carbon dioxide process and being dismantled. On completion of dismantling, the component was thoroughly water washed, chemically cleaned, and transported for on-pile erection in PFBR.

10.5.2   Sodium Water Reaction Test Rig

SOWART (Figure 10.105) was constructed to study the phenomenon of self-wastage and adjacent tube wastage with respect to an SG tube material in the case of a sodium water reaction. In addition, this facility was also a test bed for testing and development of different types of in-sodium and in-argon hydrogen sensors. SOWART is housed in the 23 m tall low bay of Engineering Hall-III. This facility consists of a sodium storage tank with 10 tons of sodium holdup, a cold leg containing a 20 m³/h EM pump,

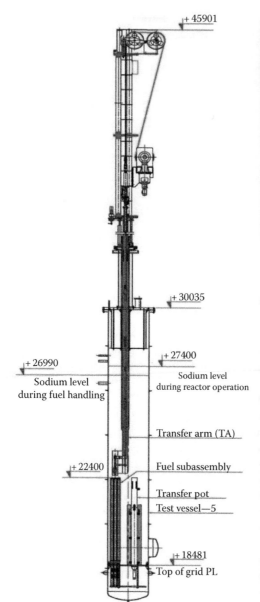

Transfer arm: schematic

Transfer arm inside TV-5

FIGURE 10.104 TA and its components.

a heater vessel with 150 kW immersion heaters, a large air-cooled cold trap, plugging indicator, sampler, a hot leg containing a heater vessel with 150 kW immersion heaters, expansion tank, test sections, and hydrogen meter section. Apart from this, the facility has a main heater exchanger of 380 kW capacity and a sodium to air heater exchanger of 150 kW capacity.

The facility was commissioned in 2001, and the cumulative operating hours of the test facility is 30,000 h. The maximum operating temperature of the test facility is 525°C, and the rated sodium flow is 10 m³/h. The material of construction of this test facility is austenitic stainless steel, grade 316. In PFBR, a single tube wall with a thickness of 2.3 mm separates low-pressure liquid sodium from high-pressure water/steam in SG. An accidental leakage of water into sodium causes sodium water reaction leading to damage of the leaking tube itself (self-wastage) and damage on its adjacent

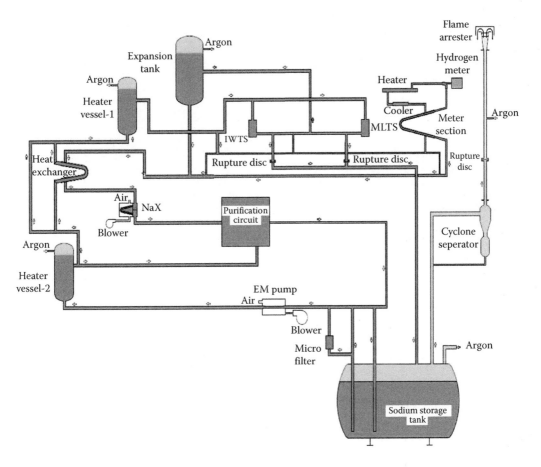

FIGURE 10.105 SOWART loop (schematic).

tube (impingement wastage). A few tests carried out in SOWART to understand self-wastage and impingement wastages are highlighted in the following.

10.5.2.1 Self-Wastage Experiments

Self-wastage studies were conducted in sodium water reaction test rig to understand the wastage phenomenon. Experiments of modified 9Cr–1Mo PFBR tube specimens were conducted with steam at 450°C/17 MPa and sodium at 450°C. Steam was injected into flowing sodium through a pin hole of 0.1 mm diameter in the leak simulator at a leak rate of 23 mg/s. On completion of the experiment, the tube specimen was removed for analysis (Figure 10.106). The tests have revealed the following:

- 0.94 mm thickness underwent self-wastage in 18 min.
- Average self-wastage rate is 0.00087 mm/s.
- For 2¼Cr–1Mo self-wastage as mentioned in literature is 0.001 mm/s, and in comparison, it is observed that modified 9Cr–1Mo is 1.2 times more resistant to self-wastage.

10.5.2.2 Impingement Wastage Experiments

Leaking tubes of modified 9Cr–1Mo/9Cr–1Mo or nickel having a calibrated pinhole leak are used for injecting steam into sodium. This tube is inserted into the test section from the bottom. One end of the leak simulator tube is dummied, and the other end is connected to the steam system. The target tube assembly, which has a target tube of modified 9Cr–1Mo, is fixed in the test section from top as shown in Figure 10.107. Reaction jet from the leaking tube produces wastage on

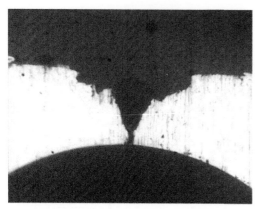
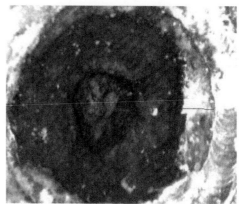

FIGURE 10.106 Self-wastage test results (photo).

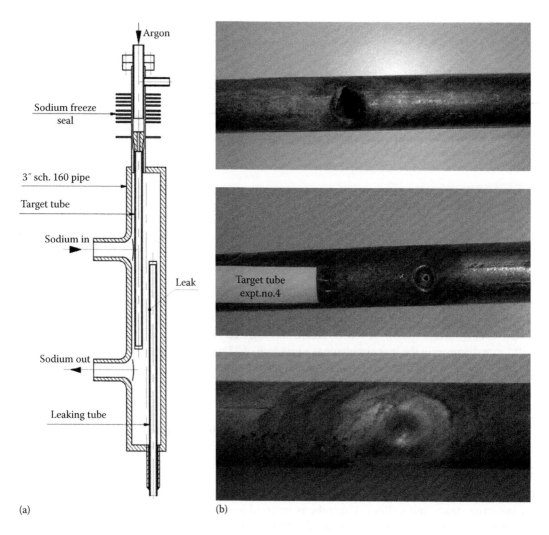

(a)

(b)

FIGURE 10.107 Details and results of impingement wastage studies: (a) test section—schematic and (b) impingement wastage on tube specimens.

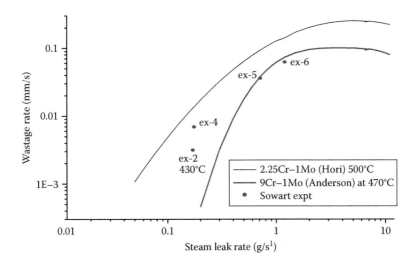

FIGURE 10.108 Impingement wastage rate of different tube materials.

the surface of the target tube. Impingement wastage studies were carried out for steam leak rates from 100 to 800 mg/s. The tests revealed the following observation.

The ratio of maximum depth of penetration to the time required, that is, wastage rate, is determined for all the cases. It is found that wastage rate is increasing as the steam leak rate. The wastage rate is compared with correlations, for 2¼Cr–1Mo steel and 9Cr–1Mo steel, as mentioned before. It is observed from experiments 2 and 4 that at lower temperature, 430°C, relative wastage resistance of 9Cr–1Mo is found about two times higher than that of higher temperature, 500°C. Anderson's observation (based on experiments at 470°C) is that at higher leak rate (>0.5 g/s), wastage resistance of 9Cr–1Mo is very less (close to 2¼Cr–1Mo) and lower leak rate (<0.2 g/s) wastage resistance is much higher (Figure 10.108). But in most of the other literatures, the average relative resistance of 9Cr–1Mo is reported to be 1.2–1.5. In all the experiments, a relative wastage resistance greater than 1.96 could be obtained. Even at a higher leak rate (>0.7 g/s), a wastage resistance greater than 2.2 was observed. Though at higher leaks wastage resistance obtained is close to that predicted by the Anderson correlation for 9Cr–1Mo, at lower leaks wastage resistance is found much lower than the predicted value. However, it was always found higher than the wastage resistance of 2¼Cr–1Mo steel.

10.5.3 STEAM GENERATOR TEST FACILITY

SGTF was constructed and operated to validate the design of once-through SG of PFBR (Figure 10.109) as well as for future fast breeder reactors (FBRs). The facility is housed in a separate building with a height of 38 m. It consists of a sodium storage tank with 18 tons of sodium holdup. An SG of 5.5 MWt with 19 tubes, 23 m long without weld joints and with one expansion bend; a furnace oil-fired heater (5.7 MW) to heat the sodium; and a 170 m³/h annular linear induction EM pump for sodium circulation, cold trap, plugging indicator, etc., are the major components of the facility. Auxiliary circuits such as cover gas system, SG leak detection system, sodium water reaction products discharge circuit, and conventional steam water system are also part of the test facility. The facility (Figure 10.110) was commissioned in 2003, and the cumulative operating hours of the test facility is 40,000 h. The maximum operating temperature of the test facility is 530°C, and the maximum sodium flow rate is 170 m³/h. The material of construction of this test facility is austenitic stainless steel, grade 316 LN for the sodium systems and modified 9Cr–1Mo for the SG. Experiments carried out in the facility are described in the following.

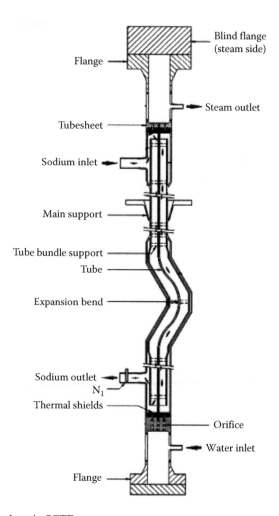

FIGURE 10.109   SG mock-up in SGTF.

10.5.3.1   Assessment of Heat Transfer Performance of Steam Generator

Sufficient heat transfer area margin is provided in SG to account for the uncertainties in two-phase flow heat transfer correlations. Experiments were carried out to assess the actual margin. At rated power of 5.5 MWt and rated steam temperature and pressure conditions, the sodium temperature to SG was only 516°C against the expected 525°C as per design. With power increased further, 6.09 MWt heat transfer could be achieved at 525°C. This corresponds to a heat transfer area margin of 12.25%.

10.5.3.2   Performance Assessment of SG Thermal Baffles

To minimize the normal temperature gradient across SG tube sheets and protect the tube sheets from thermal shocks during plant transients, thermal baffle assemblies are provided for both the top and bottom tube sheets. Loss of feed water flow transient results in maximum thermal shock, and the SG cold end bottom tube sheet is the most affected. The incident was simulated numerically and by experiments. Tests showed that when the average rate of temperature rise of sodium is 62°C/min, the thermal shield mitigated this to 22.5°C/min at the bottom tube sheet (Figure 10.111). Experiments proved that the thermal baffles absorb the thermal shock adequately and protect the tube sheets as intended.

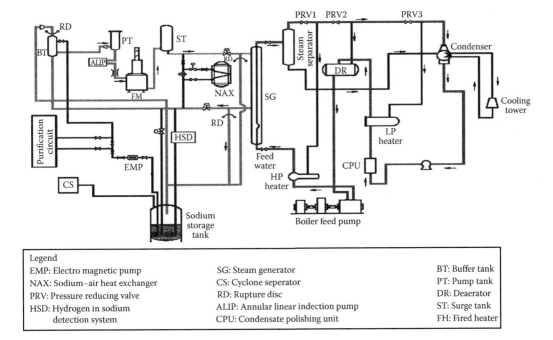

FIGURE 10.110 SGTF loop (schematic).

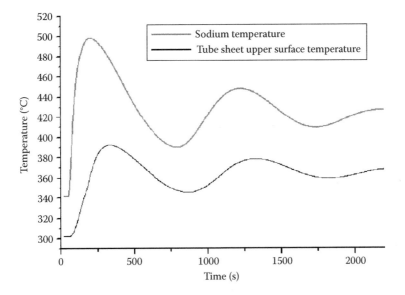

FIGURE 10.111 Bottom tube sheet temperature variations followed by loss of feed water flow to SG (obtained by experimental simulations).

10.5.3.3 Two-Phase Flow Instability Studies

Once-through SGs are susceptible to water/steam flow instabilities due to two-phase flows in the same tube. Mapping the zones of instability is required to quickly pass through these zones during start-up and power raise. Experiments were carried out at the start-up power of 20% of the nominal power (1.1 MWt). Flow instability was observed at steam pressure below 140 kg/cm². At the threshold of instability, outlet steam temperature oscillations were found to vary up to 60°C.

At higher power levels, instability was not observed even at steam pressures as low as 100 kg/cm². Experiments were also carried out to assess the flow instability in SG during start-up similar to PFBR. The aim was to find out the sodium flow corresponding to the minimum target power with which start-up can be done without flow instabilities. It has been observed that the plant start-up is absolutely stable when the target power is above 75%. Below the 75% target power, instability was observed when the SG steam outlet condition is near to dry saturation.

10.5.3.4 Experimental Evaluation of Hydrogen Flux

Hydrogen produced due to water side corrosion of SG tubes and diffusing to sodium is a major impurity in the secondary sodium system of FBR and has to be quantified for fixing the capacity of cold trap. Hydrogen flux for 2.25Cr–1Mo steel is available in literature, and the steady-state value is 1.8–2.2×10^{-7} gH/m² s. No data are available in literature for hydrogen flux for PFBR SG material of mod 9Cr–1Mo steel. Experiments were carried out in the model SG (Figure 10.112), and the value determined is 1.25×10^{-7} gH/m² s.

10.5.3.5 Feasibility of Using Acoustic Sensors for Steam Generator Tube Leak Detection

For the detection of tube failure in SGs, acoustic method wherein capturing the noise generated due to the evolution of hydrogen serves as a supplementary leak monitoring technique is utilized. Studies have been carried out by injecting argon into SG sodium through calibrated orifices and observing the response of acoustic sensors (Figure 10.113) located at various elevations on SG shell. Different signal processing techniques were successfully used for acoustic leak detection. Based on the results, an online leak detection system using acoustic technique is being developed for SGTF SG. This will be further extended to PFBR.

10.5.3.6 Stress Evaluation during Operation of Steam Generator with Plugged Tube

In case of a tube leak in SG of FBR, the leaking tube will be isolated by plugging and operation will be resumed. Isolation of tubes results in deviation of temperature distribution in the affected and nearby tubes and hence the associated stresses. Tests were conducted by plugging one tube at a time,

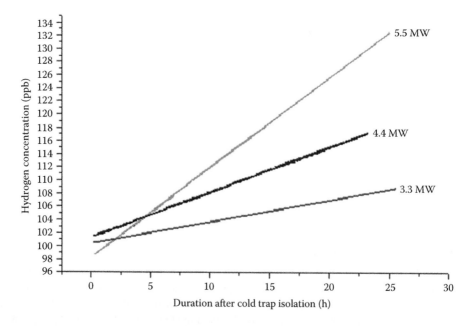

FIGURE 10.112 Hydrogen flux variation with power.

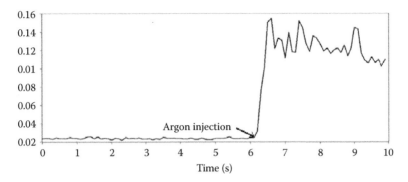

FIGURE 10.113 Variation of acoustic sensor output on argon injection.

and temperature distribution of tubes was recorded and analyzed. Results indicate that a single tube plugging does not significantly add to the thermomechanical loading on SG. These observations are in agreement with the analytical predictions.

10.5.3.7 SG Endurance Test

SG was operated at rated steam temperature and pressure (17.2 MPa and 493°C) for more than 5000 h. It was also operated at the nominal power of 5.5 MWt continuously for 28 days. SG was in operation close to 10,000 h cumulatively. The operation of SG without any incident has demonstrated the reliability and robustness of the design.

10.5.4 SADHANA Test Facility

This facility (Figure 10.114) was constructed and operated to study the thermal hydraulic behavior of safety grade decay heat removal (SGDHR) system (Figure 10.115) of PFBR. This is a 1:22 scale model with a 355 kW thermal power capacity. Richardson number, the nondimensional number characterizing buoyancy forces, was simulated for the studies. SADHANA facility is housed in the high bay of Engineering Hall-III. This facility is connected to the sodium storage tank of LCTR. The sodium holdup of the loop is 2.7 tons. This facility consists of a test vessel with immersion heaters of 450 kW capacity simulating the main vessel of PFBR with decay heat from core, a decay heat exchanger (DHX) inside the test vessel, an air heat exchanger (AHX) for heat removal at an elevation of 19 m from the test vessel, a chimney for inducing the natural draft connected to the AHX, and an expansion tank. Sodium filling and circulation are by an annular linear induction pump of 5 m³/h capacity. The circulating loop consists of a sodium to air heat exchanger, an economizer exchanger, and a smaller heater vessel.

The facility was commissioned in 2009, and the cumulative operating hours of the test facility is 4000 h. The maximum operating temperature of the test facility is 550°C. The material of construction of this test facility is austenitic stainless steel, grade 316 L. The highlights of the experiments carried out in the facility are given in the following.

10.5.4.1 Estimation of Heat Transport Capability

At a sodium pool temperature of 550°C, the sodium mass flow rate induced in the secondary loop was 1.63 kg/s with a DHX outlet temperature of 520°C and AHX outlet temperature of 317°C. For this condition, the power transported was 425 kW from the sodium pool. The nominal capacity of the loop is 355 kW, and at 550°C sodium pool temperature, secondary loop removes 19.4% more power than its nominal capacity. The relation between sodium mass flow rate induced in secondary sodium system with respect to the mean sodium temperature difference between hot leg and cold leg was established as $0.544W = 0.0913\Delta T$.

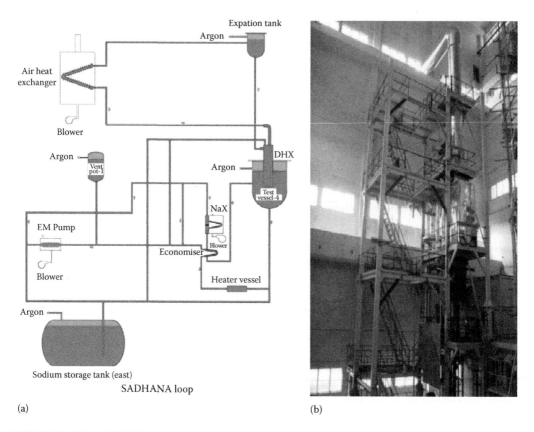

FIGURE 10.114   SADHANA test facility: (a) flow sheet and (b) photograph.

10.5.4.2 Transient Experiments: Consequences of Sudden Opening of AHX Outlet Damper

The response of SGDHR system when the AHX dampers are suddenly open has been studied and found that the system will be fully functional in around 510 s after the initiation of opening the dampers (Figure 10.116). The time required to open the damper is 70 s. It has been observed that from the initial sodium flow, the time taken to achieve the steady-state sodium flow after opening the damper was practically independent of the sodium pool temperature. The rate of rise of nondimensional sodium flow and power transported is approximately the same for all the cases of experiment conducted. The establishment of sodium flow, air flow, and various temperatures after the sudden opening of the damper was smooth and has not indicated major oscillations.

10.5.4.3 Heat Transport Performance with Level Drop of Sodium in Primary Pool

Experiments have been conducted by reducing the sodium level, which leads to a reduction of 88% of the original primary sodium flow entry area to DHX simulating the primary sodium leak condition. The results obtained from these studies have indicated that 88% reduction in sodium level causes 2% reduction in secondary flow and 5% reduction in power transported by the system (Figure 10.117).

10.5.4.4 Evaluation of Heat Transport with Partial Opening of AHX Outlet Damper

Heat transport experiments were conducted simulating outlet damper failure event in the SGDHR system. The heat transport capability of the system is low at lower damper opening and increasing with increase in damper opening. At 50% damper opening, the heat transport capability of the system is stabilized and no appreciable change has been observed further. A similar trend was observed for flow in secondary sodium loop.

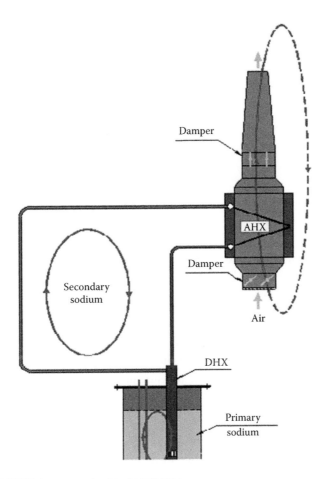

FIGURE 10.115   SGDHR loop conceived in SADHANA.

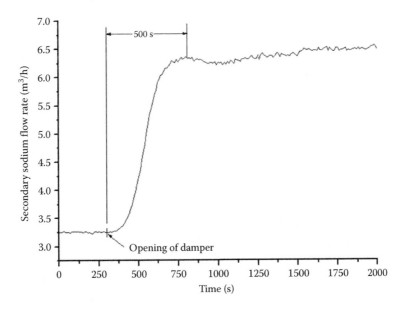

FIGURE 10.116   Development of sodium flow when the AHX dampers opened with a sodium pool temperature of 550°C.

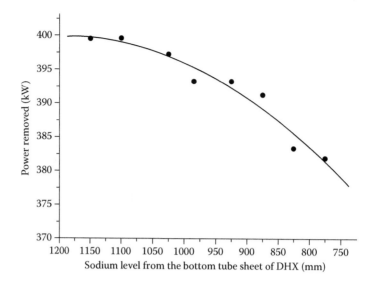

FIGURE 10.117 Power removal by SADHANA loop during level drop studies.

10.5.4.5 Evaluation of Resistance Offered by the AHX Tube Bundle for Naturally Induced Air Flow

Experiments were conducted to establish the pressure drop characteristics of finned tube bundle in AHX and the damper provided at the AHX outlet. The pressure drop offered by the damper is found to be 14.62 Pa, which is 14.4% of the thermally induced draft of 101.8 Pa. The characteristics of variation of pressure drop with respect to the air velocity become insignificant compared to the total pressure drop of the air flowing path when the damper opening is more than 50%. At 50% damper opening case, the pressure drop offered by the AHX tube bundle is velocities (Re < 3500) compared to higher air velocities (Re > 3500). The pressure drop coefficient of damper exponentially reduces with increase in percentage of damper opening.

10.5.5 SAMRAT WATER LOOP (1/4 SCALE WATER MODEL OF PFBR)

SAMRAT was constructed for various experimental requirements such as velocity measurements at IHX inlet window, free level fluctuation studies, control plug and IHX flow distribution studies, gas entrainment studies, secondary pump trip studies, thermal stratification studies, thermal striping studies, and SGDHR studies. This experimental model (Figure 10.118) was set up in Engineering Hall-II in the year 2003. This is a 1/4 scale model of PFBR primary circuit based on a two-loop design of PFBR. This loop has three water storage tanks with a total capacity of 50 m³ with provision of heaters inside the tanks. Two pumps, each of 1200 m³/h capacity at 10 bar, supply water flow during studies requiring high flow rates. Plate-type heat exchangers are provided to control the water temperature. Rod-type heaters are provided in the core of the model to simulate core heat generation. Pneumatic valves with control systems are provided to facilitate easy operation of the loop. Dedicated control room is available with data acquisition system to acquire data during experiments.

In this facility, both hot and cold plena were modeled. Perspex windows are provided for visualization in the cold pool. Core includes blanket and storage zone along with fuel zone. Individual core SA is also modeled. Velocity measurements were carried out by hot-film anemometer (HFA) and propeller anemometer, and for temperature measurement, K-type thermocouples were used (1, 0.5, and 0.25 mm diameter mineral insulated). Free level oscillation was measured by conductivity probes.

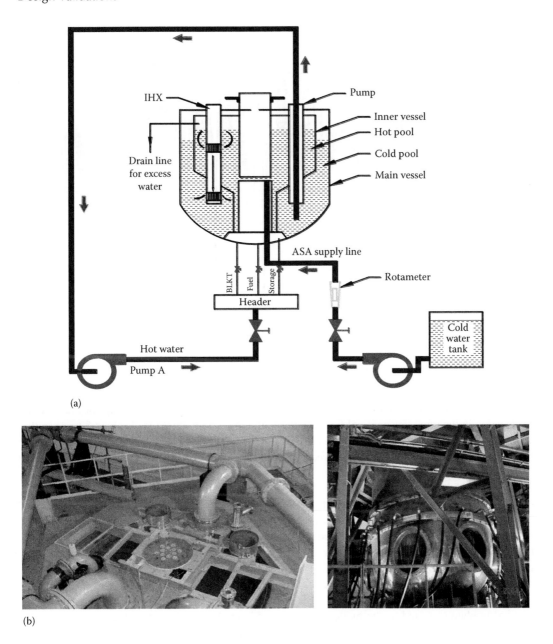

(a)

(b)

FIGURE 10.118 SAMRAT water loop: (a) flow sheet and (b) some photos.

10.5.5.1 Free Level Fluctuation Studies

Due to free level fluctuations, the components immersed in sodium will be subjected temperature fluctuations that can lead to thermal fatigue of the immersed components. Hence, it is required to assess the characteristics of free level fluctuations to understand their effect on the reactor components. The experimental study carried out in the SAMRAT model with Froude number similitude determined the magnitude and frequency of the level fluctuations in the free surface of the pool. The free level fluctuations were measured by means of conductance probes. The time domain data obtained from the experiment are analyzed statistically. It is found from the analysis of data that the free level fluctuations are

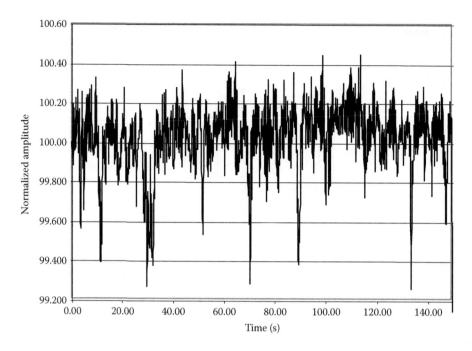

FIGURE 10.119 Time series plot in the vicinity of control plug (100% flow).

not uniform throughout the hot pool and that the ripple height and the fluctuation frequency depend on the location in the pool. For nominal flow condition, the maximum amplitude is about 82 mm in the vicinity of the control plug in reactor case as shown in Figure 10.119. The predominant frequencies of fluctuations for the reactor case vary between 0.25 and 1.6 Hz.

10.5.5.2 Gas Entrainment Studies

One of the issues to be resolved in LMFBR is argon cover gas entrainment problem from free liquid sodium surface. The entrained cover gas may hinder the normal reactor operation. High free surface velocity along with the presence of various immersed components in the hot pool is the cause of gas entrainment from the free surface. To reduce the free surface velocity and hence gas entrainment, ring-type baffle plates were considered. Initially, the optimum geometry of the baffle plate was arrived through numerical analysis using PHOENICS and experiments were conducted in the 1/4 scale water model of PFBR primary circuit with selected baffle plate geometry. The baffle plate as tested in the model is shown in Figure 10.120. Based on the numerical study previously discussed, the selected configuration of baffle plate was fabricated and installed in the SAMRAT model.

Parametric studies were carried out by varying the location of the baffle plate. For each experimental run, velocity was measured at the free surface to find out the maximum magnitude. Visual observation of the free surface was carried out to find out the occurrence of vortices. The pump suction side region was observed from the perspex window provided at the main vessel of the cold pool side to find out the gas entrainment. The free surface textures at various experimental conditions are shown in Figure 10.121.

10.5.5.3 Safety Grade Decay Heat Removal Studies

DHX–core–hot pool interaction during SGDHR condition has been assessed from primary pool thermal hydraulic studies carried out in 1/4 scale model of PFBR primary circuit using water as

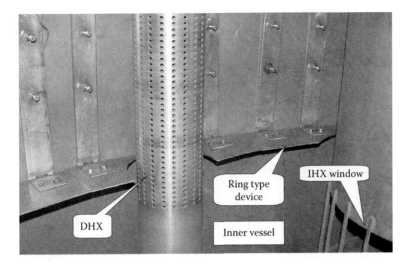

FIGURE 10.120 Baffle plate introduced in the inner vessel mock-up in SAMRAT.

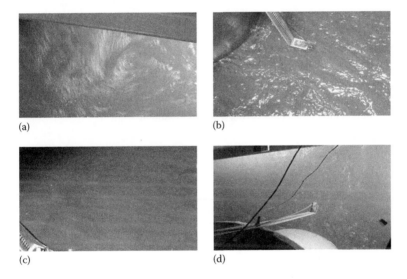

FIGURE 10.121 Free surface textures at various experimental conditions: (a) flow pattern at free surface without any baffle plate, (b) flow pattern at free surface (baffle plate fixed below 240 mm from free surface), (c) flow pattern at free surface (baffle plate fixed below 315 mm from free surface), and (d) flow pattern at free surface (baffle plate fixed below 360 mm from free surface).

simulant. In this model, immersion-type rod heaters were fixed inside the fuel SAs and storage SAs to represent the generation of decay heat in the prototype after reactor shutdown. In the secondary side of DHX, water was circulated with the aid of an external pump to simulate the heat removal by DHX. The heat removal rate by DHX for various secondary-side flow rate was estimated from earlier experiments. The shell side of DHX and the whole primary pool is cooled by natural convection.

The experimental results demonstrated the core coolability under SGDHR operating condition by establishment of natural circulation in the primary pool. Figure 10.122 shows the plot for the possible natural circulation path in the primary pool during SGDHR condition.

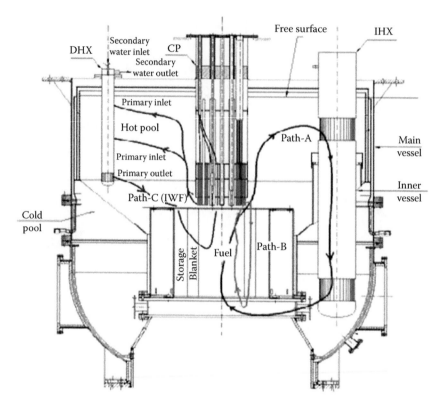

FIGURE 10.122   Natural circulation path during SGDHR condition.

10.A   APPENDIX A: 23-PARAMETER CHABOCHE VISCOPLASTIC MODEL

- *Viscoplastic flow rate*

$$\varepsilon_i^{vp} = G_i^P = \frac{3/2\langle\sigma_v/K(R)\rangle^n \cdot \exp\left(\alpha\cdot\langle\sigma_v/K(R)\rangle^{n+1}\right)\cdot\left(\sigma_i' - X_i'\right)}{J(\sigma - X)} \tag{10.A.1}$$

where

$$\sigma_v = J(\sigma - X) - \alpha_R\cdot R - k \tag{10.A.2}$$

$$K(R) = K_o + \alpha_k\cdot R \tag{10.A.3}$$

where k is the initial yield stress with $R(0) = 0$.

$$\langle u \rangle = u\cdot H(u)$$

where H is Heaveside's step function.

- *Nonlinear kinematic hardening*

$$X_i = X_{1i} + X_{2i} \tag{10.A.4}$$

$$X_{1i} = G_i^{x_1} = \frac{2}{3}c_1 a_1 \varepsilon_i^{vp} - c_1 \varphi(p) X_{1i}\cdot p - b_1\cdot\left|I\left(X_1\right)\right|^{(m_1-1)}\cdot X_{1i} \tag{10.A.4a}$$

$$X_{2i} = G_i^{x_2} = \frac{2}{3}c_2 a_2 \varepsilon_i^{vp} - c_2 \varphi(p) X_{2i} \cdot p - b_2 \cdot \left| I\left(X_2\right) \right|^{(m_2 - 1)} \cdot X_{2i}$$
(10.A.4b)

where
p is accumulated plastic strain: $p = [2/3\varepsilon^{vp} : \varepsilon^{vp}]^{1/2}$
$I(X)$ is the second invariant of $X : I = [2/3X : X]^{1/2}$

The function $\varphi(p)$ is defined by

$$\varphi = \varphi_\infty + (1 - \varphi_\infty)\exp(-bp)$$
(10.A.5)

- *Isotropic hardening*

$$R = G^R = b(Q - R)p + \tau |Q - R|^m \cdot sgn(Q_R - R)$$
(10.A.6)

- *Plastic strain memory*

$$Q = G^Q = 2\mu(Q_{max} - Q)q \quad \text{with } Q(0) = Q_o$$
(10.A.7)

q is the internal variable corresponding to the radius of the memory surface F and ξ is its center:

$$F = I(\varepsilon^{vp} - \xi) - q \leq 0$$

where

$$I = \left[\frac{2}{3}(\varepsilon^{vp} - \xi) : (\varepsilon^{vp} - \xi)\right]^{1/2}$$

$$q = G^q = \eta H(F)\langle n : n^*\rangle p$$
(10.A.8)

$$\xi_i = G_i^\xi = \frac{2}{3}(1 - \eta) \cdot H(F) \cdot \langle n : n^*\rangle \cdot p \cdot n_i$$
(10.A.9)

with

$$n_i = \frac{3/2\left(\sigma_i' - x_i'\right)}{J(\sigma - x)}$$

$$n_i^* = \frac{\sqrt{2/3} \cdot \left(\varepsilon_i^{vp} - \xi_i\right)}{I(\varepsilon^{vp} - \xi)}$$

$$Q^R = Q - Q_R^* \left\{ 1 - \left[\frac{(Q_{max} - Q)}{Q_{max}}\right]^2 \right\}$$
(10.A.10)

α, n, k, c_1, a_1, c_2, a_2, β_1, β_2, m_1, m_2, φ, b, τ, m, K_o, α_k, α_R, μ, η, Q_{max}, Q_o, and Q_R^* are the 23 material parameters that are functions of temperature. These parameters are determined for SS 316 LN in the temperature range of 293–873 K, which are presented in Table 10.2.

10.B   APPENDIX B: 20-PARAMETER VISCOPLASTIC MODEL FOR 9CR–1MO STEEL

- *Viscoplastic strain rate*

$$\varepsilon = \frac{\sqrt{3}}{2} \cdot p \cdot n$$

$$p = \left(\frac{\sigma_v}{K}\right)^n = \left[\frac{2}{3} \cdot \varepsilon \cdot \varepsilon\right]^{1/2}$$

$$n = \frac{(S-X)}{|S-X|}$$

$$\sigma_v = J(\sigma - X) - k$$

$$J(\sigma - X) = \left[\frac{3/2}{|S-X|}\right]^{1/2}$$

$$|S-X| = \left[\left(S_{ij} - X_{ij}\right) \cdot \left(S_{ij} - X_{ij}\right)\right]^{1/2}$$

$$K = K_o + \alpha \cdot R$$

$$R = R_1 + R_2$$

$$k_{ul} = u \cdot H(u)$$

where H is Heaveside's step function.

- *Kinematic hardening variables*

$$X_i = \frac{2}{3} \cdot c_i \cdot a_i \cdot \varepsilon_i - c_i \cdot p \cdot X_i - \beta_i \left[I(X_i)\right]_i^{(m-1)} X_i$$

$$X = \sum X_i \quad (i = 1, 3)$$

- *Isotropic hardening variables*

$$R_i = b_i \cdot (Q_i - R_i) \cdot p \quad (I = 1, 2)$$

$$J(X) = (2/3|X|^{1/2})$$

n, k, K, α, a_1, c_1, a_2, c_2, a_3, c_3, β_1, β_2, β_3, m_1, m_2, m_3, b_1, b_2, Q_1, and Q_2 are the 20 material parameters which are functions of temperature.

REFERENCES

10.1   Chaboche, J.L., Nouailhas, D. (1989). A unified constitutive model for cyclic viscoplasticity and its application to various stainless steels. *Trans. ASME J. Eng. Mater. Technol.*, 111, 424–430.

10.2   Chellapandi, P., Ramesh, S. (1997). 20-Parameter viscoplastic model for modeling 9Cr–1Mo steel for SFR applications. Appeared in Creep–fatigue damage design aspects specific to SFR development. IAEA, Vienna, IAEA-Tecdoc-115, pp. 185–192.

10.3   Chaboche, J.L. (1986). EVPCYCL: A finite element program in cyclic viscoplasticity. *La Recherche Ae Ârospatiale*, 2, 91–112.

10.4   (2010). *ABAQUS Version 6.10*. Dassault Systèmes Simulia Corp., Providence, RI.

10.5 SYSTUS: Multiphysics tool structural analysis. http://www.esi-group.com/software-services/virtual-environment/cfd-multiphysics/systus.

10.6 Chandonneret, M. (1989). Tests de validation de Codes de Calcul on Viscoplastique. Report of ONERA, Châtillon, France.

10.7 Nouilhas, D. (1987). A viscoplastic modelling to stainless steel behaviour. *Proceedings of the Second International Conference on Constitutive Laws for Engineering Materials, Theory and Applications*, University of Arizona, Tucson, AZ, Report: T.P. No. 1987-1.

10.8 Contesti, E., Cailletand, G., Levaillant, C. (1987). Creep damage in 17-12 SPH stainless steel notched specimens: Metallographical study and numerical modelling. *ASME J. Pressure Vessel Technol.*, 109, 228.

10.9 Cabrillat, M.T., Gatt, J.M. (1993). Evaluation of thermal ratchetting on axisymmetric thin shells at the free level of sodium inelastic analysis. *Transactions of the 12th International Conference on Structural Mechanics in Reactor Technology (SMiRT-12)*, Stuttgart, Germany, August 1993, Vol. E05, pp. 155–159.

10.10 Olschewski, J., Sievert, R., Qi, W., Bertran, A. (1993). Prediction of inelastic response of a Ni-based superalloy under thermal-mechanical cyclic loading. *Transactions of the 12th International Conference on Structural Mechanics in Reactor Technology (SMiRT-12)*, Stuttgart, Germany, Vol. L, p. 17.

10.11 CEA. (2003). CAST3M—User manual. http://www-cast3m.cea.fr/cast3m/.

10.12 (2002). RCC-MR: Appendix A16. Guide for leak before break analysis and defect assessment, AFCEN.

10.13 Hooton, D.G., Bretherton, I., Jacques, S. (2003). Application of the sigma-d (σ_d) method for the estimation of creep crack incubation at austenitic weld boundaries. *Proceedings of the Second International Conference on Integrity of High Temperature Welds*, London, U.K., November 10–12, 2003, pp. 425–434.

10.14 Fujita, K. et al. (1993). Study on flow induced vibration of a flexible weir due to fluid discharge effect of weir stiffness. *PVP-258, Flow Induced Vibration and Fluid Structure Interaction*, ASME, Denver, CO, pp. 143–150.

10.15 Chellapandi, P., Chetal, S.C., Raj, B. (2008). Investigation on buckling of FBR vessels under seismic loading with fluid structure interactions. *J. Nucl. Eng. Des.*, 238, 3208–3217.

10.16 Bolotin, V.V. (1964). *The Dynamic Stability of Elastic Systems*. Holden-Day, Inc., San Francisco, CA.

10.17 Hsu, C.S. (1963). On the parametric excitation of dynamic system having multiple degrees of freedom. *ASME J. Appl. Mech.*, 30, 367–372.

10.18 Siva Srinivas, K., Chellapandi, P. (2014). Investigation of parametric instability in elastic structures. PhD thesis. HBNI, Department of Atomic Energy, Mumbai, India.

10.19 Brochard, D., Buland, P., Hammami, L., Gantenbein, F. (1987). FBR core seismic analysis. *SMIRT-9*, Lausanne, Switzerland, Vol. E, pp. 33–42.

10.20 Brochard, B., Buland, P., Gantenbein, F., Gibert, R.J. (1987). Seismic analysis of LMFBR cores—Mockup RAPSODIE. *Transactions of the Ninth International Conference on Structural Mechanics in Reactor Technology (SMiRT-9)*, Lausanne, Switzerland, Vol. E, pp. 33–42.

10.21 Gauvain, J., Hartelli, A. (1982). Numerical integration of the vibratory motion equations of a bundle of flexible beam type structures separated by small gaps. *AIHETA, Sixth Congresso Nazionale*, Genova, Italy.

10.22 Preumont, A., Parent, J. (1983). Fluid coupling effects in LMFBR core seismic analysis. *Transactions of the Seventh International Conference on Structural Mechanics in Reactor Technology (SMiRT-7)*, Chicago, IL, Vol. E06, pp. 275–283.

10.23 Preumont, A., Kunsch, A., Parent, J. (1986). Fluid coupling coefficients in an array of hexagonal prisms. *Nucl. Eng. Des.*, 2, 51–59.

10.24 Aida, Y., Niwa, H., Kawamura, Y., Kobayashi, T., Kurihara, C., Nakamura, H., Toyoda, Y., Hagiwara, Y. (1993). Experimental and analytical studies on vertical response of FBR core components during seismic events. *Transactions of the 12th International Conference on Structural Mechanics in Reactor Technology (SMiRT-12)*, Stuttgart, Germany, Vol. E13, pp. 375–380.

10.25 Clough, R.W., Clough, D.P. (1977). Seismic response of flexible cylindrical tanks. *Transactions of the Fifth International Conference on Structural Mechanics in Reactor Technology (SMiRT-4)*, San Francisco, CA, Vol. K/1.

10.26 Housner, G.W. (1969). Finite element analysis of fluids in containers subjected to acceleration. Nuclear Reactors and Earthquakes, TID 70211. U.S. Atomic Energy Commission, Washington, DC.

10.27 Veletsos, A.S., Yang, T.Y. (1976). Dynamics of fixed base liquid storage tanks. *U.S.–Japan Seminar for Earthquake Engineering Research with Emphasis on Lifeline Systems*, Tokyo, Japan.

10.28 Fujimoto, S., Yamoto, S., Shimizu, H., Murakami, T., Sakurai, A., Kurihara, C., Mashiko, Y. (1985). Experimental and analytical study on seismic design of a large pool type LMFBR in Japan. *Transactions of Eighth International Conference on Structural Mechanics in Reactor Technology (SMiRT-8)*, Brussels, Belgium, Vol. EK1, pp. 315–320.

10.29 Housner, C.H. (1957). Dynamic pressure on accelerated fluid container. *Bull. Seismol. Soc. Am.*, 47(1), 15–35.

10.30 Athiannan, K. (1998). Investigation of buckling of thin shells. PhD thesis, Department of Applied Mechanics, IIT Madras, Chennai, India.

10.31 Bose, M.R.S.C., Thomas, G., Palaninathan, R., Damodaran, S.P., Chellapandi, P. (2001). Buckling investigations on a nuclear reactor inner vessel model. *Exp. Mech.* (*SAGE Publications*), 41(2), 144–150.

10.32 Lenoir, G., Dallongeville, M., Goldstein, S., Vidard, M. (1981). Thermal hydraulics of the annular spaces in roof slab penetrations of liquid sodium cooled fast breeder reactor. *Proceedings of the Sixth International Conference on Structural Mechanics in Reactor Technology*, Paris, France, Vol. E, pp. 1–6.

10.33 Satpathy, K., Velusamy, K., Patnaik, B.S.V., Chellapandi, P. (2013). Numerical simulation of liquid fall induced gas entrainment and its mitigation. *Int. J. Heat Mass Transfer*, 60, 392–405.

10.34 Govindha Rasu, N. (2013). Investigations of entrance flow and partial flow blockages in fuel subassemblies of fast breeder reactor. PhD thesis, Homi Bhabha National Institute, Mumbai, India.

10.35 Olive, J., Jolas, P. (1990). Internal blockage in a fissile super-phenix type of subassembly: The SCARLET experiments and their interpretation by the CAFCA-NA3 code. *Nucl. Energy*, 29, 287–293.

10.36 Velusamy, K., Raviprasan, G.R., Nema, V., Meikandamurthy, C., Selvaraj, P., Chellapandi, P., Vaidyanathan, G., Chetal, S.C. (2010). Computational fluid dynamic investigations and experimental validation of frozen seal sodium valve assembly of a fast reactor. *Ann. Nucl. Energy*, 37(11), 1423–1434.

10.37 Turner, W.D., Elrod, P.C., Siman-Tov, I.I. (1978). HEATING5: An IBM 360 heat conduction code. Report, Oak Ridge National Lab., TN, USA, ORNL/CSD/TM-15.

10.38 Wakamatsu, M., Nei, H., Hashiguchi, K. (1995). Attenuation of temperature fluctuations in thermal striping. *Nucl. Sci. Technol.*, 32(8), 752–762.

10.39 Tenchine, D. et al. (2012). Status of CATHARE code for sodium cooled fast reactors. *Nucl. Eng. Des.*, 245, 140–152.

10.40 Tenchine, D. et al. (2013). International benchmark on the natural convection test in Phenix reactor. *Nucl. Eng. Des.*, 258, 189–198.

10.41 Surle, F., Berger, R. (1994). The CORMORAN programme: a computational and experimental study of temperature fluctuations associated with a stratified liquid metal flow in a rectangular cavity. *International Atomic Energy Agency Meeting on Correlation between Material Properties and Thermohydraulics Conditions in LMFBRs*, Aix-en-Provence, France.

10.42 Trio Code.

10.43 Kimura, N., Nishimura, M., Kamide, H. (2002). Study on convective mixing for thermal striping phenomena (experimental analyses on mixing process in parallel triple-jet and comparisons between numerical methods). *JSME Int. J. Ser. B*, 45(3), 592–599.

10.44 Kimura, N., Miyakoshi, H., Kamide, H. (2003). Experimental study on thermal striping phenomena for a fast reactor-transfer characteristics of temperature fluctuation from fluid to structure. *Proceedings of the Sixth ASME–JSME Thermal Engineering Joint Conference*, Hawaii Island, HI, TED-AJ03-159.

10.45 Kimura, N., Igarashi, M., Kamide, H. (2001). Investigation of convective mixing of triple jet—Evaluation of turbulent quantities using particle image velocimetry and direct numerical simulation. *Proceedings of the Eighth International Symposium on Flow Modeling and Turbulence Measurements*, Tokyo, Japan, pp. 651–658.

10.46 Nam, H.Y., Kim, J.M. (2004). Thermal striping experimental data. Internal Report, LMR/IOC-ST-002-04-Rev.0/04. KAERI, Daejeon, South Korea.

10.47 Ushijima, S., Tanaka, N., Moriya, S. (1990). Turbulence measurements and calculation of non-isothermal coaxial jets. *Nucl. Eng. Des.*, 122, 85–94.

10.48 Nishimura, M., Tokuhiro, A., Kimura, N., Kamide, H. (2000). Numerical study on mixing of oscillating quasi-planar jets with low Reynolds number turbulent stress and heat flux equation models. *Nucl. Eng. Des.*, 202, 77–95.

10.49 Menter, F.R. (1994). Two equation eddy-viscosity turbulence models for engineering applications. *AIAA J.*, 32, 1598–1604.

10.50 Chen, H.C., Patel, V.C. (1988). Near-wall turbulence models for complex flows including separation. *AIAA J.*, 26, 641–648.

10.51 Durbin, P.A. (1995). Separated flow computations with the k-ε-υ^2 model. *AIAA J.*, 33, 659–664.

10.52 Coste, P., Quéméré, P., Roubin, P., Emonot, P., Tanaka, M., Kamide, H. (2008). Large eddy simulation of highly-fluctuational temperature and velocity fields observed in a mixing-T experiment. *Nucl. Technol.*, 164, 76–88.

10.53 Kimura, N., Kamide, H., Emonot, P., Nagasawa, K. (2007). Study on thermal striping phenomena in triple-parallel jet. Investigation on non-stationary heat transfer characteristics based on numerical simulation. *NURETH-12*, Pittsburgh, PA.

10.54 Lorenz, J.J. et al. (1974). Peripheral flow visualization studies with a 91 element bundle. *Trans. Am. Nucl. Soc.*, 17, 416–417.

10.55 Lorenz, J.J., Ginsberg, T., Morris, R.A. (1973). Experimental mixing studies and velocity measurements with a simulated 91 element LMFBR fuel assembly. Technical Memorandum, Gesellschaft fur Kernforschung MbH, ANL-CT-74-09, 13-38, Karlsruhe, Germany.

10.56 Basant et al. (2008). Thermal hydraulics within fuel subassembly of an FBR core. Indira Gandhi Centre for Atomic Research, (IGCAR), India. IGCAR-ZNPL Collaborative Project Report.

10.57 Paliwal, P.U., Parthasarathy, K., Velusamy, T., Sundararajan, P., Chellapandi, P. (2012). Characterization of cellular convection of argon in top shield penetrations of pool type liquid metal fast reactors, *Nucl. Eng. Des.*, 250, 207–218, September 2012.

10.58 Hemnath, M.G., Meikandamurthy, C., Ramakrishna, V., Rajan, K.K., Vaidyanathan, G. (2007). Cellular convection in vertical annuli of fast breeder reactors. *Ann. Nucl. Energy*, 34, 679–686.

11 Design Analysis and Methods

11.1 INTRODUCTION

A robust design is achieved by validating the basic design with numerical simulations and experimental verifications. Design of a fast reactor requires detailed studies on reactor physics, thermal hydraulics, and structural mechanics of core and assembly components. In this chapter, various concepts of core, along with methodology for calculating flux and power distributions, and various coefficients of reactivity are discussed. Also, the numerical simulation and experimental validation of problems associated with core thermal hydraulics and structural mechanics of core and assembly components in a sodium-cooled fast reactor (SFR) are brought out.

11.2 REACTOR PHYSICS

The core is constituted by certain volume fractions of the fuel, coolant, and structural materials, which are chosen to obtain the targeted linear pin power, core burnup, fuel pin diameter, etc. The core physics design starts with the determination of the fuel enrichment necessary to keep the reactor critical at full power with sufficient excess reactivity for burnup compensation. The main outputs from the calculations are (1) neutron flux and power distributions across the core, (2) breeding ratio achievable for the fuel/blanket configuration, (3) refueling scheme for the core and blankets to ensure maximum availability of the reactor and highest possible burnup for each of the fuel and blanket subassembly (SA), (5) Doppler reactivity, and (6) delayed neutron fractions. The fuel management scheme is decided to achieve targeted peak burnup for which the fuel pins are designed. Another important part of the physics design is the determination of the number, layout, and enrichment of absorber rods for reactivity control during normal operation as well as in accident situations. This section provides the details of the calculations to finalize the design.

11.2.1 HOMOGENEOUS AND HETEROGENEOUS CORES

In the homogeneous concept [11.1], the fuel SAs are arranged together constituting the core of the reactor, which is surrounded both radially and axially with blanket SAs (Figure 11.1). This configuration, for a given core size and fuel volumetric fraction, gives the lowest uniform enrichment and critical mass. The core size and fuel volumetric fraction are related to the total reactor power and the chosen fuel pin diameter. Thus, for a given reactor power and fuel pin diameter, the homogeneous core concept gives minimum fissile inventory. However, in a large-sized homogeneous core, the sodium void coefficient is positive. This is because sodium voiding results in harder neutron spectrum that leads to increased fast fission in ^{238}U and high η, the number of neutrons produced per neutron absorbed [11.2,11.3]. It also increases neutron leakage resulting in smaller sodium void coefficient. However, the former two effects (high fast fission and high η) dominate over the neutron leakage effect and hence the void coefficient is positive. In order to increase the leakage effect to reduce the positive sodium void effect, heterogeneous core concept is evolved.

In heterogeneous concept, in addition to the normal outer blanket SAs, blanket SAs are also introduced in the core, which is designated as internal blanket. These types of core with internal blankets are often called heterogeneous or parfait reactors. It can be seen that there is a great

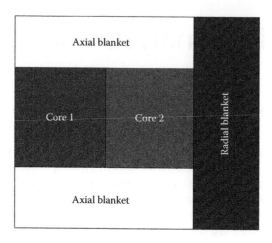

FIGURE 11.1   Conventional homogeneous core.

variety of heterogeneous cores depending upon the arrangement of internal blankets in the core region. The heterogeneous reactor configuration increases the breeding ratio (hence lower the doubling time) but improves the safety characteristics of the reactor. Neutron leakage to the blanket regions is increased and hence sodium void coefficient is reduced. The use of internal blanket pushes out the fissile material to regions of lower neutronics importance and hence the enrichment and fissile inventory gets increased. Higher enrichment results in harder neutron spectrum that leads to low sodium void effect in the fissile region due to high neutron leakage and less fast fission (enrichment being high). Thus, the increase in the neutron leakage component and the reduction in the spectrum component lead to overall reduction of the sodium void reactivity. It is this feature that makes heterogeneous reactors particularly attractive from the safety point of view. In the internal blanket, there is lack of fission neutrons and hence the neutron spectrum is much softer than the fissile region. It leads to two opposite effects: the chance of fertile material to undergo fission (fast fission effect) is reduced but the chance to capture neutrons is increased. Since the latter effect is dominant, the net effect is an increase in breeding ratio of the reactor [11.4].

The blanket region usually separates the core regions from each other. A basic characteristic of heterogeneous cores is then the degree of neutronic coupling between different zones. The coupling constant K_{ij} of the jth zone to the ith zone is the expectation value that a neutron in the jth zone will give rise to its next-generation neutron in the ith zone. As the magnitude of K_{ij} increases, the core tends to behave like a homogeneous core. For smaller value of K_{ij}, the coupling reduces between the fuel zones, and the power distribution becomes unstable in the sense that small variation in the enrichment or absorber rod positions gives rise to large variation in the fractional power produced by a zone. The typical power distributions of homogeneous and heterogeneous cores are given in Figure 11.2. This is very important for the design of control system. Loosely coupled cores require separate control systems for each fuel zone. Large flux tilts with absorber rod movement and complex flux shape change with time are possible in large decoupled cores. Thus, controlling the flux shapes in a large radially heterogeneous core requires more in-core instrumentation.

The advantage of heterogeneous design is the requirement of only one enrichment for the different fuel zones, whereas different enrichment zones are required in the homogeneous design for adequate power flattening. But the fissile inventory is much more in heterogeneous core. The presence of internal blankets leads to more complex sodium flow regulations for a desired outlet temperature uniformity. Further, in a heterogeneous core, there can be appreciable buildup of fuel material in the internal blankets, as the reactor operates, leading to much larger power swing from the fuel zones

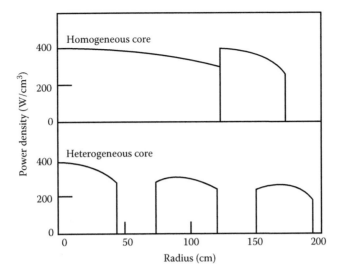

FIGURE 11.2 Typical power distributions in core.

to the blanket region than in the homogeneous core. As a consequence, unless variable flow devices are considered, large overcooling of the blanket SAs at the beginning of life becomes necessary. Thus, for the same maximum clad hot spot temperature, the mixed mean core outlet temperature in the heterogeneous core is lower than in the homogeneous case, thereby affecting the thermal efficiency.

11.2.1.1 Various Heterogeneous Models

A variety of heterogeneous arrangements have been studied internationally, each having its own advantages and disadvantages [11.4]. They can be broadly classified as modular or island type, annular or radial parfait type, axial parfait type, stepped cores, salt and pepper geometry, and seeded SAs. In the modular or island type, the fuel SAs are grouped into smaller modules in a large core, each module surrounded by one or two rows of blanket SAs (Figure 11.3). When there are only six or seven modules, the arrangement is also called *cartwheel* type. This concept results in large leakage of neutrons to blanket SAs and hence a high breeding ratio and

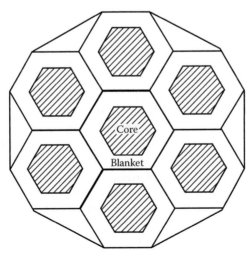

FIGURE 11.3 Heterogeneous core: cartwheel type.

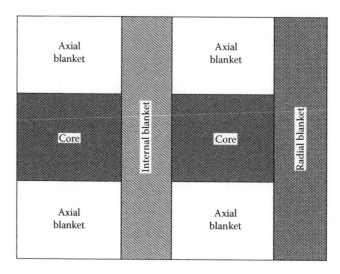

FIGURE 11.4   Annular heterogeneity.

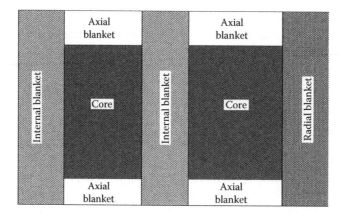

FIGURE 11.5   Bull's eye arrangement.

low sodium void coefficient. However, in this arrangement, each module with its associated blanket tends to act like a small independent reactor and may have to be independently controlled. In annular or radial parfait type, the blanket SAs are arranged in the form of rings separating rows of fissile SAs. An arrangement in which the central portion of the core contains blanket material is called *bull's eye* type. Figures 11.4 and 11.5 show two such annular arrangements. One or two annular blanket rings are considered sufficient. In the annular and modular types, there can be substantial power swings from the fuel to the blanket portion of the core as the reactor operates. As a consequence, the blanket SAs have to be overcooled at the beginning of life leading a thermal penalty and loss of efficiency. In one of the studies, a thermal penalty of 10–20 K is reported. This advantage can be avoided if arrangements to regulate the flow rate through the blanket SAs during burnup can be provided. In axial parfait type (Figure 11.6), the fertile pellets are introduced to a height of about 30 cm at the center of each SA, thus making them axially heterogeneous. The problem of overcooling of blanket SAs is absent in the axial parfait design. However, studies show that the breeding ratio is smaller and the sodium void reactivity gain is larger than for the radial parfait cores. In a stepped core, the active core height in the central SAs is considerably smaller than that in the peripheral SAs (Figure 11.7). Such stepped cores

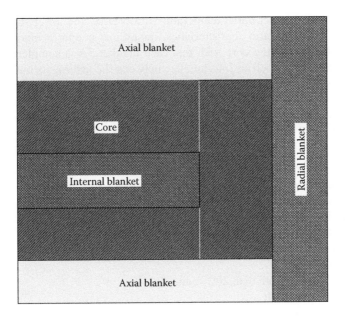

FIGURE 11.6 Axially heterogeneous.

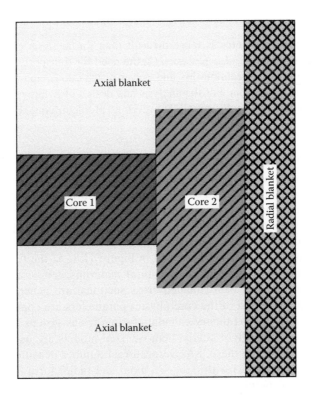

FIGURE 11.7 Stepped core arrangement.

have uniform enrichment, better outlet temperature uniformity, and lower sodium void reactivity gain but poorer breeding ratio characteristics. In salt and pepper geometry, single blanket SAs are scattered throughout the core such that each blanket SA is completely surrounded by fuel SAs. Such cores have properties intermediate between the heterogeneous core and conventional homogeneous core. This arrangement provides high breeding ratio and low sodium void reactivity gain. Using seeded SAs for the whole core is not favored due to excessive inventory requirements. Heterogeneous cores can have the following typical features (specific to a design), which distinguish them from homogeneous cores:

1. Lower sodium void reactivity ($2 vs. $7)
2. Higher breeding ratio (1.35–1.40 compared to less than 1.30)
3. Higher doubling times (18 years vs. 14 years)
4. Higher fissile inventories (30% higher and more)
5. Larger core sizes (more than 100 additional assemblies)
6. Lower Doppler coefficients (e.g., −0.005 vs. −0.009)
7. Higher enrichments
8. Lower damage fluxes (more than 20% reduction)
9. Reduced control rod worth
10. For the same number of orificing zones, higher clad midwall temperatures
11. Higher fuel compaction reactivity (e.g., 0.0058 $\Delta k/k$ vs. 0.0042 $\Delta k/k$)
12. Localized reactivity addition gives rise to localized power peaks
13. Less sensitivity of doubling times and sodium void reactivity to changes in fuel pin diameters
14. Greater sensitivity of power peaking and sodium void reactivity to small changes in enrichment split
15. Lower total inelastic strain in cladding (2% vs. 3%: general limit for any cladding)

The analysis of heterogeneous cores is more difficult than the analysis of homogeneous cores because of (1) neutron decoupling of core zones, (2) the need for transport calculations instead of diffusion calculations to obtain accurate flux and power distributions, and (3) the significance of gamma heating in internal blankets. For the analysis and design of heterogeneous cores, we need modified design approaches and analysis tools. However, no problems could be identified so far that would affect the feasibility of heterogeneous cores.

11.2.1.2 Calculation Methodology

Geometrically, the core is an arrangement of hexagonal fuel and control SAs in a triangular lattice with axial heterogenities [11.5]. For the accurate determination of several reactor physics parameters, 3D computations are necessary. Further, the partial insertion of control SA distorts the flux and fission rate distributions considerably. This distortion can be presented accurately only through 3D simulation with triangular or hexagonal meshes in the horizontal plane. The step-by-step calculation sequences are indicated in Figure 11.8. At the first step, knowing the volume fractions of the fuel, coolant, and structural materials, temperatures, and fuel data, the temperature-dependent data (self-shielded cross sections) are generated from the multi-group cross-section library. Most of the core physics parameters are computed by performing 3D neutron diffusion or transport theory calculations in various groups by dividing core into burnup zones axially and radially. Usually, homogenized models are used for the neutronics calculations (Figure 11.9). From these, SA-averaged fuel number densities and burnup values are derived. From the flux shapes within an SA, axial and radial form factors are computed. Refueling simulation is accomplished by using a refueling code, which updates the burnup data of the assemblies refueled.

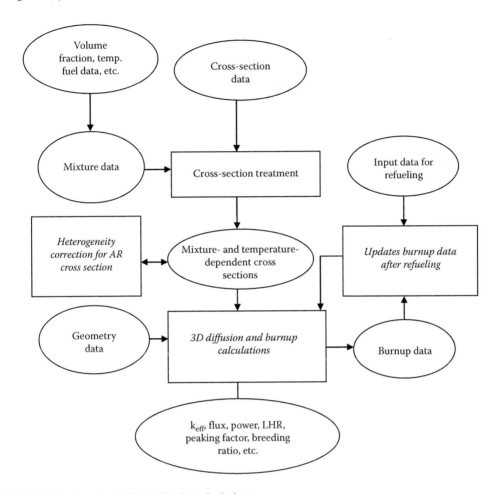

FIGURE 11.8 Flowchart of core physics calculations.

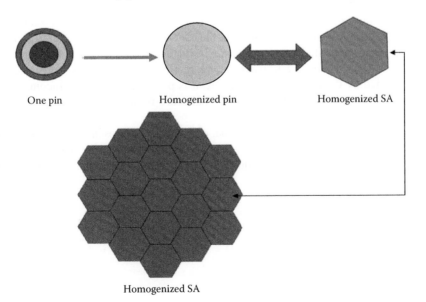

FIGURE 11.9 Homogenized core model.

11.2.1.3 Multigroup Diffusion Equation and Method of Solution

The steady-state multigroup diffusion equation is given as

$$-\nabla \cdot D_g \nabla \phi_g + \Sigma_{Rg} \phi_g = \sum_{\substack{g'=1 \\ g' \neq g}}^{G} \Sigma_{sg' \to g} \phi_{g'} + \chi_g \sum_{g'=1}^{G} \left(v\Sigma_f \right)_{g'} \phi_{g'} + S_g \tag{11.1}$$

where
  D_g is the diffusion coefficient
  Σ_{Rg} is the total removal cross section defined as

$$\Sigma_{Rg} = \Sigma_{ag} + \Sigma_{sg}$$

where
  Σ_{ag} is the absorption cross section
  Σ_{sg} is the out-scattering cross section
  $\Sigma_{sg' \to g}$ is the scattering transfer from group g' to g
  χ_g is the fraction of fission neutrons born in group g
  $\sum_{g'=1}^{G} \left(v\Sigma_f \right)_{g'}$ is the production cross section due to neutrons in all groups
  G is the total number of energy groups
  λ is the eigenvalue or k-effective
  S_g is the external source

If the external source is present, we solve an inhomogeneous equation, while we solve a homogeneous equation in the absence of external source. The flux and current are continuous within the solution domain. Appropriate boundary conditions are specified on the external boundary surfaces. In finite difference solution of diffusion equation, the reactor domain is divided into small subregions. In each subregion, the material properties are assumed to be uniform. If the flux values are estimated at the center points of the subregions called center mesh finite differencing scheme. Generally, the code like FARCOB [11.5] generates the mesh structure in X–Y plane with six triangles per hexagon. The flat surface of the hexagonal cross section of SA is assumed to be parallel to X-axis. Lines joining the centers of triangles parallel to X- and Y-axes form the mesh lines. Thus, the six triangles of a hexagon are distributed in four rows and three columns as mesh lines. Axial mesh distributions, radial and axial zones, and material distribution are specified by the user. In special cases, it is possible to consider incomplete rings of SA in the core periphery. Other than full core, half-core rotational and reflective symmetries are considered.

In the center mesh finite differencing scheme, each mesh is represented by the symbol i. Let j represent its three neighboring meshes in X–Y plane and let k denote the neighboring meshes in Z directions. For mesh i, the diffusion equation for neutron energy group g, after differencing becomes,

$$\sum_{j=1}^{3} \bar{D}_{ij}^g \frac{\left(\phi_i^g - \phi_j^g \right)}{a} hd_i + \sum_{k=1}^{2} \bar{D}_{ik}^g \left(\frac{\phi_i^g - \phi_k^g}{(d_i + d_k)/2} \right) A_i + \left[\sum_{a,i}^g + \sum_{s,i}^g \right] \phi_i^g V_i$$

$$= \sum_{g^i} \left(\sum_{s,i}^{g' \to g} + \frac{\chi^g v \sum_{f,i}^{g'}}{k_{eff}} \right) \phi_i^{g'} V_i \tag{11.2}$$

where

  ϕ's denote the fluxes at triangle centers

  D is the diffusion coefficient

  χ^g is the fission neutron yield fraction in group g

  Σs are the macroscopic cross sections

  k_{eff} is the neutron multiplication factor.

  h is the triangle side

  $a = h/\sqrt{3}$ is the distance between triangle mesh centers

  $A_i = \sqrt{3}\, h^2/4$ is the area of a triangle mesh

  d_i is the mesh height in z-direction

  $V_i = A_i d_i$ is the volume of 3D mesh i

$$\bar{D}^g_{i,j} = \frac{2}{\left(1/D^g_i\right) + \left(1/D^g_j\right)}$$

$$\bar{D}^g_{i,k} = \frac{d_i + d_k}{\left(d_i/D^g_i\right) + \left(d_k/D^g_k\right)}$$

In the X–Y plane, on reflective boundaries if for mesh i the neighbor j happens to be across the reflective boundary, then $\phi_i - \phi_j$ term reduces to zero. Reflective boundary in z-direction leads to similar cancellation. For half triangles appearing on X boundary limits with reflective boundary condition, only the volume and surfaces appearing within the problem geometry are considered. Thus, reflective boundary is automatically imposed just on the X boundary line. Zero incoming neutron current boundary condition is incorporated in the following way by introducing the parameter Γ. Let ϕ's be the flux at the boundary surface of mesh i. The parameter Γ is defined by

$$-\frac{D(\phi_s - \phi_i)}{h/2} = \Gamma\phi_s$$

$$\phi_s = \frac{D\phi_i}{D + (h\Gamma/2)} = \frac{\phi_i}{1 + (h\Gamma/2D)}$$

Assuming the current J to be constant in the interval mesh center i to the mesh surface, current can be computed in terms of ϕ_i:

$$J = \frac{D d\phi}{dr} = \frac{2D}{h}\left\{\phi_i - \frac{D\phi_i}{D + (h\Gamma/2)}\right\}$$

11.2.1.4 Power Distribution

Flux calculations are performed and neutron multiplication factor (k_{eff}) is obtained. Generally, convergence criteria for fission source at every point are 0.0001 and for multiplication factor 0.000001. Power distribution is obtained for each triangular mesh and summed up for each hexagonal SA. For power estimation, fission, neutron capture, and slowing down powers in each material are computed. For core power normalization, peak pin power in each SA is calculated. Fluxes in adjacent triangular meshes are interpolated to get flux at any point within the SA of interest. Then, knowing the spatial location of each pin of that SA, peak power pin and its location are identified. Due to the interpolation scheme, peak pin powers are estimated to a better accuracy than by using only the mesh-averaged fluxes. Peak power ratings in different enrichment zones and radial blanket are determined. Reactor power can be normalized based on an allowed peak rating or a given reactor power. Reaction rates in different material zones and breeding ratio in different core zones are also estimated. For performing burnup calculations, the average power per unit length of fuel/blanket in each burnup zone and one group capture, fission, and (n, 2n) cross sections in each zone are also computed.

11.2.1.5   Burnup Model

The core is divided into different radial and axial burnup zones. The user can make these selections arbitrarily. Fuel and blanket number densities of burnable SA are kept track of for every axial hexagonal mesh of an SA (the fuel number densities in the six triangular meshes of an SA are the same). So, the user has freedom to change radial and axial burnup zone for a fresh burnup step or after refueling. This option is very useful when the number of SA in core and their configuration change after every cycle. To get macroscopic cross sections of a given axial position of SA, microscopic cross sections are weighted with respective number of densities. Nonburnable number densities are kept constant. Microscopic cross sections are generally not corrected for self-shielding effect change due to change in heavy isotopic concentration. The burnup equations are a set of coupled linear first-order equations that are solved similar to other fuel burnup routines, for example, the method followed in LWR-WIMS [11.5].

For the ith isotope,

$$
\frac{dN_i}{dt} = -\lambda_i N_i - \sigma_i \phi N_i + \sum_k \delta_{i,j(k)} \alpha_{ki} \sigma_{c,k} \phi N_k
$$

$$
+ \sum_k \delta_{i,l(k)} \beta_{ki} \lambda_k N_k + \sum_k y_{k,i} \sigma_{f,k} \phi N_k + \sigma_{n,2n} \phi N_m \tag{11.3}
$$

where

N_i is the concentration of isotope of mass A and atomic number Z
λ_i is the decay half-life of the ith isotope
ϕ is one group total neutron flux
σ_i is one group absorption cross section of isotope i
$\sigma_{c,k}$ is one group capture cross section of isotope k
$\sigma_{f,k}$ is one group fission cross section of isotope k
δ is the Dirac delta function
N_m is the concentration of isotope with mass = A + 1 and atomic number Z
j(k), l(k) are identification of all products by capture and decay, respectively, from isotope k
α_{ki}, β_{ki} are product fractions by capture or decay
$y_{k,i}$ is the fission yield of product i from the fission of k

For lumped fission product, only the fifth term on the right-hand side is present and this term is absent for actinides. Both fourth-order Runge–Kutta and trapezoidal rule are used to solve the burnup equations depending on the option. Burnup is performed with constant power. So, the flux level for burnup is calculated at every step from

$$
\phi = \frac{P_F}{V_F \sum_i \sigma_{f,i} \phi_i N_i E_i} \tag{11.4}
$$

where

P_F is the linear heat rating
V_F is the volume per unit length
E_i is the energy yield per fission

The burnup equations can be written as

$$\frac{d\vec{N}}{dt} = [B]\vec{N} \tag{11.5}$$

where [B] matrix consists of the coefficient of $\vec{N}$ on the right-hand side of the detailed Equation 11.3 for N_i. For example, trapezoidal integration gives (neglecting Δt^2 term and higher)

$$N_i(\Delta t) = N_i(t) + \Delta t \sum_k B_{ik}(t)N(t) \tag{11.6}$$

For short half-life isotopes like N_p-239, the equilibrium concentration can be obtained without detailed solution by putting $dN_i/dt = 0$. Otherwise, the time step size required for accuracy will be too small for N_p-239. Generally, a step size of 0.5 day is used. The trapezoidal rule is accurate up to Δt^2 while fourth-order Runge–Kutta method is accurate up to Δt^4. It is also possible to select the time step size for an accuracy of <0.01% in the change ΔN_i for any isotope.

11.2.1.6 Breeding Ratio

Breeding ratio for a reactor is defined as the ratio of fissile mass produced to the fissile mass destroyed [11.1]. It can be expressed as

$$BR = \frac{FP}{FD}$$
$$\tag{11.7}$$
$$BR = \frac{(FEOC + FD \quad FBOC)}{FD}$$

where
   FP is fissile produced in a cycle
   FD is fissile destroyed in a cycle
   FEOC is fissile mass at the end of a cycle
   FBOC is fissile at the beginning of a cycle

The fissile destroyed per cycle (FD) can be calculated as:

$$FD = \text{Fissile destroyed per year (FD)} = E \times FF \times (1 + \alpha),$$

where
   E is the energy produced in MWd per cycle
   FF is the fissile mass (kg) destroyed by fission for producing 1 MWd
   α is the capture to fission ratio for the fissile material

11.2.1.7 Doubling Time Calculation

Reactor doubling time (RDT) is usually calculated by using the following expression:

$$RDT(y) = \frac{M_0}{M_g} \tag{11.8}$$

where
   M_0 is the initial fissile inventory
   M_g is the fissile mass gained per year

The calculation of M_g requires the solution of the burnup equation for the given power level and cycle length. However, an approximation to the M_g can give a simple formula for RDT, which will correctly illustrate various parameters influencing it. The standard expression used for computing RDT [11.1] is

$$RDT(y) = \frac{2.7 M_0}{GPf(1+\alpha)} \qquad (11.9)$$

where
   M_0 is the initial fissile inventory
   G is the breeding gain
   P is the reactor power in MWt
   f is the load factor
   α is the σ_c / σ_f for the fissile material

This expression does not account for the fission in fertile nuclides. In reality, fertile nuclides like U-238, Pu-240, and Pu-242 also contribute for power production through fast fission. This effect can be incorporated in the RDT computation by using the concept of Pu-239 equivalence.

11.2.1.8 Power and Temperature Coefficient Calculation

Temperature coefficient is the change of reactivity with temperature assuming temperatures throughout the core change by the same amount. Similarly, power coefficient is the change in reactivity when the reactor power goes from one steady state to another asymptotic state. Static reactivity coefficient ρ is the combination of many reactivity contributions arising from change in temperature. The various effects that contribute to the reactivity coefficients are the Doppler effect, fuel and clad axial expansion, coolant expansion, core radial expansion or bowing, control rod drive line expansion, vessel expansion, and grid plate expansion. Here, the Doppler and fuel axial expansion feedbacks are calculated as a function of change in average fuel temperatures and their corresponding Doppler constant and fuel removal worth. With axial thermal expansion, axial boundary movement of the core also contributes to the axial expansion feedback. Similarly, clad axial expansion is calculated based on the change in average clad temperatures and its removal worth. Coolant expansion reactivity is calculated based on the bulk coolant expansion/voiding and its removal worth. Core radial thermal expansion is determined based on the coolant temperature at the spacer pad location and its feedback is calculated from the fuel and clad removal worth. Similarly, boundary movement of core 1 to core 2 and core 2 to the radial blanket also contributes to the core radial expansion feedback. Grid plate and vessel expansion is determined based on the change in inlet coolant temperature and their feedbacks are calculated from the corresponding fuel and clad removal worth. Control rod drive line expansion feedback is calculated based on the outlet coolant temperature.

$\delta\rho$ is the net change in reactivity when the power goes from the nominal value P_0 to P_1. Net change reactivity is a combination of different reactivity effects as explained in the following:

$$\delta\rho = \delta\rho_D + \delta\rho_{fax} + \delta\rho_{cax\text{-}expn} + \delta\rho_{Na} + \delta\rho_{rad} + \delta\rho_{rd} + \delta\rho_g + \delta\rho_v \qquad (11.10)$$

where
   $\delta\rho_D$ is the reactivity change due to Doppler effect
   $\delta\rho_{fax}$ is the reactivity change due to fuel axial expansion
   $\delta\rho_{cax\text{-}expn}$ is the reactivity change due to clad axial expansion
   $\delta\rho_{Na}$ is the reactivity change due to bulk coolant expansion
   $\delta\rho_{rad}$ is the reactivity change due to core radial expansion
   $\delta\rho_{rd}$ is the reactivity change due to control rod drive line expansion
   $\delta\rho_g$ is the reactivity change due to grid plate expansion
   $\delta\rho_v$ is the reactivity change due to vessel expansion

11.2.1.9 Radiation Damage

The structural materials undergo radiation damage due to interaction of neutrons. The interaction of incident neutrons leads to the transfer of energy to the lattice atoms producing primary recoil or primary knock-on atoms (PKAs) [11.6]. Their passage through the lattice site creates a cascade of knock-on atoms depending upon their energy spectrum. This process is continued till the PKA comes to rest in the lattice as interstitial. The net result will be the production of point defects (vacancies and interstitials) and their clusters in the crystal lattice. The total time required for this chain of events is about 10^{-11} s. Neutron absorption by (n, α) and (n, p) interactions create atoms of helium and hydrogen and other transmutation products within the crystals. In many cases, helium is the most important. To understand and quantify the radiation damage in materials, it is necessary to know, in detail, how neutrons interact with the material that produces the PKAs and, later on, their passage and the associated effects in the crystal lattice. The simple model used to describe the energy transfer of PKAs with the atoms is by the hard-sphere collisions. They also lose energy by interacting with electrons and Coulomb field of nearby atoms.

A useful way to characterize the extent of the irradiation a piece of material has received is to specify the average number of times an atom has been displaced (dpa) from its lattice site. The total number of displacements per atoms is calculated as [11.7]

$$D = \int_0^T \sum_g \sigma_{dg} \phi_g(t) \, dt \qquad (11.10.1)$$

where
   T is the length of time of irradiation
   σ_{dg} is the displacement cross section in the neutron energy group g
   ϕ_g is the neutron flux in the group g

The various important physical models [11.8] used to arrive at the displacement cross section are discussed as follows.

11.2.1.9.1 Kinchin–Pease Model

The number of displaced atoms $\nu(E)$ generated by a PKA of energy E is estimated by Kinchin and Pease as

$$\nu(E) = \begin{cases} 0 & 0 < E < E_d \\ 1 & E_d < E < 2E_d \\ \dfrac{E}{2E_d}; & 2E_d < E < E_c \\ \dfrac{E_c}{2E_d} & E_c < E < \infty \end{cases} \qquad (11.10.2)$$

At energies above E_c, the energy loss is only electron excitation, whereas they slow down entirely by hard-core elastic scattering for energies below E_c. An atom receiving the energy greater than the displacement threshold E_d is permanently displaced, while stuck atoms receiving energy less than E_d will eventually return to a lattice site. In this model, crystal lattice effects were ignored.

11.2.1.9.2 Half-Nelson Model

Nelson proposed a semiempirical modification of the Kinchin–Pease model by introducing a number of corrections as

$$v(E) = \frac{\alpha \beta(E) W(E) E}{\gamma(E) E_f} \tag{11.10c}$$

Here, α is the factor introduced to allow for realistic atomic scattering of the hard-core approximation. The factor $\beta(E)$ is to account for the defect recombination within the cascade, but this is usually ignored and is taken as 1. The factor $W(E)$ is the fraction of the initial PKA energy dissipated in elastic collisions and is estimated using the stopping power theory of Lindhard et al. But the electronic energy losses were confined to the PKA itself. The cascade was regarded as terminated by the formation of focused replacement sequences; thus, the focusing energy E_f replaces the displacement threshold E_d. The factor $\gamma(E)$, which corresponds to the factor 2 in the Kinchin–Pease model, increases at higher energies.

11.2.1.9.3 NRT Model

The secondary displacement model of Norgett, Robinson, and Torrens (NRT) is generally used in fast reactors to estimate the number of displacements per atom (dpa) by the PKAs. The damage energy or the kinetic energy absorbed by the material when it is exposed to high energy neutrons is used to estimate the number displacements (Frenkel pairs) produced by a given PKA as

$$v(E)_{NRT} = \frac{0.8 T_d}{2 E_d} \tag{11.10d}$$

where T_d is the damage energy that is obtained from the energy partitioning theory of Lindhard et al. It is to be noted that interatomic potentials are required to characterize the partition of energy of recoiling PKA between various electronic and nuclear energy loss mechanisms. The factor 0.8 was determined from binary collision model to account for realistic scattering. For a given PKA, a range of elastic and inelastic collisions may contribute to the damage energy. To compute PKA recoil spectra, it is important to know the damage energy accurately.

To compute the radiation damage due to neutron irradiation, the basic parameter needed is the displacement cross section σ_d, which is dependent upon the energy and the interaction cross sections of the incident neutrons, PKA energy spectrum, and the probability of producing the secondary knock-on atoms. Mathematically, it can be expressed as follows:

$$\sigma_d(E) = \sum_i \sigma_i(E) \int_{T_{min}}^{T_{max}} K(E,T)_i \, v_{NRT}(T) dT \tag{11.10e}$$

where
  E is the energy of incident neutron, energy of PKA
  $\sigma_i(E)$ is the neutron cross section for the reaction i
  $K(E,T)$ is the neutron-atom energy transfer kernel
  T_{min} and T_{max} are the minimum and upper energy of recoil atoms
  $v_{NRT}(T)$ is the secondary displacement function computed using the NRT model

The $\sigma_i(E)$ can be calculated from the ENDF/B [11.9] libraries using NJOY [11.10] code system. The SPECTER code [11.11], which is very simple and very fast, is widely used for estimating radiation damage of cross sections by using the in-built data based on the older version of the ENDF/B library.

11.3 THERMAL HYDRAULICS

The design of SFR and its core is controlled by the dominance of thermal loads. In order to optimize the design of core and coolant components from the considerations of safety and economy, robust design and safety criteria are postulated. In order to ensure such criteria, detailed analysis is often necessary. Simplified and detailed analysis methods that will usually be carried out to ensure that the specified design and safety criteria are respected by the design are presented in this section.

11.3.1 CORE THERMAL HYDRAULICS

At no point, the fuel center line temperature should exceed its melting point taking into account the uncertainties in predicting the operating temperature of the reactor under normal power level and flow distribution. Similarly, the clad hot spot temperature should not exceed the melting point of the clad material, and bulk temperature of the coolant should be lower than its boiling point. Appropriate factors of safety are applied on these limits depending upon the operating and transient conditions (details are presented in Chapter 3). The main objective of thermal design and analysis is to ensure that the specified temperature limits are not exceeded in the respective materials in the core. Thermal analysis should include the following aspects with increase in fuel burnup:

- Reduction in the melting point and thermal conductivity of the fuel with burn-up and also the bond gas (between fuel and clad) purity and its thermal conductivity change with burn-up
- Flow reduction in the SA due to change in configuration of the pin and SA
- Any phase transition in the fuel
- Change in the stochiometry of fuel
- Fuel restructuring, if any
- Chemical state of fission products in the fuel
- Differential swelling and creep between fuel and clad
- Heat transfer coefficient between flowing coolant and clad surface

The thermal design is performed through analytical or simplified numerical models to arrive at the preliminary design. Subsequently, detailed analysis is carried out to confirm/validate the design. At the beginning of life under high temperature condition, analytical model is simple since the fuel is as fabricated with a gap from clad. The salient features of both analysis steps are presented as follows.

11.3.1.1 Determination of Averaged Temperatures: Analytical Method

For preliminary analysis, a singular three-density region model is proposed. The temperature profile from the center line to the bulk coolant must be arrived at by solving the heat conduction equations with appropriate boundary conditions. Subsequently, the peak temperatures (hot spot temperatures) of clad are estimated, taking into account all uncertainties. This is called hot spot analysis.

11.3.1.1.1 Fuel Temperature Profile

Considering the fuel as a heat-generating ceramic cylinder, we can derive the temperature profile from the one-dimension heat conduction equation:

$$\frac{d^2T}{dr^2} + \frac{1}{r}\frac{dT}{dr} = -\frac{Q}{K}$$

where Q is the volumetric heat generation in the fuel

Boundary conditions:

$$\frac{dT}{dr} = 0 \quad \text{at } r = 0$$

$$T = T_s \quad \text{at } r = R_f$$

where
  T_s is the fuel surface temperature
  R_f is the radius of the fuel

Integrating and eliminating the integral constants,

$$\int_{T_s}^{T(r)} KdT = \frac{Q}{4}\left(R_f^2 - r^2\right)$$

11.3.1.1.2 Heat Transfer in the Fuel-Clad Gap

$$\Delta T = \left(T - T_{co}\right) = \frac{\chi}{2\pi R_f h_g}$$

$$h_g = \frac{CK_s P_{fc}}{H\sqrt{\delta_{eff}}} + \frac{K_m}{\left(\delta_f + \delta_c\right) + \left(g_f + g_c\right) + G}$$

$$\delta_{eff} = \sqrt{\frac{\delta_f^2 + \delta_c^2}{2}}$$

where
  C is the empirical constant (m^{-1})
  $K_s = 2K_f K_c/(K_f + K_c)$
  H is Meyer's hardness of the soft material
  P_{fc} is contact pressure

The h_g is solved for open and closed gaps. When the gap is open, the $P_{fc} = 0$ and the first term vanishes, and when the gap is closed, $G = 0$. The effect of roughness (δ_f, δ_c) and the temperature jump distances (g_f, g_c) are still in the equation even after the gap closure.

11.3.1.1.3 Temperature Drop in the Clad

The heat flux passing through the clad may be expressed as

$$q = -K_c \frac{dT}{dr}$$

where
  q is the heat flux (W/m^2)
  K_c is the thermal conductivity of cladding W/mK

This is expressed in terms of linear power.

$$\therefore \chi = -K_c 2\pi r \frac{dT}{dr}$$

K_c can be assumed constant through the clad thickness.

$$\Delta T = T_{ci} - T_{co} = \frac{\chi}{2\pi K_c} \ln \frac{R_{co}}{R_{ci}}$$

11.3.1.1.4 Temperature Drop in the Film (Clad to Coolant)
The heat transfer from the clad OD to the bulk coolant is expressed as

$$q = h(T_{co} - T_b) = \frac{\chi}{2\pi R_{co}}$$

where
    h is the heat transfer coefficient (W/m²°C)
    T_{co} is the outer cladding temperature
    T_b is the coolant bulk temperature (°C)
    R_{co} is the outer radius of the cladding (m)

11.3.1.2 Determination of Peak Temperatures: Hot Spot Analysis
Three methods for analysis are deterministic method, statistical method, and semistatistical method.

11.3.1.2.1 Deterministic Method
This is the oldest and most conservative method. The inner surface clad temperature is given by

$$T_c = T_i + \Delta T_1 + \Delta T_2 + \Delta T_3$$

where
    T_i is the coolant inlet temperature
    ΔT_1 is the temperature of coolant between point of entry of channel and point of maximum design value of T_c
    ΔT_2 is the film temperature drop
    ΔT_3 is the clad temperature drop

For each of the design variables, there exists some pessimistic values and if T_c is calculated using these values, the designer can be quite sure that nowhere else will the cladding temperature be superior to this critical temperature.

11.3.1.2.2 Statistical Method
This is an optimistic method since all variables that appear in calculation are not randomly distributed. Here, it is assumed that the variables follow the law of standard statistical distribution.

11.3.1.2.3 Semistatistical Method
In this method, the variables that cause the hot spot temperature are separated into two principal groups, that is, variables of statistical origin and nonstatistical origin. Exact values of the variables of nonstatistical origin are not known in advance and they are not subject to random occurrence.

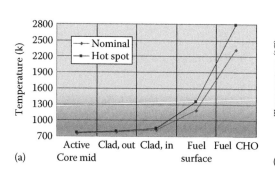
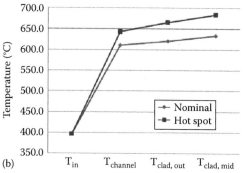

FIGURE 11.10   Fuel (a) pin and (b) clad midwall temperatures: nominal and hot spot values.

Dimensions, fuel density, fuel composition, flow distribution, and local flux perturbations are considered to be statistical in nature. Macroscopic flux distribution, heat transfer, and thermal power evaluation are treated as nonstatistical parameters. Fuel pin nominal and hot spot temperature at 400 W/cm for a typical reactor operating condition are shown in Figure 11.10a, and the clad hot spot temperature is shown in Figure 11.10b. The difference between the hot spot and nominal temperatures can be clearly seen in these figures.

11.3.1.3   Determination of Allowable LHR

Nominal power to melt is determined using conventional 1D heat transfer calculation and finding clad hot spot temperature (T_{CL}) and equating it to its melting point as shown in the following:

$$\text{Nominal: } T_{CL} = T_{Inlet} + 0.5 * \Delta T_{Channel} + \Delta T_{Film} + \Delta T_{Clad} + \Delta T_{Gap} + \Delta T_{Fuel}$$

Hot spot factors are superimposed on the various ΔTs and the power to melt is obtained, which is the design safety limit. Hot spot T_{CL} is estimated as shown in the following:

$$\text{Hot spot: } T_{CL} = T_{Inlet} + 0.5 \Delta T_{Channel} f_{channel} + \Delta T_{Film} f_{film} + \Delta T_{Clad} f_{clad} + \Delta T_{Gap} f_{gap} + \Delta T_{Fuel} f_{fuel}$$

Applying a power margin of 10% on linear heat rate (LHR) and its associated uncertainties and overshoot amounting to 5%, the allowable LHR is determined, as shown in Figure 11.11.

11.3.1.4   Flow and Temperature Distributions: Subchannel Analysis

By solving 1D forms of mass, momentum, and energy balance equations, the mean temperature of sodium, clad, and fuel at any cross section can be determined. Since the geometry of the fuel SA is highly complicated, it is impossible to obtain exact solutions due to the crossflow components of velocity generated by the wire wrap, which disturbs the 1D nature of the flow. The flow channels in an SA are also coupled to one another. There is mass, momentum, and energy exchange among the various channels. Hence, it is not possible to analyze one channel independently of the others. Also, there is flow bypass through the outermost row of fuel pins and hexcan wall. These features lead to large temperature variations along circumferential and radial directions. The inter-SA heat transfer effects also significantly influence the temperature profile of SAs located in the core periphery. Knowledge of steady and transient temperature distributions is essential for life assessment of SAs and hence to arrive at the target burnup. Bowing behavior of hexcans and fuel pins depends strongly on their temperature. In safety-related transients, onset of local coolant boiling (with its consequence on reactivity) depends on the local clad temperature. Online monitoring of fuel, clad, or coolant temperatures within SAs is not practical. Hence, temperature distribution in a fuel SA has to be determined by multidimensional computations backed up with suitable experiments.

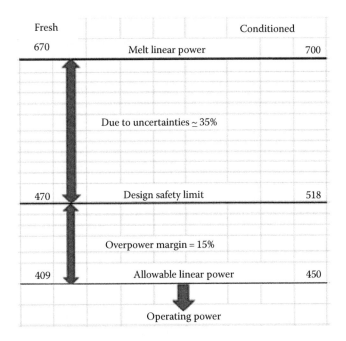

Fresh   Conditioned

670   Melt linear power   700

Due to uncertainties ~ 35%

470   Design safety limit   518

Overpower margin = 15%

409   Allowable linear power   450

Operating power

FIGURE 11.11 Basis for determining peak linear heat rate.

There are two approaches for this determination: (1) subchannel-based thermal hydraulic analysis and (2) computational fluid dynamics (CFD)-based thermal hydraulic analysis.

In a subchannel analysis, the temperature, pressure, and velocity in a subchannel are averaged, and a representative thermal hydraulic condition specifies the state of a subchannel. Flow and temperature distributions in the core are obtained by modeling and solving the conservation equations in a subchannel. Therefore, it is essential to model the inter-subchannel heat transfer between the adjacent subchannels as accurately as possible to enhance the predictability of a subchannel analysis model. One of the models adopted in SFR core design is followed in ENERGY code [11.12]. In this model (Figure 11.12a), at any axial plane, the SA is divided into two control volumes. The unknowns are two axial velocity components (U_P and U_C), one circumferential velocity ($U\theta$), two temperatures (T_P and T_C), and one pressure. The subchannel mass, momentum, and energy equations are solved for U_C and U_P, T_C, T_P, and pressure.

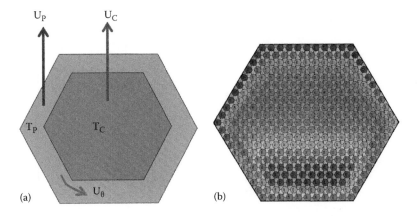

FIGURE 11.12 (a) Subchannel model adopted in ENERGY code. (b) Subchannel temperatures in a blanket SA (CADET simulation).

It may be highlighted that T_C and T_P have distribution at any cross section depending on the radial power profile. Subchannels exchange momentum and energy. But the effect of mass exchange is accounted by eddy viscosity (ε), which is taken from experiments. The value of Uθ is also taken from experiments. In general, Uθ and ε are functions of Reynolds number, pitch to diameter ratio, and helical pitch of the wire wrap. However, such models are not suitable to study natural convection, transient effects, and other conditions out of the range of experimental data for which Uθ and ε are available. There are also several other advanced subchannel analysis codes such as SUPERENERGY [11.13] and CADET [11.14] that can simulate multiple SAs of the core and inter-SA heat transfer. A typical temperature field in a blanket SA is shown in Figure 11.12b. It can be seen that due to flow bypass in the peripheral subchannels, the sodium temperature is less than that in the internal subchannels. Although the flow rates in the central subchannels are identical, there is large dissymmetry in the sodium temperature due to skewed power distribution in a blanket SA located in the core periphery.

11.3.1.5 Flow and Temperature Distributions: CFD Analysis

In the CFD-based approach, 3D forms of conservation equations of mass, momentum, and energy with suitable turbulence model are solved in the SA. The flow and temperature fields around each and every fuel pin and helical wire wrap are resolved in the computational mesh. Generation of boundary mesh for the fuel pin with helical wire wrap is one of the challenging tasks in this approach. The number of mesh points required for a realistic simulation of flow and temperature features demands a large computational time and memory. The SFR core consists of SAs that house a large number of fuel pins. The pins are provided with helically wound spacer wires to promote inter-subchannel mixing and reduce temperature asymmetry. The wires also reduce flow bypass between peripheral row of pins and hexcan wall. The pins being tightly packed as a bundle form thin winding flow channels. The traditional design methodology uses codes employing empirical information related to mixing, effective porosity, and heat transfer. With the advent of cheaper computing resources, methodologies capable of doing detailed analysis have emerged. CFD is one such tool, which is capable of revealing detailed flow/temperature fields, local hot spots, and heat transfer characteristics. The main challenge in adopting CFD-based thermal hydraulic design for SA is generation of structured mesh for the entire length of fuel pin bundle and solving the CFD equations in the large number of mesh [11.15]. The CFD analysis performed for a 19-pin bundle is shown in Figure 11.13,

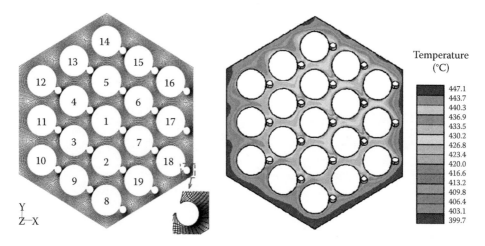

FIGURE 11.13 **(See color insert.)** CFD mesh for a 19-pin bundle (left) and sodium temperature field predicted by CFD calculations (right).

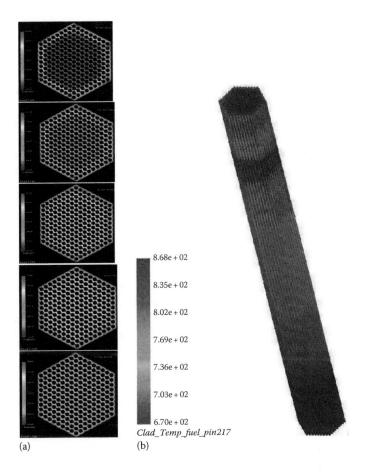

8.68e + 02

8.35e + 02

8.02e + 02

7.69e + 02

7.36e + 02

7.03e + 02

6.70e + 02

Clad_Temp_fuel_pin217

(a)   (b)

FIGURE 11.14 **(See color insert.)** (a) Sodium temperature distribution in various cross sections of the SA and (b) clad temperature in the entire bundle.

and also a 217-pin fuel bundle with 7 spacer wire pitch is depicted in Figure 11.14 [11.16]. As sodium flow upward in the SA, its temperature gradually increases due to heat addition. Unlike a subchannel model, CFD models predict detailed temperature variations within the subchannels (Figure 11.13). The main limitation of the CFD model is huge computational memory and time. For example, thermal hydraulic simulation of a 217-pin bundle for 7 helical pitch length required the usage of an 84-node parallel computing system at IGCAR and demanded a run time of about 24 h for each simulation. These simulations brought out that (1) the circumferential velocity in the peripheral subchannels is not uniform all around as assumed in the subchannel models and (2) the fully developed friction factor and the Nusselt number exhibit prominent spatial oscillations that are well correlated to the position of the helical wire in the peripheral subchannels. The entrance regime in the wire-wrap rod bundles is found to consist of two regions: (1) local development, which is completed within an axial length of ~20 times hydraulic diameter, and (2) global development requiring a longer length due to flow redistribution among the various subchannels. Further, the circumferential variation in the Nusselt number is found to be as high as 70%, which cannot be predicted by the subchannel models. The hot channel factor and hot spot factor are well predicted in the CFD-based models. Also, these models are capable of predicting flow redistributions when the flow regime changes from laminar to turbulent and flow is affected by buoyancy, especially under natural convection conditions.

11.3.2 Thermal Hydraulics of Sodium Pools

11.3.2.1 Design Criteria

11.3.2.1.1 Temperature Asymmetry in Cold Pool

During plant transients associated with one secondary loop, the temperature in the cold pool associated with the affected loop becomes hotter or colder than that of the unaffected loop. This leads to circumferential temperature difference in the cold pool structures, namely, main vessel and grid plate. The flow coast down characteristics of the secondary sodium pump and cold pool capacity need to be arrived at such that the circumferential temperature difference (temperature asymmetry) is less than a value from local thermal buckling and functional considerations (typically 30 K) in the reactor assembly shells. Similarly, the flow in the main vessel cooling system is to be distributed to respect this temperature asymmetry limit.

11.3.2.1.2 Free Level Fluctuation

Due to large sodium free surface area interfacing with argon, the hot pool free surface fluctuates. The height of the inner vessel above mean hot pool sodium level should be adequate to avoid overflow of hot pool sodium to cold pool. The amplitude of the fluctuation has to be minimum so that the height of the inner vessel can be minimum. When the level fluctuates, structures partly immersed in hot pool and partly exposed to cover gas see alternating temperatures. From high cycle fatigue considerations, the amplitude of fluctuations has to be less than a specified value to limit the high cycle fatigue damage (typically 50 mm).

11.3.2.1.3 Free Surface Velocity in the Pool

To avoid stratification risk in the pool, the velocity in the bottom part of the pool has to be high. At the same time, the velocity at the free surface has to be limited to a certain value (typically 0.5 m/s) to avoid gas entrainment. The entrainment of gas has to be limited to the extent that it does not cause any reactivity changes in the core. Continuous limited flow of argon through the core is not a serious concern. But segregation of minute bubbles in grid plate to agglomerate into a larger bubble and its sudden movement into the core are to be avoided.

11.3.2.1.4 High Cycle Temperature Fluctuations

The extent of stratification in the hot pool is a function of ratio of buoyancy forces to inertial forces. The stronger the inertial forces, the lower the axial temperature gradient. The stratification interface normally oscillates. Based on detailed structural mechanics calculations, it has been established that the axial temperature gradient has to be limited to <300 K/m. From thermal striping considerations, the peak to peak temperature fluctuations on the structures have to be limited to thermal striping limits specified in Section 11.4.2.

11.3.2.1.5 Heat Loss to Top Shield

The heat loss to the top shield has to be minimum to minimize the heat load on the top shield cooling circuit. By the provision of thermal shield, the direct radiation heat load and argon convection heat load are reduced by ~50%. But this in turn increases the bulk temperature of argon cover gas. Increase in the bulk temperature will enhance cellular convection and the associated temperature asymmetry. The bulk cover gas temperature will also affect the axial temperature gradient in the structures. Cellular convection of argon in the narrow gap penetrations of the top shield has to be managed to have temperature asymmetry less than a specified value to limit the resulting tilting of hanging components from the top shield. The annular gap size and cooling conditions of the top shield are to be optimized to respect this limit.

11.3.2.2 Analysis

11.3.2.2.1 Thermal Stratification: Design and Analysis Guidelines

The generally adopted design solution to avoid the thermal stratification effects during steady-state operating conditions of the reactor is to provide some devices that are able to break the stratification interfaces. These devices offer good mixing among the varying temperature streams, before they reach IHX primary inlet windows in the hot pool. Detailed multidimensional CFD studies aid in arriving at the design configuration of such devices. One such device adopted in the design of Indian prototype fast breeder reactor (PFBR) is a porous cylindrical skirt provided just below the control plug [11.17]. The effect of this skirt is to increase the sodium velocity entering into the hot pool and thereby promoting good mixing and avoiding stratification. In the reference design, hot sodium from fuel SAs flows directly to the IHX inlet forming low velocity stratified layers at the bottom of the hot pool as shown in Figure 11.15a. On introduction of the porous skirt below the control plug, sodium velocity in the bottom zone of hot pool is increased and thereby stratification is avoided (Figure 11.15b). However, provision of this skirt increases the pressure below the control plug bottom. This increased pressure increases the flow entering the control plug through the annular passages in the absorber rod mechanisms. Large flow in the control plug is of concern from transient temperature (cold shock) seen by the control plug parts during a reactor SCRAM. Ideally, the flow entering the control plug has to be proportional to its volume in the hot pool. This can be achieved by choosing proper combination of annular clearances between the shroud tubes and respective absorber rods drive mechanisms and the perforations in the shroud tubes. However, care has to be exercised to ensure that adequate clearances are available for facilitating smooth drop of absorber rods.

There are other detrimental effects due to the provision of the skirt. The skirt increases the radial velocity of sodium stream with which it enters the hot pool. This will result in increased velocity of sodium at the sodium free surface, which can be harmful from gas entrainment considerations. The skirt diverts sodium flow exiting from various SAs in the radial direction. This causes the sodium streams exiting out of SAs located in the periphery of the core to be masked by the sodium stream from those located in the central region. Therefore, the positioning of thermocouples in the core

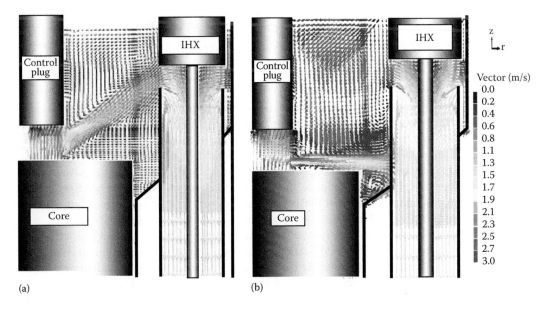

(a)   (b)

FIGURE 11.15 (a) Stratified flow in hot pool and (b) stratification eliminated.

temperature monitoring instrumentation should be very carefully decided to satisfy the SA plugging detection requirements [11.18]. Another effect that is caused by the skirt is the nonuniform velocity profile of sodium at the entry to the intermediate heat exchangers, which is critical from flow-induced vibration of IHX tubes. Detailed experimental and theoretical studies have to be carried out in selecting the proper design configuration for the cylindrical skirt. Thus, the porosity of the skirt has to be optimized considering these factors. Typical porosity value of the skirt is 10%. The skirt may not be effective in avoiding stratification in the pool at all power levels of operation of the reactor. If the low power operation is envisaged by reducing the flow (Q) proportionally to maintain the temperature rise across the core (ΔT), the Richardson number ($\Delta T/Q^2$) increases, which promotes stratification. To avoid this, a strategy can be adopted wherein the core flow can be reduced only up to a level so as to keep the Richardson number more or less unchanged, that is, the value of ($\Delta T/Q^2$) remains nearly the same for all the power levels.

11.3.2.2.2 Thermal Stratification: Numerical Simulation

Thermal effects cannot be adequately represented by water model tests, and construction and operation of large size sodium experimental facilities are also prohibitively expensive and time consuming. Hence, CFD analysis evolves as an essential approach for the prediction of temperature distributions in the reactor structures. In the CFD computations, turbulence model is a critical issue, especially for stratified flow conditions. Most of the turbulence models have been developed mainly for the forced convective flows. Identification of suitable turbulence model for stratified flow (buoyancy dominated), geometrical regimes for the computation, and establishing the optimum mesh (with respect to computational time, memory, and desired accuracy) are challenges in the CFD analysis for SFR applications. Fundamental studies [11.19,11.20] on thermal stratification show the capability of the computational approach to estimate the time-averaged vertical temperature gradients and interface position correctly, making use of standard k-ε-based turbulent models. Additionally, 3D large eddy simulation (LES) modeling calculations have been shown [11.19] to provide simulations of turbulent fluctuations and global instabilities. Recent studies by Abraham et al. [11.21] indicate that standard high Reynolds number k-ε turbulence model is capable of predicting thermal stratification interface accurately in MONJU reactor hot pool. The study brought out the need for detailed modeling of perforations in vertical shell structures for accurate prediction of stratification movement. In this field, more work is necessary to assess to what extent time-averaged gradient characteristics are representative of instantaneous conditions and to validate more precisely the prediction of the frequency and amplitude of temperature fluctuations.

11.3.2.2.3 Thermal Stratification: Experimental Simulation

Experimental studies play a vital role in the validation of the computational modeling methods and also to study certain specific problems concerning transient and multidimensional effects involved in the phenomena. In the experimental simulations of thermal stratification in the hot pool, certain nondimensional numbers should be respected between the model and prototype. The Richardson number (Ri), which represents the ratio between the buoyancy and inertia forces involved in the flow, and the Reynolds number to simulate the flow fluctuations are important nondimensional numbers need to be simulated correctly. Further, the ratio of heat transfer by convection to that by conduction should be preserved for which the Peclet number should be nearly equal.

During transient conditions, there exists a critical Richardson number Ri_c beyond which the flow pattern in the hot plenum undergoes modifications in comparison to that under nominal conditions [11.22,11.23]. This is illustrated in Figure 11.16. Figure 11.16a shows the flow configuration when Ri < Ri_c (Ri_c is typically ~2.6), where the inertial effects are large compared to buoyancy effects. Hence, sodium emerging from core is able to move upward to produce strong recirculation in the upper part of the hot pool as well as in the cavity between core and inner vessel, thus facilitating good mixing without any stratification. Figure 11.16b depicts the flow pattern when Ri > Ri_c, that is, the buoyancy effects are dominant compared to inertial effects, where the cold sodium from the

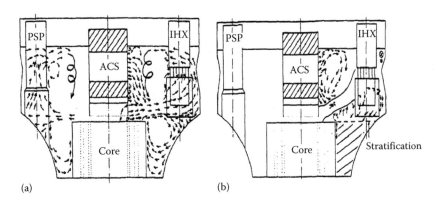

FIGURE 11.16 Stratification mechanisms in hot pool: (a) inertia driven and (b) buoyancy driven (PSP, primary sodium pump; ACS, above core structure).

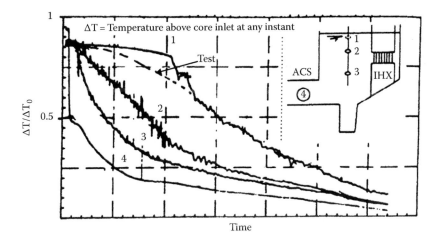

FIGURE 11.17 Hot plenum stratification after a SCRAM at 23% power in SPX1.

peripheral SAs separates from the hot sodium of fuel SAs and occupies the cavity. This leads to inadequate mixing leading to thermal stratification.

The transient thermal-hydraulic behavior of hot pool can also be characterized by means of Ri. In the same way as for steady-state conditions, fully turbulent flow in the model must be ensured by imposing a low Reynolds distortion. Comparative studies were performed in SPX1 geometry [11.23,11.24] between mock-up tests and reactor measurements to simulate the transient temperature evolutions at various locations in the hot pool following a reactor scram. During this transient, the hot pool temperature in the vicinity of core top reduces rapidly (location 4 in Figure 11.17) compared to locations far from the core (location 1). This has been simulated correctly in the model studies where Richardson number similitude is satisfied.

11.3.2.2.4 Thermal Striping: Design and Analysis Guidelines

As the complete theoretical or experimental simulation of thermal striping phenomenon in a fast reactor system is difficult, the usual methodology followed in design is a combination of simplified global analysis and detailed local analysis approach. After a first step using global thermal-hydraulic models to identify the areas where thermal striping may occur, the characteristics of the fluctuation have generally been estimated from model tests using sodium as the working fluid. Based on global 3D thermal hydraulic studies carried out for the primary sodium circuit, the localized zones

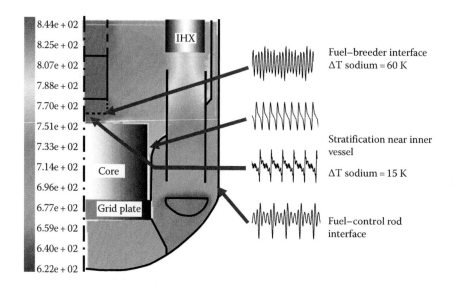

FIGURE 11.18 Predicted temperature fluctuations in primary circuit of PFBR.

prone to thermal striping identified in a pool-type SFR are: (a) fuel–breeder interface around the lattice plate, (b) fuel–breeder interface around the core cover plate, (c) bottom location of absorber rod drive mechanism where fuel and control SA sodium flows interact, and (d) main vessel near IHX outlet [11.17]. These localized zones have to be further considered for the detailed prediction of flow and temperature oscillations. With the velocity and temperature values prescribed for the jet streams, the oscillations in the mixing layer region (interface) can be predicted at different locations. For conservatism, the influence of global flow field in the domain caused by other flow sources in the circuit can be neglected. Further, for numerical prediction of the fluctuating temperature field, a simplified method can be devised by approximating the local geometry in each case to an equivalent 2D domain, and direct numerical simulation (DNS) calculations can be performed [11.17]. The temporal oscillations and peak-to-peak sodium temperature fluctuations (ΔT) at critical locations predicted for PFBR are summarized in Figure 11.18. Three-dimensional simulations would definitely lead to more accurate results compared to the 2D predictions. In this case, the 2D simulation is a compromise for the easy handling of the large geometrical domain of hot pool regions prone to thermal striping.

11.3.2.2.5 Thermal Striping: Numerical Simulation

Thermal hydraulics of thermal striping phenomenon is governed by many physical phenomena, with each one having its own difficulties and limitations. There are difficulties with the experimental quantification and modeling of turbulent mixing (temporal and spatial multiscale phenomena). The oscillating temperature field in sodium is caused by jet instabilities and turbulence. The size distribution of turbulent eddies and their life spans depend on the velocity and hydraulic diameter of issuing jets. Therefore, the experimental simulation needs to be carried out on equal scale models with exact simulation of flow velocity. Meanwhile, on the theoretical modeling of this phenomenon, progress has been made with the advanced turbulence models: LES, very large eddy simulation (VLES), and detached eddy simulation (DES). But these models are still in the development and validation stages. The qualification of these models is extremely difficult as a result of measurement difficulties and uncertainty in the validation of the numerical and experimental results. The success of the numerical simulation of thermal striping depends on how accurately the turbulent scales responsible for the oscillating flow and temperature field are modeled. These methods require fine resolution of geometry to capture the formation and movement of eddies. Enormous computational resources are required for these calculations. Performing such calculations for the entire hot plenum of an SFR is very difficult. In this context, a localized analysis approach is a better alternative. However, presently, there is a trend toward solving large numerical models that current computers can barely handle to simulate the entire range of fluctuations. On the heat transfer aspect, there are still difficulties in measuring and modeling the heat exchange between the fluid and the wall. In addition, there are theoretical problems associated with the lack of understanding of the boundary layer phenomena. Computational studies [11.25,11.26] show that simulations using high-order numerical schemes and LES turbulence models provide promising results in the prediction of temperature fluctuations in SFR systems. Menant and Villand [11.26] have predicted the temperature fluctuations in a square section duct with T-junction using the computer code TRIO-VF employing the LES model. The studies indicated that the amplitude of temperature fluctuation in the structure is about 50% of that in the fluid.

11.3.2.2.6 Thermal Striping: Experimental Simulation

In view of the difficulties associated with the complete modeling of the striping phenomenon, the usual approach followed is a combination of numerical and experimental simulations. Thermal hydraulics studies cover basic understanding of the phenomenon [11.27], establishing experimental simulation principles, assessments based on tests with water and air [11.28], and attenuation characteristics of thermal striping on the metal wall [11.29]. Experimental studies are generally performed for the validation of numerical models. International scenario in the understanding of thermal striping is very encouraging. France has reported many important observations related to thermal striping in PHENIX and SPX-1 [11.30] reactors and carried out experiments in air and sodium (Figure 11.19) to understand thermal striping during mixing of nonisothermal axisymmetric jets in a pool. The amplitude of temperature fluctuation is seen to exhibit a nonmonotonic variation with respect to axial distance. It is clear that a structure located away from the jets at about five times the hydraulic diameter would be subjected to maximum thermal striping damage. This location is found to be independent of the fluid.

In the United Kingdom, Betts et al. [11.31] carried out elaborate experimental research to understand if air or water could be used to simulate sodium in thermal striping studies. They found that air can be used to predict the temperature fluctuations in the fluid if the Reynolds number is of the order of 10^6. However, the boundary layer attenuation and the resulting temperature fluctuations in

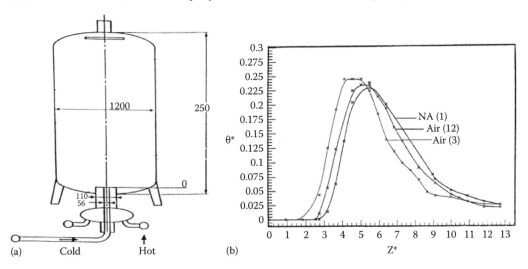

FIGURE 11.19 Sodium and air experiments on mixing of coaxial nonisothermal jets: (a) experimental setup and (b) variation of temperature fluctuation along the axis.

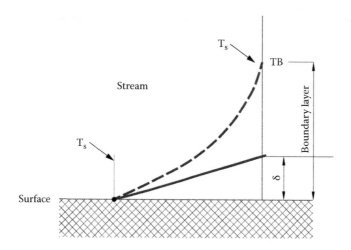

FIGURE 11.20 Model to estimate boundary layer attenuation.

the structures could not be accurately predicted by air/water tests. This is because of the large difference in the heat transfer coefficient of sodium and air. To overcome this difficulty, Wakamatsu and Hashiguchi [11.29] proposed an equivalent boundary layer model for the determination of boundary layer attenuation and the resultant temperature fluctuations in the structures (Figure 11.20). In this model, an effective stagnant sodium layer is assumed to shield the structure. The thickness of this shield is determined such that the conductive resistance of this shield is equal to the convective resistance of flowing sodium. Thermal striping phenomenon has been simulated in water tests through a dedicated test setup (Figure 11.21), developed at Indira Gandhi Centre for Atomic Research (IGCAR) [11.32]. This setup simulates thermal hydraulics in the vicinity of core cover

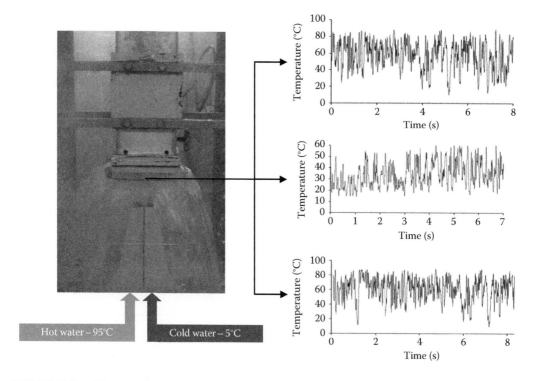

FIGURE 11.21 Water test facility to simulate thermal striping (IGCAR).

plate, located below the control plug just above the core. Temperatures and flow rates of hot and cold water varied over wide ranges. A maximum ΔT of 90 K is possible in the setup.

11.3.2.2.7 Free Surface Fluctuation: Design and Analysis Guidelines

Free level fluctuation is a strong function of the vertical velocity component in the hot pool. It has a strong bearing on the phenomenon of argon gas entrainment into the pool. In order to minimize the free surface gas entrainment in the hot pool, horizontal baffle devices attached to the inner vessel are found to be very effective by Banerjee et al. [11.33] through detailed water model tests. Provision of the horizontal baffle is found to reduce the free level fluctuations very effectively.

11.3.2.2.8 Free Level Fluctuations: Numerical Simulations

Prediction of the level fluctuation by CFD studies calls for 3D, transient studies coupled with complex modeling features for predicting free surface profiles. Experimental simulation is the favored option to understand this phenomenon. For the numerical prediction of these types of flows with moving gas–liquid interface, numerical methods such as height function method and volume of fluid (VOF) method [11.34] have been developed and incorporated in commercial CFD computer codes. In order to capture the free level fluctuations of frequency less than 1 Hz, the simulations have to be carried out also with smaller time steps. The computational domain being a large pool, the number of computational cells will be large. Solving a 3D transient problem of this nature is a computational challenge.

11.3.2.2.9 Free Level Fluctuations: Experimental Simulations

The free level fluctuations are governed by the balance between the inertial, gravitational, and frictional forces. The appropriate nondimensional number that governs the inertial force to gravitational force is Froude number (Fr) and the number that governs the inertial force to frictional force is Reynolds number (Re). These nondimensional numbers can be respected by water models. Since no heat transfer is involved, poor thermal conductivity of water is not a constraint. The effect of Re is secondary once the flow is well in the turbulent regime and hence distortion in the Re can be permitted. Hence, water experiments on geometrically similar scaled down models, respecting Fr number similarity, are carried out and the measured free level fluctuations are extrapolated for the reactor.

The experimental studies on a ¼ reactor model [11.35] reveal that the free level fluctuations are not uniform throughout the hot pool and that the ripple height and the fluctuation frequency depend on the location in the pool. For nominal flow condition, the maximum amplitude is about 82 mm in the vicinity of the control plug extrapolated to prototype conditions. The predominant frequencies of fluctuations for the reactor vary between 0.25 and 1.6 Hz. Using these data, the transient temperature fluctuations in partially submerged structures are determined by solving the transient heat conduction equation. It shall be highlighted that structural temperature fluctuations cannot be predicted by water experiments.

The free level fluctuations in the pool should be quantified for choosing the margins for avoiding the gas entrainments into the heat exchangers immersed in the hot pool. Apart from this, the free level fluctuations cause special type of structural damage called *thermal ratcheting* on thin shells in the vicinity of level fluctuations. Since the weir ensures maintaining a constant sodium free level in the vicinity of main vessel irrespective of the sodium flow rate, there is no possibility of level fluctuations in the main vessel (Figure 11.22).

11.3.2.2.10 Cellular Convection: Design and Analysis Guidelines

In fast breeder test reactor (FBTR), cellular convection formed between the reactor vessel and the large rotating shield plug led to the uneven expansion and hence the tilting of reactor vessel. In order to overcome this problem, a comparatively lighter helium gas is being injected from the top of the penetration to suppress the cellular convection.

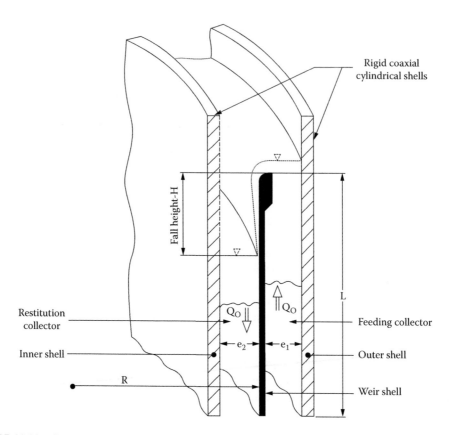

FIGURE 11.22   Schematic of weir flow system.

The tilting of components caused by cellular convection has to be managed through proper care taken in the design. The tilting of control plug causes the thermocouples of the core temperature monitoring instrumentation to get offset with respect to the center of the SA. This needs to be considered in deciding the normal location of thermocouple with respect to the SA. The tilting of pump assembly due to cellular convection needs to be considered in the design of the spherical seat support. Wherever the cellular convection effects are unmanageable, proper design provisions have to be made for arresting the convection by the use of anticonvection barriers. Thus, cellular convection is an important aspect to be considered in the design to meet the functional requirements of various systems and mechanical design of various components. It is seen that by a strong cooling of the roof slab shell, the effects of cellular convection can be effectively controlled [11.36].

11.3.2.2.11   Cellular Convection: Numerical Simulation

As the cellular convection is a geometry-dependent and boundary condition–dependent phenomenon, involving radiative heat transfer in evaporation and condensation of sodium vapor and natural convection of sodium mist–laden argon cover gas, the experimental studies have to be carried out on full-scale models [11.37]. The approach adopted in design is a combination of experimental and theoretical analyses. Experimental studies are used for the validation of theoretical models and design predictions are made using the validated theoretical models [11.38]. Cellular convection in the penetrations and resulting asymmetric temperature distribution in the shells have to be predicted by a conjugate thermal hydraulic model considering natural convection of argon, heat conduction in the multiple shells, radiative heat exchange among the shells, and forced convection cooling boundary conditions simultaneously. Suitable turbulence models (depending on the Rayleigh number) with special approaches recommended for the simulation of natural convective flows have to be adopted

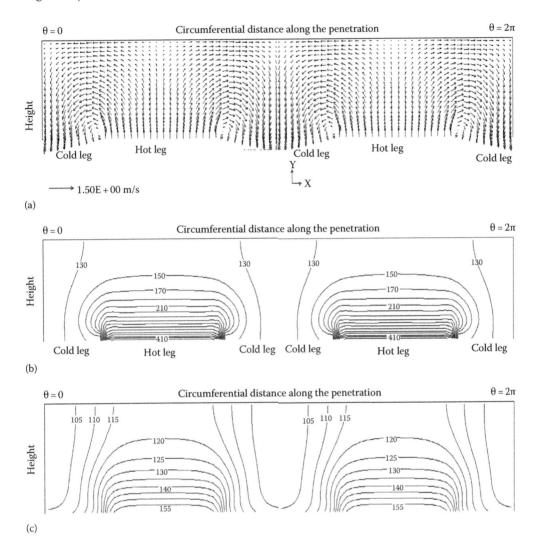

FIGURE 11.23 Important cellular convective parameters. (a) Velocity field in the developed annulus, (b) temperature distributions (°C) in the argon gas, and (c) temperature distribution (°C) in roof slab shell.

in these predictions [11.39]. The cellular convection velocity pattern and corresponding temperature pattern in the annulus predicted for a typical component penetration are shown in Figure 11.23 [11.17]. It is seen that hot argon enters the annulus from bottom at two locations, heats up the shells, gets cooled, and eventually leaves the annulus at two locations (Figure 11.23a). As a consequence of this, the maximum circumferential temperature difference in argon is 270°C (Figure 11.23b) and the same in roof slab shell is 45°C (Figure 11.23c).

11.3.2.2.12 Cellular Convection: Experimental Simulations

This aspect has been elaborated in Section 10.4.8. Here, the important aspects are highlighted for the sake of comprehensive presentation. Cellular convection of argon was studied in COBA experimental test facility by Hemnath et al. [11.40]. The COBA test vessel has two parts: the lower vessel and the upper vessel. Liquid sodium is contained in the lower vessel and an annulus of 20 mm gap width is formed in the upper vessel as shown in Figure 10.96. The upper vessel height is 1000 mm and it simulates the penetration of a pump. Experiments have been conducted at various axial temperature gradients and the conditions for onset of natural convection have been determined.

11.4   STRUCTURAL MECHANICS ANALYSIS OF SPECIAL PROBLEMS RELEVANT TO SFR

A few specific situations in SFR that cause temperature fluctuations in sodium and associated effects on the adjoining structural wall surfaces were addressed in Chapter 9. Such situations are sodium free level oscillations, thermal stratifications, and thermal striping. Apart from these, there are some failure modes specific to thin-walled shell structures. Among them, fluid elastic instability due to coupling of certain vibration modes associated with structures and fluids and buckling of thin shells subject to seismic loadings need special analytical treatments. In this part of the chapter, analysis methods adopted for deriving (1) allowable temperature fluctuations (thermal striping limits are quantified for a typical SFR as a case study and similar approach can be followed for any source of temperature fluctuations such as thermal stratification and free level oscillations), (2) fluid elastic instability regimes for thin shells (a thermal baffle incorporated in the main vessel cooling system is taken for a case study), and (3) buckling of thin shells subject to seismic loadings (main vessel, inner vessel, and thermal baffles are taken for the case study) are presented.

11.4.1   Generic Methodology to Derive Allowable Temperature Fluctuations

SFR reactor assembly structures are generally cylindrical shells with several longitudinal and circumferential seam welds. These geometries can be idealized as plates. Such assumption is valid for the thin cylindrical shells characterized by large diameter to thickness ratio (D/h). However, for a smaller D/h ratio, the plate assumption yields conservative results [11.41]. Immediate effect of imposed temperature fluctuations is generation of thermal stress fluctuations across the structural wall thickness, and on the long run, such fluctuations cause structural damage, namely, high cycle fatigue damage. To determine the damage, application of basic fracture mechanics concept is one effective approach. As per this, the plate has an initial part-through crack of size, typically 0.1 mm, which cannot be detected by the existing nondestructive evaluation (NDE) techniques and further the cracks grow due to cyclic thermal stresses. In the design stage, a maximum allowable crack size is taken as 0.5 mm. It is postulated that the damage (D) is accumulated due to creep (during hold time), low fatigue cycles (cycles involving high strain ranges) as well as high fatigue cycles (cycles involving low strain ranges). Thus, the plate can have any crack size within the range of 0.1–0.5 mm, depending upon the value of D: 0.1 for D = 0 and 0.5 mm for D = 1. Under such situations, the procedure recommended in RCC-MR: Appendix A16 [11.42] for computing the accumulated creep–fatigue damage for the structures having crack-like defect is applied [11.43]. The important aspects are presented in the following subsection.

11.4.1.1   Main Steps to Derive the Permissible Range of Temperature Fluctuations

The thickness of the idealized plate geometry is h and the size of the part-through wall crack is a. The plate is subjected to random surface temperature history. The plate has accumulated certain values of fatigue and creep damage (V and W, respectively) due to normal and design basis load cycles. With these idealizations, the temperature, stress, and stress intensity distributions are computed analytically. Subsequently, the permissible temperature range (ΔT_p) is computed as a function of the combined effect of V and W (D_{eff}). The following steps are followed to derive such relation:

- D_{eff} is computed using creep–fatigue damage interaction diagram for the specified material of construction (Figure 11.24 for the austenitic stainless steels, generally used for the reactor assembly components).
- An equivalent crack length (a_{equ}) corresponding to D_{eff} is computed as indicated in Figure 11.24.

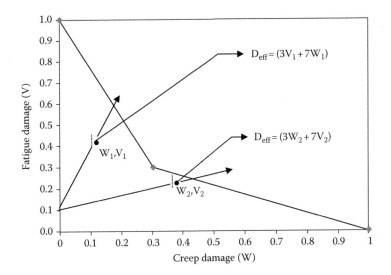

FIGURE 11.24 Creep–fatigue interaction diagram for SS 316 LN.

- The temperature distribution across the structural wall of the plate, subjected to specified random temperature history (expressed as a power spectral density [PSD]), is determined using frequency response function method [11.44].
- The thermal stress range ($\Delta\sigma_f$) corresponding to applied temperature fluctuation (ΔT_s) and subsequently, stress intensity factor range (ΔK_f) are computed as a function of frequency 'f'.
- ΔK_{rms} and subsequently ΔK_{max} are computed using idealized PSD.
- Knowing allowable $\Delta K_{allowable}$ for the given material at the specified temperature ($\Delta K_{threshold}$), permissible ΔT_p is computed as $\Delta T_p = \Delta K_{allowable}/\Delta K_{max}$.

The derivations of expressions for ΔT_s, $\Delta\sigma_f$, and ΔK_{max} are given as follows.

11.4.1.2 Transient Temperature Distribution

Temperature fluctuations in the fluid (ΔT_f) as well as on the surface (ΔT_s) are assumed to be in phase as per the following equations:

$$\Delta T_f = \Delta T_{fm} e^{i\omega t} \tag{11.11}$$

$$\Delta T_s = \Delta T_{sm} e^{i\omega t} \tag{11.12}$$

where

ΔT_{fm} and ΔT_{sm} are the maximum temperature difference (range) in the fluid and at the metal surface, respectively

ω is the frequency of oscillations expressed in radians/s, which is equal to $2\pi f$

f is the frequency expressed in cycles/s (Hz)

The back face is adiabatic. The temperature distribution across the plate thickness δ is expressed as

$$\Delta T(x,t) = \Delta T_f A(\omega,x) \tag{11.13}$$

where temperature response function: $A(\omega, x) = (P(\omega,x) + i \cdot Q(\omega,x))/(R + iS)$.

The functions P and Q are obtained so that Equation 11.13 satisfies the governing differential Equation 11.14 and adiabatic boundary condition defined by Equation 11.15:

$$\frac{\partial^2 T}{\partial x^2} = \frac{1}{k}\frac{\partial T}{\partial t} \tag{11.14}$$

where k is the thermal diffusivity of structural material.

$$\text{Adiabatic wall condition:} \frac{\partial T}{\partial x} = 0 \quad \text{at } x = \delta \tag{11.15}$$

The following analytical expressions for P and Q satisfy Equations 11.14 and 11.15:

$$\begin{aligned}
P(\omega,x) &= \cos\lambda(\delta-x)\cos h\lambda(\delta-x) \\
Q(\omega,x) &= \sin\lambda(\delta-x)\sin h\lambda(\delta-x)
\end{aligned} \tag{11.16}$$

$$\lambda = \sqrt{\frac{\omega}{2k}}$$

The analytical expressions for R and S are obtained by satisfying the following convective heat transfer boundary condition (11.17) on the surface facing the fluid temperature variations. h is the heat transfer coefficient of fluid on the metal surface.

$$-K\frac{\partial T}{\partial x} = h(\Delta T_f - \Delta T_s) \tag{11.17}$$

$$-K\Delta T_{fm}e^{i\omega t}\frac{\frac{dP}{dx}(0)+i\frac{dQ}{dx}(0)}{(R+iS)} = he^{i\omega t}\left[\Delta T_{fm} - \Delta T_{fm}\frac{P(0)+iQ(0)}{(R+iS)}\right] \tag{11.18}$$

Simplification of Equation 11.18 yields

$$-\frac{K}{h}\frac{\frac{dP}{dx}(0)+i\frac{dQ}{dx}(0)}{(R+iS)} = 1 - \frac{P(0)+iQ(0)}{R+iS} \tag{11.19}$$

$$\left[P(0)-\frac{K}{h}\frac{dP}{dx}(0)\right] + i\left[Q(0)-\frac{K}{h}\frac{dQ}{dx}(0)\right] = R+iS \tag{11.20}$$

From Equation 11.20, the following expressions for R and S are obtained:

$$R = P(0) - \frac{K}{h}\frac{dP}{dx}(0)$$

$$S = Q(0) - \frac{K}{h}\frac{dP}{dx}(0) \tag{11.21}$$

From Equation 11.16,

$$P(0) = \cos\lambda\delta\cos h\lambda\delta \quad \text{and} \quad Q(0) = \sin\lambda\delta \cdot \sin h\lambda\delta \tag{11.22}$$

Differentiation of Equation 11.20 and putting x = 0,

$$\frac{dP}{dx}(0) = -\lambda\left[\cos\lambda\delta \cdot \sin h\lambda\delta - \sin\lambda\delta\cos h\lambda\delta\right]$$
$$\frac{dQ}{dx}(0) = -\lambda\left[\sin\lambda\delta \cdot \cos h\lambda\delta + \cos\lambda\delta \cdot \sin h\lambda\delta\right] \tag{11.23}$$

Substituting Equations 11.22 and 11.23 into Equation 11.21 yields

$$R = \cos\lambda\delta\cos h\lambda\delta + \frac{K}{h}\lambda\left[\cos\lambda\delta \cdot \sin h\lambda\delta - \sin\lambda\delta \cdot \cosh\lambda\delta\right]$$
$$S = \sin\lambda\delta \cdot \sin h\lambda\delta + \frac{K}{h}\lambda\left[\sin\lambda\delta \cdot \cos h\lambda\delta + \cos\lambda\delta \cdot \sin h\lambda\delta\right] \tag{11.24}$$

$$\text{If } h \to \infty, \quad R = \cos\lambda\delta \cdot \cos h\delta \quad \text{and} \quad S = \sin\lambda\delta \cdot \sin h\lambda\delta \tag{11.25}$$

The expressions are matching with solutions given by Jones et al. [11.44] for the case $h \to \infty$. The attenuation of peak temperature on the metal wall surface is given by

$$\frac{\Delta T_{sm}}{\Delta T_{fm}} = \alpha = \sqrt{\frac{\left(P^2 + Q^2\right)}{\left(R^2 + S^2\right)}} \tag{11.26}$$

For a typical practical case: plate thickness (δ) = 30×10^{-3} m, heat transfer coefficient of sodium (h) = 40,000 W/(m^2 K), and thermal diffusivity of austenitic stainless steel (K) = 4.76×10^{-6} m^2/s at 820 K, the attenuation factor (α) is plotted as a function of frequency (f) in Hz using Equation 11.26 in Figure 11.25.

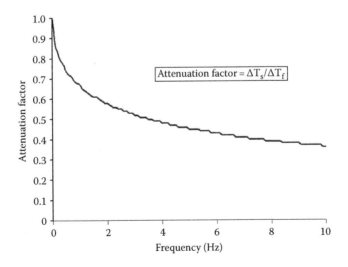

FIGURE 11.25   Temperature attenuation on the metal wall surface.

11.4.1.3 Transient Thermal Stress Distribution

It is conservatively assumed that the plate is fully constrained. The axial stress range induced at any distance x from surface is expressed as

$$\Delta\sigma(x,t) = \left[\frac{E\alpha\Delta T_f}{(1-\upsilon)} \right] \cdot A(\omega,x) \tag{11.27}$$

where E, α, and υ are Young's modulus, coefficient of thermal expansion, and Poisson's ratio, respectively, for the structural material. The peak-to-peak stress variation at any location measured from the front face is expressed as a function of ω in the following equation:

$$\Delta\sigma(x,\omega) = \left[\frac{E\alpha\Delta T_f}{(1-\upsilon)} \right] \frac{\sqrt{\left(P^2+Q^2\right)}}{\sqrt{\left(R^2+S^2\right)}} \tag{11.28}$$

From this equation, the stress range per unit ΔT_f is expressed as

$$\Delta\sigma(x,\omega) = \left[\frac{E\alpha}{(1-\upsilon)} \right] \frac{\sqrt{\left(P^2+Q^2\right)}}{\sqrt{\left(R^2+S^2\right)}} \tag{11.29}$$

The stress decay has been plotted as a function of frequency of fluctuation in Figure 11.26, using Equation 11.29 across the plate thickness (δ) = 30 × 10⁻³ m. Other parameters used are heat transfer coefficient (h) → ∝ W/(m² K), thermal diffusivity (k) = 4.76 × 10⁻⁶ m²/s, Young's modulus (E) = 1.63 × 10⁵ MPa, υ = 0.3, and α = 20.0 × 10⁻⁶/K and ΔT_{fm} of 20 K. From Figure 11.26, it is seen that, in the case of constant temperature on the wall surface (frequency = 0), the stress is constant across the thickness. When the frequency increases from 0 to 1 Hz, the stress decays rapidly across the thickness, which reaches nearly saturation at the frequency of about 1 Hz. Hence at higher frequencies, the stresses are concentrated in the vicinity of surface without significant penetration,

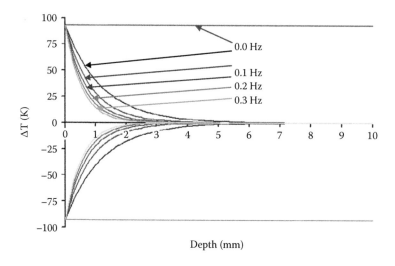

FIGURE 11.26 Decay of stress across the structural wall thickness.

implying that thermal striping can only initiate cracks and does not have potential to cause growth. This physical understanding can be derived from Figure 11.26.

11.4.1.4 Determination of ΔK_I

For the mode I field, the crack tip stress intensity factor corresponding to a part through crack a is determined using the Breckner weight function as

$$\Delta K_I(x,t) = \int_0^a \Delta\sigma(x,t)M(x)dx \qquad (11.30)$$

where the weight function M(x) for an edge cracked plate is given by

$$M(x) = \sqrt{2\pi a} \left\{ \left(\frac{1-x}{a}\right)^{-1/2} + m_1\left(\frac{1-x}{a}\right)^{1/2} + m_2\left(\frac{1-x}{a}\right)^{3/2} \right\} \qquad (11.31)$$

where m_1 and m_2 are the polynomials independent of x:

$$m_1 = 0.6147 + 17.1844\left(\frac{a}{h}\right)^2 + 8.7822\left(\frac{a}{h}\right)^6$$

$$m_2 = 0.2502 + 3.2889\left(\frac{a}{h}\right)^2 + 70.0444\left(\frac{a}{h}\right)^6$$

Using Equation 11.27 for $0 \leq a \leq h/2$

$$\Delta K_I(a,t) = -\left[\frac{E\alpha\Delta T_f}{(1-\nu)}\right]e^{i\omega t}\int_0^a A(x,\omega)M(x)dx$$

$$\Delta K_I(a,\omega) = -\left[\frac{E\alpha\Delta T_f}{(1-\nu)}\right]\sqrt{\left(I_1^2 + I_2^2\right)} \qquad (11.32)$$

where $I_1 = \int_0^a P(x)\,M(x)\,dx$ and $I_2 = \int_0^a Q(x)\,M(x)\,dx$

The stress intensity factor (SIF) per unit ΔT_f is expressed as

$$\Delta K_I(a,\omega) = -\left[\frac{E\alpha}{(1-\nu)}\right]\sqrt{\left(I_1^2 + I_2^2\right)} \qquad (11.33)$$

In Figure 11.27, ΔK_I is shown as a function of crack size a (0.1–5 mm) for three frequencies, 0.0625, 1.0, and 6.25 Hz, for the plate of thickness of 10 mm. The SIF increases monotonically for frequencies less than 1 Hz and for frequencies higher than 1 Hz, it increases only up to a certain crack size. After reaching a maximum value, SIF decreases. This sort of drooping behavior indicates that there is a possibility of crack arrest under high frequency thermal fluctuations. More pronounced drooping characteristics can be seen in the case of cylinders, which are presented in Ref. 11.41. This ensures that the assumption of plate geometry yields conservative thermal striping limits, which are preferred in the design stage.

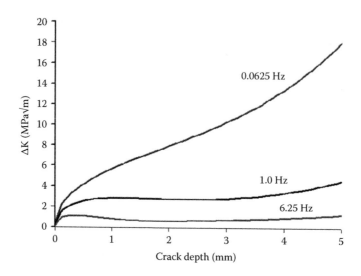

FIGURE 11.27   Stress intensity fluctuations due to thermal striping.

11.4.2   Establishing Thermal Striping Limits: A Case Study

The temperature histories are defined as PSD, from which radiation monitoring system (RMS) values can be derived by integration. A few typical PSD shapes extracted from published literature information are depicted in Figures 11.28 and 11.29: spectra obtained from a rig at AEA technology in Figure 11.28 [11.44], time history and associated spectra published by Risely in Figure 11.29 [11.45].

The methodology dealt in the previous section is used for deriving permissible temperature fluctuations (peak-to-peak temperature range) for a typical SFR, having the following parameters:

- Material = SS 316 LN
- Plate thickness, $h = 5$–30 mm
- Cutoff frequency, $f = 10$ Hz
- Plant life (40 years with 85% LF), $T = 3 \times 10^5$ h
- Effective damage, $D_{eff} = 0$–1.0

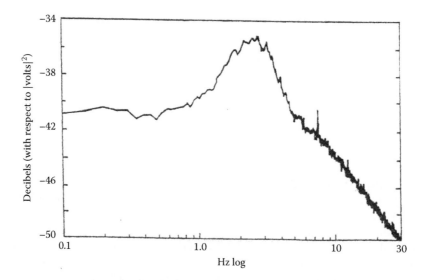

FIGURE 11.28   PSD of a typical thermal striping (PFR).

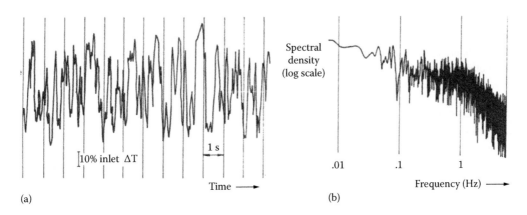

(a)

(b)

FIGURE 11.29 A few typical PSD shapes of temperature fluctuations (Risely).

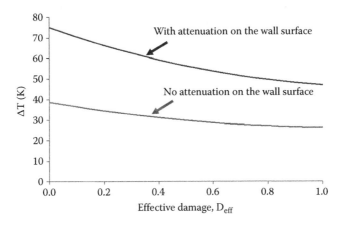

FIGURE 11.30 Permissible thermal striping limits for the structures, made of SS 316 LN. (a) Typical temperature variations and (b) typical frequency content.

- Maximum metal surface temperature = 820 K
- The material properties at 820 K:
  - Young's modulus, $E = 1.49 \times 10^5$ MPa
  - Density, $\rho = 7739$ kg/m^3
  - Specific heat, $C_p = 582$ J/kg K
  - Thermal conductivity, $\lambda = 21.54$ W/m K
  - Thermal expansion coefficient, $\alpha = 20.4 \times 10^{-6}$/K

Permissible thermal striping limits (ΔT_p) are derived as a function for two cases: (1) there is no thermal attenuation on the surface wall temperature, which has been implemented by setting $h = \alpha$, and (2) there is thermal attenuation depending upon the frequency for $h = 40{,}000$ W/m^2/K. The results are shown in Figure 11.30. It is noted from this figure that the ΔT_p reduces significantly with the increase in creep–fatigue damage: 39–26 K under the assumption of no attenuation and 75–46 K with attenuation. This also indicates that the effect of attenuation due to high frequency components of temperature spectrum is very significant.

11.4.3 INVESTIGATION OF FLUID ELASTIC INSTABILITY OF THIN SHELLS

For illustrating the principle, the weir shell of main vessel weir cooling system is taken as for the case study. The main vessel cooling system has already been described in Section 11.3. The fluid elastic instability phenomenon is also described in Chapter 9. Instability regimes are expressed as the critical combinations of associated parameters: weir flow rate and fall height in the present case. Such charts are derived numerically and compared with the experimental data. Detailed investigations including formulations, solutions, and benchmark studies are reported in [11.46]. Some important aspects are presented in the following sections.

11.4.3.1   Idealization of Weir Flow System

There are two coaxial thermal baffles: the outer one attached to the main vessel called weir shell and the inner one attached to the weir shell. The cold sodium flows upward in the annular space between the main vessel and weir shell and flows downward in annular space between the inner and outer baffle to join the cold pool. Thus, the fluid space occupied in the space between the main vessel and weir shell is termed as *feeding collector* and the annular space between the weir and inner baffle is termed as *restitution collector*. The sodium after flowing over the weir shell falls along the weir over a distance H, called *fall height*, and reaches the free surface of the restitution collector after an interval τ, called *delay time*. The idealized weir flow system thus consists of a weir shell placed in between two rigid coaxial cylindrical shells. The idealized weir flow system considered for the present analysis is shown schematically in Figure 11.31.

11.4.3.2   Fluid Elastic Instability Mechanism

Both inner baffle and weir shell could vibrate naturally with various modes. The vibration mode that has the lowest natural frequency corresponds to a shell mode with a specific circumferential wave number (n), generally lying in the range of 4–10 Hz. If the dynamic forces developed in the

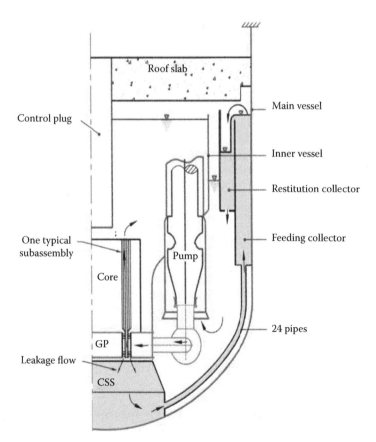

FIGURE 11.31   Main vessel weir cooling circuit.

feeding and restitution collectors could be derived from the shell displacements due to fluid and structure interaction, then the resulting vibration could attain a state of instability, termed as fluid elastic instability. In the present case, the displacement of the weir shell can grow exponentially depending upon the inherent damping mechanism available in the fluid-structure system. The fluid elastic instability that affects the weir shell is mainly due to the coupling of sloshing mode of the liquid free levels that are associated with the feeding and restitution collectors with the shell vibration modes. The mechanism is illustrated schematically in Figure 11.32. Let us assume that the weir shell is perturbed to vibrate with one of the natural frequencies of wave number n. These vibratory motions can generate wavy motions of the liquid free surfaces. The phase difference between the free surface motions depend on the weir flow rate and delay time of the liquid falling from the weir shell. If the kinetic energy imparted by the liquid falling from the weir crest on the free surface of the restitution collector is greater than the energy dissipated due to structural and fluid damping, then any perturbation introduced on the shell will develop to cause instability. For the given weir flow configuration, this situation arises under some critical combinations of flow rate and fall height. The weir flow rate, fall height, and delay time have azimuthal and temporal variations. This is the important aspect responsible for the instability since the dynamic forces on the liquid free surfaces of the feeding and restitution collectors are due to azimuthal and temporal

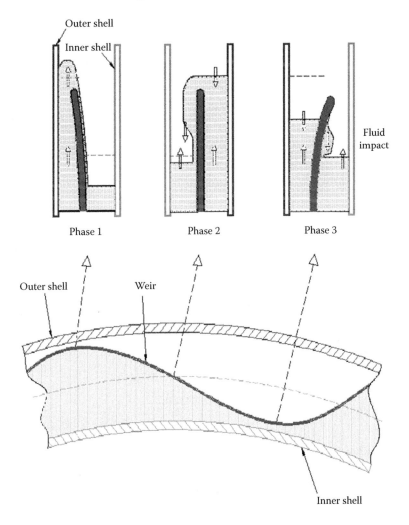

FIGURE 11.32   Weir instability phenomenon.

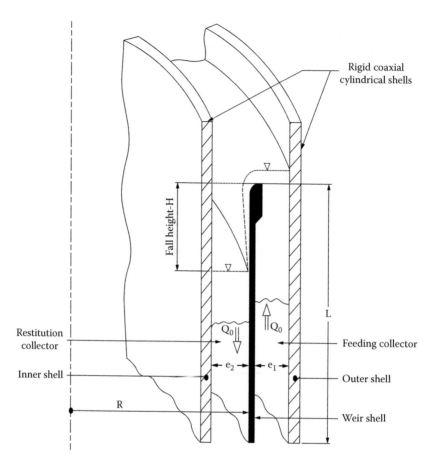

FIGURE 11.33   Schematic sketch of weir flow system.

flow variations. These are inertial forces. Apart from this, impact force due to momentum transferred by the fluid falling from the weir crest contributes to the dynamic forces on the liquid free
surface of the restitution collector (Figure 11.33). Analytical expressions for the fluid forces are
derived as follows.

11.4.3.3 Inertia Force on the Liquid Free Surface

The relation between pressure and free level displacement z is

$$p(\theta,t) = \rho g z(\theta,t) \tag{11.34}$$

$$\left(\frac{d^2 p(\theta,t)}{dt^2} \right) = \frac{\rho g d^2 z(\theta,t)}{dt^2} \tag{11.35}$$

Expressing the free level acceleration in terms of flow rate per unit length q,

$$\frac{d^2 z(\theta,t)}{dt^2} = \frac{dv}{dt} = \left(\frac{1}{e} \right) \frac{dq(\theta,t)}{dt} \tag{11.36}$$

Substituting Equation 11.36 into 11.35,

$$\frac{d^2p(\theta,t)}{dt^2} = \left(\frac{\rho g}{e}\right)\frac{dq(\theta,t)}{dt} \tag{11.37}$$

Using relation: $d^2p/d^2t = -\omega^2p$, Equation 11.37 is rewritten as

$$p(\theta,t) = -\left(\frac{\rho g}{e/\omega^2}\right)\left(\frac{dq(\theta,t)}{dt}\right) \tag{11.38}$$

Force on the free surface per unit length of circumference:

$$f(\theta,t) = \left(\frac{\rho g}{\omega^2}\right)\left(\frac{dq_1(\theta,t)}{dt}\right) \tag{11.39}$$

For feeding collector: $f_1(\theta,t) = \left(\frac{\rho g}{\omega^2}\right)\left(\frac{dq_1(\theta,t)}{dt}\right) \tag{11.40}$

For restitution collector: $f_2(\theta,t) = \left(\frac{\rho g}{\omega^2}\right)\left(\frac{dq_2(\theta,t)}{dt}\right) \tag{11.41}$

The liquid falling on the restitution collector at any time t is nothing but the liquid that was discharged by feeding collector, τ seconds earlier; τ being the time needed for the liquid to travel over the fall height H. Further flow direction of q_2 is opposite to that of q_1. Accordingly,

$$q_2(\theta,t) = -q_1(\theta,t-\tau) \tag{11.42}$$

Further, the overflow rate can be related to the elevation of the free surface w.r.t. weir crest level, $z_1(\theta,t)$ using the following equation:

$$q_1(\theta,t) = -k\cdot\sqrt{g}\cdot\left[z_1(\theta,t)\right]^{3/2} = 0 \quad \text{if } z_1 < 0 \tag{11.43}$$

The weir crest velocity (dx_s/dt) affects the overflow rate due to friction between the weir crest surface and the liquid. To account for this, Equation 11.43 is modified as follows:

$$q_1(\theta,t) = -\left\{k\cdot\sqrt{g}\cdot z_1^{3/2} - z_1\cdot\frac{dx_s}{dt}\right\} = 0 \quad \text{if } z_1 < 0 \tag{11.44}$$

$$\frac{dq_1(\theta,t)}{dt} = -\left\{k\cdot\sqrt{g}\cdot\frac{3}{2}\cdot z_1^{1/2}\cdot\frac{dz_1}{dt} - z_1\cdot\frac{d^2x_s}{d^2t} - \frac{dz_1}{dt}\cdot\frac{dx_s}{dt}\right\} = 0 \quad \text{if } z_1 < 0 \tag{11.45}$$

Substitution of Equations 11.42 and 11.45 in Equations 11.40 and 11.41 yields the following equations for f_1 and f_2 for feeding and restitution collectors, respectively:

$$f_1(\theta,t) = \left(\frac{-\rho g}{\omega^2}\right)\left\{k \cdot \sqrt{g} \cdot \frac{3}{2} \cdot z_1^{1/2} \cdot \frac{dz_1}{dt} - z_1 \cdot \frac{d^2x_s}{d^2t} - \frac{dz_1}{dt} \cdot \frac{dx_s}{dt}\right\} \tag{11.46}$$

$$f_2(\theta,t) = \left(\frac{\rho g}{\omega^2}\right)\left\{k \cdot \sqrt{g} \cdot \frac{3}{2} \cdot z_1^{1/2} \cdot \frac{dz_1}{dt} - z_1 \cdot \frac{d^2x_s}{d^2t} - \frac{dz_1}{dt} \cdot \frac{dx_s}{dt}\right\}\Big|_{t=t-\tau} \tag{11.47}$$

Impulse force on the free surface of restitution collector
  The rate of momentum of falling liquid, I, is given by

$$I = q_2(\theta,t) \cdot v_f = -q_1(\theta,t-\tau) \cdot v_f(\theta,t) \tag{11.48}$$

Equation 11.48 represents the force generated on free surface of restitution per unit length of circumference:

$$f_3(\theta,t) = -\rho \cdot q_1(\theta,t-\tau) \cdot v_f(\theta,t)$$

Substituting Equation 11.44,

$$f_3(\theta,t) = \rho \cdot \left\{k \cdot \sqrt{g} \cdot z_1^{3/2} - z_1 \cdot \frac{dx_s}{dt}\right\}\Big|_{t=t-\tau} v_f(\theta,t) \tag{11.49}$$

11.4.3.4 Formulation of Dynamic Equilibrium Equations

Based on the natural frequencies and associated mode shapes for the sloshing modes coupled by the weir shell, a system of dynamic equilibrium equations is formulated. Natural frequency analysis is carried out taking into account fluid structure interaction. Since the characteristics of the free level fluctuations depend weakly on the fluid compressibility, an acoustic medium governed by the linear wave propagation equation is considered in which fluid is defined by two parameters: density (ρ) and sound velocity (c). The governing partial differential equation is

$$\Delta^2 p - \left(\frac{1}{c^2}\right) \cdot \frac{d^2p}{d^2t} = 0 \tag{11.50}$$

with the following boundary conditions:

$$\text{On weir shell:} \frac{\partial p}{\partial n} = -\rho \cdot n \cdot \frac{\partial^2 x_s}{\partial^2 t} \tag{11.51}$$

$$\text{On free surface:} \frac{\partial p}{\partial z} = -\left(\frac{1}{g}\right) \cdot \frac{\partial^2 z}{\partial^2 t} \tag{11.52}$$

Based on these equations, variational functionals L_f and L_s are written for fluids and free surfaces, respectively. Including the functional for shell L_m, the general functional L is by a suitable combination of the associated functionals as

$$L = L_m + \left(\frac{1}{\omega^2}\right) \cdot L_f + L_s \qquad (11.53)$$

The term $(1/\omega^2)$ is necessary to identify the coupling terms in the fluid and other functionals. In order to obtain the result in the standard form $(L_1 - \omega^2 L_2)$, a new variable π, which is equal to $-p/\omega^2$, is introduced in the formulation. Thus, the functionals have the variables x_s, p, π, and z, which have harmonic variations over circumference. Typically for x_s, $x_s = \Sigma x_{s0} \cos(n\theta)$. Hence, x_{s0}, p_0, π_0, and z_0 become nodal variables in the finite element discretization. Introducing appropriate shape functions and minimizing the functional L, the following matrix expression is obtained in a standard Eigen value problem:

$$\left[M - \left(\frac{1}{\omega^2}\right)K\right]\{U\} = \{0\} \qquad (11.54)$$

$\{U\}$ vector contains modal variables: X, P, Π, and Z corresponding to x_s, p, π, and z, respectively. The natural frequencies $(\omega_{n,m})$ and mode shapes $(X_{mnm}, P_{n,m}, \Pi_{n,m}, \text{ and } Z_{n,m})$ are obtained by solving Equation 11.54 for each selected harmonic and each harmonic has m number of modes.

11.4.3.5   Uncoupled Dynamic Equilibrium Equations

Following the modal superposition principles, the dynamic equilibrium equations are uncoupled. The uncoupled equations for n > 0 are

$$M_{n,m}\left\{\frac{d^2\alpha_{n,m}(t)}{dt^2} + 2\xi\omega_{n,m}\frac{d\alpha_{n,m}(t)}{dt} + \omega^2_{n,m}\alpha(t)\right\} = F_{1n,m} + F_{2n,m} + F_{3n,m} \qquad (11.55)$$

The expressions for the force vectors need to be transformed as follows:

$$F_{1n,m} = Z_{1n,m}\int f_{1n,m}\cos(n\theta)\cdot R \cdot d\theta \qquad (11.56)$$

$$F_{2n,m} = Z_{2n,m}\int f_{2n,m}\cos(n\theta)\cdot R \cdot d\theta \qquad (11.57)$$

$$F_{3n,m} = Z_{3n,m}\int f_{3n,m}\cos(n\theta)\cdot R \cdot d\theta \qquad (11.58)$$

Using Equations 11.46, 11.47, and 11.49 and substituting $\Pi_{n,m} = (-g \cdot Z_{n,m}/\omega^2)$,

$$F_{1n,m} = \Pi_{1n,m}(\rho R)\int\left\{k \cdot \sqrt{g} \cdot \frac{3}{2} \cdot z_1^{1/2} \cdot \frac{dz_1}{dt} - z_1 \cdot \frac{d^2 x_s}{d^2 t} - \frac{dz_1}{dt} \cdot \frac{dx_s}{dt}\right\}\cos(n\theta)\cdot d\theta \qquad (11.59)$$

$$F_{2n,m} = -\Pi_{2n,m}(\rho R) \int \left\{ k \cdot \sqrt{g} \cdot \frac{3}{2} \cdot z_1^{1/2} \cdot \frac{dz_1}{dt} - z_1 \cdot \frac{d^2 x_s}{d^2 t} - \frac{dz_1}{dt} \cdot \frac{dx_s}{dt} \right\} \Big|_{t=t-\tau} \cos(n\theta) \cdot d\theta \quad (11.60)$$

$$F_{3n,m} = Z_{2n,m}(\rho R) \int \left\{ k \cdot \sqrt{g} \cdot z_1^{3/2} - z_1 \cdot \frac{dx_s}{dt} \right\} \Big|_{t=t-\tau} v_f(\theta,t) \cdot \cos(n\theta) \cdot d\theta \quad (11.61)$$

The free level and weir shell displacements are expressed as

$$z_1 = h_1(t) + \sum_n \sum_m \alpha_{n,m}(t) \cdot Z_{1n,m} \cdot \cos(n\theta) \quad (11.62)$$

$$z_2 = h_2(t) + \sum_n \sum_m \alpha_{n,m} \cdot Z_{2n,m} \cdot \cos(n\theta) \quad (11.63)$$

$$x_s = \sum_n \sum_m \alpha_{n,m} X_{n,m} \cdot \cos(n\theta) \quad (11.64)$$

11.4.3.6 Governing Equations for Mean Values (n = 0)

Requirement of flow continuity in the collectors yields (Figure 11.34)

$$2\pi e_1 \frac{dh_1(t)}{dt} + \int q_1(\theta,t) \cdot d\theta = 2\pi q_0(t) \quad (11.65)$$

$$-2\pi e_2 \frac{dh_2(t)}{dt} + \int q_1(\theta,t-\tau) \cdot d\theta = 2\pi q_0(t) \quad (11.66)$$

Fall height, impact velocity, and delay time relations:

$$v dv = \left(\frac{g - fv^2}{2d_e} \right) dz \quad (11.67)$$

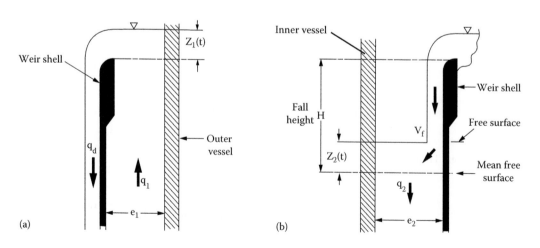

FIGURE 11.34 Flow paths in (a) feeding and (b) restitution collectors.

Replacing d_e by $(4v_0\delta/v)$ and subsequently $(8gv_0\delta/f)$ by v_∞^3 in Equation 11.67,

$$vdv = \frac{g}{v_\infty^3}\left[v_\infty^3 - v^3\right]dz \qquad (11.68)$$

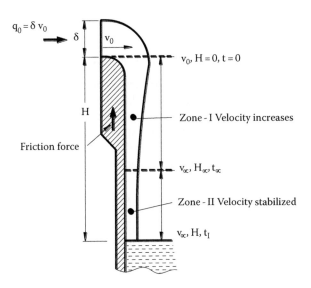

$q_0 = \delta\, v_0$

δ v_0

$v_0, H = 0, t = 0$

H

Zone - I Velocity increases

Friction force

$v_\infty, H_\infty, t_\infty$

Zone - II Velocity stabilized

v_∞, H, t_I

Solution for zone I ($H \leq H_\infty$):

$$\int \frac{vdv}{\left(v_\infty^3 - v^3\right)} = \left(\frac{g}{v_\infty^3}\right)H \qquad (11.69)$$

Integration of Equation 11.69 yields the relation between fall height and impact velocity:

$$H = \left(\frac{v_\infty^2}{3g}\right)\left\{\ln\left[\left(\sqrt{\frac{a^2 + av + v^2}{(a-v)}}\right)\right] - \sqrt{3}\cdot\tan^{-1}\left[\frac{(2v+a)}{(\sqrt{3}a)}\right]\right\} \qquad (11.70)$$

Knowing $\tau = \int dz/v$ and integrating, relation between τ and v is written as

$$\tau = \left(\frac{v_\infty}{3g}\right)\left\{\ln\left[\sqrt{\frac{\left(a^2 + av + v^2\right)}{(a-v)}}\right] + \sqrt{3}\cdot\tan^{-1}\left[\frac{(2v+a)}{(\sqrt{3}a)}\right]\right\} \qquad (11.71)$$

It is to be noted that when $v = v_\infty$, $\tau = \tau_\infty$.

Solution for zone II ($H > H_\infty$):

$$v = v_\infty \qquad (11.72)$$

$$\tau = \tau_\infty + \frac{\left(H - H_\infty\right)}{v_\infty} \qquad (11.73)$$

11.4.3.7 Numerical Solution

The modal parameters $\omega_{n,m}$, X_{mnm}, $P_{n,m}$, $\Pi_{n,m}$, $Z_{n,m}$, and $M_{n,m}$ are obtained through free vibration analysis. Using these parameters, Equations 11.59 through 11.61 are numerically integrated over the circumference to obtain the force vectors. The resulting nonlinear equilibrium equations are integrated using Newmark-β method with the $\beta = \frac{1}{4}$. The minimum time step (Δt) selected is $(2\pi/\omega_{max})/20$, which is found to provide converged solutions.

11.4.4 SOLUTION OF BENCHMARK PROBLEM FOR FLUID ELASTIC INSTABILITY

The fluid elastic instability of a 1/15 scale weir-coolant system that simulates the thermal baffle of the Japanese DFBR is analyzed to compare the test data presented by Fujita et al. [11.47]. For the analysis, water is used in place of sodium. Both free vibration analysis and dynamic response analysis are carried out, and they are compared with the experimental data. For details, refer to Section 10.2.4. The test results confirm that the numerical simulation techniques have the capability to solve the complicated fluid–structure interaction problems accurately.

11.4.5 BUCKLING ANALYSIS OF THIN SHELLS UNDER SEISMIC LOADINGS

The analysis is carried out in three broad steps: (1) natural frequency analysis, (2) seismic response analysis, and (3) buckling analysis, using the validated structural mechanics software [11.48]. Natural vibration and seismic response analyses are performed for the reactor assembly components incorporating the essential internals and sodium, taking into account fluid–structure interaction effects. Subsequently, seismic analysis is carried out in time domain to determine the realistic pressure distributions at any time step. As far as buckling analysis is concerned, the cylindrical portion of main vessel (prone to buckle under shear stress), toroidal portion of inner vessel (prone to buckle by meridional compressive stress), and upper cylindrical portions of thermal baffles (prone to buckle under hoop compressive stress) are the critical parts in the reactor assembly. Elastic–plastic buckling analysis is carried out combining the loads acting under normal operating conditions with the dynamic loads generated during a design basis earthquake, that is, safe shutdown earthquake (SSE). Accordingly, the seismic forces and pressure distributions determined through seismic analysis are applied on the shells for the buckling analysis. The buckling strength reductions due to geometrical imperfections introduced during the manufacturing stages are estimated from experimentally established correlations recommended in Ref. 11.49. Finally, the design adequacy is checked by further applying appropriate factors of safety specified in the design codes (e.g., RCC-MR [11.50]).

The finite element model (FEM) consists of structures (main vessel, inner vessel, outer and inner thermal baffles, core support structure [CSS], grid plate, core, control plug, top shield, and support skirt), hot and cold sodium pools, sodium in the feeding and restitution collectors, sodium–shell interfaces, and sodium free level boundaries for predicting sloshing. The FE mesh developed for the analysis of reactor assembly of a typical SFR is shown in Figure 11.35. More details on the FE mesh generation and seismic analysis are presented in Ref. 11.48. Natural vibration analysis is performed to determine the natural frequencies up to 50 Hz. Three mode shapes depicting (1) rocking of inner vessel along with CSS, grid plate and core about the triple point of the main vessel (1.23 Hz), (2) swaying of core along with inner vessel about grid plate (2.8 Hz), and (3) swaying of main vessel along with thermal baffles about the reactor assembly support (5.2 Hz) are shown in Figure 11.36. These modes have the dominant modal masses of 1370, 534, and 1785 tons, respectively. The modes having frequencies 1.23 and 2.8 Hz can contribute significantly in generating high seismic forces on the inner vessel and the mode having 5.2 Hz has significance from the point of imposing forces on main vessel and thermal baffles, as seen from the respective modal displacements. Subsequently, seismic response analysis is performed to derive pressure distributions and forces.

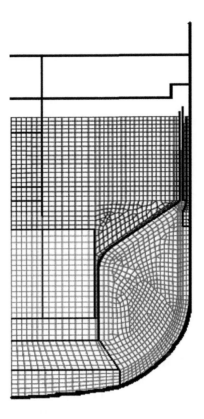

FIGURE 11.35   FEM.

11.4.5.1   Pressure Distributions on 3D Geometries

The dynamic pressure distributions generated over the vessel surfaces during an earthquake are responsible for buckling. In order to have a feel of distributions, a few typical distributions extracted from the seismic response analysis for the expanded time histories corresponding to the excitation in the X-direction are also shown in Figure 11.37. The inner and outer baffles are subjected to high dynamic pressures due to high added mass effects of the fluid confined in the narrow annular spaces between the shells. The hydrostatic pressure p_H acting on the surface that is in contact with sodium is symmetrically distributed. The pressure $\rho_{it}(\theta)$ generated in any node i at any time step t, during horizontal excitation, is superimposed as follows:

$$\text{Pressure: } P(\theta) = \rho_H + \rho_{it}(\theta) \tag{11.74}$$

$$P_H = \rho \cdot q \cdot H_i \quad \text{and} \quad \rho_{it}(\theta) = P_{ni}(t) \cdot \cos(\theta_i) \tag{11.75}$$

where
   ρ is the density of sodium at operating temperature (880 kg/m^3)
   g is the acceleration due to gravity (9.81 m/s^2)
   H_i is the height of the node i w.r.t. free level of sodium
   θ_i is the angle in radian of the node i w.r.t. X-axis
   $P_{ni}(t)$ is the pressure derived from the results of axisymmetric Fourier analysis at the specified
      time t corresponding to the originating node "n_i"

The node "n" and i, which lie in the same horizontal plane at the elevation H_i, are indicated in Figure 11.38.

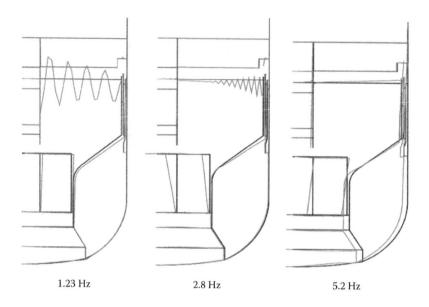

1.23 Hz   2.8 Hz   5.2 Hz

FIGURE 11.36   Critical buckling mode shapes for the reactor assembly vessels.

11.4.5.2   Force Distributions on 3D Geometries

In the FE model of main vessel, its connecting structures—CSS, thermal baffles, and roof slab—are not included. However, both static and seismic forces are applied appropriately along the edges at which these structures get isolated. The static and seismic forces acting on the CSS support skirt transmitted to the triple point are important for the buckling of main vessel. The static load applied at the triple point (W_{TP}) is 400 tons for the half symmetric model. For the inner vessel, the static loads are 6.9 tons on each IHX standpipe and 10.8 tons for each pump standpipe, which are applied uniformly along the upper edges of standpipes. The distributions of static and seismic forces acting on the main vessel and inner vessel are indicated in Figure 11.39.

11.4.5.3   Elastoplastic Buckling Analysis

Assuming that the initial geometries are ideal and there is no geometrical imperfection, the critical buckling load factors are determined, taking into account elastic–plastic deformation behavior of structural materials at the operating temperature. The important material data required is the average tensile curve. The curve recommended for austenitic stainless steel SS 316 LN in RCC-MR: Appendix Z can be adopted for the analysis. Incremental elastic–plastic analysis generally followed as situations are briefly described in the following.

Three-dimensional finite element meshes of main vessel, inner vessel, and thermal baffle generated for the buckling analysis are shown in Figure 11.40. Analysis is performed at each time step to determine the critical buckling loads, and finally, the minimum value is chosen among them. At every time step, incremental analysis is performed by applying the computed pressure distribution and other lumped forces. Analysis is carried out in two phases: incremental elastic–plastic analysis in the first phase and determination of critical buckling loads in the second phase. Here, two factors are introduced: one on the applied load combinations (F_L) and another is critical buckling load factor (F_B). Analysis provides a definite value of F_B for each F_L. That is, several sets of F_L and F_B are available at the end of the analysis. That value of F_L corresponding to F_B = unity is the critical buckling load factor. When the F_L is gradually increased in the elastic–plastic analysis, the associated F_L obtained from the buckling analysis decreases. The analysis can be terminated when the

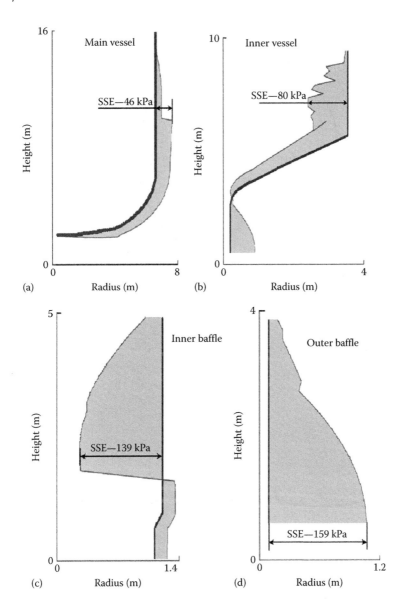

FIGURE 11.37 Pressure distributions at critical time steps. (a) Main vessel, (b) inner vessel, (c) inner baffle, and (d) outer baffle.

F_B is equal to unity. The critical buckling factor is generally obtained by plotting curves F_B versus F_L. A typical curve is shown in Figure 11.41 for the main vessel, where it is found that when applied load factor is 3.2, the buckling load factor is unity. Hence, 3.2 is the critical buckling load factor for the main vessel.

The minimum buckling factors thus obtained at every time steps, for the critical shell structures in a typical SFR, are given in Table 11.1. From the table, the minimum buckling load factors are extracted: 3.2 for main vessel, 1.9 for inner vessel, 3.2 for inner baffle, and 3.0 for outer baffle. The elastic–plastic deformations and the buckled mode shapes for the main vessel, inner vessel, and thermal baffles, which have yielded the possibly lowest load factors, are shown in Figure 11.42. Shear buckling mode shape of main vessel, asymmetrical shell buckling mode shapes of inner vessel, and thermal baffle can be seen clearly in these figures.

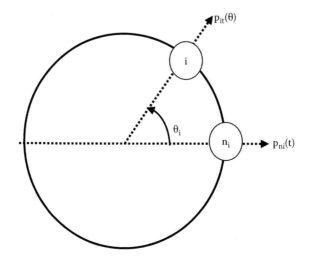

FIGURE 11.38   Position of nodes n and i.

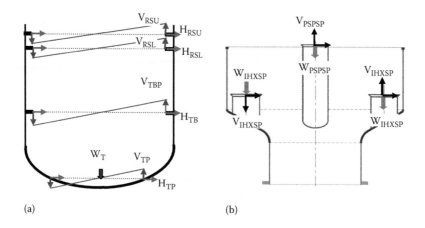

FIGURE 11.39   Distribution of forces acting on (a) main vessel and (b) inner vessel.

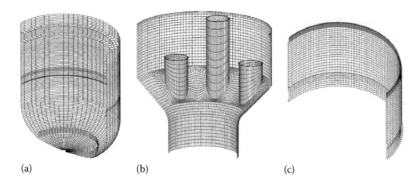

FIGURE 11.40   FEMs of SFR vessels for buckling analysis: (a) main vessel, (b) inner vessel, and (c) thermal baffle.

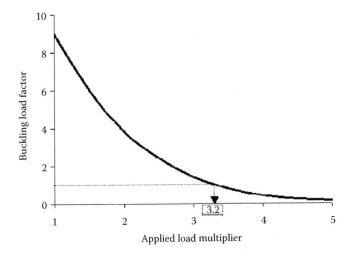

FIGURE 11.41 Critical buckling load factor for the main vessel.

TABLE 11.1
Critical Buckling Load Factors for the Vessels under SSE

S. No.	Component	X-Direction			Y-Direction		
		c	n	e	c	n	e
1	Main vessel	4.3	3.4	3.2	3.6	3.7	3.6
2	Inner vessel	1.9	2.7	2.1	2.0	1.9	1.9
3	Outer baffle	4.7	5.0	3.8	3.8	4.0	3.2
4	Inner baffle	5.6	5.8	3.8	4.4	4.2	3.0

Note: c, compressed time history; n, normal time history; e, expanded time history.

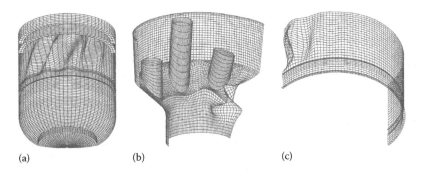

(a)   (b)   (c)

FIGURE 11.42 Elastoplastic buckled mode shapes of thin vessels of SFR: (a) main vessel, (b) inner vessel, and (c) thermal baffle.

11.4.5.4   Effects of Initial Geometrical Imperfections

The buckling strength is reduced significantly for the thin shells with geometrical deviations. The geometrical imperfections are expressed as *form tolerance*, which is the maximum radial deviation (δ) achieved during the manufacturing stage (see the sketch given as follows).

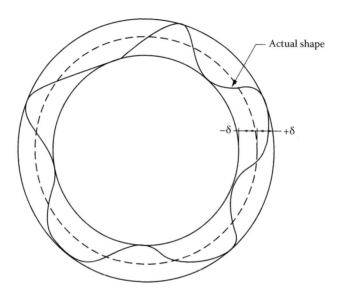

Definition of form tolerance

The buckling strength reduction factors (η) are expressed as the function of normalized tolerance (χ) in the recommendation made by Japanese design rules based on the outcome of extensive experiments conducted in Japan [11.49] as

$$\eta = \frac{1}{\left(1 + 0.19\chi^{0.65}\right)} \tag{11.76}$$

where χ is equal to $2\delta/h$ for shear force (applicable for main vessel undergoing shear buckling) and $4\delta/h$ for bending moment (applicable for inner vessel and thermal baffles, undergoing shell type of buckling). If δ is restricted to $\leq\frac{1}{2}$ of structural wall thickness (h), then the computed η values are 0.85 corresponding to $2\delta/h$ equal to 1 for main vessel and 0.8 corresponding to $4\delta/h$ equal to 2 for inner vessel and thermal baffles. Applying these factors, the minimum buckling load factors are arrived at.

11.4.5.5   Design Check

It is well known that the experimental data on buckling loads have several uncertainties, and hence, there is a need to apply high factor of safety. In this respect, the design codes specify minimum factors of safety on the computed critical buckling load. For example, the factors of safety refined in RCC-MR [11.50] are 2.5 for level A and 1.3 for level D loadings. Accordingly, the design basis loads should be less than the respective critical buckling loads divided by appropriate factor of safety for the critical components of SFR.

REFERENCES

11.1 Walter, A.E., Todd, D.R., Tsvetkov, P.V. (eds.) (2012). *Fast Spectrum Reactors.* Springer, New York, ISBN 978-1-4419-9571-1.

11.2 Duderstadt, J.J., Hamilton, L.J. (1976). *Nuclear Reactor Analysis.* John Wiley & Sons, New York, ISBN 0-471-22363-8.

11.3 Stacey, W.M. (2007). *Nuclear Reactor Physics.* Wiley-VCH Verlag Gmbh & Co. KGaA, Weinheim, Germany. ISBN 978-527-40679-1.

11.4 Barthold, W.P. et al. (1977). Potential and limitations of heterogeneous concept. Report-NEACRP-L-182, USA. June 1977. https://www.oecd-nea.org/science/docs/1977/neacrp-l-1977-182.pdf

11.5 Mohanakrishnan, P. (2008). Development and validation of a fast reactor core burnup code-FARCOB. *Ann. Nucl. Energy,* 35, 158–166.

11.6 Norgett, M.J., Robinson, M.T., Torrens, I.M. (1975). A proposed method of calculating displacement dose rates. *Nucl. Eng. Design,* 33, 50–54.

11.7 Was, G.S. (2007). *Fundamentals of Radiation Material Science.* Springer Verlag, Berlin, Germany, ISBN 978-3-540-49471-3.

11.8 Judd, A.M. (1981). *Fast Breeder Reactors: An Engineering Introduction.* Pergamon Pree Ltd., Headington Hill Hall, Oxford, U.K.

11.9 IAEA. https://www-nds.iaea.org/exfor/endf.htm.

11.10 MacFarlane, R.E., Muir, D.W. (1994). The NJOY nuclear data processing system, Version 91. Report-LA-12740M. Los Alamos National Laboratory, Los Alamos, NM.

11.11 Greenwood, L.R., Smither, R.K. (1985). SPECTER: Neutron damage calculations for material irradiations. ANL-FPP/TM-197, Argonne National Laboratory, Argonne, Illinois, USA.

11.12 Khan, E.U., Rohsenow, W.M., Sonin, A.A., Todreas, N.E. (1975). A porous body model for predicting temperature distribution in wire-wrapped fuel rod assemblies. *Nucl. Eng. Design,* 35, 1–12.

11.13 Chen, B., Todreas, N.E. (1975). Prediction of coolant temperature field in a breeder reactor including inter-assembly heat transfer. Report COO-2245-20TR. Massachusetts Institute of Technology, Massachusetts, CA.

11.14 Valentin, B. (2000). The thermal hydraulics of a pin bundle with an helical wire wrap spacer: Modeling and qualification for a new subassembly concept. LMFR core thermal hydraulics status and prospects, IAEA, Vienna. IAEA TECDOC-1157, ISSN 1011-4289.

11.15 Govindha Rasu, N., Velusamy, K., Sundararajan, T., Chellapandi, P. (2014). Simultaneous development of flow and temperature fields in wire-wrapped fuel pin bundles of sodium cooled fast reactor. *Nucl. Eng. Design,* 267, 44–60.

11.16 Basant, G. (2009). Thermal hydraulics within fuel subassembly of an FBR core. IGCAR-ZNPL collaborative project report, Zeus Numerix Pvt. Ltd., IIT, Mumbai, India.

11.17 Velusamy, K., Chellapandi, P., Chetal, S.C., Raj, B. (2010). Challenges in pool hydraulic design of Indian prototype fast breeder reactor. *Sadhana,* 35(2), 97–128.

11.18 Maity, R.K., Velusamy, K., Selvaraj, P., Chellapandi, P. (2011). Computational fluid dynamic investigations of partial blockage detection by core-temperature monitoring system of a sodium cooled fast reactor. *Nucl. Eng. Design,* 241, 4994–5008.

11.19 Surle, F. et al. (1993). Comparison between sodium stratification tests on the CORMORAN Model and TRIO-VF computation. *Proceedings of Sixth International Topical Meeting on Nuclear Reactor Thermal Hydraulics.*

11.20 Iritani, Y. et al. (1991). Development of advanced numerical simulation of thermal stratification by highly-accurate numerical method and experiments. *Proceedings of International Conference on Fast Reactor and Related Fuel Cycles.*

11.21 Abraham, J. et al. (2011). Computational fluid dynamic investigations of thermal stratification in the hot pool of Monju reactor and comparison with measured data. *14th International Topical Meeting on Nuclear Reactor Thermal Hydraulics, NURETH-14.*

11.22 Azarian, G., Astegiano, J.C., Tenchine, D., Lacroix, M., Vidard, M. (1990). Sodium thermal-hydraulics in the pool LMFBR primary vessel. *Nucl. Eng. Design,* 124, 417–430.

11.23 Astegiano, J.C. et al. (1981). EFR primary system thermal-hydraulics—Status on R&D and design studies. *Proceedings of International Conference Fast Reactor and Related Fuel Cycles.*

11.24 Francois, G., Azarian, G., Astegiano, J.C., Lacroix, C., Poet, G. (1990). Assessment of thermal-hydraulic characteristics of the primary circuit. *Nucl. Sci. Eng.,* 106, 55–63.

11.25 Muramatsu, T. (1993). Intensity and frequency evaluation of sodium temperature fluctuation related to thermal striping phenomena based on numerical methods. *Proceedings of Fifth International Symposium on Refined Flow Modelling and Turbulence Measurements.*

11.26 Menant, B., Villand, M. (1994). Thermal fluctuations induced in conducting wall by mixing sodium jets: An application of TRIO-VF using large eddy simulation modeling. *IAEA Specialist Meeting on Correlation between Material Properties and Thermohydraulics Conditions in Liquid Metal-Cooled Fast Reactors (LMFRs)*, Aix-en-Provence, France.

11.27 Tokuhiro, A., Kimura, N. (1999). An experimental investigation on thermal striping mixing phenomena of a vertical non-buoyant jet with two adjacent buoyant jets as measured by ultrasound Doppler velocimetry. *Nucl. Eng. Design*, 188, 49–73.

11.28 Moriya, S. et al. (1988). Thermal striping in coaxial jets of sodium, water and air. *Proceedings of Fourth International Conference on Liquid Metal Engineering and Technology*, Avignon, France.

11.29 Wakamatsu, H.N., Hashiguchi, K. (1995). Attenuation of temperature fluctuations in thermal striping. *J. Nucl. Sci. Technol.*, 32(8), 752–762.

11.30 Gelineau, O., Sperandio, M. (1994). Thermal fluctuation problems encountered in LMFBRs. *IAEA-IWGFR/90, Specialistic Meeting on Correlation between Material Properties and Thermohydraulics Conditions in LMFBRs*, Aix-en-Provence, France.

11.31 Betts, C. et al. (1983). Thermal striping in liquid metal cooled fast breeder reactors. *Second International Topical Manufacturing on Nuclear Reactor Thermal Hydraulics, NURETH-2*, Santa Barbara, CA, Vol. 2, pp. 1292–1301.

11.32 Chellapandi, P. et al. (2009). Thermal striping limits for components of sodium cooled fast spectrum reactors. *Nucl. Eng. Design*, 239, 2754–2765.

11.33 Banerjee, I. et al. (2013). Development of gas entrainment mitigation devices for PFBR hot pool. *Nucl. Eng. Design*, 258, 258–265.

11.34 Hirt, C.W., Nichols, B.D. (1981). Volume of fluid (VOF) method for the dynamics of free boundaries. *J. Comput. Phys.*, 39, 201–225.

11.35 Laxman, D. et al. (2004). Free level fluctuations study in ¼ scale reactor assembly model of PFBR. *NUTHOS-6*, Nara, Japan.

11.36 Velusamy, K. et al. (1998). Natural convection in narrow component penetrations of PFBR roof slab. *J. Heat Mass Transfer*, 20, 1–11.

11.37 Baldasari, J.P. et al. (1984). Open azimuthal thermosyphon in annular space—Comparisons of experimental and numerical results. *Liquid Metal Engineering and Technology*, BNES, London, U.K., pp. 463–467.

11.38 Francois, G., Azarian, G. (1989). SUPER PHENIX reactor block thermalhydraulic behaviour comparison between calculations and experimental results. *Proceedings of the 10th International Conference on Structural Mechanics in Reactor Technology*, Lyon, France, pp. 37–42.

11.39 Paliwal, P. et al. (2012). Characterization of cellular convection of argon in top shield penetrations of pool type liquid metal fast reactors. *Nucl. Eng. Design*, 250, 207–218.

11.40 Hemnath, M.G. et al. (2007). Cellular convection in vertical annuli of fast breeder reactors. *Ann. Nucl. Energy*, 34, 679–686.

11.41 Jones, I.S. (1997). The frequency response model of thermal striping for cylindrical geometries. *Fatig. Fract. Eng. Mater. Struct.*, 20(6), 871–882.

11.42 RCC-MR Appendix A16. (2002). Guide for leak before break analysis and defect assessment, AFCEN, France.

11.43 Chellapandi, P., Chetal, S.C., Raj, B. (2009). Thermal striping limits for components of sodium cooled fast spectrum reactors. *Int. J. Nucl. Eng. Design*, 239, 2754–2765.

11.44 Jones, I.S., Lewis, M.W.J. (1994). A frequency response method for calculating stress intensity factors due to thermal striping loads. *Fatig. Fract. Eng. Mater. Struct.*, 17(6), 709–720.

11.45 Clayton, A.M., Irvine, N.M. (1987). Structural assessment techniques for thermal striping. *J. Pressure Vessel Technol*, 109(3), 305–309. doi:10.1115/1.3264869, August 1,1987.

11.46 Chellapandi, P., Chetal, S.C., Raj, B. (2008). Investigation of fluid-elastic instability of weir shell in a pool type fast breeder reactor. Advances in vibration engineering. *Sci. J. Vibration Inst. India*, 7(2), 111–126.

11.47 Fujita, K. et al. (1993). Study on flow induced vibration of a flexible weir due to fluid discharge effect of weir stiffness. *PVP-258, Flow Induced Vibration and Fluid Structure Interaction*. ASME PVP 258, pp. 143–150, New York, USA.

11.48 Chellapandi, P., Chetal, S.C., Raj, B. (2008). Investigation on buckling of FBR vessels under seismic loading with fluid structure interactions. *J. Nucl. Eng. Design*, 238, 3208–3217.

11.49 Akiyama, H. (1997). *Seismic Resistance of FBR Components Influenced by Buckling*. Kajma Institute Publishing, Japan.

11.50 RCC-MR. (2007). Design and construction rules for mechanical components for FBR nuclear islands. Section I, Subsection B, AFCEN, Paris, France, Vol. 1.

Section III

Safety

Section III

12 Safety Principles and Philosophy

12.1 INTRODUCTION

In the early stages of fast reactor development, the potential for accidents in liquid sodium–cooled fast reactors (SFRs) was considered as a major safety issue, especially for severe accidents. Severe accidents occur when the reactor is "unprotected," which means that the reactor protection systems fail to scram the reactor. The consequences could include fuel melting, fuel relocation, and power excursions, up to and including energetic disassembly of the reactor core and the potential failure of the containment building. For fast reactors, these accidents are typically in one of three categories: loss of cooling (unprotected loss of flow [ULOF]), loss of normal heat removal (unprotected loss of heat sink [ULOHS]), and reactivity addition (unprotected transient overpower [UTOP]). Such events were the major focus of the licensing reactor. This combination of circumstances made the reactor designers to study and develop more advanced fast neutron reactors and concepts, including the concept of "inherent" safety, that is, design choices and features that do not require an active system to function, but are the result of inherent fundamental physical processes such as thermal expansion or gravity, which do not have a probability of failure.

The following are the fundamental requirements for the safe operation of a fast neutron reactor (or any reactor), which in turn are the goals of the inherent safety principles:

1. Avoiding large increases in core reactivity
2. Maintaining cooling of the reactor core
3. Preventing rearrangement of the fuel that would result in large energetic events

Each of these areas required significant research to acquire the basic understanding needed to successfully develop the inherent safety principles that could be adopted in the design of a fast reactor. During the 1980s and 1990s, many research and design efforts adopted one or more inherent safety features in developing new reactor designs. As a result, it became possible to design a fast reactor with safety characteristics that in some cases far exceeded the earlier designs, depending on the design philosophy and choices made using the inherent safety principles. For example, one can use inherent safety features to replace active engineered safety systems representing an aggressive nontraditional approach, which would provide greater reliability and lower the probability that a particular accident scenario would occur, or one could use inherent safety features in addition to the active engineered safety systems, providing much greater protection while maintaining the traditional design philosophy of "defense in depth" with diverse and redundant safety systems. Either approach could be successfully used, although the benefits may be different in each case.

In this chapter, safety principles and associated philosophy are introduced under three subsections: inherent and engineered safety features, operation simplicity, and radioactivity release. There are numerous papers written in these topics [12.1–12.6]. Readers are requested to read them before studying this chapter.

12.2 INHERENT AND ENGINEERED SAFETY FEATURES

Negative reactivity feedbacks during power operation can result in inherent safety of a reactor as against engineered safety. There has been continuous attempt in the nuclear community to enhance negative reactivity feedbacks in the reactor core to ensure enhanced safety. After the Chernobyl accident, persistent attention is being given in nuclear design to make the coolant void coefficient negative or to enhance the negative reactivity feedbacks from any other source. In case of SFR, the neutron spectrum is hard and the coolant void coefficient is positive. There are many ways to reduce the magnitude of the positive void coefficient and make it zero void coefficient or negative. In the Russian design of BN-800 reactors, axial blankets are replaced with a sodium plenum. It increases neutron leakage and makes the void coefficient negative. There is some more suggestion available in the literature like replacing the fertile material such as U-238 with Th-232 or replacing the fissile material Pu-239 with U-235 [12.7]. In all the previously mentioned methods, neutron economy gets affected either by loosing neutrons through leakage or by producing neutrons with reduced importance. If more neutrons undergo leakage, fewer neutrons are available for breeding.

12.2.1 GOVERNING PARAMETERS

One of the goals of inherent safety is to provide an inherent capability to prevent large increases in the reactivity of the reactor core or significant mismatch between the core power and the available coolant flow that could cause power generation to exceed the reactor cooling capability. As is widely known, the nuclear fission process in a fast reactor is affected by a number of physical phenomena that change core reactivity. The major parameters affecting core reactivity include both nuclear physics and materials phenomena and are explained in the following paragraphs.

12.2.1.1 Coolant and Pool-Type Concept

Sodium is used as a coolant in SFR. A large margin between the normal operating sodium temperature and the boiling point of sodium can accommodate significant temperature rise in the event of mismatch of heat generation and heat removal. High thermal conductivity, low viscosity, and large difference between the temperature of hot sodium at 820 K and ambient air at 310 K, coupled with a significant variation of sodium density with temperature, permit decay heat removal through the natural convection mode. The pool-type concept provides large thermal inertia and hence more time to the operator to act in case of exigencies during reactor operation.

12.2.1.2 Fuel Doppler Coefficient

The fuel Doppler coefficient is the net increase in the neutron absorption cross section caused by an increase in fuel temperature, introducing negative reactivity feedback, or the decrease in neutron absorption cross section caused by a decrease in fuel temperature causing positive reactivity feedback. Different fuel types have different Doppler coefficients, for example, the Doppler coefficient for oxide fuel is more negative than for metallic fuel. However, the Doppler feedback is affected by other inherent characteristics of the fuel, including thermophysical properties such as thermal conductivity since it affects the steady-state fuel temperature and the magnitude of temperature changes during transient conditions.

12.2.1.3 Coolant Density

Increasing the coolant temperature reduces the coolant density, mainly reducing the moderating and reflecting effects of the sodium coolant. The reactivity effect of the coolant density varies across the core, being more positive in the core interior where the moderating effect is dominant and negative at the core boundaries where the leakage effect is more important. Depending on the design choices affecting the relative importance of moderation and leakage, the reactivity feedback from increases in coolant temperature (reduction in coolant density) can be either positive or negative. Large reactor cores with a similar core height and diameter are usually dominated overall by the moderating effect, so that increasing the coolant temperature causes a positive reactivity feedback, while small cores where the height is much smaller than the diameter can be dominated by the leakage effect and increasing coolant temperature results in a negative reactivity feedback.

12.2.1.4 Fuel Axial Expansion

An increase in the fuel temperature causes the fuel to expand axially based on the thermal expansion coefficient of either the fuel or cladding, or both, depending on the chemical and stress conditions at the fuel/cladding interface. Similarly, a decrease in fuel temperature will cause the fuel to contract axially, resulting in positive reactivity feedback. The reactivity feedback from fuel expansion or contraction depends on the thermal expansion coefficient(s) and the change in fuel and/or cladding temperature during transient conditions.

12.2.1.5 Core Radial Expansion

Fast reactor cores have a significant neutron leakage fraction due to the neutron mean free path length, resulting in a large gradient of fuel reactivity worth at the edges of the core and making the reactivity of the core sensitive to changes in core geometry. If core subassemblies move outward in the radial direction, increasing the effective diameter of the core, a negative reactivity feedback is generated. Conversely, if the core subassemblies move inward, a positive reactivity feedback results.

12.2.1.6 Control Rod Driveline Expansion

The control rod driveline expansion is the relative motion between the core and the control rods caused by a change in temperature of the control rod drivelines. The drivelines are normally located in the coolant outlet, or hot, plenum and respond to changes in the core outlet temperature. An increase in the control rod driveline temperature will cause the control rods to move further into the core, introducing negative reactivity feedback. A decrease in the control rod driveline temperature will cause the control rods to move further out of the core, causing positive reactivity feedback. Depending on the reactor design, there can be other components of reactivity feedback, such as in a pool plant where the control rods are suspended from the closure deck, and the control rod position in the core is determined not only by the temperature of the control rod driveline but also by the vessel wall temperature since the core is supported by the vessel wall. All such effects must be considered to determine all of the inherent reactivity feedback mechanisms for evaluating their ability to compensate for conditions in accident situations.

12.2.2 Ways to Enhance Inherent Safety

Before finding out the possible ways to enhance the negative reactivity, there is a necessity to understand the role of the reactivity feedback during both the power rise and drops. If the magnitude of negative reactivity is high, that is, if the power coefficient is high, then the built-in excess reactivity needs to be high to counterbalance the negative reactivity while taking the reactor from zero power to nominal power. If the excess reactivity is high, the severity of the unprotected transient overpower accident (UTOPA) is also high, due to higher control rod worth. If the power coefficient is low, then the power reactivity decrement, the reactivity required to bring the reactor from nominal power to zero power, is less. The reactor will go to a safe shutdown with smaller negative reactivity feedback, and the severity of unprotected loss of flow accidents (ULOFA) is less. Thus, a higher magnitude of power coefficient does not necessarily ensure safe shutdown in all accidents. The basic aim is to enhance the negative reactivity that would effectively take the reactor asymptotically to another steady state during UTOPA as well as to take the reactor to a safe shutdown during ULOFA.

When the power goes from one steady state to another, the quasi-static reactivity balance [12.8] equation between those asymptotic states is written as follows:

$$\Delta\rho = (P-1)A + (P/F-1)B + \delta T_i C + \Delta\rho_{ext} \qquad (12.1)$$

where

coefficient A is a varying function of power (P)
B and C is a varying function of power/flow (P/F)
δT_i is the change in inlet coolant temperature
$\Delta\rho_{ext}$ is the external reactivity

Based on our earlier studies [12.3], reactivity coefficients A and B are

$$A = \left\{ \left[\frac{\delta\rho_D}{\delta T} + \Delta k_f^{j,i}\alpha_f + \Delta z\alpha_f C^i \right]\left(\frac{1}{h_{sc}} + \frac{1}{h_{fs}} \right) + \Delta k_s^{j,i}\alpha_s \frac{1}{h_{sc}} \right\} q^{j,i} \qquad (12.2)$$

$$B = \left[\frac{\delta\rho_D}{\delta T} + \Delta k_f^{j,i}\alpha_f + \Delta z\alpha_f C^i + \Delta k_s^{j,i}\alpha_s + 3\Delta k_c^{j,i}\alpha_{Na} \right]\sum_{k=1}^{j} \frac{\Delta z}{C_c\rho A_f}\frac{q^{k,i}}{v(i)}$$

$$+ \left[\begin{array}{l} 2\alpha_s\Delta k_f^{j,i}W^j + 2\alpha_s\Delta k_s^{j,i}W^j + 2R_1(R^{i+1}-R^i)\alpha_s W^j \dfrac{E_j}{2R_1+1} \\[2mm] +2R_c(R^{i+1}-R^i)\alpha_s W^j \dfrac{D_j}{2R_c+1} \end{array} \right] \sum_{k=1}^{jsp} \frac{\Delta z}{C_c\rho A_f}\frac{q^{k,i}}{v(i)} \qquad (12.3)$$

From reactivity coefficients A and B, it is possible to determine the change in P/F value (ratio of normalized power to normalized flow) and also the change in the outlet coolant temperature of the final asymptotic states. They are

$$\frac{P}{F} = 1 + \frac{A}{B} \quad \text{and} \quad \delta T_{out} = \Delta T_c \frac{\delta P}{F} = \frac{A}{B}\Delta T_c \qquad (12.4)$$

Coefficient A is contributed by the Doppler worth, the axial boundary movement worth, and the thermophysical properties such as the linear expansion coefficient and the effective heat transfer h_{fs}. Reactivity coefficient A is basically the amount of reactivity that is vested from the fuel side. Power/flow coefficient B is contributed by the Doppler component, fuel and clad axial and radial expansions, and fuel axial and radial boundary movement components as a function of coolant temperature rise. Basically, power/flow coefficient B is the amount of reactivity vested from the coolant side. Rise in coolant temperature at the spacer location contributes to the flowering reactivity feedback that enhances the negative reactivity feedback. For a given reactivity or flow perturbation, reactivity feedbacks are such a way that change in the P/F value of final asymptotic states and δT_{out} (change in outlet coolant temperature) should be small. In order to satisfy the earlier condition, the value of B should be more than A.

Based on the experience gained in accident analyses, the following methodologies are recommended to enhance the negative feedback effectively.

12.2.2.1 Addition of Moderating Material

An innovative way of improving negative reactivity feedback in liquid metal fast breeder reactor (LMFBR) is addition of moderating material [12.9,12.10]. Addition of moderating material degrades the spectrum but improves the Doppler feedback. This can be done in metal fuel FBR, which has a high breeding ratio, and hence degradation of the spectrum will not result in high penalty of neutron economy. This is more effective for a medium-sized (500–700 MWe) metal FBR than a large core (>1000 MWe) since in the large core the spectrum is softer with a lower breeding ratio and will lead to more neutronic economic penalty. Also, the Doppler feedback is available in a large core due to the soft spectrum.

Addition of moderating material is possible either as addition of diluents in the coolant or as replacing some fuel pins with pins having moderating properties. One such study has been made by replacing some fuel pins with ZrH [12.10]. Parametric studies were carried out with this change. Results of the study show that the substitution of the ZrH pin with 3%–5% of fuel pins markedly improves the reactivity coefficients. In those cores the Doppler coefficient improved two to three times, and the value was more even compared to a similar mixed oxide (MOX) core. The sodium void reactivity also had favorable value compared with the MOX and the metallic fuel core. The substitution of ZrH brings unfavorable effects for the breeding ratio, the burnup reactivity loss, and the peaking factor. However, these values for the core substituted by 3% ZrH were still better than or the same as those of conventional MOX fuel core.

The influence of hydrogenous materials on Doppler and sodium void reactivity has also been verified experimentally. A series of critical experiments were conducted in the fast critical assembly (FCA) of Japan Atomic Energy Research Institute (JAERI) to confirm the effect of hydrogenous materials on the Doppler and the sodium void reactivities and to check the applicability of design codes for fast reactors with hydrogenous materials [12.9]. The experimental results clearly showed that the hydrogenous material significantly improves the Doppler and sodium void reactivities.

12.2.2.2 Increasing the Thermal Conductivity of Fuel

This argument is more relevant to metallic fuel. The thermal conductivity of metal fuel is very high as compared to the oxide fuel. Thermal conductivity may be further enhanced by choosing a suitable alloying material. The thermal conductivity of U–Pu–Al is about 80–150 W/m/K [12.11] as compared to 20 W/m/K of U–Pu–Zr fuel. A higher thermal conductivity of fuel materials results in lower operating temperatures. Since the neutron spectrum is hard with metallic fuel, the Doppler constant is small. With a smaller steady-state operating temperature, overall the negative Doppler feedback is small in case of metal fuels. This reduces the excess reactivity and the control rod worth. So, the severity of the UTOP is reduced. A smaller Doppler feedback in metal fuel is one of the reasons for smaller power reactivity decrement (reactivity required to bring the reactor to zero power). In case of ULOFA, the overall Doppler feedback is either negative or zero due to a smaller steady-state temperature. So, the reactor goes to safe shutdown without much coolant boiling.

Since the fuel operating temperature is small in high thermal conductivity fuel, when there is ULOFA, there is a possibility that the fuel temperature increases and introduces negative reactivity and it takes the reactor to a safe shutdown. The expansion coefficient of U–Pu–Al is found to be comparatively high. So, this also may enhance the negative reactivity feedback for a reactivity accident. This may come back as a positive contribution during ULOFA in oxide fuel reactors. But in metal fuel since the difference in temperature between different operating levels is considerably small and the temperature difference between fuel and coolant temperatures is small, the average fuel temperature is expected to increase and the fuel axial expansion feedback is expected to give a negative reactivity contribution during ULOFA.

12.2.2.3 Fuel Extrusion Phenomenon: An Inherent Safety under UTOPA

If the external reactivity insertion rate is high, then there is fuel melting and molten fuel motion within the pin from the central region to the axial periphery of the core. This preclad failure in pin fuel motion is called fuel squirting in Europe and it has been called fuel extrusion in the United States. Extrusion of molten fuel introduces negative reactivity feedback, and this negative reactivity feedback takes the reactor to a new steady state or to a subcritical state.

12.2.3 Inherent Safety in Decay Heat Removal

The primary coolant system in modern sodium-cooled fast reactors can easily be configured to provide natural circulation shutdown heat removal. The capability to remove shutdown decay heat with natural circulation provides a means to maintain reactor component temperatures at acceptable levels even in the event of loss of all off-site and emergency on-site power supplies. As a result, natural circulation heat removal can play an important role in the overall inherent safety approach. Natural circulation flow arises due to the effect of gravity on a continuous fluid with a density difference along the elevation. Heavy fluid sinks to displace lighter fluid. Buoyancy-induced flow can be established when a fluid is heated, decreasing its density, at an axial position below the elevation at which the fluid is cooled, increasing its density. In a 1D model, flow occurs when the buoyancy force is great enough to overcome form, friction, and shear losses. The natural circulation flow rate is regulated by the balance between the buoyancy force and the flow-related pressure losses. When the buoyancy force is provided by a thermally driven density difference, the fluid flow rate will be determined by the fluid properties, the elevation difference between the heat sink and the heat source, and the temperature difference in the fluid between the heat source and the heat sink. Liquid sodium and its alloys are excellent fluids for natural circulation heat removal because of their thermophysical properties. Due primarily to its high thermal conductivity, liquid sodium is capable of very high convective heat transfer rates, even at the modest fluid velocities characteristic of natural circulation. This tends to minimize the temperature differences between the heat source and the fluid and between the fluid and the heat sink and to reduce the overall source-to-sink temperature difference required for natural circulation cooling. Sodium-cooled fast reactors can be configured to promote natural circulation shutdown heat removal. The key design parameters are (1) provision for a relatively free-flowing fluid natural circulation path and (2) provision for sufficient elevation difference between the heat source and the heat sink.

In the primary coolant circuit, natural circulation flow may be established along the same flow path used for normal operation. Along this path, coolant is heated in the reactor, rises to the hot plenum, and flows through the intermediate heat exchangers (IHXs) to the cold plenum and back to the reactor core. In accidents or emergency shutdown conditions in which no heat is removed in the IHX, heat can be removed by independent heat exchangers in series with the IHX for loop-type reactor designs. Alternatively, for pool-type reactor designs, direct reactor auxiliary cooling system (DRACS) heat exchangers may be located high in the cold pool. The primary coolant chilled in the DRACS heat exchangers falls near the bottom of the cold pool, where it enters the primary coolant pump inlet and travels back to the reactor. For either loop- or pool-type reactor configuration, primary coolant natural circulation carries heat from the reactor to the IHX or auxiliary heat exchangers. Heat transferred at the IHX may be removed by natural circulation in the intermediate coolant loop, if sufficient elevation of the ultimate heat sink is provided. This mode of shutdown heat removal has been demonstrated in the loop-type FFTF reactor [12.12]. If the IHX path is not available, heat may be removed through the auxiliary heat exchangers to a second natural circulation loop. The working fluid in this second loop is often specified as the sodium–potassium alloy (NaK). The NaK loop carries the heat through piping to a second heat exchanger located at a high elevation outside the containment building, where heat is rejected to environmental air. The relatively low melting point of NaK minimizes the potential for freezing in the secondary loop. Shutdown heat removal through such an auxiliary cooling system has been demonstrated in the pool-type EBR-II reactor.

12.2.4 ENGINEERED SAFETY FEATURES

General safety features of the plant can be broadly classified as inherent safety features and engineered safety features. Inherent safety features are available because of the chosen concept of the reactor system, choice of coolant, and core characteristics. Specific inherent features are the pool concept, negative reactivity coefficients, efficient sodium coolant, and easy natural convection. Some of the important engineered safety features are multiple radial entry sleeves for subassemblies, inertia on primary pump, emergency power supplies for the primary sodium pumps, core monitoring instrumentation, reactor shutdown systems, safety grade decay heat removal system (SGDHRS), protection against sodium–water reaction and external sodium leaks, and various design provisions to prevent transient overpower events and escape of radioactivity to the environment following the BDBE of core disruptive accident (CDA).

The core is configured with adequate shielding to limit radioactivity of secondary sodium and also to reduce the neutron dose on the structural components such as the grid plate, core cover plate, and main vessel, ensuring low material property degradation on account of radiation. The main vessel is made of highly ductile stainless steel material for which leak before break criteria applies. The incorporation of a safety vessel around the main vessel ensures continued core cooling under all conditions. The main vessel has no nozzle penetrations and thus offers high structural reliability. A simple shape permits in-service inspection of the vessel welds to assess its structural reliability.

The temperature and power coefficient of reactivity are designed to be negative. So in the event of disturbances in primary and secondary sodium flow or feedwater flow, the reactor stabilizes to a new power level even without the corrective action of the operator. The reactivity loss due to fuel burnup is small in FBR due to breeding of fuel and low absorption cross section of fission products (FPs) to fast neutrons. There is no xenon poisoning. The frequency of control rod operation is less (~2–3 times per shift) during steady-state operation. Thus, there is no need for automatic power regulation. This avoids the problems that may arise by failure of the automatic power regulation system.

Core monitoring is done by functionally diverse sensors. Neutron detectors are provided to monitor the power and provide signals for safety action by SCRAM on parameters like linear power, period, and reactivity. These parameters provide protection against transient overpower, transient undercooling, and anomalous reactivity addition events. On detection of any abnormality in the reactor, shutdown is assured by two independent, fast-acting shutdown systems.

12.2.5 Overall Perception on Inherent Safety Parameters

Any discussion of the reactivity coefficient should always consider the importance of buildup time. The time constant of reactivity feedback should be low. As it is known to the nuclear physics community, the feedbacks that come from the fuel (such as Doppler) are prompt, that is, with smaller time constant. Any other feedback that arises due to the core radial expansion, control rod driveline expansion, and vessel expansion may be effective after a few seconds to hundreds of seconds, that is, their time constants are large. So, the time constant is expected to be small in order to get effective feedback instantaneously.

In the case of LMFBR designed with oxide as a basic fuel material, the neutron spectrum is soft and the sodium void coefficient is comparatively small as compared to the reactor designed with metal as a basic fuel material. In oxide fuel, the Doppler feedback gives a pronounced negative feedback, which is also a prompt feedback. Among the reactivity component, Doppler gives a dominant contribution (almost three-fourths). But in the case of metal fuel though, there is a Doppler feedback that is not pronounced as it is in the oxide fuel. There are works carried out [12.9] to enhance the prompt coefficient by introducing diluent subassemblies into the reactor and soften the spectrum. The reactor core design choices can have a strong influence on the ability to use these inherent reactivity feedback mechanisms to favorably compensate for undesirable changes in core power and temperature during accident conditions. It is important to recognize that the reactivity feedback mechanisms can act in either a beneficial or detrimental manner, depending on what is required to keep the reactor power and temperatures within limits, depending on the accident conditions. For example, some accident conditions require a response to an inadvertent introduction of positive reactivity in the core, as occurs in the UTOP event. In this case, larger feedback from fuel Doppler can be beneficial. However, in other accidents, it is necessary to terminate the fission process and reduce core power to decay heat level, as in the ULOF and ULOHS, and a lower feedback from the fuel Doppler is preferred. It is also important to recognize that the change in core reactivity is the result of all the reactivity feedback mechanisms, not any individual one. Proper incorporation of inherent reactivity feedback in the reactor design requires a balanced approach that considers all potential accident conditions to achieve the best overall performance.

12.3 OPERATION SIMPLICITY

In many respects, fast breeder reactors are similar to the power reactors in operation at present. However, the core of a fast breeder has to be much more compact than that of a light water reactor (LWR). Plutonium or more highly enriched uranium is used as fuel, the fuel elements are smaller in diameter, and they are clad with stainless steel instead of zircaloy. Since water rapidly decelerates the fast-moving neutrons produced during fission to less than the energy level required for breeding, it cannot be used in fast breeders. Thus, in a fast breeder reactor we have to remove a large amount of heat from a small volume of fuel and at the same time use a coolant that does not reduce the "neutron energy unacceptably."

Production of energy in the core of the fast breeder is intense compared with thermal reactors, and therefore, the coolant must have very good heat transfer properties. For fast breeders using a liquid metal cooling system, sodium is the selected coolant since it can remove heat effectively from the compact reactor core and remains in the liquid state over a fairly broad temperature range. Sodium exhibits the best combination of required characteristics as compared with other possible coolants, including excellent heat transfer properties, a low pumping power requirement, low system pressure requirements (one can use virtually atmospheric pressure), the ability to absorb considerable energy under emergency conditions (due to its operation well below the boiling point), a tendency to react with or dissolve (and thereby retain) many FPs that may be released into the coolant through fuel element failure, and finally, good neutronic properties. The unfavorable characteristics of sodium are its chemical reactivity with air and water, its activation under irradiation, its optical opacity, and its slight neutron decelerating and absorption properties. But these disadvantages are considered in practice to be outweighed by the merits of sodium as a coolant.

12.3.1 BURNUP

Conventional light water reactors require excess reactivity on burnup in order to operate between refueling. Consumable absorbers are used to decrease it. Nonetheless, several percents of reactivity must be reserved on control rods or reactivity must be compensated by a boron absorber mixed into the coolant. Taking into account these measures and uncertainty factors, the excess reactivity in light water reactors is about 10%. A fast reactor requires much lower excess reactivity to compensate the temperature effect. The highest excess reactivity is required to compensate fuel burnup. With the metal-fueled reactor, the maximum excess reactivity can be designed to be less than the delayed neutron fraction (β). Such a burnup section can be obtained in compositions with enhanced-density fuel, for example, uranium nitride (UN) and heavy metal coolant (lead).

12.3.2 SYSTEM SAFETY

In the fast reactor, as the core is operating closer to the ambient pressure, it dramatically reduces the danger of a loss-of-coolant accident. The entire reactor core, heat exchangers, and primary cooling pumps are immersed in a pool of liquid sodium or lead, making a loss of primary coolant extremely unlikely. The coolant loops are designed to allow for cooling through natural convection, meaning that in the case of a power loss or unexpected reactor shutdown, the heat from the reactor core would be sufficient to keep the coolant circulating even if the primary cooling pumps were to fail.

Decay heat generation from short-lived FPs and actinides is comparable in both cases, starting at a high level and decreasing with time elapsed after shutdown. The high volume of the liquid sodium primary coolant in the pool configuration is designed to absorb decay heat without reaching the fuel melting temperature. The primary sodium pumps are designed with flywheels so they will coast down slowly. This coast down further aids core cooling upon shutdown. If the primary cooling loop were to be somehow suddenly stopped, or if the control rods were suddenly removed, the metal fuel can melt as accidentally demonstrated in EBR-I; however, the melting fuel is then extruded up the steel fuel cladding tubes and out of the active core region leading to permanent reactor shutdown and no further fission heat generation or fuel melting. With metal fuel, the cladding is not breached and no radioactivity is released even in extreme overpower transients.

12.3.3 PASSIVE SAFETY FEATURES

Fast reactors also have passive safety advantages as compared with conventional light water reactors. The fuel and cladding are designed such that when they expand due to increased temperatures, more neutrons would be able to escape the core, thus reducing the rate of the fission chain reaction. In other words, an increase in the core temperature will act as a feedback mechanism that decreases the core power. This attribute is known as a negative temperature coefficient of reactivity. Most LWRs also have negative reactivity coefficients; however in fast reactors, this effect is strong enough to stop the reactor from reaching core damage without external action from operators or safety systems. This was demonstrated in a series of safety tests on the prototype reactors, which are covered in an elaborate manner in Chapter 20.

All reactivity coefficients except the sodium expansion effect are negative for fast reactors. Therefore, the reactor operation is stable for any bounded disturbances. The burnup reactivity changes are small due to breeding of fuel and low absorption cross section of FPs to fast neutrons. There is no xenon poisoning as in thermal reactors. There is no need for automatic power regulation. This avoids the problems that may arise by failure of the power regulation system. Inlet temperature and power coefficients of reactivity are negative so that any off-normal increase in temperature and/or power leads to reduction in reactivity and consequent reduction in power. The expansion of coolant and structural steel results in small positive reactivity that is compensated by negative and prompt reactivity effects like Doppler and fuel expansion. There is also negative reactivity feedback from the grid plate, spacer pad, and differential control rod expansion that tends to decrease the reactor power for transient undercooling incidents. The dynamic power coefficients computed are negative, which assure the stability of the reactor.

12.4 RADIOACTIVITY RELEASE

In a typical medium- to large-sized pool-type SFR, the CDA is a beyond-design-basis accident (BDBA) resulting from the mismatch of power produced and power removed from the reactor and the shutdown system not responding on demand, typically under conditions of either ULOF or unprotected transient overpower events. The consequent thermal energy release has an equivalent mechanical work potential, usually in the range of megajoules during which high temperatures and high pressure are reached. Though the accident is BDBA, the reactor containment building (RCB) is provided to mitigate the consequences of the CDA and ensure that the dose rate at the site boundary is within the prescribed limit and habitability in the control room is not jeopardize. Two steps are involved in estimating the source term. First, one calculates the radioactive source in RCB and, second, the source term is evaluated outside RCB. Evaluation of these terms is discussed.

12.4.1 RADIOACTIVE SOURCE IN RCB

To evaluate the source term in RCB, one requires basic information, such as the FP inventory in the core, extent of core damage in CDA, FPs reaching the cover gas, and subsequent release into the RCB through top shield leaks [12.13]. Another pathway is through the main vessel melt. However, the damage evaluation studies performed for CDA show that the main vessel and top shield remain intact. So, the only path available is through top shield leaks. The FP inventory depends on the core burnup. The inventory of radiologically important nuclides is given in Table 12.1. At the end of CDA, the core will be partly in the molten state and the remaining part will be in vaporized state. For example, CDA analysis for PFBR [12.14] shows that at the end of CDA, 94% of the core is in the molten state and 20% is in the vaporized state. One has to find the fraction of FP that reaches the

TABLE 12.1

Core Inventories of Fission Products and Source Term in RCB

Isotope	Half-Life	PFBR (Bq)	OECD (Bq)	SNR-300 (Bq)	EFR (Bq)
I-131	8.02 days	1.48E+18	1.33E+18	7.40E+17	1.48E+17
I-132	2.30 h	1.96E+18	1.76E+18	9.80E+17	1.96E+17
I-133	20.80 h	2.54E+18	2.29E+18	1.27E+18	2.54E+17
I-134	52.50 months	2.53E+18	2.28E+18	1.27E+18	2.53E+17
I-135	6.57 h	2.23E+18	2.01E+18	1.12E+18	2.23E+17
Cs-134	754.50 days	2.82E+14	2.28E+14	1.41E+14	2.82E+13
Cs-137	30.07 years	8.42E+16	6.82E+16	4.21E+16	8.42E+15
Rb-88	17.78 months	5.22E+17	4.23E+17	2.61E+17	5.22E+16
Ru-103	39.26 days	2.41E+18	9.64E+16	4.10E+16	9.64E+16
Ru-106	373.59 days	1.06E+18	4.24E+16	1.80E+16	4.24E+16
Sr-89	50.53 days	7.12E+17	7.83E+16	1.21E+16	2.85E+16
Sr-90	28.79 years	2.88E+16	3.17E+15	4.90E+14	1.15E+15
Ce-141	32.50 days	2.11E+18	8.44E+16	3.59E+16	8.44E+16
Ce-144	284.89 days	1.05E+18	4.20E+16	1.79E+16	4.20E+16
Te-131m	30.00 h	1.63E+17	2.61E+16	8.15E+16	1.63E+16
Te-132	3.20 days	1.88E+18	3.01E+17	9.40E+17	1.88E+17
Ba-140	12.75 days	1.93E+18	2.12E+17	3.28E+16	7.72E+16
Zr-95	64.02 days	1.76E+18	7.04E+16	2.99E+16	7.04E+16
La-140	1.68 days	1.96E+18	7.84E+16	3.33E+16	7.84E+16
Kr-85m	4.48 h	2.26E+17	2.03E+17	2.26E+17	2.26E+17
Kr-87	76.30 months	4.08E+17	3.67E+17	4.08E+17	4.08E+17
Kr-88	2.84 h	4.94E+17	4.45E+17	4.94E+17	4.94E+17
Kr-85	10.70 years	4.69E+15	4.22E+15	4.69E+15	4.69E+15
Xe-133	5.24 days	2.55E+18	2.30E+18	2.55E+18	2.55E+18
Xe-135	9.14 h	2.66E+18	2.39E+18	2.66E+18	2.66E+18
Pu-239	24110 years	4.51E+15	4.51E+13	7.67E+13	4.51E+11

cover gas and subsequently into the RCB. A large amount of R&D effort has been put worldwide on this topic. During CDA, the core bubble expands, and then for the FP to reach the argon cover gas, the FP must pass through the sodium pool. Much of the radioactive source could be removed during the transport through sodium. A schematic of the radioactive sources in different regions of the plant under CDA scenario is given in Figure 12.1. Kress et al. [12.15,12.16] identified the following possible removable mechanisms:

- During condensation of fuel vapor onto the bubble/sodium interface, onto the structures in the bubble, and onto the entrained liquid droplets
- During the agglomeration and removal into the sodium
- FP diffusion to and reaction with sodium

Several models [12.17–12.19] have been developed for the previously mentioned phenomena. However, these have not been verified experimentally. Keeping all the uncertainties involved in the previously mentioned phenomena Organization for Economic Co-operation and Development/ Nuclear Energy Agency (OECD/NEA) recommended [12.20] the source term for RCB as given in Table 12.2. The results of several experiments conducted in several countries on the LMFBR source term during the last 20 years are summarized as follows.

The fission gas release to the cover gas is definite and total, but with some time delay. The exact location of the release from the core would not be known. The volatile FP release fraction

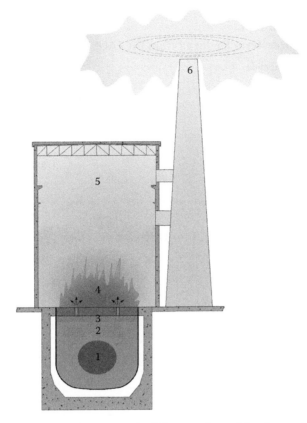

FIGURE 12.1 Schematic of radioactive sources in different regions of the plant under the CDA scenario.

TABLE 12.2
Source Term in RCB

Isotope	Release Fraction into RCB	Isotope	Release Fraction into RCB
I-131	1.00E−01	Ce-144	1.00E−01
I-132	1.00E−01	Te-131M	4.00E−02
I-133	1.00E−01	Te-132	4.00E−02
I-134	1.00E−01	Ba-140	4.00E−02
I-135	1.00E−01	Zr-95	1.00E−04
Cs-134	1.00E−01	La-140	1.00E−01
Cs-137	4.00E−02	Kr-85m	1.00E−01
Rb-88	4.00E−02	Kr-87	1.00E+00
Ru-103	4.00E−02	Kr-88	1.00E+00
Ru-106	4.00E−02	Kr-85	1.00E+00
Sr-89	4.00E−02	Xe-133	1.00E+00
Sr-90	4.00E−02	Xe-135	1.00E+00
Ce-141	1.00E−01	Pu-239	1.00E+00

depends on their physicochemical properties and evolution of thermodynamic conditions during the development of the accident. The release fractions from the core to the RCB for the 26 nuclides considered are given in Table 12.3. The values for iodine, cesium, tellurium, and fuel are taken from the values deduced from FAUST experiments [12.21]. For nonvolatile nuclides, the release fractions should be much lower. From the results of the FAUST experimental

TABLE 12.3

Source Term Deduced from the FAUST Experiment

Radionuclide	Fraction
Xe, Kr	1
I, Br	0.1
Cs	1.00E−04
Te	—
SrO	1.00E−04
UO$_2$	1.00E−04

TABLE 12.4

Source Term in RCB for Different Reactors

Radionuclide	OECD[a]	SNR-300[a]	EFR[b]	Monju[a]
Xe, Kr	0.9	1	1	1
I	0.9	0.5	0.1	0.1
Cs	0.81	0.5	0.1	0.1
Te	0.16	0.5	0.1	0.1
SrO	0.11	0.017		
Fuel	0.01	0.017	1.00E−04	2.00E−03

[a] Energetic.
[b] Nonenergetic.

program conducted in Germany [12.21], a realistic instantaneous source term is arrived and is given in Table 12.3. Based on the several experimental results, the source term estimated for different reactors [12.21] is given in Tables 12.1 and 12.4. It has also been observed that for volatile FPs, the iodine and cesium vapor contribution to the source term is negligible compared to the iodine and cesium combined with the liquid sodium aerosol contribution. For nonvolatile radionuclides, the source term depends on the pool height above the core and on the obstacles present in the plenum.

The OECD report on the nuclear aerosols has also discussed the shape parameters in detail. The resulting environmental releases are only marginally higher and given in Table 12.5. The effective dose at the site boundary due to this release is increased by only 1.7 mSv. The cumulative exposure due to the deposited activity due to Cs-137 and I-131 in the first 24 h increases only by 0.18 mSv. Hence, the total dose is still below the permitted limits.

12.4.2 Environmental Source Term

The physical processes involved in estimating the releases to the environment are quite complex. A number of codes have been developed to estimate the releases [12.22–12.25]. They essentially solve a nonlinear integral equation for the aerosol number concentration in the RCB taking into account the various physical processes such as agglomeration, sedimentation, thermophoresis, diffusiophoresis, and turbulent impaction. Where calculation of any of the processes is not possible, parameters from experiments are used to estimate the effect of the process. Fission gas release to the environment depends on the pressure time history in the RCB. The summary of radioactivity release scenario is depicted in Figure 12.2 and associated details are given in Table 12.6. The background calculation details are presented in the following paragraphs.

TABLE 12.5
Environmental Release for α = 0.15

Isotope	Core Inventory (Bq)	Environmental Release (Bq)	Release Fraction
I-131	1.48E+18	2.05E+13	1.39E–05
I-132	1.96E+18	2.72E+13	1.39E–05
I-133	2.54E+18	3.52E+13	1.39E–05
I-134	2.53E+18	3.51E+13	1.39E–05
I-135	2.23E+18	3.09E+13	1.39E–05
Cs-134	5.22E+17	8.96E+12	1.72E–05
Cs-137	2.41E+18	4.62E+12	1.92E–06
Rb-88	1.06E+18	2.03E+12	1.92E–06
Ru-103	7.12E+17	3.65E+12	5.13E–06
Ru-106	2.88E+16	1.48E+11	5.13E–06
Sr-89	2.11E+18	6.07E+12	2.88E–06
Sr-90	1.05E+18	3.02E+12	2.88E–06
Ce-141	1.63E+17	1.24E+12	7.59E–06
Ce-144	1.88E+18	1.43E+13	7.59E–06
Te-131M	1.93E+18	8.19E+12	4.24E–06
Te-132	1.76E+18	5.21E+12	2.96E–06
Ba-140	1.96E+18	6.01E+12	3.07E–06
Zr-95	4.51E+15	2.27E+07	5.03E–09
La-140	2.82E+14	1.63E+10	5.77E–05
Kr-85m	8.42E+16	4.86E+12	5.77E–05
Kr-87	2.26E+17	1.30E+15	5.77E–03
Kr-88	4.08E+17	2.35E+15	5.77E–03
Kr-85	4.94E+17	2.85E+15	5.77E–03
Xe-133	4.69E+15	2.71E+13	5.77E–03
Xe-135	2.55E+18	1.47E+16	5.77E–03
Pu-239	2.66E+18	1.53E+16	5.77E–03

12.4.3 MODEL FOR COMPUTATION

As mentioned earlier, much of the radioactivity will either reside on aerosols of sodium oxide or form their own aerosols. In the second stage, it involves computing the evolution of the aerosol size distribution using the following equation [12.26]:

$$\frac{dn(x,t)}{dt} = \left\{ \frac{1}{2} \int_0^x K(x',x-x')n(x',t)n(x-x',t)dx' - n(x,t) \int_0^\infty K(x,x')n(x',t)dx' - n(x,t)R(x) + S(x,t) \right\}$$

(12.5)

where

K(x, x′) is collision kernel predicting the probability of collision between two particles of volume
x and x′ resulting from Brownian motion, gravitational settling, and turbulent gas motion
x is the volume of the particle with radius r
x′ is the volume of the particle with radius r′
n(x, t) is the aerosol number distribution function
R(x) is the removal rate of particles produced by gravitational settling to the floor, diffusion to
the walls, etc.
S(x, t) is the source function rate of particles to the RCB

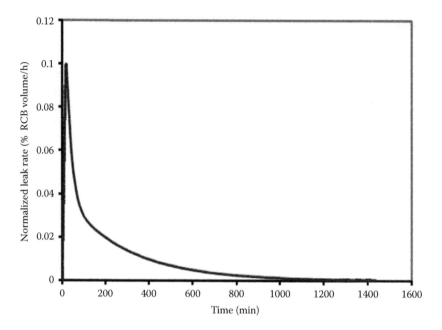

FIGURE 12.2   Leak rate versus time from the RCB.

TABLE 12.6
Radioactive Source Term Evaluation Details

Region No.	Source Terms	Methods	Code	Activity Bq[a]
1	Molten/vaporized mixture of fuel structure and coolant	Solution of a coupled set of differential equations	ORIGEN 2.0	2.9×10^{20}
2	Retention of radioactive isotopes in sodium pool	Theory/experiments	FAUST experiments	4.3×10^{19}
3	Fission gas and sodium aerosol accumulated in cover gas	Theory/experiments	FAUST experiments and ORIGEN 2.0	3.3×10^{19}
4	Passing of sodium with radioactive nuclides through penetration	D^2 law for droplet diameter with distance	NACOM and SOFIRE	2.1×10^{19}
5	Sodium aerosol and fission gas	Sodium aerosol diffusion	OECD report	8.0×10^{18}
6	Sodium aerosol and fission gas	Atmospheric dispersion	Standard double Gaussian plume model	3.7×10^{16}

[a] Indicative values given for 500 MWe PFBR.

In Equation 12.5, the first term represents agglomeration of aerosols to size x, the second term represents the removal of aerosols of size x, and the last term is the source term. There have been a number of computer codes developed [12.25,12.27–12.29] in different laboratories of the world to solve Equation 12.5. The form of the collision kernel K(x, x′) depends on the coagulation mechanisms such as Brownian motion of the particles, gravitational settling, and turbulent gas motion represented by rate kernels. These rate kernels depend on the viscosity of RCB air, gas mean free path in the RCB, temperature of RCB air, gravitational constant, Cunningham slip correction factor, aerosol material density, RCB air density, coagulation shape factor, dynamic shape factor, and particle–particle collision efficiency. These are determined by backfitting experimental measurements. Thus, the method of solution not only involves solving the nonlinear equation but also requires many experimentally determined parameters. Different codes are hardly in agreement in predicting the results for many measurements of aerosol concentrations [12.30]. Here, a simplified model is used. In this model [12.31], the assumption is made that within minutes of the release, the agglomeration process would proceed to reach a stage that could be used to describe the evolution of the aerosol concentration of an average size through the following equation:

$$\frac{dn(t)}{dt} - n(t)R \tag{12.6}$$

where $R = P_R + G_R + T_R$ and P_R, G_R, and T_R refer to the wall plating rate, gravitational sedimentation rate, and thermophoretic removal rate. Given the situation that the source is assumed to be instantaneous, this assumption is reasonable. The removal rates, P_R, G_R, and T_R, are given by Silberberg [12.15]

$$P_R = \frac{kTA_w\alpha^{1/3}}{6\pi\eta\Delta\chi r} \tag{12.7}$$

$$G_R = \frac{2g\rho_m A_f\alpha^{1/3}r^2}{9\eta V\chi} \tag{12.8}$$

$$T_R = \frac{3\eta A_w\nabla Tk_T}{2\rho_g VT} \tag{12.9}$$

where
   A_w is the deposition (wall) area
   V is the containment volume
   Δ is the parameter representing a distance over which diffusion occurs
   A_f is the available floor area
   k_T is the constant that depends on particle and gas properties
   ∇T is the temperature gradient usually taken as $(T_{gas}-T_{wall})/\Delta_T$ where Δ_T is the deposition distance parameter

If the leak rate of RCB air into the environment is L, then the fraction of mass of aerosol that would leak out of RCB in a period T is given by

$$f = \int_0^\tau L(t)\,dt\,e^{-(P_R+G_R+T_R)t} \tag{12.10}$$

The leak rate depends on the pressure rise in the RCB due to burning of sodium expelled into the RCB during the accident. The leak rate as a function of time has been estimated from the time dependence of pressure calculated from sodium fire in the RCB [12.32]. The integral f is obtained as

$$f = \frac{1}{R}\sum_{i=0}^N L(t_i,t_{i+1})(e^{-Rt_i} - e^{-Rt_{i-1}}) \tag{12.11}$$

Here, $L(t_i, t_{i+1})$ represents the leak rate during the time interval (t_i, t_{i+1}). The summation is carried over t hours, which is divided into N intervals. The radioactivity carried by aerosols into the environment is now given by

$$A_L = D_L fA \tag{12.12}$$

where

    "A" represents the radioactivity of the nuclide released into the RCB
    D_L represents a reduction factor due to aerosol deposition through leakage paths

Aerosol deposition in leakage paths such as cracks and penetrations through containment walls is an important process of removal of aerosols [12.33,12.34]. Levenson and Rahn have given possible attenuation of aerosols through leakage paths to be in the range of 1–100 in the context of LWRs. Fission gases, iodine, and cesium escape into the RCB in molecular form and get mixed uniformly with RCB air. Iodine will settle on sodium oxide aerosols [12.35]. For aerosols D_L is taken as 0.1 and for fission gases as 1.0. It may be noted that the removal rate for fission gases is zero.

12.4.4 Dose Estimates

Radioactivity released to the environment during an accident and radioactive materials that are bottled up inside the RCB are the major sources of exposures for the plant personnel as well as the general public. The activity that is bottled up within the RCB is the major contributor for the dose to the plant personnel, whereas the radioactivity that is released to the atmosphere is the major source of exposure for the general public beyond the site boundary. Normally, the dose estimates from radioactivity released to the environment have to be calculated in the case of a maximally exposed individual at the site boundary, that is, 1.5 km away from the reactor and as a function of distance from the RCB. The integrated dose rate for 24 h after accidents at different locations is shown in Figure 12.3. The pathways of exposure are

- Cloud gamma dose due to fission product noble gases (FPNGs)
- Inhalation dose due to particulate and vapors in the environment
- Exposure to ground deposited activity

As followed in PHWRs [12.36], the ingestion dose to the thyroid from the intake of radioiodine through the grass–cow–milk route has been excluded since sufficient time is available for intervention.

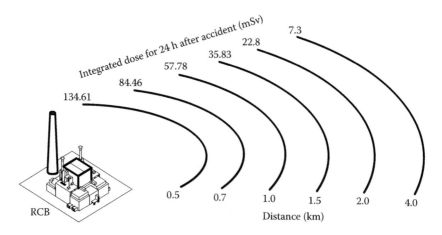

FIGURE 12.3 Atmospheric dispersion and dose rate at different distances.

12.4.5   DOSE LIMITS UNDER ACCIDENT CONDITIONS

Certain dose limits are specified under design basis accident (DBA) condition, for example, 100 mSv for the whole body and 500 mSv to a single organ (the thyroid) for the maximally exposed individual at the site boundary as per Indian regulation [12.37]. Higher dose limits are allowed for BDBA (250 mSv for the whole body and 2500 mSv for a child thyroid as per Indian regulation).

The source term that has to be used for the calculation of doses is taken from the previous section. Under a CDA event, radioactive materials from the containment building get released into the atmosphere through two pathways. The first pathway is the leakage of radionuclides through the normal ventilation route via stack. The second pathway is due to the leakage from the RCB, which meets a design target of not more than 0.1% volume of RCB per hour, for a period of about 24 h. This release is assumed to be at the ground level. In the case of fast reactors, the radioactive sodium from the coolant (which has two radionuclides, Na-22 and Na-24) is also released into the reactor containment building during CDA.

The model used for the calculation of concentration is the standard double Gaussian plume model [12.38]. The radionuclides are assumed to be released from the containment building to the atmosphere at ground level through leaks from the containment. The reactor containment building is surrounded by buildings on all sides, covering up to a height of about 30 m. A fraction of the radioactivity that leaks under the CDA will be into these buildings before reaching the environment. However, in the calculation the release is assumed to occur directly into the environment. Three pathways of exposure mentioned in the introduction are considered. Dose estimation is done for the adult group for all the pathways considered.

The stack discharges occur only for a period of 60 s. The height of the stack is 95 m. For the estimation of doses through stack discharges, two pathways of exposures are relevant: gamma exposure from the FPNG plume and inhalation exposure from the other major FPs. The most unfavorable meteorological conditions are assumed to prevail during the release of the radionuclide in order to maximize the dose estimates as per the recommendations of atomic energy regulatory board (AERB) [12.39]. The atmospheric stability category, the wind speed, the wind direction, and the other meteorological parameters depend on the release of the radionuclide.

REFERENCES

12.1   Cahalan, J. et al. (1990). Performance of metal and oxide fuels during accidents in a large liquid metal cooled reactor. *International Fast Reactor Safety Meeting*, Snowbird, UT.

12.2   Wade, D.C., Chang, Y.I. (1988). The integral fast reactor concept. Physics of operation and safety. *Nucl. Sci. Eng.*, 100, 507–524.

12.3   Sathiyasheela, T., Riyas, A., Sukanya, R., Mohanakrishnan, P., Chetal, S.C. (2013). Inherent safety aspects of metal fuelled FBR. *Nucl. Eng. Design.*, 265, 1149–1158.

12.4   Wade, D.C., Wigeland, R.A., Hill, D.J. (1997). The safety of the IFR. *Prog. Nucl. Energy*, 31(1–2), 63–82.

12.5   Wigeland, R., Cahalan, J. (2009). Fast reactor fuel type and safety performance. *Proceedings of Global*, Paris, France, September 6–11, 2009.

12.6   Royl, P. et al. (1990). Influence of metal and oxide fuel behavior on the ULOF accident in 3500 MWth heterogeneous LMR xores and comparison with other large cores. *International Fast Reactor Safety Meeting*, Snowbird, Utah, August 12–16, 1990.

12.7   Walter, A.E., Reynolds, A.B. (1981). *Fast Breeder Reactors*. Pergamon Press, New York.

12.8   Wade, D.C., Chang, Y.I. (1988). The integral fast reactor concept: Physics of operation and safety. *Nucl. Sci. Eng.*, 100, 507–524.

12.9   Tsujimoto, K. et al. (2001). Improvement of reactivity coefficients of metallic fuel LMFBR by adding moderating material. *Ann. Nucl. Energy*, 28, 831–855.

12.10 Merk, B., Fridman, E., Weiß, F.-P. (2011). On the use of a moderation layer to improve the safety behavior in sodium cooled fast reactors. *Ann. Nucl. Energy*, 38(5), 921–929.

12.11 Ishizu, T. et al. (2010). Study of the self controllability for the fast reactor core with high thermal conductivity fuel. *J. Nucl. Sci. Technol.*, 47, 684.

12.12 Beaver, T.R. et al. (1982). Transient testing of the FFTF for decay heat removal by natural convection. *Proceedings of the LMFBR Safety Topical Meeting*, European Nuclear Society, Lyon, France, Vol. II, pp. 525–534.

12.13 Indira Gandhi Centre for Atomic Research. (2001). Fission product inventories, Department of Atomic Energy. Note: PFBR/01115/DN/1081, Rev-A, India.

12.14 ULOF and CDA. (2008). Analysis of PFBR with reactivity worths based on ABBN-93 cross-section set. RPD/SAS/177/R-1.

12.15 Silberberg, M. (ed.) (1979). State of the art report on nuclear aerosols in reactor safety. A state of the art report by a group of exports of the NEA Committee on the Safety of Nuclear Installation, OECD.

12.16 Kress, T.S. et al. (1977). Source term assessment concepts for LMFBRs, aerosol release and transport analytical program. ORNL-NUREG-TM-124.

12.17 Ozisik, M.N., Kress, T.S. (1977). Effects of internal circulation velocity and non-condensible gas on vapour condensation from a rising bubble. *ANS Transactions*, 27, 551.

12.18 Theofanous, T.G., Fauske, H.K. (1973). The effects of non-condensibles on the rate of sodium vapour condensation from a single rising HCDA bubble. *Nucl. Technol.*, 9, 132.

12.19 Kennedy, M.F., Reynolds, A.B. (1973). Methods for calculating vapour and fuel transport to the secondary containment in an LMFBR accident. *Nucl. Technol.*, 20, 149.

12.20 Kress, T.S. et al. (1977). Source term assessment concepts for LMFBRs. Aerosol release and transport analytical program. ORNL-NUREG-TM-124.

12.21 Balard, F., Carluec, B. (1996). Evaluation of the LMFBR cover gas source terms and synthesis of the associated R&D. *Technical Committee Meeting on Evaluation of Radioactive Materials and Sodium Fires*, IWGFR/92, O-arai, Japan.

12.22 Berthoud, G. et al. (1988). Experiments on LMFBR aerosol source terms after severe accident. *Nucl. Technol.*, 81, 257.

12.23 Dunbar, I.H. (1985). Aerosol behaviour codes: Development, inter-comparison and applications. In *Proceedings of CSNI Specialists Meeting on Nuclear Aerosols in Reactor Safety*, Schikarski, W. O., Schock, W. (Eds.), Kernforschungszentrum Karlsruhe, KfK 3800, Federal Republic of Germany.

12.24 Beonio-Brocchieri, F. et al. (1988). Nuclear aerosol codes. *Nucl. Technol.*, 81, 193.

12.25 Gieske, J.A., Lee, K.W., Reed, L.D. (1978). HAARM-3: User manual. BMI-NULEC-1991.

12.26 Balard, F., Carluec, B. (1996). Evaluation of the LMFBR cover gas source terms and synthesis of the associated R&D. *Technical Committee Meeting on Evaluation of Radioactive Materials and Sodium Fires*, IWGFR/92, O-arai, Japan.

12.27 Hubner, R.S. et al. (1973). HAA-3 user report. AI-AEC-13038.

12.28 Obata, H. (1981). ABC-3C user's manual. CRC-081-001.

12.29 Bunz, H. (1983). A computer code for determining behaviour of contained nuclear aerosols. Pardiseko IV. KfK-3545.

12.30 Schutz, W., Minges, J., Haenscheid, W. (1985). Investigations on bubble behaviour and aerosol retention in case of a LMFBR core disruptive accident. In *Proceedings of CSNI Specialists Meeting on Nuclear Aerosols in Reactor Safety*, Schikarski, W.O., Schock, W. (eds.), KfK-3800.

12.31 Indira Gandhi Centre for Atomic Research. (2004). Release of radioactivity into environment under core disruptive accident, Department of Atomic Energy. Note: PFBR/01115/DN/1018/R-C, India.

12.32 Indira Gandhi Centre for Atomic Research. (2002). Sodium Release and Pressure Loading on reactor Containment Building during a CDA, Department of Atomic Energy. PFBR/31050/DN/1018, Rev-C, India.

12.33 Huang, T.C. (1977). Aerosol attenuation in leakage paths. *Trans. Am. Nucl. Society*, 26, 338.

12.34 Nelson, C.T. (1976). Some potential reductions in the release of radioactivity under LMFBR accident conditions. *Proceedings of International ANS/ENS Conference on Fast Reactor Safety*, Chicago, Illinois, US, October 5–8, 1976.

12.35 Baurmas et al. (1970). Behaviour of iodine in the presence of sodium oxide aerosols. *Proceedings of 11th AEC Air Cleaning Conference*, Federal Building, Richland, Washington, DC, Vol. 1, August 31–September 3, 1976.

12.36 Nuclear Power Corporation of India Limited. (1999). Accident dose computation for the PHWR station at Kaiga Site. PSAR chapter, Department of Atomic Energy, India.

12.37 Atomic Energy Regulatory Board. (1992). AERB safety guide on intervention levels and derived investigation levels for an off-site radiation emergency. SG/HS-1, India.

12.38 Hukoo, R.K., Bapat, V.N., Sitaraman, V. (1988). Manual on emergency dose evaluation, BARC-1412.

12.39 AERB Safety Guide (Draft). (1999). SG/S-5, India.

13 Safety Criteria and Basis

13.1 INTRODUCTION

A nuclear reactor performs various operating conditions that are well covered in the design. They are called *design basis conditions* as defined in Table 13.1. Design requirements under these conditions are specified to assure the plant availability and integrity of the fuel elements and mechanical components. With this, a set of design criteria are established (see Table 13.2 for the core integrity and Table 13.3 for hot and cold pool components of the reactor assembly). Safety demonstration requires not only meeting these design criteria during normal and transient conditions but also meeting the criteria relevant to the safe shutdown state following transient events. The consolidated composition of such criteria is termed as safety criteria. A few examples are given in Table 13.4. Thus, the safety criteria describe the design requirements for structures, systems, and components important to safety that shall be met for safe operation and also for the prevention or mitigation of the consequences of events that could jeopardize safety. Figure 13.1 shows the relationship between the categories, events, and safety targets [13.1].

The safety criteria are broadly classified into two categories: the first category covers the general safety requirements that are applicable to all types of reactors, while the second category covers requirements that are reactor system specific. Aspects covered under general safety principles are defense-in-depth requirements, safety functions, radiological dose limits to plant personnel and public, site boundary radiological requirements, emergency preparedness, etc.

This chapter presents the generic features of fast reactors (FRs) as well as specific safety issues related to sodium to be addressed in the safety criteria. International Atomic Energy Agency (IAEA), and other international safety standards and highlights safety criteria specified for typical pool sodium fast reactor (SFR) systems and components (as a case study) and evolving trends consequent to the post–Fukushima accident.

13.2 GENERIC FEATURES OF FAST REACTORS TO BE ADDRESSED IN THE SAFETY CRITERIA

The three important characteristics of the FR core, which are claimed to be disadvantages for the FR system compared to the pressurized water reactor (PWR), are (1) high power density (the specific power is 300–600 MW/m^3 in FR, about five times higher than that of PWR), (2) short lifetime of prompt neutrons (about 4.5×10^{-7} s in FR and 2.5×10^{-5} s in PWR), and (3) a small number of delayed neutrons (the overall fraction of delayed neutrons is 0.35% in FR and 0.6%–0.5% for PWR depending upon the burnup). A critical assessment of these three parameters is given in the following.

The high power density signifies a low thermal capacity of the core, and hence, the consequences of the disturbances cause rapid temperature changes, which in turn actuate the fast and continuous feedback by the Doppler and fuel expansion effects (inherent safety features in FR). Further, the changes in the status of the core would be known by the temperature measurements at the core outlet with very small time delay, so that at the right time, countermeasures are possible (engineered safety features). Temperature monitoring is carried out, in addition to the neutron flux monitoring. Because of these safety features, a high power density is not of a great concern in FR. Further, a high power density means small core dimensions, calling for short distance travel of the control and safety rods and resulting in shorter shutdown times. Hence, a high power density of the FR core

TABLE 13.1

Examples of Plant and Safety Criteria

S. No.	Category	Frequency of Occurrence
1	Normal and planned operations	>1
2	Anticipated operational occurrences	$>10^{-2}$/reactor-year
3	Events of low frequency (emergency)	10^{-2}–10^{-4}/reactor-year
4	Events of very low frequency (accidents)	10^{-4}–10^{-6}/reactor-year

TABLE 13.2

Examples of Design Criteria for Core Integrity

Category	Cladding Limits	Fuel Limits
1—Normal	Nonopen clad failure: MCT* < 700°C	No melting
2—Incidental	No clad failure, except due to random effect. MCT < 700°C and 700°C < MCT < 740°C during 2 h, and 740°C < MCT < 780°C during 10 min.	No melting
3—Accidental	No systematic (i.e., large number of) pin failures nor large increase in leak rate.	No melting
4—Hypothetical	"Reversible local sodium boiling" criteria.	Any predictable local fuel melting to be shown acceptable (simultaneous clad failure and fuel melting of the same pin excluded)
DEC	Coolability of damaged core; propagation of a pin bundle blockage must be prevented.	Severe accident admitted complying with the containment measures; no recriticality
Residual risk	Core collapse excluded by design.	Energetic severe accident excluded by design

MCT*, maximum clad temperature for an oxide fuel.

TABLE 13.3

Design Criteria for Hot and Cold Pool Components of the Reactor Assembly

Category	Cold Pool Components	Hot Pool Components (°C)
1—Normal	Maximum temperature is core inlet temperature	<545
2—Incidental	<520°C	<545
3—Accidental	<530°C	<545
4—Hypothetical	<630°C	<880
DEC	<630°C	—
Residual risk	—	—

is demonstrated as a safety-oriented advantage. The negative consequence of a high power density is that during the primary pump pipe rupture event without shutdown or with the total loss of the coolant, high thermal gradients would result. However, any sudden stoppage of the coolant flow is physically impossible due to the high mass moment of inertia of the coolant and pumps. Further, any mismatch between the power and coolant flow would soon initiate a fast scram. Moreover, the simultaneous failure of pumps and scram is an extremely improbable event. In spite of these, once the initiation of sodium boiling occurs, it may lead to a rapid power increase and ultimately to the destruction of the core. Hence, a highly reliable shutdown system is considered to be an essential requirement for the FR system to prevent coolant boiling.

TABLE 13.4
Examples of Safety Criteria

Category	Frequency	Plant Criteria	Safety Criteria
1	>1/year	High availability	Radiological release ALARA
2	>10^{-2}/year	Able to return to power at short term after rectification	Radiological release lower than the limit
3	>10^{-4}/year	Able to restart after inspection and repair	Radiological release lower than the limit
4	>10^{-6}/year	Plant restart not required	To maintain core coolability, limited change in core geometry
DEC	>10^{-7}/year	Loss of plant investment	Releases lower than the targets (no need for off-site provisions)
Residual risk	<10^{-7}/year	—	No "cliff edge" effect

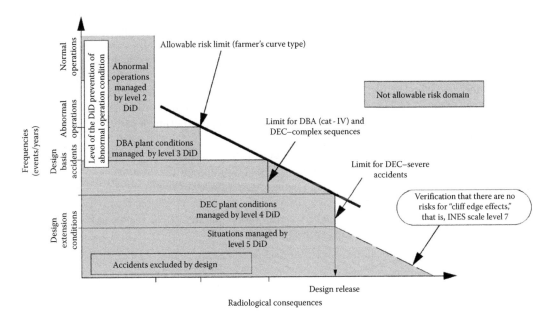

FIGURE 13.1 Relationship between categories and safety targets. (From Rouault, J. et al., Sodium fast reactor design: Fuels, neutronics, thermal-hydraulics, structural mechanics and safety, *Handbook of Nuclear Engineering*, 2010, Springer, New York, USA, Vol. 4, pp. 2321–2710, Chapter 21.)

The effective neutron lifetime in FRs, which is the mean time between two consecutive neutron generations, is shorter than in PWRs. The neutron suffers only a few collisions after its generation till it gets absorbed by another fissile or fertile atom. Such collisions take place with relatively high z nuclei, and hence, the energy loss from their birth to absorption is relatively small. Under these conditions, if the kinetics of the reactor had been defined by prompt neutrons, it would not have been possible to control the chain reaction with a mechanically regulated apparatus. However, the control in all reactors is based on the time constants of delayed neutrons and the reactivity disturbances are related to the fraction of delayed neutrons in reactor dynamic analyses. Since the overall fraction of delayed neutrons is small for FRs (0.35% for a typical FR and 0.6%–0.5% for a PWR depending upon the burnup), concern is expressed on the stability of the FR core. This is only a misconception: due to the benefits derived from fast neutrons (lower cross sections), the FRs are more stable even with a small number of delayed neutron fractions. That is, the perturbations in the operating parameters that could cause instability, for example, variations in inlet temperature, coolant flow, and power, have a considerably less influence on the reactivity (about 0.0015$/K w.r.t. temperature in FRs compared to 0.01–0.1$/K for PWRs). Apart from this, as mentioned earlier, a shorter shutdown time is easily possible in view of the short core height (<1 s in a typical FR). With this short drop time, the failure of all coolant circulation pumps (a pessimistic event) would initiate shutdown that will certainly take place well in advance so that bulk coolant boiling does not take place. Hence, a relatively shorter neutron lifetime and a small fraction of the delayed neutron fraction do not jeopardize core safety.

The earlier conclusion has been demonstrated theoretically by solving the reactor kinetic differential equations with the consideration of the delayed neutrons and is represented in Figure 13.2 [13.2]. In this figure, the periods are shown on the ordinates. In these periods, the power of the unregulated reactor increases by a factor e, if the appropriate reactivity is inserted without considering any feedback effects; controlled reactors are practically not distinguishable from each other from the reactor period point of view, as long as the reactivity is lower than about 1$. The periods are so long that control is quite possible with technically simple mechanical regulating units. With the reactivity higher than 1$, prompt neutrons determine the time behavior of the reactor ("superprompt criticality"). The periods are too short in both reactor types, in order to be able to interfere effectively with a fast shutdown. Thus, every reactor must be designed so that a superprompt critical condition cannot occur, and in all cases, this can be considered as a hypothetical event. If a superprompt critical excursion is arbitrarily postulated, it is shown that it is limited highly by feedback effects. In this case, the Doppler coefficient plays a prominent role, which depends on the neutron absorption capability of U-238, available in the fuel material, which naturally increases as the fuel temperature increases. The theory has been verified through an experiment called "SEFOR" conducted in 1963 in the United States [13.3].

There are only a few initiators that can introduce positive reactivity in the fast breeder reactor (FBR) core. The major ones among them are (1) gas entrainment, (2) oil entry, and (3) coolant voiding. Regarding gas entrainment, though isolated gas bubbles of relatively small sizes distributed in the core may be a possible scenario, they do not cause unacceptable reactivity variations. It is impossible to conceive a scenario, where a very large quantity of gas bubble (a few hundreds of liters) passing through core, that can cause significant reactivity change. Hence, gas entrainment

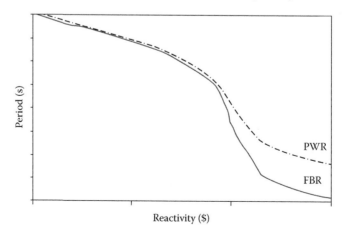

FIGURE 13.2 Reactor period versus reactivity. (From Vossebrecker, V.H., Special safety related thermal and neutron physics characteristics of sodium cooled fast breeder reactors, Warme Band 86, Heft 1, 1999.)

is not an issue in FR safety. However, as a defense in depth, design features such as mechanical seals for intermediate heat exchangers (IHXs) and porous plates attached to the reactor internals in the vicinity of sodium-free levels are incorporated to avoid/mitigate the gas entrainment mechanism effects. Regarding oil entry effects, the quantity of oil used especially for the pump bearing is minimized to a large extent such that accidental oil leak is not of a great concern to the safety. Again, as a defense in depth, a few design features (e.g., oil catch pots below the oil tanks to prevent the oil falling into the reactor sodium, oil-free bearings based on ferrofluids) are incorporated for the future design of pumps to completely eliminate the oil entry issue. With reference to coolant boiling, the sodium void coefficient is positive only in the central part of a relatively large core and there is a risk of uncontrolled power excursion. This can be safely handled with some design features such as short core height, lesser sodium volume fraction, and introduction of moderator materials.

There are some scenarios that are critical for the safety of thermal reactors that do not take place in FR. One example is the reactivity loss due to fuel burnup. This is small in SFR due to the breeding of fuel and low absorption cross section of fission products to fast neutrons. In thermal reactors, due to the higher absorption cross section of fission products, a large negative reactivity is added, especially by Xe-135 (half-life of 9.2 h). Within a short period after the shutdown, the reactivity increases considerably due to the decay of Xe-135. Since its effective cross section, in contrast to the fission cross section, is very much smaller in the fast spectrum than in the thermal spectrum, this effect does not practically play any role in the FBR. Thus, there is no xenon poisoning. The frequency of control rod operation is low (~2–3 times per shift) during steady-state operation. Thus, there is no need for automatic power regulation. This avoids the problems that may arise due to failure of the automatic power regulation system, which is one of the parameters that affect the reliability of the shutdown system.

13.3 SAFETY ISSUES RELATED TO SODIUM TO BE ADDRESSED IN THE SAFETY CRITERIA

In the 1960s of the last century, the technology of sodium as an FR coolant was well developed and brought to a commercial level. Production of sodium, meeting the reactor standard requirements, was mastered in the plants; sensors, instruments, and systems were developed for controlling sodium parameters required for successful operation of the reactors. Sodium has several advantages as a coolant for FRs. The large margin between the operating temperatures (~820 K) and the boiling point of sodium (1170 K) can accommodate a significant temperature rise in the event of mismatch of heat generation and heat removal. High thermal conductivity, low viscosity, and large difference between the temperature of hot sodium at 820 K and ambient air at 310 K, coupled with a significant variation of sodium density with temperature, permit decay heat removal through the natural convection mode. The pool-type concept provides a large thermal inertia, and hence, more time is available to the operator to act in case of exigencies during reactor operation. The temperature and power coefficient of reactivity are negative. So in the event of disturbances in primary and secondary sodium flow or feedwater flow, the reactor stabilizes to a new power level even without the corrective action of the operator.

In spite of the previously mentioned positive features, sodium leaks cannot be completely ruled out [13.4]. Some examples of sodium fire scenarios experienced during operation and maintenance are as follows: (1) the Almeria accident, which occurred in 1986, (2) at the Rapsodie reactor in France during the removal of sodium residues from the drain tanks, and (3) damage of building structures at both Rapsodie plant and FAUST facility in Forschungszentrum Karlsruhe (FZK), Germany. Sodium leaks from the radioactive circuit as well as in the steam generators cause a long outage of the reactor and a negative effect on further operation of the reactor. Safe disposition of huge amounts of radioactive sodium during the decommissioning stage would call for high investment and operational cost and complex technologies. Further, the opaqueness of sodium poses challenging technological problems for in-service inspection (ISI) and repair. Difficulties have been experienced during fuel-handling operations. This has been shown to be an important operational aspect, as demonstrated by the experience gathered from fuel-handling incidents, which occurred in fast breeder test reactors (FBTRs) in India as well as in Joyo. In the FR core, any small geometrical variations (e.g., due to internal/external excitations) could introduce significant reactivity variations. Sodium boiling may result in voids that could introduce a positive void coefficient in the SFR. Hence, special provisions should be made in core design and appropriate safety criteria should be developed to ensure reactor safety. In terms of reliability of items important to safety, with the absence of adequate long-term commercial operating experience (~400 reactor-years for SFR, specifically in France, the United Kingdom, Germany, former Soviet Union and Russia, Japan, India, China, and the United States), they have to be enhanced in a step-by-step manner with the accumulated experience and reliability demonstrated.

13.4 IAEA AND OTHER INTERNATIONAL SAFETY STANDARDS

Robust safety criteria are available for thermal reactors, for example, IAEA's safety series 50-C-D Rev D, 1988, and reissued as IAEA NS-R-1, 2000 [13.5]. IAEA is subsequently revising the NS-R-1 standard, fundamental safety principle documents, and safety assessment of facilities. Draft safety requirements have been released as DS-414 in 2009, which was finally issued as IAEA SSR 2/1 [13.6]. Revision of safety criteria is one of the current activities of high priority of Generation IV International Forum (GIF), an international forum working for the development of Gen IV reactors. Figure 13.3 depicts the hierarchy of safety standards [13.7]. In 1990, the LMFBR Safety Criteria and Guidelines for Consideration in the Design of Future Plants have been issued (Report EUR 12669 EN) by the Commission of the European Community [13.8].

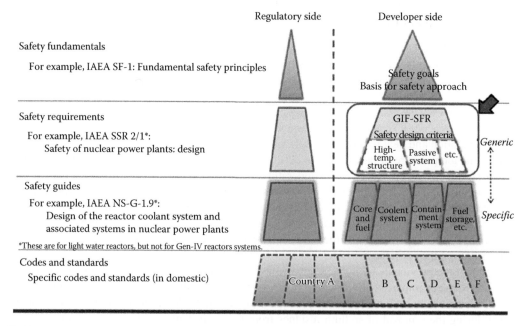

FIGURE 13.3 Hierarchy of safety standards. (From Nakai, R., Safety design criteria (SDC) for Gen-IV sodium-cooled fast reactor, *GIF-IAEA/INPRO SDC Workshop*, IAEA, Vienna, Austria, February 26–27, 2013.)

13.5 SAFETY CRITERIA FOR SFR: A FEW HIGHLIGHTS

13.5.1 Requirements for Plant Design

A few essential aspects to be taken in plant design are highlighted in the following paragraphs. The following sections have been extracted from Ref. 13.6 and reproduced as a ready reference to the readers.

13.5.1.1 Application of Defense in Depth

Application of the defense-in-depth concept throughout design and operation provides a graded protection against a wide variety of transients, anticipated operational occurrences, and accidents, including those resulting from equipment failure or human action within the plant and events that originate outside the plant. It provides a series of defense (inherent features, equipment, and procedures) aimed at preventing harmful effects to people or the environment and ensuring appropriate protection and mitigation in the event that prevention fails. The independent effectiveness, that is, the capability to achieve the specified objective of each of the different levels of defense, is a necessary element of defense in depth.

There are five levels of defense in depth (Table 13.5). The aim of the first level of defense is to prevent deviations from normal operation and to prevent system failures. This leads to the requirement that the plant be soundly and conservatively sited, designed, constructed, maintained, and operated in accordance with appropriate quality levels and engineering practices. To meet these objectives, careful attention is paid to the selection of appropriate design codes and materials and to the control of fabrication of components and of plant construction as well as plant commissioning. The aim of the second level of defense is to detect and intercept deviations from normal operational states in order to prevent anticipated operational occurrences from escalating to accident conditions. This is in recognition of the fact that some postulated initiating events (PIEs) are likely to occur over the service lifetime of a nuclear power plant, despite the care taken to prevent them. This level necessitates the provision of specific systems/features as determined in the safety analysis and the definition of operating procedures to prevent or minimize damage from such PIEs. For the third level of defense, it is assumed that, although very unlikely, the escalation of certain anticipated operational occurrences or PIEs may not be arrested by a preceding level and a more serious event may develop. These unlikely events are anticipated in the design basis for the plant, and inherent safety features, fail-safe design for the provisions of the previous levels, and additional equipment and procedures are provided to control their consequences and to achieve stable and acceptable plant states following such events. This leads to the requirement that inherent and/or engineered safety features be provided that are capable of leading the plant first to a controlled state, and subsequently to a safe shutdown state, and maintaining at least one barrier for the confinement of radioactive material.

TABLE 13.5
Five Levels of Defense in Depth

Level 1	Level 2	Level 3	Level 4	Level 5
	Plant states (considered in design)			
Normal operation	AOO	DBA	DEC	Off-site emergency response (out of the design)
	Operational states		Accident conditions	
Normal operation	Anticipated operational occurrences	Design basis accidents	Design extension conditions (including severe accident conditions)	

The aim of the fourth level of defense is to address consequences of beyond design basis accidents (BDBAs) with a very low probability of occurrence and to ensure that radioactive releases are kept as low as practicable. The most important objective of this level is the protection of the confinement function while achieving the management and mitigation of severe accident consequences. This may be achieved by taking account of available design margins, complementary measures, and procedures to prevent accident progression and by mitigation of the consequences of selected severe plant conditions, in addition to accident management procedures. The fifth and final level of defense is aimed at mitigation of the radiological consequences of potential releases of radioactive materials that may result from severe plant conditions. This requires the provision of an adequately equipped emergency control center and plans for the on-site and off-site emergency response.

13.5.1.2   Safety Classification

All structures, systems, and components, including software for instrumentation and control (I&C) that are important to safety, should be first identified and then classified on the basis of their function and significance with regard to safety. They should be designed, constructed, and maintained such that their quality and reliability are commensurate with this classification. The applicable codes and standards for design, manufacture, inspection, testing, and ISI of all these structures, systems, and components should be identified. The method for classifying the safety significance of structures, systems, and components should primarily be based on deterministic methods, complemented where appropriate by probabilistic methods and engineering judgment, with account taken of factors such as the safety function(s) to be performed by the item, the consequences of failure to perform its function, the probability that the item will be called upon to perform a safety function, and the time following a design basis event (DBE) at which it will be called upon to operate. Appropriately designed interfaces should be provided between structures, systems, and components of different safety classes to ensure that any failure in a system classified in a lower safety class will not propagate to a system classified in a higher safety class. For example, if a fluid system is interconnected with another fluid system that operates at a higher pressure, then it should be designed to withstand the higher pressure, or provision should be made to preclude the design pressure of the system operating at the lower pressure from being exceeded, on the assumption that a single failure occurs. Ref. 13.9 provides guidelines and illustrations for the classifications.

The safety classes have to be linked to design and construction codes. The civil structures and mechanical systems and components have been classified in three safety classes—class 1, class 2, and class 3—as design and construction codes are readily available corresponding to three safety classes defined earlier. Class 1E is assigned for safety and safety-related electrical equipment corresponding to three safety classes. Two design and construction classes—class 1 and class 2—are assigned to I&C systems and equipment and they correspond, respectively, to safety class 1 and 2 and safety class 3. Two design and construction classes—class 2 and class 3—are assigned to safety-related civil structures. In addition, nonnuclear safety (NNS) is assigned to structures, systems, components, and equipment that are not associated with any of the safety functions. Figure 13.4 depicts the classification of reactor assembly components of a typical pool-type reactor (prototype fast breeder reactor [PFBR]).

13.5.1.3   Identification of Common Cause Failures

The potential for common cause failures of items important to safety should be considered to determine where the principles of diversity, redundancy, and independence should be applied to achieve the necessary reliability.

13.5.1.4   Application of Single Failure Criterion

Single failure is a random failure, which results in the loss of capability of a component to perform its intended safety function. Consequential failures resulting from a single random occurrence

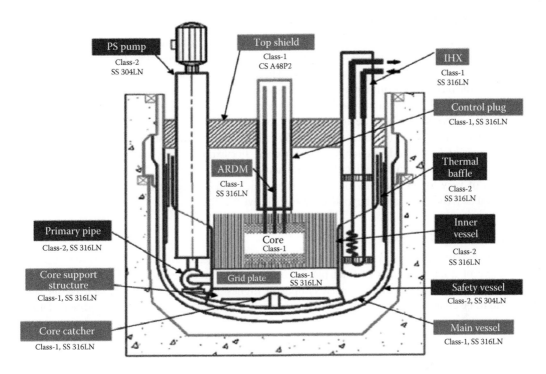

FIGURE 13.4 Safety classification of reactor assembly components of SFR.

are considered to be part of the single failure. The single failure criterion should be applied to each safety group incorporated in the plant design. For a safety group, the system level should be identified. The single failure criterion should be applied at system levels for specified systems. To test compliance of the plant with the single failure criterion, the pertinent safety group should be analyzed in the following way. A single failure (and all its consequential failures) should be assumed in turn to occur for each element of the safety group until all possible failures have been analyzed. The analyses of each pertinent safety group should then be conducted in turn until all safety groups and all failures have been considered. At no point in the single failure analysis is more than one random failure assumed to occur. Spurious action should be considered as one mode of failure when applying the concept to a safety group or system. Compliance with the criterion should be considered to have been achieved when each safety group has been shown to perform its safety function when the previous analyses are applied. Potentially harmful consequences of PIEs for the safety group are assumed to occur and the worst permissible configuration of safety systems performing the necessary safety function is assumed, taking into account maintenance, testing, inspection, and repair and allowable equipment outage times. Noncompliance with the single failure criterion should be exceptional and should be clearly justified in the safety analysis covering the situations such as very rare PIE or very improbable consequence of PIE or very low consequences of PIE or withdrawal from service for limited periods of certain components for purposes of maintenance, repair, or periodic testing. In the single failure analysis, it may not be necessary to assume the failure of a passive component designed, manufactured, inspected, and maintained in service to an extremely high quality, provided that it remains unaffected by the PIE. However, when it is assumed that a passive component does not fail, such an analytical approach should be justified, taking into account the loads and environmental conditions as well as the total period of time after the initiating event for which functioning of the component is necessary.

13.5.1.5   Identification of Design Basis Events and Beyond-Design-Basis Events

In designing the plant, it shall be recognized that challenges to all levels of defense in depth may occur and design measures shall be provided to ensure that the required safety functions are accomplished and the safety objectives can be met. These challenges stem from the DBE, which are selected on the basis of deterministic or probabilistic considerations. A combination of the two independent events, each having a low probability, is normally not considered to occur simultaneously. DBEs, which form the basis of design of a plant, are classified into different categories to include normal operation, operational transients, and PIE. DBE can be categorized on the basis of their consequences and expected frequency of occurrence. For example, the categories may be normal operation, operational transients, incidents, and accidents.

Events of very low probability of occurrence, which can lead to severe accidents and are not considered as DBEs, are called beyond-design-basis events (BDBEs).

13.5.1.6   Design Extension Condition

Scenarios making a relatively large contribution to causing core damage, scenarios that are representative of a large number of final plant damage states, credible scenarios with potential for significant releases, and credible scenarios imposing highest loads on containment (at least one such scenario to be considered) should be considered in selecting the design extension condition (DEC).

13.5.1.7   Practical Elimination of Conditions

In the IAEA SSR 2/1 [13.6], the expression "practical elimination" is defined as

> The possibility of certain conditions occurring is considered to have been practically eliminated if it is physically impossible for the conditions to occur or if the conditions can be considered with a high level of confidence to be extremely unlikely to arise.

13.5.1.8   Fail-Safe Design

The principle of fail-safe design should be considered and incorporated into the design of systems and components important to safety for the plant as appropriate and to the extent possible (if such a system or component fails, plant systems should be designed to pass into a safe state with no necessity for any action to be initiated).

13.5.1.9   Safety Support Systems

The systems that provide support to a safety system should be considered a safety support system and should be classified accordingly (either full or in part). Their reliability, redundancy, diversity, and independence and the provision of features for isolation and for testing of functional capability should be commensurate with the reliability of the system that is supported. Safety support systems necessary to maintain the plant in a safe state may include the supply of electricity, cooling water, and compressed air or other gases.

13.5.1.10   Provision for In-Service Testing, Maintenance,
Repair, Inspection, and Monitoring

Structures, systems, and components important to safety should be designed to be calibrated, tested, maintained, repaired or replaced, inspected, and monitored with respect to their functional capability over the lifetime of the nuclear power plant. The plant layout should be such that these activities are facilitated and can be performed to standards commensurate with the importance of the safety functions to be performed, with no significant reduction in system availability and without undue exposure of the site personnel to radiation. If the structures, systems, and components important to safety cannot be designed to be tested, inspected, or monitored to the extent desirable, then the other proven alternative and/or indirect methods, such as surveillance of reference items, or verified and validated computational methods should be specified and conservative safety margins or other appropriate precautions should be applied to compensate for potential undiscovered failures.

13.5.1.11   Equipment Qualification

A qualification procedure should be adopted to confirm that the items important to safety are capable of meeting, throughout their design operational lives, the demands for performing their functions while being subject to environmental conditions (vibration, temperature, electromagnetic interference, irradiation, humidity, or any likely combination thereof) prevailing at the time of need. The environmental conditions to be considered should include the variations expected in normal operation, anticipated operational occurrences, and DBEs. In the qualification program, consideration should be given to aging effects caused by various environmental factors (such as vibration, irradiation, and extreme temperature) over the expected lifetime of the equipment. Where the equipment is subject to external natural events and is needed to perform a safety function in or following such an event, the qualification program should replicate as far as practicable the conditions imposed on the equipment by the natural phenomenon, either by test or by analysis or by a combination of both. In addition, any unusual environmental conditions that can reasonably be anticipated and could arise from specific operational states, such as in periodic testing of the containment leak rate, should be included in the qualification program. To the extent possible, equipment (such as certain instrumentation) that must operate in a BDBE should be shown, with reasonable confidence, to be capable of achieving the design intent.

13.5.1.12   Materials

To ensure satisfactory performance during normal operation and accident conditions, only codified or proven materials or materials for which experience exists for structures, components, etc., should be selected based on considerations, among others, like (1) irradiation damage, (2) activation and corrosion, (3) creep and fatigue, (4) erosion, (5) compatibility with other interacting materials, (6) thermal effects, and (7) resistance to brittle fracture. The current state-of-the-art developments in material research and behavior phenomena should be an essential input for design updates.

13.5.1.13   Aging

The design should provide appropriate margins in the design for all structures, systems, and components important to safety so as to take into account relevant aging and wear-out mechanisms and potential age-related degradation, in order to ensure the capability of the structure, system, or component to perform the necessary safety function throughout its design life. Aging and wear-out effects in all normal operating conditions, testing, maintenance, maintenance outages, and plant states in a PIE and post-PIE should also be taken into account. Provision should also be made for monitoring, testing, sampling, and inspection, to assess aging mechanisms expected at the design stage (through appropriate surveillance coupons) and to identify unanticipated behavior or degradation that may occur in service. Design operating life of the plant and its components should be specified. In cases where the design life of the equipment/component is less than the design life of the plant, mid-term in situ replacement of the equipment may be warranted. Adequate provisions should be made in the design particularly for the in-core equipment and to facilitate such replacements.

13.5.1.14   Sodium Aerosol Deposition

In order to minimize sodium aerosol deposition in the clearances of different equipment on the roof of the reactor, the roof slab should be maintained at temperatures above the sodium melting point (warm roof concept).

13.5.1.15 Sharing of Structures, Systems, and Components between Reactors

Structures, systems, and components important to safety should generally not be shared between two reactors in nuclear power plants. If in exceptional cases such structures, systems, and components important to safety are shared between two reactors, it should be demonstrated that all safety requirements are met for all reactors under all operational states (including maintenance) and in DBEs. In the event of a BDBE involving one of the reactors, an orderly shutdown, cooling down, and removal of residual heat should be achievable for the other reactor(s).

13.5.1.16 Interactions between the Electrical Power Grid and the Plant

In the design of the plant, account should be taken of power grid–plant interactions, including the independence of and number of power supply lines to the plant, in relation to the necessary reliability of the power supply to plant systems important to safety.

13.5.1.17 Decommissioning

At the design stage, special consideration should be given to the incorporation of features that will facilitate the decommissioning and dismantling of the plant. Attention should be directed to keep the exposures of personnel and the public during decommissioning within limits prescribed by the regulatory body and to ensure adequate protection of the environment from radioactive contamination. The decommissioning aspect should be considered at the design stage itself including the choice of materials, such that eventual quantities of radioactive waste are minimized and decontamination is facilitated. For example, the use of stellite should be minimized for hardfacing in the primary sodium system, and the access capabilities and the facilities necessary for storing radioactive waste generated in both operation and decommissioning of the plant should be ensured.

13.5.1.18 Safety Analysis

A safety analysis of the plant design, applying deterministic methods, should be performed, demonstrating that the overall plant design ensures that radiation doses and releases are within the prescribed and acceptable limits for each DBE category. In addition, to complement the deterministic safety analysis, probabilistic safety analysis should also be performed. Computer programs, analytical methods, and plant models used in the safety analysis should be verified and validated, and adequate consideration should be given to uncertainties.

13.5.1.19 Deterministic Approach

The deterministic safety analysis should include (1) confirmation that operational limits and conditions are in compliance with the assumptions and intent of the design for the normal operation of the plant, (2) categorization of the PIEs that are appropriate for the design and site of the plant, (3) analysis and evaluation of event sequences that result from PIEs, (4) comparison of the results of the analysis with radiological acceptance criteria and design limits, (5) establishment and confirmation of the design basis, and (6) demonstration that the management of anticipated operational occurrences and DBEs is possible by automatic response of safety systems in combination with prescribed actions of the operator. The applicability of the analytical assumptions, methods, and degree of conservatism used should be verified. The safety analysis of the plant design should be updated with regard to significant changes in plant configuration, operational experience, and advances in technical knowledge and understanding of physical phenomena and should be consistent with the current or "as-built" state.

13.5.1.20 Probabilistic Approach

There are three levels of PSA: Level 1 PSA, Level 2 PSA, and Level 3 PSA. Level 1 PSA should be carried out to demonstrate that a balanced design has been achieved such that no particular feature or PIE makes a disproportionately large or significantly uncertain contribution to the overall risk. The outcome of Level 1 PSA is quantification of the core damage frequency. Level 2 PSA quantifies the various levels of the radiation release frequency. The outcome of Level 3 PSA is quantification of risk to the public. This means that Level 3 PSA will provide a systematic analysis to give confidence that the design will comply with the general safety objectives and to assess the adequacy of plant emergency procedures.

13.5.2 REQUIREMENTS FOR THE MAIN REACTOR SYSTEMS: A CASE STUDY

For this case study, a 500 MWe capacity PFBR is considered. Important aspects of the safety criteria and basis are highlighted in this section.

13.5.2.1 Core

Design safety limits (DSLs) should be established for fuel, clad, and coolant for all categories of DBEs. The reactor core and associated coolant, control and protection systems, and other safety-related systems should be designed with appropriate safety margins to ensure that the specified DSLs are not exceeded during all operational states and DBE. In case of safe shutdown earthquake (SSE), the external reactivity inserted due to the relative movement of fuel and absorber rod subassemblies should be below the prescribed limits to ensure the capability to shut down, thereby avoiding any process parameters to exceed the applicable DSL. The core subassemblies should be designed such that they maintain their integrity for their design lifetime with deformations/displacements less than the specified design limits. The fuel pins should be designed so as to ensure their integrity under irradiation up to the design burnup with clad failure less than the specified design limit. The fuel pin failure rate (gas leak failures) during operation should be minimized (reduction of activity of cover gas circuits). There should be systems for the location and removal of failed fuel (delayed neutron leak failure) from the reactor (avoiding fission product contamination of sodium). The reactor core and associated coolant, core support, and control elements (including their drive mechanisms) should be so designed that the total power coefficient, the prompt power coefficient, and the total temperature coefficient should be negative throughout the reactor life, for all operational states and accident conditions, taking into account all core loading configurations and irradiation effects.

The design of core and coolant components should be such that local coolant boiling is avoided and vibrations are within acceptable limits under all DBEs. In case of any exceptions, it should be highlighted and demonstrated that the net reactivity inserted is negative. The inlet and outlet paths of subassembly should be designed to prevent coolant blockages, which can cause the DSL to be exceeded. Provision should be made in the design to prevent wrong loading of subassemblies on the grid plate causing less coolant flow than that required to remove the heat generated. Under operational states, the power to flow mismatch in a fuel subassembly should be detected within an appropriate time so that safety limits are not exceeded. Appropriate corrective safety action is taken so that DSLs are not exceeded. The hold-down of all subassemblies during all operating condition and anticipated events should be ensured. Hydraulic or other unintentional lifting of any core subassembly from the grid plate should not be possible in any DBE. Ejection of absorber rods should be prevented when connected to their drive mechanisms under normal operating conditions and DBEs. Ejection of absorber rods, disconnected from their drive mechanisms in the shutdown conditions, should not be possible for any DBE. Sufficient shielding should be provided surrounding the core to limit radiation damage to maintain integrity of irreplaceable life-limiting reactor components (like grid plate, core support structure), to limit activation of the components for which maintenance may be required, and to limit activation of secondary sodium.

13.5.2.2   Reactor Assembly

The reactor assembly components should be designed and constructed with the highest quality with respect to materials, design standards, inspection, and fabrication. There should be no penetrations in the walls of the main vessel below the sodium-free level. Sodium from the main vessel should not get drained out below a prescribed safe limit due to any leak in the primary sodium circuit. To meet this requirement, the main vessel could be enveloped with a safety vessel to contain the sodium in case of leak from the main vessel in such a way that it is possible to maintain sufficient coolant flow through the core to remove decay heat. The intervessel space between the main and safety vessels should be maintained under inert gas atmosphere. Design should ensure that the failure of the load-bearing member does not have a cascading effect. Reliable systems should be provided for the detection of sodium leak into the space between the main and safety vessels. The cover gas system should be designed so as to prevent air ingress into the primary sodium and leakage of radioactivity in the reactor containment building (RCB). For structural integrity considerations of the main vessel, external/internal pressure should be maintained within the safe limits with application of the single failure criteria in pressure regulating systems. Water should not be used as a cooling medium for the top shields of the reactor assembly containing sodium.

The main vessel should be designed so that flaws are very unlikely to be initiated, and any flaws that are initiated would propagate in a manner that leads to leak before break. The main vessel design should take into account the expected end-of-life properties affected by erosion, creep, fatigue, the chemical environment, the radiation environment, and aging. The main vessel should be designed, manufactured, and arranged in such a way that it is possible, throughout the service lifetime of the plant, to carry out at appropriate intervals adequate inspections and demonstrate the absence of unacceptable defect or of safety significant deterioration. During normal operating and design basis fault conditions, possible geometrical variations of the core should be limited to an extent that the resulting reactivity increase is within the capability of the trip and shutdown systems. Distortions that might impede safe reactor shutdown should be prevented. Structural failure leading to geometrical variations with consequences beyond the capability of the shutdown system should be prevented.

Geometrical variations of the core and associated structures caused by increasing the power to flow ratio or increasing coolant inlet temperatures should contribute to overall negative reactivity changes. The cross section of individual sodium feed pipes should be limited to reduce the core flow change in case of pipe failure. The core support system should incorporate a high structural design margin with redundant features. The propagation of failure of the core support and grid plate should be limited by design features. Design provisions should be made for the top shield design to minimize the deposition of sodium aerosols in the narrow gaps for smooth functioning of the mechanisms.

13.5.2.3   Reactor Shutdown System

At least two reliable, independent, automatic, fast-acting shutdown systems should be provided for the reactor, operating as far as possible on diverse principles. At least one of the shutdown systems should be so designed that it meets all functional requirements even in the case of postulated core deformation. The failure probability of each shutdown system should be $<10^{-3}$ per demand in addition to the requirement that the combined failure frequency of the two shutdown systems is $<10^{-7}/$ reactor-year. With all absorber rods inside the core, the minimum shutdown margin (SDM) should be more than 1.0\$. The worth of all the absorber rods in the core should be sufficient to provide enough SDM for the fuel loading errors, namely, the replacement of the absorber rod of maximum worth with a fresh fuel subassembly of maximum enrichment. The minimum SDM under the withdrawal of two absorber rods of maximum worth (total) including antishadowing effects or failure of any one shutdown system along with the failure of the maximum worth rod of the working shutdown system should be more than 1\$ at the cold shutdown state for all PIEs. The maximum reactivity worth of an absorber rod together with its maximum withdrawal speed should be designed such that DSLs are not exceeded in case of its withdrawal. The reactivity worth, speed of action, and delay in actuation of each shutdown system should be such that during all operational states and DBEs of the reactor, the reactor is rendered sufficiently subcritical and maintained subcritical. Provision should be available to measure the drop time of all the shutoff rods. All equipment in the shutdown system should be designed such that its failure modes will not result in an unsafe condition. The design should be such that each shutdown system can be actuated manually from the main and backup control rooms. The design should be such that it is not possible for an operator to prevent a safe automatic action from taking place. The control logic of the absorber rods and their drive mechanisms should be designed to prevent unintended movement in the directions, which add positive reactivity.

13.5.2.4 Heat Transport Systems

The various coolant systems and associated control, protection, and auxiliary systems should be designed to have sufficient capacity with adequate margins and redundancy to remove the heat from the core and transport it to the ultimate heat sink, without exceeding the specified limits for fuel coolant and structures, under all operational states of the reactor and DBEs. The primary coolant circuit should be designed to prevent ingress of moderator or gas into it. In case of reactivity due to gas or moderator, the likelihood of which should be remote, the resulting reactivity change should be within the capabilities of the shutdown systems. There should be an intermediate heat transport circuit between the primary sodium and steam–water circuit to prevent carryover of hydrogenous materials and reaction products (hydrogen, sodium hydroxide, etc.) into the core in case of a sodium–water/steam reaction in steam generators. A pressure differential should be maintained in an IHX between the primary and secondary sodium coolants (except during repair of respective circuits), such that in case of any leakage, sodium would flow from the secondary to the more radioactive primary circuit. The reactor cover gas should be provided with monitors to detect the entry of air or oil ingress from sodium pump auxiliaries (mechanical seals and bearing) into sodium.

The heat transport components important to safety should be independent, redundant, and physically separate from each other to prevent common cause and cross-linked failures. The design for the previously mentioned components should include considerations of degradation of material properties (e.g., effect of sodium, temperature, and irradiation), transients, residual stresses, and flaw size. Provisions should be made for all sodium-containing systems to detect and, to the extent practical, identify the location of the leak at an early stage. Inadvertent draining of sodium through the auxiliary circuit like sodium purification should be prevented. Systems should be provided to monitor and maintain purity of the coolant and cover gas within allowable limits, which should be based on the consideration of corrosion, fouling, plugging of passages, radioactivity concentration, detection of sodium–water/steam reaction, etc. The heating systems and their controls should be appropriately designed with suitable redundancy to assure that the temperature distribution and rate of change in the temperature of the items are maintained within design limits, assuming a single failure. The effect of nonavailability of the heating power should be considered in an analysis for all states of the reactor even when there is no core decay heat.

At least two diverse decay heat removal systems should be provided to transfer heat from the core to ultimate heat sinks, such that specified DSLs are respected for all DBEs. The nonavailability of the decay heat removal function should be $<10^{-7}$/reactor-year. There should be one decay heat removal system with heat rejected to the ultimate heat sink as air from the reactor pool. This system of heat rejection from the reactor pool should have adequate diversity and redundancy in the design of heat exchangers and circulation of sodium and ultimate heat sink medium. Each of these circuits should be independent for the decay heat removal path. The consequences of the total postulated failure of one of the diverse subsystems should be considered and consequences found acceptable. For the ultimate heat sink as air, the design should take into account the effect of cyclonic and severe weather conditions. Provisions should be made for removing decay heat in auto mode. Provision should be made in the design for periodic testing of decay heat removal systems to ensure their performance at their designed capacities. All active components of decay heat exchanger (DHX) should be testable during reactor operation. The safety grade decay heat removal (SGDHR) system should remain functional after the postulated whole core accident.

Irrespective of the safety classification of the sodium system and components, all sodium piping and components should be designed for both operating basis earthquake (OBE) and SSE. Auxiliary cooling circuits (like reactor assembly top shield cooling, reactor vault concrete cooling) should be designed with sufficient capacity and redundancy to remove the heat from the sources to the ultimate heat sinks without exceeding the specified limits during all the operational conditions and DBEs. Sodium pumps should be seismically qualified and should be functional, during and after SSE. There should be adequate provisions to limit overspeeding of the pump. Continuous monitoring of appropriate parameters should be done to detect any breach in the reactor coolant boundary.

13.5.2.5 Sodium–Water Reactions: Prevention/Mitigation of Effects

Steam generators and associated circuits should be designed, constructed, and operated so as to minimize the probability of water/steam leakages and to limit the consequences resulting from sodium–water/steam reactions. Systems should be provided to detect at an early stage at all temperatures, rapidly and reliably, the presence of small water leakages into sodium and to initiate actions to stop the sodium/water reactions, so as to limit progressive increase in the size of the leak and the damage. Appropriate pressure-relieving systems like rupture disks should be provided to minimize the consequences of large sodium/water reactions. The possibility of any effect propagating to the primary side (e.g., through IHX and DHX) should be investigated and precluded. Spreading of corrosive substances in the secondary circuit and the water/steam system, following a leak, should be minimized. The release of sodium–water reaction products should not endanger parts of the plant having a safety function.

13.5.2.6 Core Subassembly Handling and Storage

Core loading errors during handling of core components should be prevented by design and administrative procedures. There should be a provision to assess the subcritical reactivity of the core, and any unacceptable reduction in the subcritical margin due to core loading errors should give alarms and should automatically stop the handling machine operations. Criticality in fresh and irradiated fuel assembly, storage should be prevented by physical systems or processes. k_{eff} under worst conditions should not exceed 0.9. The storage capacity of spent fuel storage bay should be such that during the life of the plant there should be enough unused capacity to permit full discharge of one reactor core. Adequate cooling arrangement should be made during handling and storage of subassemblies. Provisions should be made to monitor the coolant temperatures. Arrangements should be made in the design to control chemistry and radioactivity in the cooling medium of the spent fuel storage bay. Provision should be made in design to prevent draining of the liquid cooling medium from the spent fuel storage bay.

The handling and storage systems for irradiated fuel should be designed (1) to permit adequate heat removal under all operational states and DBEs; (2) to permit inspection of irradiated fuel; (3) to prevent the dropping of spent fuel in transit; (4) to prevent unacceptable handling stresses on the fuel elements or fuel SA; (5) to prevent the inadvertent dropping of heavy objects such as spent fuel casks, cranes, or other potentially damaging objects on the fuel assemblies; (6) to permit safe storage of suspect or damaged fuel elements or fuel SA; (7) to adequately identify individual fuel modules; and (8) to ensure that adequate operating and accounting procedures can be implemented to prevent any loss of fuel.

For reactors using a water pool system for fuel storage, the design should provide the means (1) for controlling the chemistry and activity of water in which irradiated fuel is handled or stored, (2) for monitoring and controlling the water level in the fuel storage pool and for detecting leakage, (3) for preventing emptying of the pool in the event of a pipe break in cooling/cleaning systems, and (4) for detecting leakages and tracking down leaky locations. Sufficient administrative controls should be planned and counterchecking the arrangements made, for example, computer surveillance of the entire handling and storage system. All fuel-handling machines and storage facilities should have sufficient shielding to limit radiation exposure of the operating personnel.

13.5.2.7 Plant Layout
The layout should be such that the redundant safety systems, along with their instrumentation and support features have sufficient physical separation, so that an accident (typically fire, flooding and earthquake) in one system does not jeopardize the availability of the other systems. The layout of the controlled area should be such that the personnel do not have to pass through areas of high radiation or contamination to gain access to areas of lower radiation or contamination. The layout should give consideration to segregating equipment containing high-activity materials. Process subsystems should be arranged to minimize the number of points where radioactive materials can deposit. The backup control room should be located sufficiently away from the main control room such as to ensure accessibility and habitability to at least one of them at all times. There should be a clear passage from the main control room to the backup control room. The layout should be such that the sodium and water/steam lines are separated as far as possible. Wherever not possible, adequate barriers should be put in place in between them. Consideration should be given to localization of any fire above the top shield (reactor deck) by provision of an appropriate barrier. The layout should ensure that the consequences of earthquake from the risk of sodium fire and damages to machines handling subassemblies between RCB and the fuel building should have minimum effects.

13.5.2.8 Electrical Power Systems
Electric power supply systems should comprise off-site and on-site power systems including emergency power supply systems. These systems should be designed, installed, tested, and operated to permit functioning of structures, systems, and components important to safety during normal operation, anticipated operational occurrences, and accident conditions. Design should ensure surveillance, testability, and maintainability of all standard components in both the on-site and off-site power supply systems. Electric power from the transmission network to the on-site electric distribution system should be supplied by two physically independent circuits. These should be designed and located so as to minimize the probability of their simultaneous failure during normal operation and under accident conditions. A switchyard common to both circuits is acceptable.

Design should be such that the possibility of coincident or simultaneous loss of power, from both off-site and on-site power systems, termed as station blackout (SBO), is highly unlikely to occur. It should, however, be demonstrated by analysis that batteries and other built-in design and operating provisions should ensure that the specified fuel, cladding, coolant, component design limits and containment integrity are maintained during the anticipated "period" of the SBO. A detailed analysis of power supply system availability considering power failure statistics should be done to assess the previously mentioned period. The duration of the SBO should be assumed taking into account national and international experiences.

The emergency power supply system consists of diesel generators, rectifiers/inverters, batteries, and associated switchgear. It should be ensured that the emergency power supply is able to supply the necessary power in any operational state, accident conditions, and severe accidents or DEC on the assumption of the coincidental loss of off-site power. The capacity of the batteries should be such as to ensure power supply control room panel indications, I&C related to safety systems, and emergency lighting for the period of the SBO.

13.5.2.9 In-Service Inspection
The structures, systems, and components important to safety should be designed to be able to be tested, inspected, or monitored. Any exemption should be justified on sound basis. ISI should meet the requirements of applicable codes for sodium-cooled reactors (e.g., ASME Section XI, Division 3).

Far from the ISI point of view, allied aspects such as adequate clearances for equipment and personnel access; adoption of suitable geometries; radiation protection; arrangements for removal, storage, and installation of parts, handling facilities; and provisions for repair/replacement of parts/components, decontamination, and weld configurations should be adopted appropriately. For the material surveillance program, suitable arrangements for location and removal of specimens (drawn from the materials actually used for the construction) should be made to study the effects of operating conditions such as irradiation, exposure to high temperature, and effects of fluids.

13.5.2.10 Sodium Fire Protection

Diverse sodium fire detection systems should be provided in the plant. These systems should be designed to survive the sodium fire. Provisions should be made for testability and replaceability of this system. Design should provide for draining or fast dumping of sodium from the circuits in case of leaks to limit the amount of sodium that can leak. In order to limit the sodium fire, barriers should be provided to prevent fire from spreading. Sodium systems should be protected against the consequences of DBEs of other components and systems. Insulation materials of sodium-bearing vessels and piping systems should be chosen ensuring the compatibility with the material of construction. Their interaction with sodium does not aggravate the consequences of a sodium leak. The leak can be detected by the installed devices before it increases to an unacceptable size. The effects of possible interaction between sodium and concrete and hence hydrogen generation due to this reaction should be considered in the design. Sodium-resistant concrete or steel liners should be provided on concrete as appropriate. Adequate storage of appropriate extinguishers should be provided at appropriate locations, considering the layout of sodium circuits. The release of sodium aerosols at the site boundary should be less than the prescribed limits for all design basis sodium leaks. Design analysis of design basis sodium fires should take into account mechanical, thermal, and chemical effects on items important to safety. Sodium fire detection and sodium firefighting system should be at the required location before receipt of sodium at the site during construction.

All the sodium piping within the RCB should have double-walled construction and filled with inert gas in the interspace or the piping should be in inerted cells/casings. Secondary sodium circuits and intermediate sodium circuits of decay heat removal system should have provisions for reducing the quantity of sodium coming out in air and catching fire, such as design satisfying leak before break, leak collection trays, and injection of inert gas in Na/air heat exchanger. Sodium leak detection should be provided, which can reliably detect and locate leaks at an early stage.

13.5.2.11 Instrumentation and Control

Sufficient instrumentation should be provided to monitor plant variables and systems over the respective ranges for normal operation, anticipated operational occurrences, accident conditions, and severe plant conditions in order to ensure that adequate information can be obtained on the status of the plant. Appropriate instrumentation, with adequate redundancy and diversity, should be provided for measuring the parameters related to the fission process and the integrity of the reactor core, the reactor cooling system, and the containment system and for obtaining any plant information required for the reliable and safe operation of the plant. Instrumentation should also be provided to evaluate the accident scenario including that required for the management of severe accidents. Appropriate controls should be provided to maintain safety-related variables within prescribed ranges and annunciate if they exceed their set limits for reactor trip parameters. During the shutdown state, the core should be monitored by appropriate instrumentation including thermocouples and neutron detectors. Provision should be made for special instrumentation and/or auxiliary neutron sources to ensure safety during initial loading of the core, subcritical monitoring in the fuel-handling state, and routine approach to criticality. The minimum count rate in the neutron detectors arising from core neutrons should be more than the prescribed limits in the shutdown state. When different neutron detector systems are used to monitor the core, in different ranges of neutron flux levels, there should be adequate overlap between any two systems covering adjacent ranges. Two systems—cover gas radioactivity and delayed neutrons in sodium—should be provided for failed fuel detection. Operation of the plant with failed fuel (gas failure) should not interfere with the detection of subsequent failed fuel. Fuel and blanket subassembly coolant outlet temperatures should be monitored by thermocouples to the extent feasible. If not feasible, offline flow measurement should be adopted to ensure that there is no coolant flow blockage. Two or more thermocouples should be used to measure the coolant outlet temperatures of fuel and blanket subassemblies. An adequate number of level probes should be provided to monitor the sodium levels in hot and cold pools of the main vessel. Provision should be provided to verify the subassembly top positions prior to the start of subassembly handling (rotatable plug movement). Control rod positions for different insertion levels should be measured and indicated. The in and out conditions of safety rods should be measured and indicated. Systems for detection of water leaks into sodium should be provided. Sodium leaks should be detected by diverse instrumentation systems, which can reliably detect and locate leaks at an early stage to allow initiation of corrective action (e.g., drainage of loops). Seismic sensors should be installed for data collection on the seismic event in order to assess the severity of the earthquakes. Seismic sensors should initiate alarms at specified levels of severity of an earthquake. In areas where inert gas bearing circuits (argon/nitrogen) are present, provision for monitoring the reduction of the oxygen level should be available to facilitate the safety of personnel entering these areas.

Design and layout of the instrumentation system should be such as to permit periodic testing, calibration, and preventive maintenance in order to detect and rectify faults and failures of instruments and their components. A control room should be provided from where the plant can be safely operated in all its operational states and can be brought and maintained in the safe state even under accident conditions. The control room design should provide appropriate measures to enable prevention of unauthorized access to items important to safety. The backup control room should be provided in the plant with sufficient I&C equipment, so that the reactor can be placed and maintained in a safe shutdown state, to ensure decay heat removal and monitor essential plant data, should there be nonavailability of the main control room. The backup control room should be physically and electrically separate from the control room. It should be ensured that the same event should not result in simultaneous failure of safety functions in both the control room and backup control room. Reliable air conditioning and ventilation system should be provided to the control room, backup control room, and instrumentation room housing safety-related instrumentation.

Independent verification and validation (IV&V) for computer-based systems is essential. The entire development should be subject to an appropriate quality assurance program. The level of reliability necessary should be commensurate with the safety importance of the system.

13.5.2.12 Plant Protection System

The plant protection system should be designed to initiate automatically the operation of appropriate systems including the reactor shutdown system in order to ensure that DSLs are not exceeded as a result of anticipated operational occurrences, to sense accident conditions, and to initiate the operation of systems required to mitigate the consequences of such accidents and be capable of overriding unsafe actions of the control systems. The protection system should be automatically initiated. The reactor should be safe even if manual action is not taken for the first 30 min. Credit for operator action in <30 min should be justified. The design, however, should be such that an operator can initiate protection system functions and can perform necessary actions to deal with circumstances that might prejudice the maintenance of the plant in a safe state but cannot negate correct protection system action at any time. The protection system should be designed to ensure that the effects of normal operation, anticipated operational occurrences, and DBEs on redundant channels do not result in loss of its function. Plant protection system design should provide diversity, independence, and physical separation between redundant channels to ensure that no single failure results in loss of protection function. The effects of natural phenomena and postulated accident conditions on any channel should not result in loss of protection system function. The plant protection system should be designed to fail into a safe state if conditions such as disconnection of the system, loss of energy (e.g., electric power and instrument air), or postulated adverse environments (e.g., extreme heat or cold, fire, pressure, steam, water, sodium reaction products, and radiation) are experienced. If signals are used in common by both the protection system and any control system, appropriate separation (such as by adequate decoupling) should be ensured and it should be demonstrated that all safety requirements of protection system are fulfilled.

13.5.2.13 Containment

A containment system should be provided to ensure or contribute to the achievement of the safety functions: confinement of radioactive substances in operational states and in accident conditions, protection of the reactor against external natural and human-induced events, and radiation shielding in operational states and in accident conditions. The strength of the containment structure, including access openings and penetrations and isolation valves, should be calculated with sufficient margins of safety on the basis of the potential internal overpressures, underpressures, and temperatures, dynamic effects such as missile impacts caused by the event, reaction forces anticipated to arise as a result of design basis accidents, and end-of-life properties affected by aging. The effects of other potential energy sources, including possible chemical and radiolytic reactions, should also be considered in design. The design basis of the containment should include a postulated whole core accident resulting in at least one credible scenario imposing the highest loads on the containment. The structural design of the containment should consider the thermal stresses arising from the calculated temperature transients and spatial temperature profiles within the structure during postulated accident conditions as well as during normal operation. Provision for maintaining the integrity of the containment in the event of a severe accident should be considered. The design should include measures to avoid failure of the containment during severe accident sequences that can challenge the containment integrity. Potential containment degradation caused by molten fuel should be prevented. Systems to control the release of fission products from containment to the environment following DBEs should be provided. The containment wall should provide sufficient shielding to permit postaccident site occupancy requirements especially control room and backup control room habitability. Within the containment, the use of water should be minimized. In case water is used, multiple barriers to contain the leak and monitoring provisions to detect the leak should be available. Waterlines should be physically separated from the sodium lines. Protection should also be made to prevent leaks under postulated accident conditions. A number of flanged joints should be minimized and should have provision for leakage detection, collection, monitoring, and drainage to a safe location. The use of oils should also be minimized. The containment should be maintained under negative pressure during all operational states of the reactor except during special operations like transportation of the IHX from the RCB in the reactor shutdown state.

The containment system should be designed so that the prescribed maximum leakage rate is not exceeded in design basis accidents throughout the service life of the plant. The design leakage rate, apart from satisfying the calculated radiological consequence limits for accidents, should be kept to a minimum in keeping with the as low as reasonably achievable (ALARA) principle. For postulated whole core accident, the dose rate at the boundary should be within the highest category DSL. The containment structure and equipment and components affecting the leak tightness of the containment system should be designed and constructed so that the leak rate can be tested at the design pressure after all penetrations have been installed. Adequate consideration should be given to the capability to control the progression of severe accidents and any leakage of radioactive materials from the containment in the event of a severe accident. The number of penetrations through the containment should be kept to a practical minimum. All penetrations through the containment and other components forming the containment pressure boundary should meet the same design requirements as the containment structure itself. If resilient seals (such as elastomeric seals or electrical cable penetrations) or expansion bellows are used with penetrations, they should be designed to have the capability for leak testing at the containment design pressure, independent of the determination of the leak rate of the containment as a whole, to demonstrate their continued integrity over the lifetime of the plant.

At least two adequately diverse parameters (value or logic) should be provided for the detection of the containment isolation signal. Each line that penetrates the containment as part of the reactor coolant boundary or that is connected directly to the containment atmosphere should be designed to fail to safety such that it will be automatically and reliably sealable in the event of a design basis accident in which the leak tightness of the containment is essential to preventing radioactive releases to the environment that exceed prescribed limits. These lines should be fitted with at least two adequate containment isolation valves arranged in series (normally with one outside and the other inside the containment, but other arrangements may be acceptable depending on the design), and each valve should be capable of being reliably and independently actuated. Isolation valves should be located as close to the containment as is practicable. Any exceptions should be justified. Containment isolation should be achievable on the assumption of a single failure.

13.5.2.14 Design for Radiation Protection

Suitable provision should be made in the design and layout of the plant to minimize exposure and contamination from all sources. Such provision should include adequate design of structures, systems, and components in terms of minimizing exposure during maintenance and inspection, shielding from direct and scattered radiation, ventilation and filtration for control of airborne radioactive materials, limiting the radioactive corrosion products by proper specification of materials, means of monitoring, control of access to the plant, and suitable decontamination facilities. The shielding design should be such that radiation levels in operating areas do not exceed the prescribed limits and should facilitate maintenance and inspection so as to minimize exposure of maintenance personnel. The ALARA principle should be applied. The shielding design should take into account possible streaming paths. Full account should be taken of the buildup of radiation levels in areas of personnel occupancy and generation of radioactive materials and wastes with time. The plant layout and procedures should provide for the control of access to radiation areas and areas of potential contamination and for minimizing contamination from the movement of radioactive materials and personnel within the plant. The plant layout should provide for efficient operation, inspection, maintenance, and replacement as necessary to minimize radiation exposure. A decontamination facility should be provided for both personnel and equipment. It should have an adequate margin to provide for anticipated off-normal conditions. The facility should have provision for handling any radioactive waste arising from decontamination.

13.5.2.15 Radioactive Waste Management

Adequate systems should be provided to treat radioactive liquid and gaseous effluents in order to keep the quantities and concentrations of radioactive discharges within prescribed limits. The ALARA principle should be applied. Adequate systems should be provided for the handling of radioactive wastes and for storing them safely on the site for a period of time consistent with the availability of the disposal route on the site.

13.5.2.16 Control of Releases of Radioactive Liquids to the Environment

The plant should include suitable means to control the release of radioactive liquids to the environment so as to conform to the ALARA principle and to ensure that emissions and concentrations remain within prescribed limits. For handling large components like the primary sodium pump and IHX, flasks should be designed to protect the personnel from radiation. The dose criteria outside the flask should be fixed considering the frequency of their handling. A suitable decontamination procedure should be established and followed for reduction of dose to the personnel.

13.6 EVOLVING TRENDS [13.10]

13.6.1 EVOLVING SAFETY APPROACH

The safety approach has been evolved to address the following issues:

- Fundamental orientation related to defense in depth
- Relationship among plant states, probabilistic and deterministic approaches
- Utilization of passive safety features
- Prevention of the cliff edge effect
- Containment function
- Provision against hazards
- Nonradiological and chemical risks

The previously mentioned aspects are detailed in the following paragraphs.

13.6.1.1 Fundamental Orientation Related to Defense in Depth

The future reactor systems should have well-balanced safety throughout whole range of accident conditions and highly reliable system with very low probability of accidents, with enhanced measures against severe accidents. In order to ensure safety for normal operation, anticipated operational occurrences, and design basis accidents, the feedback of "operation/accident experience" and "maintenance/repair experience" should be taken into account. The safety margins should be increased and the ISI technology should be elaborated. To ensuring safety in DECs, practical measures should be taken to prevent severe accident occurrences and/or mitigate consequences. Due considerations should be given on the potential sources causing common mode failures. Passive design measures should be introduced by utilizing the favorable safety features of the SFR system. Designs should be such that accident progress will be slow enough for systems to respond and appropriate actions needed to mitigate the consequences to proceed.

13.6.1.2 Relationship among Plant States: Probabilistic and Deterministic Approaches

Consideration should be given to the effective functioning of design measures for each defense-in-depth level, so that a specific event will not be a dominant factor. The identification/selection of design basis accidents and DECs shall be based on the combined use of (1) the deterministic approach based on fundamental characteristics of the reactor system supplemented by probabilistic analysis as needed and (2) operation experience and external event experience and licensing experience. The application of the probabilistic safety assessment from the beginning and throughout the design phases is encouraged to estimate the effectiveness of design measures.

13.6.1.3 Utilization of Passive Safety Features

Provisions of well-balanced design measures should be made by utilizing an appropriate combination of active and passive safety systems for enhancing safety against wide-ranging DBA/DEC. The capability of a passive safety system is expected even under the DEC although fine control might be limited. For DBA, it is important to well characterize the safety features of structures, systems, and components (including inherent characteristics). The reliability of proven technologies (safety systems with adequate redundancy and diversity) should be enhanced. For DEC, it is possible to ensure diversity with different operation principles without further multiplexing measures already applied for DBA. The use of passive safety and inherent safety features should allow termination of accidents or mitigation of consequences of a DEC, even in postulated failure of active safety systems.

13.6.1.4 Prevention of the Cliff Edge Effect

Severe accidents that could lead to a significant and sudden radioactive release due to a possible cliff edge effect, not reasonably manageable by design improvement, shall be practically eliminated by appropriate provisions. Severe accidents that are determined to be practically eliminated should be restricted to those that are not deemed physically impossible as determined by deterministic and probabilistic considerations. Safety demonstrations of practically eliminated situations shall be robust and based on deterministic and probabilistic analyses that address uncertainties and cover a large spectrum of events.

13.6.1.5 Containment Function

Containment should be designed so that it can withstand postulated severe accidents with core degradation. Safety provisions required to mitigate consequences of core degradation and to retain the degraded core materials should be built in. For radiological confinement, design provisions related to the confinement function should be enhanced, as far as reasonably achievable, and confinement measures must take into account a source term whatever the origin of the radioactive material in the plant (e.g., core and spent fuel storage).

13.6.1.6 Provision against Hazards

The possible combination of external and internal hazards should be considered in order to (1) improve the robustness of the power plant safety, (2) confirm that consequences of degraded plant situations induced by extreme hazards are acceptable, and (3) define equipments that need to be strengthened to resist extreme natural hazards beyond the reference used for plant design. As hazards are the potential common cause failures that can impact several structures, systems, and components, each safety function should rely on the appropriate diversification and physical separation to enhance redundancy.

13.6.1.7 Nonradiological and Chemical Risks

Nonradiological and chemical risks, introduced by the system features and processes, have to be reduced to as low as reasonably achievable, with the objectives to limit the impact on the outside of the plant area and to protect the health of workers and the public. Nonradiological and chemical risks must be considered in order to minimize the risk of nuclear power plant damage and to prevent simultaneous radioactive and toxic chemical releases in the environment, as cumulative consequences of an accident.

13.6.2 IDENTIFICATION OF DECs

For the future SFR, it is required to provide built-in measures to achieve practical elimination of any significant release of radioactive materials into the environment under DECs. DECs are identified based on SFR fundamental characteristics, engineering judgment, and probabilistic safety assessments. Safety studies provide information to identify DECs from the viewpoints of significance of the consequence of accidents, weak points in the current design, and degree of uncertainty. Measures for DECs are those for "prevention of core damage" and for "mitigation to ensure containment function." To determine the measures, the time margin to core damage for each DECs needs to be considered.

SFR DEC events can be grouped into two categories based on the characteristics of an SFR and probabilistic safety assessment studies: (1) failure to shut down the reactor following an off-normal initiating event (e.g., loss of flow with failure to scram, overpower transient with failure to scram, and loss of main heat removal with failure to scram) and (2) inability to remove heat from the core following an initiating event (e.g., loss of coolant flow caused by disruption in flow paths of the decay heat removal circuit, loss of the primary coolant level when the core becomes uncovered, and long-term loss of the heat sink with scram).

13.6.3 IDENTIFICATION OF DESIGN MEASURES

For the failure of reactor shutdown events, the design needs to prevent such events from damaging the core and mitigate the consequences of core damage to minimize the load on the containment function. The time margin leading to core damage is relatively small. Significant mechanical energy release, derived from prompt criticality in a core disruptive accident, has the potential to breach the reactor vessel (RV) and to significantly affect the integrity of the containment. In order to prevent core damage, the design may make use of passive or inherent reactor shutdown capabilities. Restricting generated energy and retaining/cooling of the damaged core will reduce the potential load on the containment function. The total reactivity feedback and molten fuel discharge are key issues.

For the loss of heat removal events, the design should provide a means to prevent core damage or loss of containment function by maintaining the sodium coolant level for core cooling and ensuring decay heat removal even under conditions with or without core damage. Compared to loss of shutdown events, there is generally more time prior to core damage so that a variety of diverse measures might be provided depending on the circumstances of the event. The degree of core damage may vary depending on the time margin to fuel failure after losing the decay heat removal function. Similar design approaches that address the loss of heat removal events may also be applied for a spent fuel storage pool using sodium, which might be located outside of the containment.

The coolant level in the RV should be maintained above the reactor core to achieve core cooling. This can be achieved by (1) appropriate layout of the primary piping, (2) a guard vessel (GV) in case the leak is in the RV, or (3) another boundary and/or a refill of reactor coolant, and (4) the RV and GV should be designed, manufactured, installed, maintained, and inspected to have the highest level of reliability. Thus, any leakage from the RV/GV is highly unlikely to occur. A leakage from the RV/GV will remain as a DEC, if the consequent leak (first from the RV and then from the GV) and/or common cause leak (common cause failure of the RV and the GV) cannot be practically eliminated.

To maintain the sodium coolant level, the conditions to clarify "practical elimination" are to be respected, that is, prevention of dependent failure and common cause failure. With respect to the "prevention of dependent failure" condition, the GV should withstand thermal loads due to a sodium leak from the RV, mechanical loads from earthquakes while retaining leaked sodium for a long time, and any interference with a failed RV (even considering thermal expansion, vibration by earthquake, and so on). With respect to the "prevention of common cause failure" condition, separate the support structures of the RV and GV to the extent practicable, or prevent the failure of common parts of the support structures. Prevent common cause defects in manufacturing and sufficient margins against extremely severe earthquakes should be ensured.

The heat sink is an essential part of the core cooling functions regardless whether the reactor core is intact or damaged. To ensure decay heat removal (heat sink), the heat sink should be maintained by design and recovering/adding the heat sink in accident management. An alternative means of cooling can prevent damages to the reactor core and loss of the containment function. The plant conditions depend on the time margin before core damage. If the time margin is sufficiently long to implement an alternative cooling system, or heat can be removed using natural phenomena (e.g., radiation and thermal conduction), core damage conditions will be practically eliminated. The longer the time margin is, the more varied the design choices of alternative cooling systems can be provided. A safety implication of longer time margin is illustrated in Figure 13.5.

To ensure the integrity of SFR containment, in addition to conventional loading conditions, events such as sodium combustion, sodium concrete reaction, debris–concrete interaction, and combustion of accumulated hydrogen, which have the potential to load or otherwise threaten the integrity of the containment, must be prevented or mitigated.

13.6.4 Practical Elimination of Accident Situations

The "practical elimination" of accident situations is a matter of judgment and each type of sequence must be assessed separately, taking into account the uncertainties due to the limited knowledge of some physical phenomena and the cost of implementation. This judgment cannot be solely based

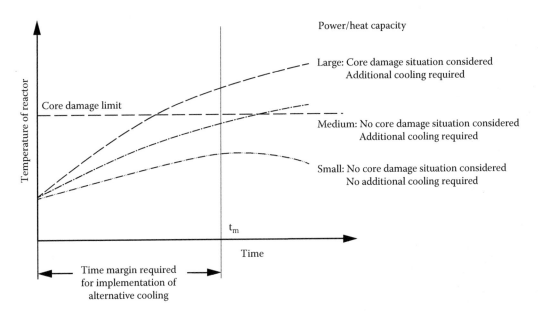

FIGURE 13.5 Ensuring decay heat removal: heat sink. (From Kubo, S., Safety design approach based on Gen-IV SFR, *GIF-IAEA/INPRO SDC Workshop*, IAEA, Vienna, Austria, February 26–27, 2013.)

on the probabilistic exclusion criteria but should be combined with careful deterministic assessment of all potential mechanisms leading to potentially large radioactive releases. In the case of SFRs, the first design objective is to make such situations physically impossible. Mitigation of the consequences of some accident situation should be excluded by design where feasible, because the implementation of additional mitigation devices, or the R&D necessary for demonstrating their effectiveness, may be prohibitively expensive or difficult to prove effective under DECs. However, for situations that are physically possible, the design process has to consider, within economic and physical constraints, all situations independent of their probability. In compliance with defense in depth, "practical elimination" is acceptable only for a limited number of situations. The "practical elimination" of some accident situation requires implementation of independent reliable design features and/or operational procedures and a robust demonstration of their efficiency, for example, combination of active and passive systems, inherent characteristics, and operating procedures to verify the efficiency of protection devices (e.g., needs ISI). For example, the failure of decay heat removal systems could be "practically eliminated" by (1) implementation of robust, redundant, independent, and diverse systems, (2) minimization of their failures due to common causes or hazards, (3) a large capability to operate in natural convection, (4) a long grace time to operate on systems to allow for their recovery after failure, and (5) demonstration complemented by probabilistic assessments.

13.6.5 Lessons Learned from Fukushima Dai-Ichi NPP Accident

Robustness in power supplies, cooling functions, heat transportation systems including final heat sink, instrumentation to identify the status of the reactor core and containment vessel, and independency and diversity of safety systems are the key requirements learned from accidents. Apart from these, enhancing passive safety functions and protection measures with adequate margins are also to be highlighted. It has become essential to evaluate safety margins against severe plant conditions and extreme external hazards. Specific to the SFR system, attention should be given to flooding in buildings with sodium equipment.

REFERENCES

13.1 Rouault, J. et al. (2010). Sodium fast reactor design: Fuels, neutronics, thermal-hydraulics, structural mechanics and safety. *Handbook of Nuclear Engineering*, Springer, New York, USA, Vol. 4, pp. 2321–2710, Chapter 21.

13.2 Vossebrecker, V.H. (1999). Special safety related thermal and neutron physics characteristics of sodium cooled fast breeder reactors. Warme Band 86, Heft 1.

13.3 Guidez, J., Martin, L., Chetal, S.C., Chellapandi, P., Raj, B. (2008). Lessons learned from sodium cooled fast reactor operation and their ramification for future reactors with respect to enhanced safety and reliability. *Nucl. Technol.*, 164, 207–220, November 2008.

13.4 McKeehan, E.R. (1970). Design and testing of the SEFOR fast reactivity excursion device (FRED). General Electric Co., Sunnyvale, CA. Breeder Reactor Development Operation, US Atomic Energy Commission (AEC), US. doi: 10.2172/4050547.

13.5 IAEA. (2000). Safety of nuclear power plants: Design requirements. IAEA Safety Standards Series No. NS-R-1, IAEA, Vienna.

13.6 IAEA. (2009). Safety of nuclear power plants: Design. Revision of the IAEA Safety Standards Series No. NS-R-1, IAEA, Vienna.

13.7 Nakai, R. (2013). Safety design criteria (SDC) for Gen-IV sodium-cooled fast reactor. *GIF-IAEA/INPRO SDC Workshop*, IAEA, Vienna, Austria, February 26–27, 2013.

13.8 Office for Official Publication of the European Communities. (1990). LMFBR safety criteria and guidelines for consideration in the design of future plants. Report of Commission of the European Communities Office for Official Publications of the European Communities, EUR 12669 EN.

13.9 Mohanakrishnan, P. (2010). Safety criteria and guidelines for design of fast breeder reactors. IGCAR internal report, India, International report, PFBR/0123/, March 2010.

13.10 Kubo, S. (2013). Safety design approach based on Gen-IV SFR. *GIF-IAEA/INPRO SDC Workshop*, IAEA, Vienna, Austria, February 2013.

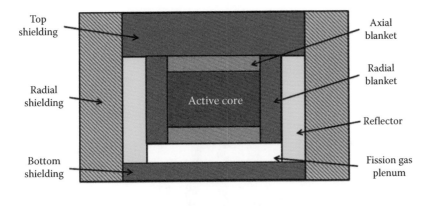

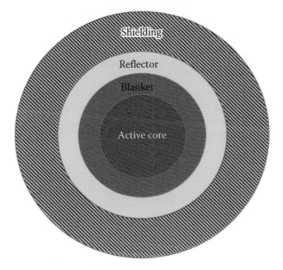

FIGURE 2.1 FSR core configuration.

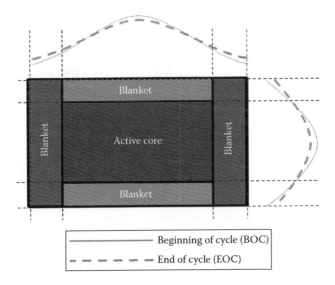

FIGURE 2.2 Power distribution in the core including blankets.

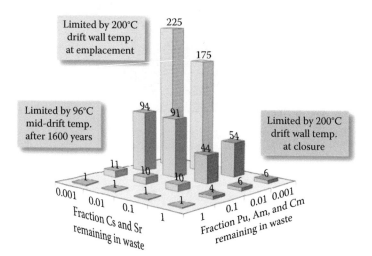

FIGURE 4.6 Storage space required for the nuclear waste as a function of MA. (From Raj, B. et al., Assessment of compatibility of a system with fast reactors with sustainability requirements and paths to its development, IAEA-CN-176-05-11, FR09, Kyoto, Japan, 2009.)

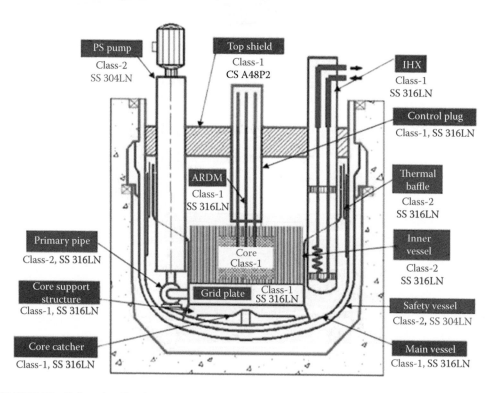

FIGURE 7.23 Safety classifications and materials of SFR reactor assembly components.

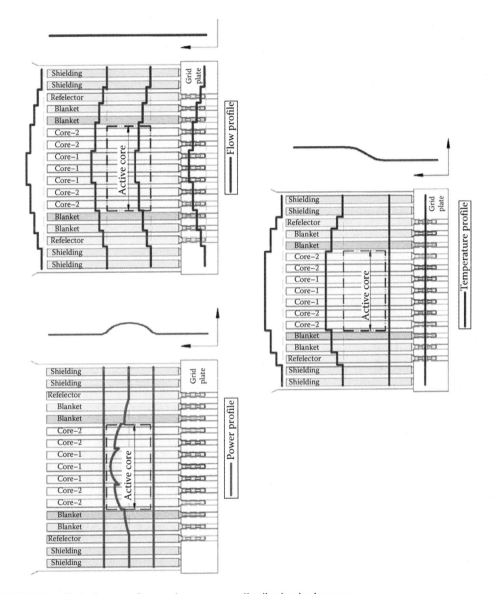

FIGURE 8.9 Typical power, flow, and temperature distribution in the core.

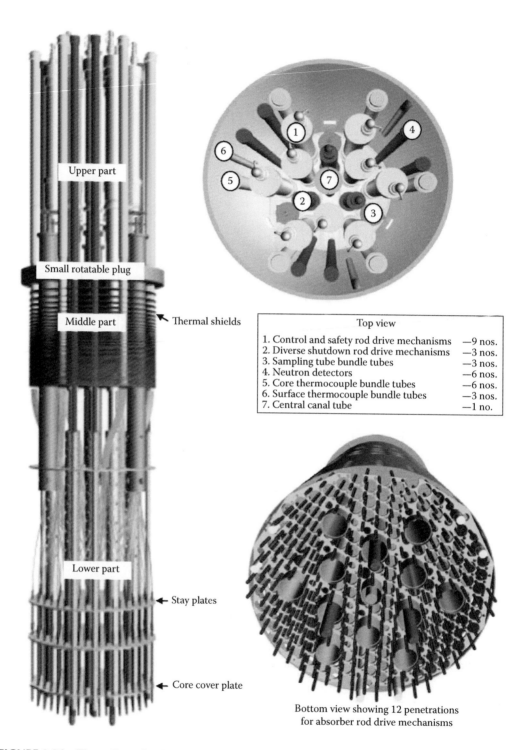

Upper part

Small rotatable plug

Middle part ◄ Thermal shields

Lower part

◄ Stay plates

◄ Core cover plate

Top view	
1. Control and safety rod drive mechanisms	—9 nos.
2. Diverse shutdown rod drive mechanisms	—3 nos.
3. Sampling tube bundle tubes	—3 nos.
4. Neutron detectors	—6 nos.
5. Core thermocouple bundle tubes	—6 nos.
6. Surface thermocouple bundle tubes	—3 nos.
7. Central canal tube	—1 no.

Bottom view showing 12 penetrations
for absorber rod drive mechanisms

FIGURE 8.24 Three-dimensional views of a typical control plug.

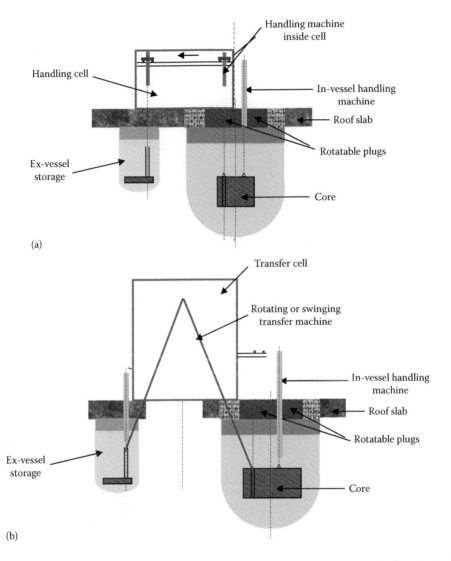

FIGURE 8.69 Cell transfer concept: (a) using fixed cell and (b) using rotating or swinging transfer lock inside transfer cell.

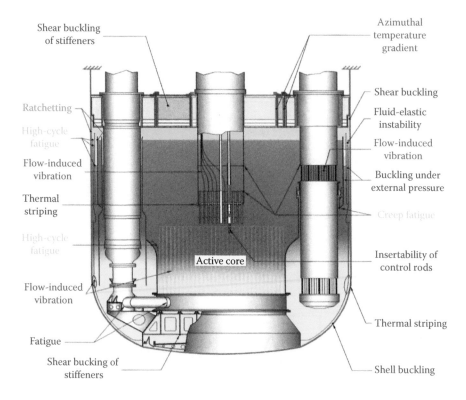

FIGURE 9.1 Failure modes considered in design.

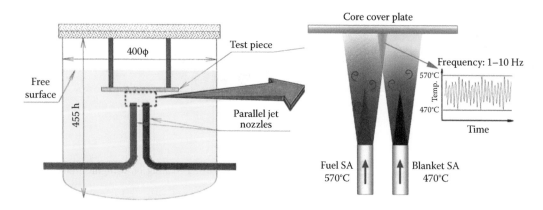

FIGURE 9.2 Thermal striping phenomenon near core cover plate.

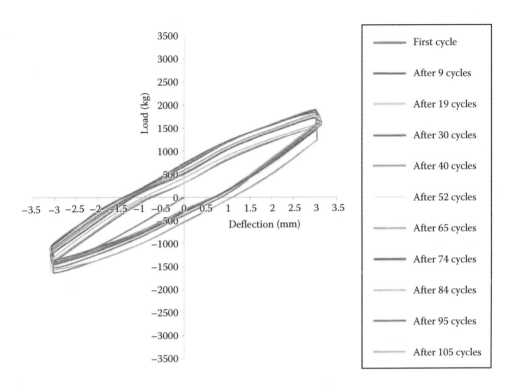

FIGURE 10.23 Load versus deflection (with 1 h hold time) plot of the plate at 873 K.

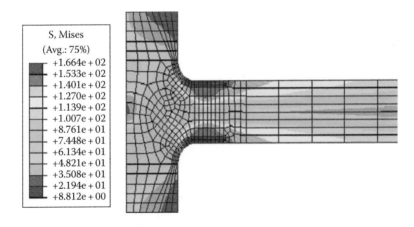

FIGURE 10.25 Stress distribution after creep relaxation (primary + secondary stress).

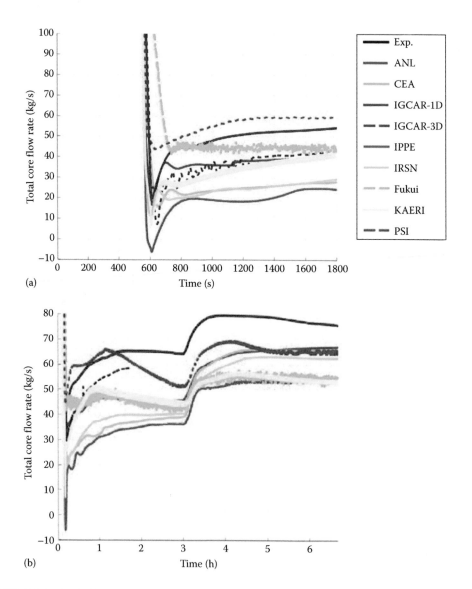

FIGURE 10.71 Prediction of natural convection evolution in the core by different participants: (a) represents the zoomed view at the initial time period of figure and (b) for more understanding.

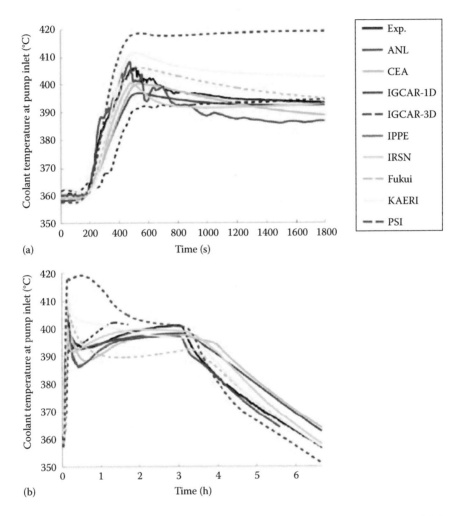

FIGURE 10.72 Evolution of core inlet temperature: (a) represents the zoomed view at the initial time period of figure and (b) for more understanding.

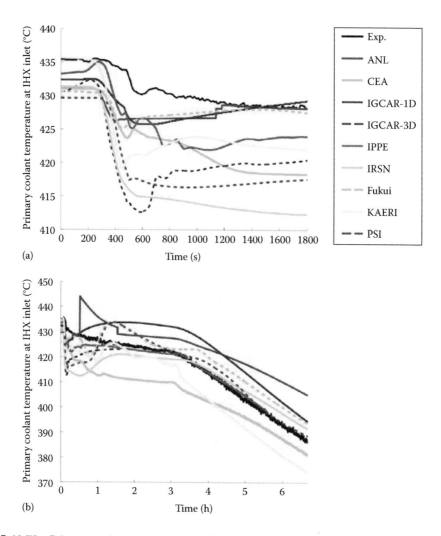

FIGURE 10.73 Primary coolant temperature predicted at heat exchanger inlet: (a) represents the zoomed view at the initial time period of figure and (b) for more understanding.

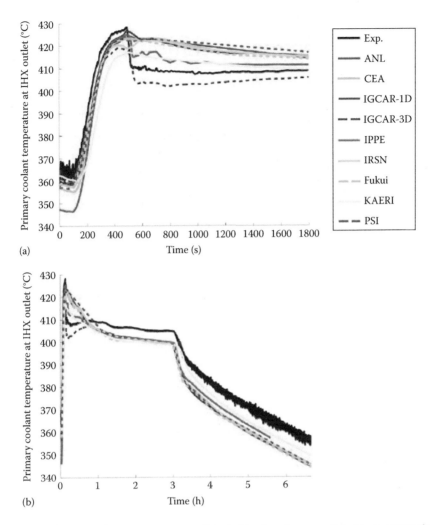

FIGURE 10.74 Primary coolant temperature predicted at heat exchanger outlet: (a) represents the zoomed view at the initial time period of figure and (b) for more understanding.

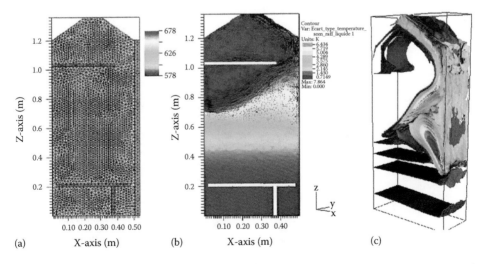

FIGURE 10.79 TRIO_U calculation of CORMORAN sodium experiment: (a) mesh, (b) velocity and temperature, and (c) temperature fluctuation.

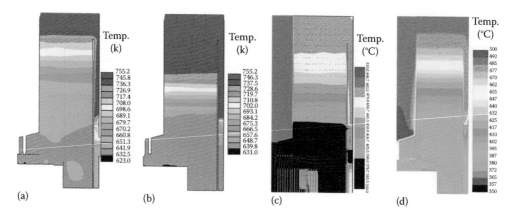

FIGURE 10.81 Transient temperature field in MONJU hot pool at 10 min: (a) India (sharp edge), (b) India (round edge), (c) Russia (round edge), and (d) China (sharp edge).

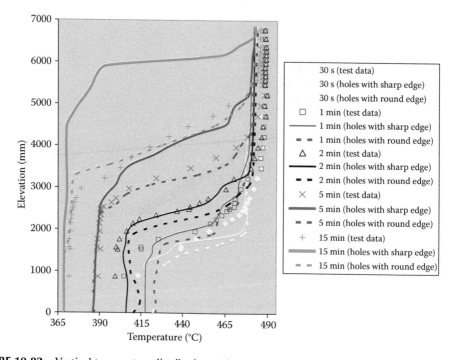

FIGURE 10.82 Vertical temperature distribution at thermocouple rack position.

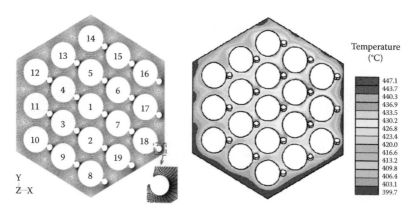

FIGURE 11.13 CFD mesh for a 19-pin bundle (left) and sodium temperature field predicted by CFD calculations (right).

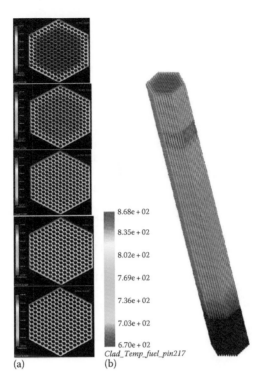

8.68e + 02

8.35e + 02

8.02e + 02

7.69e + 02

7.36e + 02

7.03e + 02

6.70e + 02

Clad_Temp_fuel_pin217

(a)   (b)

FIGURE 11.14 (a) Sodium temperature distribution in various cross sections of the SA and (b) clad temperature in the entire bundle.

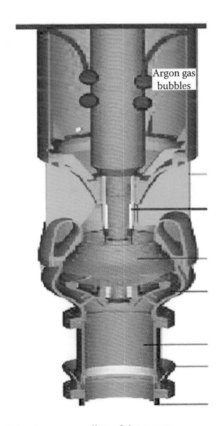

Argon gas bubbles

FIGURE 15.8 Gas entrainment due to overspeeding of the pump.

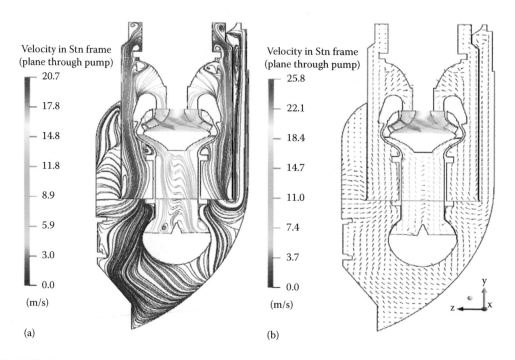

FIGURE 15.10 Flow path and sodium velocity in the cold pool. (a) Contour and (b) vector.

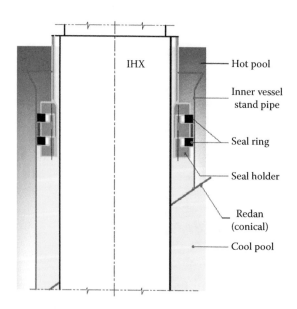

FIGURE 15.12 Mechanical seal for intermediate heat exchanger (IHX).

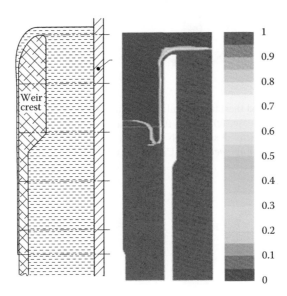

FIGURE 15.13 Weir profiling.

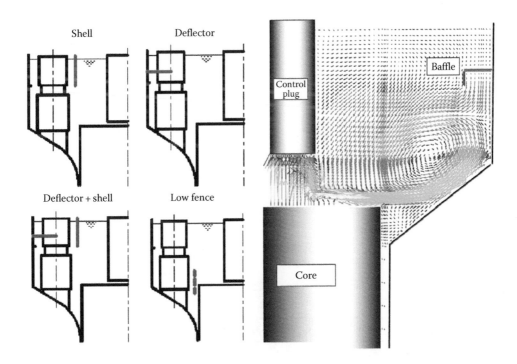

FIGURE 15.14 Baffle plates for mitigation of gas entrainment in the hot pool.

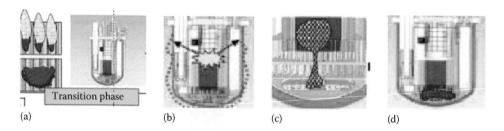

FIGURE 15.17 Mechanical and thermal hydraulic consequences of CDA. (a) Melting and vaporization of core, (b) mechanical energy release, (c) molten fuel relocation, and (d) postaccident heat removal. (From Rouault, J. et al., Sodium fast reactor design: Fuels, neutronics, thermal-hydraulics, structural mechanics and safety, in *Handbook of Nuclear Engineering*, D. Cacuci (ed.), Vol. 4, Chapter 21, Springer, New York, 2010, pp. 2321–2710.)

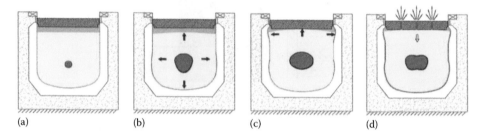

FIGURE 15.18 Mechanical consequences of CDA. (a) Initial state: 0 ms, (b) vessel pulldown: 0–50 ms, (c) slug impact: 100–150 ms, and (d) final state: 150–900 ms.

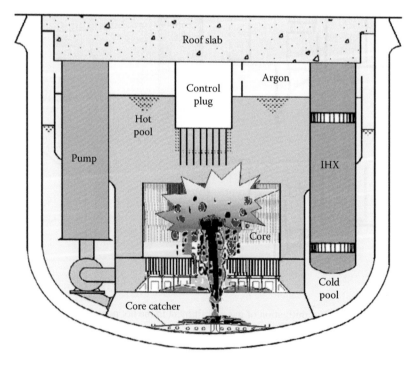

FIGURE 15.19 Molten fuel settlement in the core catcher: schematic.

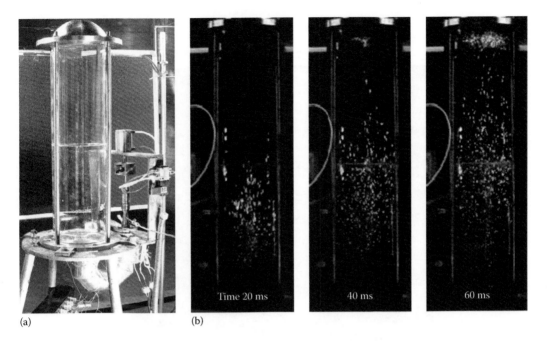

FIGURE 16.5 Sodium spray fire scenario: (a) facility and (b) some experimental images.

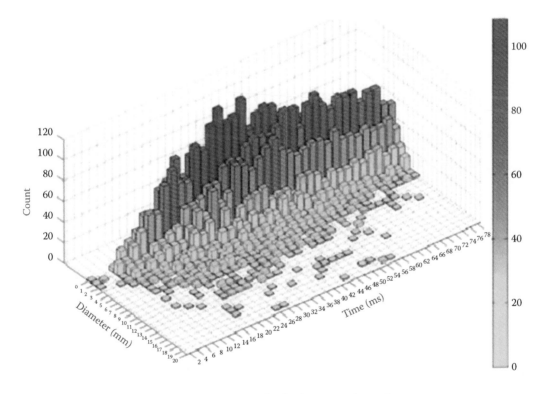

FIGURE 16.6 Sodium spray fire: flame diameter distribution—experimental.

FIGURE 16.7 Droplet burning scenario.

(a)   (b)

FIGURE 16.22 Depiction of sodium–concrete interaction scenario with limestone concrete (a) Test setup and (b) sodium concrete interaction.

FIGURE 16.23 Typical test on flowing sodium–concrete interaction: (a) specimen, (b) fire scenario, and (c) after test.

FIGURE 20.11 Demonstration of integrity of safety grade decay heat removal pipe under simulated sodium fire. (From Chellapandi, P., Overview of Indian FBR programme, *International Workshop on Prevention and Mitigation of Severe Accidents in SFR*, Tsuruga, Japan, June 11–13, 2012.)

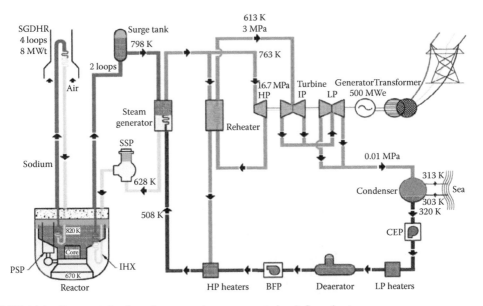

FIGURE 21.1 Prototype fast breeder reactor heat transport circuit flow sheet.

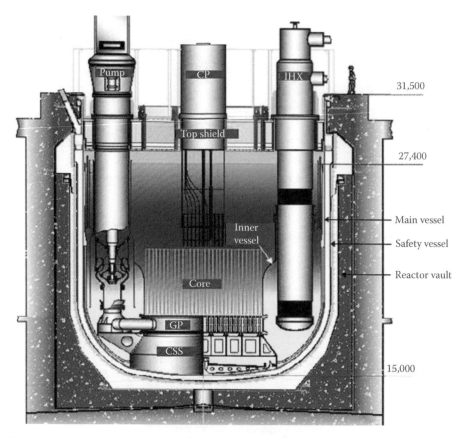

FIGURE 21.2 Schematic of prototype fast breeder reactor assembly.

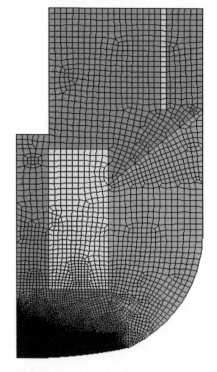

FIGURE 21.33 Computational mesh.

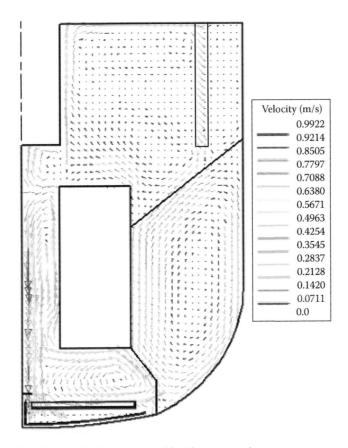

Velocity (m/s)
0.9922
0.9214
0.8505
0.7797
0.7088
0.6380
0.5671
0.4963
0.4254
0.3545
0.2837
0.2128
0.1420
0.0711
0.0

FIGURE 21.34 Sodium flow paths during postaccident heat removal.

FIGURE 22.21 Construction stages of RCB in PFBR.

FIGURE 22.22 Construction stages of reactor vault in PFBR.

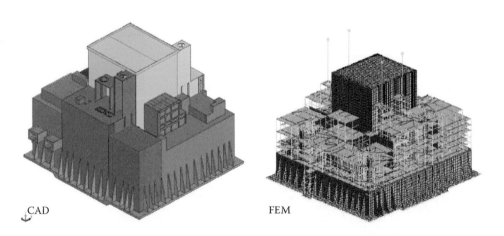

CAD

FEM

FIGURE 22.23 Computer applications for design and development of nuclear island in PFBR.

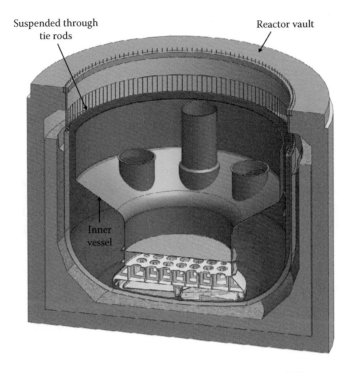

FIGURE 23.1 Reactor assembly components erected on the reactor vault in SFRs.

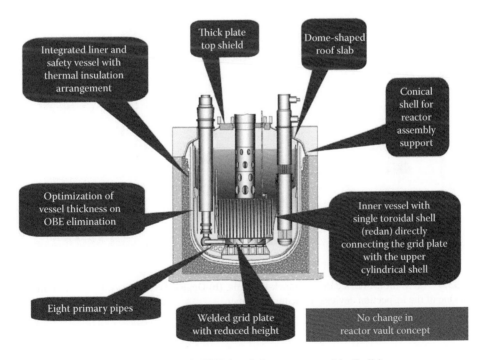

FIGURE 28.2 Improvements introduced in FBR-1 and -2 reactor assembly (India).

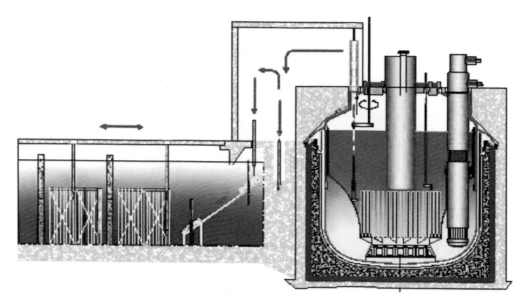

FIGURE 28.3 Fuel handling scheme conceived in FBR-1 and -2 (India).

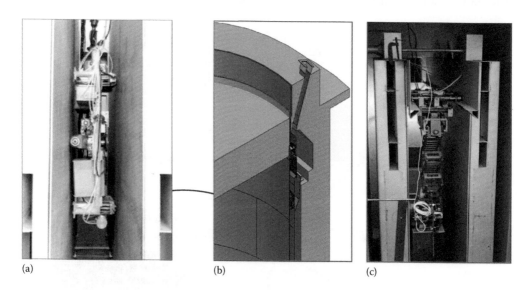

(a)   (b)   (c)

FIGURE 33.5 (a) Main ISI device during mock-up trials. (b) Depiction of the main ISI device in the interspace. (c) One of the inspection devices in the test chamber.

14 Event Analysis

14.1 INTRODUCTION

The design of various systems and equipment of a plant should ensure their robust performance under steady-state as well as transient conditions. The various plant states as defined in International Atomic Energy Agency (IAEA) [14.1] are presented in Table 14.1. The steady-state conditions are selected focusing mainly on the optimum performance of the plant. The transient conditions are identified to ensure plant safety as well as component safety. Plant dynamic analyses are performed to demonstrate safety. Important aspects that would emerge from the analysis are (1) method of prevention and mitigation, (2) response time, (3) reliability, and (4) functional performance. Various operating parameters and procedures for the plant are also worked out based on dynamic analysis. All these targets can be achieved only through dynamic simulation of the entire plant. The degree interdependence of various systems and components in achieving the ultimate design objectives can also be assessed only through such simulations.

After the successful design and construction of a power plant, the impact of various operational as well as event conditions on the plant needs to be investigated. This investigation should include the study of consequences of transients in an individual event basis as well as in a collective basis considering all occurrences during the lifetime of the plant. This is important because an event, when it occurs in an isolated manner, may not be damaging for the plant. But many such occurrences happening over a period of time will damage the plant in a cumulative manner. Plant licensing requires demonstration of safe management of impacts of routine operational events as well as unlikely and extremely unlikely events. In this regard, dynamic simulation is essential to satisfy the licensing requirements of the plant as it can demonstrate safety of the plant under various anticipated events.

An important aspect essential for complete demonstration of technology is the fault-free operation of the plant, which requires trained and qualified manpower to be available for plant operation. If we examine the operating history of power plants built in the world, several disturbances have been observed during the initial period of operation after its commissioning, with the lack of trained manpower being the major reason. Thus, dynamic simulation of the whole plant is an essential component of design evolution, operating specification, as well as safety demonstration of the plant. With the development of computer technology, this gap can be filled through training power plant operators using simulators. A large number of these simulators are of the type called full scope [14.2]. Full-scope simulators incorporate a detailed process as well as instrumentation and control (I&C) modeling of all the important systems of the plant with which the operator interacts in the actual control room. The simulator includes replica operating consoles of the power plant control room. In these simulators, the responses of the simulated unit are identical in time and indications received in the simulated control room are similar to actual conditions. Experienced operators can also be effectively retrained on these simulators because a variety of malfunctions and accident conditions can be simulated that the operator may not have experienced in the real plant. So, the operators can be effectively trained to tackle all possible scenarios of the real plant. The dynamic responses of the plant essential for developing the simulator software are derived from plant dynamic studies. The full-scope training simulator is detailed in Chapter 34.

From the previously mentioned aspects, it is obvious that dynamic simulation of the plant is essential to demonstrate/achieve safe operation of the plant. This chapter deals with the plant dynamic analysis of various design basis events. Accordingly, this chapter covers the definition

TABLE 14.1

Plant States as per IAEA

Operational States		Accident Conditions	
Normal operation	Anticipated operational occurrences	Within design basis accidents	Beyond design basis accidents
		Design basis accidents[a]	Severe accidents[b]
			Accident management

Source: IAEA, Safety of nuclear power plants: Design, IAEA Safety Standards Series No. NS-R-1, IAEA, Vienna, Austria, 2000.

[a] Accident conditions that are not design basis accidents as explicitly considered but which are encompassed by them.

[b] Beyond design basis accidents without significant core degradation.

and basis of categorization of events, typical examples of various categories of events, and various objectives of the plant dynamic analysis. Further, through a case study on a 500 MWe prototype fast breeder reactor (PFBR), some important results and the safety and practical implications for reactor operation are highlighted. However, readers should understand that the philosophy, strategy, and definition followed for event analysis are more or less the same for all kinds of reactors. Moreover, any safety guides, IAEA, for example, attempt to harmonize them to all kind of reactors with proper interpretation of the terminologies.

14.2 CATEGORIZATION OF EVENTS: BASIS, DEFINITION, AND EXPLANATION

The demonstration of the adequacy of the design with the safety objectives is made through the analysis of three kinds of conditions [14.3]:

1. *Design basis conditions*: The safety design of the plant mainly results from the analysis of these conditions. It must be shown that the consequences of accidents occurring in previous situations are matching the objectives targeted in terms of radiological releases and radiation protection. Moreover, the estimated frequencies of incidents and accidents determine the acceptable consequences for each situation postulated. And it must be checked that the risk of whole core degradation initiated by the initiating events (IEs) is very low.

2. *Design extension conditions (DECs)*: Complex sequences, limiting events, and severe accident are evaluated despite their low occurrence frequency. The consequences of these accidents are analyzed, and their consequences in the environment have to be demonstrated to be lower than the limiting release targets.

3. *Residual risk (RR) situations*: The consequences of these situations are not analyzed. If these situations are not demonstrated to be physically impossible, the prevention measures regarding their occurrence have to be demonstrated to be sufficient to "practically eliminate" them.

Figure 14.1 describes the approach that is proposed for the evolving designs in order to determine and analyze its relevant operating conditions.

14.2.1 DESIGN BASIS CONDITIONS

A design basis condition is a plant condition resulting from the combination of a normal operating condition (category 1) and of an IE belonging to the design basis area, that is, classified from category 2 to category 4 (according to its probability). IEs may arise due to component failure, operator errors,

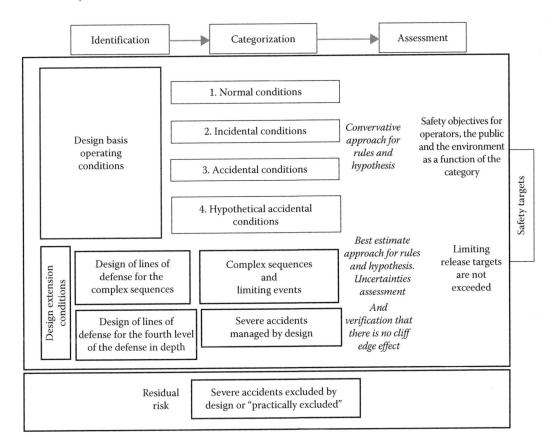

FIGURE 14.1 General approach for the safety-related design and assessment.

and internal or external hazards. Their consequences affect plant behavior. The design basis conditions are grouped in four categories on the basis of the expected occurrence frequency of the corresponding IEs. The definition of the categories made for European fast reactors is as follows:

- Normal operating conditions (category 1) are plant conditions planned and required. They include special conditions such as tests during commissioning and start-up, partial loading, shutdown states, handling states, and partial unavailability for inspection, test, maintenance, and repair. The decommissioning conditions are not included in the safety analysis of the operating plant; they will be specifically analyzed in good time. Nevertheless, considerations concerning the decommissioning have to be made. The goal of the safety analysis of normal operating conditions is to verify that their consequences on the staff and the public are as low as reasonably achievable (ALARA) and in any case lower than the corresponding release criteria.
- Category 2 conditions are operating conditions not planned but expected to occur more times during the life of the plant (mean occurrence frequency estimated greater than 10^{-2} per year). The plant shall be able to return to power in a short term after fault rectification. The goal of the safety analysis of category 2 operating conditions is to verify that their consequences on the staff and the public are ALARA and in any case lower than the corresponding release criteria.
- Category 3 conditions are operating conditions not expected to occur during the life of the plant (mean occurrence frequency between 10^{-4} per year and 10^{-2} per year) but after which plant restarting, by possible repair, is required for investment cost guarantee. The goal of

the safety analysis of category 3 operating conditions is to verify that their consequences on the public are lower than the corresponding release targets.

- Category 4 conditions are operating conditions after which plant restart is not required because there is no expectation of occurrence during the plant life. The consequences of an operating condition must not exceed category 4 limits with a mean value of their frequency higher than 10^{-7} per year. The goal of the safety analysis of category 4 operating conditions is to verify that their consequences on the public are lower than the corresponding release targets.

For the normal operating conditions (category 1), the definition and the studies of these conditions are aimed to specify the variation range of the main physical parameters featuring reactor operation. These parameters deal mainly with the protection of the physical barriers and also permit to check the integrity of the barriers (measurement of radio-contamination, tightness of barriers, etc.). These conditions include the different states of the reactors and the definition of the transients permitting to switch from one state to the other. The definition of these conditions is a starting point for the elaboration of the operating procedures (instrumentation and control system and human actions) and for the sizing of the threshold, triggering the limitation actions permitting to maintain the reactor in normal operation. Moreover, the transient studies in normal operating conditions permit to assess the performance of the reactor and the sizing of components regarding cyclic loadings. Finally, the systematic investigation of normal operating conditions permits to identify the enveloping situations to retain as an initial state to combine with an IE for the building of the operating situations. While considering each IE, the most unfavorable state must be retained.

For the incidental and accidental operating conditions (categories 2–4), the study of those conditions or, more precisely, of their representative enveloping situations is aimed to size the systems permitting their control and to avoid that they lead to nonacceptable consequences toward the facility and the surrounding. The acceptability criteria of the consequences deal with general safety objectives and, more practically, they deal with the decoupling criteria. Actually, these criteria permit to uncouple and perform separately the thermal hydraulic and neutronic transient calculations from the release calculations. The decoupling criteria will be elaborated for each category of operating situations. They deal generally with maximal admissible loadings (thermal, mechanical, and power density) of the fuel and of the two first safety barriers. These criteria are defined to bring out a margin up to the physical resistance of the fuel and of the barriers or at least (for the highest category) to limited damages.

Once the various situations have been defined, the transient calculations of the sequences corresponding to the operating situations will be performed considering conservative assumptions that are supposed to lead to appropriate design margins. The deterministic approach relies on the pertinent selection of the conservative assumptions in order to cover the various uncertainties and the possible lack of exhaustivity of the analysis. Prior to applying the penalizing value on the key physical parameters in the calculation, such parameters are supposed to be identified properly. Moreover, an aggravating failure will be taken into account in the studies of the operating situations. Finally, the end of the incidental and accidental sequences investigated must correspond to a safe final state, which is either a controlled state or the safe shutdown state. A few examples of design basis events are provided in Table 14.2.

14.2.2 DESIGN EXTENSION CONDITIONS

DECs are not defined on the basis of their occurrence frequency, but they are postulated to be bounding cases resulting from risks specific to the design or the process. Two kinds of DECs are considered: the situations for which the consequences have to be demonstrated to be limited and the severe accidents. The goal of the safety analysis of DEC is to verify that their consequences

TABLE 14.2
Typical Design Basis Events and Associated Categories

S. No.	Category	Events
1	2	Spurious reactor trip
2	2	Fuel pin failure
3	2	Inadvertent withdrawal of an absorber rod
4	2	Power failure
5	2	Circulating pump trip
6	2	Gas entrainment in the core
7	2	Minor sodium leaks
8	2	Errors in loading of sub assembly (SA)
9	2	Conventional fire
10	2	Small sodium–water reaction
11	3	Seizure of circulating pump
12	3	Operating base earthquake
13	3	Blockage of fuel subassembly
14	3	Large steam generator leak
15	3	Station blackout for short duration
16	3	Failure of core component handling machine
17	3	Large leak of cover gas
18	4	Leakage or rupture of primary pump to grid plate connection pipe
19	4	Primary sodium vessel leak
20	4	Dropping of loads
21	4	Station blackout for long duration
22	4	Safe shutdown earthquake

on the public are lower than the limiting release targets. In the safety approach developed in the European Utility Requirements (EUR), complex sequences are DECs that are not covered by the safety analysis of category 2, 3, and 4 operating conditions, but the occurrence frequency of which is not demonstrated to be sufficiently low (i.e., well below the mean value of 10^{-7}/sequence/year/plant). In the EFR safety approach, the complex sequences are complemented by limiting events defined for licensing purposes [14.3]. They are bounding cases resulting from risks specific to the design or the process. The consequences of complex sequences and limiting events are investigated, which can lead to the enhancement of the design in order to show that core damage is prevented and that the limiting release targets are not exceeded. Severe accidents are considered in order to verify that there is no "cliff edge" effect on the consequences even for very hypothetical conditions. The goal of the analysis of severe accidents is to prove the efficiency of the containment measures for limiting the consequences of core damage accidents. The radiological consequences shall be lower than the limiting release targets.

For the DECs, it should be noted that these conditions include the complex sequences, the limiting events, and the managed severe accidents. The study of these conditions must show that the release target of the fourth category of the design basis situations is reached. These situations are built without considering the rules applied for the definition of the design basis situations. The study of such situations relies on realistic hypothesis:

- The initial state of the reactor is the nominal state.
- All the systems are available except those postulated unavailable due to the accident considered.

- The instrumentation and control system operates correctly.
- The physical calculations are performed with realistic hypothesis (best-estimate calculations plus uncertainties assessment).
- No additional aggravating failure is considered.

However, the uncertainties on physical parameters able to induce "cliff edge" effects will be assessed with particular attention. The final state of the accident must correspond to a safe state ensuring durably the core subcriticality, the decay heat removal, and the radiological releases fulfilling the acceptance criteria.

14.2.3 Residual Risk Situations

These are accident conditions for which prevention regarding their occurrence is such that the analysis of their consequences is not required by the safety demonstration. On the other hand, the adequacy of the prevention of these accident conditions has to be demonstrated. Such a demonstration may be performed using probabilistic assessment. In this case, the goal is to show that accident conditions for which the consequences may exceed the limiting release targets have a mean frequency well below the threshold, as, for example, 10^{-7}/event/year/plant. For RR situations, the practical elimination of such situations must be shown according to a case by case analysis on the basis of the physical impossibility to obtain the situations considered, taking into account the design and safety provisions including ultimate systems, or by a combination of deterministic and probabilistic arguments. The adequacy of the redundancy level of the systems fulfilling safety functions and more generally of the robustness of the safety design will be analyzed considering the following:

- The aggravating failure assumption for the safety analysis of the design basis operating conditions
- The combination of IEs according to a cause–effect reasoning
- The single failure criterion

14.3 METHODOLOGY OF ANALYSIS

The motivation behind the analysis of various plant states is the demonstration of the adequacy of various safety provisions of the plant under design basis events and establishment of management guidelines under severe accident conditions. Depending on the objectives of the analysis, the methodology also varies. There are three ways of analyzing anticipated operational occurrences and design basis accidents for the safety demonstration of the plant: (1) the use of conservative computer codes with conservative initial and boundary conditions (conservative analysis), (2) use of best-estimate computer codes combined with conservative initial and boundary conditions (combined analysis), and (3) use of best-estimate computer codes with conservative and/or realistic input data but coupled with an evaluation of the uncertainties in the calculation results. The uncertainties associated with input data and the uncertainties associated with models in the best-estimate computer code are accounted for in the uncertainty evaluation. A deterministic analysis approach with conservative initial and boundary conditions is generally applied in the case of DBE for the demonstration of plant safety. This approach may not always represent the conservative scenario. There is a possibility for some important phenomenon to get masked by the predicted conservative scenario. The use of best-estimate analysis together with an evaluation of the uncertainties is increasing today. For scenarios with large margins to the acceptance criteria, it is appropriate for simplicity, and therefore economy, to use a conservative analysis. In situations where margins are low, the best-estimate analysis is essential to quantify conservatism. For beyond design basis accidents, the best-estimate analysis is performed, together with an evaluation of the uncertainties associated with the relevant phenomena. However, in determining accident management procedures, an uncertainty analysis is not usually performed.

The conservative assumptions made in the deterministic analysis should address the following [14.4]:

- It should be considered that the IE occurs at an unfavorable condition with respect to initial reactor conditions including power level, decay heat level, reactivity conditions, reactor coolant system temperature, pressure, and inventory.
- Credit of control systems in mitigating the effects of the IE should not be considered. Operation of control systems should be considered if their functioning aggravates the consequences of the event.
- Failure of all nonsafety systems should be considered in the manner that causes the most severe effects for the postulated initiating events (PIEs).
- The worst single failure should be considered to occur in the operation of the safety systems essential for mitigating the effects of the IE. Further, starting of only a minimum number of redundant trains should only be considered.
- The safety systems should be assumed to operate at their minimum performance levels.
- If the consequences on any structure, system, or component exceed the limit during the accident for which the designer did not prove full operability, they should be assumed to be unavailable.
- The actions of the plant staff to prevent or mitigate the accident should only be considered if it can be shown that there is sufficient time for them to carry out the requested actions, ample information is available for event diagnosis, adequate written procedures are available, and sufficient training has been provided.

Even though analysis of the entire plant is desired, depending on the scenario of various events, there is a possibility for simplification in the analysis by neglecting the modeling of certain parts of the plant. For example, during events originating from the core such as reactivity transients, the balance of the plant has negligible impact on the consequences. Therefore, the analysis of such events can be carried out without the models for the balance of the plant. However, in case of events originating from the balance of the plant, in order to assess the impact of the same on the reactor core, a complete modeling of the entire plant becomes essential. During whole core events such as core disruptive accident (CDA), the transient times are much smaller than the circulation time of the coolant in the circuit. Therefore, these studies can be performed with detailed modeling of the core alone without the models for the rest of the plant.

Dynamic simulation of a system comprising several components involves formulation and solution of a set of governing equations. The governing equations are formulated based on the conservation principle. These conservation equations generally form a set of partial differential equations in three special and one time coordinates. For a system involving single-phase fluid flow, three conservation equations representing the balance of mass, momentum, and energy are to be considered. For a two-phase system, the number of such equations is twice that for the single-phase system. In several situations, the complexity of the problem can be simplified by eliminating one or two space coordinates. Most of the transient phenomena happening during normal operations and IEs falling under the design basis category can be analyzed using 1D system dynamics models as there are no serious multidimensional phenomena involved during these conditions. Multidimensional spatial simulation of the entire plant is beyond the practical reach today. In situations where multidimensional process features become important to be simulated, such as those involving stratified flow conditions in pools during natural convection conditions, multidimensional modeling becomes essential. Even though some special 1D models [14.5] can take care of such effects, for accurate simulation of such phenomena, hybrid models by combining 1D and multidimensional models together are to be adopted.

The simulation of power generated in the reactor core requires the modeling of both fission power and decay power. For a reactor core of medium size, the neutronics can be decoupled in space in time. The point kinetics equation is applicable in such cases. For a large-sized reactor core, space kinetics will have to be considered. Various reactivity feedback effects due to Doppler, coolant expansion, fuel expansion, structural expansion, core radial expansion, control rod expansion, etc., have to be considered in the model. For simulating severe accident conditions, feedback effects due to coolant voiding, in-pin molten fuel motion, and core slumping also have to be considered.

Computer codes developed for water reactors are not suitable for analyzing the transient phenomenon in liquid metal–cooled fast reactors. Many transient thermal hydraulic analysis codes have been developed in different countries for their fast reactor systems. In the United States, several specialized codes have been developed for different plants, such as IANUS for fast flux test facility (FFTF), DEMO for clinch river breeder reactor project (CRBRP) [14.6], and NATDEMO for experimental breeder reactor (EBR-II) [14.7]. Brookhaven National Laboratory (BNL) has developed a general purpose code NALAP [14.8] for sodium-cooled fast reactors based on the RELAP code used worldwide for water reactor applications. Another general purpose code SSC-L has been developed for loop-type fast reactors [14.9]. For the safety analysis of severe accidents, Argonne National Laboratory has developed the SAS4A code, which has detailed models for thermal hydraulic, neutronic, and mechanical processes associated with severe accidents to simulate the behavior of the reactor core, coolant, fuel elements, and structural components under accident conditions [14.10]. In Russia, the Institute of Physics and Power Engineering (IPPE) developed the GRIF code [14.11] for the dynamic analysis of fast reactors, which has models for simulating severe accidents. Durham developed MELANI for modeling the prototype fast reactor in the United Kingdom [14.12]. CATHARE is the French code for pressurized water reactor (PWR) safety analysis. It has been used for light water reactor concepts. The CATHARE code for sodium reactors is a new development in this field [14.13]. Korea Atomic Energy Research Institution (KAERI) developed the SSC-K code for analysis of the Korea advanced liquid metal reactor (KALIMER), which is a pool-type fast reactor [14.14] based on the SSC-L code. In India, the Indira Gandhi Center for Atomic Research (IGCAR) has developed computer codes DYNAM and plant dynamics (DYANA-P) for analyzing the dynamic behavior of loop-type fast breeder test reactor (FBTR) [14.15] and pool-type PFBR [14.16]. DHDYN is a computer code developed with the multizone modeling approach for sodium pools. This code is used for studying the transient behavior of PFBR under decay heat removal conditions. A computer code system, KALDIS [14.17], is used for severe accident analysis of PFBR. For the predisassembly phase, a computer code, PREDIS, has been developed, forming part of KALDIS. Super-COPD [14.18] is a Japanese code developed for sodium fast reactor (SFR). This code has been validated against start-up tests carried out in the Monju reactor. Similarly, NETFLOW++ [14.19] is a code developed for the prediction of natural convective flow behavior in sodium systems. Japan Atomic Energy Agency (JAEA) developed SIMMER-III and SIMMER-IV for the 3D neutronics–thermohydraulics simulation of core disruptive accidents in SFR [14.20]. THACOS [14.21] is a computer code being developed in China for the dynamic simulation of a commercial demonstration fast reactor (CDFR) plant. Recently, the capability of the TRACE code [14.22] has been extended to predict 1D and 2D sodium boiling situations. Details of formulation, solution method, and validation of various codes are provided in Chapter 17.

14.4 APPLICATION OF PLANT DYNAMICS STUDY

Plant dynamic studies find application in several areas of nuclear power plant design and licensing: design of the plant protection system, thermomechanical design of systems and components, schemes of various plant operations, safety demonstration of the plant, and analysis of severe accidents. Some of the typical case studies are presented later with respect to the Indian design of 500 MWe PFBR.

14.4.1 DESIGN OF PLANT PROTECTION SYSTEM

Plant protection against various design basis events (DBEs) that can occur in the plant is ensured by reactor trip based on the values of a certain selected set of plant parameters known as safety control rod accelerated motion (SCRAM) parameters, crossing their preset threshold levels. In order to limit the consequences of various DBEs within the specified design safety limits (DSLs), an adequate number of SCRAM parameters are required in the plant protection system. Meanwhile, a large number of such parameters selected in the plant protection system affect the availability of the plant due to their spurious actuation. Therefore, there is a need for optimizing the list. The plant protection system should have reliable instruments to generate signals for the timely shutdown of the plant based on the evolution of SCRAM parameters. In order to design the instrumentation and other systems of the plant protection system, the maximum permissible delay for the actuation of these systems needs to be determined. The permissible delay time should be such that the consequences following various design basis events are limited within the specified safety criteria.

The most important measurements made in the plant are neutron flux (Φ), sodium temperature at the core inlet (θ_{RI}), central subassembly coolant outlet temperature (θ_{CSA}), individual subassembly coolant outlet temperature (θ_I), core flow (Q_{PP}), and delayed neutron detector (DND) flux. Some of the important signals derived from these measurements are reactor power (LinP), period (τ_N), reactivity (ρ), power to flow ratio (P/Q), group mean of SA sodium outlet temperature (θ_M), deviation of individual SA sodium outlet temperature from the expected value ($\delta\theta_I$), mean core temperature rise ($\Delta\theta_M$), and coolant temperature rise in the central SA ($\Delta\theta_{CSAM}$). Overpower events can be detected by the power, reactivity or reactor period, and temperature rise in the core. Undercooling events can be detected by the power to flow ratio, outlet temperature rise in the subassemblies, and reactivity. The DND detects fuel clad failures. Based on the safety criteria, an optimum list of SCRAM parameters arrived at for PFBR is given in Table 14.3.

All the electronically processed analog signals are susceptible to variations due to drift, temperature variation, supply fluctuations, accuracy of calibration, nonlinearity, and so on. Similarly, set point values are also susceptible to variation. The net impact of these effects is the shifting of the threshold values of SCRAM parameters from their intended set levels. Such deviations have to be considered in the analysis while evaluating the required response time for the instruments. The methodology adopted for estimating the response time for pump speed measurement is shown in Figure 14.2. The main utility of the pump speed parameter is to provide protection to the plant against pump seizure events. A consequence on the reactor core during a pump seizure event without the activation of the plant protection system is shown with square markers in Figure 14.2. Now, the plant protection system should be triggered at a time to limit the consequences within the DSLs

TABLE 14.3

List of Reactor SCRAM Parameters

S. No.	Parameter	Description of Parameter
1	τ	Reactor period for an "e"-fold increase
2	Lin P	Reactor power in linear channel
3	ρ	Net reactivity
4	DND	Neutron flux at delayed neutron detectors
5	P/Q	Power to flow ratio
6	θ_{CSAM}	Central subassembly sodium outlet temperature
7	$\Delta\theta_M$	Core temperature rise
8	$\Delta\theta_{CSAM}$	Central SA temperature rise
9	$\delta\theta_I$	Deviation from an expected value of a subassembly sodium temperature
10	θ_{RI}	Reactor inlet sodium temperature
11	N_P	Primary sodium pump speed

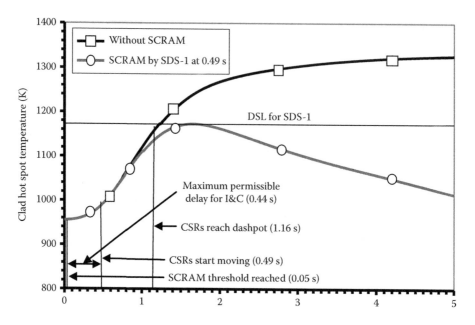

FIGURE 14.2 Permissible delays during one primary pump seizure.

specified for this event as shown with circular markers in Figure 14.2. A similar study has to be performed for all the enveloping events and response time arrived at for safety instrumentation.

14.4.2 THERMOMECHANICAL DESIGN OF COMPONENTS

An important outcome of event analysis is the transient evolution of temperatures at various critical locations in the plant. This information is useful to assess the structural damage due to fatigue and creep, thereby demonstrating the viability of various operating procedures. Some of the important structures (Figure 14.3) whose designs are governed by transient thermal loading are discussed in the following.

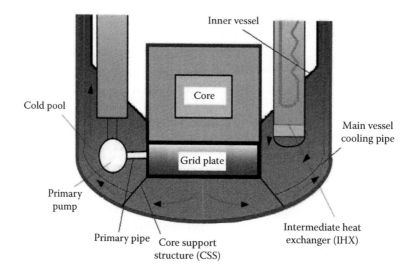

FIGURE 14.3   Important cold pool structures.

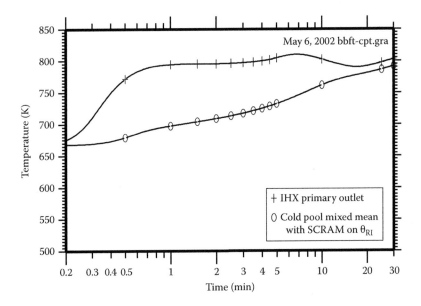

FIGURE 14.4   Transient following loss of steam water system on cold pool structures.

- *Cold pool structures*: During transients, due to the large thermal capacity of the cold pool, its temperature would be evolving slowly compared to the temperature of the sodium stream exiting the intermediate heat exchangers (IHXs). The primary pipe, grid plate shell, main vessel cooling pipes, and core support structure would all be affected by the difference between the IHX primary outlet and cold pool mixed mean temperatures. DBEs influencing these temperatures in a major way and enveloping almost all the other DBEs are trip of one primary sodium pump, trip of one secondary sodium pump, and loss of the steam–water system. The evolution of cold pool temperatures is shown in Figure 14.4 for the loss of the steam–water system. It can be seen that the cold pool structures are subjected to high temperatures in the creep regime following loss of the steam–water system.
- *Tube sheets of IHX*: IHX tube sheets are subjected to thermal loads during one primary pump trip, one secondary pump trip, one secondary pump acceleration, and IHX sleeve valve closure events, due to the difference between the primary and secondary sodium temperatures at these locations. The typical evolution of the cold end temperature during one secondary pump trip is shown in Figure 14.5, which affects the bottom tube sheet. Figure 14.6 depicts the evolution of hot pool temperatures under the primary pump trip, which affects the top tube sheet. In a similar manner, thermal shocks for other critical structures like steam generator (SG) tube sheets, hot pool structures, and steam headers during various design basis events have to be predicted for their detailed thermomechanical analysis.

14.4.3   STUDIES FOR DECIDING PLANT OPERATING STRATEGY: A CASE STUDY CARRIED OUT FOR PFBR

It is important to decide the strategy for various plant operations such that the constraints imposed by the design of various components are respected, operator action involved is simple, and thermomechanical loading on various structures is minimum. One of the important plant operations that requires a careful execution strategy is the reactor start-up, which is expected about 1000 times in the lifetime of the plant. The plant dynamics studies aid in working out a simple manual operating scheme for the start-up by satisfying various constraints imposed by (1) the design of

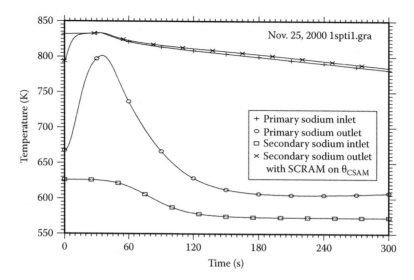

FIGURE 14.5   Transient following one secondary pump trip.

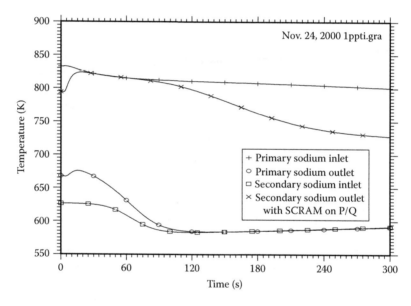

FIGURE 14.6   Transient following one primary pump trip.

components and (2) thermomechanical loading on various structures within permissible levels. The plant start-up operation, which is planned to be totally manual, comprises three major activities: (1) termination of the decay heat removal system and establishment of the regular heat sink, (2) approach to criticality in the reactor core, and (3) power raising in the reactor core by lifting the control and safety rods.

   The activity of power raising through the control rod movement is the longest operation in start-up. During this operation, it is necessary to keep the heating up rate of the hot pool below the permissible rate to avoid thermal shocks to various components. The reactivity additions during the start-up procedure should be such that the total reactivity is not more than the alarm threshold of the reactivity trip parameter. A simplified procedure with a minimum time delay between two control rod raise operations and with control rod raise by a definite distance (less than a limiting distance) every time has been worked out to raise the reactor power. This "wait and raise" mode of the start-up

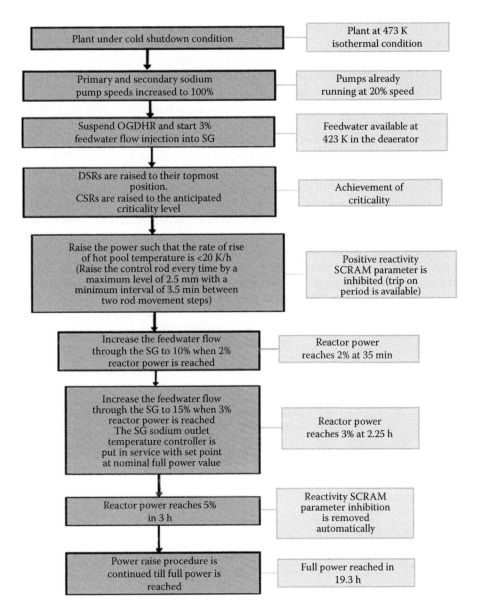

FIGURE 14.7 Reactor start-up procedure.

procedure to begin the equilibrium cycle core configuration is elaborated in Figure 14.7. The scheme respects the constraints of thermomechanical loading on the structure and minimum operator actions.

14.4.4 Demonstration of Plant Safety

Transient analyses of all the enveloping DBEs have to be carried out to predict the thermal hydraulic and neutronic behavior of the plant and hence to demonstrate the safety of the plant under design basis events. Apart from the events, mild overpower transients result due to overcooling caused by the events of increase in secondary sodium flow or feedwater flow in one or both loops. One or both primary pumps' acceleration results in a mild negative reactivity feedback. The effects are noticeable only when the reactor is operating at low power than while it is operating at nominal power. Therefore, these events have to be analyzed from the lowest possible initial steady-state condition envisaged for the reactor. At this power level, the primary sodium, secondary sodium, and feedwater flows are at their lowest levels.

There could be a need to analyze some events at low-power condition, which are acceleration of primary, secondary, and feedwater pumps and continuous withdrawal of one control rod at various initial power conditions. It has been seen from the analysis that some events may require automatic SCRAM, since the fuel, clad, and coolant temperatures cross respective safety limits without any safety actions.

Double-ended guillotine rupture of one of the four primary pipes is considered as a category 4 event in spite of the fact that the structural integrity of the pipe is ensured under all the conditions with a comfortable factor of safety. Safety demonstration of the plant during this event is very important from the point of view of justifying the selection of two loop designs for the reactor. The scenario following the pipe rupture event is shown in Figure 14.8 [14.23]. The core temperature evolution following this event is shown in Figure 14.9. Following pipe rupture, due to the low resistance encountered by the primary pumps against which they operate, pumps enter into the cavitation mode of operation. Therefore, net positive suction head (NPSH) and cavitation mode characteristics of the pump are essential to be simulated for the prediction of consequences of the event. Other important features of the system dynamics model—single-phase incompressible flow of sodium and single pressure point assumption for the grid plate—have been verified through detailed transient

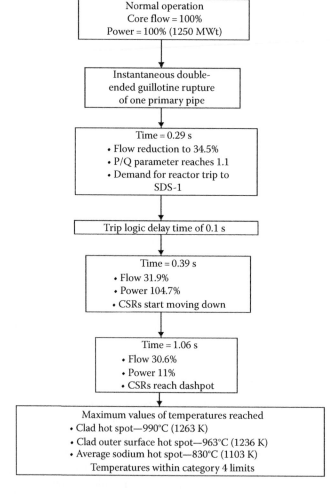

FIGURE 14.8 Scenario following primary pipe rupture event in PFBR.

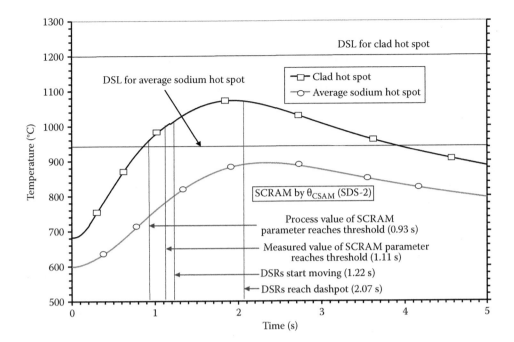

FIGURE 14.9 Temperature evolution following primary pipe rupture event.

pressure analysis and 3D hydraulic analysis of the grid plate. It has been established that there is no flow redistribution among various SAs and no void formation during the event due to transient pressure effects.

14.5 SUMMARY

The results obtained from event analysis are checked with the DSLs to ensure the integrity of the fuel, fuel cladding, and coolant during all the design basis events covering category I to category IV. Any events postulated beyond this DBEs will be categorized under accident conditions, which will be covered in Chapter 15.

REFERENCES

14.1   IAEA. (2000). Safety of nuclear power plants: Design. IAEA Safety Standards Series No. NS-R-1. IAEA, Vienna, Austria.

14.2   IAEA. (1998). Selection, specification, design and use of various nuclear power plant training simulators. IAEA-TECDOC-995. IAEA, Vienna, Austria.

14.3   Rouault, J., Chellepandi, P., Raj, B., Dufour, P., Latge, C. et al. (2010). Sodium fast reactor design: Fuels, neutronics, thermal-hydraulics, structural mechanics and safety. In *Handbook of Nuclear Engineering*, D. Cacuci (ed.), Vol. 4, Chapter 21, pp. 2321–2710. Springer, New York.

14.4   IAEA. (2001). Safety assessment and verification for nuclear power plants. Safety Standards Series No. NS-G-1.2. IAEA, Vienna, Austria.

14.5   Haihua, Z., Ling, Z., Hongbin, Z. (2014). Simulation of thermal stratification in BWR suppression pools with one dimensional modeling method. *Ann. Nucl. Energy*, 63, 533–540.

14.6   Albright, D.C., Bari, R.A. (1978). Primary pipe rupture accident analysis for clinch river breeder reactor. *Nucl. Technol.*, 39, 225–257.

14.7   Mohr, D., Feldman, E.E. (1981). A dynamic behavior of the EBR-II plant during natural convection with NATDEMO code. In *Decay Heat Removal and Natural Convection in FBRs*, A.K. Agrawal, J.G. Guppy (eds.). Hemisphere Publications, New York.

14.8 Martin, B.A., Agrawal, A.K., Albright, D.C., Epel, L.G., Maise, G. (1975). NALAP an LMFBR system transient code. Report No. BNL-50457. Brookhaven National Laboratory, Upton, NY.

14.9 Khatib Rahbar, M., Guppy, J.G., Cerbone, R.J. (1978). An advanced thermo hydraulic simulation code for transients in LMFBRs (SSC-L code). Report No. BNL-NUREG-50773. Brookhaven National Laboratory, Upton, NY.

14.10 Cahalan, J.E., Wei, T. (1990). Modeling developments for the SAS4A and SASSYS computer codes. *Proceedings of the International Meeting on Fast Reactor Safety*, American Nuclear Society, Snowbird, UT.

14.11 Chvetsov, I., Volkov, A. (1998). 3-D thermal hydraulic analysis of transient heat removal from fast reactor core using immersion coolers. *Proceedings of the IAEA Technical Committee Meeting on Methods and Codes for Calculations of Thermal Hydraulic Parameters for Fuel, Absorber pins and Assemblies of LMBFR with Traditional and Burner Cores*, Obninsk, Russia, July 27–31, 1998.

14.12 Durham, M. E. (1976). Influence of reactor design on establishment of natural circulation in a pool-type LMFBR. *J. Br. Nucl. Energy Soc.*, 15, 305–310.

14.13 Geffraye, G., Antoni, O., Farvacque, M., Kadri, D., Lavialle, G., Rameau, B., Ruby, A. (2011). CATHARE 2 V2.5 2: A single version for various applications. *Nucl. Eng. Des.*, 241, 4456–4463.

14.14 Chang, W.P., Kwon, Y.M., Lee, Y.B., Hahn, D. (2002). Model development for analysis of the Korea advanced liquid metal reactor. *Nucl. Eng. Des.*, 217, 63–80.

14.15 Vaidyanathan, G., Kasinathan, N., Velusamy, K. (2010). Dynamic model of fast breeder test reactor. *Ann. Nucl. Energy*, 37, 450–462.

14.16 Natesan, K., Kasinathan, N., Velusamy, K., Selvaraj, P., Chellapandi, P., Chetal, S.C. (2011). Dynamic simulation of accidental closure of intermediate heat exchanger isolation valve in a pool type LMFBR. *Ann. Nucl. Energy*, 38, 748–756.

14.17 Harish, R., Sathiyasheela, T., Srinivasan, G.S., Singh, O.P. (1999). KALDIS: A computer code system for core disruptive accident analysis of fast reactors. Report No. IGC-208. Indira Gandhi Center for Atomic Research, Kalpakkam, India.

14.18 Yamada, F. et al. (2009). Validation of plant dynamics analysis code super-COPD by MONJU startup tests. *Proceedings of the International Conference on Fast Reactors and Related Fuel Cycles: Challenges and Opportunities*, Kyoto, Japan.

14.19 Mochizuki, H. (2010). Development of the plant dynamics analysis code NETFLOW++. *Nucl. Eng. Des.*, 240, 577–587.

14.20 Tobita, Y., Kondo, Sa., Yamano, H., Fujita, S., Morita, K., Maschek, W., Coste, P., Pigny, S., Louvet, J., Cadiou, T. (2003). SIMMER-III: A computer program for LMFR core disruptive accident analysis— Version 3. A model summary and program description. Report JNC TN9400 2003-071. Japan Nuclear Cycle Development Institute, Ibaraski, Japan.

14.21 Hu, B.X., Wu, Y.W., Tian, W.X., Su, G.H., Qiu, S.Z. (2013). Development of a transient thermal-hydraulic code for analysis of China demonstration fast reactor. *Ann. Nucl. Energy*, 55, 302–311.

14.22 Chenu, A., Mikityuk, K., Chawla, R. (2009). One and two dimensional simulations of sodium boiling under loss-of flow conditions in a pin bundle with the TRACE code. *Proceedings of 13th International Conference on Nuclear Reactor Thermal Hydraulics*, Kanazawa City, Japan.

14.23 Natesan, K., Kasinathan, N., Velusamy, K., Selvaraj, P., Chellapandi, P., Chetal, S.C. (2006). Thermal hydraulic investigations of primary coolant pipe rupture in an LMFBR. *Nucl. Eng. Des.*, 236, 1165–1178.

15 Severe Accident Analysis

15.1 INTRODUCTION

The assessment of various energy generating options by worldwide experts concludes that the sodium-cooled fast reactors (SFRs) have high potential for providing affordable energy in 10–20 years in the countries mastering this technology. The conclusion has been derived after rigorous discussions involving comprehensively all the issues such as economy, safety, and environment. In the domain of safety, the concerns are expressed since the 1950s particularly referring to the risk of an uncontrolled power excursion in case of large-sized fast reactor systems and positive sodium void effects in case of SFRs. In the context of advanced reactor concepts, for example, Gen IV systems, treatment of severe accidents in the design is one of the key issues of R&D plans [15.1]. This requires complete understanding of various scenarios and the associated phenomena in the allied domains of science, engineering, and technology that can be hypothesized for robust safety demonstration. Accident progression in the fast reactor is considerably different from that in the thermal reactor. The thermal reactor core is optimized with respect to the fuel to moderator ratio for just optimum moderation, due to which any motion of the fuel material will lead to negative reactivity under loss of moderation. In the fast reactor, the core is not in optimum reactivity configuration. This means that any motion of fuel, depending upon core compaction and core expansion or dispersion, could introduce either positive or negative reactivity, respectively. If there is fuel melting, there will be core compaction due to the downward motion of molten fuel or fuel slumping, which will lead to large positive reactivity addition. This in turn will result in a superprompt critical excursion and release of a lot of thermal energy followed by mechanical consequences. The severe accident scenario in the fast reactor is defined based on this physics.

For such accident to take place, at least two or more low probability failures must take place in sequence, for example, a large reactivity insertion event coupled with complete failure of the plant protection system. Those accidents, which involve degradation/melting of the whole core, are termed as severe accidents or core disruptive accident (CDA) in the fast reactor context. CDA is a less probable event, and hence, it is referred to as a hypothetical accident, even though it may be mechanistically defined. Despite its low occurrence frequency, such an accident is to be analyzed to demonstrate that its consequences result in yielding lower radioactivity release than the specified limits to the environment.

This chapter brings out a few potential initiating events (IEs) that could result in a CDA with special attention to flow blockages in the core. Following this, accident scenarios are addressed: unprotected loss of flow (ULOF), unprotected transient overpower (UTOP), and loss of heat sink (LOHS). Then, thermal as well as mechanical energy releases and their consequences; postaccident heat removal (PAHR) aspects, mechanisms, design, and safety requirements; radioactivity source term evaluations, radioactivity release paths; and finally, some important major and minor events, description and feedbacks are presented in sequence.

15.2 INITIATING EVENTS

The transients get initiated in the core when there is a mismatch between heat generation and heat removal. These transients are called anticipated transients without SCRAM (ATWS) in a general term. There are basically three ATWS, the IEs that can lead to a CDA in SFRs. They are explained in the following.

15.2.1 INITIATING EVENTS LEADING TO UTOPA

The transients can be caused by excess reactivity addition, during which there is more heat generation than heat removal. These are the IEs leading to unprotected transient overpower accident (UTOPA). The physical phenomena of these IEs are uncontrolled withdrawal of control rods or a large gas entrainment into the center portion of the active core. To counterbalance the effects of these IEs, only the Doppler feedback could be efficient. Figure 15.1 shows the consequence of a gas bubble leading to a

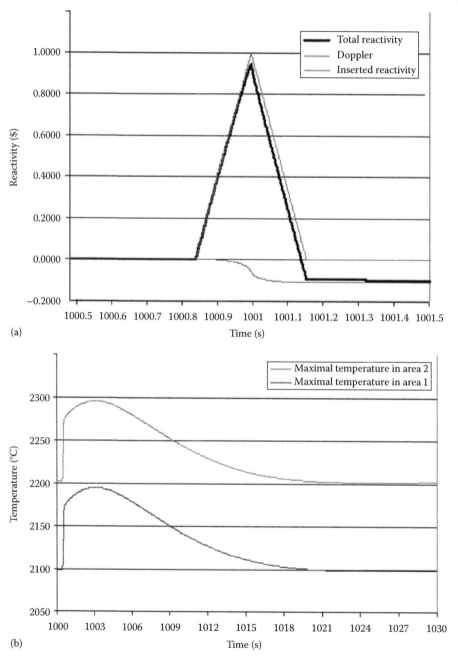

FIGURE 15.1   Example of a fast UTOP. (a) Reactivity variation and (b) temperature variation. (From Rouault, J. et al., Sodium fast reactor design: Fuels, neutronics, thermal-hydraulics, structural mechanics and safety, in *Handbook of Nuclear Engineering*, D. Cacuci (ed.), Vol. 4, Chapter 21, Springer, New York, 2010, pp. 2321–2710.)

reactivity insertion of 1 $ in 0.15 s and then coming back to normal in 0.15 s, the maximal temperatures in two areas of the studied core and the variation of the different components of the reactivity [15.2].

15.2.2 Initiating Events Leading to ULOFA

The transients can also take place due to flow starvation, during which heat removal is less than heat generation. These are the IEs for the unprotected loss of flow accident (ULOFA). ULOFA could arise from loss of power to the primary coolant pump or pump seizure or ruptures of primary pipes, etc., causing a flow coastdown and a large mismatch in the core power and coolant flow (P/F ratio). Depending on the timing and magnitude of the reactivity feedback, the reduction in power could be sufficient to prevent excessive temperatures and coolant boiling. An example of primary pump coast-down to 28% of nominal flow without actuating the control rods is illustrated in Figure 15.2 [15.2].

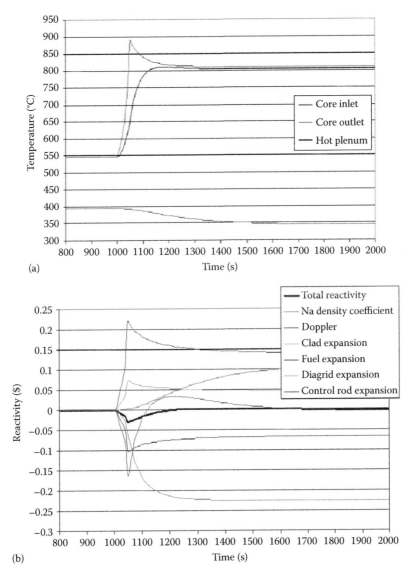

FIGURE 15.2 Example of a ULOF. (a) Reactivity variation and (b) temperature variation. (From Rouault, J. et al., Sodium fast reactor design: Fuels, neutronics, thermal-hydraulics, structural mechanics and safety, in *Handbook of Nuclear Engineering*, D. Cacuci (ed.), Vol. 4, Chapter 21, Springer, New York, 2010, pp. 2321–2710.)

This figure presents the evolution of temperatures in the hot plenum, the core inlet and outlet temperatures, and the influences of various reactivity feedbacks. The Doppler is the first reactivity feedback opposing the effect of the decrease in the sodium coolant density. After a few minutes, this Doppler reactivity feedback is positive because of the power decrease, and the reactivity feedback is due to the fuel, the structure, and the control rod expansion counterbalancing the sodium void and other positive reactivity feedbacks. Consequently, the occurrence of coolant boiling depends on the design provisions such as the inertia of the pumps, flow halving time, and core reactivity feedback coefficients.

15.2.3   Initiating Events Leading to ULOHS

The loss of main heat sink without SCRAM is characterized by the loss of the ability to reject heat at the steam generator, such as sudden loss of generator load and increasing core temperature. The occurrence of an unprotected loss of heat sink (ULOHS) causes a rise in the core temperature that yields a negative reactivity feedback. The main challenge is to avoid high temperature for the structures over a prolonged period that could cause severe thermomechanical damages, resulting in loss of the core support function of the associated structures.

15.2.4   Flow Blockages in the Core

The fast breeder reactor (FBR) core is very compact in the sense that only a minimum coolant area is permitted to remove the nuclear heat, to preserve the fast spectrum characteristics. This is also the reason for selecting liquid metal as the coolant for the fast spectrum reactor. The heat generation per unit length of the fuel pin is much higher than that generated in fuel pins of the thermal reactor. Hence, the absence of coolant even in a localized region can cause significant damage to the cladding due to local hot spots. Thus, flow blockage is a critical phenomenon to be investigated in the SFR in depth (Figure 15.3) [15.3]. The sources of flow blockage are fragments from failed fuel, foreign materials left over during fabrication, oil ingress into the primary sodium circuit, spacer wire snapping, chemical products during operation, and excessive pin deformation due to swelling. The presence of a wire-type/grid-type spacer is also another source of flow blockage. The possibility of flow blockage is more pronounced in the case of the grid spacer concept than the wire wrap spacer concept. In order to detect any such blockages, several provisions are incorporated in the core design. A sufficient number of neutron detectors are set up to monitor the reactor power. The sodium temperatures at the fuel subassembly outlet are monitored by providing thermocouples in each subassembly outlet. The central subassembly

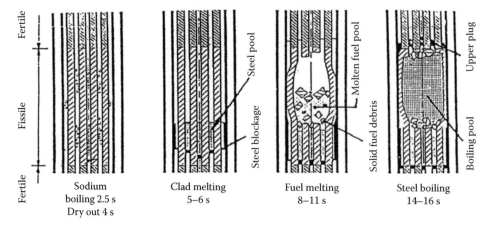

FIGURE 15.3 Flow blockage scenario in a subassembly—SCARABEE. (From Livolant, M. et al., SCARABEE: A test reactor and programme to study fuel melting and propagation in connexion with local faults, objectives and results, *International Fast Reactor Safety Meeting*, Snowbird, UT, 1990, Session 2, Vol. II.)

sodium outlet temperature is monitored generally by fast-response thermocouples. These will indirectly indicate the sodium flow blockages. Failed fuel detection is done by monitoring the cover gas activity and delayed neutrons in the primary sodium entering the IHX. Delayed neutron detector (DND) by primary sodium sampling is envisaged to SCRAM the reactor. Provisions would be incorporated to ensure at least two SCRAM parameters for all events affecting the core.

15.2.4.1 Investigation of Partial Blockage

In order to arrive at the extent of flow blockage that can be permitted, that is, the specified design safety limits that would be respected, an analysis is carried out [15.4]. Though it is carried out for prototype fast breeder reactor (PFBR) as the reference case, the conclusions appear to be generic. Out of many results, one important result is highlighted in Figure 15.4. This figure shows "the rate of flow reduction versus extent of flow reduction" to check whether the $\delta\theta$ SCRAM parameter (Table 14.3) could limit the maximum clad hot spot below 1073 K. In this figure, the effective regions, where the appropriate safety criterion is met, are indicated. From this figure, important conclusions are arrived at. The extent of inlet flow reduction less than 32% at any rate of flow reduction is tolerable. The extent of flow reduction greater than 95% is not tolerable for any rate of flow reduction. For the extent of flow reduction between 60% and 95%, the rate of flow reduction less than 5%/s is tolerable, and for the extent of flow reduction between 32% and 60%, the tolerable rate of flow reduction varies in between 6%/s and 5%/s, respectively.

15.2.4.2 Investigation of Total Instantaneous Blockage

Under the assumption of total instantaneous blockage (TIB), a single subassembly melting progresses at the most to the neighboring six subassemblies. However, there is a concern that if the hot fuel settles at the bottom, the main vessel wall may melt and give way to the sodium and the molten fuel to spill out of the primary containment. Also, there is concern for recriticality if the fuel settles down in a more reactive configuration. Therefore, a core catcher below the grid plate is provided. This would collect the molten fuel and debris in a suitably dispersed manner to avoid criticality and ensure long-term cooling. However, the amount of molten fuel released in the melting of seven subassemblies is relatively small compared to the actual mass of fuel that could make the critical configuration.

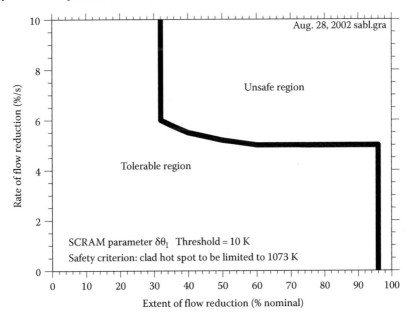

FIGURE 15.4 Tolerable limits for partial inlet blockage at the subassembly.

FIGURE 15.5 Blockage adaptors at the SA top.

For PFBR, the total fuel mass of all seven subassemblies is 0.3 ton, compared to 1.0 ton of fuel in the most reactive configuration required to achieve recriticality. Therefore, the slumped molten fuel of seven subassemblies does not have any recriticality potential. The core catcher located below the core support structure can collect the debris without creating any opportunity that molten fuel can have direct contact with the main vessel bottom. Further, the debris settling on the core catcher can be cooled by natural convection. Toward facilitating cooling in case of PFBRs, a chimney is provided at the center to aid in the natural convection flow of sodium. Preliminary calculations have shown that the maximum temperature of the core catcher is 1014 K and the temperature of the core catcher bottom plate is less than 923 K [15.5].

Now let us discuss the chances of occurrence of a TIB. In this respect, it is worth to know the generic conclusions derived from several international experiences [15.6]. The rate of blockage would be very small. Transition from local boiling to bulk boiling would take hours. Fission gas release does not affect cooling of neighboring pins. The fast propagation of local faults does not occur. With the availability of DND signals, it is possible to SCRAM the reactor within 20–30 s. There is no concern related to fuel–coolant interaction in case of TIB. The reactor shutdown system and decay heat removal capability are not impaired. Radial melt through is a slow process. There are several features added in the reactor to avoid TIB. The coolant enters into the subassembly in the radial direction rather than the axial entry with multiple holes at two elevations, discriminators at the bottom of the subassembly to avoid wrong loading, and thorough check of each and every subassembly before entry into the reactor. In the sleeves of the grid plate and in the feet of all the core subassemblies, multiple holes are provided. Apart from these, the total blockage at the outlet of a subassembly is ruled out by providing an adapter, which ensures an alternate path for flow through the side gaps (Figure 15.5). Thus, TIB is a beyond design basis event (BDBE). In case TIB is not detected, the situation may lead to a CDA.

15.2.5 GENERATION OF SODIUM VOIDS IN THE CORE

One of the major consequences of abnormal void fraction in liquid sodium is that it inserts a positive reactivity effect into the core. Sodium boiling in the core can create local sodium vapor pockets.

Apart from these, there are several mechanisms to replace sodium, thereby creating a void in the core. The void may lead to severe accident only if there is a massive introduction of gas/void into the core. The following are the major sources of gas entrainment in SFRs:

- Continuous gas production in the core: in sodium (neon and tritium), in fuel pins (Xe, Kr), in control rods (He), and in the outer B_4C neutronic shielding (He).
- Gas production due to reactions between Na and oil/elastomer products.
- Cavitation-induced local and stable sodium vapor pockets (relatively small).
- Pressurized gas leak from auxiliary circuits (nitrogen leak from purification units, detection devices, etc.).
- Entrainment of argon gas from the hot pool and transfer to the cold pool during sodium overflow on the weir (Figure 15.6). This happens if the weir crest is not properly profiled or under a weir shell instability condition that happens under some critical combination of sodium flow rate and fall height (Section 10.2.4).
- Gas entrainment by vortex-induced mechanisms in the hot pool and sodium-free surface (Figure 15.7): (a) liquid fall, (b) vortex activity, (c) drain-type vortex, and (d) free surface shearing phenomenon.

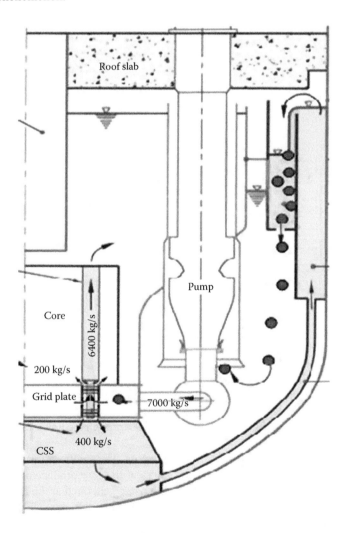

FIGURE 15.6   Gas entrainment through weir flow.

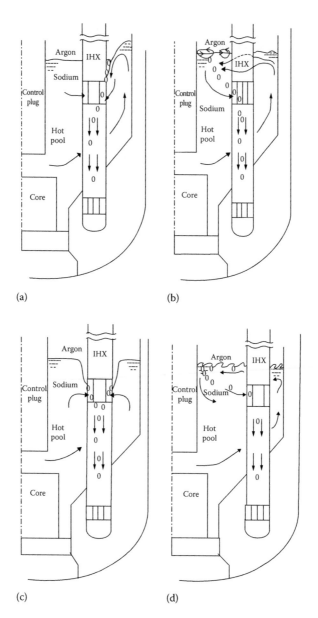

FIGURE 15.7 Gas entrainment mechanisms associated with sodium hot pool free level. (a) Liquid fall (Large upward velocity), (b) vortex activity at free surface (Large horizontal velocity), (c) drain-type vortex (Large drain velocity), and (d) free surface shearing (Large horizontal velocity).

- Overspeeding of primary pumps can cause a high risk of inlet pump draining and argon entrainment (Figure 15.8).
- Differential dissolution of argon in hot and cold pools through the hot pool and pump shell.

Sodium voids are critical IEs particularly for large reactors. The smaller the core, the lower is the void effect. The sodium voids generated in the periphery of the core produce negative reactivity in the core since neutron losses are higher. However, if they are generated at the core central portion, they introduce positive reactivity to the core. Typical total void coefficients for international reactors are 1.8 \$ ± 1.35 \$ for Phénix, 5.3 \$ ± 1.60 \$ for Superphénix, 5.8–6.2 \$ for EFR [15.7], and

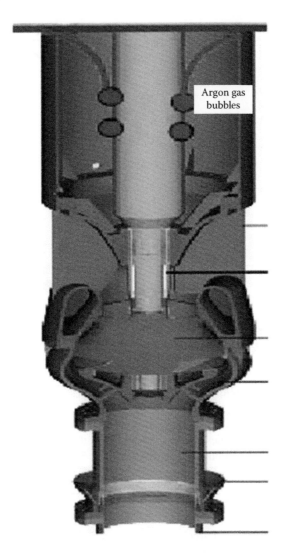

FIGURE 15.8 **(See color insert.)** Gas entrainment due to overspeeding of the pump.

2.7 $ for PFBR [15.8]. Hence, the design has to minimize the gas sources or limit the gas accumulation zones in liquid sodium, for example, elimination of argon pocket seals in IHX, use of purger subassemblies to prevent gas flow through positive reactivity zones in the core, avoiding oil sources by opting for magnetic bearings to replace the mechanical bearing in the primary pump rotor, and introduction of baffles to mitigate gas entrainments into the sodium pool. It is also preferable to have the provision for the void fraction measurement in the liquid sodium with adequate accuracy to eliminate false alarms.

15.2.5.1 Gas Entrainment Mitigation Mechanisms Introduced in PFBR: An Illustration

Extensive experimental and numerical investigations were carried out to mitigate the gas entrainment risks in PFBR. Basic experiments were conducted to characterize gas entrainment in the hot pool free surface. Further, efforts were put in for the numerical prediction of free surface velocities and numerical assessment of gas entrainment mitigation mechanisms introduced in the hot pool. The baffle plates incorporated in the hot pool to mitigate gas entrainment were validated through

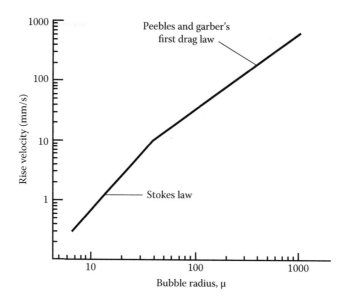

FIGURE 15.9 Terminal velocity of argon gas bubble in liquid sodium.

simulated experiments on a ¼ scale water model of primary system (called Scaled Model of Reactor Assembly for Thermal hydraulics [SAMRAT]). Details can be seen in Ref. 15.9. A few specific aspects are highlighted in the following:

- To quantify the differential dissolution of argon in hot and cold pools at 820 and 670 K, respectively, terminal velocities of argon bubbles from the hot pool to cold sodium pool are estimated that for the argon bubble size ranging from 50 to 100 μm, the terminal velocity of argon bubbles would be ~0.02 m/s (Figure 15.9), compared to the sodium velocity of ~0.5 m/s in the cold pool (Figure 15.10). Hence, the argon bubbles are likely to be carried by sodium.
- Gas dissolution studies through the sodium around the primary pump indicate that the upper bound mass flow rate of argon entering the grid plate is 0.5 g/s and the void fraction of argon in the grid plate is 1.5×10^{-5}. The void fraction increases to ~8.5×10^{-5} at the subassembly top. The associated reactivity change is found to be negligible.
- Purger subassemblies are provided in the grid plate at six locations to purge the gas accumulating beneath the grid plate through peripheral locations, so that they do not add any positive reactivity effects (Figure 15.11).
- Mechanical seals to significantly minimize the sodium flow bypass from the hot pool to the cold pool (Figure 15.12).
- Proper profile of the weir shell of the main vessel cooling circuit and qualification through experimental hydraulic simulations (Figure 15.13).
- To prevent entrainment of argon gas into IHX inlet windows, sufficient sodium head is provided above the IHX inlet window in the hot pool.
- To prevent gas entrainment in the cold pool, the depth of the vortex around the pump shaft (0.4 m) is kept shorter than the submergence depth of the hydrostatic bearing (3 m).
- To mitigate gas entrainment in the hot pool, baffle plates are placed in the control plug and inner vessel. The locations and sizes of baffles are finalized based on computational fluid dynamics (CFD) simulations (Figure 15.14). The design was also validated with the experimental simulation in SAMRAT (Figure 15.15) [15.10].

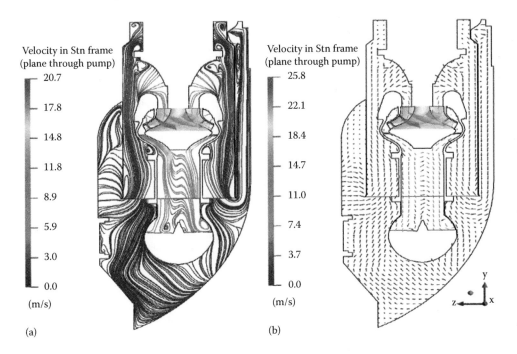

FIGURE 15.10   (See color insert.) Flow path and sodium velocity in the cold pool. (a) Contour and (b) vector.

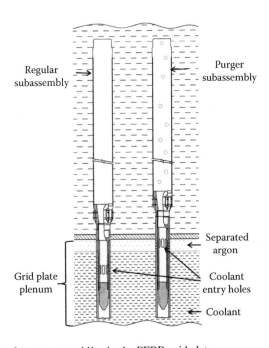

FIGURE 15.11   Location of purger assemblies in the PFBR grid plate.

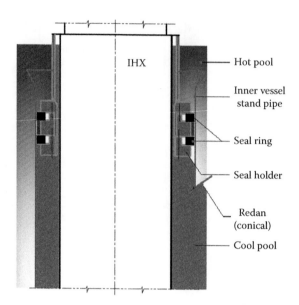

FIGURE 15.12 **(See color insert.)** Mechanical seal for intermediate heat exchanger (IHX).

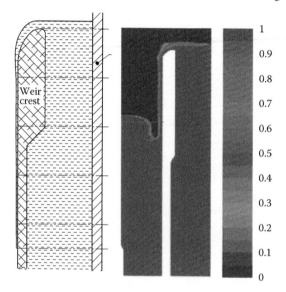

FIGURE 15.13 **(See color insert.)** Weir profiling.

15.3 SEVERE ACCIDENT SCENARIOS

The IEs described in the previous section can lead to a severe accident. However, several international studies with in-depth modeling and supporting experiments indicate that these transients will be benign and do not lead to severe accident. In spite of this, on a hypothetical basis, one can postulate scenarios of severe consequences staring with IEs. The accident initiating transients described in the previous section are either reactivity based or flow based. Among them, ULOF is considered to be the most severe IE that can lead to increase in power and temperatures and ultimately to severe accident. Four characteristics of large liquid metal fast breeder reactor (LMFBR) cores—positive sodium void worth, core geometry not in its most reactive configuration, large margins to coolant boiling, and high fuel reactivity worth in case of core slumping—have led to the choice of ULOFA

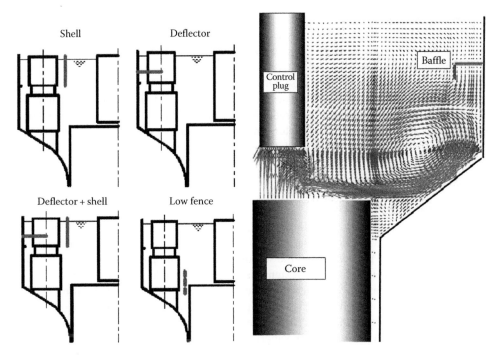

FIGURE 15.14 **(See color insert.)** Baffle plates for mitigation of gas entrainment in the hot pool.

Normalized height of baffle plate above core top (H_n)	V_{max} at free surface (m/s) (measured)	V_{max} at free surface (m/s) (num.)	Visual observation
Without baffle plate	0.56	0.50	Vortex activated entrainment from free surface was observed
0.808	0.30	0.33	Free surface turbulence is reduced but vortex activated entrainment from free surface was observed
0.748	0.25	0.30	The free surface is calm and no vortex activated gas entrainment was observed

FIGURE 15.15 Validation of gas entrainment mitigation baffles in SAMRAT.

as the principal scenario for study by the international community. However, there are strong experimental evidences that suggest that due to the in-pin motion of molten fuel (fuel squirting), UTOP transient will not lead to accident but can take the rector to a higher power level.

The CDA scenario is described exhaustively starting with the IE, through various stages, until the estimation of mechanical energy release. Different physical phenomena are dominant at various stages of the accident and also the timescales involved are different. Therefore, accident progression is analyzed deterministically in different phases using the cause and effect phenomenology, adopting a conservative approach where deterministic analysis is not possible. These phases are the predisassembly phase, transition phase, and disassembly phase and are described in the following.

15.3.1 Severe Accident Phases

15.3.1.1 Predisassembly Phase

This phase involves temperature rise in the fuel, clad, and coolant. In case of ULOF, the first scenario is sodium void formation at the upper portion of the active core region and subsequent propagation to the central portion. When the sodium voids have propagated to the central portion, high positive reactivity is inserted to the extent that the reactor could reach a superprompt critical condition with simultaneous fuel temperature rise. Subsequent phenomenon could be melting of fuel in the central portion with further rise in temperatures. The predisassembly phase terminates with the initiation of boiling of the fuel. Thus, the end of this phase is the onset of conditions for the generation of high internal pressures from vaporized core materials and consequent hydrodynamic core disassembly.

For the given power and configuration, the transient behavior of the reactor depends on the fuel. For the oxide core, the temperature difference between the fuel and sodium is very high compared to the metal-fueled core. Accordingly, the drop of the average fuel temperature is much higher than that in the metallic core. Because of this, there will be high positive reactivity addition due to Doppler feedback for oxide fuelled core and in case of metallic fuelled core, the reactivity effect is insignificant (it may be slightly negative). Thus, the metallic core could march toward the subcritical state. On the contrary, the reactor with the oxide core can become subcritical, even when the sodium void coefficient is positive by design, provided some passive shutdown system is incorporated in the core as an ultimate shutdown system. This system should act without any support of active elements, once the coolant temperature exceeds certain safety limits by introducing automatically an adequate quantity of strong neutron absorber material.

The predisassembly analysis has to predict the transient response from accident initiation to the point of neutronic shutdown with essentially intact geometry and a gradual core meltdown scenario. The computations are generally performed deterministically for core neutronics, reactivity feedbacks, thermal hydraulics, sodium boiling, fuel pin failure, fuel slumping, fuel relocation, and fuel–coolant interactions.

15.3.1.2 Transition Phase

At the end of the predisassembly phase, the reactor can become subcritical if there are sufficient negative reactivity feedbacks. If the negative reactivity feedbacks are insufficient, the fuel and clad will melt and form a molten pool. This is called the transition phase, since the fuel attains gradual transition from solid to liquid phase and since this phase is sandwiched between the predisassembly and disassembly phases. The core will boil until a permanently cooled subcritical configuration could be established, or criticality conditions can recur. This phase is sandwiched between the predisassembly and disassembly phase (third phase), in a conventional sense. However, current advancement in CDA analysis treats the entire phenomenon as a transition from IE to mechanical consequences. This kind of analysis involves extensive multiphysics modeling with solid, liquid, and vapor fields and is grouped into density and energy components. It is possible to address a finer behavior of molten fuel relocations in various pathways. This would provide realistic energy

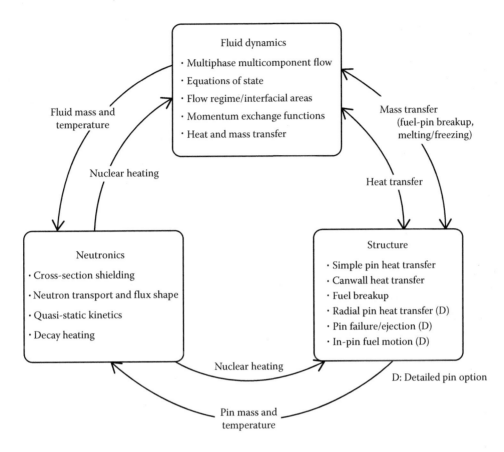

FIGURE 15.16 A comprehensive scheme of transition phase modeling.

releases. It is possible to study the effects of design improvements to prevent recriticality, for example. In a typical modeling of this phase in the SIMMER code [15.11], neutron kinetics is modeled by an improved quasi-static method, in which a space- and time-dependent neutron transport equation is factorized into a shape function that represents the neutron flux distribution but changes only slowly with time and an amplitude function that accounts for the time evolution of the reactor power. The flux shape calculations are based on a multigroup transport theory. The reactivity and other kinetics parameters are calculated from the neutron flux and macroscopic cross sections. Then, the amplitude equation is solved to determine the reactor power. A schematic of a more comprehensive transition phase modeling is given in Figure 15.16.

15.3.1.3 Disassembly Phase

The predisassembly phase provides the initial conditions for the disassembly phase. Once the fuel starts dispersing upon vaporization, its displacement feedback dominates and other reactivity feedbacks except the Doppler can be ignored. The core loses its integrity. This phase lasts till the reactor attains subcriticality due to fuel dispersal. The associated timescales are short (of the order of milliseconds). The rate of reactivity addition due to fuel slumping and sodium voiding determines to a large extent the accident scenario in disassembly phase and the ultimate energy release. The other factors that have major influence in the disassembly phase are the fuel temperature distribution and net reactivity in the core at the end of the predisassembly phase. In the disassembly phase, the high reactivity addition rate from fuel slumping leads to an energetic excursion, with the release of a lot of thermal energy. The reactivity excursion gets controlled by reactivity feedbacks due to Doppler and outward material motion, and the core bubble becomes subcritical.

In a pessimistic approach followed for PFBR (Chapter 21), the transition phase is ignored and conservative inputs are considered. A prompt critical excursion is considered with a high reactivity addition rate. The power pulse generates high temperatures, vaporizes the molten fuel, or liquefies the solid fuel present. High temperatures lead to high fuel vapor pressures. The pressure gradient disassembles the core. In this disassembly configuration, a high negative reactivity is created due to Doppler and material displacements. This overrides the input reactivity ramp and the core becomes subcritical within a very short time, typically 10–20 ms. The power pulse accumulates a few thousand MJ of thermal energy. With the addition of such high thermal energy in the disassembly phase, the core will become a mixture of molten fuel and steel along with the pressurized vapor of fuel, steel, and sodium.

15.3.2 Consequences of CDA

The vapor portion of the core that results in at the end of the disassembly phase is called "core bubble." The core bubble is neither in thermal equilibrium with the surrounding sodium (in view of its high temperature compared to the sodium temperature) nor in mechanical equilibrium with the sodium (in view of its high pressure compared to the sodium pressure). Hence, the state of the core bubble marches toward attaining thermal as well as mechanical equilibrium. Toward attaining mechanical equilibrium, the bubble exerts pressure on the sodium and thereby expands rapidly generating pressure/shock waves in the sodium. These pressure/shock waves are responsible for mechanical consequences, which are elaborated in the following section.

Toward attaining thermal equilibrium, it transfers the heat to the surrounding sodium and gets cooled; in fact it gets quenched. The core bubble vapor, after getting quenched by the surrounding sodium, gets converted into the liquid phase and mixes with the bulk of molten fuel and steel. In view of its higher density, the molten fuel and steel move downward, melting the structures on the path of travel, and ultimately settle on the core catcher in the form of core debris. Subsequently, the decay heat should be removed over a prolonged period by proper postaccident decay heat removal mechanisms. This is the thermal consequence of CDA. Apart from these, there is a radiological consequence. The fission gas in the core bubble mixes with the cover gas and ejects to the reactor containment building (RCB) through the top shield penetrations. This is the major radioactive source term for radioactivity release from the stack.

A typical CDA scenario caused by ULOFA is illustrated in Figure 15.17. The ULOFA leads to melting of almost the whole core, with simultaneous generation of the vapor phase consisting of fuel, coolant, and structural materials. The heat generated due to power excursion could melt the core to the extent of generating saturated liquid and vapor phases (Figure 15.17a). The mechanical energy release is due to the expansion of the vapor phase from the initial pressure (P_o) to the final pressure (P_f) equal to the ambient condition prevailing in the reactor. During this rapid expansion of core bubble in the liquid sodium environment, the pressure waves are generated, which deform

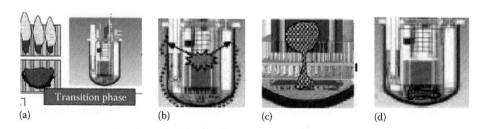

(a) (b) (c) (d)

FIGURE 15.17 **(See color insert.)** Mechanical and thermal hydraulic consequences of CDA. (a) Melting and vaporization of core, (b) mechanical energy release, (c) molten fuel relocation, and (d) postaccident heat removal. (From Rouault, J. et al., Sodium fast reactor design: Fuels, neutronics, thermal-hydraulics, structural mechanics and safety, in *Handbook of Nuclear Engineering*, D. Cacuci (ed.), Vol. 4, Chapter 21, Springer, New York, 2010, pp. 2321–2710.)

the shell structures surrounding the core (Figure 15.17b). Subsequent to completion of mechanical work, the two-phase bubble condenses due to the heat exchange with sodium pool, which is at 855 K, and starts moving downward melting the support structures, such as the grid plate and core support structure forming solid debris (Figure 15.17c). Finally, the debris settles on the core catcher and subsequently it will be cooled by natural convection (Figure 15.17d). Thus, there are mechanical, thermal, and radiological consequences of a CDA. These are elaborated in the following sections.

15.4 MECHANICAL ENERGY RELEASE AND CONSEQUENCES

The mechanical consequences are studied with the objective of ensuring the structural integrity of the primary containment and RCB. The main vessel along with the top shield forms the primary containment in SFRs. Toward assessing the structural integrity of the primary containment, in the beginning, conservative estimation was made on mechanical energy release. In 1956, Bethe and Tait analyzed reactivity excursions terminated by fuel vapor expansions. In this analysis, coherent core compaction was hypothesized and then a gravity-driven collapse of the core and a hydrodynamic core disassembly were assumed. Hence, this model yielded high energy release. It also assumes no Doppler feedback and no delayed neutrons [15.12–15.14]. In fact for the EBR-II, a Bethe–Tait analysis was used to demonstrate the primary system containment capability. At Argonne National Laboratory (ANL), during FFTF licensing, Jackson [15.15] extended the phenomenological perspective to include the mechanistic treatment of the accident initiation phase, coolant boiling at normal power, coolant-voided fuel disruption, spatially incoherent core melting in the transition phase, tracking of material relocation and interaction with structures, blockage formation and remelting, core boil up and dispersal, consideration of recriticality during transition to whole core melting, and ultimate core dispersal to a coolable, subcritical state. Over the years, the understanding of such phenomenon has improved considerably, and lesser and lesser energy release values have been estimated. Table 15.1 shows the progression in the trend of energy release analysis from examples of various reactors, indicating lower and lower values for energy release, as reflected by the reducing value of the ratio of CDA energy release (MJ) to reactor power (MWt). Currently, a stage has come where insignificant energy release has been envisaged in CDA as indicated for EFR with in-pin fuel motion (Table 15.1). One simplified and idealized scenario for the mechanical consequences of a CDA is illustrated later.

The core bubble contains a highly pressurized two-phase mixture, which generates pressure waves. In view of its high pressure, the bubble expands, and in this process, the liquid phase is converted to the vapor phase. An immediate effect of pressure waves is plastic deformation of surrounding structures, which offer resistance for propagation of pressure waves. Due to the presence of the cover gas space above the sodium-free level, there is less resistance for the movement of liquid in the upward direction, and hence, a portion of sodium above the bubble is accelerated upward. As

TABLE 15.1
CDA Energy Release for Reactors

Reactor	Power, P (MWt)	Mechanical Energy, W (MJ)	Ratio, W/P
SPX-1	3000	800	0.270
SPX-2	3500	110	0.030
BN 800	2100	50	0.024
DFBR	1600	50	0.031
EFR	3600	150	0.040
PFBR	1250	100	0.080

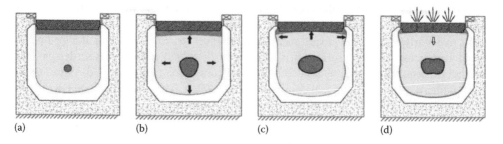

(a)   (b)   (c)   (d)

FIGURE 15.18   (See color insert.) Mechanical consequences of CDA. (a) Initial state: 0 ms, (b) vessel pulldown: 0–50 ms, (c) slug impact: 100–150 ms, and (d) final state: 150–900 ms.

a result, a net force develops in the downward direction loading and the pulldown of the main vessel. This, in turn, produces a compressive force on the reactor vault. The accelerated sodium continues to move upward for a certain period (50 ms typical) during which there are no significant mechanical deformations, till sodium impacts on the top shield. Once sodium impacts on the top shield, which is termed as "sodium slug impact," the kinetic energy of moving sodium is converted into pressure energy. Consequently, the pressure in the cover gas as well as in sodium increases steeply, producing (1) further overall plastic deformation on the main vessel, (2) large local deformation on the main vessel near the top shield junction in the form of bulging, and (3) impact force on the top shield in the upward direction. The slug impact phenomenon (from the start of impact till stabilization of vessel deformation) occurs in 100–150 ms (typical).

During the slug impact, the bolts of top shield components elongate and the seals in the annular gaps of the top shield may fail. As a result, sodium may fill the top shield penetrations. Subsequently, sodium leaks to the RCB. In the quasi-static condition when the sodium leak phenomenon occurs, the core bubble pressure drops. This is mainly due to the cooling of the bubble by the surrounding subcooled sodium, which has high heat capacity, while the volume of the core bubble remains unchanged. The quasi-static condition prevails in 150–900 ms (typical). The leaked sodium burns and causes temperature and pressure rise in the RCB and the RCB is designed for this pressure. Beyond 900 ms, the mechanical consequences are terminated except for the creep of the main vessel bottom portion, which is at a high temperature, depending upon the heat removal provisions made in the design. Figure 15.18 depicts the previously mentioned scenarios.

15.5   POSTACCIDENT HEAT REMOVAL

The formation of a molten mixture of fuel and steel is the end of the transient phase of CDA. The subsequent phenomena involved are (1) relocation of molten fuel, (2) molten fuel–coolant interaction, (3) core debris settlement behavior, and (4) core catcher and long-term coolability aspects. Figure 15.19 depicts the core debris relocation and settlement scenario. These are explained in the following sections.

15.5.1   RELOCATION OF MOLTEN FUEL

The core debris generated by quenching of the molten core debris in sodium is expected to travel downward. Before the particles settle on the main vessel/core catcher, the debris have to cross the lower axial blanket, grid plate, and core support structure. The debris may reach the foot of the subassembly and are then collected between the sleeves of the grid plate. If the mass of the debris is large enough, they might spread and deposit on the peripheral region of the grid plate. An analysis of downward streaming from the core and freezing of the molten fuel/steel mixture indicated that at least one-third of the fuel will reach the grid plate in a very short time period (several seconds) [15.16]. This fraction would be higher if the flow through the control rod channels is high or if the

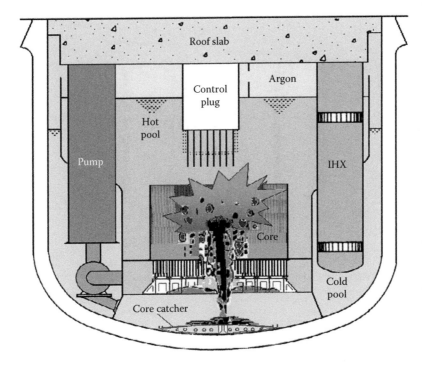

FIGURE 15.19 **(See color insert.)** Molten fuel settlement in the core catcher: schematic.

remaining fuel mass becomes critical and remelts the frozen fuel plug. The debris could melt the lower plate of the grid plate or open a way through the sleeves of the grid plate. The French study [15.17] for superphenix (SPX-1) has indicated that if the debris are deposited on the grid plate 100 s after the accident, the time needed to melt the lower plate of the grid plate would be 1600 s. At the same time, incipient sodium boiling below the grid plate is expected. Apart from the axial melting through the lower assembly structure, the downward relocation of molten fuel and steel may also occur by streaming through the coolant channels in the lower assembly structure. Streaming is a much more rapid mechanism for relocation. Molten fuel may not penetrate far in the pin cladding structure without plugging, because of the small channel flow area/cladding surface area and due to the significant potential for steel melting. However, large coolant paths existing below the active core region facilitate the steaming process.

15.5.2 MOLTEN FUEL–COOLANT INTERACTIONS

The molten metals of fuel and steel emerge from the grid plate and core at a highly superheated condition and have a higher tendency to travel downward. In the process of travel, they lose the heat and solidify partly/completely. In view of high velocity and temperature of the liquid (typically about 5000 K) jet penetrating the liquid sodium, which is at much lower temperatures (typically 700 K), several thermal hydraulics phenomena take place. The most complicated one is jet instabilities as depicted in Figure 15.20. Depending upon the interactions, the geometrical and dispersion characteristics of debris settling on the core catcher plate would vary. From the heat transfer point of view, even spreading and formation of a uniform layer is desirable. It has been established from studies of fuel–coolant interactions that molten fuel and steel tend to fragment and form solid particulates upon quenching in sodium. The size distribution of these particles significantly affects the ultimate coolability of the debris bed. In-pile tests in transient reactor test facility (TREAT) [15.18] have shown that the major fraction of debris (70%–80%) has diameters <1000 μm. The average size of a fuel particle is ~200 μm and that of steel particles is ~400 μm.

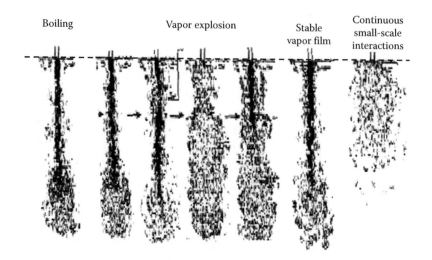

FIGURE 15.20   Molten fuel jet and coolant interactions.

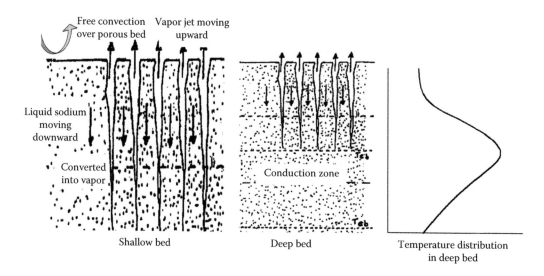

FIGURE 15.21   Heat transfer scenario from porous debris bed.

In all experiments, the difference in appearance between metallic particles and UO_2 particles has been observed. The metal particles are generally rounded, whereas UO_2 particles are irregular and of smaller size. The fragmentation is primarily due to thermal and mechanical interactions including hydraulic forces and thermal stress cracking. Particles of bigger size offer lesser hydraulic resistance for the same porosity of the bed. Particle size distribution is a strong function of the defragmentation process and coolant to fuel material combination. Apart from the particle size distribution, the settlement behavior is also an important factor from heat transfer considerations. In this respect, two configurations can be conceived: shallow bed and deep bed. This categorization depends on the penetrating depth of the coolant. Depending on the amount of vapor jet generated, dry out condition occurs (Figure 15.21). Under dry out, vapor jet movement and hence sodium movement could be stopped. Dry out depends on the decay power density, porosity, particle size, etc.

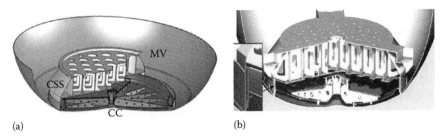

FIGURE 15.22   In-vessel core catchers in international SFRs: (a) PFBR and (b) European fast reactor (EFR).

15.5.3   Core Catcher Concepts

In the event of a large accumulation of core debris (10%–100%) at the bottom of the main vessel, the bed would be in a molten state. If the steel and fuel separate due to density differences, then no steel boiling is expected. For a homogeneous mixture of fuel and steel, the maximum layer temperature can be 3200 K. Therefore, steel boiling may occur and this would increase upward heat transfer and sodium boiling leading to further pressurization of the main vessel. The debris bed in contact with the bottom of the main vessel would penetrate through the wall. The vessel may fail due to structural melting and creep rupture. Characteristic failure time of 12–60 min has been reported for melt through failure if heat removal from the bottom of the vessel is by thermal radiation only. High pressure exerted on the vessel would cause structural failure prior to melting. Conditions following failure of the main vessel and safety vessel are very serious. It can lead to failure of the liner resulting in chemical attack of concrete by core debris and sodium resulting in degradation of the structure, hydrogen production, and pressurization of containment. In view of the previously mentioned serious consequences, there is a need for accommodating the core debris in a safe manner. This is accomplished by placing a core catcher for collection of debris in the sodium plenum under the core support structure. Schematic sketches of in-vessel core catcher conceived in PFBR and EFR are shown in Figure 15.22.

The core catcher with heat removal capability from the bottom could increase the coolability range to a value from 55% to 100% of the core fuel [15.19]. The overall conclusion of the in-vessel retention study is that a partial core meltdown involving up to approximately 50% of the core can be accommodated within the main vessel provided a heat sink is available. One important advantage for a pool-type SFR is that the cooling capabilities of the sodium inventories can still be used if the main vessel remains intact. The thermal capacity of the in-vessel sodium between the normal operating temperature and sodium boiling point is substantial and provides a period of several hours before significant sodium boiling occurs. For a number of core melt accident sequences, the decay heat removal loop may still be functioning, leading to the satisfactory mode of heat rejection. Natural convection occurs in internally heated molten pools, with hot fluid rising and colder fluid moving downward. The onset of motion in a fluid occurs when buoyancy forces caused by temperature gradients exceed the viscous forces. Inertial forces also affect the transport process after flow develops.

15.5.4   Postaccident Heat Removal

Heat generation in the core debris settling on the core catcher following a CDA involves (1) transient heat generation associated with decreasing fission process (immediately after neutronic shutdown), (2) decay heat of the activated fuel nuclides, (3) decay heat of fission products (FPs), and (4) decay heat due to stainless steel activation. Transient heat generation is significant only for a very short duration (~10 s) after attaining subcriticality. In large power plants, considerable decay heat will be generated in the radial blanket subassemblies and such heat generation should also be included in the PAHR studies.

The PAHR performance is to be ensured to respect three criteria: (1) the creep damage on the triple point of the main vessel during the transient that is within permissible limits, (2) the sodium temperature under the debris tray that is below sodium boiling to avoid deterioration of heat transfer in the downward direction, and (3) sufficient mechanical resistance for the core catcher plates taking into account the mechanical loads and temperature reached. The PAHR study should consider three different possible scenarios: (1) The lower plenum remains closed after the passage of the debris, (2) the lower plenum is in communication with the upper plenum through a hole at the diagrid level, and (3) the lower plenum is in communication with the upper plenum through a broad hole at the diagrid level. The transfer of heat by natural convection of sodium around the debris tray and under the support plate can be evaluated under the steady-state assumption.

Further, in the innovative SFR, the PAHR capability is enhanced by incorporating features for (1) the protection against direct melt jet attack on the grid plate, (2) enlargement of coolant inventory under the core support structure for molten fuel quenching and fragmentation into small particles, (3) chimney for effective coolant circulation, (4) guide tubes for fuel debris settling to the lower plate (i.e., the debris goes down when the bed height exceeds a certain level by fluidization), and (5) multilayered debris tray for debris retention within a limited bed height of cooling and subcritical state.

In case the core debris melts through the vessel, heat transfer and interaction of the debris in the reactor cavity must be examined. Two topics are of principal interest: first, techniques that might be employed to reduce the consequences of melt through such as engineered core retention concepts (i.e., devices specifically designed to stop the flow of molten fuel debris) and, second, the interaction of sodium and core debris with a structural material below the reactor vessel. Details can be found in Ref. 15.20.

15.6 RADIOLOGICAL CONSEQUENCES

Though the CDA is beyond design basis accident (BDBA), RCB is provided to mitigate the consequences of CDA, ensuring that the dose rate at the site boundary is within the prescribed limit and habitability in the control room is not jeopardized. Two steps are involved in estimating the source term: First, one calculates the radioactive source in the RCB and, second, the source term is evaluated outside the RCB. Evaluation of the radioactive source in the RCB is discussed later. Evaluation of the environmental radioactive source term has been discussed in Section 12.4.

To evaluate the source term in the RCB, one requires the basic information, such as FP inventory in the core, extent of core damage in CDA, FPs reaching the cover gas, and subsequent release into the RCB through top shield leaks. Another pathway is through the main vessel melt. However, the damage evaluation studies performed for CDA show that the main vessel and top shield remain intact. So, the only path available is through top shield leaks. FP inventory depends on the core burnup. The inventory of radiologically important nuclides is given in Table 15.2 for various reactors. At the end of CDA, the core will be partly in the molten state and the remaining part will be in the vaporized state. For example, CDA analysis for PFBR [15.21] shows that at the end of CDA, 94% of the core is in the molten state and 20% is in the vaporized state. One has to find the fraction of FP that reaches the cover gas and subsequently into the RCB. A large amount of R&D effort has been put worldwide on this topic. During CDA, the core bubble expands, and then for the FP to reach the argon cover gas, the FP must pass through the sodium pool. Much of the radioactive source could be removed during the transport through sodium. A schematic of the radioactive sources in different regions of the plant under the CDA scenario is given in Figure 15.23. Kress et al. [15.22,15.23] identified the following possible removable mechanisms:

- During condensation of the fuel vapor onto the bubble/sodium interface, onto the structures in the bubble, and onto the entrained liquid droplets
- During the agglomeration and removal into the sodium
- FP diffusion to and reaction with sodium

TABLE 15.2

Core Inventories of Fission Products and Source Term in RCB

Isotope	Half-Life	PFBR (Bq)	OECD (Bq)	SNR-300 (Bq)	EFR (Bq)
I-131	8.02 d	1.48E+18	1.33E+18	7.40E+17	1.48E+17
I-132	2.30 h	1.96E+18	1.76E+18	9.80E+17	1.96E+17
I-133	20.80 h	2.54E+18	2.29E+18	1.27E+18	2.54E+17
I-134	52.50 m	2.53E+18	2.28E+18	1.27E+18	2.53E+17
I-135	6.57 h	2.23E+18	2.01E+18	1.12E+18	2.23E+17
Cs-134	754.50 d	2.82E+14	2.28E+14	1.41E+14	2.82E+13
Cs-137	30.07 y	8.42E+16	6.82E+16	4.21E+16	8.42E+15
Rb-88	17.78 m	5.22E+17	4.23E+17	2.61E+17	5.22E+16
Ru-103	39.26 d	2.41E+18	9.64E+16	4.10E+16	9.64E+16
Ru-106	373.59 d	1.06E+18	4.24E+16	1.80E+16	4.24E+16
Sr-89	50.53 d	7.12E+17	7.83E+16	1.21E+16	2.85E+16
Sr-90	28.79 y	2.88E+16	3.17E+15	4.90E+14	1.15E+15
Ce-141	32.50 d	2.11E+18	8.44E+16	3.59E+16	8.44E+16
Ce-144	284.89 d	1.05E+18	4.20E+16	1.79E+16	4.20E+16
Te-131m	30.00 h	1.63E+17	2.61E+16	8.15E+16	1.63E+16
Te-132	3.20 d	1.88E+18	3.01E+17	9.40E+17	1.88E+17
Ba-140	12.75 d	1.93E+18	2.12E+17	3.28E+16	7.72E+16
Zr-95	64.02 d	1.76E+18	7.04E+16	2.99E+16	7.04E+16
La-140	1.68 d	1.96E+18	7.84E+16	3.33E+16	7.84E+16
Kr-85m	4.48 h	2.26E+17	2.03E+17	2.26E+17	2.26E+17
Kr-87	76.30 m	4.08E+17	3.67E+17	4.08E+17	4.08E+17
Kr-88	2.84 h	4.94E+17	4.45E+17	4.94E+17	4.94E+17
Kr-85	10.70 y	4.69E+15	4.22E+15	4.69E+15	4.69E+15
Xe-133	5.24 d	2.55E+18	2.30E+18	2.55E+18	2.55E+18
Xe-135	9.14 h	2.66E+18	2.39E+18	2.66E+18	2.66E+18
Pu-239	24110 y	4.51E+15	4.51E+13	7.67E+13	4.51E+11

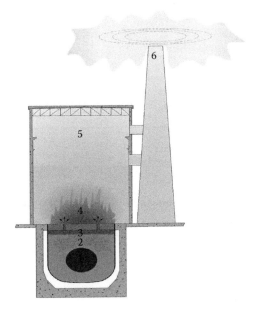

FIGURE 15.23 Schematic of radioactive sources under the CDA scenario. 1, Core; 2, Hot sodium pool; 3, Cover gas; 4, Sodium fire from ejected sodium; 5, Reactor Containment Building; 6, Stack.

Several models [15.24–15.26] have been developed for the previously mentioned phenomena. However, these have not been verified experimentally. Keeping all the uncertainties involved in the previously mentioned phenomena, Organization for Economic Co-operation and Development/ Nuclear Energy Agency (OECD/NEA) recommended [15.23] the source term for the RCB as given in Table 15.3. The results of several experiments conducted in several countries on the LMFBR source term during the last 20 years are summarized as follows.

The fission gas release to the cover gas is quite high and, for this reason, does not need more precise studies. The volatile FP release fraction depends on their physicochemical properties and evolution of thermodynamic conditions during the development of the accident. The nonvolatile FP and fuel release fraction are closely bound to the molten fuel–coolant interaction and then to the coolant decontamination factor depending on many parameters such as bubble formation and pool depth. The release fractions from the core to the RCB for the 26 nuclides considered are given in Table 15.3. The values for iodine, cesium, tellurium, and fuel are taken from the values deduced from FAUST experiments (Table 15.4). For nonvolatile nuclides, the release fractions should be much lower. From the results of the FAUST experimental program conducted in Germany [15.27], a realistic instantaneous source term is arrived and is given in Table 15.5. Based on the several experimental results, the source term estimated for different reactors [15.27] is given in Tables 15.2 and 15.5. It has also been observed that for volatile FPs, the iodine and cesium vapor contribution to

TABLE 15.3

Source Term in RCB

Isotope	Release Fraction into RCB	Isotope	Release Fraction into RCB
I-131	1.00E–01	Ce-144	1.00E–01
I-132	1.00E–01	Te-131M	4.00E–02
I-133	1.00E–01	Te-132	4.00E–02
I-134	1.00E–01	Ba-140	4.00E–02
I-135	1.00E–01	Zr-95	1.00E–04
Cs-134	1.00E–01	La-140	1.00E–01
Cs-137	4.00E–02	Kr-85m	1.00E–01
Rb-88	4.00E–02	Kr-87	1.00E+00
Ru-103	4.00E–02	Kr-88	1.00E+00
Ru-106	4.00E–02	Kr-85	1.00E+00
Sr-89	4.00E–02	Xe-133	1.00E+00
Sr-90	4.00E–02	Xe-135	1.00E+00
Ce-141	1.00E–01	Pu-239	1.00E+00

TABLE 15.4

Source Term Deduced from the FAUST Experiment

Radionuclide	Fraction
Xe, Kr	1
I, Br	0.1
Cs	1.00E–04
Te	—
SrO	1.00E–04
UO2	1.00E–04

TABLE 15.5

Source Term in RCB for Different Reactors

Radionuclide	OECD[a]	SNR-300[a]	EFR[b]	MONJU[a]
Xe, Kr	0.9	1	1	1
I	0.9	0.5	0.1	0.1
Cs	0.81	0.5	0.1	0.1
Te	0.16	0.5	0.1	0.1
SrO	0.11	0.017		
Fuel	0.01	0.017	1.00E–04	2.00E–03

[a] Energetic.
[b] Nonenergetic.

the source term is negligible compared to the iodine and cesium combined with liquid sodium aerosol contribution. For nonvolatile radionuclides, the source term depends on the pool height above the core and on the obstacles present in the plenum. From this source, the radioactive source to the environment through the stack can be evaluated as described in Section 12.4.

REFERENCES

15.1 Dufour, P., Fiorini, G.L. (2005). Context and objectives of the workshop. *Workshop on Severe Accidents for the Sodium Fast Reactor*, Cadarache, France, December 2005.

15.2 Rouault, J., Chellapandi, P., Raj, B., Dufour, P., Latge, C. et al. (2010). Sodium fast reactor design: Fuels, neutronics, thermal-hydraulics, structural mechanics and safety. In *Handbook of Nuclear Engineering*, D. Cacuci (ed.), Vol. 4, Chapter 21, pp. 2321–2710. Springer, New York.

15.3 Livolant, M., Dadillon, J., Kayser, G., Moxon, D. (1990). SCARABEE: A test reactor and programme to study fuel melting and propagation in connexion with local faults, objectives and results. *International Fast Reactor Safety Meeting*, Snowbird, UT, Session 2, Vol. II.

15.4 Maity, R.K., Velusamy, K., Selvaraj, P., Chellapandi, P. (2011). Computational fluid dynamic investigations of partial blockage detection by core-temperature monitoring system of a sodium cooled fast reactor. *Nucl. Eng. Des.*, 241, 4994–5008.

15.5 Natesan, K. (2007). Post accident heat removal analysis for PFBR. *IGCAR-CEA Seminar on Liquid Metal Fast Reactor Safety*, Kalpakkam, India, February 2007.

15.6 IAEA. (1999). Status of liquid metal cooled fast reactor technology. IAEA-TECDOC-1083. IAEA, Vienna, Austria, April 1999.

15.7 Baque, F. (2007). Feedback experience on CEA gas entrainment studies. *IGCAR-CEA Meeting on LMFBR Safety*, Kalpakkam, India, December 2007.

15.8 Riyas, A., Devan, K., Mohanakrishnan, P. (2013). Perturbation analysis of prototype fast breeder reactor equilibrium core using IGCAR and ERANOS code systems. *Nucl. Eng. Des.*, 255, 112–122.

15.9 Velusamy, K., Chellapandi, P., Chetal, S.C., Raj, B. (2010). Challenges in Pool Hydraulic Design of Indian Prototype Fast Breeder Reactor. *SADHANA*, 35(Part 2), 97–128.

15.10 Banerjee, I. et al. (2013). Development of gas entrainment mitigation devices for PFBR hot pool. *Nucl. Eng. Des.*, 258, 258–265.

15.11 Yamano, H. et al. (2008). Development of 3-D CDA analysis code: SIMMER-IV and its application to reactor case. *Nucl. Eng. Des.*, 238, 66.

15.12 Endo, H. (1986). Analysis of core meltdown for protected loss of heat sink accident in a fast breeder reactor. *BNES Conference on Science and Technology of Fast Reactor Safety*, Guernsey, U.K., Vol. 2, pp. 77–83.

15.13 Fauske, H.K. (1983). An assessment of the accident energetics potential in connection with hypothetical loss of heat-sink accidents in LMFBRs. *Trans. ANS*, 45, 366.

15.14 Theofanous, T.G., Bell, C.R. (1985). An assessment of CRBR core disruptive accident energetic. *Proceedings of the International Topical Meet on Fast Reactor Safety*, Knoxville, TN, Vol. 1, pp. 471–480.

15.15 Jackson, J.F. et al. (1974). Trends in LMFBR hypothetical accident analysis. *Proceedings of the Fast Reactor Safety Meeting*, American Nuclear Society, Beverly Hills, CA, CONF-740401-P3, pp. 1241–1264, April 1974.

15.16 Gluekler, E.L. et al. (1982). Analysis of in-vessel core debris retention in large LMFBRS. *Proceedings of the LMFBR Safety Topical Meeting*, Lyon, France.

15.17 Le, R.C., Kayer, G. (1979). An internal core catcher for a pool type LMFBR and connected studies. *Proceedings of the International Meeting on Fast Reactor Safety Technology*, Seattle, WA.

15.18 Gabor, J.D. et al. (1974). Studies and experiments on heat removal from fuel debris in sodium. *Proceedings of the Fast Reactor Safety Conference*, Beverly Hills, CA, FONF-740401, p. 823.

15.19 Gluekler, E.L., Huang, T.C., Jospeh, D. (1979). In-vessel retention of core debris in LMFBRS. *Proceeding of the International Meeting in Fast Reactor Safety Technology*, Seattle, WA.

15.20 Waltar, A.E., Todd, D.R., Tsvetkov, P.V. (2012). *Fast Spectrum Reactors*. Springer, New York, USA.

15.21 Srinivasan et al. (2008). ULOF and CDA Analysis of PFBR with reactivity worths based on ABBN-93 cross-section set. Indira Gandhi Centre for Atomic Research (IGCAR), Internal Report of IGCAR: RPD/SAS/177.

15.22 Silberberg, M. (ed.) (1979). State of the art report on nuclear aerosols in reactor safety. NEA Committee on the Safety of Nuclear Installation, OECD, Washington, DC.

15.23 Kress, T.S. et al. (1977). Source term assessment concepts for LMFBRs. Aerosol Release and Transport Analytical Program, Oak Ridge National Lab., Tennessee, USA. ORNL-NUREG-TM-124.

15.24 Ozisik, M.N., Kress, T.S. (1977). Effects of internal circulation velocity and non-condensible gas on vapour condensation from a rising bubble. *ANS Trans.*, 27, 551.

15.25 Theofanous, T.G., Fauske, H.K. (1973). The effects of non-condensibles on the rate of sodium vapour condensation from a single rising HCDA bubble. *Nucl. Technol.*, 9, 132.

15.26 Kennedy, M.F., Reynolds, A.B. (1973). Methods for calculating vapour and fuel transport to the secondary containment in an LMFBR accident. *Nucl. Technol.*, 20, 149.

15.27 Balard, F., Carluec, B. (1996). Evaluation of the LMFBR cover gas source terms and synthesis of the associated R&D. *Technical Committee Meeting on Evaluation of Radioactive Materials and Sodium Fires*, O-arai, Japan, IWGFR/92.

16 Sodium Safety

16.1 INTRODUCTION

The development of SFRs has been possible thanks to the attractive nuclear, physical, and even some chemical properties of sodium. Sodium is the most common element of alkali metals. Sodium has only one stable isotope: ^{23}Na. Neutron flux leads to the formation of radioactive isotopes: ^{24}Na (half-life is 14.98 h), inducing the necessity to wait for decay before some interventions on primary circuits, and ^{22}Na (half-life is 2.6 years), to be taken into account during the decommissioning stage. This low activation is also a very attractive property of sodium for nuclear use in sodium fast reactors (SFRs). Sodium is in the liquid state at 98°C and the boiling point is 883°C. This wide range in the liquid state at atmospheric pressure is responsible for the high thermal inertia of the sodium system, a favorable safety feature. The density of sodium is always less than that of water. It has a value of around 850 kg/m^3 at 400°C. The density of the liquid phase is higher than the solid phase (volume expansion is about 2.7% during solidification). Due to this characteristic, it is necessary to follow specific procedures for melting sodium in storage vessels or containers. The viscosity of sodium at 400°C (310 Pa·s) is of the same order as the viscosity of water at 100°C (280 Pa·s). Due to such similarity between density and viscosity of sodium and water, it is possible to carry out experimental studies with water to simulate hydraulics of sodium components. The thermal conductivity of sodium is very high: about 76.6 W/m/K at 573 K. Comparatively, the conductivity of water varies from 0.6 W/m/K at 20°C to 0.465 W/m/K at 350°C (at a pressure of 150 bars), whereas the conductivity of sodium is between 100 and 150 times higher at atmospheric pressure. In view of its high boiling point, sodium is considered as not very volatile. The fact that sodium is not very volatile has several consequences: in normal operation, evaporation rapidly attains an equilibrium level (condensation = vaporization). Consequently, in various gas plenums, and particularly in the main vessel, the mass transfer toward the colder roof of the slab is rather limited, particularly in the presence of argon (due to its low thermal conductivity). Nevertheless, the SFR operational feedback shows that sodium aerosols are deposited in the upper structures or narrow gaps. Hence, it is necessary to take care of the vapor and aerosol traps in the cover gas to prevent any related issues. Due to this low volatility, the sodium flames are very short and the heat produced by the fire is rather low: thus, it is possible to extinguish the fire by spreading a powder mixture of Na carbonate, Li carbonate, and graphite. Sodium, like all metals, has a very low electrical resistivity. These attractive conduction properties are widely used in sodium technology: instrumentation, level probes, flow measurements, electromagnetic pumps, and leak detection. The speed of sound in sodium varies little with temperature. Sound waves therefore propagate very well through sodium. This property is widely used in all metrology and visualization techniques in sodium and compensates the opacity in sodium for in-service inspection operations. The influence of temperature on sonic velocity is high. Using this, the temperature could be also deduced from the sound velocity in sodium. Since sodium has a tendency to lose its external electron, it will have very significant reducing characteristics, as the case with all alkali metals. It reacts exothermically with water, potentially with violence, as a function of local conditions. This reaction with water producing sodium hydroxide and hydrogen gas is strongly exothermic (162 kJ/mol of Na) and extremely fast. For these reasons, sodium–water reaction (SWR) that can occur in steam generator (SG) units is considered as an important safety issue and several measures are developed and implemented to mitigate this event. Nevertheless, this reaction with water is favorably used for the development of cleaning processes for structural material wetted with sodium during handling operations and, moreover, for the conversion of large amounts of sodium into sodium hydroxide at the end of the reactor operation, during the decommissioning phase. Solid sodium quickly oxidizes in air and liquid sodium burns in air over melting point, that is, 98°C, if it is spread out in air and over 140°C in other cases; it forms sodium peroxide Na_2O_2 or, with limited oxygen, the oxide Na_2O.

Due to its high chemical reactivity, extreme care is required in handling elemental/metallic liquid or solid sodium: it must be stored or used in the liquid phase in an inert (oxygen and moisture-free) atmosphere such as nitrogen or argon. Further, specific means to reduce the occurrence of sodium leaks and detection systems have to be developed. In case of leak detection with insulated electrical wire (commonly used), the leaked sodium would be in contact with the wire mounted on the pipe and generates a warning. For mitigating sodium fires, several approaches have been developed: minimize the oxygen available for combustion (e.g., nitrogen injection), specific catchers (e.g., leak collection trays [LCTs]), and dedicated volume apportionments.

Sodium contains various nonradioactive impurities either present from the start or introduced during operation. In the primary circuit, there is essentially one source of discontinuous contamination by oxygen and hydrogen at the start of the cycle and also at the end of the fuel/component handling operations. There is another continuous source of hydrogen coming by permeation from the secondary circuit. In the intermediate loops, there is essentially one continuous source of hydrogen, mainly due to aqueous corrosion in SG units. In the event of SWR in SG, there is one discontinuous source of hydrogen, sodium hydroxide, Na_2O, and NaH crystallized products, which is also the case after contamination by air and moisture, following handling operation. Thus, oxygen and hydrogen constitute the major contaminants and their solubilities in sodium are very low. Their solubility becomes insignificant near the melting temperature of sodium (this is a property specific to sodium, in comparison with other liquid metals, used as coolants). Oxygen contributes to corrosion of steel, and the corroded activation products are transported from the core toward the components and mainly to the intermediate heat exchangers (IHX), resulting in contamination. All the elements that constitute the steel (iron, Cr, Ni, Mn, C) are susceptible to dissolution in sodium. Even if it is generally considered that corrosion is only significant for temperatures higher than 540°C and for oxygen contents of the order of 10 ppm, SFR requires to be operated with an oxygen content below 3 ppm in normal operation to limit corrosion. Stress corrosion cracking phenomena can occur in the presence of aqueous soda, when the component, wetted by a residual film of sodium, is filled with air and moisture (which reacts with the film of sodium). Over 80°C for ferritic steels and over 110°C for austenitic steels, transgranular cracking can occur. When preheating the facilities, prior to filling with sodium, such temperatures are unavoidable. Therefore, it is necessary to avoid any presence of residual aqueous soda on components, gaps, etc., by appropriate rinsing and drying processes.

Because steels tend to exchange nonmetallic elements such as carbon and nitrogen with the liquid metal, the effects of these elements have to be taken into account for the evaluation of the corrosion behavior of steels in liquid sodium and for mass transfer calculations, but also for the potential evolution of mechanical properties. At high temperatures, the diffusion rate of carbon in steel is fast enough to cause changes in carbon concentrations in cladding materials. These evolutions have generally marginal effects on the corrosion behavior of steels at 600°C but could be much more important at higher temperatures. In intermediate loops, hydrogen content has to be maintained as low as possible (<0.1 ppm), in order to allow a fast detection of water ingress. Moreover, since an SWR induces a very high hydrogen content, hydrogen embrittlement on the structural materials should be avoided by an efficient sodium purification process. The detection can be carried out with hydrogen meters; the diffusion type with Ni membrane is generally used (mass spectrometer and electrochemical cell have been developed for the prototype fast breeder reactor [PFBR]).

Two main sodium purification processes have been developed with regard to oxygen and hydrogen control: cold trapping and hot trapping. Cold trapping by crystallization of Na_2O and NaH lowers the sodium temperature below the saturation temperature, thus creating optimal conditions for Na_2O and/or NaH nucleation and growth on a steel packing distributed in an auxiliary cooled vessel. Cold trapping is the worldwide process used for sodium purification in the SFR due to its undeniable advantages: oxygen and hydrogen are the two most important impurities trapped. Highest efficiency and capacity can be achieved with (1) optimized designs of cold traps, (2) the ability of the cold trap to be regenerated by extracting solely packing, and/or (3) in situ appropriate chemical process. Hot trapping or *getter* operation is based on the capacity of the chosen material to oxidize, thereby absorbing oxygen, when it is placed in the sodium containing oxygen. This process is generally chosen for small sodium volumes to be purified and when the risk of Na_2O dissolution by loss of the cooling function in the cold trap is unacceptable. As an example, the $Zr_{0.87}$–$Ti_{0.13}$ alloy has been qualified for hot trapping in an irradiation loop for Phénix. For hydrogen trapping, *hydride* traps (i.e., yttrium) could also be set up, but due to their very low potential capacity and reversibility, their application for large sodium volumes has never been foreseen. The monitoring of sodium quality can be performed by numerous techniques, for example, plugging meters, hydrogen meter, and oxygen meter.

Having briefed on some various important aspects of sodium, specific safety issues concerned with its violent chemical reaction with air and water are addressed in this chapter. Accordingly, this chapter addresses sodium fire scenarios, SWRs, sodium–concrete interactions, sodium fire mitigation effects, and innovative fire extinguishers. For more details on the science and technology aspects of sodium, readers are requested to refer to [16.1].

16.2 SODIUM FIRE

16.2.1 Sources of Sodium Leaks in SFR

Radioactive primary sodium leaks under normal operating conditions are not possible, since all the primary sodium lines within the reactor containment building (RCB) as well as passing through the RCB are double enveloped, filled with inert gas and periodical monitoring provisions. Sodium leaks from other components, piping, and systems particularly in the secondary circuit are possible. There are several sodium leaks reported in the literature and the details of these leaks are presented in Section 27.1. The sodium leak that occurred in a Japanese SFR (Monju) in 1995 [16.2] is worth highlighting. About 1000 kg of sodium leaked from a thermowell damaged by flow-induced vibration. The scenario of the leak event and consequences is shown in Figure 16.1. The BN-600 reactor experienced numerous sodium fires in the initial years of reactor operation, primarily due to pipe and weld cracking [16.3]. The largest primary sodium leak occurred in 1993,

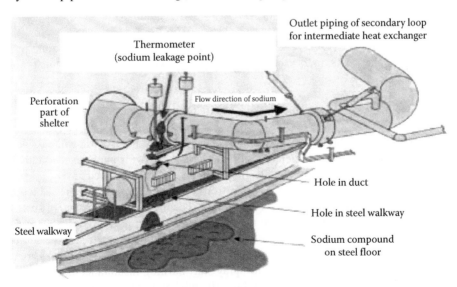

FIGURE 16.1 Sodium leak incident in Monju (SFR in Japan).

with about 1000 L of sodium that leaked from a cracked pipeline in the neighborhood of a weld joint. From the experience of sodium leaks, it is noted that the leak before break concept could be justified and the maximum quantity of leak is not more than 2 tons [16.4]. This also provides an experimental basis for the design of LCTs and the quantity of fire extinguisher materials required. For PFBR, this value is about 2 tons. Apart from these, sodium leaks are postulated under hypothetical events for the safety study. This scenario has been explained in Chapter 15. To highlight, a limited quantity of primary sodium is ejected out through top shield penetrations, when the rotatable plugs are lifted up under sodium slug impact during a core disruptive accident (CDA). This sodium provides the source term for the sodium fire and consequent design basis temperature and pressure loadings to the RCB.

16.2.2 Sodium Fire Scenarios and Consequences

The equations governing the chemistry of sodium fire are expressed in the following forms. Sodium oxide is formed by

$$4Na + O_2 \rightarrow 2Na_2O \quad \Delta H = -104 \text{ kcal/mol}$$

Sodium oxide (Na_2O) is a white powder that melts at 1193 K and dissociates (in the liquid form) at 2100 K. Hence, Na_2O cannot exist as a gas at temperatures much above 2100 K. Sodium peroxide is represented by the following reaction:

$$2Na + O_2 \rightarrow Na_2O_2 \quad \Delta H = -124 \text{ kcal/mol}$$

Na_2O_2 is a yellow-white powder that melts at 733 K with some decomposition. It decomposes at 930 K. During sodium combustion, the presence of water vapor results in the formation of NaOH via the following reactions:

$$2Na + 2H_2O \rightarrow 2NaOH + H_2 \quad \Delta H = -90 \text{ kcal/mol}$$
$$Na_2O + H_2O \rightarrow 2NaOH$$
$$2Na_2O_2 + 2H_2O \rightarrow 4NaOH + O_2$$
$$Na_2O_2 + H_2 \rightarrow 2NaOH \quad \Delta H = -80 \text{ kcal/mol}$$

Sodium hydroxide is a white powder that melts at 591.4 K and boils at 1663 K. Sodium hydroxide can react with atmospheric CO_2 to form sodium carbonate. Sodium carbonate is a white powder that melts at 1124 K:

$$2NaOH + CO_2 \rightarrow Na_2CO_3 + H_2O$$

All these reactions result in fires characterized by very low flames and a dense white oxide smoke, which is nontoxic but can cause irritation to respiratory organs. Apart from these, if sodium is burnt in oxygen under pressure, sodium superoxide, that is, NaO_2, will be produced.

In SFR situations, sodium leaks are possible from the cracked pipe and the leaked sodium spreads on the floor in the form of a pool (Figure 16.2). The fire from the surface of such sodium pool is termed as "pool fire." Typically, the mean burning rates for pool fire are about 6 kg/m²/h for surface combustion and 40 kg/m²/h for vapor-phase combustion. Sometimes, the sodium leak from the cracked pipe emerges like a spray in the form of fine droplets when the cracks are too narrow and sharp. The sodium droplets from the spray start burning instantaneously in the ambient atmosphere. This scenario is termed as spray fire. The pool and spray fire scenarios are depicted schematically in Figure 16.3. Between these two, the spray fire is generally considered to be more

FIGURE 16.2 Pool fire scenario: schematic.

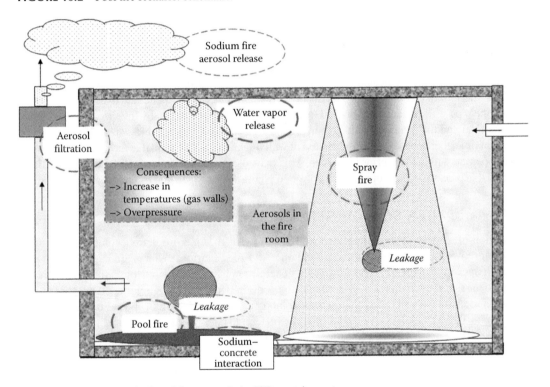

FIGURE 16.3 Sodium leak and fire scenario in SFR containment.

aggressive in terms of effective burning, burning rate, and resulting thermal and pressure loads to the containment. This is because the spray fire ignites in a highly divided state with a large number of smaller droplets, which provide more surface area for combustion. On the other hand, complete burning of the pool is not possible, since burning is limited to a certain depth of the pool surface. However, the burning time is longer for the pool fire. Further, thermal effects on the structural materials (particularly the concrete floor) are much greater in view of direct contact and longer burning time [16.5].

From the design point of view, the sodium fire, apart from the temperature and pressure rise, generates a dense white smoke and sodium aerosols. The smoke generated during sodium fire obscures vision and complicates fire fighting activities. In pool fire scenario, around 30% of the reacted sodium can be released as aerosols, which can be even higher in spray fires. The aerosols formed as a result of sodium–air combustion—sodium oxide (Na_2O), sodium dioxide (Na_2O_2), and sodium hydroxide (NaOH)—pose a health hazard to the personnel and to the general public if the release exceeds certain specified limits (2 mg/m^3 for 8 h). They may also cause structural damage to concrete buildings due to deposition on floors, walls, and ceilings.

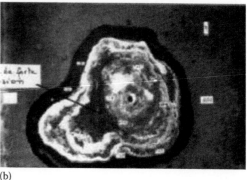

(a)   (b)

FIGURE 16.4 (a) Sodium–insulation interaction and (b) corrosion of steel pipe. (From Latge, C., *Interaction Sodium and Materials, Fast Reactor Science and Technology*, CNEA, Bariloche, Argentina, October 2012.)

Experiences have also indicated that very small leaks could not be detected effectively by the current sensor design and technologies. The undetected small leaks would penetrate over the interspace between the pipe wall surface and thermal insulation. The sodium chemically interacts with the insulation material and the reaction product generated is highly corrosive to stainless steel. If they are not detected in time, there could be significant loss of wall thickness to the extent that the pipe may develop a larger opening without any prior indication. This may also challenge leak before break justification. Figure 16.4 depicts the nature of a typical corroded piping wall surface due to such reaction [16.6].

16.2.3 International Studies on Sodium Spray Fires: Highlights

Combustion of a spray is complicated to study because of the interdroplet phenomena associated with it. Moreover, several studies are reported on pool fire scenarios. Readers are requested to read a review paper by Newman, which discusses general combustion behavior of sodium extensively [16.5]. He carried out a number of experiments on sodium pool fires and gave a good account of the physical and chemical processes that govern combustion of liquid sodium. Based on his observations in a fairly deep pool, he concluded that the different stages of the burning behavior—surface combustion and vapor-phase combustion—were attributable to the pool temperature. Hence, the focus in this section is on spray fire scenarios.

A single-droplet combustion was investigated initially to understand the basic concepts of spray combustion. The theories of single-droplet combustion were then expanded to create spray combustion theories. The experimental works on sodium fire have shown that oxygen concentration, spray velocity, and droplet diameter are few parameters that could affect sodium spray combustion. Richard et al. [16.7] studied experimentally the combustion of a single droplet with diameters ranging from 1000 to 3000 μm under controlled conditions using a movie camera. They observed the vapor-phase combustion of sodium separated from the droplet by a visible dead space. They found that the burning of sodium drops follows essentially the D^2 law that has been established for the combustion of hydrocarbon fuel droplets, that is, the square of the droplet diameter decreases linearly with time during combustion. The burning rate was found to increase linearly with the initial droplet diameter. They also reported that the dependence of the burning rate with the initial temperature was little and varied more significantly with the mole fraction of oxygen, which confirmed that the burning rate was controlled by oxygen diffusion to the droplet surface. Krolikowski et al. [16.8] developed a model to predict the burning rate and burning temperature of a single sodium droplet moving through air. The model was based on a diffusion-controlled vapor-phase combustion process. The reaction rate of an individual droplet was correlated to the burning rate of spray by means of a quasi-steady-state approach and an averaging technique. They assumed that each droplet in the spray is of equal size and distributed evenly across available space. The theoretical predictions were compared with their experimental results. It was found that the theory correctly predicted the experimentally observed variations with respect to oxygen content, spray velocity, and particle size. Morewitz et al. [16.9] studied experimentally sodium droplet combustion for different droplets of size varying from 0.209 to 0.731 cm with different fall distances and reported the mass burning rates for the same. Droplets of size greater than 0.8 cm were found to break up into smaller satellite droplets, and as the droplet size increased for the given falling height, it was reported that the combustion efficiency decreased, that is, a lot of unburnt sodium was found in the quenched droplets. Tsai [16.10] developed a computer code called NACOM to analyze the sodium spray fire. This code uses a single-droplet vapor-phase combustion theory, which is well established for combustion of hydro-carbon fuel droplets. The burning rate of spray is estimated by summing up the burning rates of all the individual droplets of various sizes. The size variation of the droplets is accommodated by including a droplet size distribution function called Nukiyama–Tanasawa distribution function. Also, a preignition model is included in the code to estimate the burning rate of the droplet before it attains the vapor-phase combustion.

Kawabe et al. [16.11] studied experimentally the sodium spray combustion to measure the burning rate, pressure, and temperature rise transients in the containment building. The experiments were conducted in a 2 m^3 capacity closed vessel and about 0.4 kg of sodium was sprayed. The oxygen concentration was varied from 0 to 0.21 mol fraction in their experiments. It was shown that the peak pressure rise was a strong function of the oxygen mole fraction. The measured results were compared with the analytical model based on a vapor-phase combustion mechanism in which the heat transfer from the flame to the droplet is the combustion rate–determining process. Okano and Yamaguchi [16.12] investigated numerically the combustion of a free falling single liquid sodium droplet in a steady airflow (forced convection) using the computational fluid dynamics (CFD) code COMET with a four-step global sodium reaction mechanism leading to the formation of Na_2O, Na_2O_2, and NaOH through the thermodynamic equilibrium approach. They obtained a flame temperature of about 1700 K and identified different regions in which Na_2O or NaOH is formed when a sodium drop burns. Makino and Fukada [16.13] carried out experimental studies on the ignition and combustion behavior of a falling sodium droplet. They identified the effect of parameters such as droplet temperature, droplet size, oxygen concentration, and relative speed of the droplet on the ignition delay of a sodium droplet. They verified that the D^2 law is valid for the burning sodium droplet. Yuasa [16.14] also investigated experimentally the ignition and burning behavior of a single droplet of sodium. He reported that the ignition temperature increases with the increase in the moisture content of air. The ignition temperature can be as high as 300°C for moist air (water vapor pressure of 20 mm Hg) when compared to the ignition temperature of 200°C for dry air. He concluded that the presence of the nonporous sodium hydroxide layer was responsible for the delayed ignition in moist air conditions. Overall, his work was focused on the sensitivity of the ignition point of sodium for different ambient conditions.

Apart from the previously mentioned aspects, there has been significant research activity, mainly aimed at developing a numerical model for the sodium fire scenario. Several sodium fire codes and facilities have been developed in various countries and had been used to validate experimental results, which are described in Chapters 17 and 18, respectively.

16.2.4 Sodium Fire Studies for PFBR: A Specific Case Study

A few interesting fundamental experimental studies that have been carried out to investigate the effect of small sodium leaks, sodium–insulation interaction and sodium droplet burning, particle distribution during sodium spray fire, and oxygen concentration in sodium burning are presented in the following sections.

In case of sodium, leaks from a very small crack have a lot of uncertainties, in the sense that they may leak and then stop for a while and again may restart. This leak may be too small for reliable detection. The important consequence is the applicability of the leak before break concept for sodium pipes under such smaller leaks (see Section 16.2.2). In studying such small leaks, a dedicated innovative experimental setup consisting of a stainless steel vessel with a 0.3 mm diameter pinhole was employed. The results indicated that the leak rate is a function of temperature, pressure, and sodium purity. Leak rates are found to be random and the average leak rates are in good agreement with the theoretically calculated values. Conditions for unplugging have been established in terms of temperature and pressure. The condition for unplugging of sodium was found to be above 290°C in conjunction with 4 bar pressure [16.15]. Further experiments are focused on the interaction of sodium and thermal insulation to investigate the corrosion of piping following a small sodium leak event.

A series of experiments were conducted in a unique facility with small sodium inventories (2–5 g) to understand the spray fire scenario and validate containment building pressure rise under CDA resultant fire and combined fire (spray followed by pool) scenarios [16.16]. The facility consists of a highly instrumented quartz cylindrical chamber and associated systems and some typical small-scale sodium spray fire scenarios are given in Figure 16.5. The particle size distribution of an ejected sodium droplet from a nozzle of 1.6 mm was measured by processing the recorded images of a high-speed camera. A typical particle size distribution is shown in Figure 16.6. The variation in flame diameter of burning sodium droplets with time was measured and used as input for modeling the spray fire scenario. The lifetime of a single burning droplet was measured by processing the image captured during combustion (Figure 16.7). Sodium oxide aerosol produced during sodium combustion was allowed to settle down and analyzed for its size distribution using the laser scattering technique, as shown in Figure 16.8.

In order to determine the heat load accurately, heat transfer to the ambient atmosphere is computed from the temperature, size distribution, and lifetime of the film emanating from a single particle. The integrated heat from individual particles knowing the particle size distribution will provide a realistic thermal load to the containment. From this value and knowing the maximum possible heat that can be generated from the total mass of the sodium burnt, the burning efficiency of a spray fire can be calculated. With this, a numerical analysis has been carried out to understand the single sodium droplet combustion behavior in air and hence to

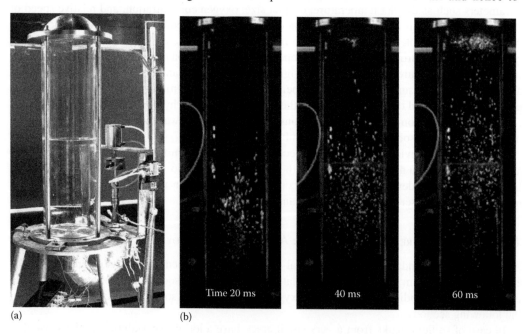

(a)   (b)

FIGURE 16.5   (See color insert.) Sodium spray fire scenario: (a) facility and (b) some experimental images.

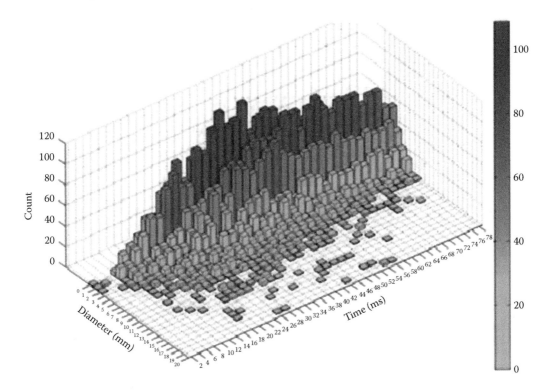

FIGURE 16.6 **(See color insert.)** Sodium spray fire: flame diameter distribution—experimental.

FIGURE 16.7 **(See color insert.)** Droplet burning scenario.

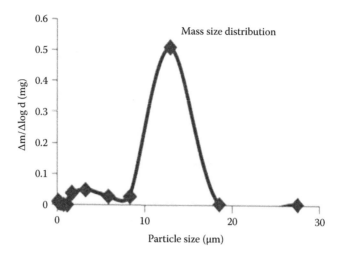

FIGURE 16.8 Sodium aerosol distribution (spray fire).

develop a sodium spray fire model. The two stages of droplet combustion have been considered. The preignition stage has been modeled based on a temperature-dependent surface oxidation process. It has been found that the droplet size, initial temperature, velocity, and atmospheric oxygen concentration are major parameters influencing the ignition delay time of the droplet. The ignition delay time of the droplet of various sizes has been estimated using a computer code and compared with the experimental results available in literature. The steady-state burning behavior of a droplet has been analyzed based on the diffusion-controlled vapor-phase combustion process. The major parameters affecting the burning rate of the droplet are droplet diameter, velocity, and ambient oxygen concentration. The burning rate of the droplet is reduced drastically when the ambient oxygen concentration falls below 4 mol%, which is also experimentally observed. Below this concentration level, vapor-phase combustion is no longer valid since the estimated flame temperature approaches the melting point of Na_2O and boiling point of sodium. The maximum flame temperature estimated was in the range of 2100–1800 K, and it is not affected by the size and velocity of the droplet. The results of this study are being used to model spray combustion, with appropriate consideration for size and spatial distribution of droplets in the spray and interaction between the individual droplets. More details of numerical simulation and benchmark results can be found in [16.17].

Adopting the previously mentioned numerical simulation, a benchmark computation was carried out to quantify the sodium combustion characteristics of two typical droplet sizes (1 and 5 mm). The numerical simulation involves both preignition and vapor combustion phases. The evolution of droplet sizes and their temperatures in these phases are shown in Figure 16.9. The temperature evolution of the flame resulting from droplet combustion is quantified in Figure 16.10. During the combustion process, the droplet size decreases in view of loss of mass being burnt; the flame diameter increases and the oxygen concentration decreases in close vicinity of the flame surface. The computed ratio of the flame diameter and droplet diameter is shown as a function of oxygen concentration in Figure 16.11. The fraction of heat transported to the containment atmosphere forms an important input for computing the temperature and pressure rise in the RCB. Accordingly, the heat transported to the unburnt droplet (radially inward) as well as to the atmosphere (radially inward) is determined and presented in Figure 16.12 during the preignition phase and Figure 16.13 during the vapor combustion phase. The ratio of heat transported to the atmosphere and the heat generated by complete burning of the droplet presented in Figure 16.14 shows that such ratio (~0.75%) is independent of the droplet diameter [16.18].

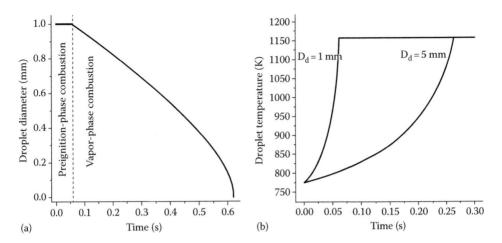

FIGURE 16.9 Sodium droplet burning phases, shrinking core, and temperature evolution. (a) Evolution of droplet size and (b) evolution of droplet temperature.

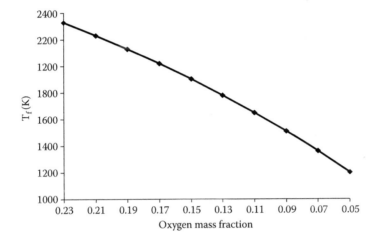

FIGURE 16.10 Evolution of flame temperature and adjoining oxygen concentration.

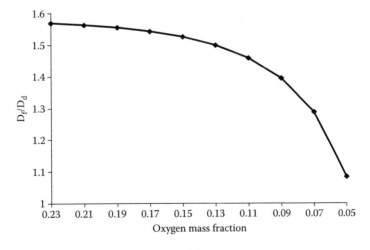

FIGURE 16.11 Evolution of flame diameter and adjoining oxygen concentration.

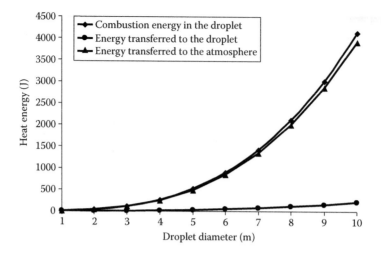

FIGURE 16.12   Heat balance during sodium droplet combustion (preignition phase).

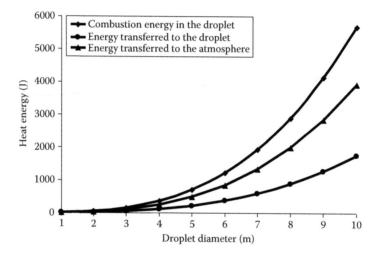

FIGURE 16.13   Heat balance during sodium droplet combustion (vapor phase).

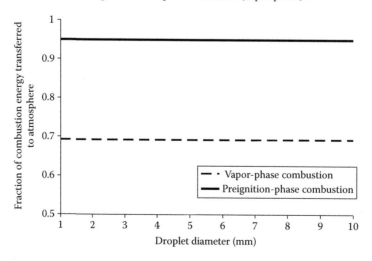

FIGURE 16.14   Heat energy transferred to containment atmosphere.

16.3 SODIUM–WATER INTERACTION

In an SG of SFR employing the steam–water system for power conversion, steam/water at high pressure and sodium at low pressure are separated by a steel tube wall. Any small crack or hole in the tube would result in SWR, which produces heat and other reaction products, mainly $NaOH$, Na_2O, and NaH. These products cause corrosion and erosion of nearby tubes leading to further escalation of water/steam leak rates. Exothermic chemical reactions occur between sodium and water:

$$Na + H_2O(l) \rightarrow NaOH + \frac{1}{2}H_2 \quad \Delta H_{298}^\circ = -35.2 \text{ kcal/mol}$$

$$Na + H_2O(g) \rightarrow NaOH + \frac{1}{2}H_2 \quad \Delta H_{298}^\circ = -45.0 \text{ kcal/mol}$$

$$Na + \frac{1}{2}H_2 \rightarrow NaH \quad \Delta H_{298}^\circ = -13.7 \text{ kcal/mol}$$

$$Na + NaOH \rightarrow Na_2O + \frac{1}{2}H_2 \quad \Delta H_{298}^\circ = +2.6 \text{ kcal/mol}$$

$$2Na + H_2O(l) \rightarrow Na_2O + H_2 \quad \Delta H_{298}^\circ = -31.1 \text{ kcal/mol}$$

$$2Na + NaOH \rightarrow Na_2O + NaH \quad \Delta H_{298}^\circ = -11.1 \text{ kcal/mol}$$

From the previous reactions, it can be estimated that about 1.5 mol of hydrogen gas gets released if all the reactions occur simultaneously. This energy release leads to rise in temperature and instantaneous pressure buildup in a confined vessel, deciding the design limits for the vessel. Apart from the enthalpy of the reaction, hydrogen gas that evolved as a product of the reaction also builds up pressure and has the potential to explode if it comes in contact with oxygen and may lead to severe damage of equipment and personnel. All these factors increase the safety concern for the design of a sodium–water heat exchanger in fast reactors. Therefore, several engineered safety features are provided for prevention, detection, and mitigation of SWR in fast reactors. Generally in SGs, the probable zones of failure are tube to tube sheet weld joints. A small crack in the tube at the tube to tube support may be developed due to fretting corrosion caused by flow-induced vibration. In all these cases, steam or liquid water under high pressure starts leaking into the sodium side, which is the low-pressure side of the heat exchanger, at high velocities like a jet depending on the size of the failure. Several minor SWRs have been reported in literature [16.19]. Some notable ones are detailed in the following paragraph.

In the French SFR Phénix SGs, four leaks occurred after 9 years of operation. All leaks were found in the reheater modules at the butt welds on the hottest part of the tube. The reason for failure is thermal fatigue. In the start-up sequence, the reheater was fed with water via a turbine bypass. As temperature rose, water changed to steam. This has caused major thermal stress particularly around the weld beads where extra thickness was present. In the first leak alone, approximately 8 m^3 of sodium entered in the steam circuit. Modifications made after the incident are as follows: Excess thickness at weld point is reduced, a maximum excess thickness of welds is specified at the design stage itself, a new start-up procedure was made to eliminate any possibility of liquid water reaching the reheater stages, a second nitrogen system is added to each stage, and draining of sodium was included in the procedure. All 36 reheater modules have been replaced by new modules. Major SWRs also occurred in PFR. PFR was commissioned in 1973. The three evaporator units were first filled with water in 1974, and since that time, there had been 33 leak events in which a total of about 75 small leaks had been found. All of the leaks had been in the tube to tube sheet welds resulting in steam entering the low-pressure gas space above sodium. The reason for failure is stress corrosion cracking from the water side due to chlorine atmosphere. The most important thing has been the presence of the hard and stressed tube to tube sheet weld that had not had any postweld heat treatment. The main solution adopted to obtain reliable operation of the existing evaporators is to insert sleeves in the tubes to cover the problem welds.

16.3.1 Classification of Leaks in SG and Their Effects

Depending on the leak rate and its effect, the SWR is classified into four types: microleaks, small leaks, intermediate leaks, and large leaks. Microleaks are those that emerge from a leak size ~0.07 mm with a leak rate <0.1 g/s. Only self-wastage occurs during the microleaks. The phenomenon of self-wastage is shown in Figure 16.15a. Any leaks <0.01 g/s may disappear due to plugging by SWR products. Small leaks are defined as the leak rate that lies in the range of 0.1–10 g/s and leak size in the range of 0.07–1 mm. For these leak rates, some secondary tube failures may be possible in a few minutes due to impingement wastage of a tube that is directly opposite to the leaking tube; damage to the adjacent tubes is possible due to corrosion and erosion (Figure 16.15b). The possible temperature ranges of reaction zones are 300°C–500°C at the center of the flame (unreacted steam), 1200°C–1400°C at the center of the reaction zone, 900°C–1000°C at the outside sodium-rich zone, and 1000°C–1200°C at the tube surfaces. Intermediate leaks are defined as the leak rate that lies in the range of 10–2000 g/s and leak size in the range of 1–7 mm. Wastage of multiple tubes could occur due to overheating. Generally, the pressure rise of the system is relatively slow. Large leaks are defined when the leak rate >2 kg/s with a leak hole diameter >7 mm. The pressure rise due to hydrogen generation is the major effect. Surge tank and rupture disks are utilized to reduce the pressure.

A large SWR is of great concern to safety and reliability of the reactors. In the large leak SWR, the major effect is increased pressure in the secondary sodium system and components (Figure 16.16). The pressure increase at the IHX tube sheet is more critical as it is the boundary that separates the

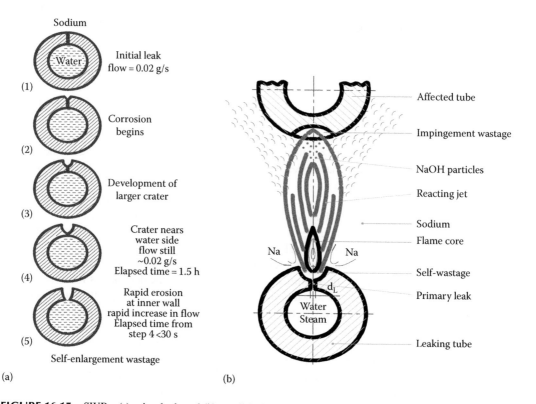

FIGURE 16.15 SWRs: (a) microleak and (b) small leak.

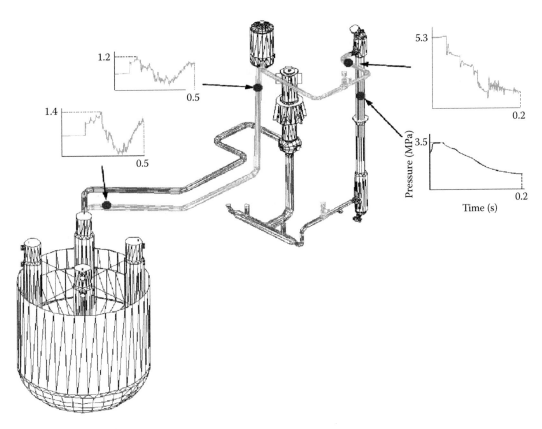

FIGURE 16.16 Pressure wave propagation under large leak SWR.

radioactive primary and secondary sodium. Failure of the IHX tube sheet will lead to secondary sodium entering into the primary sodium system leading to possible core neutron flux fluctuations. A surge tank is provided on the upstream of the SG to mitigate the initial pressure spike. Further, rupture disks are provided at both the sodium inlet and outlet of the SG. When the sodium pressure at the rupture disk goes above the set pressure, the rupture disk will burst and relieve the system pressure. In fact, the design loads for SG shell, sodium piping, piping supports and snubbers, and IHX tube to tube sheet thickness are decided by the transient pressure and forces developed during SWR in the shell and subsequently at various pipe bend locations.

Large leak SWR experiments are carried out by different countries (France, Federal Republic of Germany (FRG), the United Kingdom, the United States, USSR, and Japan) for validating their computer codes (PLEXUS, POOL+HEINKO, FLOOD, TRANSWRAP-II, SWACS) and also to confirm their design against large SWRs [16.20]. Either a full-scale or scale-down model SG is used in these experiments. The general conclusion derived from these experiments is that the pressure rise is the major effect and the wastage is not as severe as in the case of small leak. There is no leak propagation due to the fast movement of the sodium–hydrogen/steam interface. For the medium leak range, experiments are carried out by Power Reactor and Nuclear Fuel Development Corporation (PNC), Japan, for the Monju reactor. The simultaneous wastage on many tubes and also the temperature rise in the tubes are predicted by these experiments. Overheating experiments in these leak range are carried out by INTERATOM, commissariat à l'énergie atomique et aux énergies alternatives (CEA), and United Kingdom Atomic Energy Authority (UKAEA) [16.21]. The main conclusions from these experiments are only for leak rates above 80 g/s where overheating failure is possible. An uncooled tube will take minimum 4 s to fail, and in case of cooled tubes, it takes minimum 20 s. In India, the Sodium–Water Reaction Test Facility (SOWART) was

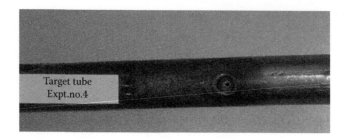

FIGURE 16.17 Impingement wastage on target tube. (From Kishore, S. et al., *Nucl. Eng. Des.*, 243, 49, 2012.)

constructed to study various aspects of the SWR phenomenon [16.22]. This test facility is equipped with both hydrogen in sodium detectors and hydrogen in argon detectors. The steam system can produce superheated steam of 17.2 MPa at 753 K. Leak simulators of mod. 9Cr–1Mo having a calibrated pinhole leak were used for injecting steam into sodium. A target tube of mod. 9Cr–1Mo is fixed into the impingement wastage test section from the top in such a way that the reaction jet from the leaking tube hits on the target tube and produces wastage on its surface. Six self-wastage experiments and nine impingement wastage experiments were carried out to understand the behavior of leak propagation and data generation. Figure 16.17 depicts the photograph of one typical self-wasted tube. The wastage resistance of mod. 9Cr–1Mo steel is experimentally validated and found to be better than that of 2.25Cr–1Mo steel, as seen in Figure 16.18 [16.22].

Wastage tests for mod. 9Cr–1Mo steels were performed by Korea Atomic Energy Research Institute (KAERI) in a small leak SWR test facility [16.23]. Test specimens were exposed to small leaks of steam, issuing into stagnant sodium in the temperature range of 400°C and 450°C. The pressure and temperature conditions of steam were 150 kg/cm² and 350°C, respectively. Circular-type defects were used in these tests whose diameter ranged from 200 to 400 μm. The wastage data obtained in the stagnant sodium systems are shown in Figure 16.19. It showed that the tube spacing and sodium temperature have significant effects on the tube material wastage rate. Wastage rate increased as the leak rate increased. Comparative experimental evaluation of enlargement rates in two different materials—2.25Cr–1Mo and mod. 9Cr–1Mo steels—was also performed. Enlargement rate was found to be slightly faster in the 2.25Cr–1Mo steel than in the mod. 9Cr–1Mo steel.

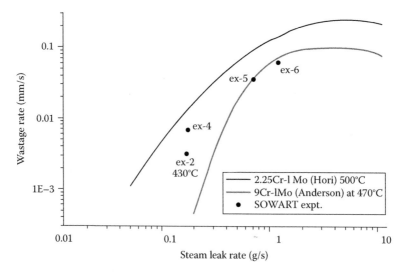

FIGURE 16.18 Comparison of wastage rate with literature data. (From Kishore, S. et al., *Nucl. Eng. Des.*, 243, 49, 2012.)

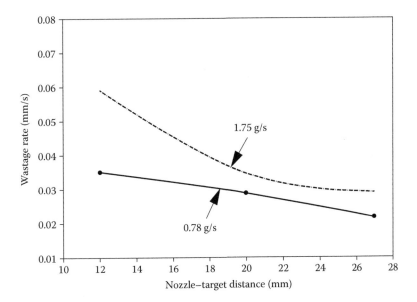

FIGURE 16.19 Experimental prediction of wastage rate in mod. 9Cr–1Mo. (From Jeong, J.-Y. et al., Wastage behavior of modified 9Cr-1Mo steel tube material by sodium–water reaction, *Transactions of the Korean Nuclear Society Autumn Meeting*, Gyeongju, Korea, 2009.)

To visualize the mechanism of sodium–steam interaction and the characteristics of subsequent temperature and pressure rise, an experimental setup has been installed at Indira Gandhi Centre for Atomic Energy (IGCAR). This setup consists of a water injection system and a cubical test chamber as shown in Figure 16.20a. The test chamber is provided with a heater-mounted cup to create a sodium pool, inert gas purging system, gas monitoring system, and a piezoelectric-type transient pressure sensor with a range of 0–500 psi at 10 mV/psi sensitivity. The water injection system is connected to a water supply tank through a solenoid valve and a needle-type release nozzle. The nozzle is initially placed above the sodium pool and will be immersed during the run by a piston

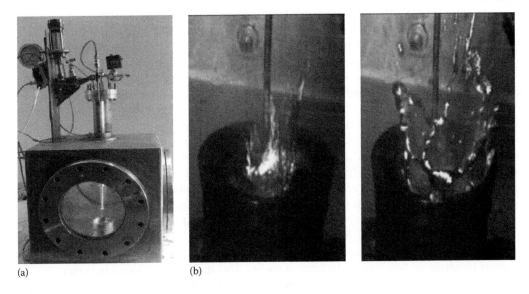

(a) (b)

FIGURE 16.20 Benchmark test facility for simulating the SWR scenario: (a) test setup and (b) typical SWR scenario.

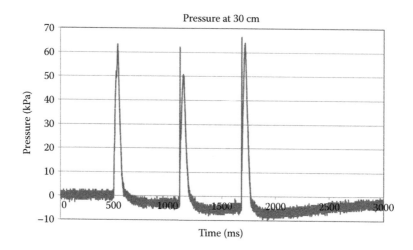

FIGURE 16.21 Pressure pulses generated during an SWR test.

valve–type arrangement and a pneumatic actuator. This prevents the chance of injection line blockage by sodium. Few experiments were conducted by injecting water at about 6 mL/min into 10 g of sodium pool at 350°C. The reaction was captured with a high-speed camera and the scenario is shown in Figure 16.20b. Transient pressure pulse in the interaction chamber was monitored and indicated in Figure 16.21. This shows that the pressure rise is cyclic in nature with short-duration pulses, rather than monotonic behavior.

16.3.2 DESIGN PROVISIONS TO PREVENT/MITIGATE THE EFFECTS OF SWR IN SG

Once an SWR is confirmed through the leak detection system, immediately a series of automatic/manual actions are taken to stop the SWR. Once identified, the SWR can be stopped in the small leak range itself. However, as a defense in depth, a large leak SWR is considered as one of the design basis events, and necessary design provisions are made to reduce the damage to the system components. Prior to the under sodium leak event in the PFR superheater in February 1987, all countries agreed that one double-ended guillotine (DEG) rupture in 1 ms would be a conservative assumption for the design basis leak (DBL) [16.24]. The under sodium leak event in the PFR superheater has changed the scenario [16.25]. During this leak event a total of 40 tubes failed, each one equivalent to a DEG failure. The 39 secondary failures occurred due to overheating in about 10 s. This incident has led to a review of the DBL. Because of the complexity of the processes, it is very difficult to predict the evolution of an SG leak event in a deterministic way due to many random possibilities. In order to limit overheating, sodium needs to be taken away from the reaction site. This necessitates two rupture disks, one at the top and one at the bottom. Simultaneous failure of both rupture disks will result in sodium moving out of the reaction zone in both directions. Because of the fast movement of sodium out of the reaction zone, there is no time to overheat additional tubes; thus, additional tube failures are suppressed. As the SG size increases, the DEG of a single tube is not sufficient to push the sodium away from the reaction site. A study has been carried out with the SWEPT computer code to estimate the number of simultaneous DEG failures required to ensure rupture of both rupture disks, pushing sodium away from the reaction site. From the study performed for PFBR [16.26], it has been found that the DBL is an instantaneous DEG failure of three tubes at the top of the SG. However, it is possible to predict an upper bound on the maximum possible pressures in the SG, IHX, and other secondary sodium circuit components and pipelines by assuming simultaneous failure of all the tubes in the SG. This approach is followed in France.

Required design provisions have been incorporated for PFBR in India to reduce damage to the components. High-strength ferritic steel (modified 9Cr–1Mo) is chosen for SFR SGs under design, instead of austenitic stainless steel (prone to stress corrosion cracking in caustic environment).

The most critical portions of the sodium–water boundary, that is, tube welds, are provided with additional features like long seamless tubes to reduce the number of tube to tube sheet welds, avoiding tube to tube joining, welding of tube to tube sheet with internal bore welding machine, stringent quality check for each joint, and postweld heat treatment of individual tube to tube sheet joint. The design also has been validated through experiments for flow-induced vibration of the tube bundle. A detailed study of the effects of SWR due to various water/steam leak rates has been carried out. Appropriate leak detection systems to positively identify SWR have been incorporated. Necessary automatic safety actions have been provided to quickly terminate SWR.

16.4   SODIUM–CONCRETE INTERACTION

Concrete structures in SFR are largely used as foundation, support, radiation shielding, and containment with some specific features [16.27]. The interaction of accidentally spilled liquid sodium at elevated temperature of 550°C and above can cause deterioration of concrete in several ways. The contact between hot sodium and concrete could result in (1) reaction between sodium and water vapor released by concrete, producing hydrogen, (2) reaction between sodium and carbon dioxide (T > 650°C) that damages the concrete, and (3) exothermal reactions between concrete solid components (e.g., SiO_2) with sodium or sodium oxide and hydroxide. Thus, hydrogen production, concrete damage, and heat release are of great concern in SFRs. The degradation mechanisms may be dominated by thermal and chemical effects, elevated concrete pore pressure and fluid flow rates, and thermal instability and chemical instability of concrete when exposed to hot liquid sodium [16.28]. Hence, studies are required to characterize sodium–concrete interactions, develop inert concretes and concrete protection methodology with a liner or metallic coating or painting.

Marcherta [16.29] observed that concrete exposed to elevated temperatures undergoes thermally induced relative displacement between the cement paste and aggregate resulting in the rupture of the bond between these materials. Various researchers have studied liquid sodium and concrete interactions either in air or in inert argon atmosphere to simulate various accident scenarios in sodium-cooled fast breeder reactors. Sodium exposure experiments are equipped with oxygen monitor, hydrogen monitor, hygrometer, pressure transducers, and thermocouples. Bae et al. [16.30] reported that the possibility of a hydrogen explosion reaction cannot be neglected, because the lowest flammable limit of hydrogen is 4.0 mol%, whereas the experimental maximum concentration reached up to 31 mol% in several cases. The possibility of an alkali–aggregate reaction in limestone concrete cannot be neglected as limestone aggregates may contain impurities like amorphous silica. Dolomite with magnesium carbonate may participate in alkali carbonate reaction leading to various disintegration mechanisms in concrete. The alkali–aggregate reaction can be described as a chemical reaction between reactive components of an aggregate (amorphous silica or carbonates like dolomite) and alkali hydroxides present in the cement [16.31]. In the case of limestone concrete and liquid–sodium interactions, the alkali is directly available as sodium hydroxide, as a product of the interaction between sodium at elevated temperature and moisture present in the concrete. The expansive products of an alkali–aggregate reaction can cause early distress and reduction in serviceability of concrete structures [16.32]. Different theories are proposed by researchers for the expansion caused by the alkali–aggregate reaction. The theory of imbibition (absorption of fluid by a solid or colloid that causes swelling) or osmotic pressure, the theory of ion diffusion, the theory of crystallization pressure, and the theory of gel dispersion are some explanations given for the expansion caused by the alkali–aggregate reaction [16.31–16.33]. According to Das et al. [16.28], when hot liquid sodium at high temperature interacts with concrete, heat is transferred to the concrete through conduction. In an inert argon atmosphere, heat is transported to the argon by radiation and convection. Hot sodium and concrete interactions can lead to various endothermic and exothermic reactions, resulting in dehydration and erosion of concrete along with the production of hydrogen gas. Degradation of concrete strength and accumulation of hydrogen gas associated with hydrogen burning could challenge the integrity of the structure due to overpressurization. It could possibly

TABLE 16.1

Reaction Products of Sodium with Different Types of Aggregates

Types of Aggregates	Major Components	Major Reaction Products with Sodium
Basalt	SiO_2, Al_2O_3	Na_2SiO_3, $NaAlO_2$
Magnetite	Fe_3O_4	Na_2O, FeO, Fe
Limestone	$CaCO_3$, $CaMg(CO_3)_2$	CaO, MgO, Na_2CO_3, C
Granite	SiO_2, CaO	Na_2SiO_3, Na_2CaSiO_4
Serpentine	$Mg_3SiO_2O_5(OH)_4$	MgO, Na_2SiO_3
Greywacke	SiO_2	Na_2SiO_3

lead to the release of radioactive materials that are extremely harmful. Thus, it is necessary to understand sodium–concrete reactions to predict thermal energy, hydrogen gas release, and degradation of concrete. Bae et al. [16.30] divide the sodium and concrete interactions into three parts: (1) sodium and free water in concrete, (2) sodium and chemically bound water in concrete, and (3) sodium and concrete aggregates. All the previously mentioned interactions include the initiation, propagation, and termination phases. When hot sodium comes in contact with concrete, evaporable water gets released from the concrete and is driven to the concrete surface by pressure generated during heating up of concrete. The porosity of concrete increases with increase in temperature and transport of water vapor occurs by Darcian flow [16.34]. The reaction of this water vapor with sodium results in the formation of liquid sodium hydroxide, solid sodium monoxide, and gaseous hydrogen. Even though a number of chemical reactions are involved in sodium–concrete interactions, nearly 78% of the reaction products and heat are produced from primary sodium and water interactions (as per the below equation):

$$Na(l) + H_2O(g) \rightarrow NaOH(l) + H_2(g) \quad +185\ kJ.$$

Chemically bound water from concrete is released over a range of 200°C–700°C and is transported to the surface by Darcian flow resulting in the same interactions as in the case of free water. Table 16.1 shows the products of sodium–concrete interactions for various types of aggregates [16.27]. Studies by Premila et al. [16.35] revealed that fused NaOH did not disintegrate the limestone concrete below 300°C and the effect of sodium was only visible at elevated temperatures. Midinfrared spectroscopy study was used to conclude that at 800°C, lowering of symmetry in the calcite structure occurs due to sodium interactions, which is absent in limestone concrete subjected to thermal exposure alone. The severity of damage of limestone concrete is dictated primarily by compressive strength and hardness of concrete along with the intensity and duration of sodium fire.

During the sodium leak incident in the secondary sodium circuit, the sodium will be ejected out under pressure and may interact with the underlying structural concrete. In order to prevent the degradation of concrete, a sacrificial layer needs to be used in the SG building. However, the depth of the layer has to be suggested for different orientations (vertical walls and flooring). Engineering-scale experiments conducted by Fritzke and Schultheiss [16.36] on cylindrical concrete specimens of 250 mm diameter and 300 mm height exposing 6.2 kg of sodium at 550°C simulating sodium pools on the floor due to accident revealed that maximum depths of penetration of sodium were observed to be about 50 mm. Tests conducted at IGCAR (Figure 16.22) revealed similar conclusions [16.37]. Hence, a sacrificial limestone concrete layer on the structural concrete with a minimum thickness of 50 mm can protect the structural concrete from damage from attack of hot liquid sodium pool for 30 min of exposure.

To study the effect of sodium interaction on impinging, sloping, and stagnant surfaces, an experiment was conducted at IGCAR with limestone concrete block (Figure 16.23a) embedded

(a)   (b)

FIGURE 16.22   (See color insert.) Depiction of sodium–concrete interaction scenario with limestone concrete. (a) Test setup and (b) sodium concrete interaction.

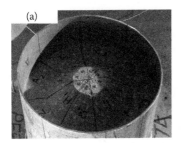

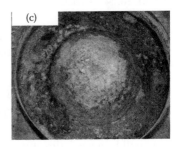

FIGURE 16.23   (See color insert.) Typical test on flowing sodium–concrete interaction: (a) specimen, (b) fire scenario, and (c) after test.

with thermowells at various locations to obtain temperature distribution within the block. Sodium injected at 500°C onto the impact zone of concrete block was observed to burn in the form of spray and then later as pool (Figure 16.23b). The sodium-exposed concrete block is shown in Figure 16.23c. The peak temperatures at the impact, sloping, and stagnant zones were observed to be 840°C, 780°C, and 900°C, respectively [16.38]. The temperature distribution in a typical specimen at various depths is shown in Figure 16.24. Maximum temperatures in the block at 5, 25, and 45 mm depth from the surface are observed to be 115°C, 80°C, and 60°C, respectively. It was also observed that the sloping zone was less affected by sodium than other zones due to less contact time. Preliminary posttest analysis indicated no noticeable visual damage on the specimen.

Fundamental investigations on chemical reactions of liquid sodium with concrete were carried out using thermal analysis experiments and many researches, and it is reported that for various types of aggregates, the threshold temperatures varied only in a small range (697–820 K), whereas the enthalpies of the reactions varied vastly (ranging between 250 and 2172 J/g). A study by Chasanov and Staahl [16.39] on the immersion of dolomite concrete in a pool of hot liquid sodium at 500°C showed that vibrated concrete specimens performed better than hand-rodded specimens. In addition, simultaneous liberation of heat, gases, and aerosols poses a multitude of hazards to health and the working environment of plant operators. Formation of reaction products such as elemental carbon and hydrated sodium silicate was observed during the interaction.

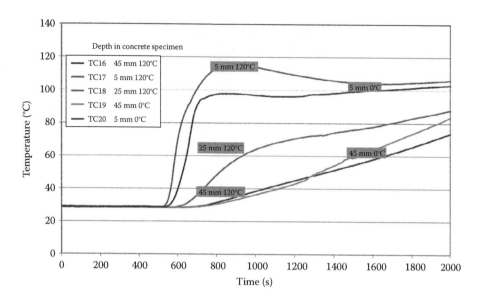

FIGURE 16.24 Concrete temperatures across the depth (sodium impact).

When liquid sodium comes in contact with concrete, the following reactions take place. Hot liquid sodium reacts with constituents of limestone concrete to form sodium carbonate and carbon causing discoloring [16.40]:

$$4Na(l) + 3CaCO_3(s) \rightarrow 2Na_2CO_3(s) + 3CaO(s) + C(s) + 512 \text{ kJ}$$
$$4Na(l) + 3MgCO_3(s) \rightarrow 2Na_2CO_3(s) + 3MgO(s) + C(s) + 727 \text{ kJ}$$

There will be a few indirect reactions among the products and reactants forming CO_2, etc:

$$CaCO_3(s) \rightarrow CaO(s) + CO_2(g) \quad (\text{at } 890°C)$$
$$MgCO_3(s) \rightarrow MgO(s) + CO_2(g) \quad (\text{at } 400°C - 540°C)$$

Steep thermal gradients exist due to the poor thermal conductivity of concrete, and the solid constituents locally experience thermal stresses because of (1) differential and anisotropic thermal expansion, (2) dilation induced by phase transitions, and (3) escalation of pore pressure driven by generation of steam and carbon dioxide:

$$CaMg(CO_3)_2(s) + 2NaOH(l) \rightarrow Mg(OH)_2(s) + CaCO_3(s) + Na_2CO_3(s)$$
$$xSiO_2(s) + yNaOH/Na_2O \rightarrow aNa_2O \cdot bSiO_2 \cdot cH_2O$$
$$xCaO \cdot ySiO_2 \cdot zH_2O \rightarrow xCaO \cdot ySiO_2 \cdot (z-n)H_2O + nH_2O(g)$$
$$4Na(l) + 3CaO \cdot SiO_2(s) \rightarrow 2Na_2O \cdot SiO_2(s) + 3CaO(s) + Si(s)$$

Water released will further contact the liquid sodium and generate hydrogen:

$$Na(l) + H_2O(g) \rightarrow NaOH(l) + \frac{1}{2}H_2(g) + 185 \text{ kJ}$$
$$NaOH(l) + Na(l) \rightarrow Na_2O(s) + \frac{1}{2}H_2(g)$$
$$Na(l) + H_2O(g) \rightarrow Na_2O(s) + H_2(g) + 128 \text{ kJ}$$

Carbon released by the reaction of liquid sodium with carbon dioxide reacts with hydrogen generated during the previously mentioned reaction, producing methane gas:

$$4Na(l) + CO_2(g) \rightarrow 2Na_2O(s) + C(s)$$

$$\left[C\right]_{Na} + \left[H\right]_{Na} \rightarrow CH_4(g)$$

16.4.1 Design Provisions

Many design provisions are usually made in SFRs to prevent sodium–concrete interactions: (1) design and construction of the sloping floor with a sodium collection pit, which helps in minimizing the mass of sodium accumulating on the floor, its contact area with the concrete surface, and exposure period; (2) installation of LCTs below the strategic pipelines and critical equipment/vessels containing liquid sodium, which will suppress the fire and avoid physical contact between the concrete surface and sodium; (3) provision of a sacrificial layer on the concrete used in the SG building; and (4) provision of steel/refractory stainless steel liners over the entire surface of the building (e.g., Monju reactor in Japan).

Reviews of earlier research papers related to the interaction of concrete with sodium fire by Parida et al. [16.37] and Schultheiss [16.40] have revealed that limestone concrete is more resistant to sodium attack than basalt and quartz. The results of preliminary experiments at 500°C on the interaction of sodium and concrete specimens containing limestone as an aggregate (Chasanov and Staahl [16.39]) concluded that the partial dehydration of concrete before immersion in sodium eliminated the cracking and spelling in the specimens. Casselman [16.41] carried out a series of experiments on sodium–concrete interaction under neutral atmosphere and indicated that the sodium temperature plays an important role at the interfaces of sodium and concrete. However, concrete thickness does not appear as an important parameter. Noumowéa et al. [16.42] concluded that concrete containing limestone aggregates could be used in applications involving elevated temperatures.

16.5 SODIUM FIRE MITIGATION

Burning sodium as a spray or as a pool of large surface area or high temperature may build up pressure and cause building damage. Sodium metal and sodium fume corrosion products arising from combustion can impair equipment severely. The fumes from a sodium fire of comparatively small size quickly reduce visibility to a yard or less. The caustic and irritating nature of the combustion products makes it unpleasant to breathe. The fire should be extinguished as early as possible in order to protect the personnel as well as equipment nearby.

Sodium fires can be mitigated by two methods: active and passive. Active methods include application of fire extinguishing materials or inert gas onto the fire, whereas passive methods are those that cause self-extinguishment of fire by draining sodium into inert vaults.

16.5.1 Passive Methods

In secondary heat transfer circuits of FBRs, leakage of liquid sodium from the pipelines is postulated as one of the design basis accidents. The leaked sodium can undergo combustion in spray, pool, or combined form, generating a dense white smoke and flame leading to a rise in the ambient gas temperature, causing a potential hazard to the plant and operating personnel. To mitigate the thermal and chemical consequences of sodium fire, LCTs are designed as a passive fire protection system. The sloping top cover of LCT facilitates sodium drain into a holdup vessel

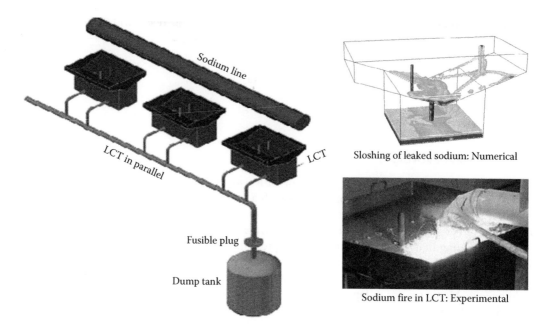

FIGURE 16.25   Sodium collection and burning scenario in a typical LCT.

through drainpipes, where fire is mitigated by oxygen starvation. A typical sodium LCT and the method of sodium collection are indicated in Figure 16.25 [16.43,16.44]. These LCTs are strategically arranged under the secondary sodium piping to collect leaking sodium into a holdup vessel and subsequently to a sodium dump tank connected through a low-melting fusible plug as shown in Figure 16.25. Experiments conducted at IGCAR on the qualification of a leak collection system indicated that proper drainage of liquid sodium into the dump tank occurred with minimum sodium burning [16.45].

16.5.2   ACTIVE METHODS

16.5.2.1   Fire Mitigation by Nitrogen Injection

A conventional method is to isolate the system combined with nitrogen flooding. Steel catch pans are also provided to prevent the interaction of concrete with sodium. Space isolation basically involves the prevention of oxygen into the cell in which sodium is burning and thus allows the oxygen concentration to fall below the limit that will not sustain the sodium–oxygen reaction. Also nitrogen is manually injected at a sufficiently high rate to pressurize the cell in order to prevent the inflow of air. Generally, 1%–2% oxygen in nitrogen is maintained in such systems, which varies in different countries since many others report that an average of 4%–5% oxygen in nitrogen can be used in such systems to mitigate the sodium fire due to insufficient oxygen for combustion reaction. The percentage of sodium burnt as a function of oxygen concentration controlled by the injection of nitrogen was studied at IGCAR [16.46]. It was observed that ignition of sodium fire at 500°C occurs only with a minimum oxygen concentration of 4%. The percentage of sodium combustion with oxygen concentration is given in Figure 16.26.

16.5.2.2   Sodium Fire Extinguisher Powder

Another active fire protection method involves the use of extinguishers to mitigate sodium fire, that is, the application of dry powder–based extinguishers on the fire. Sodium pool fires can be extinguished by the application of powders, which blanket the sodium fire and extinguish it by preventing oxygen supply. They are applied by scoop/shovel or by injecting fluidized powder under nitrogen

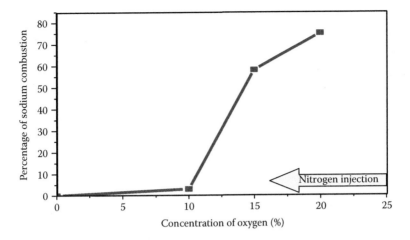

FIGURE 16.26   Sodium fire mitigation through nitrogen injection.

over the burning sodium or by using an extinguisher with an applicator [16.47]. The extinguishing powders shall have the following characteristics for better functioning [16.48]:

- Density should be less than 0.8 g/cm^3.
- They should be stable and nonhygroscopic.
- They should be free flowing and compatible with liquid sodium.
- The ratio of the quantity of required powder to sodium should be small.
- They should rapidly cool the metal to reduce the possibility of re-ignition.
- They should be inexpensive, available, and easily disposable.
- Reaction with sodium (if any) must be nonexothermic/nonviolent.

Various sodium fire extinguishers reported for extinguishing sodium fires are (1) sodium bicarbonates (dry chemical powder), (2) sodium carbonates (soda ash), (3) calcium carbonate (technical grade), (4) vermiculite, (5) expandable graphite (Graphex CK 23, Markalina [mixture of carbonates]), (6) ternary eutectic chloride (TEC), (7) intumescence-based flame retardant (IGCAR), and (8) carbon microsphere (CMS) (IGCAR). The comparison among various fire extinguishers is shown in Table 16.2.

TABLE 16.2
Comparison between Commercially Available Sodium Fire Extinguishers

Particulars	Sodium Bicarbonate	Sodium Carbonate	Vermiculite Large	Small	Ternary Eutectic Chloride	Graphex CK 23
Average ratio of the weight of powder to the weight of sodium	1.1	5.5	0.8	1.1	2.9	—
Average rate of cooling	10	2	3	16	18.5	—
Cost	Costly	Cheap but drying is essential.	Cheap		Costly	Costly
Mode of application	By showel and extinguisher	Showel only.	By showel only		By showel and extinguishers	—
Storage	Airtight container	Polythene bag.	Any container		Air tight container	—
Disposability	Difficult	Easy.	Easy		Easy	Easy
Free flowing	Nicely	Nicely.	Nicely		Nicely	Nicely

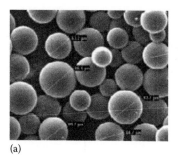 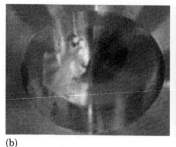 

(a)   (b)

FIGURE 16.27   (a) CMS and (b) sodium fire extinguishment using CMS.

Among the few suitable options, sodium bicarbonate (DCP) is the best available powder, in spite of the difficulty in disposal. The ratio of the weight of powder to the weight of sodium is extremely small. It rapidly cools the burning metal. Next to this ranks vermiculite, which is cheap and easy to store, but cannot be used in standard extinguishers. The cooling rate is also slow, since it only covers the fire but does not stop combustion completely. Calcium carbonate can be ranked number three. It is cheap and available freely, but due to the difficulty in application, it cannot be ranked higher. TEC is not suitable for extinguishing large sodium fires as these give rise to flaming and sputtering. TEC is found to absorb moisture on exposure to atmosphere. A sodium bicarbonate–based extinguisher (dry chemical powder) is commercially popular, but with the requirement of a large quantity and difficulty in removal and disposal, innovative powders are being developed at IGCAR [16.49]. Intumescent materials expand 200 times its volume under the action of heat and cover sodium fire, extinguishing it by preventing oxygen supply. They are of low density and environment friendly and facilitate easy removal and disposal. CMS blankets sodium fire and extinguishes it by preventing oxygen supply. They possess high thermal stability and chemical inertness and are nonhygroscopic, nontoxic, and capable of being directed onto fires from conventional nozzles. A CMS-based innovative sodium fire extinguisher was tested on a small-scale sodium fire (Figure 16.26). The sodium metal could be easily recovered once the fire is extinguished. A scanning electron microscope (SEM) image of CMSs is shown in Figure 16.27.

REFERENCES

16.1   Rouault, J., Chellapandi, P., Raj, B., Dufour, P., Latge, C. et al. (2010). Sodium fast reactor design: Fuels, neutronics, thermal-hydraulics, structural mechanics and safety. In *Handbook of Nuclear Engineering*, Cacuci, D.G. (ed.), Springer, U.S., Vol. 4, pp. 2321–2710, Chapter 21.

16.2   Tsuruga, F., Kobayashi, H. (1995). Leakage of sodium coolant from secondary cooling loop in prototype fast breeder reactor MONJU. Failure Knowledge Database. Tokyo Institute of Technology, Tokyo, Japan.

16.3   Olivier, T.J. (2007). Metal fire implications for advanced reactors, part 1: Literature review, SANDIA report, SAND2007-6332, October 2007.

16.4   Guidez, J., Martin, L., Chetal, S.C., Chellapandi, P., Raj, B. (November 2008). Lessons learned from sodium cooled fast reactor operation and their ramification for future reactors with respect to enhanced safety and reliability. *Nucl. Technol.*, 164, 207–220.

16.5   Newman, R.N. (1983). The ignition and burning behaviour of sodium metal in air. *Prog. Nucl. Energy*, 12, 119–147.

16.6   Latge, C. (2012). Interaction sodium and materials. *Fast Reactor Science and Technology*, CNEA, Bariloche, Argentina, October 2012.

16.7   Richard, J.R., Delbourgo, R., Laffitte, P. (1969). Spontaneous ignition and combustion of sodium droplets in various oxidizing atmospheres at atmospheric pressure. *Twelfth Symposium (International) on Combustion*, The combustion institute, pp. 39–48.

16.8   Krolikowski, T.S., Lebowitz, L., Wilson, R.E., Cassulo, J.C. (1969). The reaction of a molten sodium spray with air in an enclosed volume: Part 2—Theoretical model. *Nucl. Sci. Eng.*, 38, 161–166.

16.9 Morewitz, H.A., Johnson, R.P., Nelson, C.T. (1977). Experiments on sodium fires and aerosols. *Nucl. Eng. Des.*, 42, 123–135.

16.10 Tsai, S.S. (1980). The NACOM code for analysis of postulated sodium fires in LMFBRs, NUREG/CR-1405, March 1980.

16.11 Kawabe, R., Suzuoki, A., Minato, A. (1982). A study on sodium spray combustion. *Nucl. Eng. Des.*, 75, 49–56.

16.12 Okano, Y., Yamaguchi, A. (2003). Numerical simulation of free-falling liquid sodium droplet combustion. *Ann. Nucl. Energy*, 30, 1863–1878.

16.13 Makino, A., Fukada, H. (2005). Ignition and combustion of a falling, single sodium droplet. *Proc. Combust. Inst.*, 30, 2047–2054.

16.14 Yuasa, S. (1985). Spontaneous ignition of sodium in dry and moist air streams. *Twentieth International Symposium on Combustion*, University of Michigan, Ann Arbor, Michigan, USA, pp. 1869–1876.

16.15 Avinash, Ch.S.S.S. et al. (2013). Experimental study on sodium leaks through small openings. *Fluid Mechanics and Fluid Power*, NCFMFP-2013 NIT Hamirpur, India.

16.16 Ponraju, D. et al. (2013). Experimental and numerical simulation of sodium safety in SFR. *International Conference on Fast Reactors and Related Fuel Cycles: Safe Technologies and Sustainable Scenarios (FR13)*, Paris, France.

16.17 Muthu Saravanan, S. (2011). Numerical investigation and characterization of single sodium droplet. MTech thesis report. HBNI, Mumbai, India.

16.18 Muthu Saravanan, S., Mangarjuna Rao, P., Nashine, B.K., Chellapandi, P. (2011). Numerical investigation of sodium droplet ignition in the atmospheric air. *Int. J. Nucl. Energy Sci. Technol.*, 6(4), 284–297.

16.19 (1996). Fast reactor fuel failures and SG leaks: Transient and accident analysis approaches. IAEA-TECDOC-908. IAEA, Vienna, Austria.

16.20 Hori, M. (1980). Sodium water reaction in SG of LMFBR. *Atom. Energy Rev.*, 18(3), 707–778.

16.21 Ruloff, G., Hubner, R. (1990). SNR-300 SG accident philosophy—Assessment due to new understanding in sodium water reaction, IAEA, IWGFR/78. *SG Failure and Failure Propagation Experience*, Aix-en-Provence, France, pp. 62–70.

16.22 Kishore, S., Ashok Kumar, A., Chandramouli, S., Nashine, B.K., Rajan, K.K., Kalyanasundaram, P., Chetal, S.C. (2012). An experimental study on impingement wastage of mod 9Cr1Mo steel due to sodium water reaction. *Nucl. Eng. Des.*, 243, 49–55.

16.23 Jeong, J.-Y., Kim, J.-M., Kim, T.-J., Choi, J.-H., Ki, B.-H. (2009). Wastage behavior of modified 9Cr-1Mo steel tube material by sodium–water reaction. *Transactions of the Korean Nuclear Society Autumn Meeting*, Gyeongju, Korea.

16.24 (1983). Summary report, General conclusions and recommendations. IWGFR/50. IAEA, The Hague, the Netherlands, p. 6.

16.25 Currie, R. et al. (1990). The under sodium leak in the PFR super heater 2 in February 1987, IAEA, IWGFR/78. *SG Failure and Failure Propagation Experience*, Aix-en-Provence, France, pp. 107–132.

16.26 Selvaraj, P., Vaidyanathan, G., Chetal, S.C. (1996). Review of design basis accident for large leak sodium-water reaction for PFBR. *Fourth International Conference on Nuclear Engineering*, New Orleans, LA.

16.27 Mohammed Haneefa, K., Santhanam, M., Parida, F.C. (2013). Review of concrete performance at elevated temperature and hot sodium exposure applications in nuclear industry. *Nucl. Eng. Des.*, 258, 76–88.

16.28 Das, S.K., Sharma, A.K., Parida, F.C., Kashinathan, N. (2009). Experimental study on thermo chemical phenomena during interactions and limestone concrete with liquid sodium under inert atmosphere. *Construct. Build. Mater.*, 23, 179–188.

16.29 Marcherta, A.H. (1984). Thermo-mechanical analysis of concrete in LFMBR programs. *Nucl. Eng. Des.*, 82, 47–62.

16.30 Bae, J.H., Shin, M.S., Min, B.H., Kim, S.M. (1998). Experimental study on sodium–concrete reactions. *J. Korean Nucl. Soc.*, 30, 568–580.

16.31 Diaz, E.G., Riche, J., Bulteel, D., Vernet, C. (2006). Mechanism of damage for alkali–silica reactions. *Cement Concr. Res.*, 36, 395–400.

16.32 Mo, X., Fournier, B. (2007). Investigation of structural properties associated with alkali–silica reaction by means of macro and micro structural analysis. *Mater. Charact.*, 58, 179–189.

16.33 Chatterji, S. (2005). Chemistry of alkali–silica reaction and testing of aggregates. *Cement Concr. Compos.*, 27, 778–795.

16.34 Chawla, T.C., Pederesen, D.R. (1985). A review of modeling concepts for sodium–concrete reactions and a model for liquid sodium transport to the unreacted concrete surface. *Nucl. Eng. Des.*, 88, 85–91.

16.35 Premila, M., Sivasubramanian, K., Amarendra, G., Sundar, C.S. (2008). Thermo-chemical degradation of limestone aggregate concrete of exposure of sodium fire. *J. Nucl. Mater.*, 375, 263–269.

16.36 Fritzke, H.W., Schultheiss, G.F. (1983). An experiment study on sodium concrete interaction on mitigating protective layers. *Seventh International Conference on Structural Mechanics in Reactor Technology (SMiRT)*, Chicago, IL, pp. 135–142.

16.37 Parida, F.C. et al. (2006). Sodium exposure tests on limestone concrete used as sacrificial protection layer in FBR. *Proceedings of the International Conference on Nuclear Engineering (ICONE 14)*, Miami, FL, July 2006.

16.38 Chellapandi, P. et. al. (2014). Studies on mitigation of sodium fire. *IGC Newsletter*, 99, 15–18, January 2014.

16.39 Chasanov, M.G., Staahl, G.E. (1977). High temperature sodium-concrete interactions. *J. Nucl. Mater.*, 66, 217–220.

16.40 Schultheiss, G.F. (1983). Investigation of sodium concrete interaction and the effect of different by products. Report No. GKSS 83/E/59.

16.41 Casselman, C. (1981). Consequences of interaction between sodium and concrete. *Nucl. Eng. Des.*, 68, 207–212.

16.42 Noumowéa, A. et al. (2009). Thermo-mechanical characteristics of concrete at elevated temperatures up to 310°C. *Nucl. Eng. Des.*, 239, 470–476.

16.43 Diwakar, S.V., Mangarjuna Rao, P., Kasinathan, N., Das, S.K., Sundararajan, T. (2011). Numerical prediction of fire extinguishment characteristics of sodium leak collection tray in a fast breeder reactor. *Nucl. Eng. Des.*, 241, 5189–5202.

16.44 Nashine, B.K. et al. (2013). Qualification of leak collection system. IGC annual report.

16.45 Kim, B.H., Jeong, J.Y., Choi, J.H., Kim, T.J., Nam, H.Y. (2008). Analytical study of sodium fire characteristics under sodium leak accidents, Korea Atomic Energy Research Institute, Daejeon, Korea, pp. 305–353.

16.46 Pradeep, A. et al. (2013). Sodium fire extinguishment by nitrogen injection. Internal report.

16.47 Ballif, J.L. et al. (1979). Liquid metals fire control engineering handbook. HEDL-TME-79-17, UC-41.

16.48 Ponraju, D. et al. (2011). Specifications for sodium fire extinguisher. Internal report.

16.49 Snehalatha, V. et al. (2013). Small scale sodium fire extinguishment. Carbon microspheres. Internal report.

17 Computer Codes and Validation

17.1 INTRODUCTION

Most of the severe accidents cannot be experimentally simulated in a real scale mainly due to several safety aspects that need to be taken care of. Hence, traditionally, numerical simulations are inevitable in all the phases of safety analysis, starting from the initiating event. The analysis requirements are derived from the safety criteria defined for a specific SFR. In this chapter, a few computer codes widely used for safety analysis as well as sodium fire studies are highlighted in this chapter. Apart from these, some salient numerical results of unprotected loss of flow accident (ULOFA) analysis performed through an international benchmark on a Russian reactor (BN-800) under the International Atomic Energy Agency (IAEA) framework are presented. The purpose of this chapter is to expose the readers a few important codes and modes of validations. It is difficult to provide a complete spectrum of several codes and validation studies performed by various countries. However, extensive literature is listed, and further, more details provided on a system of computer codes employed for the safety analysis of Indian SFR prototype fast breeder reactor (PFBR) are presented as a case study.

17.2 COMPUTER CODES FOR SEVERE ACCIDENT ANALYSIS

17.2.1 INTERNATIONAL CODES

As explained in the preceding chapters, the accident propagates through various phases: predisassembly phase, transition phase, disassembly phase, and mechanical energy release/system response phase. Computer code developments over the past four decades have gone through all these phases, with thorough verifications both within pile and out-of-pile experiments. In this context, readers should be aware of the three-code series—SAS, VENUS, and Sn, implicit, multifield, multicomponent, Eulerian recriticality (SIMMER)—which are briefly explained in the following sections.

17.2.1.1 SAS Code Series

The SAS series of codes was developed in the 1970s at the Argonne National Laboratory (ANL) in the United States. The first code in the series is SAS1A [17.1]. Subsequently, SAS2A [17.2], SAS2B, SAS3A [17.3], SAS3D [17.4], and SAS4A [17.5] were developed. The last version is SAS4A, which was widely used in the United States, Europe, and Japan. The SAS1A code started with a single pin representation in a channel. The different coolant channels are not coupled thermohydraulically. It performs neutronics calculations by the point kinetics model taking into account reactivity feedbacks from Doppler, fuel and clad axial expansion feedback, coolant density feedback, and radial feedback. The sodium boiling model is a two-phase slug ejection and single-bubble model. Fuel dynamics and clad deformation models are also included. All the calculations are performed for the fresh pins only. The next version of the SAS series is SAS2A. This code is a multichannel code, which can couple thermohydraulics between the different channels. The primary loop model is included in this version. The code was also improved to include transient fuel pin mechanics (fission gas pressure, effect of fission gases on clad deformation, etc.). The sodium boiling model is a two-phase slip flow slug ejection multiple-bubble mode. The SAS3 series of codes

489

(SAS3A–SAS3D) were further improved to calculate fuel–coolant interaction (FCI), FCI-driven fuel motion and voiding, and primary and secondary loops for thermohydraulics. In the SAS4A code, new features were incorporated: a balance of the force model to calculate the fuel axial expansion and coupling of the clad motion and sodium boiling. The sodium boiling model was improved to treat a variable coolant flow cross section to couple with pin mechanics. A unified molten fuel motion model was also incorporated. All the SAS codes use the point kinetics model for neutronics. The space–time kinetics model may be required when there is a large-scale fuel and clad motion. The recent SAS4A modeling includes extension of the boiling model to treat sudden fission gas release upon pin failure, fuel deformation to handle advanced cladding materials, and metallic fuel modeling capability to the fuel relocation model [17.6]. Various evolutionary features of the SAS series are summarized in Table 17.1.

17.2.1.2 VENUS Series

As explained in Chapter 15, after the predisassembly analysis, the disassembly phase is analyzed, skipping the transition phase. However, in order to take off the transition phase scenarios, certain specific assumptions are made in the disassembly phase so as to yield conservative results, in terms of energy release. The most widely used computer code for the disassembly phase is VENUS, developed by ANL [17.7]. VENUS is a 2D coupled neutronics–hydrodynamics program that calculates the dynamic behavior of an liquid metal fast breeder reactor (LMFBR) during a prompt critical disassembly excursion. This code has been used extensively in LMFBR disassembly energetic simulations. The model employed in the code utilizes the space-independent neutronics, 2D (R–Z) Lagrangian hydrodynamics and an energy density–dependent equation of state. Reactivity feedbacks due to Doppler broadening and material motion are explicitly taken into account. The model can analyze both high- and low-density systems. The density changes could be explicitly calculated, which in turn allows the use of an accurate density-dependent equation of state. VENUS II is an improved version of VENUS [17.8]; particularly VENUS II uses a more accurate equation of state developed by ANL. Reactivity insertion along with reactivity feedbacks is simulated by a point kinetics module which produces power and energy. Energy production leads to an increase in rise of temperature in the fuel system. Heat transfer is ignored due to the short time frame (e.g., the transient lasts only about 10 ms for PFBR). Temperature rise is computed through specific heat, ignoring vaporized fuel. Temperature gives rise to pressure through a density-dependent equation of state. The density changes affect pressure enormously and depend on the reduced volume of fuel during the transient. Reduced volume is the ratio of volume at a given temperature to that at a critical point. Reduced volume varies in the transient and hence the density and the pressure generated for varying temperatures. Fuel motion is computed through 2D hydrodynamics, which in turn affects the densities influencing pressure generation. The rise in temperature in the system gives a large negative Doppler feedback. Fuel motion and displacement give a large negative reactivity feedback. The negative feedback due to Doppler and fuel displacement counteract with the external reactivity in addition due to fuel slumping, and the reactor becomes subcritical at the end of the transient. The VENUS II code has been validated against Kiwi-TNT, SNAPTRAN-2, and SNAPTRAN-3 reactor disassembly experiments [17.9].

A number of disassembly computer codes were developed by successive improvements of the original method implemented in VENUS, for example, the SCHAMBETA code developed by Korea [17.10]. This code is more relevant to metal-fueled SFRs. Accordingly, improvements were made to the equation of state used to estimate the pressures. Thus, the VENUS code developed at ANL is the first major step followed by the advanced mechanistic code series, SIMMER, for example.

17.2.1.3 SIMMER Series

Analysis of the transition phase of core disruptive accident (CDA) requires treatment of the multi-phase multicomponent thermohydraulics coupled with space- and energy-dependent neutron kinetics. Such a comprehensive modeling is carried out in the SIMMER series of codes. The code has gone through four versions: SIMMER [17.11], SIMMER II [17.12], SIMMER III [17.13], and, the

TABLE 17.1

Evolution of the SAS Series of Codes

Physical Mechanism/Modeling	SAS1A	SAS2A	SAS3A	SAS3D	SAS4A
Geometry of calculations	R–Z single fuel pin in each radial ring. No coupling between radial channels	Multichannel code—one pin calculation in each channel with flow coupling between radial channels	Multichannel code—one pin calculation in each channel with flow coupling between radial channels	Multichannel code—one pin calculation in each channel with flow coupling between radial channels	Multichannel code—one pin calculation in each channel with flow coupling between radial channels
Neutronics	Point kinetics	Point kinetics	Point kinetics	Point kinetics	Point kinetics
Steady-state conditions	Supplied as input or calculated for fresh pins	Supplied as input for either fresh fuel or irradiated fuel	Steady-state fuel pin characterization calculated	Steady-state fuel pin characterization calculated	Steady-state fuel pin characterization calculated
Heat transfer module	Transient fuel pin heat transfer for fresh fuel	Transient fuel pin heat transfer for irradiated fuel	Transient fuel pin heat transfer for irradiated fuel	Transient fuel pin heat transfer for irradiated fuel	Transient fuel pin heat transfer for irradiated fuel
Feedback models	1. Doppler 2. Axial expansion of fuel, clad, and structure 3. Coolant density and void 4. Radial expansion of structure	1. Doppler 2. Axial expansion of fuel, clad, and structure 3. Coolant density and void 4. Radial expansion of structure 5. Fuel slumping	1. Doppler 2. Axial expansion of fuel, clad, and structure 3. Coolant density and void 4. Radial expansion of structure 5. Fuel slumping	1. Doppler 2. Axial expansion of fuel, clad, and structure 3. Coolant density and void 4. Radial expansion of structure 5. Fuel slumping	1. Doppler 2. Balance of force model for axial expansion of fuel 3. Coolant density and void 4. Radial expansion of structure 5. Fuel and steel relocation
Sodium void model	Combination of annular slip two-phase flow and single-bubble slug ejection model	Multiple-bubble slug ejection coolant voiding model, treats voiding due to release of fission gas also	Multiple-bubble sodium boiling model with moving film treatment	Multiple-bubble sodium boiling model with moving film treatment	Multiple-bubble boiling model with variable coolant flow cross-sectional treatment to allow consistent coupling with pin mechanics and clad motion models
Fuel pin mechanics	No	Fuel pin mechanics calculated in transient	Fuel pin mechanics calculated in transient	Fuel pin mechanics calculated in transient	Fuel pin mechanics calculated in transient
Fuel deformation model and fuel pin behavior	Elastic–plastic cladding and elastic-fuel deformation model for fresh fuel	Elastic–plastic cladding and elastic-fuel deformation model for fresh and irradiated fuel	Elastic–plastic cladding and elastic-fuel deformation model for fresh and irradiated fuel	Elastic–plastic cladding and elastic-fuel deformation model for fresh and irradiated fuel	Elastic–plastic cladding and elastic-fuel deformation model for fresh and irradiated fuel

(Continued)

TABLE 17.1 (*Continued*)
Evolution of the SAS Series of Codes

Physical Mechanism/ Modeling	SAS1A	SAS2A	SAS3A	SAS3D	SAS4A
Fuel dynamics model	Partially molten fuel motion model in intact clad.	Partially molten fuel motion model in intact clad.	Cladding and fuel motion models for voided assemblies in a loss of flow accident.	Cladding and fuel motion models for voided assemblies in a loss of flow accident.	Unified molten cavity model connecting axial meshes. Includes effects of fission gas release from the fuel upon melting. Fuel motion in intact pins and fuel and steel motion in voided and disrupted coolant channels; calculations can also continue beyond coolant voiding and after clad motion has begun.
Fuel–coolant interaction	No	No	FCI model for transient overpower conditions	FCI model for transient overpower conditions	FCI-driven voiding and fuel motion for treatment of in-pin and out-of-pin fuel motion
Primary and intermediate loop model	No	Thermohydraulic model for primary loop	Thermohydraulic model for primary loop	Thermohydraulic model of primary and intermediate heat transport systems. Limited capability of handling transients for intermediate heat transport system	Thermohydraulic model of primary and intermediate heat transport systems. Limited capability of handling transients, natural circulation in the primary and intermediate loops, loss of heat sink cases, and pipe rupture
Cladding motion	No	No	No	Clad motion model	Clad motion coupled with the sodium voiding model for realistic assessment
Input to disassembly calculations	Supplies output as input to the weak explosion code MARS	Supplies output as input to the weak explosion code VENUS	Supplies output as input to the weak explosion code SIMMER	Supplies output as input to the weak explosion code SIMMER	Supplies output as input to the weak explosion code SIMMER

latest, SIMMER IV [17.14]. In the SIMMER series, all the LMFBR core materials are modeled in the structure (fuel pins and subassembly [SA] duct walls), liquid (liquid materials and solid particles), and vapor (mixture of vapor species) fields, together with the mass momentum and energy exchange among them.

In the first version of the code (SIMMER), neutronics and fluid dynamics behavior during a core disruptive accident are coupled in 2D cylindrical (R, Z) coordinates. Either the time-dependent multigroup diffusion equation or the neutron transport equation is solved by a quasi-static method to predict neutron behavior. By comparison with critical experiments, it has been shown that the enhanced accuracy of transport theory calculations is important when voided regions develop in a disrupted core as anticipated for the transition phase. Hydrodynamics is calculated in SIMMER by defining three fields—structure field, liquid field, and vapor field. Core materials are referred to as components and two types of components are defined—density components and energy components. Density components are used to follow material motion and energy components are used to predict material temperatures.

SIMMER II, an updated version of the SIMMER code, has six basic components: fertile fuel, fissile fuel, steel, sodium, absorber material, and fission gas. Each field, however, keeps track of more components than this. The structure field consists of solid fuel, cladding and structure, and unreleased fission gas. Also, when a material is frozen, it is removed from the liquid field and added to the structure field. The liquid field consists of a uniform mixture of all molten materials. In addition, the liquid field contains solid particles of fuel and steel which may be flowing with the liquid. The vapor field contains a uniform mixture of all vapors and inert gases. The conservation equations are solved for each field in Eulerian coordinates, together with the appropriate equations of state. A single velocity is calculated for each fluid field. Within a fluid field, all components move with the velocity of the field. The structure field remains stationary and thus has no momentum equation. The field does, however, influence the motion of the fluid fields. For the structure and liquid fields, separate energy equations are solved for each component. Only one energy field is used for the vapor field since all gases are assumed to be mixed and at the same temperature. The difference between density components and energy components results from the need to keep track of fertile and fissile material separately in the mass flow equation. This need arises because fuel of different fissile fractions must be mixed. The fertile and fissile components are assumed to be intimately mixed in each field, however, so that they are always at the same temperature. Therefore, only a single fuel component is needed for the energy equation. Coupling between the fields and components is taken into account in the conservation equations. Phase transitions are modeled with considerable complexity. Mass transfer between fields is included in the mass conservation equation. Melting and freezing are calculated by comparison of material energies with solidus and liquidus energies. Vaporization and condensation are calculated using a nonequilibrium conduction-limited phase transition model in which the phase transition rates are proportional to the difference between saturation and actual temperatures of the components. A multicomponent vaporization/condensation model is used to treat the effects of noncondensable and multicomponent interference. Momentum coupling is accounted for by drag force terms and by momentum exchange with mass transfer between fields, incorporated through the momentum balance equation. The drag correlations are in a simple form applicable for two-phase dispersed flow which is satisfactory for a wide range of events postulated for CDA. Rapid compression of fluids involving steep pressure gradients is handled by the pseudo-viscous pressure method. Energy coupling is accounted through the energy balance equation. Energy transfer through convection and radiation between fields or between components within a field, energy exchange with mass transfer, and energy transfer due to drag heating are denoted by various terms in the momentum balance equation. All of the energy from PdV work (expansion or compression) appears in the vapor field energy equation. The SIMMER II code had limitations that only two moving fields (liquid mixture and vapor mixture) and only a single-flow regime (dispersed droplet flow) are treated.

SIMMER III is the improved version of SIMMER II to more accurately and efficiently handle the cross sections. Fluid dynamics was further improved on the lines of the AFDM code, which considered the fundamental fluid dynamics methods for a three-velocity-field convection algorithm coupled with multiple flow regime modeling (SIMMER II is a two-field and single-flow regime code). Up to SIMMER III, calculations were done in 2D. SIIMER IV was upgraded again to perform 3D calculations for neutronics, space–time kinetics, and thermal hydraulics. Thus, the 3D version has been called SIMMER IV. An international effort is to use SIMMER IV for the practical reactor.

17.2.2 Computer Codes for Severe Accident Analysis in India

A computer code system, KALDIS, is used for CDA analysis [17.15]. KALDIS is constituted by two codes: PREDIS and VENUS II. For the predisassembly phase, PREDIS was developed as a part of KALDIS. The processes modeled in the code are core neutronics, transient thermal hydraulics, reactivity feedbacks like Doppler, fuel and clad axial expansion, coolant expansion, spacer pad, grid plate, main vessel and differential control rod expansion, coolant boiling, and clad and fuel melting and slumping. For efficiency in calculations, each fuel SA is represented by one pin. Also, all SAs in a radial ring are represented by one SA. This is a conservative approach because the boiling or melting in an SA or ring of SAs is instantaneous, while in actual situations, boiling or melting will spread with finite pace, using the point kinetics model. A detailed temperature distribution in the fuel, clad, and coolant is calculated using either the lumped model of heat transfer or the exact heat conduction model. The transfer time delays are accounted implicitly in the code. The sodium boiling model is developed based on the formulation of the SAS1A [17.1]. On fuel melting, fuel slumping is modeled conservatively to give a positive reactivity. It is assumed that when one-third of the fuel pin melts, the upper part slumps down and gives a positive reactivity. The accelerated fall of the fuel slumping is used to calculate the reactivity addition rates. For the disassembly phase, the VENUS II code, which also forms part of the KALDIS computer code system, is used. In this code, point kinetics is used for power calculations, and using the 2D reactivity worth and Doppler coefficient distribution, the calculations of displacement and Doppler reactivity feedbacks are performed. Calculations are stopped once the multiplication factor of the reactor is 0.98 or 0.95 due to outward material movement of the core.

The calculations of the work energy potential of the thermal energy released in the disassembly phase are determined using the straightforward methodology available for isentropic expansion of fuel vapor up to one atmospheric pressure or up to the upper cover gas volume. This algorithm has been made a part of the VENUS II code.

17.2.3 IAEA Benchmark on BN-800: Validation of Severe Accident Codes

For unprotected loss of flow accident (LOFA), the code is tested up to the onset of boiling against the BN-800 IAEA coordinated research project (CRP) benchmark [17.16]. A comparison of a few important parameters is provided in Table 17.2. It is found that onset of boiling and its location, the sodium void reactivity, and the net reactivity are predicted nearly the same by all countries, even though some differences are noted in the Doppler reactivity. Specifically, the Indian code PREDIS is also validated under the IAEA CRP benchmark [17.17] and further against the fast breeder test reactor (FBTR) reactivity transient [17.18].

17.3 COMPUTER CODES FOR THE MECHANICAL CONSEQUENCES

17.3.1 International Computer Codes

Analysis for mechanical consequences is very complex, since it involves calculations for fast fluid transients, large fluid and structural displacements and deformations, and strong nonlinear fluid–structure interactions. The problem is further complicated by (1) simultaneous treatment of both geometrical and material nonlinearities of structures, (2) geometrical discontinuities like sharp

TABLE 17.2

Comparison of Results for the BN-800 IAEA CRP Benchmark

Parameter	Germany	France	Japan	Russia	India
Time (s)	17.96	18.93	18.96	16.72	17.60
Channel no.	5/1	5/1	5/1	5/1	5/1
Axial position from core bottom (cm)	84–90	95	87–94	85	84
Normalized power	0.66	0.63	0.63	0.71	0.71
Net reactivity ($)	−0.17	−0.183	−0.183	−0.135	−0.147
Doppler reactivity	0.026	−0.005	−0.004	+0.039	+0.027
Fuel axial expansion reactivity ($)	−0.003	+0.015	+0.014	+0.017	+0.020
Sodium reactivity ($)	−0.207	−0.205	−0.205	−0.188	−0.223

corners and irregularities present in the fluid region in the vicinity of structures, (3) a wide spectrum of phenomena such as flow through perforated structures and multidimensional sliding interfaces, and (4) complicated flow of fluid and gas phases around the core structures given earlier and other internals. In the past 40 years, substantial research effort has been devoted to the development of numerical methods and computer codes for analyzing SFR primary system response. Most codes currently in use employ a mixed Lagrangian and Eulerian approach to analyze the fluid transients, in conjunction with the Lagrangian method to calculate structural responses. Since there are numerous finite element structural dynamics programs available for solving complicated structures with material and geometrical nonlinearities, current research mainly focuses on the development of effective solution methods to treat fast fluid transients and fluid–structure interactions.

It is known that in the Lagrangian hydrodynamic approach, the mesh used to compute the coolant motion moves with the coolant and the governing equations do not have the transport terms. Because of this, computer code development for containment analysis in the late 1960s and early 1970s almost exclusively utilized the Lagrangian mesh for fluid calculation. The first Lagrangian containment computer program is REXCO-H [17.19] developed at ANL. Later, a number of Lagrangian codes have evolved and have been reported in the literature, including REXCO-HEP [17.20], ASTARTE of the United Kingdom Atomic Energy Authority (UKAEA) [17.21], ARES of Interatom Germany [17.22], and SIRIUS [17.23] of Commissariat à l'énergie atomique et aux énergies alternatives (CEA), France. Although Lagrangian computer codes have been used for analyzing primary containment response under CDA, their analyses are still limited due to the difficulties in treating excessive grid distortions, flow around corners and irregularities, and outflow boundary conditions. To deal with such situations, codes that utilize a Eulerian method for the description of the coolant, in conjunction with a Lagrangian method for the treatment of the structure, were developed. In the Eulerian hydrodynamic analysis, the mesh used for the description of coolant motion is fixed in space. Consequently, such a mesh is ideal for treating excessive material distortions and outflow boundary conditions. The first Eulerian containment code is ICECO [17.24], developed in 1975. Subsequently, a Eulerian containment code, SEURBNUK-2 [17.25], was developed jointly by Euratom, Belgonucleaire, and the United Kingdom for the European communities. Ref. 17.26 describes PISCES-2 DELK, a Eulerian version of the PISCES code, which has also been applied to safety problems related to FBR containment. In a similar line, a Eulerian–Lagrangian coupled code, CASSIOPEE [17.27], was developed at CEA, France. The basic difficulty with a pure Eulerian code is the complexity involved in modeling the fluid–fluid and fluid–structure interfaces. To eliminate the disadvantages and still maintain the advantages of both the Lagrangian and Eulerian methods, coupled Lagrangian–Eulerian techniques were employed. Computer codes EURDYN [17.28], ALICE [17.29], and PLEXUS [17.30] are of this category. The techniques used in these codes for calculating the coolant motion are based on the "arbitrary Lagrangian and Eulerian (ALE)" coordinate system, originally suggested by Hirt et al. [17.31]. As per ALE, in the region where the coolant is expected to move extensively, the vertices of the fluid mesh can be made to move in an optimum manner, so that excessive mesh distortions can be completely eliminated by using a continuous rezoning process. Also, at the fluid–structure interface, the vertices of the fluid mesh can be made to move with the structural nodal points to simplify the computational procedure and to avoid irregular cell calculations. Further, in the ALE algorithm, it is easy to treat internal thin shells, perforated structures, curved reactor bottom, and highly distorted core gas bubble.

The ALICE code was used for the analysis of the Clinch River Breeder Reactor Project (CRBRP), while the PLEXUS code was used for the analysis of the European fast reactor (EFR). Commercial demonstration fast reactor (CDFR) and SNR-300 projects were analyzed using the SEURBNUK code, while the SIRUS, CASSIOPPE, and SEURBNUK/EURDYN codes were employed for SPX-1 and SPX-2. For the analyses for SFRs in India, an in-house computer code called FUSTIN was developed. It is an axisymmetric finite element code that solves a set of governing differential equations of fluid, structure, and fluid–structure interaction dynamics written in the ALE coordinate system. Mathematical modeling details of the FUSTIN code are described in [17.32].

17.3.2 Validation of Codes

17.3.2.1 COVA Series [17.33]

The UKAEA and Joint Research Centre, Euratom, established in 1973 a collaborative experimental program called COVA (code validation) for carrying out a series of small-scale (the maximum diameter of the vessel is 560 mm and height varied over the range of 700–1120 mm), well-instrumented tests aimed at providing high-quality data on stresses, strains, and loads when a well-characterized energy source is released within a fluid in a containment vessel. In the United Kingdom, these data were used to validate ASTARTE and SEURBNUK, which are used in studies of the response of the FBR primary containment system in the event of a CDA. Parallel but complementary programs using identical experiments were carried out at three collaborative sites (AWRE, Foulness and AEEW, Winfrith, in the United Kingdom and JRC, Ispra). Theoretical groups at JRC–Ispra, AEEW–Winfrith, and AWRE–Aldermaston collaborated in the design of the tests, analysis of the experimental data, and development of various codes.

The series of experiments, starting with simple tests of rigid cylindrical tanks partly filled with water, increase in complexity in such a way that only one new feature is introduced at a time. The test sequences covered topics relevant to both loop- and pool-type geometries. The parameters are (a) the height/diameter ratio to cover pool-type and loop-type reactor geometries, (b) size of the cover gas volume, and (c) explosive charge to vary both pressure and energy release. Design features investigated include (a) the rigid or deformable main vessel, (b) inner vessel of variable thickness and shapes, (c) grid plate, (d) neutron shields, and (e) core support structure and control plug.

17.3.2.2 CONT Benchmark Exercise [17.34]

The Commission of the European Communities (CONT) benchmark calculation exercise is a project defined by commission of the european communities (CEC), Italy. Under this exercise, a simplified pool-type reactor geometry (representative of a CDFR reactor assembly: U.K. SFR) undergoing a postulated CDA was defined. The reactor roof is represented by a rigid plate fixed to the ground and the top part of the reactor vessel is clamped to the roof. This oversimplified representation of the reactor roof structures and hold-down system has been chosen in order to allow for the participation of various computer codes having different limited capabilities regarding roof modeling. For similar reasons, the representation of the core structures given earlier has not been included. The effects of the internal structures have been schematically taken into account by the inclusion of a stainless steel cylindrical inner tank surrounding the core region. Three fluids, sodium as coolant, argon as cover gas, and expanding fuel vapor as a core bubble, are present within the reactor vessel. The main calculation parameters are strains in the main vessel and internal vessel, cover gas pressure history, slug impact time, energy balance, etc. The analysis results were obtained using six computer codes: ASTARTE, CASSIOPEE, PISCES-2 DELK, SEURBNUK, SIRIUS, and SEURBNUK/EURDYN. The details of these computer codes and basic input details are provided in Table 17.3.

17.3.2.3 Validation of Computer Codes Used in French

Several experimental programs were used to validate the fast reactor containment codes CASSIOPEE, SIRIUS, and PLEXUS that have been developed by CEA for SPX reactor safety studies. The MANON, MARA, and MARS are the major programs that were carried out by CEA/DRNR. Further, the codes were validated under the APRICOT program, a code comparison exercise, sponsored by the ERDA and COVA program, carried out jointly by the UKAEA and the JRC, Ispra. Some details of the MANON, MARA, and MARS test series are described later.

17.3.2.4 MANON Program [17.35]

CEA/DRNR, Cadarache, developed the MANON test program. A spherical charge of solid, low-pressure, low-density L 54/16 explosive source developed by CEA/DRNR and CEA/DAM was used for experimental simulations. All MANON tests consist of a deformable steel cylinder filled with water and closed at the top and bottom by a rigid flat roof. The cylinder diameter is 38 cm and its total height 38 cm. About 15 firings were performed in this series. The first six used a hexogene source to design properly the instrumental device, and for the remaining tests, the L 54/16 explosive charge was used. Some of these are repeated in order to check the validity of the instrument results and to correct operation of the charged. An inner vessel of 23 cm diameter was present in one test, and in another firing, argon gas replaced water to study neighboring media.

17.3.2.5 MARA Series [17.35,17.36]

The MARA test series (up to MARA 10) on a 1/30 scale model of SPX was carried out by CEA, France, to validate the PLEXUS code. The series comprised simple vessel configurations, progressively increasing complexity with internal components and deforming roof, added step by step. The last test in the MARA 10 series is an integral model experiment that involves different kinds of deformable structures: vessel as in MARA 01 and MARA 02, core support structure and grid plate as in MARA 04, and deforming roof as in MARA 08 along with control plug and radial shield. The CASSIOPEE and SIRIUS predictions are compared to experimental data generated from MARA 01 and 02. Other MARA results involving more complex internal structures featuring the SPX are compared with the PLEXUS code.

17.3.2.6 MARS Test [17.37]

The MARS test is the most complex and detailed mock-up conducted during 1982 in CEA, Cadarache, involving a large energy release simulating a CDA. The test results were well compared with PLEXUS predictions.

17.3.2.7 Validation of REXCO Codes in the United States [17.38,17.39]

TNT was mainly used for fast flux test facility (FFTF) mock-up tests toward validating REXCO predictions. Further, the SL-I accident was used as experimental data for comparison with analytical methods. Basic tests were also carried out at the Naval Ordinance Laboratory. ALICE code analysis results of CRBRP were validated using COVA test results.

17.3.2.8 TRIG Series for the Validation of FUSTIN in India [17.32,17.40]

The FUSTIN code has been extensively validated by solving many standard published problems—MANON (CEA, Cadarache), COVA (United Kingdom), and CONT (Italy)—which include code to code comparison as well as experimental data. The international codes–ASTARTE, CASSIOPEE, PISCES-2 DELK, SEURBNUK, SIRIUS, EURDYN, and PLEXUS–are compared

TABLE 17.3
Computer Codes Participating in the CONT Benchmark

Computer code	ASTARTE-4B	CASSIOPEE	PISCES-2 DELK	SEURBNUK	SEURBNUK	SEURBNUK/ EURDYN	SIRIUS
Origin	UKAEA	CEA	PI	UKAEA–JRC	UKAEA–JRC	UKAEA–JRC	CEA
Calculator	ENEA–Bologna	ENEA–Bologna	API–Gouda	UKAEA–Risley	JRC–Ispra	JRC–Ispra	CEA–Cadarache
Fluid (compressible/inviscid)—finite element discretization							
Coordinate	Lagrangian	EL	EL	Eulerian		Eulerian	Lagrangian
Time integration	Explicit	Explicit	Explicit	Implicit		Implicit	Explicit
Number of cells	$22 \times 39 = 858$	$18 \times 30 = 540$	$21 \times 44 = 924$	$26 \times 55 = 1430$		$26 \times 55 = 1430$	$8 \times 26 = 208$
Time step (μs)	16–6	35	55	100		100	60
Structure—Thin shell model; explicit time integration							
Method	FD	FD	FD	Fn	FD	FE	FE
No. of segments	61	20	42	75	75	75	17
Time step (μs)	16–6	35	55	20	20	20	60
Fluid–structure interaction							
Type of coupling	Weak	Strong	Strong	Weak	Weak	Weak	Weak
Cover gas model	Gas bag	Gas bag	Continuum	Gas bag	Gas bag	Gas bag	Gas bag
Bubble model	Continuum	Gas bag	Gas bag	Gas bag	Gas bag	Gas bag	Gas bag

FD, finite differences; FE, finite elements; EL, Eulerian and Lagrangian.

with FUSTIN. Further, the code validation exercise has been further continued by developing indigenous facilities under the TRIG program. The purpose of the TRIG series is (1) to qualify the specially developed chemical explosive for PFBR studies based on tests involving simple geometry (TRIG-I), (2) to quantify the rupture limit for the SS 316 under actual loading condition (TRIG-I), (3) to understand the deformation and rupture behavior of the main vessel without internals (TRIG-II) and with internals (TRIG-III), and (4) to identify a possible sodium leak through the roof slab (TRIG-III). TRIG tests were compared with FUSTIN prediction. A total of eight benchmark problems are presented for validating the FUSTIN code: two numbers for structural modeling, one number for hydrodynamic modeling, five numbers for hydrodynamic and structural modeling including fluid structural interaction effects, and three numbers for validating two-phase element formulation. In all these problems, the performance of the automatic mesh description algorithm is assessed systematically by graphically displaying the mesh configurations at discrete time intervals.

17.4 RADIOACTIVE RELEASE

In Chapter 16, radioactive sources in the SFR from a CDA were presented in detail. The estimation of radioactivity release to the site boundary (public) involves a series of computations starting from the core and then transport to the cover gas, then to the RCB, and finally to the site boundary through the stack. The estimation of radioactivity materials (principally fission products, plutonium, and sodium oxide), till it reaches the stack, is carried out by each country using their in-house computer codes. Computer codes, such as ORIGEN [17.41] and RIBD [17.42,17.43] (U.S. codes), are now available to calculate detailed inventories of all fission product isotopes from various fuel isotopes released under CDA. The COMRADEX [17.44] code is a popular code available to estimate the source term. This code follows the radiological inventory through a series of chambers within the containment. The initial chamber monitored is normally the reactor vessel. This chamber then leaks to the RCB. The COMRADEX code also requires a quantitative understanding of processes and mechanisms involved in the release and transport of radioactive materials from the fuel and sodium into the chambers. The noble gases (Kr and Xe) are usually assumed to be released from both the fuel and the sodium. The release of fuel aerosols and remaining fission products is more difficult to assess. There are several paths by which some of these materials can become airborne, contributing to the radioactive aerosol source term. In CDA, they might move upward to the cover gas in a large bubble and then escape from the reactor vessel through openings in the head resulting from mechanical damage caused by the accident. The halogens (principally iodine) and some of the volatile fission products could be released by this path, and some of the fuel and solid fission products could be suspended as aerosols. Sodium leaks from the hot pool to the RCB, which is estimated through a computational model such as CACECO in the United States and SOSPIL [17.45] in India. CACECO combined with COMRADEX could provide an estimate of the percentage of these elements that escape to the RCB. The source term release and transport processes were reviewed in Ref. 17.46. It is worth mentioning that the experimentally verified mechanistic models are not available to date to evaluate the release of solids from the reactor vessel and to follow them through various barriers to the RCB. The analyses are generally based on an arbitrary source term. Typically, 1% of the solids initially in the core is released along with the fission materials.

Readers should be aware of a few popular codes for the source term evaluations for research reactor facilities: MELCOR, SCDAP, or MAAP4 [17.47–17.50]. MELCOR is a fully integrated computer code for severe accident analyses that include specific modules for (nonexplosive) core melt progression and fission product release and transport within multivolume interconnected systems. Codes such as SPARC, BUSCA, SOPHAEROS, CONTAIN, and GOTHIC allow detailed assessments to be conducted on radionuclide transport and retention in various reactor systems, such as the pool, the primary heat transport system, and the containment structure [17.51–17.54]. Codes such as MicroShield, MARMER, MERCURE, and QAD [17.55–17.58] are appropriate for evaluating direct dose rates in a wide variety of source/shielding configurations.

For the atmospheric dispersion studies, a large number of codes are available. However, codes that take into account the effects of radioactive sodium aerosol dispersions are limited. In the general context, a number of methods and techniques of modeling the atmospheric dispersion of radioactive material in the radiological risk assessment have been developed in the last 50 years. These methods can be broadly classified under Gaussian plume models, Lagrangian puff and particle models, coupled atmospheric dispersion models, and CFD-based models. The Gaussian plume models such as MACCS [17.59], COSYMA [17.60], and ADMS [17.58] are used to obtain dispersion estimates for evaluating design basis accidental releases of radioactive materials for safety analysis. They are simple with minimum inputs and depict the distribution of time-averaged concentrations typically over a period of 1 h in an approximate way using 1–3-year composite joint frequency distribution of winds and atmospheric stability class of the site. Such codes are suited mainly for plain terrain with homogeneity of wind and atmospheric conditions and provide conservative estimates. In areas of complex topography with nonhomogeneous atmospheric flow and stability conditions, Lagrangian Puff and particle dispersions like SPEEDI [17.61] of JAEA, Japan, FLEXPART of Germany [17.62], and HYSPLIT [17.63] of NOAA, United States, are employed. At the complex coastal site, the land–sea breeze and thermal internal boundary layer are two important phenomena that influence the dispersion of radioactive releases. The approach followed for PFBR constructed at the Kalpakkam site is that these effects are incorporated in the FLEXPART dispersion model by integrating with MM5/WRF prognostic atmospheric models so that the time- and space-varying meteorological information for spatial dose assessment is possible [17.64]. The codes MM5-FLEXPART and the WRF weather model are validated for the Kalpakkam site by conducting wind field experiments and SF_6 tracer dispersion experiments [17.65,17.66]. Turbulence measurements at the site are used to develop suitable short-range turbulence diffusion physics in the previously mentioned models [17.67].

17.4.1 Approach Adopted for SFR in India

An analysis was carried out for PFBR with the assumption that the radionuclides from the containment building get released into the atmosphere through two pathways. The first pathway is the leakage of RCB activity through normal ventilation via the stack for 60 s during which fission product noble gases (Kr-87, Kr-88, Kr-85, Xe-133, Xe-135), iodine (I-131, I-132, I-133, I-134, I-135), and cesium (Cs-134, Cs-137) get released. The second pathway is the ground-level release to the environment due to leak in the RCB at a rate of 0.1% volume of RCB per hour for a period of 24 h. The ground-level activity release consists mainly of I-131, I-133, I-135, Cs-134, Cs-137, Ru-103, Ru-106, Sr-89, Sr-90, Ce-141, Ce-144, Ba-140, Pu-239, Kr-87, Kr-85, and Xe-133. Further, the environmental dispersion analysis was conducted using the simple Gaussian plume model and the coupled atmosphere and particle dispersion model MM5-FLEXPART. In this approach, first, the meteorological condition of the site was simulated using the weather prediction model MM5/WRF, and then the dispersion was simulated using the predicted 3D time-varying atmospheric parameters after validating with site observations. The dose modules for cloudshine, groundshine, and inhalation are incorporated in the codes by taking into account the simulated air concentration and deposited activity outputs. The coupled model has realistically simulated the spatiotemporal atmospheric flow condition consisting of the land–sea breeze circulation and the effect fumigation by the thermal internal boundary layer on the radiological dose (cloudshine, groundshine, inhalation) at the site boundary and in the off-site distance range of up to 25 km. An intercomparison of the estimates from GPM and MM5-FLEXPART revealed that the estimates for the site boundary dose would be much lower than the values given by GPM and the values are within the stipulated limits. The codes are tested and incorporated in the Online Nuclear Emergency Decision Support System (ONERS) designed for the Kalpakkam site for radiological accidents. The Lagrangian particle dispersion model FLEXPART-WRF implemented in ONERS-DSS was validated for the Fukushima accidental releases by performing regional-scale simulations for the March 11–31, 2011, accident period using data on radiation dose and activity deposition published by MEXT, Japan. The predicted effective dose was used to assess the lifetime attributable health risk of the public in a 40 km impact zone around the Fukushima reactor. The study helped assess the time attainment of low risk for the inhabitation of people in affected areas. More details are presented in Refs. 17.64–17.67.

17.5 SODIUM FIRE CODES

A list of international codes for numerically simulating sodium fire scenarios is presented in Table 17.4. Apart from these, FEUMIX is used in France for the simulation of sodium–concrete interactions, particularly for the calculation of gas pressure, temperatures of gas and walls and mass of produced aerosols. The analysis domain includes a fire room along with connected rooms, and a simplified modeling for the sodium pool fire and water vapor release from concrete walls is employed. Apart from this, a 1D code called SORBET is also used in France for the thermohydraulic behavior simulation for a concrete block including the determination of temperatures, pressure, water vapor, and liquid contents [17.68]. The code also predicts water and carbon dioxide release from the hot side and cold side. For sodium–concrete chemical reaction simulations, the REBUS computer code is used with the objective of deriving the composition of reaction products and heat release, hydrogen generation and energy balance [17.68]. The codes NACOM and SOFIRE are currently used for SFRs in India to analyze the sodium spray and pool fire, respectively [17.69]. The predictions of the NACOM code are conservative, as it neglects the effects of droplet interaction,

TABLE 17.4
Sodium Fire Codes

Code	Phenomena	Developed by	Description
SPRAY-3A	Spray fire	Hanford Engineering Development Laboratory (HEDL)	One-dimensional code to investigate sodium spray fire based on the vapor-phase combustion model.
SOMIX	Spray fire	Atomics International (AI)	Retains the SPRAY combustion model. It models the gas circulation two-dimensionally.
SPOOL	Spray/pool fire	Argonne National Laboratory (ANL)	An integrated code for spray/pool fire based on SPRAY and SOFIRE-II.
SOFIRE-II	Pool fire	AI	One-dimensional pool fire code based on the surface combustion model.
NACOM	Spray fire	Brookhaven National Laboratory (BNL)	One-dimensional code based on vapor-phase combustion.
CONTAIN-LMR	Spray/pool fire	Sandia National Laboratory (SNL)	Two-dimensional code to investigate sodium pool/spray fire.
PULSAR	Spray fire	Laboratory for Aerosol Physics and Filtration Technique, Germany	Two-dimensional code designed to calculate the thermodynamic consequences and the release of aerosols from burning sprayed sodium in confined atmosphere.
SOFIA-II	Spray fire	Energy Research Laboratory (ERL), Japan	Two-dimensional code based on a vapor phase and combustion to calculate pressure and temperature transients.
ASSCOPS	Spray/pool fire	Japan Nuclear Cycle Development Institute (JNC)	An integrated code for spray/pool fire based on SPRAY and SOFIRE-II. Capable of predicting aerosol generation and sodium–concrete interaction.
SPHINCS	Spray/pool fire	JNC	Multidimensional code based on lumped-mass approximation of space (zone model).
AQUA-SF	Spray/pool fire	JNC	Multidimensional numerical analysis code for sodium combustion.

nonuniformity of the gas temperature across the spray zone, and depletion of oxygen inside the spray zone in determining the spray burning rate. Also, very few details of the other codes are available in open literature, bringing the need for in-house development of reliable sodium fire codes. Accordingly, sodium droplet combustion has been modeled numerically and parameters affecting the phenomena are investigated [17.70].

REFERENCES

17.1 Carter, J.C. et al. (1970). SAS1A—A computer code for the analysis of fast reactor power and flow transients. ANL-7607.

17.2 Dunn, F.E. et al. (1974). SAS2A LMFBR accident analysis computer code. ANL-8138.

17.3 Stevenson, M.G. et al. (1974). Current status and experimental basis of the SAS LMFBR accident analysis code system. *Proceedings of the International Conference on Fast Reactor Safety*, Beverly Hills, CA.

17.4 Cahalan, J.E. et al. (1977). A preliminary users guide to version 1.0 of the SAS 3D accident analysis code. SR-239831, ANL.

17.5 Wider, H.U. et al. (1982). Status and validation of the SAS4A accident analysis code system. *Proceedings of the LMFBR Safety Topical Meeting*, ANS, Lyon-Ecully, France.

17.6 Cahalan, J.C., Wei, T.Y.C. (1990). Modeling development for the SAS4A and SASYS computer codes. *Proceedings of the International Fast Reactor Safety Meeting*, Snowbird, UT.

17.7 Sha, W.T., Hughs, T.H. (1970). VENUS: A two dimensional coupled neutronics hydrodynamics computer program for fast reactor power excursions. ANL-7701, October 1970.

17.8 Jackson, J.F., Nicholson, R.B. (1972). VENUS-II: A LMFBR disassembly program. ANL-7951.

17.9 Bott, T.F., Jackson, J.F. (1976). Experimental comparison studies with the VENUS-II computer code. *Proceedings of the International Meeting on Fast Reactor Safety and Related Physics, III*, Chicago, IL, October 1976, p. 1134.

17.10 Suk, S.-D., Hahn, D. (2002). Analysis of core disruptive accident energetics for liquid metal reactor. *J. Korean Nucl. Soc.*, 34(2), 117–131.

17.11 Bell, C.R. (1977). SIMMER-I: An Sn Implicit, Multifield, Multicomponent, Eulerian, recriticality code for LMFBR disrupted core analysis. Report LA-NUREG-6467-MS. Los Alamos Scientific Laboratory, Los Alamos, NM.

17.12 Smith, L.L. et al. (1978). SIMMER-II: A computer program for LMFBR disrupted core analysis, Vol. 2. NUREG/CR-0453, LA-7515-M.

17.13 Kondo, S. et al. (1999). Current status and validation of the SIMMER-III LMFR safety analysis of core. *Proceedings of the Seventh International Conference on Nuclear Engineering (ICONE-7)*, Tokyo, Japan, pp. 19–23.

17.14 Yamano, H. et al. (2009). A three-dimensional neutronics-thermohydraulics simulation of core disruptive accident in sodium-cooled fast reactor. *Nucl. Eng. Des.*, 239, 1673–1681.

17.15 Harish, R. et al. (1999). KALDIS: A computer code system for core disruptive accident analysis in fast reactors. IGCAR Report-IGC 208.

17.16 IAEA. (2000). Transient and accident analysis of a BN-800 type LMFR with near zero void effect. Final Report on an International Benchmark Programme, Supported by the International Atomic Energy Agency and the European Commission, 1994–1998. IAEA-Tecdoc-1139, IAEA, Vienna, ISSN 1011-4289. http://www-pub.iaea.org/MTCD/publications/PDF/te_1139_prn.pdf.

17.17 Om Pal, S., Harish, R. (1998). Results of transient calculations upto onset of boiling of a comparative calculation for unprotected loss of flow accident in BN-800 type reactor with near zero void reactivity coefficient. *IAEA/EC Consultancy Meeting on the Comparative Calculations for Severe Accident in BN-800 Reactor*, Obninsk, Russia.

17.18 Srinivasan, G.S., Om Pal, S. (1999). Validation of computer code PREDIS against FBTR experimental reactivity transients. IGCAR Report: RPD-SAS/FBTR/01100/CR/011.

17.19 Chang, Y.W., Gvildys, J., Fistedis, S.H. (1973). Analysis of primary containment response using a hydrodynamic elastic-plastic computer code. *Proceedings of the Second International Conference on Structural Mechanics in Reactor Technology (SMiRT)*, Berlin, Germany.

17.20 Chang, Y.M., Vildys, J.G. (1975). REXCO-HEP: A two-dimensional computer code for calculating the primary system response in fast reactors. ANL-75-19, ANL.

17.21 Cowler, M.S. (1974). ASTARTE—A 2-D Lagrangian code for unsteady compressible flow theoretical description. AWRE-44-91.

17.22 Doerbecker, K. (1972). ARES: Ein 2-Dim Rechenprogram zur Beschreibung der Kuzzeitigen Auswirkungen einer Hypothetischen Unkontrollierten Nukleren Exkursion auf Rektortank, Drhdeckel and Tankeinbanten, gezeigt am beispoel des SNR 300, Reaktortangung, Hamburg, Germany.

17.23 Blanchet, Y., Obry, P., Louvet, J. (1981). Treatment of fluid-structure interaction with the SIRIUS computer code. Paper B8/8. *Transactions of the Sixth International Conference on Structural Mechanics in Reactor Technology (SMiRT)*, Paris, France.

17.24 Wang, C.Y. (1975). ICECO—An implicit Eulerian method for calculating fluid transient in fast reactor containment. ANL-75-81. Argonne National Laboratory, Lemont, IL.

17.25 Cameron, I.G. et al. (1978). The computer code SEURBNUK-2 for fast reactor containment studies. *Comput. Phys. Commun.*, 13, 197.

17.26 Cowler, M.S., Hancock, S.L. (1979). Dynamic fluid-structure analysis of shells using the PISCES-2DELK computer code. Paper B1/6. *Transactions of the Fifth International Conference on Structural Mechanics in Reactor Technology (SMiRT)*, Berlin, Germany.

17.27 Graveleau, J.L., Louvet, P. (1979). Calculation of fluid-structure interaction for reactor safety with the CASSIOPEE code. Paper B1/7. *Transactions of the Fifth International Conference on Structural Mechanics in Reactor Technology (SMiRT)*, Berlin, Germany.

17.28 Donea, J.P., Fasoli-Stella, P., Giuliani, S., Halleux, J.P., Jones, A.V. (1980). The computer code EURDYN-1M for transient dynamic fluid-structure interaction. EUR 6751. Commission of the European Communities, Directorate-General, 'Scientific And Technical Information And Information Management', Bâtiment Jean Monnet, Luxembourg.

17.29 Wang, C.Y., Zeuch, W.R. (1982). ALICE-II: An arbitrary Lagrangian–Eulerian code for containment analysis with complex internals. *Trans. Am. Nucl. Soc.*, 41, 364.

17.30 Hoffmann, A., Lepareux, M., Jamet, P. (1986). PLEXUS: A general program for fast dynamic analysis. CEA, DEMT/86, p. 295.

17.31 Hirt, W., Amsden, A.A., Cook, J.L. (1974). An arbitrary Lagrangian–Eulerian computing method for all fluid speeds. *J. Comput. Phys.*, 14, 227–253.

17.32 Chellapandi, P., Chetal, S.C., Raj, B. (2010). Structural integrity assessment of reactor assembly components of a pool type sodium fast reactor under core disruptive accident—Part 1: Development of computer code and validations. *J. Nucl. Technol.*, 172(1), 1–15.

17.33 Hoskin, N.E., Lancefield, M.J. (1978). The COVA programme for the validation of computer codes for fast reactor containment studies. *Nucl. Eng. Des.*, 46, 1–46.

17.34 Benuzzi, A. (1987). Comparison of different LMFBR primary containment codes applied to a bench mark problems. *Nucl. Eng. Des.*, 100, 239–249.

17.35 Blanchet, Y. et al. (1981). Experimental validation of the containment codes SIRIUS and CASSIOPPE. *Transactions of SMiRT 6*, Paris, France, August 1981, Vol. B, p. B8/1.

17.36 Louvet, J. et al. (1987). MARA 10: An integral model experiments in supports of LMFBR containment analysis. *Transactions of SMiRT 9*, Lausanne, Switzerland, August 1987, Vol. E, pp. 331–337.

17.37 Cariou, Y. et al. (1997). LMR's whole core accident: Validation of the PLEXUS code by comparison with MARS test. *Transactions of SMiRT 14*, Lyon, France, August 1997, pp. 339–346.

17.38 Simpson, D.E. (1975). The hypothetical core disruptive accident. HEDL S/A-741 REV.

17.39 Romander, C.M., Cagliostro, D.J. (1979). Structural response of 1/20 scale models of the CRBR to a simulated HCDA. CRBRP Report No. 3, Rec-2, pp. 1–45.

17.40 Terminal Ballistic Research Laboratory. (2002). Investigation of mechanical consequences of a core disruptive accident in fast breeder reactor based on simulated tests on scaled down models. Collaborative Project No. TBRL/IGCAR/TRIG/1997. TBRL, Chandigarh, India.

17.41 Bell, M.J. (1973). ORIGEN—The ORNL isotope generation and depletion code. ORNL-4628. Oak Ridge National Laboratory, Oak Ridge, TN.

17.42 Gumprecht, R.O. (1968). Mathematical basis of computer code RIBD. DUN-4136. Douglas United Nuclear, Inc., Richland, WA.

17.43 Man, D.R. (1975). A user's manual for computer code RIBD II, a fission product inventory code. MEDL-TME 75-26. Hanford Engineering Development Laboratory, Richland, WA.

17.44 Spangler, G.W., Boling, M., Rhoades, W.A., Willis, C.A. (1967). Description of the COMRADEX code. AI-67-TDR 108. Rockwell International, Canoga Park, CA.

17.45 Velusamy, K., Chellapandi, P. (2007). Sodium release and design pressure for reactor containment building during a core disruptive accident. *CEA-IGCAR Technical Seminar on Liquid Metal Fast Reactor Safety Aspects Related to Severe Accidents*, IGCAR, Kalpakkam, India, February 12–16, 2007.

17.46 Reynolds, R.B., Kress, T.S. (1980). Aerosol source considerations for LMFBR core disruptive accidents. *Proceedings of the CSNI Specialists Meeting on Nuclear Aerosols in Reactor Safety*, Gatlinburg, TN.

17.47 Summers, R.M. et al. (1991). MELCOR 1.8.0: A computer code for nuclear reactor severe accident source term and risk assessment analysis. Rep. NUREG/CR-5531. Nuclear Regulatory Commission, Washington, DC.

17.48 Nuclear Regulatory Commission. (2001). SCDAP/RELAP5/MOD3.2 code manual, Vols. 1–5. Rep. NUREG/CR-6150, Rev. 2. NRC, Washington, DC.

17.49 Fauske Associates, Inc., MAAP4. (1994). Modular accident analysis program for LWR power plants, Vols. 1–4. Res. Proj. 3131-2. EPRI, Palo Alto, CA.

17.50 Nuclear Regulatory Commission. (1991a). SPARC-90: A code for calculating fission product capture in suppression pools. Rep. NUREG/CR-5765. NRC, Washington, DC.

17.51 Nuclear Regulatory Commission. (1991b). RAMSDALE, S.A., BUSCA-JUN90 reference manual, SRD R542, Safety and Reliability Directorate. UKAEA, Culcheth, U.K.

17.52 Missirlian, M., Alpy, N., Kissane, M.P. (2001). SOPHAEROS code version 2.0: Theoretical manual, IPSN Note: Technique SEMAR 00/39.

17.53 Wiles, L.E. et al. (1994). GOTHIC: Containment analysis package, Vols. 1–4. Rep. NAI 8907. Numerical Applications, Inc., Richland, WA.

17.54 Grove Software. (2003). Micro-Shield, Version 6.02, User's manual. Grove Software, Lynchburg, VA.

17.55 Devillers, C., Dupont, C. (1974). MERCURE-IV: Un programme de Monte Carlo à trois dimensions pour l'intégration de noyaux ponctuels d'atténuation enligne droite. CEA-N-1726. CEA, Paris, France.

17.56 Cain, V.R. (1977). A users manual for QAD-CG, the combinatorial geometry version of the QAD-P5A point kernel shielding code. Rep. NE007. Bechtel Power Corp., San Francisco, CA.

17.57 Kloosterman, J.L., Looserman, J.L. (1990). MARMER: A flexible point kernel shielding code. IRI-131-89-03/2. OECD, Paris, France.

17.58 Carruthers, D.J., Holroyd, R.J., Hunt, J.C.R., Weng, W.S., Robins, A.G., Apsley, D.D., Thomson, D.J., Smith, F.B. (1994). UK-ADMS—A new approach to modelling dispersion in the earth's atmospheric boundary layer. *J. Wind Eng. Ind. Aerodyn.*, 52, 139–153.

17.59 Chanin, D.I. et al. (1990). MELCOR accident consequence code system (MACCS). NUREG/CR-4691. Sandia National Laboratories, USNRC, Washington, DC.

17.60 Commission of European Communities (1991). COSYMA: A new programme package for accident consequence assessment. EUR 13028 EN. CEC, Commission of the European Communities, Directorate-General, Telecommunications, Information Industries And Innovation, L-2920 Luxembourg.

17.61 Chino, M., Ishikawa, H., Yamazawa, H. (1993). SPEEDI and WSPEEDI: Japanese emergency response system to predict radiological impacts in local and workplace areas due to a nuclear accident. *Radiat. Prot. Dosim.*, 50, 2.

17.62 Draxler, R.R., Hess, G.D. (1997). Description of the HYSPLIT-4 modeling system. NOAA Technical Memorandum ERL ARL-224.

17.63 Stohl, A. (1999). The FLEXPART particle dispersion model version 3.1. User guide, Vol. 13. Lehrstuhl für Bioklimatologie und Immissionsforschung, University of Munich, Am Hochanger, Freising, Germany, p. 85354.

17.64 Srinivas, C.V., Venkatesan, R. (2005). A simulation study of dispersion of air borne radionuclides from a nuclear power plant under a hypothetical accidental scenario at a tropical coastal site. *Atmos. Environ.*, 39, 1497–1511.

17.65 Srinivas, C.V., Venkatesan, R., Bhaskaran, R., Venkatraman, B. (2012). Round robin exercise on atmospheric flowfield modelling at Kalpakkam phase I. Radiological Safety Division, IGCAR. Report Submitted to Board of Research in Nuclear Sciences, p. 126.

17.66 Srinivas, C.V., Venkatesan, R., Somayaji, K.M., Yesubabu, V., Nagaraju, C., Chellapandi, P. (2010). Performance Evaluation of the Real-Time Atmospheric Model MM5 used in Emergency Response System. IGC-307. Indira Gandhi Centre for Atomic Research, Kalpakkam, India.

17.67 Srinivas, C.V., Venkatesan, R., Somayaji, K.M., Indira, R. (2008). A simulation study of short-range atmospheric dispersion for hypothetical air-borne effluent releases using different turbulent diffusion methods in HYSPLIT. *Air Qual. Atmos. Health*, 2, 21–28. doi 10.1007/s11869-009-0030-6.

17.68 Rigollet, L. (2010). R&D studies related to sodium risks. *Workshop on Safety AERB*, Mumbai, India, April 2010.

17.69 Ponraju, D. et al. (2013). Experimental and numerical simulation of sodium safety in SFR. *International Conference on Fast Reactors and Related Fuel Cycles: Safe Technologies and Sustainable Scenarios (FR13)*, Paris, France, March 2013.

17.70 Muthu Saravanan, S., Mangarjuna Rao, P., Nashine, B.K., Chellapandi, P. (2011). Numerical investigation of sodium droplet ignition in the atmospheric air. *Int. J. Nucl. Energy Sci. Technol.*, 6(4), 284–297.

18 Test Facilities and Programs

18.1 INTRODUCTION

With the limited operating experience of SFRs (~400 reactor years), limited number of safety experiments carried out in the test reactors (experimental breeder reactor (EBR)-II, southwest experimental fast oxide reactor (SEFOR), Rapsodie, Phénix), less number of transient test facilities (Transient Reactor Test Facility [TREAT], CABRI, SCARABEE), and with high demand for enhanced reactor safety, the development of novel test facilities to address comprehensively all the issues related to SFR safety is considered as essential for the growth of SFRs. The facilities, apart from simulating various severe accident scenarios, should generate quantum test data to validate the associated numerical simulation tools explained in Chapter 17. In this respect, new facilities are planned with extensive sensors and instrumentations (under sodium viewing, high-energy x-rays, thermal imaging, innovative imaging techniques, advanced tomography). Hence, it is also expected that the test facilities planned now will be executed under broad national collaborations. A first step toward this is the compilation of the test facilities having potential to pursue research on SFR safety and release in the form of a document by a Task Group on Advanced Reactor Experimental Facilities (TAREF) formed under nuclear energy agency (NEA) [18.1].

Various countries have initiated their program on future innovative fast reactor designs and concepts to achieve and demonstrate the Gen IV targets as close as possible. The initiatives revolve around, apart from SFR, the concepts of heavy metal–cooled fast reactors, gas-cooled fast reactors, and molten salt reactors. The alternative reactor design concepts are primarily aimed at higher cost effectiveness, safety, higher temperatures, and proliferation resistance. In the first category, the reactors under development are CFR-600 in China, advanced sodium technological reactor for industrial demonstration (ASTRID) in France, FBR1 and 2 in India, 4S and JSFR in Japan prototype gen-iv sodium-cooled fast reactor (PGSFR) in Korea, multi-purpose fast neutron research reactor (MBIR) in Russia, and power reactor innovative small module (PRISM) and traveling-wave reactor (TWR-P) in the United States. Reactors with alternative concepts include MYRRHA in Belgium, ALFRED and european lead fast reactor (ELFR) in Italy, PEACER in Korea, BREST-300 and SVBR-100 in Russia, G4M in the United States, Allegro in Europe, etc. Readers may refer to the recent IAEA publication "Status of innovative fast reactor designs and concepts" released in October 2013 for more details on reactor concepts [18.2]. Apart from this, the current status of test reactors is addressed in Part V: International SFR Program (Chapters 26 through 28) in this book. All these reactors can be effectively used to generate quantum data for future SFRs to provide satisfactory answers to almost all the safety issues by appropriately incorporating necessary features/provisions. Apart from these reactors, readers should also be aware of certain test facilities that had contributed significantly to the safety studies in the past. Some of the countries are revamping/developing several facilities to address the evolving safety issues comprehensively. In this chapter, details of a few of the test facilities that were employed in the past, currently being constructed and planned, are presented. Further, facilities addressing four broad aspects—core safety (flow blockage, ULOF, UTOP, slow power transients, sodium void propagation, etc.), molten fuel–coolant interaction (FCI) studies, post-accident decay heat removal scenario (heat transfer from debris and pool hydraulics) and sodium safety—are only brought out in this chapter. Tables 18.1 through 18.4 provide the list of such facilities under these four topics.

TABLE 18.1
Facilities for Core Safety

S. No.	Facility	Country	Status	Scope
1	SCARABEE	France	Closed	Results of the SCARABEE program with regard to material motion have high relevance to SFR accident analysis and computer code validation.
2	CABRI	France	Renovation	The experimental reactor has recently been recognized by TAREF as the most appropriate facility to address irradiated fuel behavior under incidental and accidental conditions.
3	IGR	Kazakhstan	Operational	Most experiments now are connected with nuclear reactor safety issues, although some work is conducted under the ITER project.
4	AR-1	Russia	Upgradation	The objective of the facility is to study the thermal hydraulic processes under start-up, normal operation, transient and accident conditions, flow stability, and heat transfer characteristics under liquid metal coolant boiling modes.
5	TREAT	United States	Discussion for renovation	The experimental reactor was also considered in the medium term for its relevance to severe accident issues, particularly simulating the fast power transients.
6	ACRR	United States	Operational	Though ACRR is mainly used for radiation effects studies, it has high potential for use of fuel transient testing under abnormal and/or accident conditions of SFRs (fast power transient from low initial power).

18.2 OVERVIEW OF TEST FACILITIES RELATED TO CORE SAFETY

18.2.1 SCARABEE: FRENCH FACILITY (CEA) [18.3]

SCARABEE is a high-flux reactor facility. Its center is housing a loop capable of containing one or seven pins. These pins are cooled with liquid sodium supplied by an external circuit simulating pin level thermal hydraulic conditions that prevail inside a fast reactor. The core generates neutrons that the fuel pins were subjected to again simulating the reactor conditions. The core consists of a pool-type reactor in which the assemblies are composed of plate-type fuels, made of U–Al alloy containing 26% uranium in weight. The plates are gathered into assemblies comprising a maximum of 21 plates. Four of those assemblies contain only 15 such plates, in order to allow space for in-core motion of control and safety rods. The core comprises 40 assemblies and 668 plates; some peripheral assemblies are only partly loaded, so as to adjust core reactivity. Cooling is ensured by demineralized water circulating from the bottom up. The core and its structures, together with the in-pile cell, are installed in a hall inside a 105 m³ container. The main equipment of the reactor is schematized in Figure 18.1. It involves essentially the driver core, the control and safety rods, and the core cooling system.

The SCARABEE experiments were mainly aimed at fulfilling several objectives: (1) to obtain knowledge on fuel element failure effects and fuel dynamics under its main aspects, for example, fuel pin failure thresholds, failure propagation and subsequent events, behavior of molten fuel, and consequences of contact with liquid sodium; (2) to identify the nature and behavior of detection signals at each stage; and (3) to develop one or several theoretical models to describe involved phenomena, at least up to hexcan failure. Results of the SCARABEE program with regard to material motion have high relevance to SFR accident analysis for various reasons: (1) Cladding and fuel motion phenomena were observed that influence model development; (2) parameters like wetting criteria, or the two-phase friction factor multiplier for the molten cladding sodium vapor system, were estimated quantitatively

TABLE 18.2
Facilities for Fuel–Coolant Interaction Studies

S. No.	Facility	Country	Status	Scope
1	PLINIUS-VULCANO	France	Refurbishment needed for sodium use	Corium melting facility used for LWR severe accidents study. Molten corium (50–100 kg) capability with molten metals. Potential use for core catcher design studies + need of conversion to sodium: under study.
2	PLINIUS-KROTOS	France	Refurbishment needed for sodium use	Corium–water interaction facility used for LWR severe accidents study. Molten corium (5 kg) dropped into water, triggering steam explosion; need of conversion to sodium under study.
3	SOFI	Indian	Operational	The objective of the facility is to generate data for understanding the various fuels, melt solidification and fragmentation, dispersion/relocation of debris, and settlement behavior on core catcher and to validate the numerical models for predicting the previously mentioned phenomena.
4	MELT	Japan	Operational	Out-of-pile facility for the study of molten material behavior under core disruptive accident (CDA) in SFRs. Also used for FCI in the water system. Small sodium loop available. Use of induction-heated crucible with simulating materials (melt of alumina, steel, and tin, 2300°C, 20 L capacity).
5	CAFE	United States	Operational	Core alloy flow and erosion facility. Used for study of materials flow, freezing of metallic fuel bearing melts, effects of eutectic liquefaction of channel structure (1D horizontal channel).
6	MCCI	United States	Operational	Large reactor test cell (1000 m³) for the study of LWR severe accidents issues (radioactive core melt, steam explosion, hydrogen production issues) may be used for fuel–coolant interaction and PAHR issues of SFR.
7	SURTSEY	United States	Operational	SURTSEY facility used for LWR severe accident studies. Possible use for large tests with molten materials and their interaction. Large sealed pressure vessel, 1/10th scale with pressure and temperature measurements.

by analysis of instrumentation signals and calculations; and (3) computer codes were verified through comparison between numerical predictions and test results observed during and after the experiments.

18.2.2 CABRI: French Facility [18.4]

CABRI is an experimental reactor located at the Cadarache research center and operated by CEA since 1978. Its objective is to study the fuel rod behavior under reactivity-initiated accident (RIA) conditions. CABRI (Figure 18.2) is an open pool-type research reactor composed of a driver core, 80 cm high, 60 cm length, 60 cm wide, made of 1488 UO_2 rods enriched with 6% of U-235. These rods have been specially designed to support an injection of reactivity (austenitic steel cladding, large pellet/cladding gap). In steady-state conditions, the core power is controlled through 6 hafnium control rods up to a maximum power of 25 MW. The reactor includes an experimental loop specially designed to position in the center of the driver core the instrumented test device housing the fuel rod to be tested. The CABRI facility was originally devoted to the study of fast breeder reactor fuel pin behavior subjected to a power transient simulating sodium vaporization in the core and an ejection of a control rod. Between 1978 and 2001, 59 experiments were performed on Superphénix and Phénix fuel pin types in the framework of four international programs: CABRI 1, CABRI 2, CABRI FAST, and CABRI RAFT. In 2001, IRSN and EDF, with a broad international cooperation, initiated an international program in

TABLE 18.3

Facilities for Postaccident Decay Heat Removal Studies

S. No.	Facility	Country	Status	Scope
1	KASOLA	Germany	Construction	A key feature of the facility is its flexibility with respect to a wide spectrum of thermal hydraulic experiments.
2	VERDON	France	To be operational for SFR conditions	Out-of-pile facility for studying irradiated fuel behavior and fission products release under simulated heat transient. Use of induction furnace under various atmospheres (He, H_2, steam, air) up to 2700°C. Online fission product measurements.
3	MERARG	France	Operational	Out-of-pile facility for studying irradiated fuel behavior and fission gas release under simulated heat transient. Use of induction furnace under neutral atmosphere up to 2700°C. Online fission gas release measurements.
4	Safety decay heat removal loop in natrium (SADHANA)	India	Operational	Thermal hydraulics simulation of SGDHR circuit. This 1:22 scaled model loop.
5	Postaccident thermal hydraulics (PATH)	India	Operational	Thermal hydraulic simulation through water. Core debris simulated with wood's metal.
6	Safety studies in reactor assembly (SASTRA)	India	Construction	Thermal hydraulic simulation with sodium. Core debris simulated with wood's metal.
7	AtheNa	Japan	Construction	The facility is being built to satisfy the safety design policy for severe accident measures for the in-vessel retention and protection against loss of heat removal system (LOHRS) and anticipated transient without scram (ATWS).

the CABRI facility under OECD auspices. The first two tests were performed in November 2002 in the sodium loop. They involved subjecting two highly irradiated PWR UO_2 fuels (burnup around 75 GWd/ton of metal) with advanced claddings to typical RIA power excursions. The other tests were carried out in the water loop that replaced the sodium loop to simulate thermal hydraulic conditions representative of the nominal operating PWR conditions (155 bar, 300°C). The facility was modified by CEA from 2003 to 2010, which included sodium loop dismantling, the implementation of the new pressurized water loop, and an overall facility refurbishment and a safety review.

The CABRI experimental reactor has recently been recognized by TAREF as the most appropriate facility to address irradiated fuel behavior under incidental and accidental conditions: fuel safety issues such as margins to fuel melting and deterministic pin failure and severe accident issues such as consequences of various accidents leading to fuel melting, with associated consequences and risk of critical events and energy release. It is expected that the facility may be available for testing in 2020 for fast reactor safety studies after completion of LWR dedicated safety programs.

18.2.3 IGR: KAZAKHSTAN FACILITY [18.5]

The impulse graphite reactor (IGR) was established at the National Nuclear Center of Kazakhstan. It is the one of the oldest research reactors in the world. The IGR is a unique source of neutron and gamma radiation characterized by a high dynamics of power change (Figure 18.3). The intense development of reactor engineering in the 1950s predetermined the creation of the pulse reactor for

TABLE 18.4
Facilities for Sodium Safety

S. No.	Facility	Country	Scope
1	DIADEMO	France	Development of SWR with hydrogen detection.
2	Mini sodium fire experimental (MINA)	India	This facility is meant for evaluating sodium burning rates, sodium aerosol behavior, and sodium–concrete interaction and qualifying the sodium fire fighting system and validation of combined fire (i.e., spray followed by pool) scenarios and sodium fire codes.
3	SOCA	India	The facility was built to simulate the sodium fire scenario above the top shield consequent to a CDA. This facility is meant for studying the combined effect of sodium and secondary cable fire on the integrity of important components like DHX piping.
4	SFEF	India	To carry out large-scale sodium pool/spray fire experiments to investigate sodium aerosol behavior, sodium–concrete interaction, etc. The facility is also used for qualifying actual sodium fire fighting systems and also prototype leak collection trays.
5	Sodium water reaction test rig (SOWART)	India	To study the behavior of self-wastage and leak enlargement during sodium–water reaction, to study impingement wastage, and to develop leak detection methods.
6	SAPFIRE	Japan	Test platform dedicated to the study of sodium leak accident consequences such as spray, columnar, pool-type fires, sodium concrete interaction, sodium structure interaction with chemical reaction, and aerosol behavior. Test section, 1000 m^3; 0.2 MPa, ventilation rate, 70 N m^3/min; 10 ton of sodium.
7	SWAT-1R	Japan	Facility dedicated to the study of water–sodium interaction due to pipe rupture in a steam generator and to self-wastage behavior of double-tube steam generator pipe. Tank volume, 630 L, $T_{max} = 580°C$, $P_{max} = 1.96$ MPa.
8	SWAT-3R	Japan	Facility dedicated to the investigation of the propagation of steam generator tube rupture in prototypical conditions (large-scale facility). Fifteen tons of sodium, T = 555°C, $P_{max} = 1.96$ MPa, tank volume: 10 m^3, water heaters of 3.1 and 4.8 m^3.

experimental research of nonstationary physical processes occurring in the core when high reactivity is imparted. The IGR was built in the shortest time possible and allowed to start experimental study of pulse reactor dynamics in 1961. In 1962, the studies of fuel and structure materials behavior in advanced reactor facilities including nuclear jet propulsion (NJP) began. The important technical parameters are as follows: neutron flux maximum density, 7×10^{16} n/cm² s; thermal neutron maximum fluence, 3.7×10^{16} n/cm²; and minimum half-width of impulse, 0.12 s.

The reactor runs at 1 GW in the steady mode and 10 GW in the pulsed mode. The core contains 10 kg of 90% enriched fuel (U-235). Fuel is made from uranium graphite blocks. Graphite blocks are placed in a uranium solution, and the uranium is gradually absorbed into the graphite. The reactor was first brought into operation in 1960 and was initially built to study nuclear reactor accidents. The reactor was designed to run for about 1 year, after which a major accident was to be simulated and the reactor destroyed in the accident. IGR's location was chosen so that the simulated accident would take place far from any populated area. However, during the first year of operation, it was noted that in the simulation of minor accidents, the characteristics of the reactor were such that even fairly major accidents could be simulated without destroying the reactor. Therefore, it was decided to keep the IGR in operation. The IGR is expected to operate till the end of its service life on the existing fuel itself. In addition, 7 kg of fresh fuel is stored at the reactor site. The reactor simulated an accident very similar to that at Chernobyl long before the Chernobyl accident. But because the IGR was a military research reactor, the results of the experiment were secret and were not shared with the civilian

FIGURE 18.1   SCARABEE: French facility (CEA). (From Bailly, J. et al., *Nucl. Eng. Des.*, 59, 237, 1980.)

power sector. Until 1991, the IGR was put into operation about 120–130 times/year. Since 1991, the number of experiments and tests was significantly decreased. In 1996, the IGR was brought into operation 37 times. In the first 8 months of 1997, it was used 20 times. Most experiments now are connected with nuclear reactor safety issues, although some work is conducted under the ITER project.

18.2.4   AR-1 (IPPE): Russian Test Facility [18.6]

The objective of the facility is to study the thermal hydraulic processes under start-up, normal operation, transient and accident conditions, flow stability, and heat transfer characteristics under liquid metal coolant boiling modes. The AR-1 test facility consists of two loops: test sodium loop and auxiliary loop with sodium–potassium alloy as a coolant. The purpose of upgrading the AR-1 test facility is to model coolant boiling processes that can occur in the SFR core under conditions of the ULOF beyond-design-basis accident and design-basis accident with fuel subassembly cross-sectional blockage. Accordingly, the main test parameters are chosen for simulating conditions close to the reactor: sodium velocity, sodium temperature, and heat flux from the fuel pin simulators to the coolant. The first stage of tests includes steady and unsteady tests toward investigating sodium vaporization and condensation modes in the area of the core and sodium cavity of a single test subassembly. The first series of tests is planned on sodium boiling in the single test FSA with seven fuel pin simulators. The instrumentation for measurements permits to investigate the sodium boiling process both within the bundle of fuel pin simulators and outside of it (in the sodium cavity up to the expansion tank). In steady-state tests, the sodium boiling mode is provided by means of a step-by-step increase in power

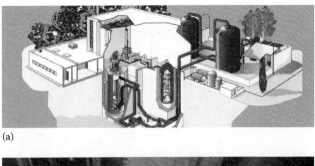

(a)

(b)

FIGURE 18.2 CABRI: the French facility. (a) CABRI facility layout and (b) CABRI open loop type reactor. (From Blanc, H. et al., The sodium CABRI loop, *International Conference on Fast Reactor Safety*, Aix-en-Provence, France, 1967.)

FIGURE 18.3 IGR: Kazakhstan facility. (From Ewell, E., *International Conference on Nonproliferation Problems*, NISNP trip report, KAZ970900, 1997, pp. 13–14.)

FIGURE 18.4 AR-1 (IPPE): Russian test facility. (From Ashurko, I.M. et al., Activities on experimental substantiation of SFR safety in accidents with sodium boiling, *International Conference on Fast Reactors and Related Fuel Cycles: Safe Technologies and Sustainable Scenarios (FR13)*, Paris, France, March 2013.)

of the heaters with the subsequent stabilization of sodium parameters at each power level. Tests are scheduled to be carried out for both sodium forced flow mode through the test subassembly and sodium natural circulation mode. In unsteady tests, it is planned to model the accidental conditions for the ULOF transient. Modeling the sodium boiling mode caused by the FSA cross-sectional blockage is provided by means of a total stop of the coolant flow rate through the test subassembly with the constant power of heaters. The program of tests envisages investigation of the degree and specific features of influence of value of the height of the sodium cavity over the core on accident transient conditions, in particular on behavior and parameters of the sodium boiling mode. The second stage of tests includes investigation of stability of sodium boiling in parallel channels under the coolant natural circulation mode. A photographic view of a typical loop of the facility is shown in Figure 18.4.

18.2.5 TREAT: U.S. DOE [18.7]

The fast reactor fuel including fuel cladding and wrappers could be subjected to short bursts of very intense, high-power radiation, posing a threat to their integrity in the core. To simulate such transients, TREAT was started at Idaho National Laboratory in 1959. It is an air-cooled test facility with uranium-impregnated graphite blocks. TREAT was used primarily to test liquid metal reactor fuel elements, initially for the EBR-II and then for the Fast Flux Test Facility (FFTF), the Clinch River Breeder Reactor Plant (CRBRP), the British Prototype Fast Reactor (PFR), and finally, the Integral Fast Reactor (IFR). TREAT was used to study fuel meltdowns, metal–water reactions, interactions between overheated fuel and coolant, and the transient behavior of fuels for high-temperature systems. Its purpose was to simulate accident conditions leading to fuel damage, including melting or vaporization in test specimens. In its steady-state mode of operation, TREAT was also used as a large neutron radiography facility and could examine assemblies up to 15 ft in length. Both oxide and metal elements were tested in dry capsules and in flowing sodium loops. The data obtained were instrumental in establishing the behavior of the fuel under off-normal and accident conditions, a necessary part of the safety analysis of various reactors. During its decades of operation, TREAT provided stress testing of nuclear fuels and rapid, high-energy neutron pulses that simulate accident conditions to facilitate the design of more durable fuels, establish performance limits, validate design codes, and help regulators define safety limits. The TREAT experimental reactor was also considered in the medium term for its relevance to severe accident issues, particularly simulating the fast power transients. The reactor was operated during 1959–1994. Throughout its about 35-year history, it has proven to be a safe, reliable, and versatile facility, compiling a distinguished

(a)   (b)

FIGURE 18.5 TREAT: U.S. DOE. (a) TREAT facility and (b) TREAT core. (From Crawford, D.C. et al., Review of experiments and results from the transient reactor test (treat) facility, *ANS Winter Meeting*, Washington, DC, November 1998.)

record of successful experiments. Now, the U.S. DOE has a proposal on resuming transient testing. Refurbishing and operating TREAT are under consideration by DOE. An overall photographic view of the reactor as well as a photo focusing on the core is shown in Figure 18.5.

18.2.6 ACRR: U.S. FACILITY [18.8]

The Annular Core Research Reactor (ACRR) facility was built in Sandia Laboratory. In ACRR, various test objects can be subjected to a mixed photon and neutron irradiation environment featuring either a very rapid pulse rate or a long-term, steady-state rate. The radiation produced at the ACRR is used for neutron scattering experiments; nondestructive testing, including neutron radiography, neutron activation analysis, and testing of materials; advanced reactor fuel development and testing; radioisotope production; basic radiation effects science; and public outreach and education. The ACRR offers several special features: a large central cavity with very little radiation gradient, despite a capability for high radiation intensity, the ability to determine with a high degree of accuracy the actual radiation dose delivered to each test article, and a limited capability to tailor the neutron energy spectrum and reduce or increase the photon intensity by selecting the appropriate interaction material to be positioned between the core and the test article. Though ACRR is mainly used for radiation effects studies, it has high potential for use of fuel transient testing under abnormal and/or accident conditions of SFRs (fast power transient from low initial power). A photographic view of the facility is shown in Figure 18.6.

18.3 OVERVIEW OF TEST FACILITIES RELATED TO MOLTEN FUEL–COOLANT INTERACTIONS

18.3.1 PLINIUS-VULCANO: FRENCH FACILITY [18.9]

PLINIUS-VULCANO facility has a furnace in which 50–100 kg of prototypic corium can be molten. The molten corium can then be poured into a test section with specific instrumentation. The VULCANO furnace used heating based on transferred plasma arc technology at the centerline of a rotating cylinder. This facility can melt a large spectrum of components from UO_2–ZrO_2 to mixtures with some metals and/or concrete decomposition products. The plasma generator gases

FIGURE 18.6   ACRR: U.S. facility. (From Walker, J.V. et al., Design and proposed utilization of the SANDIA annular core research reactor (ACRR), Sandia Laboratories, Albuquerque, NM, *International Meeting on Fast Reactor Safety Technology*, Seattle, WA, August 1979, SAND79-1646C.)

are argon and/or nitrogen plus, in some cases corium fumes. The maximum available power is around 600 kW (1000 A–600 V). Metal melting furnaces using induction heating technology are also available to add metals to the melt. It has a capacity to melt 8 kg operating with an RF generator at about 22 kHz. Thermitic reaction technology has been developed enabling the melting of some corium compositions, which works on the principle of exothermal redox reactions. Experiments have been conducted in this facility to study the spreading behavior of corium on various surfaces and the transient interaction with substrate. A photographic view of the high-frequency power generator and the furnace portion of the facility is shown in Figure 18.7.

18.3.2   PLINIUS-KROTOS: French Facility [18.10]

The PLINIUS-KROTOS facility was built and used in JRC–Ispra and recently transferred to CEA. The facility is dedicated to the study of FCIs in connection with steam explosions. Heat transfer between the hot molten core and the colder volatile water is so intense and rapid that the timescale for heat transfer is shorter than the timescale for pressure relief, leading to the formation of a shock wave. This shock wave is intensified as a result of further mixing and energy transfer as it travels through the mixture. This facility can melt about 4.5 kg of prototypic corium or 1 kg of alumina and poured in a water-filled test section. Premixing and explosion phases can be studied, which consist of four main parts: the furnace, the transfer channel, the test section, and the x-ray radioscopy system. The furnace is a water-cooled stainless steel container designed to withstand 4 MPa pressure and equipped with a three-phase cylindrical heating resistor made of tungsten. In order to avoid heat losses, the heating element is surrounded by a series of concentric reflectors and closed by circular lids made of molybdenum. The facility is developed to operate in inert atmosphere or vacuum at temperatures up to 2800°C. The x-ray radioscopy system has been specifically developed and assembled on KROTOS facility to trace the fragmentation of the melt within the coolant, allowing the three phases (water, void, and melt) to be clearly distinguished. Both spontaneous triggered explosions have been observed in KROTOS. A photographic view of a crucible within the furnace is shown in Figure 18.8.

18.3.3   SOFI: Indian Facility [18.11]

The objective of facility is to generate data for understanding the various fuels (U/UO$_2$/Zr/steel), melt solidification and fragmentation, dispersion/relocation of debris, settlement behavior on core catcher, to validate the numerical models for predicting the previously mentioned phenomena and ultimately accumulate realistic debris bed for carrying the experimental postaccident heat removal studies in SASTRA [18.18]. A schematic sketch and the front view of the facility are shown in Figure 18.9. Table 18.5 provides roadmap of sodium–fuel interaction (SOFI) studies, capacity additions, and other details comprehensively.

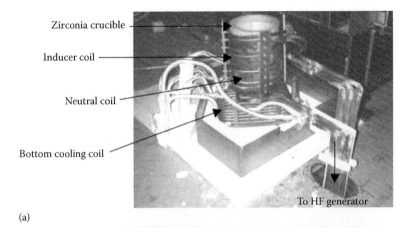

Zirconia crucible

Inducer coil

Neutral coil

Bottom cooling coil

To HF generator

(a)

(b)

FIGURE 18.7 PLINIUS-VULCANO: French facility. (a) High frequency power generator and (b) furnace portion of the facility. (From Journeau, C. et al., Oxide-metal corium–concrete interaction test in the VULCANO facility, Paper no. 7328, *Proceedings of the ICAPP 2007*, Nice, France, 2007.)

18.3.4 MELT: Japanese Facility [18.1]

The MELT facility at JAEA was built to perform out-of-pile experiments relating to the molten material behavior in the CDA. The main part of this facility is an induction heating crucible capable of generating a melt of alumina, steel, and tin, with the highest temperature of 2300°C and maximum amount of 20 L. The power of induction heating is 300 kW. This facility was established in Japan and was utilized in the investigation of FCI in the water system and the structure erosion behavior by melt jets. A small sodium loop to perform FCI experiments in the sodium system is also available. X-ray video and high-speed video are available for the visualization of multiphase transient phenomena in the test section. A schematic sketch of the facility illustrating the working principle is shown in Figure 18.10.

18.3.5 CAFÉ: U.S. Facility [18.12]

The CAFÉ facility is a part of the work sponsored by JAEA and performed at Argonne National Laboratory (ANL). The main objective of the facility is to investigate the fundamental flow and

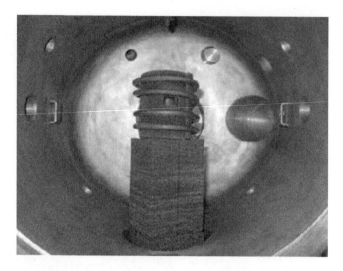

FIGURE 18.8 PLINIUS-KROTOS: French facility. (From Zabiego, M. et al., The KROTOS KFC and SERENA/KS1 tests: Experimental results and MC3D calculations, *Seventh International Conference on Multiphase Flow (ICMF 2010)*, Tampa, FL, 2010.)

FIGURE 18.9 SOFI: Indian facility. (From Das, S.K. et al., *Nucl. Eng. Des.*, 265, 1246, 2013.)

freezing behavior of uranium and uranium alloy melts in contact with metallic surfaces, including (1) dynamic changes in melt composition and properties, (2) chemical erosion/ablation of structure, (3) melting of structure and freezing of melt, and (4) flow of melts over newly solidified composi-tions. This is applicable to understanding the flow and freezing of molten fuel–cladding alloys within pin cladding, through assemblies, or in contact with the ex-core structure. Uranium and uranium–iron eutectic melts are generated by induction heating and are allowed to flow within open-sided, inclined, stainless steel troughs. The study involves investigation of uranium alloy fuels involving the formation of low-melting-point phases with iron-based (e.g., stainless steel) surfaces. A photographic view of a portion of the facility is shown in Figure 18.11.

TABLE 18.5
Details and Roadmap for SOFI at IGCAR—India

Facility Name	Sodium–Fuel Interaction (SOFI) Facility		
Commissioning	Phase I, January 2014	Phase II, End 2014	Phase III, Mid-2016
Research area	• To derive matured knowledge on induction furnace, cold crucible, high-temperature monitoring, imaging system, establishing design basis pressure for the mechanical design of the interaction vessel for large capacity systems • To generate data related to the compatibility of various fuel debris settling on the core catcher with a sacrificial layer and sodium. • To develop some special alloys	• To understand molten fuel–coolant interactions particularly with UO_2 in sodium including fuel fragmentation aspects, debris relocation behavior, and morphological characteristics of generated debris bed	• The effect of larger mass on the results derived from phase II • Further, to study the debris spread characteristic on the core catcher, heat transfer within porous debris bed • Erosion behavior of sacrificial materials during melt impingement • To generate and accumulate debris particles simulating scenarios with the use of SASTRA for performing PAHR studies
Power	50 kWe	200 kWe	Induction
Inductor	0–2 kHz	100–200 kHz	
Melt mass	About 1 kg	About 10 kg	About 20 kg
Important operation features	• Several short time duration tests • Easy for dismantling, transport, and handling • Type "C" and two-color pyrometer for temperature monitoring • In-sodium pressure transducer • Imaging of debris bed with high-power x-ray • Melt pouring into test section using a combination of remotely operated pneumatic valves	• Type "C" and two-color pyrometer for temperature monitoring • In-sodium pressure transducer • Imaging of debris bed with high-power x-ray system • Melt pouring into test section using a set of remotely operated pneumatic valves	Same as phase II

18.3.6 MCCI: U.S. Facility [18.13]

In a core melt accident, if the molten core is not retained in vessel despite severe accident mitigation actions, the core debris will relocate to the reactor cavity region and interact with the structural concrete—potentially resulting in basemat failure through erosion or overpressurization. This would result in the release of fission products into the environment. Although this is a late release event, the radiological consequences could be substantial enough to warrant an effective mitigation strategy to prevent such a release. The Melt Coolability and Concrete Interaction (MCCI) Project was dedicated to provide experimental data on this severe accident phenomena and to resolve two important accident management issues: (1) verify that molten debris that has spread on the base

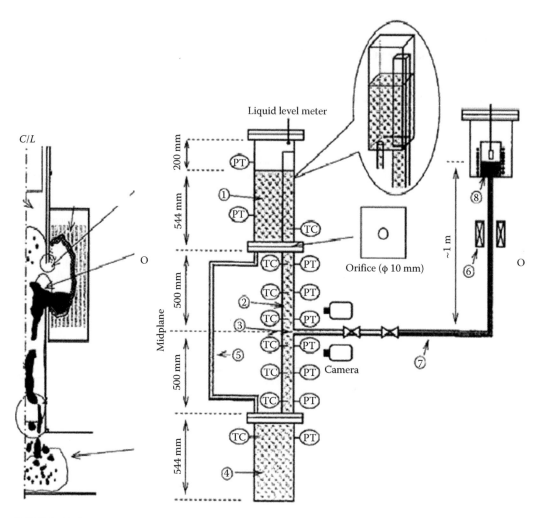

FIGURE 18.10 MELT-II: Japanese facility. (From Task Group on Advanced Reactors Experimental Facilities (TAREF), Experimental facilities for sodium fast reactor safety studies, NEA No. 6903, Nuclear Energy Agency, Organization for Economic Co-operation and Development, Paris, France, 2011.)

of the containment can be stabilized and cooled by water flooding from the top and (2) assess the 2D, long-term interaction of the molten mass with the concrete structure of the containment as the kinetics of such interaction is essential for assessing the consequences of a severe accident. To achieve these basic objectives, supporting experiments and analyses were performed at ANL. A first melt–concrete interaction test with siliceous concrete in 2003 produced unexpected results (a strong asymmetry in concrete ablation). These tests produced excellent data on axial and radial concrete ablation. A photographic view of a portion of the facility is shown in Figure 18.12.

18.3.7 SURTSEY: U.S. Facility [18.14]

The SURTSEY test facility at Sandia National Laboratories (SNL) is used to perform scaled experiments that simulate high-pressure melt ejection (HPME) accidents in a nuclear power plant (NPP). These experiments are designed to investigate melt dispersal from a reactor cavity and the resulting containment loads if the reactor pressure vessel lower head fails while the reactor coolant system is still at elevated pressures. Collectively, these phenomena are referred to as HPME from the reactor pressure vessel and direct containment heating (DCH). Integral effects tests were performed in this

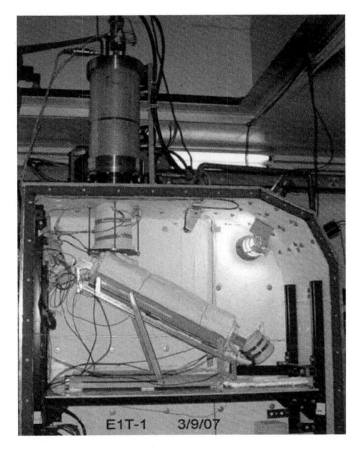

FIGURE 18.11 CAFÉ: U.S. facility. (From Farmer, M.T. et al., U.S. perspective on technology gaps and R&D needs for SFRs, *International Workshop on Prevention and Mitigation of Severe Accidents in Sodium-Cooled Fast Reactors*, Tsuruga, Japan, 2012.)

facility with a 1/10th scale model of the Calvert Cliffs nuclear power plant test vessel. The experiments investigated the effects of codispersal of water, steam, and molten core simulant materials on DCH loads under prototypic accident conditions and plant configurations. The results indicated that large amounts of coejected water reduced the DCH load by a small amount. A photographic view of the facility is shown in Figure 18.13.

18.4 TEST FACILITIES RELATED TO POSTACCIDENT HEAT REMOVAL

18.4.1 KASOLA (KIT): German Facility [18.15]

The Karlsruhe Sodium Laboratory (KASOLA) at the Institute for Neutron Physics and Reactor Technology is a versatile experimental facility to investigate flow phenomena in liquid sodium for nuclear and nonnuclear applications. A key feature of the facility is its flexibility with respect to a wide spectrum of thermal hydraulic experiments. It is planned to carry out several studies: validation and improvement of turbulent liquid metal heat transfer models in CFD tools on a limited geometric scale; development of free surface liquid metal targets for accelerator applications; development of models to describe free surface liquid metal flow; investigation of transition in convective flow patterns between forced, mixed, and free convection modes; qualification of CFD and system codes to simulate adequately the transition from the channel flow to the large plenum (collector tank); and thermal hydraulic investigations of flow patterns in fuel bundles or pool

FIGURE 18.12   MCCI: U.S. facility. (From Farmer, M.T. et al., A summary of findings from the melt coolability and concrete interaction (MCCI) program, Paper 7544, *International Congress on Advances in Nuclear Power Plants (ICAPP 07)*, Nice, France, 2007.)

configurations at prototypical or scaled heights. An in-service inspection and repair (ISIR) monitors liquid metal systems. The facility is in commissioning stage. Figure 18.14 depicts the schematic sketch of the test loops.

18.4.2   VERDON Laboratory: French Facility at CEA [18.1]

The VERDON laboratory, established at CEA Cadarache center, is composed of two high-activity cells (C4 and C5) and a glove box. The C4 cell is mostly devoted to the sample reception and to the pre-/posttest gamma scanning on a dedicated gamma spectrometry bench. The C5 cell is dedicated to the realization of the thermal hydraulic sequence with two different experimental loops (one designed for fission product (FP) release studies and the other for FP release and transport studies) and to online gamma spectrometry measurements: they are performed thanks to three complementary online gamma spectrometry stations in order to follow the FP release kinetics outside the fuel sample located in the furnace during the thermal hydraulic sequence.

The experimental loop is mainly composed of a furnace topped by an aerosol filter whose function is to trap all the FPs under the aerosol form. At the end of the loop, a May-Pack filter partly filled with zeolite (impregnated with silver) is provided in order to trap potential molecular iodine and a condenser whose function is to recover the steam of the experimental gases. Between the furnace and just before the condenser, each element of the circuit is heated at 150°C ± 20% in order to avoid any cold point leading to steam condensation along the circuit. Finally,

FIGURE 18.13 SURTSEY: U.S. facility. (From Blanchat, T.K. et al., Direct containment heating experiments at low reactor coolant system pressure in the SURTSEY test facility, Washington, DC: Division of Systems Analysis and Regulatory Effectiveness, Office of Nuclear Regulatory Research, U.S. Nuclear Regulatory, NUREG/CR-5746, SAND 99-1634, 1999.)

two safety filters prevent any residual traces of aerosol FP from going out of the high-activity cell and into the glove box where the fission and carrier gases are analyzed and stored. Inside the glove box, gas analysis can be performed either online by a microgas chromatograph or sequentially by four sampling aliquots. All gases (injected, produced, or released during the test) are directed to a 3 m³ storage vessel. The last aliquot is dedicated to gas sampling from this vessel. A photographic view of the facility is shown in Figure 18.15.

18.4.3 MERARG Facility: French Facility at CEA [18.1]

This facility is operational in a dedicated hot cell of the LECA-STAR laboratory at the CEA Cadarache center (Figure 18.16). A first version of this facility is already operational for LWR fuels. In the future, the MERARG facility could also be used to qualify fuels for Gen IV reactors. MERARG is able to reproduce temperature transients on an irradiated LWR fuel pellet. The objectives of the facility are to address the safety issue of potential FP releases by (1) studying the fission gas release mechanisms, (2) quantifying the fission gas source term in case of accidental sequences (e.g., LOCA for PWR), and (3) measuring the fission gas total inventory in the fuel sample. The facility has a maximum temperature up to 2800°C. The temperature increase rate can be controlled from 0.05°C/s up to 50°C/s. Power injection for accelerated ramps can be from 50°C/s up to 200°C/s. Circulating atmosphere in the furnace are Ar, He, and air with a Pt crucible. Fission gas release is measured by online gamma spectrometry as well as online microgas chromatography.

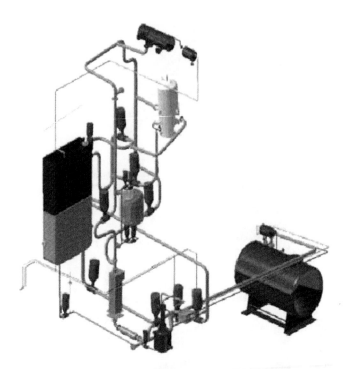

FIGURE 18.14   KASOLA (KIT): German facility. (From Hering, W. et al., Scientific program of the Karlsruhe sodium facility (KASOLA), *Proceedings of the International Conference on Fast Reactors and Related Fuel Cycles: Safe Technologies and Sustainable Scenarios (FR13)*, Paris, France, March 2013.)

FIGURE 18.15   VERDON laboratory: French facility at CEA. (From Task Group on Advanced Reactors Experimental Facilities (TAREF), Experimental facilities for sodium fast reactor safety studies, NEA No. 6903, Nuclear Energy Agency, Organization for Economic Co-operation and Development, Paris, France, 2011.)

18.4.4   SADHANA: INDIAN FACILITY AT IGCAR [18.16]

It is a 355 kW capacity sodium test facility established at IGCAR to study the thermal hydraulic behavior of the safety grade decay heat removal system of PFBR (Figure 18.17). The experimental facility contains a sodium to sodium decay heat exchanger (DHX), sodium to air heat exchanger (AHX), a test vessel containing sodium pool, chimney, and associated piping. The maximum operating temperature is 873 K and the maximum flow is 5 m³/h at 0.5 MPa. This facility is located in

FIGURE 18.16 MERARG facility: French facility at CEA. (From Task Group on Advanced Reactors Experimental Facilities (TAREF), Experimental facilities for sodium fast reactor safety studies, NEA No. 6903, Nuclear Energy Agency, Organization for Economic Co-operation and Development, Paris, France, 2011.)

FIGURE 18.17 SADHANA: Indian facility. (From Padmakumar, G. et al., *Prog. Nucl. Energy*, 66, 99, 2013.)

Engineering Hall-III. In SADHANA, the sodium in the test vessel, which simulates the hot pool of PFBRs, is heated by immersion electrical heaters. This heat is transferred to the secondary sodium through the decay heat exchanger. The secondary sodium gets circulated in the secondary loop by the buoyancy head developed in the loop due to the temperature difference in the hot and cold legs of the loop. The heat from secondary sodium circuit is rejected to the atmosphere through the air heat exchanger. A 20 m high chimney develops the airflow required to transfer the heat from the secondary sodium to the atmosphere through the AHX. This 1:22 scaled model loop is designed on the Richardson number similitude. Sodium holdup in this facility is 3 m^3.

18.4.5 PATH: INDIAN FACILITY [18.17]

The objective of the facility is to assess natural convection with decay heat exchanger units immersed in the hot pool during a postaccident heat removal condition. This facility is about 3 m diameter vessel filled with water housing the essential internals (grid plate, inner vessel, core support structure, core catcher, decay heat exchangers, etc.) and extensively instrumented to measure transient temperatures and flows within the water pool at a higher temperature (<80°C) during natural convection modes. It has core catcher plate on which any debris can be spread to simulate the realistic heat transfer process within the debris bed. This is being used at IGCAR to optimize the core catcher design for future SFRs in India. More details are presented in Table 18.5. Figure 18.18 shows a photograph of the facility along with temperature mappings at two instants, captured in a typical experiment.

18.4.6 SASTRA: INDIAN FACILITY [18.18]

This facility is being built at IGCAR, which is similar to PATH, but uses the sodium pool. Further, the facilities will have features to simulate interwrapper flow within the core subassemblies. It is planned to carry out tests with debris collected from the test series conducted in SOFI. With SASTRA, data will be generated toward source term analysis (transportation of fission products in the sodium pool and its retention ability, subsequent transportation from the cover gas to outer space

FIGURE 18.18 PATH: Indian facility. (From Gnanadhas, L. et al., *Nucl. Eng. Des.*, 241, 3839, 2011.)

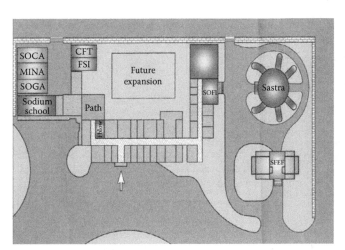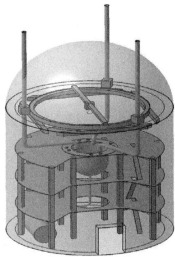

FIGURE 18.19 SASTRA layout: Indian facility. (From Chellapandi, P. et al., Overview of molten FCI studies towards SFR development, *SERENA/OECD Project Seminar and Associated Meeting*, CEA, Cadarache, France, November 2012.)

through top shield penetration). This facility will be ready by December 2015. Figure 18.19 depicts an architectural view of SASTRA.

18.4.7 AtheNa: Japanese Facility [18.19]

The facility is being built to satisfy the safety design policy for severe accident measures for in-vessel retention and protection against LOHRS and ATWS. More attention was called for LOHRS after the Fukushima accidents. The facility would also address the candidates of measures against LOHRS, alternative cooling system, and external vessel cooling and internal vessel cooling systems. The phenomena to be investigated in the facility are heat transfer and thermohydraulics in the reactor vessel, external vessel, and steam generator as well as local heat removal characteristics from the damaged core. Ultimately, the facility will help establish the criteria for the practical elimination of LOHRS. Tests will be planned to identify the design measures and design tools to be developed and validated internationally regarding SFR safety. The facility is under construction. The dimension of the facility is 130 m × 62 m × 55 m and total floor area is 11,000 m² and the sodium inventory is 260 tons (Figure 18.20).

18.5 TEST FACILITIES RELATED TO SODIUM SAFETY

18.5.1 DIADEMO: French Facility at CEA [18.20]

The major objective of this facility is to qualify the innovative sodium instrumentation as well as for thermal hydraulic performance evaluation of heat exchangers. The DIADEMO-Na loop is commissioned in the CEA at Cadarache to acquire heat exchange data to validate the design studies. The total sodium volume that can be handled in the loop is about 300 m³. A total of two test sections exist. Available dimensions for the heat exchanger mock-up are 1.2 m × 0.2 m × 0.2 m. Sodium flow rates range from 0 to 2 m³/h. Sodium temperatures in the system would range from 180°C to 550°C. The maximum sodium pressure in the system is 3 bar. There is an active purification system in which gas (nitrogen) temperatures range up to 550°C with maximum gas pressures being 100 bar. A photographic view of the portion of the test loop facility is shown in Figure 18.21.

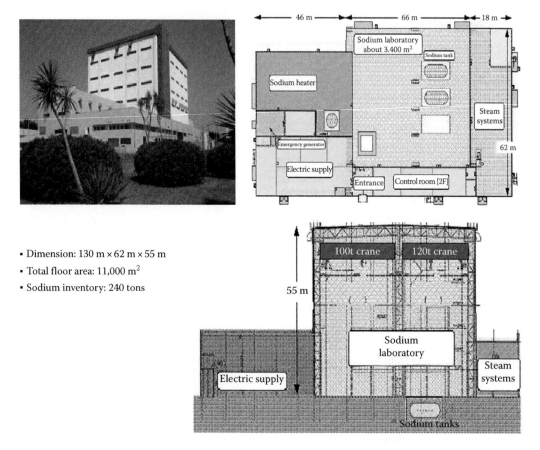

- Dimension: 130 m × 62 m × 55 m
- Total floor area: 11,000 m^2
- Sodium inventory: 240 tons

FIGURE 18.20   AtheNa: Japanese facility. (From Ohira, H. and Uto, N., Progress on fast reactor development in Japan, *46th TWG-FR Annual Meeting*, IAEA HQ, Vienna, Austria, May 2013.)

18.5.2   MINA: INDIAN FACILITY FOR THE SODIUM FIRE STUDIES [18.21]

MINA, a mini sodium experimental facility (Figure 18.22), was exclusively set up to study sodium combustion science in the field of sodium fire safety. This facility has an experimental hall of size 5.6 m × 5.4 m × 4.6 m (139 m^3 volume) and its design pressure is 4 bar at 500°C. It is fabricated with 16 mm thickness mild steel plates and the outside of these plates is strengthened by welding ISMB 150 stiffeners (i.e., I-beam) in the form of 300 mm × 300 mm size grid. These I-beam stiffeners are provided on all the six outer surfaces of the hall and the inside surface of the hall is provided with a 1.6 mm thick SS 304 L material liner, and it is bonded to the mild steel plates by plug welding at regular intervals. The experimental hall is provided with a leak-tight door of 2.0 m × 1.5 m on its northern side wall and a control room is located outside the hall. The southern side of the experimental hall is provided with a sodium loop with necessary tanks, valves, heaters, related instrumentation, and cover gas system. The facility is provided with two numbers of view ports meant for viewing/video-graphing and thermal imaging. Gas and wall temperatures of the hall are measured at 25 locations using K-type thermocouples and a strain gauge–type pressure transducer is installed to measure the gas pressure inside the hall. This facility is meant for evaluating sodium burning rates, sodium aerosol behavior, sodium–concrete interaction, qualifying sodium fire fighting system, and validation of combined fire (i.e., spray followed by pool) scenarios and sodium fire codes.

FIGURE 18.21 DIADEMO: French facility at CEA. (From Gastaldi, O. et al., Experimental platforms in support of the ASTRID program: Existing and planned facilities at CEA, *Proceedings of the Technical Meeting on Existing and Proposed Experimental Facilities for Fast Neutron Systems*, Vienna, Austria, 2013.)

18.5.3 SOCA: Indian Facility [18.21]

The Sodium Cable Fire Test Facility (SOCA) was built to simulate the sodium fire scenario above the top shield consequent to a CDA. SOCA consists of three major modules, that is, the test chamber, sodium ejection system, and exhaust gas treatment system. The schematic of the facility is shown in Figure 18.23. The test chamber is designed to withstand instantaneous burning of 10 kg sodium spray fire, that is, 10 bar pressure and 773 K temperature, and it is based on the most severe thermal consequence of limiting fire under enclosed conditions. The sodium release system of SOCA is designed to simulate sodium ejection from the annular gap of the roof slab–large rotatable plug during CDA, which consists of the sodium vessel, ring header, and argon system for pressurization. The jets of sodium are created by means of a ring header that contains equally distributed nozzles of 1.5 mm diameter along the circumference. The sodium release system is a unique design to eject sodium through multiple nozzles at a desired rate. An exhaust gas treatment system is provided to remove the harmful sodium aerosols and other toxic gaseous products from the exhaust gas before venting it into the atmosphere. This system consists of a venturi scrubber followed by two countercurrent packed bed columns connected in series. The exhaust gas treatment system is designed for the maximum gas flow rate of 0.6 m^3/s and it is based on severe conditions of limiting fire under rupture disk failure. Moreover, the system is designed to reduce the sodium aerosol concentration in the exit gas to the permissible

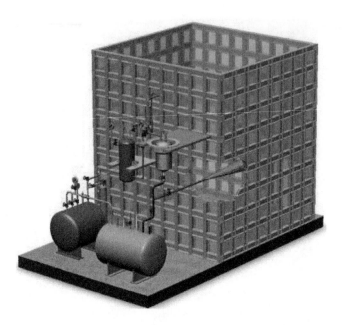

FIGURE 18.22 MINA: Indian facility for the sodium fire studies. (From Ponraju, D. et al., Experimental and numerical simulation of sodium safety in SFR, *International Conference on Fast Reactors and Related Fuel Cycles: Safe Technologies and Sustainable Scenarios (FR13)*, Paris, France, 2013.)

limit of 2 mg/m^3. This facility is meant for studying the combined effect of sodium and secondary cable fire on the integrity of important components like DHX piping.

18.5.4 SFEF: INDIAN FACILITY FOR LARGE-SCALE SODIUM FIRE STUDIES [18.21]

The Sodium Fire Experimental Facility (SFEF) consists of an experimental hall, sodium equipment hall in the ground floor, and a control room in the first floor. The experimental hall is 9 m long, 6 m wide, and 10 m high with a volume of 540 m^3. This is made up of 450 mm thick RCC floor, walls, and ceiling with a design pressure of 50 kPa (g) and temperature of 65°C. All the four walls and ceiling are provided with 50 mm thick insulation and the floor with 150 mm insulation with calcium silicate boards. In order to prevent caustic attack on the concrete and to facilitate thorough washing after conducting the experiments, the inside of the wall is completely lined with SS sheets. Personnel and material access into the experimental hall is through an airtight door of 1.8 m width and 2.1 m height in the western wall. Several leak-tight through-wall penetrations are provided for sodium piping entry, inert gas supply, thermocouples, videography, air, and aerosol sampling. The schematic of the facility is shown along with a photograph in Figure 18.24. The performance of the sodium leak collection tray provided below the sodium pipes of PFBR to minimize the hazardous effects of sodium fire is evaluated in this facility.

18.5.5 SOWART: INDIAN FACILITY [18.22]

Study on self-wastage and impingement wastage is necessary for the safe operation of steam generators. In the SOWART facility, a steam leak is simulated and the sodium–water reaction is being carried out. This facility is located at IGCAR (Figure 18.25). The sodium inventory of this facility is 10 tons. It consists of an online purification system with an air-cooled cold trap, EM pump (flat linear induction type) for sodium circulation, two heater vessels with immersion heaters, steam injection system, hydrogen sensors, microleak test section, and impingement wastage test section.

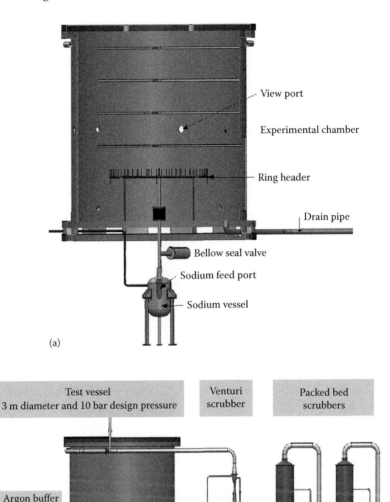

View port

Experimental chamber

Ring header

Drain pipe

Bellow seal valve

Sodium feed port

Sodium vessel

(a)

Test vessel
3 m diameter and 10 bar design pressure

Venturi
scrubber

Packed bed
scrubbers

Argon buffer
tank

(b)

FIGURE 18.23 SOCA: Indian facility. (a) Internal details of the test vessel and (b) SOCA test loop. (From Ponraju, D. et al., Experimental and numerical simulation of sodium safety in SFR, *International Conference on Fast Reactors and Related Fuel Cycles: Safe Technologies and Sustainable Scenarios (FR13)*, Paris, France, 2013.)

The maximum operating temperature is 803 K and the maximum flow is 20 m^3/h at a pressure of 0.37 MPa. This loop can be utilized for dynamic testing of components in sodium apart from the sodium–water reaction studies.

This facility was constructed to study the behavior of self-wastage and leak enlargement during sodium–water reaction, to study impingement wastage, and to develop leak detection methods. Self-wastage studies were carried out at different steam leak rates in the range of 10–50 mg/s. Model cold trap testing was completed in this test facility. Different hydrogen meters were calibrated by injecting a known quantity of hydrogen.

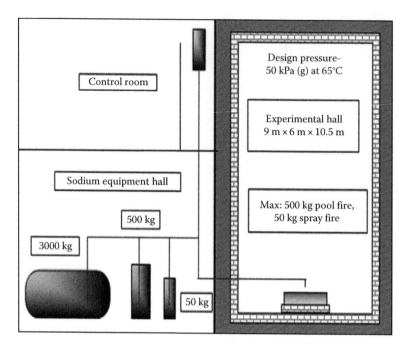

FIGURE 18.24 SFEF: Indian facility for large-scale sodium fire studies. (From Ponraju, D. et al., Experimental and numerical simulation of sodium safety in SFR, *International Conference on Fast Reactors and Related Fuel Cycles: Safe Technologies and Sustainable Scenarios (FR13)*, Paris, France, 2013.)

18.5.6 SAPFIRE: JAPANESE FACILITY [18.1]

SAPFIRE is a large-scale sodium leak, fire, and aerosols test facility constructed in 1985 (Figure 18.26). It is a test platform to investigate various phenomena in a sodium leak accident such as spray, columnar, and pool-type fire, sodium–concrete interaction, sodium–structure interaction with chemical reaction, and aerosol behavior. The volume of the test section is about 100 m³, the allowable pressure is 0.2 MPa, and the maximum ventilation rate is 70 N m³/min. The total amount of sodium in the test loop is 10 tons. It consists of three test rigs: SOLFA-1 (two-story concrete cell), SOLFA-2, and FRAT-1 (small stainless steel vessel). SOLFA-2 is 10 m in height, 3.4 m in diameter, and 200 m³ in inner volume.

18.5.7 SWAT-1R/3R: JAPANESE FACILITIES [18.23]

The prevention of a sodium–water reaction incident in SFR steam generators is essential to maintain plant reliability, but even in the case of leak, restraining leak development at lower stages could mitigate the damage to the plant. A relatively minor leak, classified as a small leak or a microleak, preceded most of the sodium–water reaction incidents that occurred in LMFBR steam generators in the past. Therefore, failure propagation should be understood to evaluate the system integrity in sodium–water reaction events. In Japan, failure propagation studies started in the 1960s using SWAT test facilities to understand the leak development phenomena. SWAT-1R is a facility dedicated to the study of water–sodium interaction due to pipe rupture in a steam generator and to self-wastage behavior of a double-tube steam generator pipe (Figure 18.27). The tank volume is 630 L, T_{max} = 580°C, P_{max} = 1.96 MPa. SWAT-3R is a large-scale facility dedicated to the investigation of the propagation of steam generator tube rupture in prototypical conditions (Figure 18.28), with 15 tons of sodium, T = 555°C; P_{max} = 1.96 MPa; tank volume, 10 m³; and water heaters of 3.1 and 4.8 m³.

The major objective of these facilities is to carry out a pipe failure propagation study toward understanding the leak development behavior and also to develop an analytical method based on the

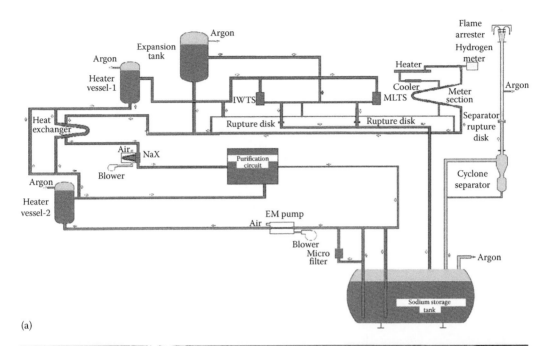

(a)

(b)

FIGURE 18.25 SOWART: Indian facility. (a) SOWART loop and (b) SOWART test facility. (From Kishore, S. et al., *Nucl. Eng. Des.*, 243, 49, 2012.)

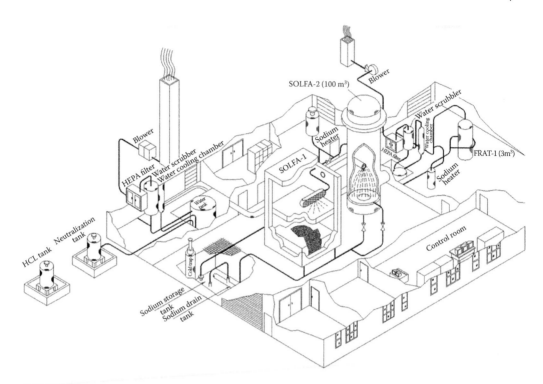

FIGURE 18.26 SAPFIRE: Japanese facility. (From Task Group on Advanced Reactors Experimental Facilities (TAREF), Experimental facilities for sodium fast reactor safety studies, NEA No. 6903, Nuclear Energy Agency, Organization for Economic Co-operation and Development, Paris, France, 2011.)

(a)

(b)

FIGURE 18.27 SWAT-1R: Japanese facilities. (a) SWAT-1R facility loop and (b) SWAT-1R facility building. (From Tanabe, H. and Wachi, E., Review on steam generator tube failure propagation study in Japan, *Proceedings of the Specialists Meeting on Steam Generator Failure and Failure Propagation Experience*, Aix-en-Provence, France, 1990.)

knowledge obtained through the experiments. Finally, this knowledge will be applied to the actual SFR steam generators to evaluate the conservatism of design basis leak (DBL). Therefore, various kinds of experimental studies in relevant fields were conducted using SWAT test facilities and the LEAP code was developed. The main experimental studies that were started in various leak ranges are (1) self-wastage tests in the microleak, (2) target wastage tests in the small leak, (3) multilube wastage tests in the intermediate leak, (4) overheating proof tests in the large leak, and (5) sequential failure propagation tests from the small leak to the large leak.

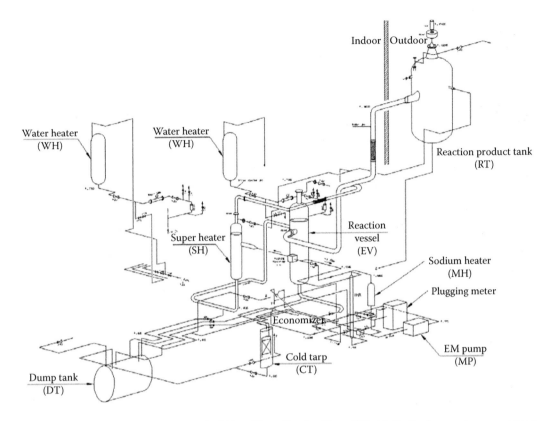

FIGURE 18.28 SWAT-3R: Japanese facilities. (From Tanabe, H. and Wachi, E., Review on steam generator tube failure propagation study in Japan, *Proceedings of the Specialists Meeting on Steam Generator Failure and Failure Propagation Experience*, Aix-en-Provence, France, 1990.)

REFERENCES

18.1 Task Group on Advanced Reactors Experimental Facilities (TAREF). (2011). Experimental facilities for sodium fast reactor safety studies. NEA No. 6903. Nuclear Energy Agency, Organization for Economic Co-operation and Development, Paris, France, ISBN: 978-92-64-99155-2.

18.2 IAEA. (2013). Status of innovative fast reactor designs and concepts: A supplement to the IAEA advanced reactors information system (ARIS). Nuclear Power Technology Development Section, Division of Nuclear Power, Department of Nuclear Energy, (TAREF) formed under NEA. http://www.iaea.org/NuclearPower/Downloadable/FR/booklet-fr-2013.pdf

18.3 Bailly, J., Tattegrain, A., Saroul, J. (1980). The SCARABEE facility—Its main characteristics and the experimental program. *Nucl. Eng. Des.*, 59, 237–255.

18.4 Blanc, H., Millot, J.P., Lions, H. (1967). The sodium CABRI loop. *International Conference on Fast Reactor Safety*, Aix-en-Provence, France.

18.5 Ewell, E. (1997). *International Conference on Nonproliferation Problems*, NISNP Trip Report, KAZ970900, pp. 13–14.

18.6 Ashurko, I.M., Sorokin, A.P., Privezentsev, V.V., Volkov, A.V., Khafizov, R.R., Ivanov, E.F. (2013). Activities on experimental substantiation of SFR safety in accidents with sodium boiling. *International Conference on Fast Reactors and Related Fuel Cycles: Safe Technologies and Sustainable Scenarios (FR13)*, Paris, France, March 2013.

18.7 Crawford, D.C., Deitrich, L.W., Holtz, R.E., Swanson, R.W., Wright, A.E. (1998). Review of experiments and results from the transient reactor test (treat) facility. *ANS Winter Meeting*, Washington, DC, November 1998.

18.8 Walker, J.V. et al. (1979). Design and proposed utilization of the SANDIA annular core research reactor (ACRR), Sandia Laboratories, Albuquerque, NM. *International Meeting on Fast Reactor Safety Technology*, Seattle, WA, August 1979, SAND79-1646C.

18.9   Journeau, C., Piluso, P., Haquet, J.-F., Saretta, S., Boccaccio, E., Bonne, J.-M. (2007). Oxide-metal corium–concrete interaction test in the VULCANO facility, Paper no. 7328. *Proceedings of the ICAPP 2007*, Nice, France.

18.10   Zabiego, M., Brayer, C., Grishchenko, D., Dajon, J.-B., Fouquart, P., Bullado, Y., Compagnon, F., Correggio, P., Haquet, J.-F., Piluso, P. (2010). The KROTOS KFC and SERENA/KS1 tests: Experimental results and MC3D calculations. *Seventh International Conference on Multiphase Flow (ICMF 2010)*, Tampa, FL.

18.11   Das, S.K. et al. (2013). Post accident heat removal: Numerical and experimental simulation, IAEA TM on fast reactor physics and technology. *Nucl. Eng. Des.*, 265, 1246–1254.

18.12   Farmer, M.T., Olivier, T., Sofu, T. (2012). U.S. perspective on technology gaps and R&D needs for SFRs. *International Workshop on Prevention and Mitigation of Severe Accidents in Sodium-Cooled Fast Reactors*, Tsuruga, Japan.

18.13   Farmer, M.T., Lomperski, S., Kilsdonk, D., Aeschlimann, S.R.W., Basu, S. (2007). A summary of findings from the melt coolability and concrete interaction (MCCI) program, Paper 7544. *International Congress on Advances in Nuclear Power Plants (ICAPP 2007)*, Nice, France.

18.14   Blanchat, T.K., Pilch, M.M., Lee, R.Y., Meyer, L., Petit, M. (1999). Direct containment heating experiments at low reactor coolant system pressure in the SURTSEY test facility, Washington, DC: Division of Systems Analysis and Regulatory Effectiveness, Office of Nuclear Regulatory Research, U.S. Nuclear Regulatory, NUREG/CR-5746, SAND 99-1634.

18.15   Hering, W., Stieglitz, R., Jianu, A., Lux, R., Onea, A.M., Homann, Ch. (2013). Scientific program of the Karlsruhe sodium facility (KASOLA). *Proceedings of the International Conference on Fast Reactors and Related Fuel Cycles: Safe Technologies and Sustainable Scenarios (FR13)*, Paris, France, March 2013.

18.16   Padmakumar, G. et al. (2013). SADHANA facility for simulation of natural convection in the SGDHR system of PFBR. *Prog. Nucl. Energy*, 66, 99–107.

18.17   Gnanadhas, L. et al. (2011). PATH—An experimental facility for natural circulation heat transfer studies related to post accident thermal hydraulics. *Nucl. Eng. Des.*, 241, 3839–3850.

18.18   Chellapandi, P., Das, S.K., Hemant Rao, E., Harvey, J., Nashine, B.K., Chetal, S.C. (2012). Overview of molten fuel coolant interaction studies towards SFR development. *SERENA/OECD Project Seminar and Associated Meeting*, CEA, Cadarache, France, November 2012.

18.19   Ohira, H., Uto, N. (2013). Progress on fast reactor development in Japan. *46th TWG-FR Annual Meeting*, IAEA HQ, Vienna, Austria, May 2013.

18.20   Gastaldi, O. et al. (2013). Experimental platforms in support of the ASTRID program: Existing and planned facilities at CEA. *Proceedings of the Technical Meeting on Existing and Proposed Experimental Facilities for Fast Neutron Systems*, Vienna, Austria.

18.21   Ponraju, D. et al. (2013). Experimental and numerical simulation of sodium safety in SFR. *International Conference on Fast Reactors and Related Fuel Cycles: Safe Technologies and Sustainable Scenarios (FR13)*, Paris, France.

18.22   Kishore, S., Ashok Kumar, A., Chandramouli, S., Nashine, B.K., Rajan, K.K., Kalyanasundaram, P., Chetal, S.C. (2012). An experimental study on impingement wastage of mod 9Cr-1Mo steel due to sodium water reaction. *Nucl. Eng. Des.*, 243, 49–55.

18.23   Tanabe, H., Wachi, E. (1990). Review on steam generator tube failure propagation study in Japan. *Proceedings of the Specialists Meeting on Steam Generator Failure and Failure Propagation Experience*, Aix-en-Provence, France.

19 Safety Experiments in Reactors

19.1 INTRODUCTION

Efficacy of most of the safety features that are incorporated in a particular reactor can be demonstrated in the reactor itself. Certain transient tests are to be conducted necessarily as a part of regulatory requirements. Apart from these, the reactors are used to generate data for validating numerical simulations of initiating events and severe accident scenarios. The performance of shutdown systems and decay heat removal systems has been assessed in almost all the test reactors, imposing certain pessimistic transients. Demonstration of performance of passive safety systems in the reactor is a very difficult exercise. Off-normal transient tests during operating tenure of the reactor are performed after careful planning and analysis. Severe transient tests simulating long station blackout conditions, for example, are conducted generally at the end-of-reactor life. These sorts of tests play a very important role in raising the confidence on the overall reactor safety among designers, regulators, and the public.

In this chapter, a few important safety experiments conducted in the reactors—Rapsodie and Phénix (France), fast breeder test reactor (FBTR) (India), BOR-60 (Russia), and fast flux test facility (FFTF) and experimental breeder reactor-II (EBR-II) (United States)—are explained, highlighting the objectives and important results/outcome.

19.2 HIGHLIGHTS OF SAFETY EXPERIMENTS

19.2.1 RAPSODIE

19.2.1.1 Natural Circulation Tests [19.1]

Tests were conducted for establishing the natural circulation in the primary and secondary circuits, raising the power up to 750 kWt and maintaining at the same level subsequently. Following the reactor scram at nominal conditions of 22.4 MWt, the transition to natural circulation in the whole installation including air coolers was investigated, simulating the total loss of electric power supply.

19.2.1.2 Reactor Transients without Scram [19.1]

The loss of flow without scram tests consisted of the shutdown of the primary circuit and secondary circuit pumps, as well as the tertiary circuit fans, and the nonoperation of the safety rods. The reactor output reached 21.2 MW (more than 50% of the rated value), while the mean coolant temperatures at the reactor inlet and outlet came to 402°C and 507°C, respectively. The principal characteristics of these processes are depicted in Figure 19.1. In this test, the maximum fissile subchannel temperature rose to 800°C, whereas the nuclear power decreased continuously without any intervention from the reactor control. A comparison of calculation results and experimental data demonstrated that the fuel residing in the core shared a state of coalescence with the fuel element cladding and expanded with the cladding upon heating up. Good agreement was observed between the calculation results and the experimental data concerning the coolant temperature at the subassembly outlet. These experiments demonstrated the inherently stable reactor behavior under a severe accident condition, including unprotected loss of flow (ULOF) situations.

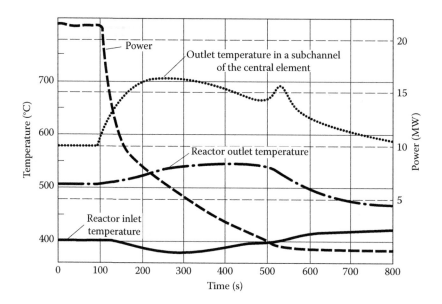

FIGURE 19.1 Rapsodie end-of-life test: characteristics during an unprotected loss of flow accident transient. (From Technical reports on design features and operating experience of experimental fast reactors, International Atomic Energy Agency, Vienna, 2013, No. NP-T-1.9.)

19.2.2 PHÉNIX

The Phénix was shut down in March 2009 after operating successfully for more than 35 years. Before the decommissioning, a series of end-of-life tests in the areas of core neutronics, fuel behavior, and thermal hydraulics were performed during May–December 2009. The control rod withdrawal test was one such serial carried out to validate the European fast reactor neutronics code system called ERANOS in predicting local power distribution due to the movement of absorber rods at nominal power. In this context, the International Atomic Energy Agency (IAEA) had initiated a collaborative research project (CRP) on Control Rod Withdrawal and Sodium Natural Circulation Tests Performed during the Phénix End-of-Life Experiments. Eight organizations from seven member states took part in the benchmark, that is, ANL, CEA, IGCAR, IPPE, IRSN, KAERI, PSI, and the University of Fukui. Each organization has performed computations and has contributed to the analysis and recommendations. Details of this CRP can be found in Ref. 19.2.

The measured results and the geometrical data required for this analysis have been supplied by the IAEA to all the participants of the aforementioned CRP. The parameters considered are absorber rod worth, criticality for four core states, and the deviation of subassembly power or sodium heating change in a critical state with respect to the reference configuration. IGCAR has simulated all the test configurations by both FARCOB and ERANOS 2.1 systems and the results computed were compared with the measured values [19.3]. In this test, two main control rods SCP-1 and SCP-4 were moved by reactivity balancing method by maintaining the total power constant and three critical core states (Steps 1–3) with distorted radial power distribution with respect to the reference state were explored (Figure 19.2). It is found that Step 2 has the maximum distortion of power distribution. The accuracy in the prediction of criticality is closely related to the errors associated with the neutron cross sections, the geometrical modeling of the core, and the calculation method. Both these code systems use 3D neutron diffusion theory for criticality calculations (33 groups in ERANOS and 26 groups in FARCOB). It is found that both codes predict the criticality for all the four critical core states. Thus, this benchmark exercise has provided an experimental validation of FARCOB code system against Phénix reactor for criticality, absorber rod worth, and radial power distribution. The computed results from these codes are close to the experimental values.

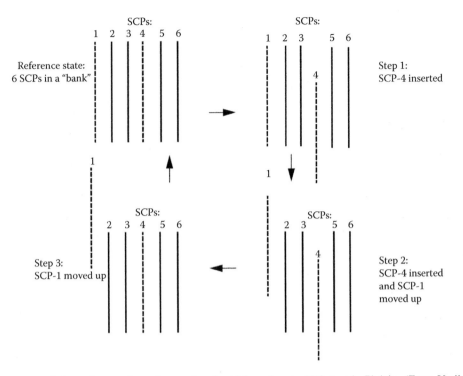

FIGURE 19.2 Schematic overview of control rod withdrawal end-of-life test in Phénix. (From Vasile, A. et al., IAEA Coordinated Research Project (CRP) on control rod withdrawal and sodium natural circulation tests performed during the PHENIX end-of-life experiments, IAEA, Vienna, Austria, September 2008.)

19.2.3 NATURAL CONVECTION TESTS IN FBTR [19.4]

FBTR is a loop-type reactor without any dedicated safety grade decay removal system as in the current designs of pool-type reactors. FBTR steam generators (SGs) are housed inside a common insulated casing. The casing is provided with trap doors at the bottom and a chimney at the top. During the normal operation, the trap doors are in closed position and when there is a demand for decay heat removal by natural convection, the trap doors are opened. As a result, ambient air that enters the casing absorbs the heat from the SG shell as well as from the walls of the casing and leaves through the chimney, which enhances the buoyancy-induced air flow. Toward demonstrating the development of natural convection in the secondary sodium loop and effective removal of heat by SG casing, a natural convection test has been carried out in FBTR.

It was decided to simulate a decay power of 180 kW. This corresponds to about 2% of the nominal operating thermal power of the reactor, viz., 10 MWt. The decay power was simulated by normal fission heat, and hence, primary sodium pumps were operating leading to primary under forced flow. On the other hand, the secondary sodium pumps were tripped and steam–water system was isolated. The trap doors of the SG casing were opened. A part of the heat loss was also from the long secondary sodium piping. Measured secondary flows in the east and west loops are depicted in Figure 19.3. It is seen that the secondary flow comes down to minimum within a few minutes and then rises as governed by the buoyancy forces. In order to simulate this event in the FBTR plant dynamics code DYNAM, the SG module of the code was replaced by a pipe thermal model to account for heat removal by natural convection of air. Secondary flows in east and west loops predicted by this model are included in Figure 19.3. It can be seen that the observed flow of secondary sodium stagnates from 15 to 75 min in the test, while in the predictions it shows an almost uniform rise. This is explained by the thermal stratification in the piping, which is not modeled effectively in the 1D code. Nevertheless, the final flows are very close. This test has

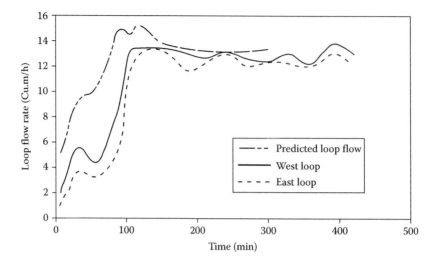

FIGURE 19.3 Secondary sodium flows in fast breeder test reactor during natural convection test at 180 kWt. (From Vaidyanathan, G. et al., *Ann. Nucl. Energy*, 37(4), 450, 2010.)

given confidence for safe decay heat removal by natural circulation of air through trap door housing as well as utilization of this code for natural convection assessment.

19.2.4 BOR-60 [19.5]

In order to demonstrate the reactor safety, three kinds of tests were performed: (1) introduction of gas into the core, (2) sodium boiling, and (3) flow blockage in the experimental fuel assembly. Tests have demonstrated robust dynamic and thermal hydraulic characteristics, thereby the presence of inherent safety properties. In the sodium boiling tests, the primary pumps were tripped and all reactor scram parameters were inhibited. The temperature of the fuel cladding exceeded 1000°C only for a short duration (Figure 19.4). For the flow blockage tests, a cell with instrumentation in the fifth

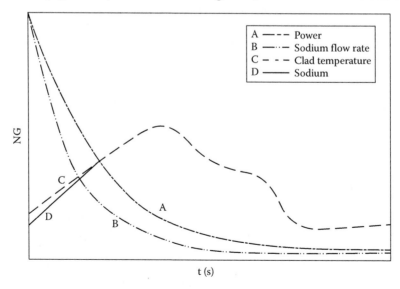

FIGURE 19.4 Transient evolution of BOR-60 core under unprotected loss of flow simulation. (A trend line is shown here. For details, refer to Ref. 19.5.) (From Gadzhiev, G.I. et al., *Atom. Energy*, 91(5), 913, 2001.)

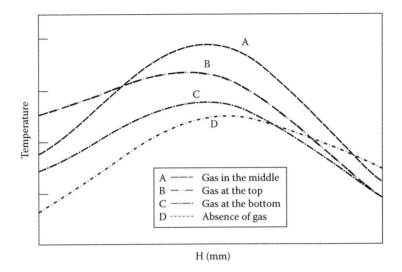

Temperature

A ---- Gas in the middle
B --- Gas at the top
C ---- Gas at the bottom
D ...... Absence of gas

H (mm)

FIGURE 19.5 Temperature distributions across the core under gas injection to the BOR-60 core. (A trend line is shown here. For details, refer to Ref. 19.5.) (From Gadzhiev, G.I. et al., *Atom. Energy*, 91(5), 913, 2001.)

row of the core was mostly used. In one of the experiments, gas was introduced continuously at the different sections along the height of the core. The signals from the neutron and temperature sensors and the noise in the signals were measured and investigated. Deviation of temperature distribution due to the introduction of gas was observed by measuring the temperature at three points at the exit from a fuel assembly. The deviation is due to the redistribution of the flow through the cell of a fuel assembly (Figure 19.5). A thermocouple was used to determine the change in the temperature of the cladding of a fuel element in the first row of a fuel assembly with different gas content. Specifically, for sodium heated by 97°C and 16% gas content, the cladding temperature increased by only 15°C, attesting to the absence of any appreciable drying of the surface. Out of three series of experiments involving gas (steam) injection, in the first two series, boiling with condensation of steam bubbles above the fuel column in the simulated fuel pins was monitored and also boiling was monitored above the fuel assembly head to correlate the boiling of the entire volume of the fuel assembly. Acoustic and temperature sensors were used in the experiments. Boiling was reached by continually increasing the reactor power. The onset of boiling was easily determined from the noise in the temperature and the neutron flux, acoustic noise, and the calculation of the balance of reactivity. The measurements of the average temperature at the top of a fuel assembly at boiling onset attest to the absence of any appreciable heating of the coolant at the location of the thermocouples. Analysis of the signals showed that the acoustic noise of boiling is nonstationary. Boiling starts with several pulsed signals with duration up to 1 ms, which within tens of milliseconds follow one another almost continuously and merge into a continuous noise signal with a duration of several tenths of a second. In a steady-boiling regime, single pulses separated by 10–20 ms are observed just like before the burst. The pulsed signals recorded with submersible sensors were about 20 dB above background. A burst with steam emission in the head is detected just as successfully by acoustic sensors above the location of boiling and at a distance of ~0.5 m from this location. However, when the steam bubbles collapse in the interior of the assembly, it becomes difficult to observe boiling using distant acoustic sensors. Periodic components with frequencies 1 and 6 Hz were found in the neutron-noise spectrum on boiling. These components correlate well with the temperature noise and the envelope of the acoustic noise. In the third series of experiments, the thermal hydraulic characteristics of the sodium flow were simulated close to the conditions in real fuel assemblies with a partially blocked flow at the entrance into the fuel bundle. A special probe, making it possible to measure the temperature in the assembly head and above the exit windows and the neutron flux density, was used to measure the parameters. The onset of boiling with power up to 14.8 MW is accompanied by increasing the high-frequency component in the range 15–20 Hz. At 20.7 MW, the noise in the low-frequency range (up to 2 Hz) increases by a factor of 10, peaks appear at frequencies 0.9 and 1.4 Hz, and the spectrum amplitude decreases somewhat in the range 10–20 Hz. The high-frequency component of the spectrum is characteristic for local boiling, when small steam bubbles start to form and collapse.

In some experiments, special devices were introduced into the core to study the behavior of fuel elements in emergency situations with a flow regulator. A demonstration experiment with a sudden decrease of the coolant flow rate for short time duration indicated the sodium temperature rise up to 1000°C and fuel-element claddings to melt was performed in such a circuit. In this experiment, even the interior case of the circuit melted, sodium filled the heat-insulating gas gap, and the temperature automatically dropped. Thus, this experiment demonstrated the prompt functioning of all safety features incorporated in the core.

19.2.5 Fast Flux Test Facility

There is no separate backup heat removal system at FFTF. Hence, dedicated tests were conducted in 1980 to establish that the reactor had sufficient natural convection heat removal capability and core cooling is assured for all accidents identified in safety analysis reports. Apart from these, a series of test conducted in July 1986 demonstrated a degree of passive safety with a modified FFTF core. The test series culminated in a test in which, with the reactor operating at 50% power, all the coolant pumps were shut down in order to evaluate the core's response to simulated loss of electric power to the pumps. These tests were conducted without reactor scram and without any operator's intervention. For the purpose of the tests, nine of the Inconel reflectors immediately outside the core were replaced with the gas expansion modules (GEMs). A GEM is essentially an inverted test tube with argon gas compressed at the top. With the pumps operating and providing full flow, a standing column of sodium is maintained in the bottom of the tube and the column of sodium keeps the argon compressed at the top of the tube and serves as a neutron reflector. When the pumps stop such as during the loss-of-flow tests, the pressure drops, the argon expands, and the sodium is driven down and outside of the tube, which effectively removes the reflector allowing more neutrons to escape from the core, adding negative reactivity and thereby tending to shut the reactor down. The need for such a large negative reactivity insertion to ensure safe shutdown upon loss of flow conditions results mainly from the need when using oxide fuels for a very rapid decrease in power and coolant temperature to prevent sodium boiling. For more details, readers are requested to see Ref. 19.6.

19.2.6 EBR-II [19.7,19.8]

The most severe safety tests were conducted in April 1986 and demonstrated that sodium-cooled fast reactor (SFR) with metallic fuel could safely accommodate unprotected loss of flow accident (ULOFA). Experiments simulating the ULOFA were conducted at full power in the presence of different pump coast-down times: active rundown, by controlling the pump speed, pump stop time 300 and 100 s, and passive coast down accompanied by the shutdown of the auxiliary electromagnetic sodium pump. The EBR-II pump system has small inertia, leading to a fast coast down. Therefore, it was decided to use the stored energy in both the pumps and the motor–generator set and to control the coast down with the magnetic clutch that couples the motor and generator. A comparison of peak temperatures demonstrated the decisive influence exerted by the rundown time of the primary pumps on the fuel element cladding and reactor coolant temperature values (Figure 19.6).

Experiments relating to the unprotected loss of heat sink (ULOHS) were also conducted at full power. The temperature placed at the reactor outlet was reduced, while the temperature placed at the reactor inlet was increased. Owing to the experimental confirmation of the fact that safety properties are inherent in the fast reactors, design work has been undertaken, aiming at ensuring the

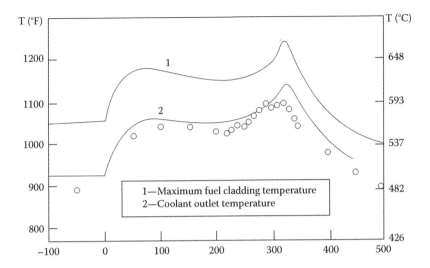

FIGURE 19.6 Characteristics of EBR-II reactor during unprotected loss of flow accident transient. (From Lahm, C.E. et al., *Nucl. Eng. Des.*, 101, 25, 1987.)

creation of conditions that favor, to the greatest extent possible, the utilization of inherent factors. It is generally known that an increase in the reactor inlet temperature and the associated reactor coolant temperature rise are accompanied by the thermal expansion of the core diameter, the bending of the grid subassembly outward from the center, and the elongation of the control rod transfer bars, thereby resulting in the introduction of negative reactivity.

The natural convection test was conducted with reactor inlet at 351°C (normal value is 371°C) and the initial reactor power and flow set at 28.5% and 32.1% of their respective full power values. The secondary pump trip circuit was bypassed in order to continue secondary intermediate heat exchanger (IHX) cooling following the loss of primary forced flow. The test was initiated by loss of power supply to primary pumps, which led to the coast down of pumps and reactor SCRAM at 1.8 s. One of the primary pumps reached to zero speed in 36.5 s and the other in 42.5 s. The experiment was simulated by the plant dynamics code NATDEMO [19.9], and the comparisons of predictions with the measured values are shown in Figure 19.7. It can be seen that immediately following the pump trip, the top-of-core coolant temperature is seen to increase. At 1.8 s, the temperature drops rapidly because of reactor SCRAM. Due to the large and rapid drop in power and the relatively slower decrease in primary flow, the core is overcooled for a short period. However, this condition shortly reverses due to the continuing decrease in the coolant flow rate and a relatively slower decrease in the neutron plus decay power, resulting in a gradual increase in coolant temperature. This increase in temperature is reversed as the convective flow develops, and a peak of about 410°C occurs at 55 s, and a slow monotonic decrease continues thereafter. The overall agreement between the measured and calculated dynamic thermal response of the coolant is excellent. The small differences in temperature levels can be accounted for by the different initial power-to-flow ratios and heat capacities of the measured and modeled subassemblies.

The important conclusions from these tests and the supporting analysis are as follows:

- Natural circulation flow can be demonstrated to be a reliable method of decay heat removal under emergency conditions in SFRs.
- The plant parameters for this type of event can be predicted with prior knowledge of the hydraulic, thermal, and neutronic characteristics of the system.
- The prediction of the detailed temperature distribution within the core, especially the local hot channel temperature rise, requires the use of a multichannel core model that accounts for flow redistribution and inter-subassembly heat transfer.

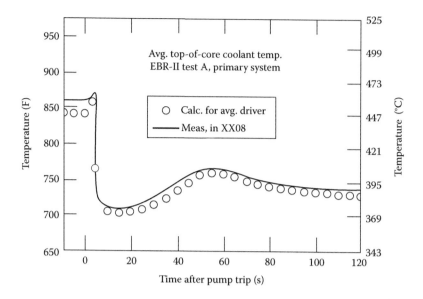

FIGURE 19.7 Evolution of core outlet temperatures: natural convection test in EBR-II. (From Singer, R.M. et al., Response of EBR-II to a complete loss of primary forced flow during power operation, *Specialists' Meeting on Decay Heat Removal and Natural Convection in FBR's*, Brookhaven National Laboratory Upton, Long Island, NY, February 1980.)

19.3 CONCLUSION

Various in-pile and end-of-life safety experiments conducted in test reactors such as BOR-60, FFTF, EBR-II, Rapsodie, FBTR, and Phénix have clearly demonstrated the SFR safety during ULOF, sodium boiling, subassembly blockage conditions, etc. During relatively long station blackout conditions, the natural convection capability of decay heat removal systems has also been confirmed. In particular, the role of inherent safety features of metallic core was realized. The data generated during these safety experiments were used even now for validating the computer codes performing safety analysis.

REFERENCES

19.1 Technical reports on design features and operating experience of experimental fast reactors, International atomic energy agency, Vienna, 2013. No. NP-T-1.9.

19.2 Vasile, A. et al. (2008). IAEA Coordinated Research Project (CRP) on control rod withdrawal and sodium natural circulation tests performed during the PHENIX end-of-life experiments. IAEA, Vienna, Austria, September 2008.

19.3 Devan, K., Alagan, M., Bachchan, A., Mohanakrishnan, P., Chellapandi, P., Chetal, S.C. (2012). A comparative study of FARCOB and ERANOS-2.1 neutronics codes in predicting the Phenix control rod withdrawal end-of-life experimental results. *Nucl. Eng. Des.*, 245, 89–98.

19.4 Vaidyanathan, G. et al. (2010). Dynamic model of fast breeder test reactor. *Ann. Nucl. Energy*, 37(4), 450–462.

19.5 Gadzhiev, G.I., Efimov, V.N., Zhemkov, Yu.I., Korol'kov, A.S., Polyakov, V.I., Shtynda, Yu.E., Revyakin, Yu.L. (2001). Some experimental work performed on the BOR-60 reactor. *Atom. Energy*, 91(5), 913–922.

19.6 Wootan, D.W., Butner, R.S., Omberg, R.P., Makenas, B.J., Nielsen, D.L. OSTI ID: 1012521. *Trans. Am. Nucl. Soc.*, 102(1), 556–557. http://www.osti.gov/scitech/biblio/1012521.

19.7 Lahm, C.E., Koenig, J.F., Betten, P.R., Bottcher, J.H., Lehto, W.K., Seidel, B.R. (1987). EBR-II driver fuel qualification for loss-of-flow and loss-of-heat-sink tests without scram. *Nucl. Eng. Des.*, 101, 25–34.

19.8   Planchon, H.P., Sackett, J.I., Golden, G.H., Sevy, R.H. (1987). Implications of the EBR-II inherent safety demonstration test. *Nucl. Eng. Des.*, 101, 75–90.

19.9   Singer, R.M., Gillette, J.L., Mohr, D., Tokar, J.V., Sullivan, J.E., Dean, E.M. (1980). Response of EBR-II to a complete loss of primary forced flow during power operation. *Specialists' Meeting on Decay Heat Removal and Natural Convection in FBR's*, Brookhaven National Laboratory Upton, Long Island, NY, February 1980.

20 Severe Accident Management

20.1 INTRODUCTION

A set of measures and actions to be taken during the evolution of a severe accident are called severe accident management (SAM), and its objectives are to establish the actions to be taken to prevent severe damage to the core, to terminate the progress of core damage once it has started, to achieve a stable and controlled state of the reactor core or core debris, to maintain the containment integrity, to minimize on-site and off-site releases, and to return the plant to a controlled safe state. SAM should cover all states of plant operation as well as selected external events, such as fires, floods, earthquakes, and extreme weather conditions that could damage a significant part of the plant. The current trend is to implement certain evolving safety criteria to existing reactors to the maximum extent possible as well as for the new designs. The new sodium-cooled fast reactor (SFR) power plant should be equipped with dedicated systems for SAM allowing to attain its safety objectives.

Most of the actions and measures to be taken under SAM are common to all types of nuclear reactors and they are not duplicated in this chapter. Moreover, there are many publications on this topic and readers are requested to see them. Regarding SFRs, only a few reactors are currently under operation and several new designs are evolving (GEN IV reactors, for example). In this chapter, the main focus is to expose the readers to the SAM-related activities published in the literature on one or two existing SFRs and the design features that are under considerations for the inclusion in the new designs.

In this chapter, the SAM measures planned in prototype fast breeder reactor (PFBR) are highlighted, specifically detailed analyses that were carried out (1) to assess the consequences of a few selected design extension conditions and (2) to estimate the additional safety margins on the integrity of the safety-related systems under postulated design extension conditions. Finally, certain design features that are under considerations for the inclusion in the new designs are presented. However, only technical aspects are explained in this chapter. Other managerial and regulatory issues are not addressed here since they are more generic to any kind of reactors.

20.2 ANALYSIS FOR THE CONSEQUENCES OF DESIGN EXTENSION CONDITIONS: PFBR CASE STUDY

This case study helps the readers to have a quantitative feel of the consequences of various design extension conditions investigated for the PFBR: SFR representing an existing reactor.

20.2.1 PRIMARY CONTAINMENT CAPACITY AGAINST CDA

The energy release under core disruptive accident (CDA) is estimated by assuming the highly pessimistic severe accident scenarios, and accordingly, mechanical energy release is arrived at. The basis of such approach is explained in Chapter 21. However, CDA with 100 MJ of mechanical energy release is considered for the analysis of mechanical consequences in PFBR. For this energy release, the structural integrity of primary containment including main vessel (MV), top shield, safety grade decay heat removal systems (SGDHRS), and reactor containment building (RCB) has been ensured.

Core catcher has been designed for withstanding the transient pressure generated during a CDA and subsequently the self-weight of core debris. The functionality, that is, cooling capability of decay heat exchangers (DHXs) immersed in hot pool, has also been ensured based on numerical and experimental simulations. The postaccident heat removal path through perforations generated in the grid plate due to melt-through of molten fuel has been confirmed by scaled down experiments. Thermal hydraulics study performed with such a perforation path has also demonstrated that the core debris settled on the core catcher could be cooled over a prolonged period without affecting the integrity of the MV. Further, analysis with higher energy release (above 100 MJ) has been performed. Results indicated that the critical issue that decides the limiting value is the fall of sodium level due to MV expansion (large permanent deformations) to a level resulting in inadequate cooling capability of DHXs. This value is found to be 500 MJ [20.1]. Experimental activities toward in-depth understanding of all associated phenomena (molten fuel coolant interactions, settlement behavior of core debris on core catcher, heat transfer characteristics, etc.) as well as for generating data for the development of computer codes are the activities of high priority in this domain. Details are presented in Ref. 20.1.

20.2.2 SEISMIC CAPACITY MARGINS FOR THE REACTOR COMPONENTS BEYOND SSE

The peak ground acceleration for design basis earthquake (safe shutdown earthquake [SSE]) is 0.156 g for PFBR. The site-dependent design basis ground motion parameters for PFBR are derived based on the deterministic methods and confirmed by probabilistic seismic hazard assessment. Design of reactor systems, components, and equipment has been qualified for 0.156 g design basis earthquake. The design criteria for all these have been respected w.r.t. functional limits, primary stress limits, factors of safety on buckling, functional performance requirements for core, shutdown systems, and primary pumps. Further, safety margins have been quantified based on the results obtained from the seismic analyses and shake table experiments. A typical analysis indicates that the reactor assembly components can withstand peak ground acceleration up to 1.41 times of the design basis peak ground acceleration. This implies that PFBR can withstand an earthquake of peak ground acceleration of 0.22 g (0.156 × 1.41) without violating any design basis safety limits. In the event of a large earthquake, which triggers the reactor SCRAM, the decay heat generated in the core is to be removed by SGDHRS. Hence, the structural integrity of these loops is ensured under seismic loading and qualified based on design rules of the class 1 piping system. Sufficient margins are also ensured for the pipe lines by adding extra snubbers/seismic arrestors, where the margins are not sufficient. Further, the safety margin of nuclear island connected building (NICB), particularly the reinforced concrete structures over SSE loads, was analyzed based on approved design methodology. By applying the total dead load and base shear of NICB, an average range of safety margin of NICB is evaluated and it is observed that there is an additional margin of 1.35 over SSE loads. More details are presented in Ref. 20.2.

20.2.3 SEQUENTIAL LEAKAGE OF MAIN AND SAFETY VESSELS

During the unlikely event of MV leak, which is one of the design basis events, MV–safety vessel (SV) interspace gap is selected such that intermediate heat exchanger (IHX) inlet windows and DHX inlet windows are sufficiently covered. Hence, during MV leak condition, decay heat removal is possible. Further, the integrity of MV and SV under an earthquake during the MV leaked condition is ensured. Under this condition, the sodium leaked from MV occupies the intervessel space, which generates large dynamic pressure causing high threat to both MV and SV [20.3]. The scenario is further extended postulating a subsequent sodium leak in SV as well. Under this condition, the IHX and DHX inlet windows are uncovered. In such a situation, decay heat is removed by sending nitrogen through biological shield cooling system or by pushing back the sodium from the SV by pressurizing with nitrogen.

A scheme for transferring the sodium from secondary loop to MV for compensating the sodium level fall in MV is proposed for PFBR. As per this scheme, the secondary loop sodium is first dumped into the secondary sodium storage tank (SSST). The sodium in the cold traps and IHXs remains there itself. Excluding this sodium, the remaining sodium is left in the storage tank after draining of the loop. Out of the 205 tons of sodium, around 144 tons can be transferred into the MV by following a systematic sequence of steps: transfer of sodium from the SSST to sodium transfer tank (STT) by electromagnetic (EM) pump, transfer of sodium from STT into argon buffer tank (ABT) by using initial sodium filling circuit and EM pump, and finally transfer from ABT into MV by primary sodium fill and drain circuit. In case of sodium leak in MV, reactor will be in shutdown condition and sodium temperature will be maintained around 200°C. If both the MV and SV are leaking, then the sodium will enter into the reactor vault. Since the reactor vault is lined with carbon steel and it is inerted with nitrogen gas, there is no concern of sodium fire. However, the reactor vault temperatures may rise, and hence, the integrity of the reactor vault concrete has to be assured. More details are presented in Ref. 20.4.

20.2.4 Multiple Failures of SGDHR Circuits

Following the prolonged station blackout (SBO), the decay heat is removed by redundant, diversified passive decay heat removal systems (SGDHRS). No external power supply is required for operating these systems except for opening of the dampers of air heat exchangers (AHXs). There are four SGDHR circuits available, out of which three circuits are sufficient to meet all design basis events (DBEs). As a postulated design extension event, multiple failures of SGDHR circuits are considered. The detailed analysis shows that even with the double failure (i.e., only two out of four circuits are available), the hot pool temperature does not cross 923 K (650°C) and the temperature starts reducing within 1 h. It is also seen that even when the SBO is combined with MV leak, two SGDHR circuits can keep the hot pool temperatures less than 923 K (650°C). Hence, the category 4 design safety limit (DSL) is met even with failure of two SGDHR circuits during SBO combined with MV leak. Hence, a minimum of two SGDHR circuits are essential at least up to 7 h, and beyond that, one circuit is adequate. More details are presented in Ref. 20.5.

20.2.5 Investigation of Sodium Freezing

It is essential to ensure that sodium does not freeze during a prolonged SBO event, when the core has only decay heat. To avoid the risk of sodium freezing, it is necessary to have a controlled cooling of sodium after shutdown of the reactor. In the case of prolonged SBO, one of the main heat removal paths is SGDHRS. This system has to be engineered properly for controlled heat removal. Toward this, two different SGDHRS operating strategies were studied: (1) controlled manipulation of AHX dampers of all the four AHXs and (2) sequential closing of AHX location of dampers one after the other. Out of these two options, the latter strategy of sequential closing of AHX dampers (thereby reducing the rate of heat removal gradually), depicted in Figure 20.1, is found to be the best one because the number of damper operations required is the minimum. By adopting this strategy, even under complete power failure condition, with the available battery power and pneumatic power, each of the dampers can be operated two to three times and the decay heat removal operation can be continued for more than 10 days without any risk of sodium freezing. Results are highlighted in Ref. 20.6.

20.2.6 Cooling of Reactor Vault and Top Shield Cooling Systems

The temperature evolutions in reactor vault, which is cooled by dedicated water cooling system, and roof slab, which is cooled by a dedicated air cooling system, are determined in the case of loss of cooling. The scenario considered is as follows: following the SBO, the reactor is shut down and decay heat is removed by three SGDHR loops, which would come into service ½ h after the

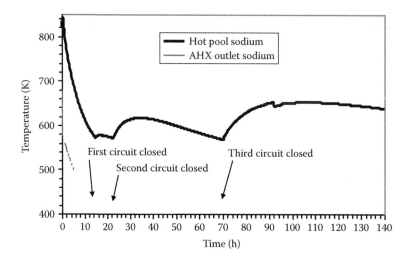

FIGURE 20.1 Safety grade decay heat removal loops operating strategy during prolonged station blackout to avoid sodium freezing.

initiation of SBO. The cooling systems for reactor vault and top shield are not available and temperature evolutions are determined [20.7]. In case of reactor vault, the temperature limits are 65°C during steady-state operation, 70°C during SGDHR operation, and 177°C during SBO condition. It is seen that the maximum temperature in the inner vault is 80°C after 10 days whereas the maximum temperature in the outer vault is 46°C. In case of prolonged SBO beyond 10 days, there is a risk of concrete temperature exceeding permissible limit. Hence, a steady-state analysis has been carried out with hot pool temperature at 200°C and 150°C. The maximum temperatures of vault are 98°C in the case of 200°C sodium temperature and 76°C in the case of 150°C sodium temperature. For the top shield, the temperature limits are as follows: temperature of the seals in the plugs ≤150°C and temperature of the shielding concrete ≤200°C (based on test data). It is seen that the temperature of the seals is the same as that of the top plate, which reaches a maximum value of 140°C at the end of 10 days. The maximum temperature attained by the shielding concrete is only 180°C [20.8].

20.2.7 Integrity of Fresh and Spent Subassembly Storage Bay

Both fresh and spent fuel storage bays and their components are designed for SSE and adequate margin exists in the design to withstand earthquakes of magnitude equal to SSE. Under extended SBO, there is no effect on fresh subassembly storage bay (FSSB), while the temperature of water rises in the case of spent fuel subassembly storage bay (SSSB). The pool water temperature reaches 65°C after 8 days/40.5 h for normal/full core unloading conditions, respectively. This is adequate to add water from makeup water tank/fire-water hydrant. The decrease in water level is less than 1 m, which meets the normal shielding requirements of ≤10 μSv/h. Apart from these, analysis has been done for verifying the minimum spacing between containers under severe earthquake beyond SSE and its impact on K_{eff} to verify margin with respect to criticality safety. Alternate bore wells for pumping water using diesel-driven pumps are planned for makeup to SSSB for long-term cooling, in case fire water is not adequate to meet the requirements. The temperature of pool water and level is monitored during extended SBO conditions. Details can be seen in Ref. 20.9.

20.2.8 Beyond Design Basis Flood Level

Taking into consideration severe cyclonic and tsunami conditions, the design basis flood levels (DBFLs) were arrived at. The finished ground level (FGL) of NICB has been decided based on the DBFL for a 1000-year return period since it houses the safety-related structures. However, the

DBFL for power island is based on a 100-year return period. During the 2004 tsunami, the water rose by a maximum of 4.714 m above the mean sea level (MSL). However, this level is lower than the DBFL for both nuclear and power islands, and hence, the site is safe in case of a future tsunami of a similar magnitude. Further, a worst tsunami postulated under a combination of high tide plus potential tsunami plus reflection wave on the sea side because of shore protection measure yields a conservative estimate of 7.33 m above the MSL. It is now considered to extend the tsunami bund along the entire stretch of the PFBR to cover the safety-related structures, for achieving higher safety margins. Details can be seen in Ref. 20.10.

20.2.9 Measures against Station Blackout by Tsunami: Japanese Approach

Readers are requested to study Ref. 20.11, which illustrates clearly the various measures taken by Japan to ensure the reactor safety against the SBO condition followed by a tsunami. Figure 20.2 illustrates such measures schematically.

20.3 IMPROVED SAFETY FEATURES FOR FUTURE SFRs [20.12]

20.3.1 Ultimate Shutdown System to Limit Core Damage

A few innovations are introduced in the shutdown systems to mitigate the extent of the core damage, in case the primary and secondary shutdown systems are not coming into action at the crucial instant when the scram is demanded. One such feature is introduction of an *ultimate shutdown system*. Such system will operate without calling for any external action, but they can operate sensing the plant parameters such as sodium flow and temperatures. The primary and secondary shutdown systems are designed to ensure that their drop time after getting scram signal is less than 1 s (typical). In case such action does not take place, the overall temperatures of fuel and coolant would rise. Depending upon the safety requirement, the ultimate shutdown system can be designed to shut down the plant to limit the core damage, that is, melting of the fuel, overheating of cladding, and sodium boiling.

In a typical ultimate shutdown system, neutron poisons such as liquid lithium/boron carbide granules are kept just above the active core by a fuse plug. On abnormal temperature rise, the fuse plug melts and the poison gets introduced in the active core zone, thus shutting down the reactor automatically (Figure 20.3). Other passive means such as (1) temperature-sensitive magnetic switch employing Curie point magnet [20.13] and (2) absorber rod–enhanced expansion device would shut down the reactor once the average sodium temperature at the core outlet exceeds a specified value [20.14]. These concepts are shown in Figure 20.4. Apart from these, a concept operating on the principle of hydraulically suspending the absorber rod could also shut the reactor automatically on loss of flow [20.15]. This concept is illustrated schematically in Figure 20.5.

20.3.2 Practical Elimination of Recriticality

The current design strategy is to rule out any possibility of energy excursions by limiting the potential sources of excessive void reactivity insertion in the initiating phase of CDA. Apart from selecting the appropriate design parameters such as maximum void reactivity coefficient, exclusion of lumped molten-fuel-pool formation by introducing an inner duct in a fuel subassembly is an innovative design (FAIDUS) conceived by JAEA for the SFR. Figure 20.6 shows an example of the early discharge of the molten fuel of about 20% during the unprotected loss of flow accident (ULOFA). In the FAIDUS, the top end of the inner duct is open, whereas the bottom end is closed and therefore it is expected that the molten fuel will be discharged from the reactor core toward the upper sodium plenum through the inner duct. The possibility of upward discharge of a high-density melt driven by coolant vapor has been confirmed experimentally by the JAEA. More details can be seen in Ref. 20.16.

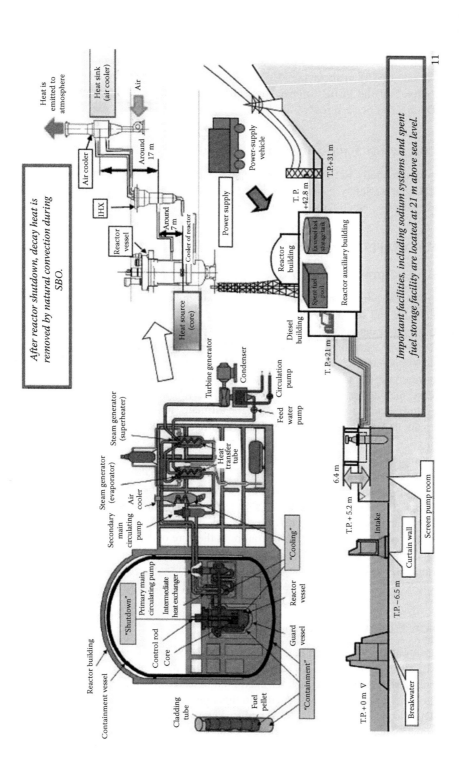

FIGURE 20.2 Measures against station blackout followed by tsunami. (From Nakai, R., Safety implication for Gen-IV SFR based on the lesson learned from the Fukushima Dai-ichi NPPs accident, Japan Atomic Energy Agency, *Proceedings of the JAEA-IAEA International Workshop on Prevention and Mitigation of Severe Accidents in Sodium-Cooled Fast Reactors*, Tsuruga, Japan, June 2012.)

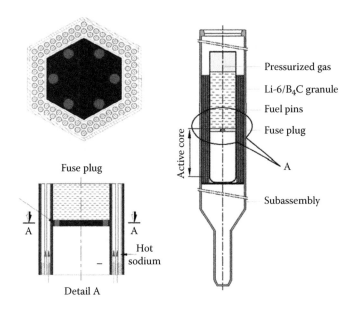

FIGURE 20.3 Ultimate shutdown system based on the flow of liquid poison or B$_4$C granules.

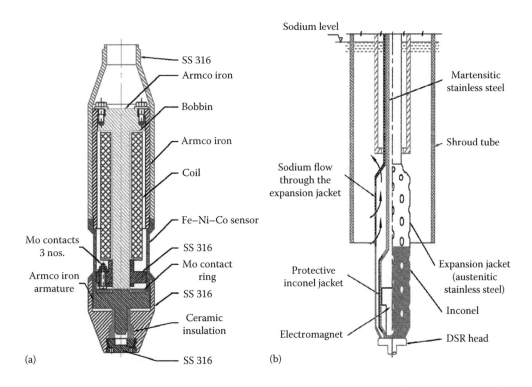

FIGURE 20.4 Ultimate shutdown system working based on temperature rise: (a) Curie point magnet. (From Bojarasky, E. et al., Inherently effective shutdown system with Curie point controlled sensor/switch unit, KfK Report 4989, May 1992.) (b) Enhanced expansion of CRDM. (From Edelmann, M., Development of passive shut-down systems for the European Fast Reactor (EFR), *Technical Committee Meeting on Absorber Materials, Control Rods and Designs of Backup Reactivity Shutdown Systems for Breakeven Cores and Burner Cores for Reducing Plutonium Stockpiles*, Obninsk, Russian Federation, July 3–7, 1995, pp. 69–79.)

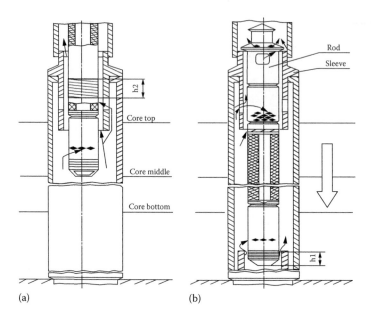

FIGURE 20.5 Hydraulically suspended control rods to prevent unprotected loss of flow accidents (ULOFA): (a) position of rods under normal condition and (b) position of rods under ULOF condition. (From Alexandrov, Yu.K. et al., Main features of the BN-800 passive shutdown rods, *Technical Committee Meeting on Absorber Materials, Control Rods and Designs of Backup Reactivity Shutdown Systems for Breakeven Cores and Burner Cores for Reducing Plutonium Stockpiles*, Obninsk, Russian Federation, July 3–7, 1995, pp. 107–112.)

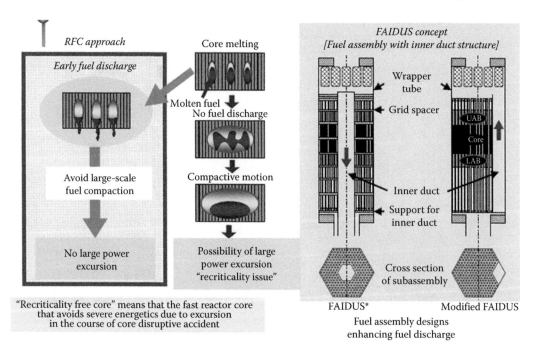

FIGURE 20.6 Fuel subassembly concept with duct structure (FAIDUS). (From Nakai, R., *Design and Assessment Approach on Advanced SFR Safety with Emphasis on CDA Issue, FR09*, ICC Kyoto, Japan, December 7–11, 2009.)

20.3.3 Maintaining Stability Core Debris on the Core Catcher under CDA

Subsequent to partial or whole core melting of core subassemblies, the molten core (debris) would be settled at the bottom of the MV. This may cause the rupture of the vessel, if it is not protected. To mitigate the damage to the main vessel in such type of events, core catcher is placed with the objective of maintaining the stable conditions within the vessel. Core catcher placed at the bottom of the MV can support the debris from whole core melting. It can maintain the fuel geometry in subcritical configuration. Hence, the core catcher does very important safety function. The robust design of core catcher involves definition of scenarios for the core meltdown relocation and settling behavior of debris within sodium. Readers can refer to Ref. 20.17 for understanding the various features that are incorporated for the effective performance of a core catcher for an innovative French SFR (Figure 20.7).

20.3.4 Postaccident Decay Heat Removal Systems

The molten fuel generated as a consequence of a CDA is relocated and settled on the core catcher by melt-through of the grid plate and core support structure. The decay heat of debris has to be removed over a long period. This requirement motivates to develop several innovative concepts for the long-term postaccident heat removal process. In the design of the core catcher conceived in the 1980s, it is postulated that the core debris generated by melting of a limited number of subassemblies only needed to be accommodated (seven subassemblies for SPX-1 and PFBR). However, in the safety criteria evolved for the innovative SFRs, the core catcher should be designed to withstand the mechanical and thermal loads corresponding to the debris generated by whole core melting. In this respect, the long-term coolability of debris is the most critical issue in SAM. Toward this, several innovative passive concepts have been proposed, particularly for the DHXs. A passive

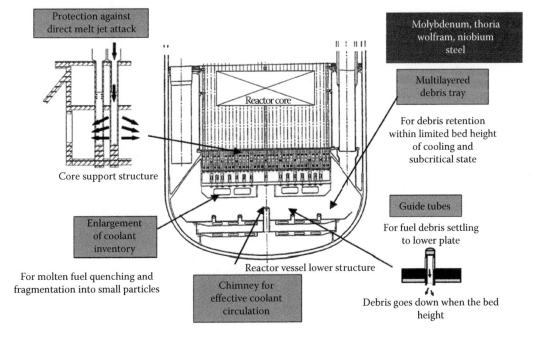

FIGURE 20.7 Various postaccident heat removal features. (From Dufour, Ph., Post accident heat removal analysis for SPX and EFR, *CEA-IGCAR Technical Seminar on Liquid Metal Fast Reactor Safety Aspects Related to Severe Accidents*, IGCAR, Kalpakkam, India, February 12–16, 2007.)

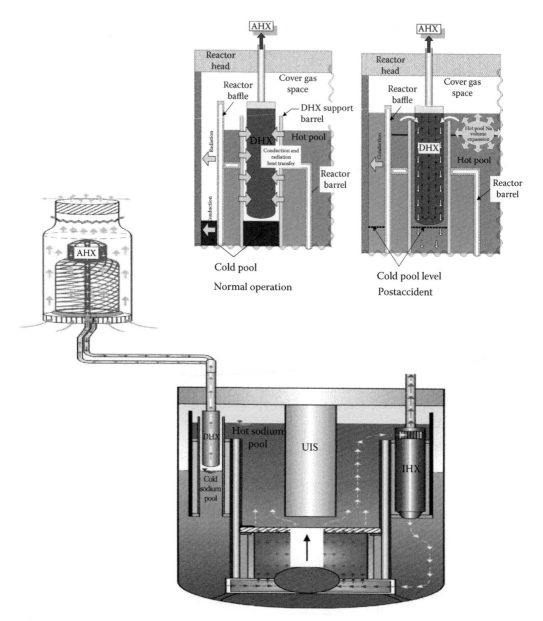

FIGURE 20.8 Passive decay heat removal concept developed by Korea. (From Yeong-il, K., Status of fast reactor technology development program in Korea, *43rd IAEA TWG-FR Meeting*, Brussels, Belgium, May 17–21, 2010.)

decay heat removal heat exchanger concept proposal by Korea is illustrated in Figure 20.8. In this concept, the DHX is penetrating the inner vessel, connecting the cold and hot pools. During normal operation, the sodium flow through this DHX is insignificant. However, when the hot pool temperature is raised monotonically, the sodium level would rise, which enhances the sodium flow through DHX during postaccident condition. The sodium after removing the heat from the core debris settled on core catcher enters into the DHX, flows down, and finally mixes with the cold pool. In this process, the heat is transported to intermediate sodium in the SGDHR circuit. More details can be seen in Ref. 20.18.

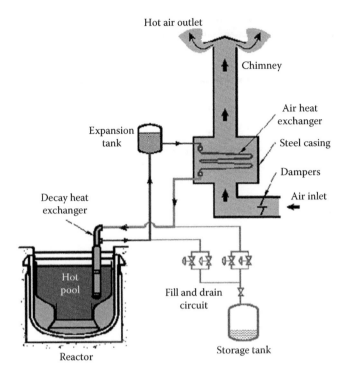

FIGURE 20.9 Decay heat removal through immersed heat exchangers.

In PFBR, the decay heat is removed through four independent SGDHR systems. Heat from sodium to AHXs placed at elevated locations will be dissipated to ambient air by natural circulation. These systems do not demand external power supply. Each loop consists of a DHX of 8 MWt capacity with tube side linked to an intermediate sodium circuit that is connected to sodium–air heat exchanger. The ultimate heat sink is air. The layout of SGDHR circuit ensures decay heat removal by natural convection in primary sodium, intermediate sodium, and air side. Two dampers of diverse design are provided at the inlet and outlet of AHX and two diverse designs of DHX and AHX are provided to enhance reliability. Figure 20.9 shows one typical SGDHR loop. In order to ensure the postaccident heat removal function, the integrity of DHX is demonstrated upon transient loadings caused by CDA. Details are presented in Chapter 21.

The sodium ejected from the MV through top shield penetrations during a CDA is the source of sodium fire. Since the SGDHR piping are penetrating through the top shield and passing through the top shield platform, the sodium fire should not cause any damage to the piping to the extent that it could affect the integrity, thereby circulation of sodium within the intermediate circuit (Figure 20.10). For PFBR, the integrity of SGDHR is ensured experimentally by carrying out a simulated sodium fire in a top shield platform mockup called SOCA. In this mockup study, a representative SGDHR pipe segment is subjected to sodium fire within a pressure vessel that simulates the top shield platform ambient conditions (Figure 20.11). More details can be found in Ref. 20.19.

For future SFRs, several innovative design concepts are proposed [20.20], for example, passive decay heat removal features such as the use of shape memory alloys for automatic opening of dampers and thermal valves which opens to enhance the flow paths during decay heat removal conditions. Placing DHXs in both hot and cold pools, decay heat removal by dedicated cooling system in SG shell (without linking the availability of steam water system), and creation of convective flow path for the sodium in the inner vessel to facilitate the heat removal, in case the temperature exceeds the limits, contribute to the postaccident heat removal function.

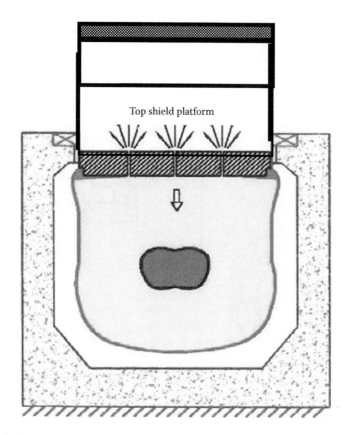

FIGURE 20.10   Sodium fire scenario within top shield platform following a core disruptive accident.

FIGURE 20.11   **(See color insert.)** Demonstration of integrity of safety grade decay heat removal pipe under simulated sodium fire. (From Chellapandi, P., Overview of Indian FBR programme, *International Workshop on Prevention and Mitigation of Severe Accidents in SFR*, Tsuruga, Japan, June 11–13, 2012.)

20.3.5 Containment Features to Withstand CDA Consequences

A specific aspect that has to be taken care in ensuring the containment integrity is the temperature and pressure rise generated in the containment subsequent to CDA due to sodium fire.

20.4 SUMMARY

Technical aspects that are associated with SAM have been explained with reference to PFBR. Sodium, safety grade decay heat circuit, shutdown system, recriticality, and core catcher bring special issues to be addressed in SFR compared to thermal reactors.

REFERENCES

20.1 Chellapandi, P., Srinivasan, G.S., Chetal, S.C. (2013). Primary containment capacity of a pool type sodium cooled fast reactor against core disruptive accident loadings. *Nucl. Eng. Des.*, 256, 178–187.

20.2 Sajish, S.D., Chellapandi, P., Chetal, S.C. (2012). Assessment of seismic capacity of 500 MWe PFBR beyond safe shutdown earthquake. *IAEA Technical Meeting on Impact of Fukushima on Current and Future FR Design*, Dresden, Germany, March 2012.

20.3 Chellapandi, P., Chetal, S.C., Raj, B. (2012). Numerical simulation of fluid-structure interaction dynamics under seismic loadings between main and safety vessels in a sodium fast reactor. *Nucl. Eng. Des.*, 253, 125–141.

20.4 Sritharan, R. (2013). Postulation of design extension condition (BDBE) of sodium leak in both main vessel and safety vessel. Internal Report: IGC/PFBR/30000/DN/1114.

20.5 Parthasarathy, U. (2013). Evolution of hot pool temperature following complete power failure event. Internal Report: IGC/PFBR/34000/DN/1140.

20.6 Natesan, K., Parthasarathy, U., Abraham, J., Velusamy, K., Selvaraj, P., Chellapandi, P. (2014). Thermal hydraulic synthesis of PFBR design in light of Fukushima accident. *New Horizons in Nuclear Reactor Thermal Hydraulics and Safety*, Mumbai, India, January 14–16, 2014, pp. 1–6.

20.7 Abraham, J. (2013). Temperature evolution in reactor vault during extended station black out. Internal Report: IGC/PFBR/21100/DN/1034.

20.8 Abraham, J. (2013). Temperature evolution in roof slab during extended station black out. Internal Report: IGC/PFBR/31310/DN/1061.

20.9 Rajan, S. (2013). Impact of Fukushima type incidents on fresh and spent fuel storage. Internal Report: IGC/PFBR/35000/DN/1037.

20.10 Satish Kumar, L. (2013). Effect of flooding on essential systems in Power Island. Internal Report: IGC/PFBR/70000/DN/1007.

20.11 Nakai, R. (2012). Safety implication for Gen-IV SFR based on the lesson learned from the Fukushima Dai-ichi NPPs accident, Japan Atomic Energy Agency. *Proceedings of the JAEA-IAEA International Workshop on Prevention and Mitigation of Severe Accidents in Sodium-Cooled Fast Reactors*, Tsuruga, Japan, June 2012.

20.12 Chellapandi, P. (2014). Thermal hydraulic synthesis of PFBR design in light of Fukushima accident. *New Horizons in Nuclear Reactor Thermal Hydraulics and Safety*, Mumbai, India, January 14–16, 2014, pp. 1–6.

20.13 Bojarasky, E., Muller, K., Reiser, H. (1992). Inherently effective shutdown system with Curie point controlled sensor/switch unit. KfK Report 4989, May 1992.

20.14 Edelmann, M. (1995). Development of passive shut-down systems for the European Fast Reactor (EFR). *Technical Committee Meeting on Absorber Materials, Control Rods and Designs of Backup Reactivity Shutdown Systems for Breakeven Cores and Burner Cores for Reducing Plutonium Stockpiles*, Obninsk, Russian Federation, July 3–7, 1995, pp. 69–79.

20.15 Alexandrov, Yu.K. et al. (1995). Main features of the BN-800 passive shutdown rods. *Technical Committee Meeting on Absorber Materials, Control Rods and Designs of Backup Reactivity Shutdown Systems for Breakeven Cores and Burner Cores for Reducing Plutonium Stockpiles*, Obninsk, Russian Federation, July 3–7, 1995, pp. 107–112.

20.16 Matsuba, K.-i. et al. (2013). Mechanism of upward fuel discharge during core disruptive accidents in sodium-cooled fast reactors. *J. Eng. Gas Turbines Power, Trans. ASME*, 135, 032901 (1–9), March 2013.

20.17 Dufour, Ph. (2007). Post accident heat removal analysis for SPX and EFR. *CEA-IGCAR Technical Seminar on Liquid Metal Fast Reactor Safety Aspects Related to Severe Accidents*, IGCAR, Kalpakkam, India, February 12–16, 2007.

20.18 Yeong-il, K. (2010). Status of fast reactor technology development program in Korea. *43rd IAEA TWG-FR Meeting*, Brussels, Belgium, May 17–21, 2010.

20.19 Chellapandi, P. (2012). Overview of Indian FBR programme. *International Workshop on Prevention and Mitigation of Severe Accidents in SFR*, Tsuruga, Japan, June 11–13, 2012.

20.20 Chellapandi, P. (2007). Indian perspective on CDA scenario for future FBRs. *Technical Seminar on LMFR Safety Aspects Related to Severe Accidents*, IGCAR, Kalpakkam, India, February 12–16, 2007.

20.21 Nakai, R. (2009). *Design and Assessment Approach on Advanced SFR Safety with Emphasis on CDA Issue, FR09*, ICC Kyoto, Japan, December 7–11, 2009.

21 Safety Analysis of PFBR
A Case Study

21.1 INTRODUCTION

Prototype fast breeder reactor (PFBR) is a 500 MWe capacity pool-type reactor with two primary loops, two secondary loops, and four steam generators (SGs) per loop. The overall flow diagram comprising primary circuit housed in the reactor assembly, secondary sodium circuit, and balance of plant is shown in Figure 21.1. The nuclear heat generated in the core is removed by circulating sodium from the cold pool at 670 K to the hot pool at 820 K. The sodium from hot pool after transporting its heat to the four intermediate heat exchangers (IHXs) mixes with the cold pool. The circulation of sodium from the cold pool to the hot pool is maintained by two primary sodium pumps (PSPs) and the flow of sodium through IHX is driven by a level difference (1.5 m of sodium) between the hot and cold pools. The heat from IHX is in turn transported to eight SGs by sodium flowing in the secondary circuit. Steam produced in SG is supplied to turbo generator. In the reactor assembly (Figure 21.2), the main vessel houses the entire primary sodium circuit including the core. Sodium is filled in the main vessel with free surfaces blanketed by argon. The inner vessel separates the hot and cold sodium pools. The reactor core consists of about 1757 subassemblies (SAs) including 181 fuel SAs. The control plug, positioned just above the core, houses mainly 12 absorber rod drive mechanisms. The top shield supports the PSPs, IHX, control plug, and fuel handling systems. PFBR uses mixed oxide with depleted uranium and about 25% Pu oxide as fuel. For the core components, 20% cold worked D9 material (15% Cr–15% Ni with Mo and Ti) is used to have better irradiation resistance. Austenitic stainless steel type 316 LN is the main structural material for the out-of-core components.

In this chapter, the safety features engineered in PFBR are presented first. Subsequently, severe accident analysis is presented, which covers broadly four aspects: (1) reactor physics analysis to estimate thermal and mechanical energy releases, (2) assessment of mechanical consequences (structural deformations, sodium release to reactor containment building [RCB], and temperature and pressure rise in RCB), (3) postaccident heat removal (PAHR) aspects, and (4) radioactivity release to site boundary. Readers are requested to read the event analysis results of PFBR highlighted in Chapter 14 for getting a comprehensive picture on the safety analysis of a sodium-cooled fast reactor (SFR), through the case study presented on PFBR.

21.2 SAFETY FEATURES INCORPORATED IN PFBR

PFBR has several safety features. These are presented in detail in Ref. 21.1. In this section, only the essential aspects and results relevant to severe accident scenarios are highlighted.

21.2.1 Negative Reactivity Coefficients

The temperature and power coefficient of reactivity are negative so that any off-normal increase in temperature or power leads to a reduction in reactivity and the consequent reduction in power. The expansion of coolant and structural steel results in small positive reactivity (+0.241 pcm/K). This is compensated by negative and prompt (time constant < 1 ms) reactivity effects like Doppler (−1.320 pcm/K) and fuel expansion (−0.236 pcm/K), which are slow (time constant ~50–100 s).

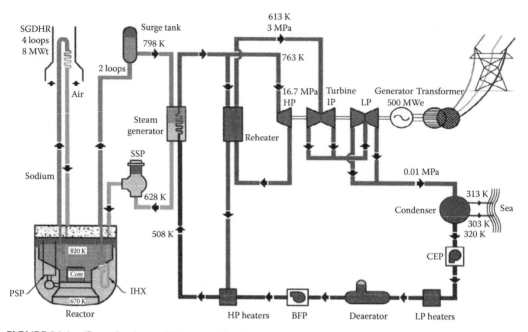

FIGURE 21.1 **(See color insert.)** Prototype fast breeder reactor heat transport circuit flow sheet.

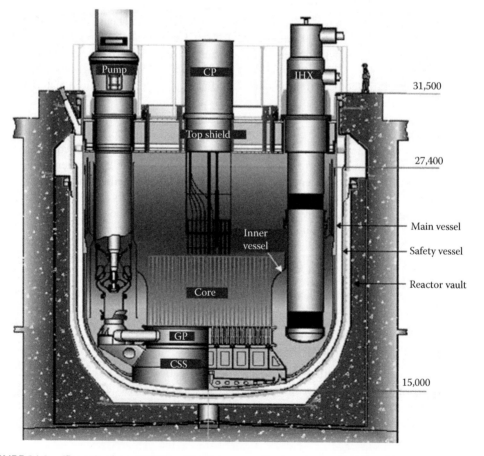

FIGURE 21.2 **(See color insert.)** Schematic of prototype fast breeder reactor assembly.

The reactivity feedback resulting from grid plate expansion, spacer pad expansion of core subassemblies (CSAs) (−0.869 pcm/K), and differential control rod expansion (−1 pcm/K) tend to shut down the reactor for transients under loss of cooling incidents.

21.2.2 CORE MONITORING

The core is monitored functionally by diverse sensors. Neutron detectors are provided to monitor the power and provide signals for safety action (SCRAM) on parameters like linear power, period, and reactivity. These parameters provide protection against transient over power, transient under cooling, and anomalous reactivity addition events. Monitoring of sodium temperature is done by providing three fast response thermocouples mounted on the central canal plug to monitor central fuel SA sodium outlet temperature (θ_{CSAM}) and used on 2/3 voting mode, which is a hardwired system. Two thermocouples each within a single thermowell are provided over the other fuel SAs to monitor the sodium outlet temperature from individual SAs (θ_i). Three thermocouples within one thermowell are provided in each of the two PSP suction sides to monitor the reactor inlet temperature (θ_{RI}). Signals are processed with hardwired electronics in 2/3 voting mode. Online computed parameters like the mean fuel SA sodium outlet temperature (θ_m), mean core sodium temperature rise ($\Delta\theta_m$), and deviation of individual SA sodium outlet temperature from an expected value ($\delta\theta_i$) are obtained from the measurements. The required computations are done using computers meant for class 1 applications in 2/3 voting mode. The central SA and coolant inlet thermocouple signals are processed through triplicated hardwired electronic circuits. This measurement is also used for deriving temperature rise across central SA parameter ($\Delta\theta_{CSAM}$).

SCRAM on various parameters takes care of core safety: θ_{CSAM} for the clad and coolant temperature limits, θ_{RI} for disturbances in secondary sodium and steam-water circuits affecting the reactor core, $\Delta\theta_{CSAM}$ for global undercooling events in the core, and $\delta\theta_i$ for local faults in SAs. Flow delivered by the PSP is measured and power to flow ratio is also monitored. Pump discharge head and speed are measured and used as trip parameters for protection against pump discharge pipe rupture and pump seizure events, respectively. Failure of fuel is detected by monitoring the cover gas fission product activity (alarm parameter) and delayed neutron detection (DND) in the primary coolant. These provisions ensure the availability of at least two diverse safety parameters as far as possible for each design basis event (DBE).

The clad rupture and consequent ejection of radioactive material from the failed fuel pin are detected by cover gas monitoring systems and delayed neutron detectors (DNDs) positioned at eight locations in the hot pool. The layout of the detectors is schematically shown in Figure 21.3. The number of DND detectors and their locations were optimized based on detailed 3D transient computational fluid dynamics simulations considering mixing and decay of delayed neutron precursors during their transport from failed SA to detector location. Based on this, the locations of eight numbers of DND, one on either side of the IHX, have been finalized, by which pin failure in any fuel SA can be detected within 1 min at any power level. Furthermore, in order to maintain a good sensitivity of detection of clad failures, operation of the reactor with direct contact of fuel and sodium is avoided by removing fuel elements with clad ruptures. This is accomplished after the localization by individual sodium sampling at the outlet of each SA (Figure 21.4). This system is called failed fuel localization module (FFLM), and there are three such modules placed in the control plug.

21.2.3 MEASURES TO PREVENT SODIUM VOIDING

Various causes of voiding are identified and ensured that they do not lead to any transient. To vent the gas entrapped in grid plate, purger SAs are provided in the design. Blockage of sodium flow to fuel SA or global flow reduction is prevented by provision of multiple radial entry holes for the coolant in the grid plate sleeve and SA foot. At the outlet of the fuel and breeder SAs, an adapter has been incorporated to provide an alternate passage for the coolant in the event of blockage at

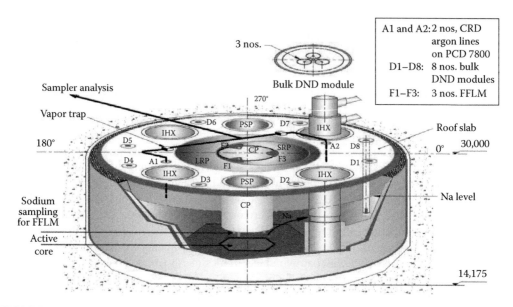

FIGURE 21.3 Clad rupture detection systems.

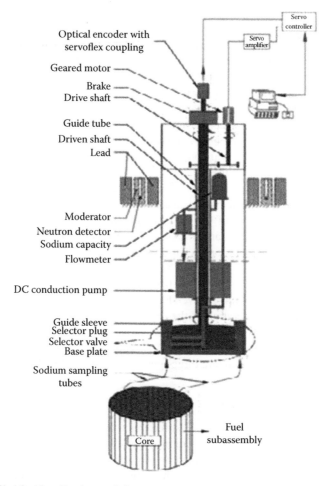

FIGURE 21.4 Failed fuel localization module.

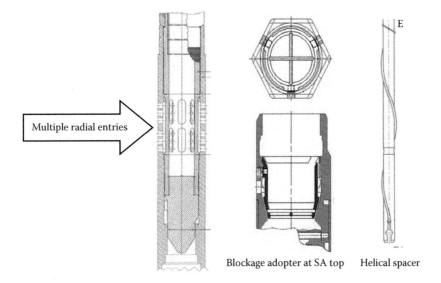

Multiple radial entries

Blockage adopter at SA top   Helical spacer

FIGURE 21.5   Design provisions at the top and bottom to prevent subassembly blockages.

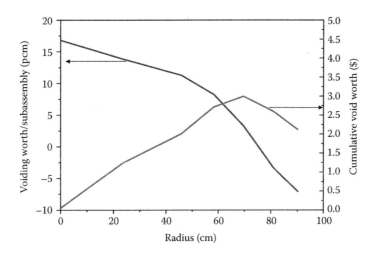

FIGURE 21.6   Sodium void reactivity effect versus core radius.

the top (Figure 21.5). High inertial flywheel (8 s flow halving time) ensures slow flow reduction for loss of power supply to the PSP. Moreover, multiple fault detection provisions like SA thermocouples and anomalous reactivity detection by reactivity meter detect sodium boiling or gas passing through the core. The whole core sodium void coefficient is 2\$ as seen in Figure 21.6. The maximum value can be 3\$ in case of central voiding, which is lesser as compared to other international values (6\$ for Japanese DFBR, >5\$ for the French SPX-1). Super prompt critical excursion experiments conducted on SFR have clearly established that such excursions are very well arrested by Doppler reactivity feedback (Section 13.2).

21.2.4   SHUTDOWN SYSTEMS

On detection of any abnormality in the reactor, shutdown is assured by two independent, fast-acting shutdown systems (SDSs). Each system consists of sensors, analog processing circuits, logics, absorber rods, and associated drive mechanisms. Reactor cold shutdown is accomplished independently by both the systems by free fall of the B_4C absorber rods even when one rod remains stuck. The response time of SDS to initiate rod drop is <200 ms and the free fall drop time of the rod is <1 s, which is sufficient to protect an incident up to 3$/s. Sufficient independence and diversity are provided in the design of sensors, analog signal processing circuits, SCRAM circuits, SCRAM logic, SCRAM switch, absorber rods, and mechanisms of the two systems. With this, the failure frequency of SDS is found to be 6.4×10^{-7}/reactor-year, which is less than the specified limit (<10^{-6}/reactor-year). The failure frequency of individual systems is 8×10^{-4} and 4.4×10^{-4}/reactor-year, which is less than the specified limit (<10^{-3}/reactor-year). In the reliability analysis, common cause failures between redundant nondiverse components/systems are accounted appropriately.

21.2.5 DECAY HEAT REMOVAL SYSTEMS

Decay heat is about 1.5% and 0.7% of nominal power 1 h and 1 day, respectively, after reactor shutdown. If off-site power is available, decay heat removal is through the normal heat transport system, that is, through SGs and steam-water system. The system is known as operation grade decay heat removal system. In case of loss of off-site power and loss of secondary circuit or steam-water circuits, the decay heat is removed through four independent safety grade decay heat removal (SGDHR) loops (Figure 21.7). The SGDHR operation is automatic. Each loop consists of a decay heat exchanger (DHX) with 8 MWt capacity with tube side linked to an intermediate sodium circuit that is connected to sodium–air heat exchanger (AHX). The ultimate heat sink is air. The layout of SGDHR circuit ensures decay heat removal by natural convection in primary sodium, intermediate sodium, and air side. Two dampers of diverse design are provided at the inlet and outlet of AHX and two diverse designs of DHX and AHX are provided to enhance reliability. Diesel and battery power is also provided to drive the primary pumps at 15% of the speed for conditions of off-site power failure and station blackout conditions as a defense in depth approach. Reliability analysis is carried out by fault tree method including common cause failure between redundant nondiverse components/systems. Passive system failures in the form of sodium leak at loop boundaries,

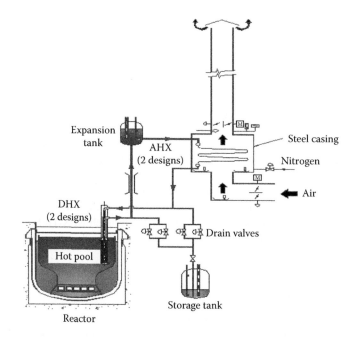

FIGURE 21.7 Safety grade decay heat removal system.

flow blockage, and freezing are considered in the analysis. The failure frequency of SGDHR function is found to be 1.5×10^{-7}/reactor-year, which practically meets the specified prescribed limit of 1×10^{-7}/reactor-year, considering the conservatism in the analysis.

21.2.6 REACTOR CONTAINMENT BUILDING

The RCB is a rectangular building of 35×40 m size and 54.5 m height above finished floor level, surrounding the primary circuit and top shield, which is provided as an ultimate safety against the radioactive fission products released during the postulated accident to ensure that the radiological dose limits for members of the public are not exceeded. Single containment, nonvented type, and reinforced cement concrete construction are the principal design features. Its safety classification is 2 and seismic categorization is 1. Toward the design of the RCB, the temperature and pressure rise in the RCB were estimated based on burning of 350 kg of sodium ejected through the top shield penetrations.

21.2.7 CORE CATCHER

Under the assumption of total instantaneous blockage (TIB), a single SA melting progresses at the most to the neighboring six SAs. However, if the hot fuel settles at the bottom, the main vessel bottom portion may melt and give way to the sodium and the molten fuel. Also, there is a concern for recriticality if the fuel settles down in a more reactive configuration. Therefore, a core catcher is provided below the core support structure (Figure 21.8) and is supported on the main vessel, covering the central area of the active core. This would collect the molten fuel and suitably disperse it and ensure long-term cooling. The amount of molten fuel released in the melting of seven SAs is only ~0.3 ton. On the other hand, calculations indicate that a minimum of 1 ton of fuel is required for recriticality. Therefore, the slumped molten fuel of seven SAs does not have any recriticality potential, and provision of core catcher is only to maintain coolable geometry. Even in the unlikely event of melting progressing to 2 rings, having 19 SAs, the molten fuel released is 0.8 ton, which is less than the minimum critical mass of 1 ton. For a radius of 3.2 m of fuel spread in the core catcher, the critical mass required is 80 tons, which is well below the total mass of the fuel in 181 SAs, which is 9.2 tons. Calculations show that the maximum temperature of sodium is 1014 K and the temperature of core catcher bottom plate is <923 K. It can withstand a temperature of 1173 K for 308 days, even with the whole core load. More details are presented on this aspect in Section 21.8.

FIGURE 21.8   Schematic sketch of core catcher of prototype fast breeder reactor.

21.3 SEVERE ACCIDENT ANALYSIS

PFBR has many inherent and engineered safety features that the probability of occurrence of a severe accident involving melting of the whole core, that is, core disruptive accident (CDA), is very low (<10^{-6}/year) that it is categorized under beyond design basis event (BDBE). The accident scenarios and mechanical and thermal consequences are described in the following sections.

21.3.1 ACCIDENT SCENARIOS AND ENERGY RELEASE

The thermal energy release depends upon the reactivity addition rate in the disassembly phase, which in turn depends upon the assumptions made on the sodium void propagation, fuel displacement/slumping characteristics, and reactivity feedback mechanisms. Apart from this, the cross-section data, the nature of temperature distributions assumed for the disrupted core, and the cross-section data employed in the analysis decide the work potential value. One of the important parameters influencing the coolant void generation/propagation is flow halving time. With lower flow halving time, the coolant voids could generate below the core top and spread rapidly to the core center, resulting in high positive reactivity rate in the disassembly phase. With higher flow halving time, the coolant boiling starts at the upper portion of the active core, which introduces negative reactivity due to the high neutron leakage. Analysis with pessimistic assumptions: shorter low halving time of 2 s, coherent core lumping, absence of feedbacks, flat temperature distribution across the core at the end of the disassembly phase, and use of conservative cross-section data (CV2M cross-section set), yields a pessimistic reactivity addition rate of 200$/s, and the associated work potential is ~1000 MJ. Analysis with optimistic assumptions: longer flow halving time of 8 s, incoherent core, presence of all feedbacks, realistic temperature distribution across the core, and use of realistic cross-section data (ABBN cross-section set), yields an optimistic reactivity addition rate of 10.5$/s, and the associated work potential is insignificant (<1 MJ). It is also found that the assumptions made on the fuel dispersion behavior have significant influence on the reactivity addition rate. If a conservative fuel slumping model is employed without considering molten fuel dispersion (incoherent core), higher energy release is possible. As per this, the active core zone is divided into three zones axially. The molten fuel from the middle one-third occupies the core lower portion and the fuel from the top one-third occupies the middle portion. This leads to a reactivity addition rate of 65$/s and a work potential of 100 MJ. In addition, the temperature distribution of the core at the end of the disassembly phase can change the work potential value significantly: assumption of a flat temperature distribution can yield the work potential of 268 MJ for the reactivity addition rate of 50$/s, compared to 100 MJ for the reactivity addition of 65$/s with the realistic temperature distributions.

The synthesis of these results motivates to investigate the mechanical consequences of a CDA over a wide range of work potentials corresponding to the reactivity addition rates ranging from 25$/s to 200$/s (Figure 21.9).

21.3.2 MECHANICAL CONSEQUENCES OF CDA

The mechanical consequences are due to rapid expansion of core bubble, releasing high pressure waves. The pressure waves impose direct loading on the surrounding structures, such as main vessel and its internals, causing large deformations. Besides, under the bubble pressure, the sodium slug above the core bubble gets accelerated upward. Once the accelerated sodium impacts at the bottom of the top shield, a high local pressure is developed. Under this pressure, the upper portion of the main vessel gets bulged; a portion of the accelerated sodium occupies the available penetrations in the top shield and the cover gas gets compressed at the peripheral region in the vicinity of the top shield bottom. At this state, the core bubble attains a shape with the maximum volume (V_{max}) condition, releasing the maximum mechanical energy, which is the expansion work ($\int P \cdot dV$) integrated over the initial volume to V_{max} with associated pressure ($P_{quasistatic}$) and the vessel has attained the

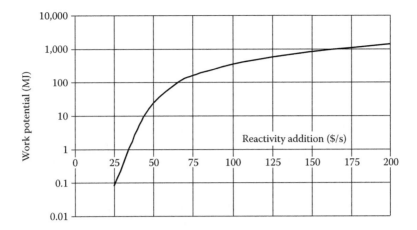

FIGURE 21.9 Pessimistic energy release under core disruptive accident in prototype fast breeder reactor.

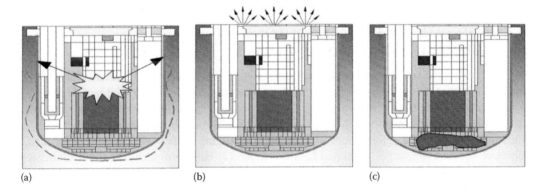

FIGURE 21.10 Important consequences of a core disruptive accident in sodium-cooled fast reactor. (a) Explosion of vaporized fuel coolant mixture, (b) ejection of sodium through top shield penetrations, and (c) settlement of core debris on the core catcher.

maximum deformed configuration (Figure 21.10a). The time duration involved to attain this state is generally <1 s. (It is also customary to define a term called *work potential* for assessing the maximum possible mechanical energy that could be released by the core bubble by assuming that the core bubble expands till its pressure is equal to the ambient pressure [P_{amb}]. P_{amb} is generally taken as a normal atmospheric pressure of one bar. Thus, the work potential quantifies the severity of a CDA.) Mechanical consequences further continue with the ejection of sodium present in the gaps out of the top shield under *pseudo-static pressure* of the core (its initial value is $P_{quasistatic}$). This pressure drops rapidly when the core bubble gets cooled by the liquid sodium itself. The event of sodium release continues till $P_{quasistatic}$ reduces to P_{amb}, which could last for 1–2 s typically (Figure 21.10b). Afterward, the sodium remaining in the top shield penetrations falls back into sodium pool by gravity. The condensation of core bubble reduces the pressure inside the vessel, promoting favorable conditions for draining of sodium in the penetrations. The ejected sodium immediately burns on the top surface of the top shield increasing temperature and pressure, which are the design basis loads for the RCB.

Subsequent to the mechanical consequences, the focus is on the decay heat removal aspects of the core, particularly the relocated molten fuel and structural materials settled on the core catcher in the form of debris after penetrating through the structures below the core (Figure 21.10c). During this event, the decay heat generated by the core debris is removed continuously till it becomes insignificant. There are many means to achieve long-term coolability of core debris, for example, incorporation of adequate number of dedicated safety grade decay heat removal heat exchangers (SGDHXs), permanently immersed in the sodium pool within the main vessel. In view of their important role, the functionality of decay heat removal mechanisms should be ensured for the long-term coolability of the core debris, for which their structural integrity has also to be assessed during the investigation of the mechanical consequences of CDA.

Based on extensive numerical and experimental simulations, the structural deformations of the main vessel along with the top shield and its internals, structural integrity of IHX and SGDHX, sodium release to the RCB, and temperature and pressure rise in the RCB consequent to sodium fire above the top shield are assessed with comfortable margins for 100 MJ of work potential. Internationally, the severe accident analysis methodologies of SFR have been critically relooked in recent years and several scenarios have been postulated. Development of advanced numerical and experimental techniques is carried out with the objective of determining realistic energy release, which is still evolving. In view of this, a parametric study on mechanical energy release values beyond 100 MJ, that is, in the range 100–1000 MJ, has been undertaken for the PFBR to assess the worst possible effects with increase in energy values. The investigation focuses on issues related to structural integrity of primary containment including RCB and postaccident cooling capability. The highlights of results are presented in the following sections. More details can be found in Refs. 21.2–21.4.

21.4 ASSESSMENT OF PRIMARY CONTAINMENT POTENTIAL: HIGHLIGHTS OF ANALYSIS

The maximum mechanical energy that can be absorbed by the main vessel is determined by the idealized model of the main vessel without any internals. Rapid expansion of core bubble confined by the liquid sodium, housed in thin vessel, having cover gas space above free level depicts complicated deformation mechanics with the fast transient fluid–structure interaction effects. There are three main phases of the deformations in sequence: (1) bulging of the vessel bottom under direct impact of the pressure waves generated by the bubble, (2) bulk movements of the sodium slug toward the top shield compressing the cover gas bulk wherein vessel deformations are insignificant, and (3) radial local bulging of the upper portion of the vessel due to sodium slug impact at the bottom of the top shield. Mechanical consequences resulting in sodium ejection to the RCB are schematically shown in Figure 21.11. The duration of each of these phases and quantum of deformation depends strongly on the work potential of the bubble.

For the analysis of mechanical consequences, a dedicated computer code called FUSTIN, which simulates several complex phenomena involved in determining the transient pressures, vessel displacements, and strains, was developed and validated thoroughly by solving relevant benchmark problems. The details of the FUSTIN code including validations were already presented in Chapter 17. Some important results obtained through the application of this code for the PFBR are presented in the following section.

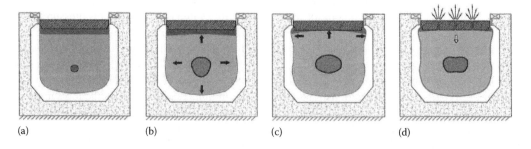

(a)   (b)   (c)   (d)

FIGURE 21.11 Mechanical consequences resulting in sodium ejection to reactor containment building under a core disruptive accident: (a) initial state—0 ms; (b) vessel pulldown—0–50 ms; (c) slug impact—100–150 ms; and (d) final state—150–900 ms.

21.4.1 IDEALIZED GEOMETRY AND LOADING DETAILS

The analysis geometry consists of main vessel, core support structure, grid plate, control plug, top shield, and its support skirt. The vessel contains sodium with argon cover gas gap of 800 mm. The masses of core, grid plate, and core support structure are distributed appropriately. The masses of thermal baffles are lumped at a junction point on the main vessel. The top shield is considered to be rigid. The argon space is treated as a single homogeneous medium obeying the polytrophic equation of state, defined as $PV\gamma = \text{constant}$. The structural material properties corresponding to the temperatures of the structures at pre-disassembly phase are considered. The bottom portion of main vessel, core support structure, grid plate, and bottom portion of the inner vessel are at 685 K. The upper portions of the main vessel and inner vessel and control plug are at high temperature (855 K). Thus, the temperatures of various portions of structures lie in the range 685–855 K. Accordingly, the true stress–true strain curves for SS 316 LN, the structural material for the reactor assembly components, as recommended in RCC-MR [21.5] at metal average temperatures 685, 773, and 855 K, are used. Readers are requested to read Ref. 21.6 for more details.

The analysis of the main vessel with only distributed/added masses, conserving the net mass of reactor assembly, provides conservative results, which is demonstrated through analysis, carried out with increasing complexities, and reported in Ref. 21.7. In this reference, basically two analyses are presented: in the first analysis, only the main vessel is considered without any internals and lumped masses, thereby the main vessel has to absorb the mechanical energy released by the core and would be subjected to maximum possible strain, and in the second analysis, the essential internals and lumped masses are considered, which is expected to provide realistic predictions. In view of getting conservative results possibly with minimum computational time, analysis is carried out for the case of main vessel without any internals. The finite element mesh of this geometry is shown along with the idealized geometry employed for the present study in Figure 21.12. The numerical results are presented in the following sections.

21.4.2 SEQUENCE OF MECHANICAL LOADINGS AND ENERGY ABSORPTIONS

Figure 21.13 depicts the sequence of energy absorbed by the vessel while core has various work potentials. A fraction of work potential absorbed by the upper portion of the vessel is compared with the fraction absorbed by the bottom portion in Figure 21.14. While at lower work potentials the upper portion of vessel as well as cover gas compression absorbs higher fractions of work potential

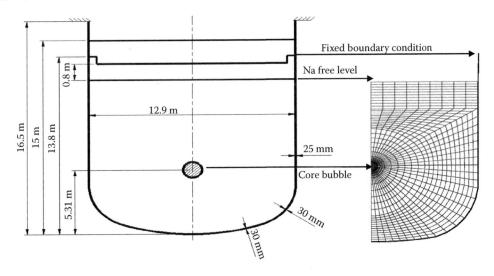

FIGURE 21.12 Geometrical idealization of main vessel without internals and finite element mesh.

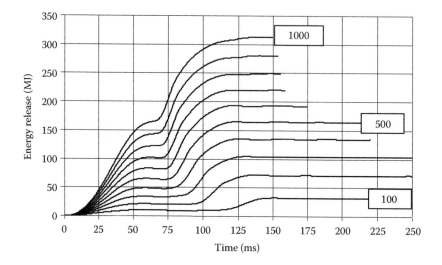

FIGURE 21.13   Strain energy absorbed by the vessel.

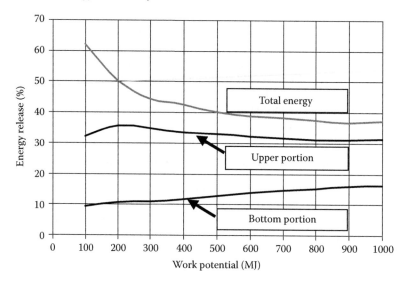

FIGURE 21.14   Energy balance in the vessel.

compared to fraction absorbed by the bottom portion, trend reverses at higher work potentials. Further, it is seen in Figure 21.14 that the net fraction of energy released by the core bubble is decreasing, while the work potential of core bubble increases and tends to stabilize (~36%). This implies that the impact effects, which cause local deformations, get saturated at high mechanical energy release and the vessel absorbs the energy uniformly, enhancing its energy absorbing potential.

21.4.3   MAIN VESSEL DEFORMATIONS

Figure 21.15 shows the radial deformation profiles along the developed length of the vessel. The absolute values of the downward displacement, radial bulging at the middle portion at the elevation of the core center, and radial bulging just below the top shield junction are quantified in this figure. The values indicate that the main vessel would have mechanical interactions with the safety vessel both at the bottom and at the top, depending upon the intervessel space. The safety vessel could contribute in load sharing subsequent to such interactions, which is not simulated in the analysis to

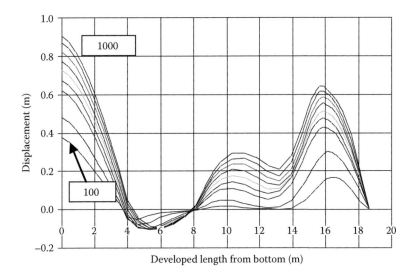

FIGURE 21.15 Radial displacement in the vessel.

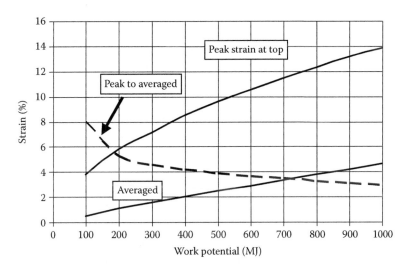

FIGURE 21.16 Membrane strains in the vessel.

preserve conservatism. The local strains at the upper portion as well as averaged strains in the vessel are presented in Figure 21.16. The ratio of peak strain to the averaged strain plotted at various work potentials shown in this figure confirms the conclusion derived from the previous subsection that the deformation becomes more uniform compared to lower work potential cases. This is a favorable feature that the energy absorbing potential is not linear and the vessel can absorb higher energy without undergoing rupture locally upon application of higher energy by the core bubble. It is also shown that the peak strain in the vessel for the work potential of 1000 MJ is 14%. From the structural integrity consideration, this strain value is acceptable for the main vessel [21.8].

21.4.4 SLUG IMPACT LOADINGS AND THEIR EFFECTS

The evolution of upward velocity values during sodium slug impact phenomenon is shown in Figure 21.17 for four specific work potentials (100, 200, 500, and 1000 MJ). From this figure, it is brought out that the loading on the top shield is gradual at low work potentials typically at

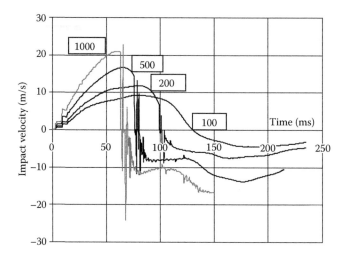

FIGURE 21.17   Sodium slug impact.

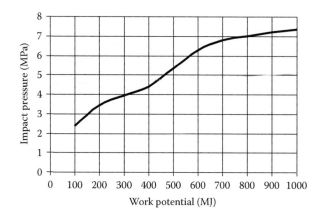

FIGURE 21.18   Peak impact pressure on top shield.

100 and 200 MJ. Figure 21.18 shows the impact pressure experienced by the top shield, which shows that at higher work potentials, the top shield is subjected to high impact pressure. The time to initiate sodium slug impact is shorter for higher work potential, which also tends to stabilize in Figure 21.18. In view of short duration of impact loadings and high mass inertia of top shield structures, it has high potential to absorb higher impact loads, and hence, the integrity of the top shield would not be of concern and do not decide the acceptable work potential.

21.5   SODIUM LEAK THROUGH TOP SHIELD AND CONTAINMENT DESIGN PRESSURE

The peak pressure developed during the sodium slug impact phenomenon causes elongation of the hold down bolts and failure of seals in the top shield, providing leak gaps for the sodium occupied in the penetration. The quasi-static pressure sustaining in the impacted sodium drives this sodium into the RCB, which occurs as far as the quasi-static pressure is higher than the ambient pressure above the top shield. To understand further, the status of core bubble at the time of impact is depicted for four work potentials (100, 200, 500, and 1000 MJ) in Figure 21.19. The status of core bubble as well as cover gas space during quasi-static phase is also depicted in Figure 21.20. From the analysis,

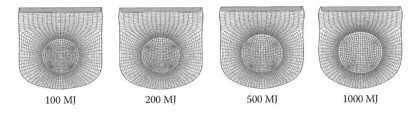

100 MJ   200 MJ   500 MJ   1000 MJ

FIGURE 21.19 Status of core bubble and cover gas space marching toward slug impact.

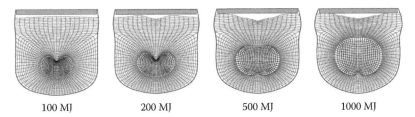

100 MJ   200 MJ   500 MJ   1000 MJ

FIGURE 21.20 Status of core bubble and cover gas space marching toward quasi-static condition.

the pressure of impacted sodium that has a tendency to eject through top shield penetrations is computed. This pressure value is in quasi-static equilibrium with the core bubble pressure, which itself would decay continuously (this pressure is termed as *quasistatic pressure [Po]*).

Starting from the initial quasi-static pressure, the decay of bubble pressure while getting cooled by the surrounding colder sodium and the resulting sodium leak rate through all the top shield penetrations are evaluated for the work potential of 100 MJ, the design basis for PFBR [21.9]. Toward this, a separate structural analysis of top shield and its components is performed, which indicates that the hold down bolts of components, such as rotatable plugs, control plug, IHX, PSP, DHX, etc., undergo plastic elongation by 0.5–1 mm. With these inputs along with Po (the starting value of quasi-static pressure), Vo (the initial volume of the core bubble [81 m³]), time constant τ (0.8 s) [21.10], entry loss coefficient (0.5), exit loss coefficient (1.0), and the 90° sharp bend loss coefficient (1.0), the sodium leak rate versus time is estimated. The sodium leak rates through various leak paths available in the top shield are shown in Figure 21.21. It is observed that the total sodium leakage is ~350 kg [21.11].

21.6 TEMPERATURE AND PRESSURE RISE IN RCB

Sodium release of 350 kg, the theoretically computed upper bound value, is taken as input for the estimation of temperature and pressure rise in RCB under CDA. Even though sodium ejection through the penetrations is a complex phenomenon, a simplified assumption is made for this part of the calculation, wherein the entire sodium is assumed to get ejected out in a horizontal direction (due to geometrical features of the penetrations) and get collected as pool over the top shield and burn. The event is analyzed as a pool fire using SOFIRE II code [21.12]. As 100% sodium monoxide, as a reaction product in diffusion-controlled sodium fires, can burn more sodium and give higher thermal consequences than 100% peroxide or any other ratios of oxides as reaction products, the same is considered for the analysis. The evolution of temperature and pressure rise is shown in Figures 21.22 and 21.23, respectively. The peak gas temperature in RCB is estimated to be about 331 K and the peak pressure rise is about 9 kPa. Based on the information available on the prediction capability of SOFIRE II code, a factor of 1.3 is applied on the pressure rise in RCB. Accordingly, the pressure rise of 11.7 kPa is considered as the maximum possible pressure rise in RCB due to a complete burning of 350 kg of sodium [21.11].

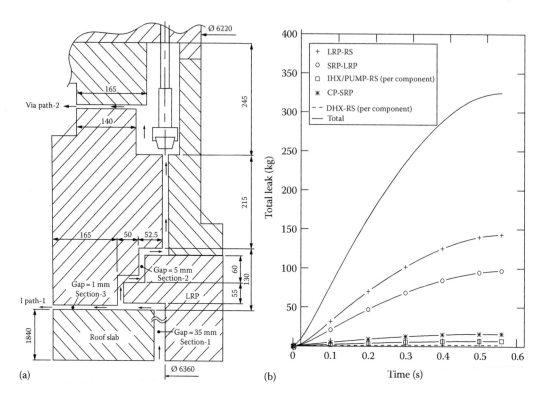

FIGURE 21.21 Various leak paths in top shield and sodium release through the leak paths. (a) Leak paths and (b) total leak vs. time.

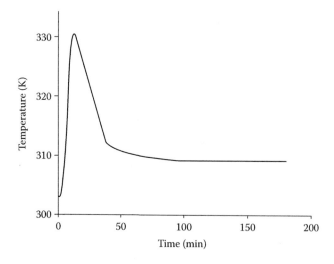

FIGURE 21.22 Temperature rise in reactor containment building.

Computation is repeated for other static pressures corresponding to various work potentials assumed and the corresponding sodium releases are presented in Figure 21.24. By postulating a conservative sodium fire scenario in RCB, temperature and pressure rises are computed for 100 MJ of work potential in Ref. 21.9. Computations are repeated to determine the pressure in RCB due to sodium fire in RCB corresponding to higher work potentials and results are presented in Figure 21.25. The results indicate that the containment loadings would attain the saturation at higher work potentials.

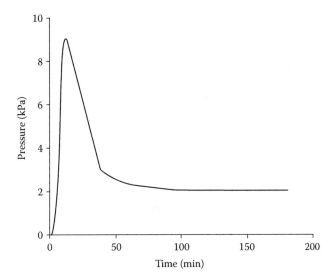

FIGURE 21.23   Pressure rise in reactor containment building.

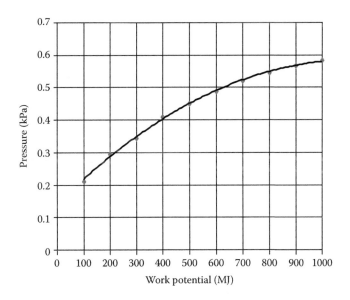

FIGURE 21.24   Quasi-static cover gas pressure.

21.7   EXPERIMENTAL SIMULATIONS

In the theoretical assessment of structural integrity, an axisymmetric analysis was performed, by which it was not possible for ensuring the structural integrity of IHX and DHXs, which are important for ensuring safe PAHR conditions. Moreover, these exchangers have very complicated geometrical features such as a large number of thin tubes and perforated tube sheets, which are difficult to model. Hence, in order to demonstrate the structural integrity of these components, an experimental route was adopted. Under this, 10 tests were conducted with LDE on 1/13 scale mock-ups that have the geometrical features essential to simulate the important phenomena. More details of the experiments can be found in Ref. 21.13.

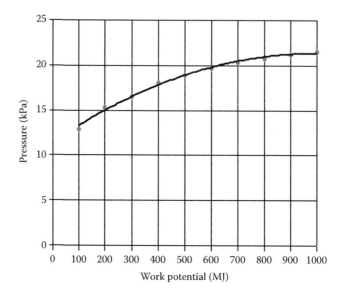

FIGURE 21.25   Pressure rise in reactor containment building.

21.7.1 MOCK-UP AND INSTRUMENTATION DETAILS

Figure 21.26 shows the geometrical details of the mock-up along with a photograph of the fabricated mock-up. The main vessel mock-up is made up of SS 304 LN by assembly of cylindrical part and torispherical dished end. The box-type core support structure model is made essentially out of standard Tee sections representing the prototype satisfactorily. The grid plate is made of two parallel perforated plates with intermediate shell. Rigidity effects of CSA sleeves are considered by increasing the thickness of the plates appropriately. The grid plate is bolted to core support structure at its periphery. Apart from this, it gets support over the core support structure at intermediate locations through spacer pads welded to core support structure. The inner vessel is an assembly of lower and upper cylindrical shells connected by conical intermediate shell, which in turn has six penetrations with cylindrical standpipes for four IHX and two pumps. The conical shell is joined with the lower

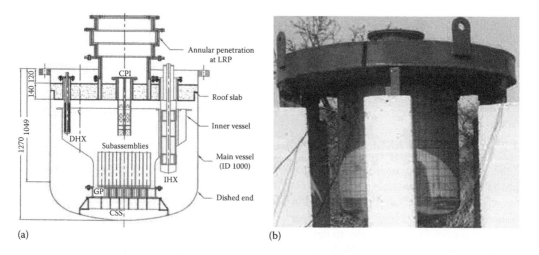

FIGURE 21.26   (a) Schematic of mockup and (b) photograph of fabricated structure.

cylindrical shell through a smooth torus-shaped shell. The inner vessel is bolted to the grid plate at its periphery. The entire core is simulated with 37 hexagonal CSA models (50 mm face to face width), which have matching rigidity characteristics as that of CSA in reactor. Each CSA is filled with the lead to simulate the mass inertia of the core. The control plug consists of outer sheath with simulated perforations and intermediate stay plates. A perforated skirt is welded to the bottom of the control plug. Other internals that do not affect the simulation are not incorporated. The top shield assembly consists of roof slab, rotatable plugs, and control plug. The roof slab is an annular box-type structure having 10 penetrations essentially to accommodate four pumps and two IHX and four DHX models. The large rotatable plug is mounted at the inner periphery of the roof slab. The small rotatable plug and the top portion of the control plug are made integral, which are mounted at the inner periphery of the large rotatable plug. The top shield components—roof slab and large and small rotatable plugs—are filled with concrete to simulate mass inertia. The components inserted through the roof slab are four IHX, four DHX, and two pumps. The pumps are simulated by a cylindrical tube with appropriate thickness to represent the rigidity characteristics. The DHX and IHX models have all the essential parts of the prototype components. The tubes are scaled down properly. They are placed only in three rows (two outermost and one innermost), and hence, only less numbers are accommodated, which will not affect the simulations. The entire reactor assembly model is supported on the reactor vault model through the cylindrical skirt. The reactor vault is represented by six steel columns embedded in the concrete columns to facilitate the photography.

The transient displacements of the main vessel, top shield, lifting of plugs, control plug, DHX, and IHX are captured through two high-speed cameras: one is digital (3000 pictures per second) and the other is conventional (5000 pictures per second). Sufficient number of strain gauges is pasted at the critical locations in the main vessel, IHX, and DHX, later to decide on the structural integrity. A few strain gauges are pasted on the cylindrical skirt to know the loads on the vault. Two accelerometers are placed on the top shield to understand the inertial loads. Two pressure transducers are placed on the top shield to measure the dynamic pressures at its bottom. The permanent deformations of various internal structures are observed only after dismantling the assembly.

21.7.2 SIMULATION OF ENERGY RELEASE

As per the similitude principle, the energy requirement for a scaled down model is E/S^3, where E is work potential for prototype (100 MJ) and S is the scale factor (13). It has been established from a series of tests on simple cylindrical shell with rigid cover and completely filled with water, the mechanical energy conversion efficiency of LDE as established (2.3 kJ/g) with a standard deviation of 0.22. For a typical 100 MJ of energy release in prototype, the energy required for the 1/13 scale model is 45 kJ, which can be released by 20 g of LDE. Taking a safety margin of 10%, 22 g is used in the trials.

21.7.3 IMPORTANT RESULTS

The transient evolution of the bottom portion of the main vessel is recorded by high-speed camera from which the displacement–time history is derived. A typical photography of bottom displacements is shown in Figure 21.27. Displacement history derived from high-speed photographic records is shown in Figure 21.28. The extrapolated values, 108 and 78 mm (8.3×13 and 6×13) are lower compared to 160 and 135 mm as per the theoretical values predicted by FUSTIN (Figure 21.15). This is very much satisfactory. Transient strains are measured at the critical locations of the upper portion. Figure 21.29 shows a typical record, indicating the maximum strain of about 1.56%. The sudden increase in strain occurs during slug impact, which is clearly depicted in the records. The strains measured at the various critical locations of the main vessel are found to be very much less than the rupture strains, and hence, the main vessel

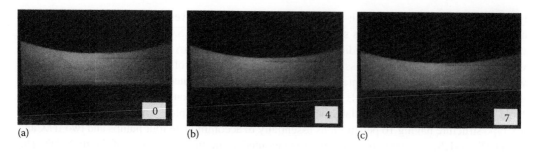

(a)   (b)   (c)

FIGURE 21.27   Evolution of main vessel bottom displacement. (a) At 0 ms, (b) at 4 ms, and (c) at 7 ms.

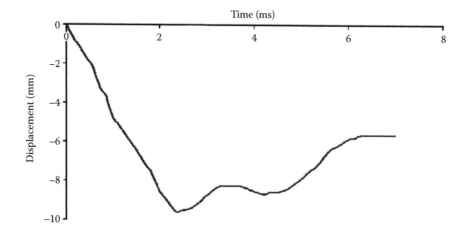

FIGURE 21.28   Displacement of main vessel bottom (measured).

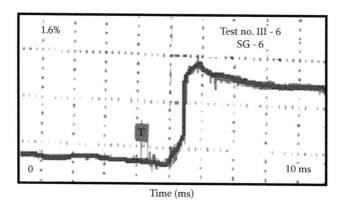

FIGURE 21.29   Hoop strain due to bulging of upper portion of main vessel.

integrity is ensured with comfortable margins. The measured strain value of 1.56% is very much close to the numerical prediction (1.6%) by FUSTIN [21.13]. Tests were repeated with increasing the charge mass and demonstrated that the main vessel can withstand a work potential of 1200 MJ [21.13]. Figure 21.30 clearly depicts the un-deformed and deformed vessels highlighting the maximum strain values (7% at the top and 9% at the bottom portions). The energy equivalent is derived by applying the appropriate scaling laws [21.14].

In order to assess the energy-absorbing potentials of the IHX and DHX, the models of these components were inspected. They could be removed easily (Figure 21.31). Eleven tests completed

(a)   (b)

FIGURE 21.30 Un-deformed and deformed main vessel after explosion ($\cong$1200 MJ): (a) before test and (b) after test.

FIGURE 21.31 Demonstration of structural integrity of heat exchangers by visual inspection.

on 1/13 scale mock-ups had demonstrated the structural integrity of IHX and DHX. It is derived from this series that the maximum mechanical energy that the PFBR can absorb to maintain the structural integrity is decided by DHX deformations, particularly to maintain the coolable geometry. This value is found to be 500 MJ. Furthermore, both rotatable plugs remain intact implying that there is no significant deformation and they were not ejected out.

In order to estimate the possible sodium that could be ejected from the top shield penetrations, five tests were conducted introducing a specific feature—aluminum ducts filled with cotton to absorb water. By knowing the difference in weight of the ducts before and after the test, water leak through each path is quantified. The water leak is measured in 1/13 scale mock-up experiment (Q_m) and extrapolated to the reactor condition (Q_p). The minimum and maximum quantities of water leaks measured in the five experiments through all the penetrations, which all simulate ($110/13^3$) MJ of energy, are 1.75 and 2.415 kg, respectively. The maximum quantity of sodium leak in the reactor estimated by extrapolation is 275 kg. This value is ~75 kg less than the theoretical prediction indicating the conservatism built-in in the numerical model. The simulation of sodium leak from water tests is depicted in Figure 21.32.

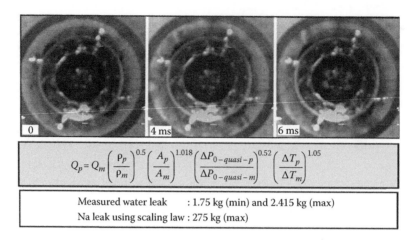

$$Q_p = Q_m \left(\frac{\rho_p}{\rho_m} \right)^{0.5} \left(\frac{A_p}{A_m} \right)^{1.018} \left(\frac{\Delta P_{0-quasi-p}}{\Delta P_{0-quasi-m}} \right)^{0.52} \left(\frac{\Delta T_p}{\Delta T_m} \right)^{1.05}$$

Measured water leak   : 1.75 kg (min) and 2.415 kg (max)
Na leak using scaling law : 275 kg (max)

FIGURE 21.32 Simulation of sodium leak corresponding to 100 MJ energy release.

21.8 POSTACCIDENT HEAT REMOVAL

A pessimistic scenario is conceived for investigating PAHR aspects following the CDA. The molten mixture of core materials (core debris) and steel moves downward. The core debris in contact with liquid sodium gets fragmented due to quenching. Thermomechanical interactions include hydraulic forces and thermal stresses which may lead to cracking. The average size of particles is assumed to be ~200 μm for fuel and 400 μm for steel. Metallic particles are generally rounded and UO$_2$ particles are irregular. The core debris settles on the core catcher positioned just below the grid plate. In case such core catcher is not incorporated, the decay heat generated within the debris bed would be transferred ultimately to the main vessel bottom. Since the main vessel is practically insulated at the outer surface in view of static nitrogen medium in contact, this may lead to melting attack to the main vessel resulting in possible failure of the main vessel itself. Thus, the core catcher serves as an in-vessel core debris retention device and provides PAHR of debris by natural circulation. The reactor core shall be designed with appropriate margins such that the core debris can be adequately cooled. In the process of settlement of molten fuel, the porous bed of 30%–40% porosity is formed over the core catcher and a large-diameter hole developed in grid plate facilitates a communication path for the sodium to flow between bottom plenum and upper plenum. The minimum time taken by the debris to reach the core catcher is 1000s. Subsequently, heat removal over a prolonged period is possible through DHX dipped in hot pool since they are integral and functional after CDA, which is the safety design requirement.

Thermal hydraulics analysis was performed employing an established CFD code to predict the natural convection flow pattern during PAHR. It is assumed that 25 MW heat is generated in the debris. The heat sink is at the DHX location. DHX sink is modeled as a function of hot pool temperature. Forty percent porosity for the debris bed with a mean particle size of 0.3 mm is assumed. The pressure drop of porous bed is modeled by Ergun's correlation. Buoyancy effect is modeled by Boussinesq approximation. The computational model is developed with about 30,000 meshes (Figure 21.33). The analysis predicts multiple loop paths (Figure 21.34). In the hot pool, sodium flows upward along the control plug wall and flows downward along the inner vessel. Sodium flows downward along the center and then flows upward along the periphery of the pore. There is a good mixing layer at the core top with the hot pool flow pattern. In the cold pool, sodium flows upward along the inner vessel and flows downward along the main vessel. The CFD analysis thus has indicated the flow pattern in pools and core support structure volume and established the capability of DHX in effectively removing decay heat of core debris following whole core meltdown. It is planned to get the transient temperature evolution following a severe core accident. More details are available in Ref. 21.15.

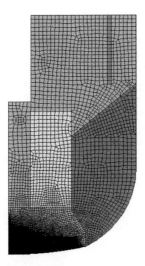

FIGURE 21.33   **(See color insert.)** Computational mesh.

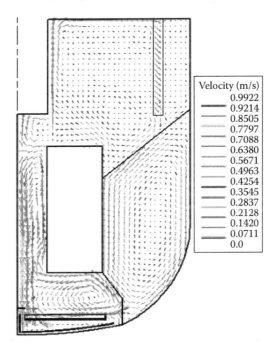

FIGURE 21.34   **(See color insert.)** Sodium flow paths during postaccident heat removal.

In order to understand the heat transfer characteristics of core debris collected on the core catcher, tests are carried out with woods metal in water. The woods metal debris was characterized by simulating the molten fuel coolant interactions and dispersion behavior on the core catcher plate (Figure 21.35a). Subsequently, tests were completed by melting the uranium and steel and subsequently pouring them into the hot sodium pool. Toward carrying out these tests with uranium and steel combinations, they are melted by induction heating with cold crucible method (Figure 21.35b). The facility is called SOFI.

In order to validate the natural convection flow patterns predicted numerically, a dedicated facility called PATH for conducting postaccident thermal hydraulics studies in the 1/20 scaled down model has been built (Figure 21.36a). With these facilities, heat removal capabilities for debris

(a)   (b)

FIGURE 21.35 Debris settled on core catcher plate (simulations): (a) Woods metal (400°C) in water and (b) U+SS (2300°C) in Na (400°C).

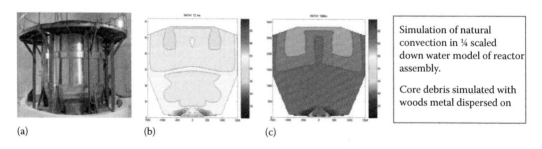

(a)   (b)   (c)

FIGURE 21.36 Heat removal from debris bed settled on core catcher (simulations with postaccident thermal hydraulics). (a) 1/20th scaled down model of PATH, (b) spatial temperature distribution at 72 h, and (c) spatial temperature distribution at 100 h.

corresponding to core melting have been studied. In this facility, the discrete temperatures are mapped so that the resulting temperature distributions can be compared directly (Figure 21.36b). Readers are requested to read the publication on the PATH facility [21.16].

21.9 SITE BOUNDARY DOSE

Source term in RCB is calculated by estimating the fission product in the core and fraction of the core damaged and considering the fractions that can come to RCB. Fission gas release to RCB is 100% of fission gas produced in the disrupted core. The volatile fission product release fraction depends upon its interaction with sodium and is taken as 10% based on international experience. Nonvolatile fission product and fuel release fraction also depends upon complex processes and is taken as 1.0×10^{-4} based on international experience. For calculating the source in the environment, a model is developed in which physical processes like agglomeration, sedimentation, thermophoresis, and diffusiophoresis are considered. It is found that radiologically important nuclides like fission gases, iodine, cesium, and other radio nuclei contribute ground release activity as 3.6×10^{16}, 1.2×10^{14}, 4.9×10^{12}, and 4.9×10^{13} Bq. Site boundary dose is calculated by considering the pathway of cloud gamma dose from fission product noble gases, inhalation dose from fission products, and external gamma dose from ground deposited activity. It has been found that, at the site boundary (1.5 km), the total calculated effective dose is 36 mSv, which includes the plume gamma (28.8 mSv), inhalation route (6.35 mSv), contribution of iodine (2.71 mSv), thyroid dose due to iodine (54.1 mSv), and external exposure from contaminated ground surfaces for 24 h (0.68 mSv). More details are available in Ref. 21.17.

21.10 SUMMARY

PFBR has adequate inherent and engineered safety features. Future reactors would have enhanced and passive safety features. For PFBR, a CDA having a work potential of 100 MJ has been considered as a BDBE. The structural integrity of primary containment has been assured due to mechanical loadings resulting from CDA. Furthermore, temperature and pressure rise in RCB following the sodium fire, as a consequence of sodium release under CDA, form the design basis loads for RCB. In order to raise the confidence on the structural integrity of primary containment as well as RCB, a parametric study on mechanical energy release values in the range 100–1000 MJ has been undertaken with some pessimistic assumptions. The analysis indicates that primary containment has high potential to withstand the transient forces generated by energy release even more than 1000 MJ. The sodium ejection into the RCB through top shield penetrations under sodium slug impact phenomenon is limited with higher energy mainly due to large dimensional changes in vessel associated with shorter transient duration. However, it is worth noting that the deformations of DHXs immersed in the sodium pool could limit the acceptable work potential. For PFBR, this value is found to be 500 MJ from the simulated experimental study. Theoretical estimate of the sodium leak has adequate conservatism when compared to the value-derived simulated test results. The computed radioactive dose values are well below the specified limit of 100 mSv applicable for a design basis accident. The CDA is a beyond design basis accident (BDBA) for PFBR. Hence, PFBR meets all the specified safety criteria with comfortable margins.

REFERENCES

21.1 Chetal, S.C., Chellapandi, P., Mohanakrishnan, C.P., Pillai, P., Puthiyavinayagam, P., Selvaraj, T.K., Shanmugam, C. et al. (2007). Safety design of prototype fast breeder reactor. *ICAPP 2007*, Nice, France, May 13–18, 2007.

21.2 Chellapandi, P., Chetal, S.C., Raj, B. (2010). Structural integrity assessment of reactor assembly components of a pool type sodium fast reactor under core disruptive accident—Part 1: Development of computer code and validations. *J. Nucl. Technol.*, 172(1), 1–15.

21.3 Chellapandi, P, Chetal, S.C., Raj, B. (2010). Structural integrity assessment of reactor assembly components of a pool type sodium fast reactor under core disruptive accident—Part 2: Analysis for a 500 MWe prototype fast breeder reactor. *J. Nucl. Technol.*, 172(1), 16–28.

21.4 Rouault, J., Chellepandi, P., Raj, B., Dufour, P., Latge, C., Paret, L., Pinto, P.L. et al. (2010). Sodium fast reactor design: Fuels, neutronics, thermal-hydraulics, structural mechanics and safety, *Handbook of Nuclear Engineering*, Cacuci, D.G (ed.), Springer, U.S., Vol. 4, pp. 2321–2710, Chapter 21.

21.5 AFCEN-Technical Appendix A3. (2002). RCC-MR Section I, Subsection Z.

21.6 Chellapandi, P., Suresh Kumar, R., Chetal, S.C., Raj, B. (2007). Numerical and experimental simulation of large elastoplastic deformations of FBR main vessel under core disruptive accident loadings. *IMPLAST, Symposium on Plasticity and Impact Mechanics*, Ruhr University, Bochum, Germany.

21.7   Chellapandi, P., Chetal, S.C., Bhoje, S.B. (2000). Effects of reactor internals on structural integrity of PFBR main vessel under CDA, ASME, New York, USA, Vol. PVP-403, pp. 161–172.

21.8   Kaguchi, H., Nakamura, T., Kubo, S. (1999). Strain limits for structural integrity assessment of fast reactors under CDA. *Proceedings of the ICONE-7*, Tokyo, Japan.

21.9   Velusamy, K., Chellapandi, P., Satpathy, K., Verma, N., Raviprasan, G.R., Rajendrakumar, M., Chetal, S.C. (2011). Fundamental approach to specify thermal and pressure loadings on containment buildings of sodium cooled fast reactors during a core disruptive accident. *Ann. Nucl. Energy*, 38, 2475–2487.

21.10 Satpathy, K., Velusamy, K., Chellapandi, P. (2007). Condensation behaviour of fuel vapour in subcooled liquid sodium during a severe accident in a fast breeder reactor. *International Conference on Modeling and Simulation*, CIT, Coimbatore, India, pp. 27–29.

21.11 Velusamy, K., Chellapandi, P. (2007). Sodium release and design pressure for reactor containment building during a core disruptive accident. *CEA-IGCAR Technical Seminar on Liquid Metal Fast Reactor Safety Aspects Related to Severe Accidents*, Kalpakkam, India.

21.12 Beiriger, P., Hopenfeld, J. (1979). SOFIRE II user report, AI-AEC-13055, Atomics International Division, Rockwell International, Canoga Park, CA.

21.13 Lal, H., Chellapandi, P. (2002). Investigation of mechanical consequences of core disruptive accident in fast breeder reactor based on simulated tests on scaled down models, Kalpakkam, India, Collaborative Project No. TBRL/IGCAR/TRIG/1997.

21.14 Wise, W.R., Proctor, J.F. (1965). Explosive containment laws for nuclear reactor vessels, NOLTR-63-140, pp. 1–109. Naval Ordnance Laboratory, White Oak, MD.

21.15 Natesan, K. (2007). Post accident heat removal analysis for PFBR. *CEA-IGCAR Technical Seminar on Liquid Metal Fast Reactor Safety Aspects Related to Severe Accidents*, Kalpakkam, India.

21.16 Gnanadhas, L. et al. (2011). PATH—An experimental facility for natural circulation heat transfer studies related to post accident thermal hydraulics. *Nucl. Eng. Des.*, 241, 3839–3850.

21.17 Indira, R., Rajagopal, V., Baskaran, R. (2007). Source term studies for prototype fast breeder reactor. *CEA-IGCAR Technical Seminar on Liquid Metal Fast Reactor Safety Aspects Related to Severe Accidents*, Kalpakkam, India.

Section IV

Construction and Commissioning

Section IV

Concepts and Connections

22 Specific Aspects of Civil Structures and Construction

22.1 INTRODUCTION

A nuclear power plant consists of various facilities, systems, components, machinery, equipment, piping, ducting, etc. They are arranged systematically within the plant layout in such a manner that the overall construction and operating costs are minimized while meeting certain safety criteria. Regarding the safety criteria, the layout should be such that the redundant safety systems, along with their instrumentation and support feature, have sufficient physical separation so that in case of any accident, one does not jeopardize the availability of other systems. The plant layout is designed to isolate all the items important to safety from unacceptable hazards by using structural features of the plant/buildings. The plant should be designed to facilitate easy access for operations, inspection calibrations, maintenance, and repairs of equipment with a view to limit the exposure of plant personnel and the spread of contaminations. Thus, in the construction of a nuclear power plant, the plant layout plays the most crucial role in facilitating simple construction, safe operation and maintenance, and access for repair.

The major buildings that constitute the layout are the reactor containment building (RCB), steam generator buildings (SGBs), fuel building (FB), decontamination building (DCB), control building (CB), turbine building (TB), electrical building (EB), rad waste building (RWB), and service building (SB). Apart from these, the layout comprises the site assembly shop, maintenance building, and switch yard. Fuel handling and storage involve all activities related to receipt, storage, and inspection before use, transfer of fresh fuel into the reactor, removal of spent fuel from the reactor, preliminary inspection of selected assemblies, and shipping to the reprocessing plant. The fuel handling activities are carried out mainly in the FB and partially in the RCB. The FB houses the fresh and spent fuel handling equipment, storage areas for fresh and irradiated spent fuel, and space for movement of fresh and spent fuel using transport casks. The secondary sodium components and pipings are located in the SGB. Within each SGB, the safety grade decay heat removal system components are located on opposite sides ensuring the availability of one system against common cause failure in case of nonavailability of the other. The layout of sodium piping is made such that the piping is nearly symmetrical and with minimum length. Each SGB houses steam generators (SGs), surge tank, secondary sodium pumps, secondary sodium storage tank, hydrogen leak detector (HLD) for each SG, one HLD for a common header, and a depressurizer tank. A reserve sodium storage tank is provided in one of the SGBs for easy access by the transport tanker when receiving sodium. The RWB houses liquid and gaseous effluent storage tanks (transient storage), solid waste storage area (temporary), and exhaust fans and filter banks for the ventilation system of all radioactive buildings. The CB houses the main control room, handling control room, computer room, local control center, and cable spreading areas. The dependency of each of the previously mentioned buildings is illustrated schematically in Figure 22.1. The location of various buildings is shown in Figure 22.2. The building blocks arranged in the plant layout of a typical medium-sized sodium fast reactors (SFR) (500 MWe capacity prototype fast breeder reactor [PFBR]) as well as a large-sized (1450 MWe capacity European fast reactor [EFR]) plant are shown in Figures 22.3 and 22.4, respectively [22.1].

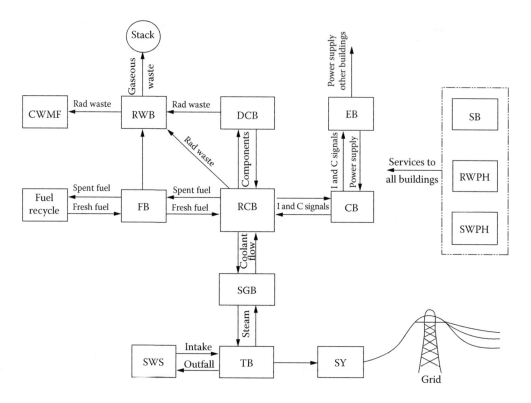

FIGURE 22.1   Functional connectivity of various buildings in the SFR nuclear island.

Specific aspects of important reactor buildings and construction details are presented in this chapter from the information derived from medium- and large-sized power plants. Particularly, three important buildings are described in detail: the RCB, including the reactor vault, FBs, and SGBs.

22.2   SPECIFIC ASPECTS OF REACTOR BUILDINGS

22.2.1   Reactor Containment Building

An RCB is conceived to house the systems and components having the potential of radioactive releases, particularly the reactor assembly including the core (Figure 22.5). The RCB is the tallest one (a height of 72.6 m in PFBRs). In SFRs, the primary sodium transfers the heat to the secondary sodium through intermediate heat exchangers. The secondary sodium inlet and outlet lines penetrate the RCB in order to reach the SGB. The main vessel of the reactor assembly is housed inside the reactor vault, which is a cylindrical concrete structure supported on the bottom base raft of the RCB. The shape of the RCB is an important aspect that has a significant impact on economy. Classically, the RCB of any nuclear power plant is of circular cylindrical shape with a dome-shaped head, which is adopted for all water reactors. Since the design pressure for the RCB is high and the dome-shaped structure has a high-pressure bearing capacity, this shape is chosen. This shape is not generally called for in the RCB of SFRs, since the design pressure is relatively low compared to the water reactor; rather a simple rectangular shape is adequate. This has enabled reductions to be made in the construction time and has yielded important advantages in installation, which include improved walkway layout for both personnel and components, an overhead crane simpler than a polar crane, easier physical separation of the electrical supply channels, simpler identification and sign posting for the operator, and maximum utilization of building volumes. A rectangular secondary containment instead of a dome-shaped one has been adopted for the superphénix (SPX-2) reactor building after comparing different versions. The concrete masses of the RCB for SPX-1 and SPX-2 are compared

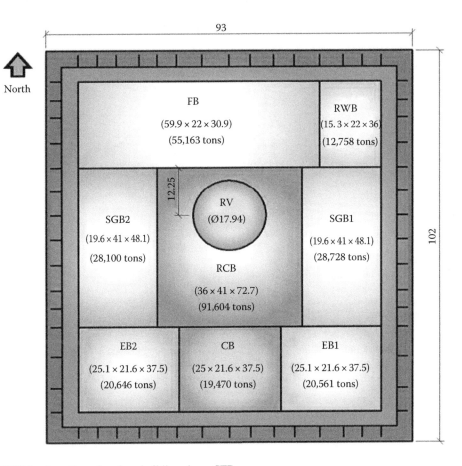

FIGURE 22.2 Location of various buildings in an SFR.

FIGURE 22.3 Plant layout of a 500 MWe capacity PFBR (architectural view).

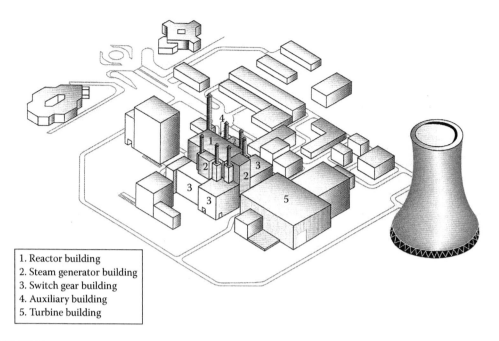

1. Reactor building
2. Steam generator building
3. Switch gear building
4. Auxiliary building
5. Turbine building

FIGURE 22.4 Layout of important buildings in a 1450 MWe European fast reactor. (From Savage, M. et al., Improvements to be made to fast reactors to enhance their competitively, *Proceedings of the Conference on Fast Breeder Systems: Experience Gained and Path to Economic Power Generation*, Richland, WA, September 13–17, 1987.)

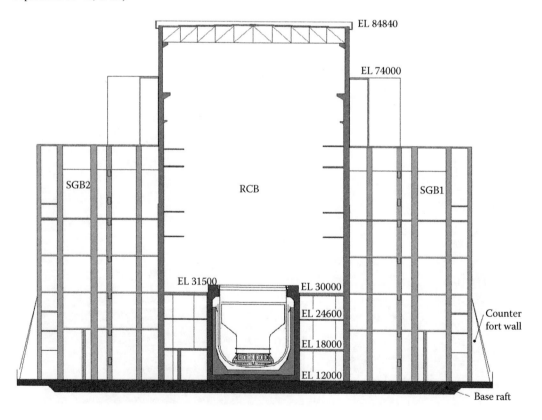

FIGURE 22.5 Location of reactor assembly in the RCB.

in Table 22.1 [22.2]. The reduction of building volumes of the SPX-2 reactor was not only due to the giving up of the round shape of the reactor building, which was also favored by the following:

- The compact arrangement of the secondary circuit equipment
- Simplified ex-reactor fuel handling system by deletion of the external fuel subassembly storage
- Reduced dimensions and a special design of the reactor vault
- A simplified decay heat removal system and a number of other design solutions

When comparing the containment complexities between a typical pool-type SFR and a typical PWR, the cross section of the containment structures for BN-600 M and vodo-vodyanoi energetichesky reaktor (VVER-1000) is shown in Figure 22.6.

TABLE 22.1
RCB Building Details of SPX-1 and SPX-2 Reactors

Details	SPX-1: 1200 MWe	SPX-2: 1500 MWe
Main building shape	Round	Rectangular
Site surface total specific	9,900 m²	6,500 m²
	8.3 m²/MW(e)	4.3 m²/MW(e)
Building volume total specific	489,900 m³	330,000 m³
	408 m³/MW(e)	222 m³/MW(e)
Concrete volume total specific	132,000 m³	94,000 m³
	110 m³/MW(e)	63 m³/MW(e)

Source: Rineiski, A.A., *Improving Economics of Fast Reactor Designs by Reducing the Amount of Plant Materials*, IAEA Vienna, 2004, ISSN 1011-4289.

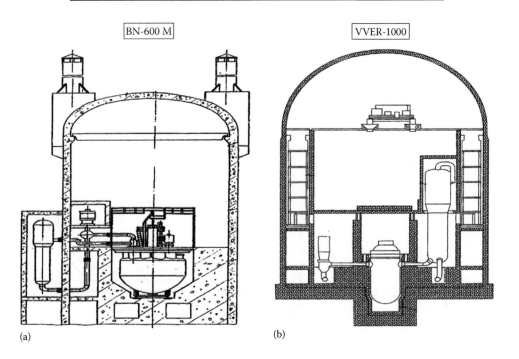

(a) (b)

FIGURE 22.6 RCB for (a) SFR and (b) PWR. (From Rineiski, A.A., *Improving Economics of Fast Reactor Designs by Reducing the Amount of Plant Materials*, IAEA Vienna, 2004, ISSN 1011-4289.)

22.2.2 Reactor Vault

The reactor assembly is supported at the top of the reactor vault. In order to support the safety vessel independently, the vault is structurally divided into two portions: the inner cylindrical structure and the outer cylindrical structure. The reactor assembly is supported by the outer wall and the safety vessel is supported by the inner wall. The double-wall concept adopted for PFBR is schematically illustrated in Figure 22.7. Both the inner and outer walls of the vault are made of reinforced concrete. In order to minimize the heat flux coming from the hot sodium pool, thereby maintaining the temperature of the concrete within acceptable values (typically 60°C under normal operating condition), the main vessel is completely surrounded by a leak-tight safety vessel with layers of highly polished stainless steel sheets to act as metallic insulation covering its outer surface. This arrangement reduces both the heat losses and the thermal cycle skin fatigue. The concrete is kept cool by the vault cooling system (generally cooled by water). The design of embedment for the reactor assembly involves very complicated features: (1) to withstand large forces developed under normal, seismic, and core disruptive accident (CDA) situations, (2) to provide an additional barrier between the vault and main vessel to take care of the eventuality of a possible leak in the biological shield cooling system, (3) to provide leak tightness required for nitrogen in the interspace between the main vessel and safety vessel, (4) to accommodate the relative seismic and thermal movement between the inner and outer walls of the reactor vault, (5) to respect the temperature limits for concrete, and (6) for constructability aspects. The material of the support structure is made of carbon steel of special grade (A48P2), and the adequacy of the strength of steel structures, tie rods, concrete thickness, and reinforcements is ensured by detailed static and dynamic stress analysis.

In the case of PFBRs, the dead load of the reactor assembly is ~4000 tons, transmitted to the upper lateral portion of the reactor vault under a normal operating condition. Apart from these, high dynamic forces in the form of axial, shear, and bending moments are developed during seismic events. The thickness of the outer wall is 1 m with the increased thickness of ~2 m in the upper portion (upper lateral) for the smooth distribution of loads to the concrete. The top edge of the reactor assembly shell is welded to a rigid box-type structure having a height of 400 mm, made of top and bottom plate flanges connected by 288 vertical radial stiffeners. This structure is supported by

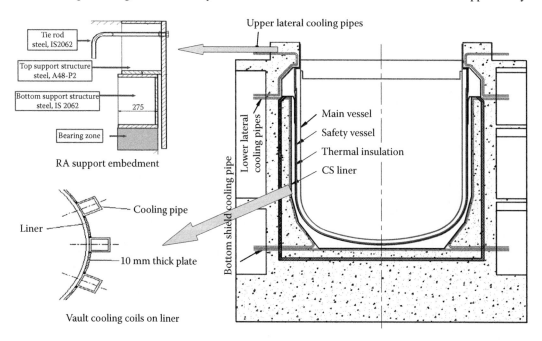

FIGURE 22.7 Reactor vault with embedment and cooling coil details.

another similar box-type structure with matching stiffeners embedded in the upper lateral region. The embedded structures are provided with 144 vertical tie rods to absorb the mechanical energy released under core disruptive accident, and 144 horizontal tie rods are incorporated to offer resistance to the shear forces developed under seismic loads. The upper lateral region of the reactor vault is provided with a double liner arrangement (1) to have an additional barrier between the vault and main vessel to take care of the eventuality of a possible leak in the biological shield cooling system, (2) to provide leak tightness required for nitrogen in the interspace between the main vessel and safety vessel, and (3) to accommodate the relative seismic and thermal movement between the inner and outer walls of the reactor vault. Vertical tie rods penetrate the inner liner at their bottom, where the circular head of the tie rod butts against the horizontal bottom surface of the inner liner and is welded to the inner liner. The top ends of the tie rods are fastened with washers and nut and are closed with end caps and seal welded to provide leak tightness for nitrogen in the interspace between the main vessel and the safety vessel. The geometrical details of the reactor vault, embedment, and cooling coil are schematically shown in Figure 22.7.

In Figure 22.8, the vault construction of a BN-600 M, a pool-type fast breeder reactor, is compared against a typical pressurized water reactor (VVER-1000). The diameter of the reactor vessel is about 12.5 m in the BN-600 M, whereas the corresponding size for VVER is ~4.5 m, since in a pool-type fast reactor, in order to restrict the radiation damage to the reactor structural materials, a large amount of in-vessel shielding is generally provided. Coupled with the large amount of liquid coolant around, the dose contribution due to core gamma is very less. However, due to activation of liquid sodium, that is, Na-24 [Na-23 + (n, γ) $\rightarrow$ Na-24, half life, 15 h] and Na-22 [Na-23 + (n, 2n) $\rightarrow$ Na-22, half life, 2.6 years], which are strong gamma emitters, shielding requirements

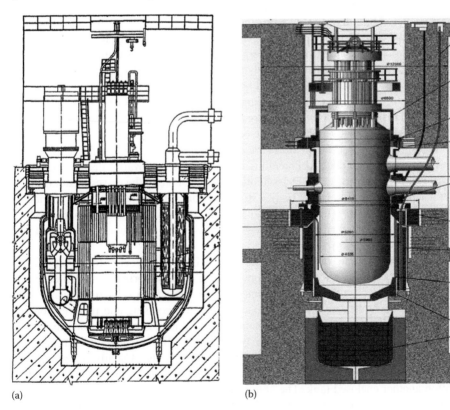

(a) (b)

FIGURE 22.8 Reactor vault supporting the reactor vessel for (a) SFR and (b) PWR. (From Rineiski, A.A., *Improving Economics of Fast Reactor Designs by Reducing the Amount of Plant Materials*, IAEA Vienna, 2004, ISSN 1011-4289.)

exist. A concrete wall of ~2 m thick meets this requirement. This concrete wall also serves the purpose of transferring the load of the reactor block to the raft. In the pressurized water reactor, in-vessel shielding is not significant, deciding the thickness of biological shielding required. For the typical VVER (1000 MWe), the shielding thickness is ~2.3 m of concrete, which is slightly higher when compared to an SFR. The size of a pool-type SFR vault is much higher compared to the size of a PWR vault (~8.7 m for PWR against ~16 m in SFR). Hence, the amount of concrete shielding is significantly less in PWRs. However, due to lesser height of the concrete vault structure in SFRs, that is, ~18 m against 27.5 m in VVER, the overall concrete requirement for the vault structure is comparable to that in VVER-type PWRs (~1800 m^3).

22.2.3 Structures Associated with Fuel Handling and Storage

During refueling of an SFR, spent fuel is replaced by fresh fuel, and the movement of the fuel assembly starting from receipt from the fuel fabrication plant, loading into the reactor to generate power, and dispatch to the reprocessing plant after unloading from the reactor involves various stages (Figure 22.9). A typical layout of the fuel handling route in PWRs is indicated in Figure 22.10. The fuel handling operations are simpler in PWRs with a fewer steps and are rigorous in SFRs. Generally, all fuel handling operations are carried out partly in the RCB and FB. The FB is linked to the RCB for the transfer of fuel assemblies. The radioactivity of spent fuel requires a sufficient shielding thickness for the concrete cell walls/floor, which is generally higher than that required based on structural considerations. To reduce the cell wall/floor thickness, high-density concrete is used instead of normal concrete. To minimize the differential movement between the fuel entry and exit points on the machine linking the two buildings under seismic conditions, both the FB and RCB are located on a common raft along with other buildings. Primary system components like the sodium pump and the intermediate heat exchangers become radioactive due to sodium sticking to its surface, induced radioactivity due to its closeness to the reactor core, as well as corrosion products depositing on the colder regions of the previously mentioned components. Hence, these components are removed using leak-tight containers called handling flasks. The height of the RCB is usually influenced by the maximum height of the handling flask used to remove a pump or intermediate heat exchanger. The weight of the flask due to its large shielding requirements is usually of the order of hundreds of tons, which directly influences the capacity of the crane required in the building and thus the civil design of the building. Efforts are being made to simplify the fuel handling system and to arrive at solutions to reduce fuel handling time. One such innovative design concept studied by the French is the direct transfer of subassemblies from the core to the ex-vessel storage locations using a fuel handling crane (Figure 22.11) [22.3]. In this concept, a concrete cell is provided in the center inside which the control plug is withdrawn maintaining leak tightness. Through the fuel handling corridor, a floor-mounted crane moves to gain access to the core subassemblies and transfers to the ex-vessel storage.

Another major characteristic of fuel handling influencing civil layout is the spent fuel storage. After the Fukushima accident, the design and operation of the bay storing spent fuel have assumed greater importance. Usually, this is a large open water pool, and the subassemblies are stored in order to reduce the decay heat to a level compatible with reprocessing and refabrication requirements. The bay is a concrete tank lined with a stainless steel liner designed for safe shutdown earthquake (SSE) loads. Leakage of water from the bay, especially if it is located underground, needs to be avoided to prevent the spread of radioactivity to the groundwater. Usually, a tank-in-tank concept is used (Figure 22.12) with an inner tank containing the bay water located above the building floor or within another enclosed tank. The annulus between the inner tank and the building floor/outer tank is monitored for leakages with remedial provisions to take care of leakages. For a twin unit layout, the fuel handling system is shared between the units. Figure 22.13 shows a comparison of the civil layout of interconnected buildings laid on a common raft for a single/twin unit.

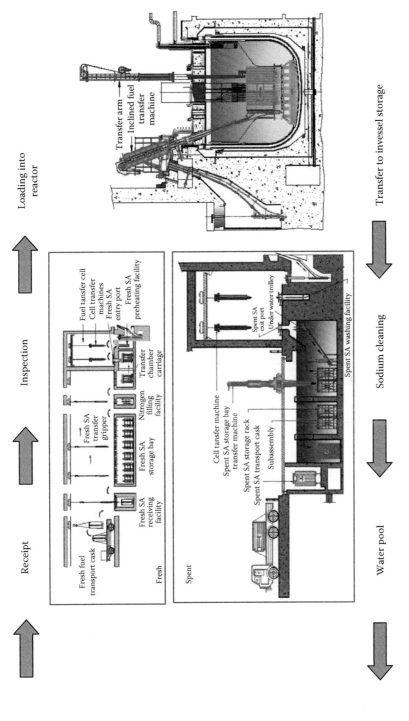

FIGURE 22.9 Buildings associated with fuel handling operations in SFR.

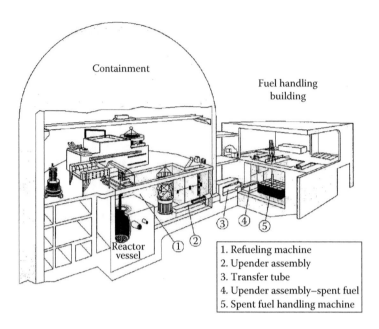

1. Refueling machine
2. Upender assembly
3. Transfer tube
4. Upender assembly–spent fuel
5. Spent fuel handling machine

FIGURE 22.10 Buildings associated with fuel handling operations in PWR.

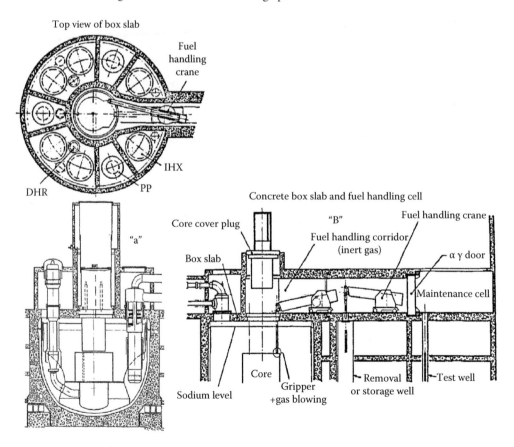

FIGURE 22.11 An innovative fuel handling scheme proposal by the French. (From Rineiski, A.A., *Improving Economics of Fast Reactor Designs by Reducing the Amount of Plant Materials*, IAEA Vienna, 2004, ISSN 1011-4289.)

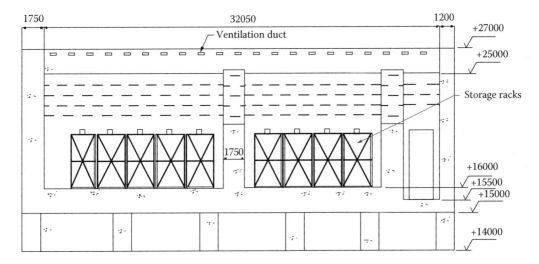

FIGURE 22.12 Typical arrangement of spent fuel storage pool.

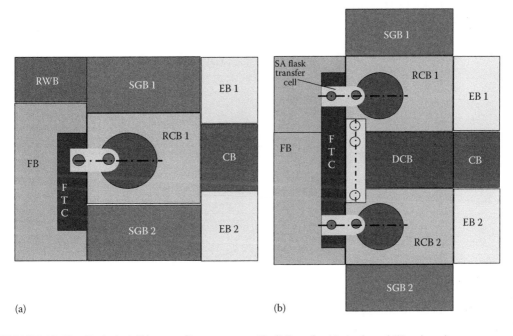

(a)   (b)

FIGURE 22.13 Typical civil layout of interconnected buildings for (a) single and (b) twin units.

22.2.4 Salient Features of a Steam Generator Building

The presence of a huge volume of sodium in the secondary sodium piping and the coexistence of sodium and water in the SG call for certain special features for the SGB. Prevention of sodium fire in case of sodium leaks and sodium–water reaction in the SG in case of water/steam leaks in the SG and/or mitigation of the effects calls for specific provisions. Some of them are provision of a sacrificial sodium-resistant concrete liner of sufficient thickness on the floors containing sodium piping/tanks, nitrogen injection, adequate thickness of walls to withstand the temperature and pressure rise in case of design basis leaks, slope for every floor to drain the leaked sodium in case of a large leak, complete isolation of water/steam systems by steel enclosures, sodium aerosol detectors at various locations to continuously monitor air for the sodium leak in

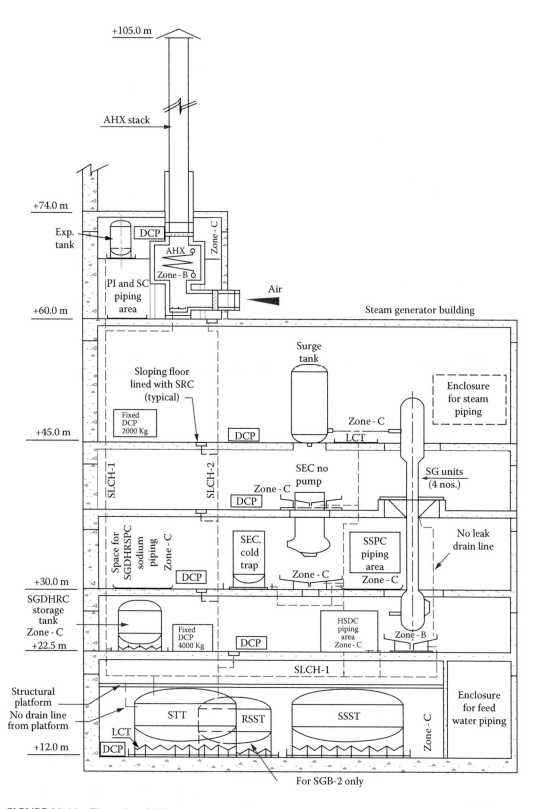

FIGURE 22.14 Fire extinguishing system in SGB.

the building, dry chemical powder (DCP) of significant volume stored in every floor for sodium fire extinguishing, leak collection trays below all sodium piping, and breakable panels to limit the pressure buildup within the building in case of any inadvertent sodium fire. Figure 22.14 shows the various fire extinguishing systems incorporated in the SGB.

In PWR/pressurized heavy water reactor (PHWR), the RCB is to be designed for the higher design pressure (~1400 kPa) (i.e., maximum pressure due to SG leak incident), whereas in SFRs, the SGs are located in a separate building (SGB); hence, the RCB is designed for a comparatively lower design pressure (~25 kPa), which helps in maintaining a smaller and simpler rectangular containment building instead of a larger and costly dome-shaped building. Unlike in PWR/PHWR, the SFR requires many auxiliary systems to support the secondary sodium system and most of them are to be housed in the SGB. In PWR/PHWR, the leakage on the shell side of the SG needs to be monitored for radioactivity (tritium), whereas there is no radioactivity release in the case of shell side leak in the SG for SFRs.

22.2.5 Nuclear Island Connected Buildings on the Base Raft

In Figure 22.1, the functional links of various buildings have been brought out. As per Figure 22.1, the RCB, FB, and SGB-1 and SGB-2 need to be structurally tied together to establish the functional links. Apart from these, other safety-related buildings—RWB, CB and EB-1 and EB-2—are also located nearby. The layout should be as compact as possible while meeting other requirements w.r.t. safety, economy and maintenance, brought out in Section 22.1. Hence, connecting them structurally has several benefits. This gives a symmetrical plant configuration of eight buildings (i.e., RCB located centrally, SGB-1, SGB-2, FB, RWB, CB, EB-1, and EB-2). Torsion effects can be minimized when structures have a more symmetrical plan configuration. Requirements of length of pipes, cables, AC, and ventilation duct are reduced.

Once buildings are interconnected, constructing them on the common base raft is the best option, having isolated the foundation for each of the buildings. Figure 22.2 shows the interconnected buildings constructed on the common base raft. This option has several advantages. The resulting increase in the area of the foundation mat has the secondary advantage of reducing peak acceleration response structural complex. They provide high stability under seismic loadings.

22.3 CHALLENGES IN CIVIL CONSTRUCTION

Construction of a nuclear plant involves many challenging activities requiring extensive pre-planning, optimization deployment of semiskilled and skilled manpower for execution of the work, adequate quality assurance checks to ensure long-term performance, and reducing the time for construction by adopting innovative practices and solutions, as illustrated in the following sections.

The first challenge in nuclear plant construction is large-scale excavation including rock blasting, provision of dewatering and earth-retaining structures, and construction of large-sized base mat or raft of the interconnected buildings. Figure 22.15 depicts the extensive excavation work that was completed for the construction of Monju (Japanese SFR) [22.4]. Prefabrication and placement of reinforcement bars with adequate spacing to ensure concrete flow and large-scale pumping of concrete at multiple points to ensure quick completion of pours are adopted. A record concreting of 5000 m³ in a single pour was carried out for the base raft of PFBR shown in Figure 22.16 (size 100 m long × 100 m width × 3.5 m thick).

Open-top and modular construction is adopted nowadays to reduce construction time. Modularization into self-supporting, freestanding modules significantly reduces the site work and hence construction time, if applied to the critical path items. This reduces on-site manpower, increases productivity and quality under factory environment, and ensures reusability of modules to similar future plants. However, this calls for additional engineering for modules,

FIGURE 22.15   Start of construction of Monju after site excavation (October 1985). (From Hayashi, A. et al., Impact of safety and licensing consideration on Monju, *Proceedings of the International Topical Meeting on Fast Reactor Safety*, American Nuclear Society, Knoxville, TN, April 21–24, 1985.)

FIGURE 22.16   Construction stage of base raft of PFBR.

additional temporary support steels, additional transportation costs, and increased lifting/rigging requirements. The overall effect is a reduction in construction time and cost. There are three levels of modularization: prefabrication, preassembly, and module assembly. Prefabrication involves joining materials to form a component part, preassembly involves joining component parts to create a subunit, and module involves assembling of subunits to create a module. Figure 22.17 shows a typical modularization and open-top construction methodology applied

FIGURE 22.17 Modular and open-top construction adopted in ABWR. (From Blewbury Energy Initiative: Nuclear Fission Energy, The role of nuclear fission power stations: Pros, cons and UK status, http://www. blewbury.co.uk/energy/fission.htm, December 2013.)

to advanced boiling water reactor (ABWR) construction [22.5]. A large-capacity outdoor crane is essential for lifting the modules and precise placement in position.

There are several challenges in the construction of a nuclear plant located in a coastal site. Seawater is used for condenser cooling as an ultimate heat sink. An intake structure is provided to draw water from the sea. Sand transportation and littoral drift are large at certain coastal locations both at the beach and at the seabed. Engineering of the intake structure is taken into consideration to avoid excessive sand entering the pump house and clogging the intake passage to condenser cooling water. Figure 22.18 shows the schematic view of an underground tunnel-type intake structure. The structure consists of a vertical intake shaft located inside the sea, vertical onshore outlet shaft, submarine tunnel linking intake and outlet shafts located below the seabed, channel leading to the pump house, and approach jetty for the offshore intake shaft provided parallel to the submarine tunnel over the sea. Boreholes are drilled at regular intervals along the intake structure to verify the geological characteristics of the seabed. Construction of the intake structure particularly a caisson, a concrete structure located a few kilometers from the plant inside the sea, is a challenging task involving excavation, rock drilling, and concreting in harsh environments and getting the required alignments between the intake and outlet shafts. Figure 22.19 shows some photographs taken during construction of the intake structure for PFBRs.

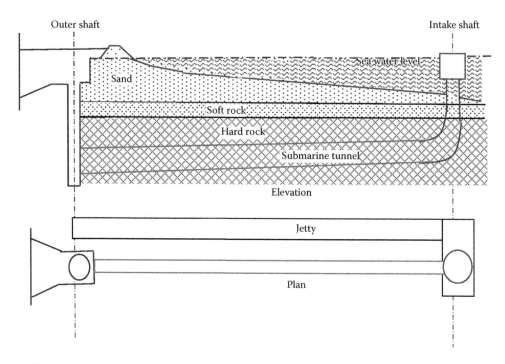

FIGURE 22.18   Schematic of underground tunnel-type intake structure.

FIGURE 22.19   Construction of intake structure for PFBR: (a) intake shaft, (b) outlet shaft, (c) caisson top view, and (d) submarine tunnel.

Nuclear power plant structures are designed for severe earthquake loadings, which results in extensive reinforcement of concrete at certain heavily loaded locations. This poses a significant challenge in prefabrication, handling, and placement in situ of reinforcement bar arrangement before concrete pouring, mock-ups to validate the concrete flow characteristics, and adequate quality assurance during pouring to ensure required quality of construction.

The construction stages of the nuclear power island are shown in Figure 22.20 for the BN-800 reactor [22.6]. Similarly a few illustrations during construction stages of the RCB and the reactor vault in the case of PFBRs are provided in Figures 22.21 and 22.22, respectively, for PFBR.

With the current and ever-growing capacity of computational power, it is essential to exploit computer applications for the design and construction. Three-dimensional (3D) solid modeling is used in contemporary facility design to provide a 3D layout of the proposed facility. It allows for greater visualization of a project and is the standard approach for plant engineering. The process of using a 3D design software starts from generating (initial design stage itself) the solid model of the components. After the solid model is completed, the 3D design software can be used to automatically generate the various plan, elevation, and detail views needed to fabricate the components. Design changes, if any, are made in the model, which gets updated automatically in the drawings. By using 3D design processes, the designer can complete the design sooner. They can also change the design more efficiently when evaluating design alternatives. An example of using 3D modeling to visualize the plant during various stages of construction of the PFBR nuclear island is shown in Figure 22.23.

FIGURE 22.20 Construction stage of the RCB and power island (BN-800). (From Nevskli, V.P. et al., *Sov. Atom. Energy*, 51, 691, 1981.)

FIGURE 22.21 **(See color insert.)** Construction stages of RCB in PFBR.

FIGURE 22.22   **(See color insert.)** Construction stages of reactor vault in PFBR.

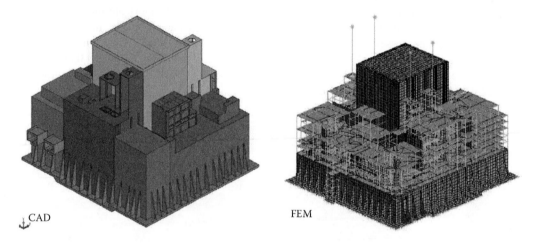

CAD   FEM

FIGURE 22.23   **(See color insert.)** Computer applications for design and development of nuclear island in PFBR.

REFERENCES

22.1   Lefevre, J.C. et al., (1998). European fast reactor: Outcome of design study. EFR Associates.
22.2   Rineiski, A.A., *Improving Economics of Fast Reactor Designs by Reducing the Amount of Plant Materials*, IAEA Vienna, 2004, ISSN 1011-4289.
22.3   Savage, M. et al. (1987). Improvements to be made to fast reactors to enhance their competitively. *Proceedings of the Conference on Fast Breeder Systems: Experience Gained and Path to Economic Power Generation*, Richland, WA, September 13–17, 1987.
22.4   Hayashi, A., Takahashi, T., Izumi, A., Yanagisawa, T. (1985). Impact of safety and licensing consideration on Monju. *Proceedings of the International Topical Meeting on Fast Reactor Safety*, American Nuclear, Knoxville, TN Society, April 21–24, 1985.
22.5   Nevskli, V.P., Malyshev, V.M., Kupnyi, V.I. (1981). Experience with the design, construction and commissioning of BN-600 reactor unit at Beloyarsk Nuclear Power Station. *Sov. Atom. Energy*, 51, 691–696, November 1981.
22.6   Blewbury Energy Initiative: Nuclear Fission Energy. (2013). The role of nuclear fission power stations: Pros, cons and UK status, December. http://www.blewbury.co.uk/energy/fission.htm.

23 Manufacturing and Erection of Mechanical Components

23.1 SPECIFIC FEATURES OF SFR COMPONENTS W.R.T. MANUFACTURING AND ERECTION

Sodium fast reactor (SFR) components, in general, are characterized as large-diameter thin-walled shell and slender structures. Tight manufacturing tolerances are specified to enhance their buckling strength as well as to have possibly minimum vessel dimensions. In the reactor assembly, the main vessel, thermal baffles, inner vessel, core support structure (CSS), and grid plate (GP) are to be positioned sequentially maintaining the coaxiality with the safety vessel, so that the core central line will be in line with the central lines of coaxial vessels: one of the requirements to facilitate smooth operation of control rods as well as facilitate accurate monitoring of sodium temperature emerging from the core subassemblies (CSAs). Further, they have to be erected accurately to maintain the annular gaps for uniform sodium flows and temperatures. During the manufacturing stage, single-side welds are unavoidable at some difficult locations particularly in the case of box-type structures. In-service inspection is difficult with the presence of sodium, and hence, stringent quality control is required in the preservice level itself. From the dimensional stability point of view, residual stresses should be kept at minimum value by adopting robust heat treatment processes and mock-up trials. It is preferable to use a minimum number of materials when considering economy and material data generation, also enhancing the reliability of performance of materials in the operation. Austenitic stainless steels, the main structural material in particular, call for careful considerations for welding without significant weld repairs and distortions. Construction experience of international fast reactors and prototype fast breeder reactor (PFBR) indicates that reactor assembly components decide the project time schedule, even though their cost is relatively small compared to civil, sodium circuits and balance of plant (BoP). There is only limited experience on manufacturing and erection of components. Apart from these, the design and manufacturing codes are still evolving. These are the major challenges in the manufacturing and erection of reactor assembly components. Figure 23.1 depicts the arrangement of reactor assembly components in a typical pool-type SFR.

23.2 MANUFACTURING AND ERECTION TOLERANCES: BASIS AND CHALLENGES

Generally, the term tolerance as used in engineering practice can be defined as the limit or the range of variation in (1) nominal dimensions and (2) geometrical profile of the components and their interfaces that can be tolerated or allowed during the manufacture of individual components, their assembly, and erection/installation of assembly. Nominal dimensions are the ideal or theoretically required dimensions of the features of components such as diameter, height, and thickness. Geometry includes ovality, roundness, straightness, perpendicularity, parallelism, profile, and concentricity. Any total limit or tolerance on a particular dimension or

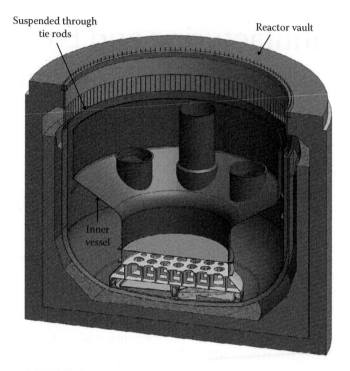

FIGURE 23.1 **(See color insert.)** Reactor assembly components erected on the reactor vault in SFRs.

geometrical feature is judicially apportioned or divided into three parts: manufacturing tolerance, assembly tolerance, and erection or installation tolerance. The tolerance design consists of the following three phases:

1. Preparation and optimization of the total limit, range, or tolerance
2. Apportioning or dividing the previously mentioned elements for manufacturing, assembly, and installation stages
3. Communication of the tolerances to industry through engineering drawings

Tolerances are specified considering various factors, such as operability or satisfactory functioning of the components and plant (also termed as functionality), inspectability of the achieved dimensions/assembly, access for in-service inspection, economy (cost of manufacture is inversely proportional to the tolerance range), and structural integrity of components and assembly under all loading conditions (related to buckling of thin shells).

As far as pool-type SFR is concerned, there are several requirements for specifying stringent tolerances. The primary sodium and all internals of the reactor assembly, such as the CSS, GP, CSAs, inner vessel, thermal baffles, and primary pipes, are housed inside the main vessel. The CSAs are self-standing structures inserted into the GP. The GP in turn is supported on the CSS. The bottom of the CSS is welded to the main vessel bottom dished end. The inner vessel is supported at the periphery of the GP. The top shield, consisting of the roof slab, large and small rotatable plugs (LRP and SRP), and control plug, forms the top closure for the main vessel. The main vessel is welded at the roof slab periphery. Thus, the main vessel is a suspended shell from the roof slab with all its internal components, the core and sodium coolant. The roof slab is supported on the reactor vault. The control plug is supported on the SRP that in turn is supported on the LRP. Similarly, the LRP is supported on the roof slab. The basis for the specification of various tolerances for the manufacture and erection of reactor assembly components to meet the major safety, functional, and interface requirements is briefly discussed in the following sections.

23.2.1 Fabrication Tolerances: Form Tolerances and Their Effects

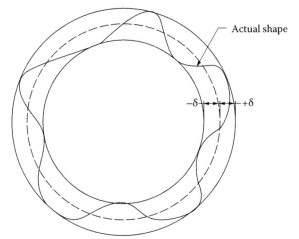

Definition of form tolerance

Large-diameter thin shells, manufactured with several welding of petals made of austenitic stainless steel, pose challenges in achieving distortion-free structures. In view of the concern on the possible sensitization of stainless steels, heat treatment to ensure dimensional stability should be carried out carefully. The tolerances achieved after completion of manufacture and inspection are defined in the adjoining sketch.

In large-diameter shells, the form tolerances are defined in circumferential and meridian directions using templates. The allowable deviations and associated template dimensions that are to be used for the measurements are defined in the manufacturing codes. Figure 23.2 illustrates the definitions of various form tolerances and measurement techniques. The geometrical imperfections are expressed in terms of the maximum radial deviation (δ) achieved during the manufacturing stage as defined in the sketch. δ is normalized w.r.t. shell wall thickness as (δ/h). Figure 23.3 presents the summary of allowable form tolerances recommended in international codes for the main vessel.

Apart from codal requirements, form tolerances are specified more stringently for meeting certain functional requirements as follows:

- To achieve enhanced buckling strength, which is influenced by the geometrical profile of the shell defined by tolerances on the radius of the shell and dished end profile and local deviations such as roundness and straightness
- To ensure uniformity of flow of the coolant for main vessel cooling, which is affected by the variation in the annular gap between the main vessel and the thermal baffle and also the horizontality of the thermal baffle weir

23.2.1.1 Effect on Buckling Strength of Thin Shells

The buckling strength reduction factors (η) are expressed as the function of normalized tolerance (δ/h) in the recommendation made by Japanese design rules based on the outcome of extensive experiments conducted in Japan [23.1] in the form of the following equation:

$$\eta = \frac{1}{\left(1+0.19\chi^{0.65}\right)} \tag{23.1}$$

where χ is equal to ($2\delta/h$) for shear force (applicable for main vessel undergoing shear buckling) and $4\delta/h$ for bending moment (applicable for inner vessel and thermal baffles undergoing shell type of

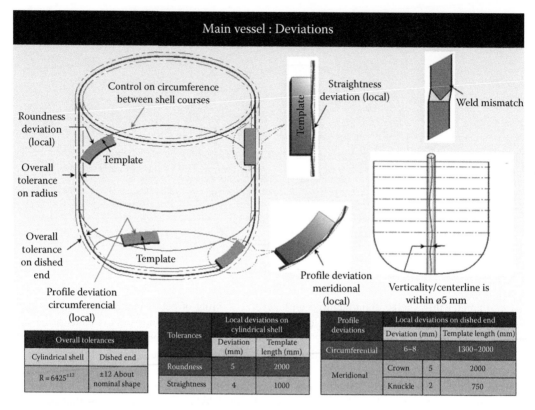

FIGURE 23.2 Form tolerances: definitions and measurements with templates.

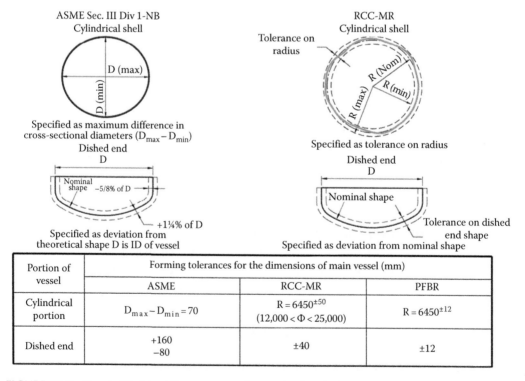

Portion of vessel	Forming tolerances for the dimensions of main vessel (mm)		
	ASME	RCC-MR	PFBR
Cylindrical portion	$D_{max} - D_{min} = 70$	$R = 6450^{\pm50}$ $(12{,}000 < \Phi < 25{,}000)$	$R = 6450^{\pm12}$
Dished end	$+160$ -80	±40	±12

FIGURE 23.3 Permissible form tolerances as per international codes.

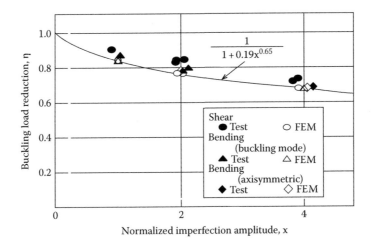

FIGURE 23.4 Effect of form tolerance on the buckling strength of thin shells. (From Akiyama, H., *Seismic Resistance of FBR Components Influenced by Buckling*, Kajma Institute Publishing, Kajma institute, Japan, 1997.)

buckling). The effect of imperfection is much severe on the buckling under bending moment compared to buckling under shear force. Application of this approach provides a net effect of imperfections on the allowable buckling strength:

$$\text{Actual buckling strength} = \eta \times \text{theoretical buckling strength}$$

The η versus χ is shown in Figure 23.4.

In order to have a feel on the effects, results of analysis carried out for PFBRs are highlighted here. Details can be seen in Ref. 23.1. The seismic forces induced on thin shells of PFBRs—main vessel, inner vessel, and thermal baffles—are computed and subsequently buckling analysis is carried out. The shear buckling of main vessel straight portion and the buckling of toroidal portion of the inner vessel and thermal baffles are found to be important. The theoretically computed load multipliers on the seismic loads that can cause buckling are 3.2 for the main vessel (shear buckling mode), 1.9 for the inner vessel (bending buckling mode), and 3.2 for the inner and 3 for the outer baffles (buckling modes). The buckling strength reduction factors ("η" values) are 0.85, corresponding to $2\delta/h$ equal to 1 for the main vessel, and 0.8, corresponding to $4\delta/h$ equal to 2 for the inner vessel and thermal baffles. Applying these factors, the minimum buckling load factors are arrived at $(3.2 \times 0.85) = 2.72$ for the main vessel, $(1.9 \times 0.8) = 1.52$ for the inner vessel, $(3.2 \times 0.8) = 2.56$ for the inner baffle, and $(3 \times 0.8) = 2.4$ for the outer baffle. The inner vessel is found to be the most critical component that buckles under seismic forces induced by a safe shutdown earthquake with a load multiplier of 1.52, which is higher than the minimum factor of safety of 1.3 required as per the design code and construction rules for mechanical components of nuclear installations (RCC-MR) (2002) for service level D conditions. Further, a parametric study was carried out by varying the imperfection δ to quantify the allowable site seismic peak ground acceleration that yields the net buckling load factor acceptable as per RCC-MR. The results are presented for the inner vessel, which is allowable peak acceleration versus δ/h in Figure 23.5. It is clear from Figure 23.5 that the vessels can accept higher seismic forces if the vessels are manufactured with possibly lower imperfections. This has significant economic benefits. Thus, the form tolerances have high impact on economy.

23.2.1.2 Effect on Diameters of Thin Vessels

There are many coaxial thin vessels that constitute reactor assembly. The inner vessel is the innermost one, which is surrounded by the inner thermal baffle, then the outer baffle, the main vessel, and finally the safety vessel. In order to avoid any mechanical interactions under seismic

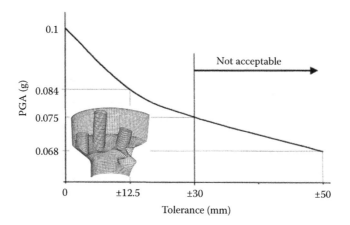

FIGURE 23.5   Effect of manufacturing tolerances on permissible seismic accelerations.

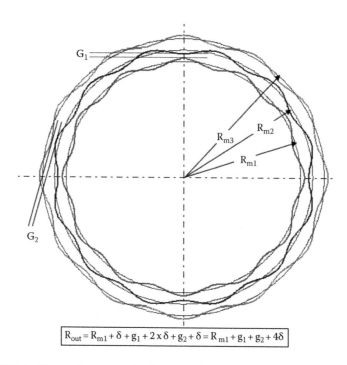

$$R_{out} = R_{m1} + \delta + g_1 + 2 \times \delta + g_2 + \delta = R_{m1} + g_1 + g_2 + 4\delta$$

FIGURE 23.6   Effect of form tolerance on the diameter of circular cylindrical shells.

loadings, minimum annular gaps are provided between the mean diameters of the shells. Further, an adequate gap should be provided between the main vessel and the safety vessel to allow free movements of the in-service inspection device. The minimum radial gap is the difference between the minimum radius of the outer vessel and the maximum radius of the inner vessel. If δ is larger, then the diameter of the outer shell should also be larger. Thus, the diameters of the vessels are governed by δ. This is made clear in the sketch shown in Figure 23.6, where three coaxial shells with mean radii of (r_{m1}, r_{m2}, and r_{m3}) are considered with the minimum radial gap requirements: g_1 between the innermost shell and the intermediate shell ($r_{m2} - r_{m1}$) and g_2 between the intermediate shell and the outer shell ($r_{m3} - r_{m2}$). The outermost radius of the shell is derived from the following: $R_{out} = R_{m1} + \delta + g_1 + 2\delta + g_2 + \delta = R_{m1} + g_1 + g_2 + 4\delta$. The last term ($4\delta$) indicates the effect of form tolerance. If $\delta = 100$ mm (say), then the outermost radius of the subsequent structure/shell required will be 400 mm, which is equal to 800 mm in diameter, thereby making a significant impact on economy.

23.2.1.3 Weld Mismatch

Another effect of δ is to produce mismatch between the shells when they are joined by welding. Construction codes specify tight tolerance on form tolerance at the weld interfaces (codal position will be presented subsequently). The discontinuity effects and forces/moments developed at the interfaces having weld mismatch are depicted in Figure 23.7. A few examples in the practical situations will be presented in subsequent sections.

23.2.2 MACHINING TOLERANCES

These aspects are discussed with reference to the GP, the component involved in extensive machining. The GP supports the reactor core and acts as a coolant distribution plenum for CSAs. It also supports the inner vessel and primary tilting mechanism. It is supported on the CSS, which is welded to the main vessel. The GP is designed for an internal pressure required to pump the coolant (0.8 MPa for PFBR). In addition, a dead load due to the core, inner vessel, and primary tilting mechanism acts on the GP (about 600 tons for PFBR). In addition, the seismic-induced forces and moments are considered in the design. Structurally, the GP is made of two plates spaced and interconnected at the periphery by a cylindrical shell and by a number of tubes called sleeves. Both the plates are perforated in an identical triangular pattern and the corresponding holes are in vertical alignment in the assembled structure. The major functional and other criteria for arriving at tolerances are primarily to facilitate easy assembly of a large number of machined parts and endure proper handling of CSAs by respective handling mechanisms; minimize leakage of sodium through bolted joints between the top/bottom plates and shell assembly and that between sleeves and plates; minimize tilting of the inner vessel supported over the GP, thereby avoiding interference at the mechanical seal between the intermediate heat exchanger and the inner vessel; and limit the core movements (flowering/compaction) during any dynamic forces, such as seismic effects. Figure 23.8 depicts various machining tolerances on the GP as an example.

23.2.2.1 Basis of Specifying Various Tolerances

For proper handling of CSAs and control rods through their respective mechanisms, the misalignment permitted between the top of the CSAs and the control rod and the gripper of the corresponding handling mechanisms is to be within certain limits. There are many factors that

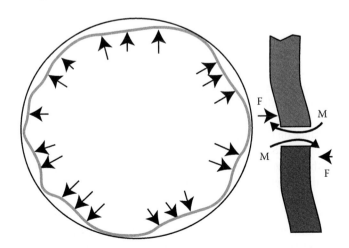

FIGURE 23.7 Forces and moments developed at the weld mismatch.

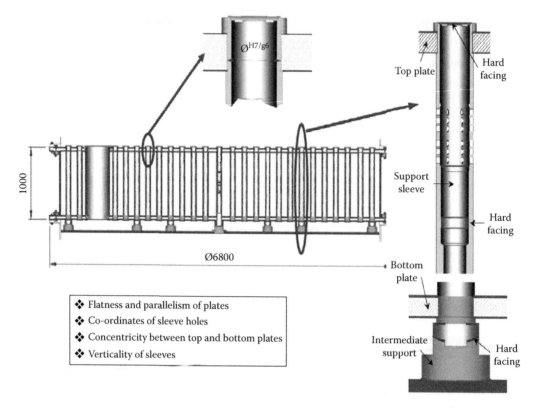

FIGURE 23.8 Important machining tolerances on the GP—an example.

contribute to this misalignment/offset. Among these, the factors that can be attributed to the GP and the core are as follows:

1. Straightness and bowing of CSAs/control rods under irradiation
2. Slopes and deflections of the GP under core and other loads
3. Position and verticality of sleeves (due to manufacturing and installation tolerances) that hold the CSAs/control rod
4. Clearance between the sleeve and foot of the CSA at the guide location

Among these, bowing takes a major portion of permitted offset. Slopes and deflections of GP are controlled in the design. The portion of the offset apportioned for manufacturing and site installation tolerances of the GP is also limited to possibly the smallest value (3 mm for PFBR), and the rest is apportioned to account for other factors. The position of the sleeve is a combination of (a) manufacturing tolerance on the position of the sleeve within the GP and (b) site installation/erection tolerances on the position and orientation of the GP. The verticality of the sleeve is composed of (a) manufacturing tolerance on the perpendicularity of the sleeve w.r.t. the bottom surface of the GP, which is supported over the CSS, and (b) site installation tolerance on the horizontality of the GP after erection, that is, the horizontality of the CSS flange face that supports the GP. The tolerances on the previously mentioned features are to be specified to limit the resulting offset/misalignment within specified limits.

In case of a bolted GP, the flanged bolted joint between top/bottom plates and shell flanges is the metal-to-metal contact joint without any sealing arrangements. This is because of the difficulty in machining large-diameter precision grooves to close tolerances for the use of metallic O-rings in sodium environment. For given flange dimensions, the factors that affect the leakage through these joints are the flatness and surface finish of the mating surfaces, namely, that of plates and shell flanges. The machining tolerances on surface flatness are to be specified to assess the variation of leak rate for various values of surface flatness and finish, and based on this, the surface flatness is specified, for which the resulting leakage should be negligible. Similarly, leakage through sleeve penetrations of the top and bottom plates is to be studied carefully, and accordingly, close clearance fit needs to be specified between the sleeves and sleeve holes in the plates, which will also facilitate easy assembly.

23.2.3 Erection Tolerances

23.2.3.1 Tolerance Requirement for Smooth Operation of Absorber Rods

Specific tolerances particularly on verticality and horizontality are written to ensure smooth reactor operation and safe shutdown functions. To achieve this, tight tolerances are to be specified on the parameters governing the insertability of absorber rods. This in turn demands accurate positioning of control and safety rod drive mechanisms (CSRDMs) and diverse safety rod drive mechanism (DSRDMs) within the control plug shroud tubes w.r.t. the central line of the GP and the verticality of the shroud tubes and horizontality of the GP. The horizontality of the control plug depends on the horizontally of the rotatable plugs and roof slab. On the whole, the horizontality of the top shield and the coaxiality and horizontality of the GP are the most important parameters. In view of the large diameters of the top shield and GP, hanging support arrangement of the main vessel supporting the GP, and the temperature gradients in these components, maintaining the specified tolerances is the most challenging task. The other most important requirement is rigidity and dimensional stability of these structures. Since these components are not independent, manufacturing tolerances, position accuracy and dimensional stability of all the components lying in the core support path and CSAs are closely interlinked, which have to be specified after thorough understanding of design requirements and constraints imposed during the manufacturing and erection process. The tolerance requirements for the insertability of absorber rods are illustrated schematically in Figure 23.9.

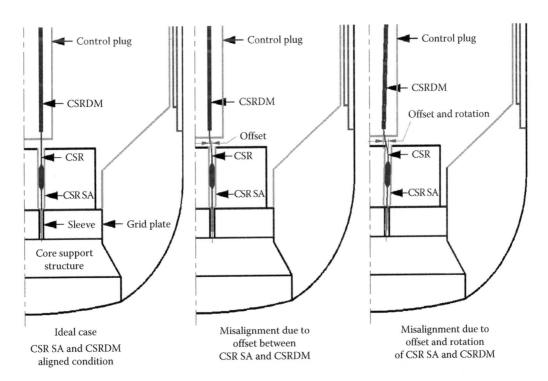

FIGURE 23.9 Basis of specification of tolerances for the insertability of absorber rods.

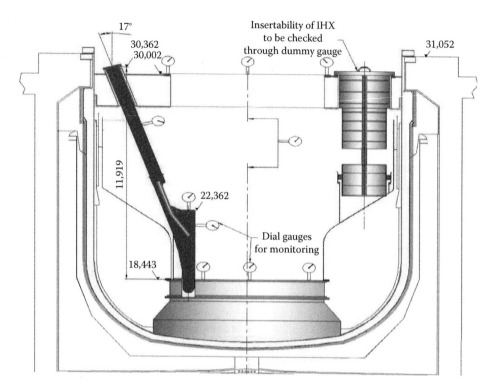

FIGURE 23.10 Erection tolerances of components penetrating the top shield.

23.2.3.2 Tolerances Required for Smooth Fuel Handling Operations

Proper functioning of the fuel handling machine is governed by the appropriate tolerances in straightness of the machine, straightness of travel of the gripper and guide tube in the machine, verticality of CSAs, interactive forces between the adjacent CSAs and between the foot of the CSA and GP sleeve due to the dimensional tolerances, alignment of the gripper of the fuel handling machine with the head of CSAs. Erection tolerances of components penetrating the top shield are specified for PFBRs in Figure 23.10.

23.2.3.3 Radial Gap Uniformity for the Smooth Operation of ISI Equipment Inspecting the Main Vessel

The tolerances that govern the gap between the main and safety vessels are form tolerances that could be achieved during the fabrication of vessels, relative elevation, verticality, and concentricity between the main and safety vessels. Figure 23.11 depicts the gap requirement for the free movement of the in-service inspection (ISI) vehicle.

23.2.3.4 Tolerance Required for Ensuring the Required Sodium Flow through Weir Shell Spaces

The horizontality of the weir shell crest is very important as illustrated in Figure 23.12. The top of the weir shell should be monitored during erection for horizontality to accomplish uniform sodium flow all around the circumference.

23.3 MANUFACTURING CODES AND PRACTICES

Different practices, standards, and guidelines are followed by countries around the world in design, manufacture, and erection of mechanical equipment. These depend on their respective strength and good engineering practices derived over the years of practice and manufacturing technological base. However, the components that form part of the reactor systems are critical to safety, and varying

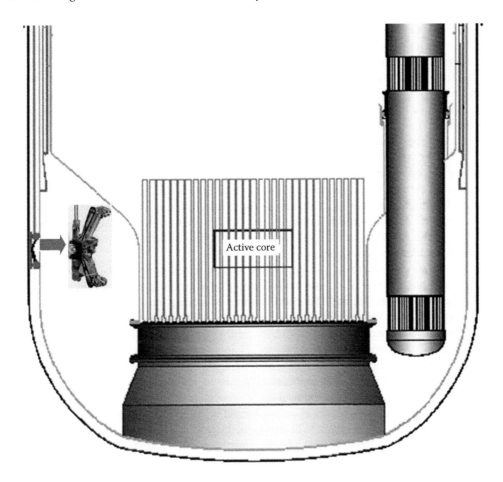

FIGURE 23.11 Movements of the ISI vehicle in the interspace between the main and safety vessels.

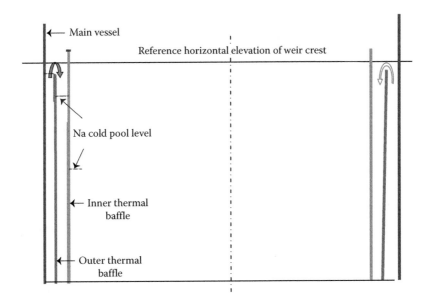

FIGURE 23.12 Requirement of horizontality for weir crest for uniformity of sodium overflow.

practices adopted during their design, manufacture, and erection may not be sufficient to ensure that the required level of margins commensurate with their safety classification. It is under this background where the international design codes such as BS, American Society of Mechanical Engineers (ASME)–Deutsches Institut für Normung (DIN), and French codes play a vital role in harmonizing and standardizing the design, manufacturing, and erection standards across the world. The adoption of these codes ensures uniformity of margins and brings the equipment on the same platform aiding better comparison and standardization and facilitating interchangeability. Further, the adoption of an international code raises the level of confidence of the purchaser and opens the market for the products from the vendors across different countries enhancing their marketability, besides aiding the regulatory practices and other statutory requirements.

Design codes provide both mandatory and nonmandatory requirements for compliance starting from classification, categorization, material selection, material properties (chemical, mechanical, physical), material qualification, design rules and analysis methodologies, fabrication and qualification procedures, nondestructive (ND) examination methods, delivery conditions, installation requirements, and in-service inspection requirements.

A well-designed component may not perform to its requirements due to either one or a combination of improper material selection, poor manufacturing or ND examination practices or noncompliance to erection requirements, etc. The manufacturing of a component involves a combination of wide-ranging operations that include machining, forming, and joining. For example, in the manufacture of reactor components, welding is an inherent process. Often difficulty is encountered in meeting the quality requirements specified for weld joints, and hence, greater care is to be exercised during edge preparation, fit-up, and execution of welding. While these steps can be progressed with caution, the difficulty is in predicting and controlling the welding distortion/shrinkage. Taking this into account, it is preferable to have built-in design provisions to avoid/minimize welding distortion. This assumes greater significance at critical locations wherein the design requirements such as horizontality, verticality, and level are to be maintained from functional requirements. Construction experience of SFR is very much limited and manufacturing codes and standards are still evolving. The manufacturing codes such as RCC-MR and ASME specify the certain requirements that are brought out in this chapter along with the basis of such requirements. This section addresses three major aspects: welds, inspection, and heat treatment.

23.3.1 Issues Related to Welds

The primary requirement in any design process is to ensure that the component performs the intended function satisfactorily throughout the targeted life under the influence of all the service loadings. In order to comply with this requirement, the designer should choose the correct materials, assess service conditions that require conversance with design and failure theories, and choose appropriate manufacturing and inspection methods. In spite of these, early failures can still occur since the welds are the weak links in the component. A specific case study on the failure experiences in worldwide SFRs indicates that about 50% failures are due to welds as illustrated in Table 23.1 [23.2].

TABLE 23.1
Weld Failures in Fast Reactors

	DFR	PFR	PX	SPX	KNK	Total
Weld failures	3	7	17	1	10	38
Total number of failures	7	19	20	2	21	69

Source: Structural Integrity of the Fast Reactor, Report by the UK Nuclear Industry, Technical Advisory Group on Structural Integrity, AEA Technology, Warrington, U.K., 1992.

In the design stage, the welds are analyzed in detail strictly respecting the applicable design code criteria. Further, in the process of licensing of power plants for extended operation, the welds are reviewed critically on nondestructive evaluation (NDE) techniques. In some situations, the defects observed in the components through NDE are analyzed critically to decide upon its acceptance. Under these situations, it is essential to apply robust design rules/assessment procedures with complete understanding of the behavior of welds under operating conditions and the factors influencing its behavior, particularly at high temperatures.

The design codes provide rules based on the vast experiences gained over the years. As ideal defect-free weld joints are not possible, it is required to specify the acceptable defects or deviations for manufacturing purposes. In this regard, the codes specify the acceptable defects and inspection requirements as per the weld category. Various codes differ in their approach and they can be broadly classified into conventional and nuclear codes. For the purpose of illustration, the scope is restricted to a typical conventional design code called ASME Section VIII Div 1 [23.3] and nuclear code ASME Section III, Subsection NH [23.4]. The French code RCC-MR provides design rules exclusively for fast breeder reactor components [23.5]. The main failure modes that are precluded once the design satisfies the code requirements are gross yielding and tensile rupture (time independent) and creep strain, creep rupture, creep, and fatigue damage (time dependent). It is assumed that the components are free from cracks at the beginning of life and the initiation of crack is the failure as per the design codes due to creep fatigue damage accumulation, apart from yielding and creep rupture.

The rules recommended in the codes are generally followed for the weld design. Accordingly, the highlights of ASME Section III, Div 1, Subsection NH [23.4] and RCC-MR [23.5] are presented.

23.3.2   Challenges in Austenitic Stainless Steel Welds

Austenitic stainless steels are commonly used structural materials in SFRs. In terms of manufacturing, mainly welding involves several difficulties. The very high coefficient of expansion associated with this material leads to high welding distortion. This material has poor thermal conductivity; therefore, a lower welding current is required (typically 25% less than carbon steel) and narrower joint preparations can be tolerated. All common welding processes can be used successfully. However, high deposition rates associated with submerged arc welding (SAW) could cause solidification cracking and possibly sensitization, unless adequate precautions are taken. The solidification strength of austenitic stainless steel can be seriously impaired by small additions of impurities such as sulfur and phosphorous; this coupled with the materials' high coefficient of expansion can cause serious solidification cracking problems. Most 304 type alloys are designed to solidify initially as delta ferrite, which has a high solubility for sulfur, transforming to austenite upon further cooling, creating an austenitic material containing tiny patches of residual delta ferrite, therefore not a true austenitic in the strict sense of the word. The filler metal often contains further additions of delta ferrite to ensure crack-free welds. The delta ferrite can transform to a very brittle phase called sigma, if heated above 550°C for very prolonged periods.

If any part of the stainless steel is heated in the range of 500°C–800°C for any reasonable time, there is a risk that the chrome will form chrome carbides (a compound formed with carbon) with any carbon present in the steel. This reduces the chrome available to provide the passive film and leads to preferential corrosion, which can be severe. This is often referred to as sensitization. Therefore, it is advisable when welding stainless steel to use low heat input and restrict the maximum interpass temperature to around 175°C, although sensitization of modern low carbon grades is unlikely.

Heat treatment and the thermal cycle caused by welding have little influence on the mechanical properties of austenitic stainless steel. Austenitic steels are not susceptible to hydrogen cracking; therefore, preheating is seldom required, except to reduce the risk of shrinkage stresses in thick sections. Postweld heat treatment is seldom required as this material as a high resistance to brittle fracture. Stress relief is carried out to reduce the risk of stress corrosion cracking (SCC); however, this is likely to cause sensitization unless a stabilized grade is used (limited stress relief can be achieved with a low temperature of around 450°C). It has good corrosion resistance, but quite severe corrosion can occur in certain environments. The right choice of welding consumable and welding technique can be crucial as the weld metal can corrode more than the parent material.

To ensure good corrosion resistance of the weld root, it must be protected from the atmosphere by an inert gas shield during welding and subsequent cooling. The gas shield should be contained around the root of the weld by a suitable means, which permit a continuous gas flow through the area. The gases used are typically argon or helium. The biggest cause of failure in the pressure plant made of stainless steel is SCC. This type of corrosion forms deep cracks in the material and is caused by the presence of chlorides in the process fluid or heating water/steam, when the material is subjected to tensile stress (this stress includes residual stress). Significant increases in nickel and also molybdenum will reduce the risk. Carbon steel tools, supports, or even sparks from grinding carbon steel can embed fragments into the surface of the stainless steel. These fragments can then rust if moistened. Therefore, it is recommended that stainless steel fabrication be carried out in a separate designated area and special stainless steel tools used where possible. The code rules for achieving sound welds are described in the following section.

23.3.3 Acceptable Weld Joints for Stainless Steel Vessels Containing Sodium

Weld joints are specified based on the component classification, that is, class 1, 2, 3, and nonnuclear category. Considering the importance of welds in the primary boundary to ensure the structural integrity of the component that has bearing on the radioactivity release to the atmosphere, only full-penetration joints are acceptable for these welds. The definition of weld joints as per RCC-MR is reproduced in Figure 23.13. Prior to undertaking production weld, the weld procedure and welder should be qualified as per purchaser's specification (drafted in line with codal requirements). The permissible weld joints in typical SFR shell structures containing primary sodium are given in Figure 23.14. The basis of such recommendations is listed as follows:

- Only full radiographable weld geometries are permitted. If the joint is not amenable for radiography, complete ultrasonic test (UT) is to be performed. If both are not possible, liquid penetrant examination (LPE) may be performed every two layers (t < 20 mm) or every three layers (t ≥ 20 mm).
- No permanent backing strip.
- Smooth transition at unequal weld joints (min 1:3).
- Acceptable weld mismatch (t ≥ 5 mm)
  - For both sides accessible, (t/10 + 1) with a maximum of 4 mm
  - For single-side access, (t/20 + 1) with a maximum of 3 mm
- Circumference control essential for welding two shell courses.
- No undercut. Welds ground flush.

Two typical design details with thermal sleeves shown in Figure 23.15 (not addressed in codes) are recommended for high-temperature applications. The diameter transition should typically be made using a standard reducer to provide increased structural flexibility and reduce the sharpness of the thermal gradient. Thus, the thermal stresses as well as discontinuity stresses and the resulting fatigue damage are minimized. The vessel supports, such as skirt, lug, or column, should be attached to the vessel shell or head using full-penetration welds. Two typical recommended configurations are shown in Figure 23.16, one with a forged ring section and another with a weld buildup. In both design concepts, clear access is provided for welding (butt weld) and also for inspecting, thus increasing the potential for achieving good overall weld quality. Further, skirt supported vessels in high-temperature applications will have a relatively high thermal gradient at the skirt to shell intersection. The local thermal gradient can be minimized by graded insulation. Another design concept often adopted is to provide slots in the skirt to increase the flexibility.

Examples	Definition of Types Welded Joints				
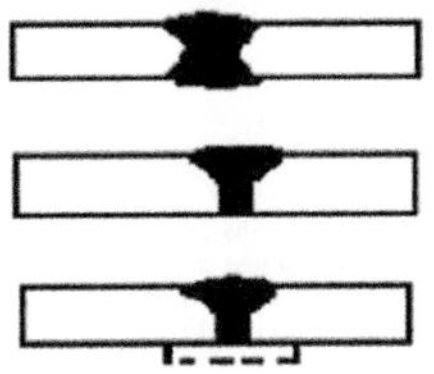	I.1	Butt welding	Full penetration	Two sides accessible	Back welding
	I.2	Butt welding	Full penetration	Two sides accessible	Gaseous back protection with or without insert
	I.3	Butt welding	Full penetration	Two sides accessible	On temporary backing strip can be inspected after removal of the strip
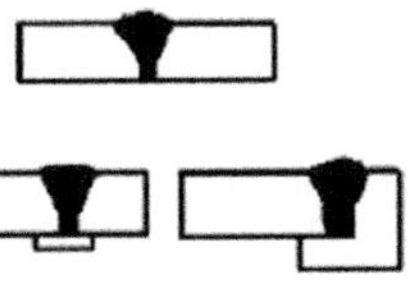	II.1	Butt welding	Full penetration	Back side inaccessible	Gaseous protection with or without insert
	II.2	Butt welding	Full penetration	Back side inaccessible	Permanent backing strip
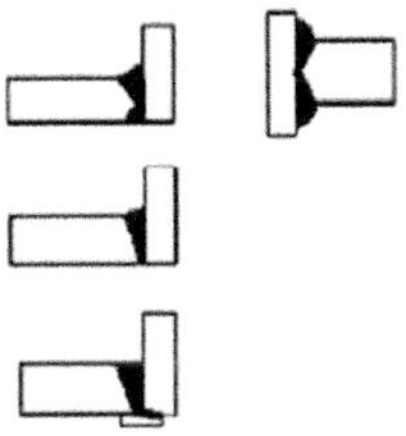	III.1	Fillet or T	Full penetration	Two sides accessible	Back weld or back machining
	III.2	Fillet or T	Full penetration	Back side inaccessible	Gaseous back protection
	III.3	Fillet or T	Full penetration	Back side inaccessible	Permanent backing strip
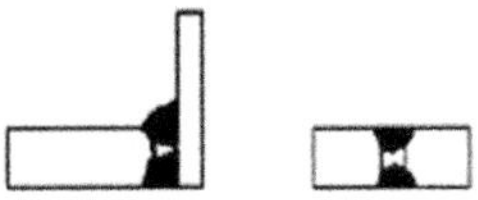	IV.1 IV.2	Fillet or T butt welding	Partial penetration	Double opening preparation	Double bead
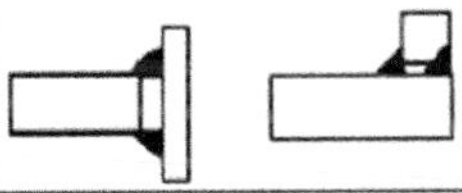	V	Fillet or T	Partial penetration or no penetration	Straight edges or single opening preparation	Double bead
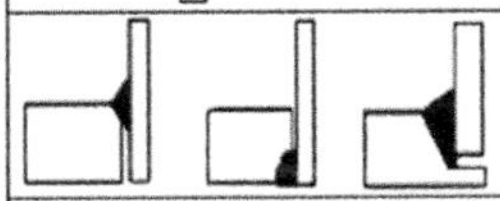	VI	Fillet or T butt welding	Partial penetration	Single opening preparation	Single bead
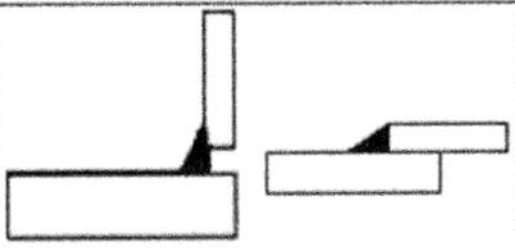	Fillet or T	No penetration	Straight edges preparation	Single bead	Single bead

FIGURE 23.13 Weld configurations as defined in RCC-MR Section III, Subsection B. (From ASME, Rules for Construction of Nuclear Power Plant Components, Subsection NB: Class 1 Components, Section III Div 1, 2007.)

23.3.4 GENERAL GUIDELINES FOR ROBUST DESIGN

The following are the general guidelines for robust design:

- Socket welds shall be avoided for application, having risk of crevice corrosion. In addition, socket welds have drawbacks, one of which is stress concentration that has detrimental effects on fatigue behavior in comparison to butt welds.
- For applications involving fatigue loading, it is important to specify the flush weld against the weld with a minimum stress concentration factor. For more details, reference may be made to Table NB 3681 (ASME Section III, Div 1 Subsection NB class 1 component) [23.6].

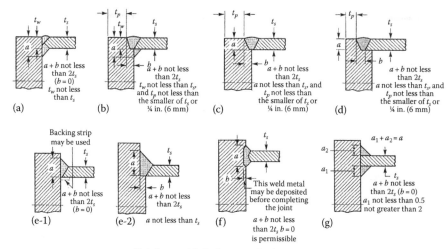

General notes:
(1) For supported tubesheets: a + b not less than 2ts, c not less than 0.7ts or 1.4tr, whichever is less.
(2) For unsupported tubesheets: a + b not less than 3ts, c not less than ts or 2tr, whichever is less.
(3) ts and tr are as defined in UG-34(b).
(4) See UW-13(e) (3) for definition of supported tubesheet.
(5) Dimension b is produced by the weld preparation and shall be verified after fit up and before welding.

FIGURE 23.14 (a–l) Acceptable corner joints for flat plates attached to shells. (From ASME, Rules for Construction of Pressure Vessels, Section VIII Div 1, 2001.)

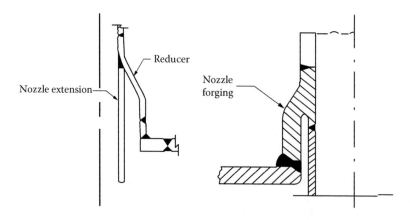

FIGURE 23.15 Nozzle welds with thermal sleeves for high-temperature applications.

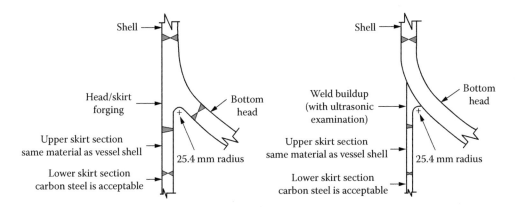

FIGURE 23.16 Typical preferred design for high-temperature vessel supports.

- For applications involving welding of two dissimilar steels having a wide difference in thermal expansion and involving fatigue loading like the example of 2.25Cr-1Mo to austenitic stainless steel piping in the power sector, especially two shift operations, it is required to incorporate an intermediate sleeve with a minimum length of 2.5/RT with an expansion coefficient intermediate between the two base materials like alloy 800 in this case.
- For heat exchanger applications involving media not compatible with each other like sodium and water in sodium-heated steam generated, it is worthwhile to have economical comparison between the internal bore butt weld of the tube to tube sheet with raised spigot, which permits volumetric inspection, prevents crevice corrosion, and allows the weld to be located in a low stress zone, and the traditional rolled and welded joint. The additional cost of an internal bore weld is likely to outweigh the material and manufacturing cost associated with the rolled and welded joint.

23.3.5 WELD MISMATCH AND CONTROL

A perfect matching of the internal or external surfaces of any two parts to be welded, before and during welding, along the entire length or circumference of the weld, is practically not possible. The degree of linear misalignment between the parts to be welded is called weld mismatch. Depending upon the mismatch, the quality of welds and forces/moments developed at the interface having the discontinuity would vary. Hence, the parts joined by welding are to be aligned, adjusted, and maintained in position during the welding operation, using processes such as jacks, clamps, bridges, tack welds, and special jigs in order to meet the tolerances given in the construction codes. Surface alignment after welding shall be such that ND examination may be properly performed. The allowable mismatch tolerances are defined as follows:

1. The centerlines of the parts to be assembled shall be aligned to within the fabrication tolerances.
2. For internal and external surface alignment tolerances between parts of identical thickness, for thickness, e, less than 5 mm, linear misalignment shall be less than 0.25e, and for thickness more than 5 mm, it shall be limited to 0.1e + 1 (with a maximum of 4 mm). If so, they can be welded directly without any correction.
3. For parts of different thicknesses, the tolerance of the offset of the centerlines of the plate thickness should respect the requirement specified earlier (2). Subsequently after welding, excess material can be removed from the thicker section to have the transition slope ≤0.25 (Figure 23.17).

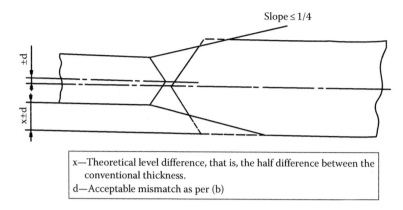

FIGURE 23.17   Alignment of parts of different thicknesses.

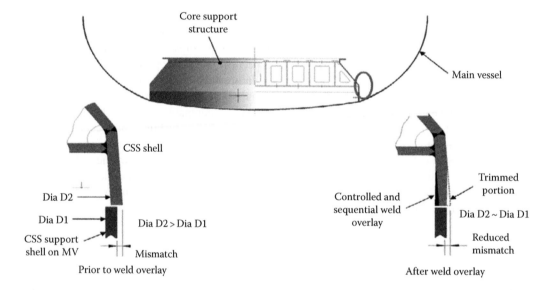

FIGURE 23.18   Mismatch correction between mating parts in a practical situation in an SFR.

In some situations, if requirement (2) and (3) are possible (this situation is illustrated in Figure 23.18), the code permits deposition of a weld metal termed as "weld overlay." This has to be done carefully respecting special inspection requirements specified in the code. In this case, the examination procedures shall cover the area where the metal is removed or where the metal is deposited. The removal of metal from the component with the thickest section shall only be permitted when this operation does not induce unacceptable stresses, for the class of component under consideration; when the stress level is unacceptable, only the deposition of the weld metal shall be permitted.

It is also possible to advantageously adopt the weld overlay technique to generate favorable distortion effects (resulting in thermomechanical effects) to facilitate joining of parts with edge mismatch to respect the codal limits (2 and 3). This is illustrated in the following case study.

Let the "bottom cylindrical shell of the CSS" be welded to the "support shell in the main vessel." The circumference of the "bottom cylindrical shell of the CSS" is to be controlled for proper weld fit-up with the "support shell in the main vessel." However, it was observed that the "bottom cylindrical shell" has taken a conical form resulting in the increase of circumference beyond the tolerance limit (Figure 23.18). Various options to overcome this problem, such as correction by mechanical means and correction by means of weld shrinkage through weld overlay on the inner surface of the

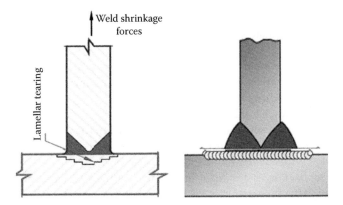

FIGURE 23.19 Weld overlay to avoid lamellar tearing.

bottom shell, were studied in detail, and the method of correction by weld overlay was adopted. Weld overlay on the inner surface of the bottom shell was carried out throughout the circumference, and the radius and circumference were measured after each layer of deposition. When the required circumference was achieved, the overlay was stopped. The outer and inner surfaces were ground smoothly to get a cylindrical shape for the bottom shell. The phenomenon of weld shrinkage has been constructively utilized here to meet the interface requirement at the bottom of the CSS. Due to this correction, the appropriate fit-up for welding could be achieved successfully.

It is worth to note that for the carbon steel items, overlay with an appropriate compatible material is sometimes deposited either at cruciform or at T junctions (which are prone to undergo lamellar tearing) to overcome/avoid lamellar tearing problems (Figure 23.19).

23.3.6 WELD INSPECTIONS

23.3.6.1 Techniques

Commonly used inspection techniques are LPE, radiography test (RT), and UT. Inspection is to be carried out at three stages: on weld edge preparation (LPE), during welding (LPE), and on finish weld (LPE and RT/UT). The acceptance criteria are that for radiography, lack of penetration, lack of fusion, crack, and fissures are not permitted. Case to case basis, gas cavities, and isolated slag inclusions may be permitted based on wall thickness. The quality of the RT is equivalent to 2-1 t as per the code. This means that the penetrameter is a device used in radiographic testing to evaluate the quality of radiographic images. The idea is to place a penetrameter on a specimen that is being radiographically tested. Demonstrating the detail of the penetrameter alongside the specimen shows that a certain level or percent of radiographic sensitivity has been achieved. Radiographic sensitivity is defined as the smallest or thinnest material change that a radiograph reveals. In Europe, penetrameters are referred to as image quality indicators (IQIs). Penetrameters come in a wide range of shapes, sizes, and configurations. They can be small metal plates containing the same- or variable-sized holes, or they can be a series of variable-diameter wires contained in plastic packets. Penetrameter wires or plates are made from the same type of material as the specimen being tested. The quality of RT specified as 2-1 t indicates that the penetrameter thickness is 2% of the weld thickness and the diameter of the hole in it is equal to the thickness of the penetrameter. The radiographic image should be able to record the previously mentioned hole in the penetrameter.

For the UT, the quality of the weld specified is based on echo amplitude. Volumetric indications (porosity, gas cavity, slag inclusion, etc.) are permitted on a case to case basis. Any nonvolumetric reflectors such as cracks and lack of fusion, lack of penetration, and undercut are not acceptable. For LPE, any linear indication >2 mm is not permissible. For more details, the readers are requested to refer to the codes in [23.6].

23.3.6.2 Time of Examination: During Execution of the Weld

- For welds at the intermediate stage, the examinations shall be performed after preparation of the surface bead.
- When it is carried out on welds exempted from volumetric examinations after completion, LPE during execution of the weld is performed with the following frequency:
  - Every three layers when e $\geq$ 20 mm
  - Every two layers when e < 20 mm
- After preparation of the back of the weld, the manufacturer is recommended to carry out an examination by magnetic particle examination or by liquid penetration examination. This examination is mandatory in the following cases:
  - For welds on backing strips when the removal of the backing strip is followed by a backing run. After preparation of the back of the weld, an examination shall be carried out by magnetic particle examination on ferritic steels.
  - When no final examination of the weld by radiographic or ultrasonic methods is specified.

23.3.6.3 Time of Inspection after Completion of the Weld

23.3.6.3.1 Volumetric Examinations

For full-penetration welds and homogeneous nonalloy and low-alloy steel buttering

- The final volumetric examination shall be carried out after the final stress-relieving heat treatment or after intermediate heat treatment if this is performed at a temperature within the limits specified for the final stress-relieving heat treatment.
- If both radiographic and ultrasonic examinations are required, the final volumetric examination after stress-relieving heat treatment shall be an ultrasonic examination. However, ultrasonic examination may be performed before heat treatment, and in that case, a second ultrasonic examination shall be carried out with the aid of shear waves only and with a single refraction angle after stress-relieving heat treatment.

23.3.6.3.2 Surface Examinations

- It shall be carried out after the final heat treatment if the thicknesses involved, the processes used, or the nature of the materials give grounds for fears that defects may occur or may be aggravated during the heat treatments.
- However, surface examination shall be carried out after the final stress-relieving heat treatment or after an intermediate heat treatment if this is performed at a temperature within the limits specified for the final stress-relieving heat treatment in the following cases:
  - Welds made with backing strips when these have been removed
  - If volumetric examination of the weld is not practicable

23.3.7 HEAT TREATMENT

Heat treatment is the controlled heating and cooling of metals to alter their physical and mechanical properties without changing the product shape. Heat treatment is sometimes done inadvertently due to manufacturing processes that either heat or cool the metal such as welding or forming. Heat treatment is often associated with increasing the strength of the material, but it can also be used to alter certain manufacturability objectives such as to improve machining, to improve formability, and to restore ductility after a cold working operation. Thus, it is a very enabling manufacturing process that can not only help other manufacturing process, but can also improve product performance by increasing strength or other desirable characteristics. Steels are particularly suitable for heat treatment, since they respond well to heat treatment. Broadly, the heat treatments followed for stainless steels could be categorized under solution annealing, stress-relieving, and dimensional stabilization. Basically, solution annealing heat treatment at 1050°C–1150°C is carried out to regain the base material properties, dissolve pre-cipitates, remove adverse effects of excessive cold work, and restructure the grains. Controlled heating and cooling to avoid inadvertent formation of undesirable precipitates during heat treatment are very important.

The stress-relieving heat treatment, for example, of Ni–Cr (Co free) hardfaced SS parts, with a maximum of up to 750°C along with appropriate hold times, is essential to relieve the residual stress in the sections introduced on account of various forming and welding operations. The stress-relieving operation, in general, helps in the removal of locked-in stresses and brings the ductility values on par with the original material. Stress-relieving heat treatment is also normally applied to formed parts with excessive cold work (>15%, RCC-MR, and >10%, ASME). However in some cases that are with excessive cold work beyond the limits specified, the need for a stress-relieving heat treatment can be dispensed after proper material testing results and with justification on the adequacy of the remaining ductility as fitness for purpose case.

The dimensional stabilization heat treatment is carried out for either the precision of machined parts or assemblies in order avoid stress-induced distortions during operation. The parts/assemblies are normally heat treated to about 50°C more than the operating temperatures with predetermined hold times dictated by the thickness of parts in the component/assembly. No metallurgical changes are induced in the structure during this heat treatment. Further, due care is to exercise control of the temperature gradient across the length of the components in all the previously mentioned heat treatments, which is very essential to minimize distortion, buildup of unfavorable stresses, etc.

For postweld heat treatment of carbon steel components, design codes such as RCC-MR, and ASME provide guidelines for the estimation of the effective thickness approach for specifying the need for heat treatment. This is particularly very useful for box structures, wherein the need for heat treatment is very essential for relieving the built-in stresses during welding. Heat treating all box structures would be unnecessary and uneconomical. The effective thickness approach is very handy in this regard, and the heat treatment temperature is in the range of 550°C–625°C. The maximum temperature of heat treatment increases with increasing alloy content.

23.3.8 Weld Strength Reduction Factors

The welds are a weak link of the structures or components, and the structural wall thickness, creep, and fatigue life of the components are significantly affected by the presence of welds. The methods adopted by the weld strength reduction factors are described in the following:

The basic allowable stress intensity S_t for the weld is reduced by J_t. Similarly, in the vicinity of a weld (defined by ±3 times the thickness to either side of the weld centerline), the creep fatigue evaluation shall utilize reduced values of the allowable number of design cycles N_d and the allowable time duration T_d. The N_d value is read from the design fatigue curve and the T_d value is read from the minimum stress to rupture curve. The design fatigue and rupture curves for welds are derived from the base material data by applying appropriate strength reduction factors: J_f for the fatigue strength reduction factor and J_r for the creep strength reduction factors. The method of application weld design procedure recommended in RCC-MR by establishing the design fatigue curve and creep rupture curves for welds is explained schematically in Figures 23.20 and 23.21, respectively. The net factor of safety that the design code RCC-MR has for accounting various factors that limit the component life is illustrated in Figure 23.22. More details can be seen in Ref. 23.7.

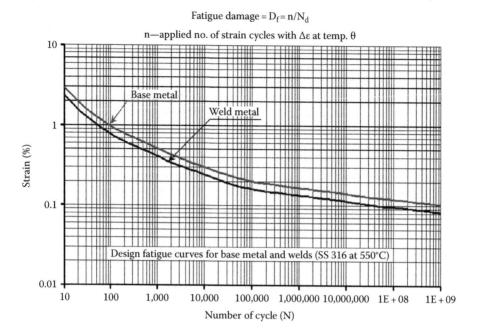

FIGURE 23.20 RCC-MR fatigue damage assessment procedure for welds. (From RCC-MR, Construction Rules for Mechanical Components of FBR Nuclear Island for class 1 components, Subsection B: Class 1 Components, AFCEN, 2007.)

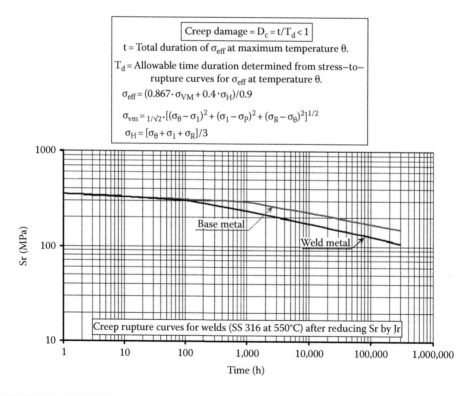

FIGURE 23.21 RCC-MR creep damage assessment procedure for welds. (From RCC-MR, Construction Rules for Mechanical Components of FBR Nuclear Island for class 1 components, Subsection B: Class 1 Components, AFCEN, 2007.)

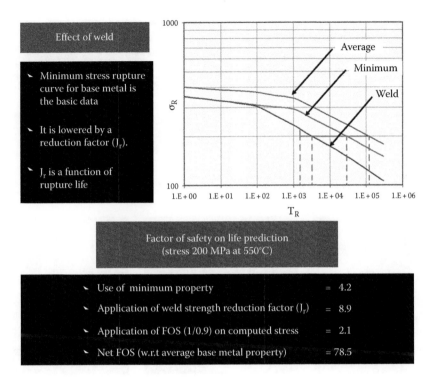

FIGURE 23.22 Net factor of safety incorporated in RCC-MR for creep design of welds.

23.3.9 LESSONS LEARNED FROM TECHNOLOGY DEVELOPMENT OF WELDED GRID PLATE: CASE STUDY

In this case study, lessons learned during the technology development exercise carried out for welded GPs are presented. The overall diameter of the GP is 6.1 m. It has more than 900 sleeves connecting the top and bottom plates and more than 700 spigots attached to the top plate. Welded joints are provided between the shell and the top and bottom plates, the sleeves and the plates, and the spigots and the top plate. Geometrical details are presented in Figure 23.23. The innovative GP concept conceived for future SFRs is the first of its kind, and hence, there is a need to demonstrate the manufacturing technology satisfying the required tight dimensional tolerances. Several challenges were faced during technology development. The important issue is distortion control. Tight distortion control measures need to be taken for the welds between (1) sleeves and plates,

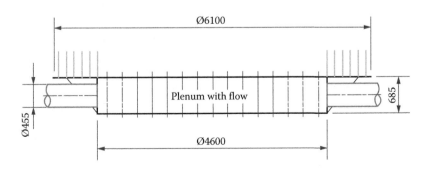

FIGURE 23.23 Geometrical details of the welded GP.

FIGURE 23.24 Manufacturing stages of the welded GP.

(2) shell and plates, (3) spigot and plate, and (4) nozzle and shell. Manufacturing stages are shown in Figure 23.24. A few challenging welds that have been accomplished after introducing some innovative ideas are shown in Figures 23.25 and 23.26 for sleeve to plate and plate to shell joints, respectively. Challenges faced and lessons learned are presented comprehensively in Table 23.2. More details can be seen in Ref. 23.8.

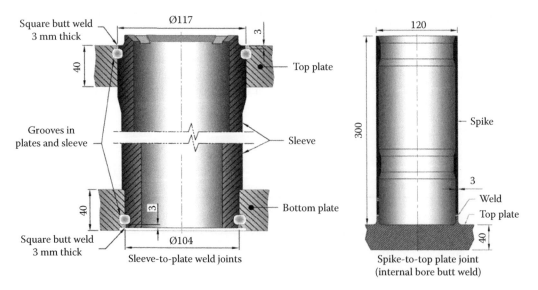

FIGURE 23.25 Challenging welded joints in the welded GP.

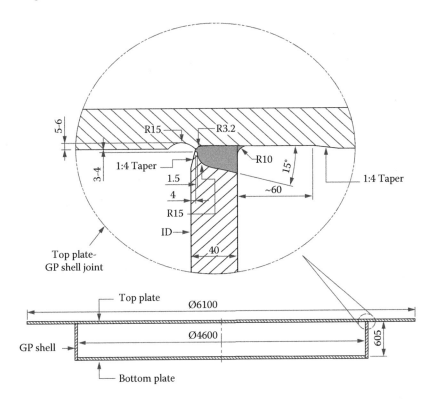

FIGURE 23.26 Typical plate to shell weld joint achieved with a groove.

TABLE 23.2
Experiences Gained during Technology Development of a Grid Plate

Challenge Faced	Solution Adopted/Lessons Learned
Distortion in the top plate while machining after long seam welding.	Full stress relieving of the plate was done before further machining.
Due to reduced height of the grid plate, the proposed double-sided welds for shell to plate joints could not be carried out.	The joint configuration is modified such that the required inside fillet radius is achieved even with one-sided welding from the outside. The new joint design is amenable for the required nondestructive testing (NDT).
Jacks were used for controlling distortion during integration welding of plates with the shell. Bowing of the top and bottom plates took place after removal of distortion control jacks. This resulted in an increase in the distance between plates at various locations. Hence, the allowance kept on the length of sleeves was not sufficient.	The final minimum thickness of the top and bottom plates relaxed from 40 to 35 mm, considering the developmental nature of the job. In the case of the grid plate for the reactor, more margins for sleeve lengths and full stress relieving will be carried out after integration welding (before removal of distortion control jacks).
Distortion of the hardfaced top conical support portion of the sleeve during welding between the sleeves and plate.	Configuration and welding process were suitably modified.
Tight tolerances on the welded assembly were specified.	Appropriate machining and welding sequences were adopted that helped to achieve the dimensions within specified limits.

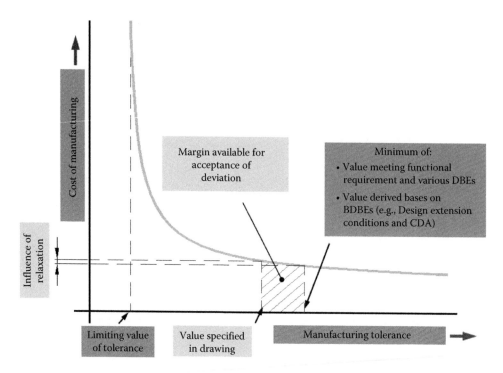

FIGURE 23.27   Manufacturing cost versus specified manufacturing tolerance.

23.4   SUMMARY

Once various tolerances (their definition, basis, and effects that could arise during manufacturing and erection stages) are understood, erection procedures can be defined appropriately. The specification of optimum tolerance is a key point in achieving economy and safety, which has been summarized in Figure 23.27. Each country or organization has its own manufacturing and erection methodologies/strategies. Hence, it is not possible to present them all. However, generalized approach and manufacturing procedures along with various tolerance limits recommended in the manufacturing codes are presented in this chapter. Further, a few illustrations are presented in Chapter 24 particularly with more detailed information for PFBRs.

REFERENCES

23.1 Akiyama, H. (1997). Seismic Resistance of FBR Components Influenced by Buckling. Kajma Institute Publishing, Japan.
23.2 Structural Integrity of the Fast Reactor. (1992). Report by the UK Nuclear Industry, Technical Advisory Group on Structural Integrity, AEA Technology, Warrington, U.K.
23.3 ASME. (2001). Rules for Construction of Pressure Vessels. Section VIII Div 1.
23.4 ASME. (2007). Rules for Construction of Nuclear Power Plant Components, Subsection NH: Class 1 Components. Section III Div 1.
23.5 RCC-MR. (2007). Construction Rules for Mechanical Components of FBR Nuclear Island for class 1 components. Subsection B: Class 1 Components. AFCEN.
23.6 ASME. (2007). Rules for Construction of Nuclear Power Plant Components, Subsection NB: Class 1 Components. Section III Div 1.
23.7 Chellapandi, P., Chetal, S.C. Chapter 9: Design against cracking in ferrous weldments. In *Weld Cracking in Ferrous Alloys*. Woodhead Publishing, Cambridge, U.K.
23.8 IGCAR Annual. (2012). Report technology development of components for innovative CFBR reactor assembly, IGCAR, India.

24 Illustrations from International SFRs

24.1 MONJU

24.1.1 MANUFACTURE OF KEY COMPONENTS

The reactor vessel of Monju is a circular cylindrical shell, made of austenitic stainless steel about 7 m in diameter, 18 m high, and 50 mm thick. Figure 24.1 shows a schematic sketch of a reactor assembly (RA). To improve the structural reliability by reducing the number of welded parts, the vessel is composed of 12 pieces of forged rings and has only circumferential welds. The shield plug had to be firm enough to support the components installed on it and to keep them properly aligned with the core internals. For this reason, the plug was manufactured with a proven thick rib structure. Since the main circulation pump in the primary system is vertical with a thin shaft about 6 m long, much consideration was given to the fabrication of rotating parts to ensure a highly controlled and balanced rotation. The balance of the shaft was tested under the same high-temperature conditions as in the actual operation. In the helical coil–type steam generator (SG), butt welds were used for both the tube-to-tube and the tube-to-tubesheet joints of the SG. An automated welding machine was developed, principally for the tubesheet welds. The shield plug, fuel handling machine, ex-vessel storage tank, and other major components were assembled in advance at the factory, and assembly tests were conducted to confirm the performance of the system prior to reassembly on site.

24.1.2 CONSTRUCTION

Construction of Monju began in October 1985 with the foundation work for the reactor building and reactor auxiliary building, which are located at the center of the plant. The construction of the reactor containment building (RCB) was completed in April 1987. The reactor vessel was installed in October 1989 (Figure 24.2). Figure 24.3 shows the core internals after installation. Construction was completed in April 1991. Functional tests were carried out between May 1991 and December 1992. Reactor criticality was achieved in April 1994. The entire construction process took ~6 years and criticality was achieved in ~3 years after carrying out various functional and start-up tests.

A great deal of reinforcement and many anchor bolts for components were installed in the reactor building, which has a complicated shape and contains the reactor vessel and other major components. Prior to construction, models of the reinforcement arrangement were built to see how the reinforcement would interact with the component basement structures and whether the fabrication sequence could be improved. The reactor cavity wall around the reactor vessel is made of serpentine concrete covered with steel plates. This steel cover is a structural member supporting the reactor vessel. It had to be installed with extreme precision, particularly with regard to remote refueling requirements. For this reason, each segment was fabricated in the factory and had its dimensions confirmed by temporary assembly before being finally assembled on site, maintaining a tolerance of 1 mm or less. The reactor vessel and internals were constructed in the following order: (1) guard vessel for the reactor vessel, (2) reactor vessel, (3) core internals, (4) top shield plug, (5) upper core structure, and (6) control rod drive mechanisms. In transportation, the guard vessel and the reactor vessel weighed ~500 tons. Both were shipped to the site and moved with rollers from the site assembly shop to the reactor vessel compartment over a period of 5 days. Installation work then

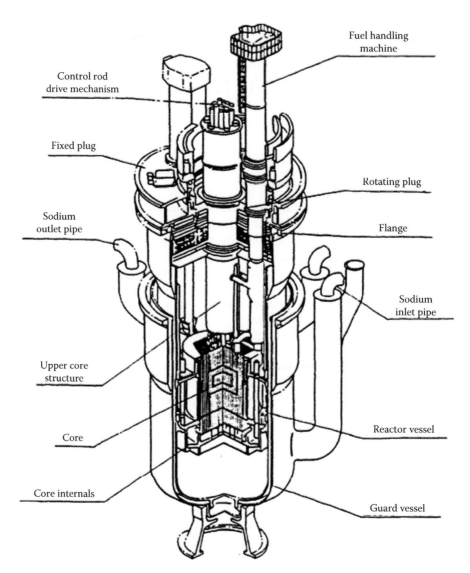

FIGURE 24.1 Schematic sketch of the RA of Monju. (From Takahashi, T. et al., *Nucl. Technol.*, 89(2), 162, 1990.)

proceeded, with special attention paid to assure both accurate installation and cleanliness of the components. Pieces of the main cooling pipe work were manufactured in the factory; the piping supports and the main cooling piping connections to the components were then welded on site. The same automatic welding technique was usually applied at the site as had been used in the factory to maintain quality. Figures 24.4 and 24.5 depict the manufacturing and erection stages of typical intermediate heat exchanger (IHX) and SG (evaporator). More details may be seen in Ref. 24.1.

24.2 SUPER PHENIX (SPX1)

The main vessel is a 21 m diameter, fabricated austenitic steel structure designed to contain the core, the internals, 3500 tons of sodium, the four primary pumps, and eight intermediate heat exchangers. The core support structure is welded to the main vessel. Diagrid is a cylindrical box structure and holds the subassemblies. The weight of the diagrid with subassemblies and the inner

FIGURE 24.2 Erection of reactor vessel of Monju. (From Takahashi, T. et al., *Nucl. Technol.*, 89(2), 162, 1990.)

FIGURE 24.3 Erection of core subassemblies in Monju. (From Takahashi, T. et al., *Nucl. Technol.*, 89(2), 162, 1990.)

vessel is transmitted to the core support structure. Eight discharge pipes from four primary pumps supply the sodium coolant to the core through the diagrid. The inner vessel is a fabricated, double-walled steel structure designed to provide a stagnant region of sodium between hot and cold sodium pools. A roof slab forms the primary containment boundary in the axially upward direction and supports primary pumps and intermediate heat exchangers. In order to fulfill the demanding requirements of containment and component support, the roof slab had to be an extremely large, annular metal box structure filled with heavy concrete shielding. The lower surface was heat insulated. The cylindrical opening in the middle of the roof slab was closed with two rotatable plugs,

(a)   (b)

FIGURE 24.4   Internals of intermediate heat exchangers: (a) assembled and (b) erected. (From Takahashi, T. et al., *Nucl. Technol.*, 89(2), 162, 1990.)

FIGURE 24.5   140 heat transfer tubes are coiled in SG (evaporator) for Monju. (From Takahashi, T. et al., *Nucl. Technol.*, 89(2), 162, 1990.)

one located off center within the other. The core cover plug, on the small rotatable plug, supports core instrumentation and safe shutdown mechanisms.

24.2.1 Construction

24.2.1.1 Reactor Assembly

The construction activity at the site started in mid-1977. Two factors that strongly affect the construction strategy and schedule are the larger dimensions of the RA as seen in Figure 24.6

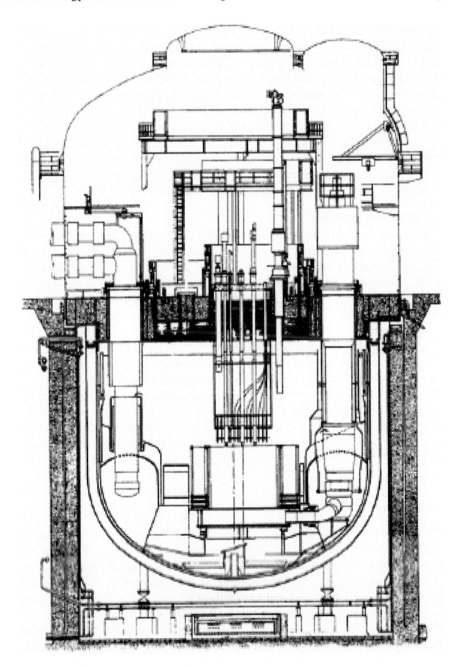

FIGURE 24.6 Schematic sketch of the RA of Super Phenix.

with the reactor vessel diameter of about 20 m and the extremely thin walled shells (25–60 mm). Considering the difficulty in handling and incompatibility with normal rail and road transportation, the equipment was built in the following three stages:

1. Parts of the various components were prefabricated at manufacturing sites where full facilities were available and in sizes compatible for rail/road transportation.
2. Individual parts were transported to the site and assembled in the special site workshop.
3. Finally, the various components were assembled, from outer to inner, in their final positions in the reactor building.

24.2.1.2 Vessels and Internals

Owing to very large diameters (21 m for reactor vessel and 22.5 m for safety vessel), the vessels are actually built in the site workshop. Shop prefabrication involved rolling of the steel petals, about 1.8 m in width, and assembling them into panels of sizes that were within the limits of rail/road transportation. The panels were assembled and vessels were built on site. Considering the difficulty posed by automatic welding of 316 SS steel with high boron content (possibility of microcracking), manual welding was opted using covered electrodes under extremely stringent operating conditions. The diagrid is an entirely different type of construction when compared to the vessels. It is a steel lattice component precision assembled with several bolts and equipped with 835 holders for receiving the core subassemblies. Considering the large machining requirements that were not available at the site workshop, the component was manufactured at the manufacturer's.

24.2.1.3 Top Closing Structure

The components of the top closing structure consisting of a roof slab, a large rotatable plug, and a core cover plug were prefabricated at the manufacturer's and assembled at the site workshop. However, the small rotatable plug was entirely assembled at the manufacturer's and transported to the site. The six inner and outer segments of roof slab were assembled at the supplier's using partially automatic, multipass welding and each segment was stress relieved. At the site workshop, these sectors were integrated by manual welding. Considering that the local thickness was less than 35 mm, no stress relieving was carried out. The total weight of the component was ~800 tons. A large rotatable plug was prefabricated in two parts at the supplier's and integrated at the site workshop. Various thick rings in the plug support arrangements were machined from thick butt-welded plates in two pieces assembled on site. Local stress relieving was carried out for these rings at the site. Considering the extreme operating conditions for core cover plug, austenitic steel construction was adopted except for support plate (at the top). Due to several penetrations, concrete shielding was not adopted for this plug and, alternatively, the required numbers of steel plates were stacked at the same level where shielding material was provided in the adjoining plugs and roof slab. The component was manufactured in two parts, assembled and instrumented at the site.

24.2.2 Site Erection

Site erection formed part of the third stage of the construction process. The components were transported from the site workshop to the transfer location near RCB by trolleys running on rails. From the transfer station, the components were vaulted into RCB through an opening in its wall using a special handling system (Figure 24.7). The handling system consisted of two stable gantries forming a rigid structure, capable of raising and moving heavy loads. The gantries were equipped with hydraulic hoists. Elevated skids on piles extended into the RCB where this overhead runway was used to install the components in the reactor pit. The sequence of the erection of components was as follows: (1) safety vessel; (2) assembly of main vessel, core catcher, core support structure, and base of thermal baffle; (3) inner vessel equipped with pump standpipes and cylindrical shell of baffle; (4) roof slab; (5) diagrid and IHX sleeves; and (6) rotatable plugs and core cover plug. Figure 24.8 shows the

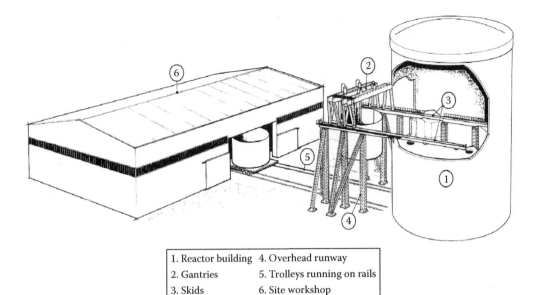

1. Reactor building	4. Overhead runway
2. Gantries	5. Trolleys running on rails
3. Skids	6. Site workshop

FIGURE 24.7 Large component handling system used in SPX. (From NERSA, The Creys-Malville power plant, Electricite de France, Direction de L' Equipement, Region D'Equiopment, Alpes-Lyon, France, 1987.)

FIGURE 24.8 Handling of safety vessel and main vessel in series for SPX. (From NERSA, The Creys-Malville power plant, Electricite de France, Direction de L' Equipement, Region D'Equiopment, Alpes-Lyon, France, 1987.)

erection of the reactor vessel followed by the erection of the safety vessel. To facilitate joining of main and safety vessels with the roof slab, both the vessels were raised using jacks provided below the safety vessel. Further, there were some intermediate stages of manufacturing and erection of control plug, IHX, primary sodium pumps, and SG, as shown in Figures 24.9 through 24.12.

More details of manufacturing and construction of SPX reactor components may be seen in Ref. 24.2.

(a)

(b)

(c)

(d)

FIGURE 24.9 Stages of manufacture and erection of SPX-control plug. (a) Breda, February 1981: Core cover plug lattice, (b) August 6, 1982: The core cover plug positioned the reactor, (c) Breda, November 1981: The core cover plug bottom plate being fitted, and (d) Breda, August 1981: Fitting guides for the cladding failure detection tubes onto the core cover plug bottom plate. (From NERSA, The Creys-Malville power plant, Electricite de France, Direction de L' Equipement, Region D'Equiopment, Alpes-Lyon, France, 1987.)

24.3   500 MWE PROTOTYPE FAST BREEDER REACTOR

24.3.1   BRIEF DETAILS OF PROTOTYPE FAST BREEDER REACTOR

Prototype fast breeder reactor (PFBR) is a 500 MWe capacity pool-type sodium-cooled fast reactor (SFR), designed and developed at Indira Gandhi Centre for Atomic Research (IGCAR), Kalpakkam in India. PFBR has two primary and two secondary loops with four SGs per loop. The temperature of the cold pool is 670 K and that of the hot pool is 820 K. The sodium from the hot pool mixes with the cold pool after transporting its heat to four IHXs. The circulation of sodium from the cold pool

FIGURE 24.10 Stages of manufacture and erection of SPX-IHX. (From NERSA, The Creys-Malville power plant, Electricite de France, Direction de L' Equipement, Region D'Equiopment, Alpes-Lyon, France, 1987.)

FIGURE 24.11 Stages of manufacture and erection of SPX-primary pump. (From NERSA, The Creys-Malville power plant, Electricite de France, Direction de L' Equipement, Region D'Equiopment, Alpes-Lyon, France, 1987.)

FIGURE 24.12 Stages of manufacture and erection of SPX-SG. (a) Steam generator being hoisted for positioning in its buildings A steam generator weighs 191 tons, (b) Bringing a steam generator into the building specially made for it, (c) Creusot-loire, March 1981: Steam generator in the workshop, and (d) January 1983: Connecting the steam generator outlets to the steam collector. (From NERSA, The Creys-Malville power plant, Electricite de France, Direction de L' Equipement, Region D'Equiopment, Alpes-Lyon, France, 1987.)

to the hot pool is maintained by two primary sodium pumps and the flow of sodium through IHX is driven by a level difference (1.5 m of sodium) between the hot and cold pools. The heat from IHX is in turn transported to eight SGs by sodium flowing in the secondary circuit. The steam produced in the SG is supplied to a turbo generator. In the RA (Figure 24.13), the main vessel houses the entire primary sodium circuit including the core. Sodium is filled in the main vessel with free surfaces blanketed by argon gas. The inner vessel separates the hot and cold sodium pools. The reactor core consists of about 1757 subassemblies including 181 fuel subassemblies. The control plug, positioned just above the core, houses mainly 12 absorber rod drive mechanisms. The top shield supports the primary sodium pumps, IHX, control plug, and fuel handling systems. PFBR uses mixed oxide with depleted uranium, 20.7% Pu oxide in the inner core and 27.7% Pu oxide in the outer core. For the core components, 20% cold worked D9 material (15% Cr–15% Ni with Mo and Ti) is used for better irradiation resistance. Austenitic stainless steel type 316 LN is the main structural material for the

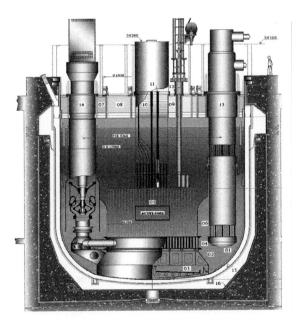

01. Main vessel
02. Core support structure
03. Core catcher
04. Grid plate
05. Core
06. Inner vessel
07. Roof slab
08. Large rotatable plug
09. Small rotatable plug
10. Control plug
11. Control and safety rod mechanism
12. In-vessel transfer machine
13. Intermediate heat exchanger
14. Primary sodium pump
15. Safety vessel
16. Reactor vault

FIGURE 24.13 Schematic sketch of the RA of PFBR.

out-of-core components and modified 9Cr-1Mo (grade 91) is chosen for SG. PFBR is designed for a plant life of 40 years with a load factor of 75%. The commissioning of PFBR has been started.

24.3.2 MANUFACTURING STRATEGY FOLLOWED

Bharatiya Nabhikiya Vidyut Nigam Limited (BHAVINI), a government company, is constructing PFBR, India's first commercial SFR project, in coordination with the IGCAR (responsible for design and R&D). Further, several major industries in the country were involved in the manufacture of the reactor components. Certain manufacturing and welding procedure qualifications, for example, relaxation of requirements for the non-load-carrying and nonsodium wetted boundaries of safety vessel and core catcher, were accepted. Testing and inspection procedures were simplified based on their importance and duty. An alternative bend test procedure was suggested for dissimilar welding at the roof slab. Procedures for the assembly of parts were simplified to ensure erection tolerance limits. However, for the main vessel, care was taken to practically avoid repair of welds or laminations. With a thorough and comprehensive analysis respecting interface requirements with other components, as-built dimensions were accepted. In certain noncritical cases (e.g., core catcher and safety vessels), the concept of *fitness for purpose* approach was adopted judiciously.

24.3.3 MANUFACTURING CHALLENGES

For the manufacturing of components, robust construction codes, standards, and methodologies are still evolving and they need to be adopted with thorough validation. The most important aspect is the specification of manufacturing tolerances, which should be written on a rational basis of functional requirements and structural integrity considerations; they should truly reflect similar industrial experiences and capabilities. Appropriate and novel mock-up trials are essential when a component is manufactured for the first time. The assembly sequences should be well understood and efficient handling schemes are to be designed and validated. It is essential that the components are manufactured practically with insignificant repairs. A few challenging issues experienced during the manufacture of PFBR components are addressed here. More details can be seen in Ref. 24.3.

24.3.3.1 Large-Diameter Thin Shell Structures

Major manufacturing challenges for the large-diameter thin vessels—main vessel, thermal baffles, inner vessel, safety vessel—are addressed here. The basic plates should not have any defects such as laminations (high-quality control is essential). For the large lengths of welds resulting during integration of individual petals, stringent control on the manufacturing deviations, such as form tolerances (<½ thickness), verticality and horizontality (<± 2 mm) and low residual stress are to be achieved without any heat treatment. High-quality welds should be achieved without any significant weld repairs. Cold forming limiting value should be <10% and close form tolerances limited to ±12 mm (≤0.2% R) as shown in Figure 24.14. These have been achieved by adopting (1) stringent dimensional control on petal level and subsequently robust weld fit up and weld sequence methodology, (2) state-of-the-art techniques for inspection and quality control, (3) numerical simulations of forming and welding procedures and innovative mock-up trials, (4) incorporating the lessons learnt from feedback experiences of various industries, and (5) elaborate technology development exercises.

24.3.3.2 Grid Plate

The grid plate accomplishes the important function of supporting the core subassemblies (SA), the inner vessel, and the in-vessel transfer position. It also serves as a plenum for distributing sufficient coolant flow to the individual SAs and also to the main vessel cooling pipes. Geometrically, the grid plate is a circular box–type structure of bolted construction. It consists of top and bottom plates and an outer cylindrical shell. The top and bottom plates are interconnected by a number of tubes called *sleeves*. Each SA rests on the conically shaped hard-faced surface (chrome nitrided) at the top of the respective sleeve. The nickel-base alloy chosen for hardfacing is highly susceptible to cracking and it was required to deposit this alloy on components of very large dimensions without any cracks. Due to the large volume of hard-face material deposits, achieving hardfacing without cracking required special technology development efforts involving designers, fabricators, and metallurgists. This helped successful hardfacing of the bottom plate of the prototype grid plate assembly. Further, the hardfacing of the inner surfaces of a very large number of sleeves (1757), in which

Component	ASME	Form tolerance on radius (mm)		
	$(ID_{Max} - ID_{Min})$	RCC-MR	PFBR	
			Specified	Achieved
Main vessel	±70	±50	±12	<±12 (during fit up) <±18 (at isolated locations)
Safety vessel	±70	±50	±12	±12 (at majority locations) ±18 (at isolated locations)
Inner vessel	±67	±50	±12	<±8 (during fit up) <±20 (at isolated locations)

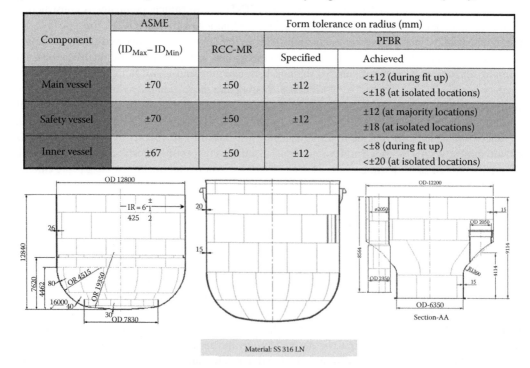

Material: SS 316 LN

FIGURE 24.14 Form tolerances specified and achieved for these vessels.

 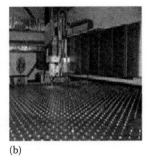 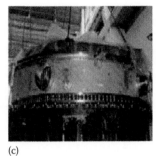

(a) (b) (c)

FIGURE 24.15 Important manufacturing stages and handling of the grid plate. (a) Hardfacing, (b) machining, and (c) handling.

the fuel subassembly rests, was another critical task that was successfully completed through an indigenous technology. Nozzle welding with least distortion, large-diameter Colmonoy hardfacing without defects, hardfacing of a large number of sleeves on the inner diameter at a depth of 500 mm were some of the accomplishments during the manufacturing of the grid plate. This was possible by adopting challenging methods in dimensional control, innovative distortion control methods, modern dimensional inspection methods, handling of plates with large percentage of perforation, and an innovative way of heat treatment of large dimensioned components and novel techniques, indigenously developed for handling/assembly of large number of items (>14,000). Figure 24.15 shows the various manufacturing stages and handling of the grid plate.

24.3.3.3 Roof Slab

A box-type structure, made of A48P2 carbon steel (similar to ASTM A516 Grade 70 with lower limits for sulfur) specified in the RCC-MR, consists of a bottom plate and a top plate with the interspace filled with concrete. This is a massive structure of about 12.9 m diameter and 1.8 m height and weighing about 650 tons. The thickness of the plate was chosen to be 30 mm with the consideration that a stress-relieving heat treatment is not mandatory as per the fabrication codes. The fabrication of the box-type structure involves dissimilar metal weld joint between A48P2 and 316 LN SS shells. A specific issue caused significant time delay in laminar tearing problem. The plates were supposed to meet a thickness ductility of 25% and ultrasonic inspection requirements. However, during fabrication of the roof slab, lamellar tearing was observed in one of the heats (batch) of this steel, though this heat (batch) had met all the specification requirements. This was overcome by buttering the plate surface by welding, as permitted by the American Society of Mechanical Engineers (ASME), and by changing the joint design. Figure 24.16 provides the details of the various weld configurations in the roof slab and the configurations selected for avoiding laminar tearing based on extensive R&D activities.

24.3.3.4 Steam Generators

An SG is the most critical component because its reliable operation decides the load factor of an SFR plant. It is made of modified 9Cr-1Mo steel. It is a tall component of about 25 m height and has about 540 tubes. The tube-to-tubesheet joint is the most critical one, since any leak in this joint can result in direct sodium–water reaction, producing severe consequences. Hence, by adopting raised spigot in-bore welding, tube-to-tubesheet joints are welded with stringent acceptance standards on dimensions and weld quality without any defects such as lack of penetration and lack of fusion, cracks, undercuts, and unacceptable porosity. The maximum concavity achieved is practically zero and maximum weld thinning is less than the permissible value of <0.2 mm. As SGs are made of grade 91 steel, this necessitates a dissimilar joint between stainless steel piping headers of grade 91 steel. A hot-wire NG-TIG welding process was employed the first time in India for the fabrication of header-to-tubesheet joint in an SG. Figure 24.17 shows the schematic of an SG as well as the tube-to-tubesheet welding joint details. Figure 24.18 shows the last manufacturing stages of the SG at the Indian industry.

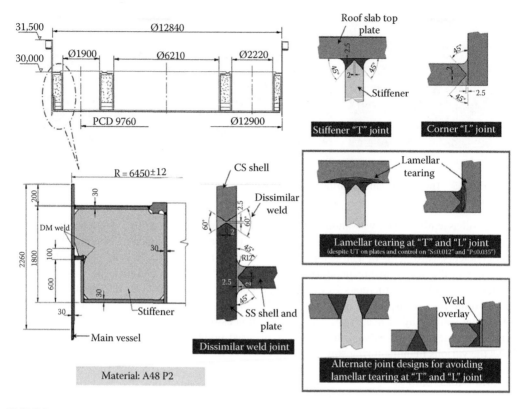

FIGURE 24.16 Improved weld configurations to overcome laminar tearing problems.

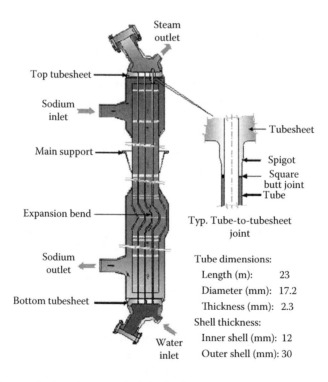

FIGURE 24.17 SG and tube-to-tubesheet weld joint details.

FIGURE 24.18 A manufacturing stage of SG at the industry.

24.3.3.5 Thermal Insulation Panels

Austenitic stainless steel plate–type insulations are provided in the form of panels around the safety vessel to reduce heat transfer to reactor vault. As these types of insulations are not commercially available, they were designed and fabricated indigenously using 0.1 mm thick sheets stacked to form panels. Dimples were provided to ensure spacing between the sheets. Details of a typical panel are shown in Figure 24.19. The manufacture and assembly of the panels were completed overcoming several challenges in thin sheets forming, achieving uniform emissivity, uniform spacing, formation of dimples without cracks, and complicated assembly sequences over the safety vessel. Innovative experiments were carried out on thermal insulation panels to confirm compliance of thermal and seismic design requirements (Figure 24.20).

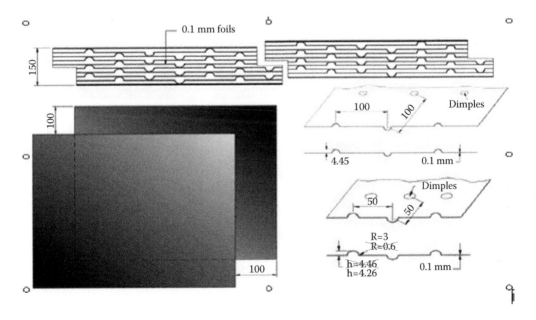

FIGURE 24.19 Geometrical features and details of the thermal insulation panel.

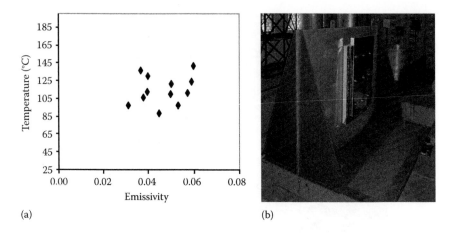

(a)   (b)

FIGURE 24.20   Qualification of insulation panels for (a) thermal (temperature vs. emissivity) and (b) seismic performances.

24.3.4 ERECTION CHALLENGES

In order to complete the project in time, it was required to carry out civil construction and equipment erection in parallel, involving state-of-the-art erection equipments and construction methodologies and optimized construction sequences. The installation of individual permanent components was completed precisely meeting the specified erection tolerances. An elegant methodology was finalized for the subsequent erection of the main vessel along with internals and top shield, respecting various erection tolerances as well as giving due considerations to schedule and economy. Transportation of thin shell

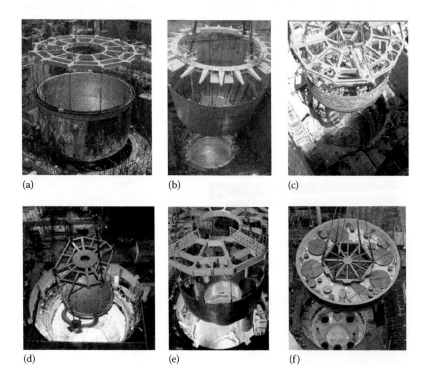

(a)   (b)   (c)

(d)   (e)   (f)

FIGURE 24.21   Erection of RA components. (a) Safety vessel, (b) main vessel, (c) thermal baffles, (d) grid plate, (e) inner vessel, and (f) roof slab.

FIGURE 24.22 Loading of the last few dummy subassemblies onto the grid plate.

FIGURE 24.23 Erection scenario of primary ramp and primary tilting mechanism.

structures from the site assembly to the support locations is another challenging activity in the construction, executed in innovative or novel ways to achieve economy without affecting safety. Erection of very large dimensioned and slender components is achieved with highly stringent dimensional accuracies on horizontality over 15 m (<±1 mm), the most challenging task completed for the first time in the country with systematically planned mock-up trials. The application of modern computer software has helped to resolve many of the assembly and construction problems. Figure 24.21 depicts the erection sequence of the RA components. Figure 24.22 shows the loading of the last few dummy core subassemblies on the grid plate. Figure 24.23 depicts the erection scenarios of fuel handling parts (primary ramp and primary tilting mechanism) into the main vessel. Figure 24.24 shows the erection stages of rotatable plugs, control plug, IHX, and turbo generator.

24.3.5 MAJOR DECISIONS TAKEN DURING THE ERECTION OF COMPONENTS

Computer simulations were extensively used based on 3-D virtual models for establishing erection sequences. Figure 24.25 shows the erection sequence established for the assembly components. Further, a full-scale mock-up was built for visualizing the complicated layout of the top shield components (Figure 24.26). This mock-up will further help to ensure the availability of space and access for adding and removing complementary shield blocks at any time and smooth operation of trailing cables without any entanglement during the rotation of plugs. Computer simulations and

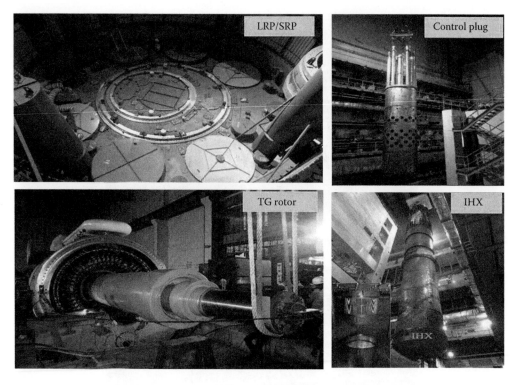

FIGURE 24.24 Erection stages of (a) rotatable plugs (LRP/SRP), (b) control plug, (c) TG rotor, and (d) IHX.

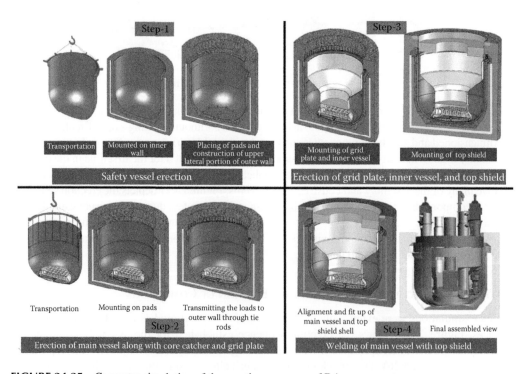

FIGURE 24.25 Computer simulation of the erection sequence of RA components.

FIGURE 24.26 Full-scale top shield layout mock-up.

mock-up trials helped to ensure good access for critical welds, to establish techniques and tools for the mismatch correction procedure and methodology and appropriate welding sequence to minimize distortion controls. Based on these, a guidelines document has been created for industries and erection agencies.

Handling scheme structures were designed, developed, and tested in novel ways to achieve minimum structural material, no welded attachments, and minimum assembly time. Many useful first-of-a-kind mock-up trials, ideal for training of crane operations, were carried out as a precaution before going for safety vessel and other component handlings. It is worth mentioning two typical full-scale mock-ups built and employed at IGCAR for the successful erection of the safety vessel (Figure 24.27) and welding of roof slab hanging shell and main vessel (Figure 24.28).

FIGURE 24.27 Safety vessel erection mock-up.

FIGURE 24.28   Mock-up for the welding of roof slab hanging shell and main vessel.

24.3.6   Important Lessons Learnt

24.3.6.1   Manufacturing of Components

Technology development prior to the start of the construction is essential for components having long manufacturing/delivery time. The important outcome of the technology development exercise undertaken for the PFBR components is presented comprehensively in Figure 24.29. Judicious choice of tolerances, number and location of welds, and inspections has to be made. Robust criteria need to be applied for the acceptance of manufacturing deviations and material compositions. Indigenous materials should be used after being qualified by the relevant manufacturing process apart from routine standards. Manufacturing drawings should be finalized after a few rounds of discussions with prospective industries giving due considerations to economy. Subsequently, the revision of manufacturing drawings should be minimized, in particular for the dimensional tolerances. Obtaining permanent components of the RA from the same manufacturer will certainly help to minimize the integration problems and also avoid

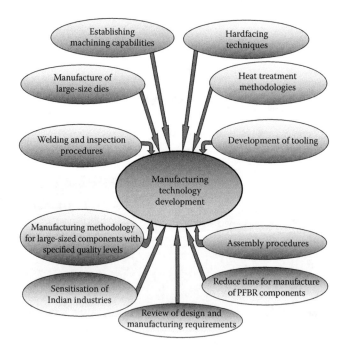

FIGURE 24.29   Outcome of the technology development exercise.

delays in the project. In case industry needs technical support, decisions should be taken quickly based on scientific input giving due considerations to international experience.

24.3.6.2 Erection of Reactor Assembly Components

Erection sequences and handling systems should be finalized after detailed discussions about the use of advanced computer software techniques. Care should be taken so that there is no revision of erection sequences and handling systems and schemes. Judicious choice of construction sequences of civil, mechanical, and electrical systems is essential. It is preferable to manufacture all of the RA components as factory-made single package items, thereby eliminating the need for a site assembly workshop. One possible strategy is schematically explained in Figure 24.30. As per this, the manufacturing of the RA components and civil construction of the reactor vault along with the safety vessel are carried out in parallel with the matching schedule so that the RA is erected without delay. Subsequently, other reactor internals kept ready in site assembly workshop will be erected. Toward achieving this, it is essential to motivate the prospective industries by attracting them with the promise of business opportunities in the long run.

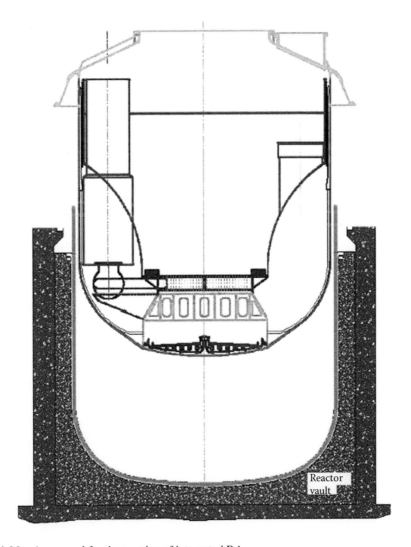

FIGURE 24.30 A proposal for the erection of integrated RA.

24.4 SUMMARY

Manufacturing large-size, thin-walled vessels made of stainless steel with tight form tolerances and machining and assembly of grid plate and steam generators with close tolerances are some of the challenging issues in the SFR. These have been successfully resolved for the PFBR as well as in other SFRs. In order to arrive at optimum manufacturing tolerances, elaborate manufacturing technology development works prior to the start of the construction may be required. Welding and pre-service inspection techniques and strategies should be finalized through adequate and appropriate mock-up trials. Judicious choice of tolerances, number and location of welds, and inspections need to be made. Robust criteria should be applied for the acceptance of manufacturing deviations and material compositions. From the experience gained through the manufacture and erection of RA components, SFR designers are in a position to recommend important guidelines and approaches to future editions of manufacturing codes/standards. The history of design, manufacturing, and erection experiences is to be documented systematically for future applications toward achieving economic competitiveness of the technology respecting the enhanced safety requirements.

REFERENCES

24.1 Takahashi, T., Yamaguchi, O., Kobori, T. (1990). Construction of the MONJU prototype fast breeder reactor. *Nucl. Technol.*, 89(2), 162–176.

24.2 NERSA. (1987). The Creys-Malville power plant. Electricite de France, Direction de L' Equipement, Region D'Equiopment, Alpes-Lyon, France.

24.3 Chellapandi, P., Chetal, S.C. (2013). Manufacture and erection of SFR components: Feedback from FBR experience. *Proceeding of International Conference on Fast Reactors and Related Fuel Cycles-Safe Technologies and Sustainable Scenarios (FR13)*, Paris, France.

25 Commissioning Issues
Various Phases and Experiences

Fast reactors are more complicated than thermal reactors in a technical respect and their exploitation demands considerable effort and time. As far as sodium fast reactors (SFRs) are concerned, the number of units is limited and almost all of them are experimental, test, and prototype reactors. Hence, the commissioning experience for power reactors is also limited. The basic objectives of the commissioning plan are (1) to achieve initial criticality and the initial demonstration run at various power levels from zero power to full power, consistent with safety requirements, and (2) to obtain necessary test and performance data (can be obtained only during the initial start-up period). In view of its high melting point (~90°C at ambient pressure), the sodium filling has to be done carefully after inerting and preheating the entire system more or less uniformly. This is a challenging and time-consuming activity. SFRs have seen a few minor sodium leaks during commissioning stage that have caused delay. In this chapter, various commissioning stages along with rich experience reported for three reactors Fast Flux Test Facility (FFTF: a loop-type reactor in the United States), Phénix (Prototype reactor in France), and BN-600 (Power reactor in Russia) are presented. Elaborate information and data are available in Refs. 25.1–25.7. In this chapter, only the salient aspects are highlighted.

25.1 FAST FLUX TEST FACILITY

25.1.1 COMMISSIONING AND POWER START-UP PROCEDURES

The key events in commissioning include start-up of the cell heat rejection systems, activation of the inert gas purification and sampling system, and preheat and fill of the secondary and primary sodium systems. The key events and start-up sequences are shown in Figure 25.1. To start hot functional testing, both the primary and secondary coolant systems and associated auxiliaries are inerted, preheated, and filled with sodium. The following are the major tests completed before inerting and preheating:

- Cell heat rejection system testing
- Inert gas purification and sampling
- Cell leak rate testing
- Fuel handling system testing
- Integrated leak rate testing, including sodium purification and sampling
- Inert preheat and filling of secondary system
- Inert preheat and filling of primary system

After filling sodium both in primary and in secondary sodium systems, flow and temperature performance testing is conducted, preparatory to fuel loading. The principal activities include confirmation of equipment performance, measurement of pipe movement during heatup of the high-temperature sodium loops, preliminary testing to verify vibrational characteristics, and sodium filtering and purification to achieve system cleanup. After hot functional testing has proven the reactor plant systems ready for radioactive operations, the reactor is incrementally loaded to criticality in its simplest configuration by substituting fueled assemblies for dummy core

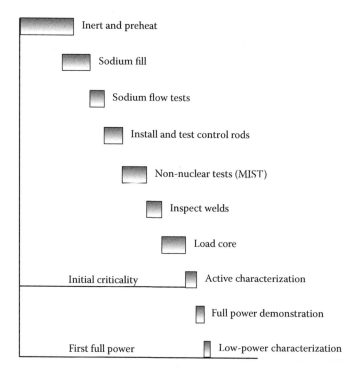

FIGURE 25.1 Various stages of the commissioning of FFTF. (From Carlisle, C.S. and Noordhoff, B.H., Fast flux test facility startup plan, HEDL-SA-1216 Rev, *ANS Reactor Operations Division Eighth Conference on Reactor Operating Experience*, Chattanooga, TN, 1977.)

elements. At that point, those low-power physics tests that must be accomplished prior to power operation are conducted. On completion of these tests, an approach to full power is planned, followed by a 48 h initial full power demonstration run to prove out the basic plant and uncover any major deficiencies. Following successful initial power demonstration, the balance of testing necessary to prove out the plant design and prepare the facility for irradiation testing is undertaken. This includes core characterization tests, natural circulation cooling tests, activation and testing of closed-loop systems, and full proof testing of auxiliary systems such as fuel handling, examination and storage, failed fuel monitoring, and sodium/cover gas chemistry and purification. Successful completion of this period of testing will bring the facility into readiness for the intended operation.

The hot functional testing is carried out at two temperatures: the major portion of the planned tests at 400°F and the rest at the maximum isothermal system temperature (MIST) achieved by pump heat and electrical power to the trace heaters. This temperature is approximately 750°F. Various steps involved till the reactor attains the first criticality are illustrated in Figure 25.1. Low-power measurements of nuclear parameters are made in an in-vessel thimble prior to initial full power to establish physics parameters of the core for comparison with calculations. Additional characterization tests and natural circulation tests are conducted following the initial 2-day full-power run. Several core adjustments and reactivity assessments follow, leading to an 8-day full-power run for core characterization purposes. Final core adjustments are made and nonstandard configuration tests performed to complete the planned acceptance test program.

25.1.2 OBSERVATIONS/DATA GATHERED DURING VARIOUS STAGES

Table 25.1 provides a summary of the observations. Each stage is elaborated in the following sections.

TABLE 25.1

Important Observations during Commissioning Tests

Activity	Observation
System cleanliness check	No debris observed except small bolt in one secondary drain pipe
	Very little core filter debris in first unit removed
	Clean sodium surface in reactor vessel
System hydraulics	Close to prediction initially
	Secondary values not changed by 472°C cycles
Flow stability	Reactor vessel surface stable
	Secondary flow oscillations (5%) observed. Caused by vortices in heat up to 427°C with simulated core assemblies (SCA)
	Caused by vortices in DHX flow dividers. Acceptable following minor control system changes
Reactor vibration	Full flow measurements on in-vessel accelerometer showed vibration below acceptance criteria
	Core assemblies
	Head-hung components
	Neutron detector thimble
Piping vibration	Within specification except for thermowells on one secondary loop, which were acceptable after local stiffeners were added.
	Primary crossover piping on pump outlet, additional supports being considered
Natural circulation flow	Excellent test result thus far
	DHX losses less than expected
	EM flowmeter calibrations successful at low flow
Thermal expansions of piping	Confirmed acceptable based on dimensional survey results
	Replacement snubbers performing satisfactorily

25.1.2.1 Cell Leak Rate Testing

Piping and equipment containing primary radioactive sodium located in individual cells in the FFTF plant require the use of a controlled inert atmosphere. These cells are lined with carbon steel to reduce the likelihood of radioactive gas leakage, and in the event of sodium leakage, to reduce interactions between sodium and air or sodium and concrete. In addition, the O_2 content is controlled by inerting with N_2 to reduce potential nitriding and corrosion of stainless steel piping/components in the event of a sodium leak. The cells are designed to withstand a normal operating pressure of 0.0049 kg/cm^2 (0.07 psig), positive or negative and to have a 0.018 kg/cm^2 (+0.25 psig), external design pressure. Prior to initial criticality, each cell had to pass a leak test to demonstrate its ability to meet normal operating and external design pressures. In addition, cell leak rates are limited by the capacity of the cell atmosphere processing system, which is designed to control the cells' O_2 content and, in the event of an accident, to remove the atmospheric contents of any given cell or combination of cells. Cells were pressurized with air from the instrument air system at pressures of up to 0.144 kg/cm^2 (2.0 psig) to facilitate the identification of leaks. Testing was conducted using various techniques, including locating personnel in cells, bubble leak tests, acoustic emission, contact ultrasonic, and olfactory senses (perfume). Some redesign and rework were required to seal leaking plug gaskets and electrical terminal boxes.

The cell leak test program was successfully completed. The significant results were as follows: (1) Cells initially designed to operate at a negative pressure of (−3.18 ± 1.90 cm W.G.) −1 1/4 ± 3/4 in. W.G. were requalified to operate at positive pressure with no adverse impact on plant or public safety. In the event of an accident, a cell (or combination of cells) was purged at a negative pressure.

(2) O_2 was injected into the N_2 inerting system to maintain <2% (technical specification limit). (3) Selected cell plugs that did not require frequent access were welded to reduce leakage.

25.1.2.2 Sodium System Inerting

All heat transport system (HTS) piping and associated process systems were inerted prior to liquid metal fill to prevent chemical reaction of the sodium with oxygen and water vapor left in the system piping from the construction period. Inerting was accomplished by using evacuation/backfill and pressurization/vent techniques. All systems were initially evacuated to 700 Pa (0.1 psia) and vacuum decay tests performed to quantify any system leakage. Argon (used as a cover gas for the FFTF) was then used to backfill the systems. Secondary HTS systems were pressurized to 380 kPa (40 psia) and vented to atmosphere three times to expedite the inerting. Leak checks at mechanical joints were performed using acoustic monitors and argon leak detectors. Final system purity was verified by using portable gas analyzers and taking gas grab samples. Purity levels of 5–20 ppm oxygen (50 ppm limit) and 30–85 ppm water vapor (100 ppm limit) were achieved using these techniques. A total system volume of approximately 1,153 m^3 (20,970 ft^3) was inerted to these purity levels in preparation for the liquid metal fill.

Significant results included the following: (1) Using acoustic monitors to identify system leakage was difficult due to other noise sources such as heating and ventilating. Using argon leak detectors proved to be much more successful. (2) Argon purity was achieved using a temporary/portable analysis technique.

25.1.2.3 Sodium Fill and Initial System

Testing prior to liquid metal fill involved preheating all sodium systems to 177°C (350°F) using permanently installed electrical resistance trace heaters (designed for a 20-year life). Permanently installed oil-fired preheaters were used to preheat dump heat exchanger (DHX) tube bundles. The reactor vessel was heated externally by circulating hot air in the annulus between the reactor vessel and the guard vessel and was heated internally by circulating hot pressurized argon in a closed cycle through the reactor internals. Both of these systems were temporarily installed to accomplish the required preheat of the reactor vessel.

Secondary system sodium fill of FFTF was started by first pressure transferring the molten sodium from railroad tank cars containing 39.7 m^3 (10,700 gal) into various plant storage vessels. (3) Secondary HTS piping was then filled by gradual pressurization of the storage tank. Secondary loop fill was started on July 4, 1978, and was completed in September 1978. The three secondary loops were filled one loop at a time using a total of 238 m^3 (63,000 gal) of sodium. The primary HTS system was filled primarily by pumped transfer. Elevated piping in the primary system was filled last using a gradually increasing vacuum at high point vents to draw the sodium into elevated piping. Primary HTS fill was completed in December 1978 requiring a total of 456 m^3 (120,700 gal) of sodium. Sodium circulation was initiated by starting the main sodium pumps on pony motor (10% of full flow) and electromagnetic pumps in the process loops. Sodium chemistry was monitored by plugging temperature indicators, and initial values indicated that excellent system purity had been maintained during the fill effort. Cold trapping to remove system impurities was started as was cover gas chromatography of the primary argon cover gas. Plant testing then focused on determining the hydraulic characteristics of the HTS system and testing the control systems for all sodium wetted components. Initial testing was performed at a system temperature of 240°C (400°F), the normal refueling temperature for FFTF. The process included vibration measurements and coast-down tests on main HTS pumps from full flow conditions, main isolation valve cycling, and further sodium chemistry measurements. When system testing was completed at the refueling temperature, the plant was raised to a MIST of 421°C (790°F) using a combination of pump work and trace heat. Three thermal cycles to the MIST temperature were performed to verify system performance at elevated temperatures. Two MIST cycles were completed in March and October of 1979 prior to initial criticality and the third cycle was completed in August of 1980. A major heating and venting air conditioning (HVAC) and insulation redesign effort was made based on the results of the MIST testing, which was successfully completed to support initial ascent to full power.

25.1.2.4 Initial Power Ascent

The FFTF initial power ascent was started by performing low power physics tests for a week to obtain baseline data for power operation. On November 20, 1980, power was raised above 1 MW for the first time and the ascent to 400 MW initiated. Power was increased in 5% (20 MW) increments with hold points for data collection and plant evaluation following each increment. Extended plant holds were conducted at 10%, 35%, 75%, and 100% power. During these hold periods, nuclear instruments were recalibrated, reactor control systems tested, physics parameters measured, and extensive shield surveys conducted as required by the ATP.

During the power ascent, the reactor was intentionally shut down (scram), a total of five times. Reactor scrams conducted from 5% and 35% power verified that natural circulation flow would initiate to remove decay heat in the event that forced circulation was lost. The FFTF design does not require a separate emergency core cooling system. It relies on natural circulation of sodium within the HTS loops for emergency core cooling. Scrams to pony motor flow were conducted from 35% and 75% power to verify that plant protective systems performed satisfactorily and to monitor the plant thermal response to a scram. One additional scram was performed from 5% power for operator training and plant maintenance.

Full power operation was achieved on December 21, 1980, when reactor power was raised to 400 MW with a core ΔT of 143°C (258°F). The plant was then shut down by slowly decreasing power while gathering additional test data. This effort spanned an interval of 32 days and paved the way for further power testing to be conducted in 1981.

Significant results were as follows: (1) The power ascent was completed in advance of the scheduled 45 days. (2) No major design changes were identified. (3) Only four unplanned scrams were experienced—all at power levels below 10%. (4) Initiation of natural circulation of sodium within the HTS loops following scram tests was much better than predicted.

25.2 PHÉNIX

The first on-site sodium deliveries leading to the first tests came during July 1971. Main sodium circuit filling took place during December 1972 and January 1973, the first criticality occurred at the end of August 1973, and power buildup started at the end of October 1973, while the first coupling to the grid took place on December 13, 1973. The importance of the Phénix start-up tests relates to the distinctive prototype character of the power plant and also to its dimensions. Several tests were evolved including sodium tests, more particularly before going through reactor criticality. This procedure enabled to foresee a relatively short delay between criticality and the first power buildup. This section will bring out the various start-up steps such as planned and important associated events. Figure 25.2 provides the summary of various stages involved in the commissioning of the reactor.

25.2.1 Sodium Tests before Core Loading

The nuclear-quality sodium was transferred from auxiliary storage tanks where it was unloaded to the primary and secondary storage tanks, the external fuel storage vessel, and the reactor (Figure 25.3). Taking into account the tight interlocking of the integrated-type reactor and secondary loops, planning and optimizing led to implementing a sodium filling operation, which was practically simultaneous, of the reactor and three secondary loops. Therefore, successive anticipated operations were (1) reactor preheating in nitrogen: structures at 150°C, (2) secondary loops filling and secondary sodium pump start-up (reactor temperature hold safety), and (3) reactor vessel filling.

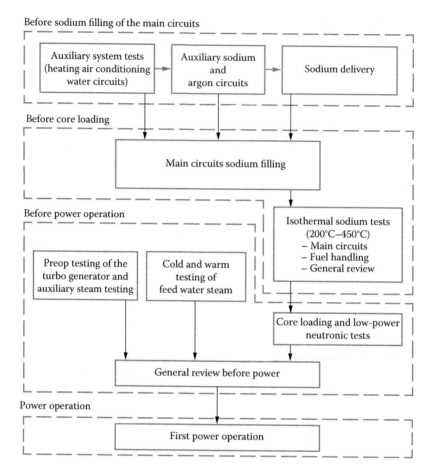

FIGURE 25.2 Phénix reactor start-up procedure. (From Carie, R., et al., Phénix startup. In: *Conference on Fast Reactor Safety*, Beverly Hills, CA. American Nuclear Society, Los Angeles, CA, pp. 1009–1020, April 2, 1974.)

After main circuits sodium filling at 150°C, the sodium temperature was raised to about 200°C through electric preheating and purified to attain the conditions needed for proper pump start-up (–180°C plugging temperature for a 200°C sodium temperature, as imposed by the manufacturer). Pump start-up and progressive increase in speed were carried out by controlling the proper operation of circuits particularly to prevent any reactor structure abnormal vibrations. Figure 25.4 shows sodium test sequencing prior to fuel loading. In subsequent isothermal operations, sodium temperature was raised to 450°C with the main goal of sodium purification and overall study of performances under temperature of the various equipments in order to identify potential problems at an early stage. Afterward, the sodium was cooled to 250°C, particularly to check the main fuel handling operations so as to learn what possible changes could be made to the initially designed systems. The next two test periods covered 350°C–450°C operations to derive the confidence on handling in case of any eventual modifications that might be called for after the first series of tests, including physical measurements (hydrogen detection, fuel failure detection). Finally, general check operations were carried out specifically to prepare for the core loading.

25.2.2 Preparatory Activities for Power Operations

These tests were concerned with the reactor (fuel loading, criticality, low-power neutronic tests), on the one hand, and with the electrical power installation (ancillary circuits, turbo-alternator, feed water circuits), on the other hand.

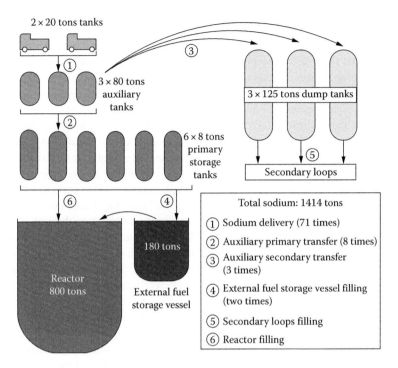

FIGURE 25.3 Sodium distribution in Phénix. (From Guillemard, B. and Le Marechal, T., PHÉNIX survey of commissioning and startup operations, *Proceedings of the International Conference Organized by BNES*, Institution of Civil Engineers, London, U.K., March 11–14, 1974.)

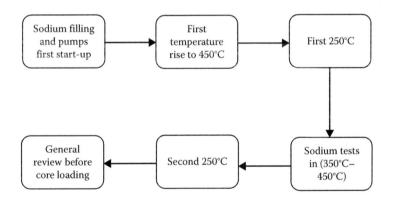

FIGURE 25.4 Sodium tests before core loading. (From Guillemard, B. and Le Marechal, T., PHÉNIX survey of commissioning and startup operations, *Proceedings of the International Conference Organized by BNES*, Institution of Civil Engineers, London, U.K., March 11–14, 1974.)

25.2.2.1 Electrical Power Installation Tests

These tests started quite early and made it possible to check the proper operation of the various ancillary circuits particularly the pumping station, while this installation was still independent from the rest of the nuclear plant. The feed water circuits were thoroughly tested to check the rated output of condensate extraction pumps and water treatment unit operations. Standby pumps as well as main feed water pumps at zero and low flow conditions were tested. Further, tests were made of the feed water circuits at nominal, 25%, and emergency flow conditions.

25.2.2.2 Low-Power Neutronic Tests

These tests took place at a temperature of primary sodium at 250°C with the secondary loops drained to ensure that there is no thermal coupling between the reactor (for handling operations) and steam generator feed water circuits that are at 150°C. These neutronic tests were performed as anticipated and their duration was mainly a function of basic fuel handling delay requirements and the configuration changes (operating configuration, handling configuration).

Fuel loading was particularly affected by difficulties in analyzing and evaluating the images delivered by ultrasonic display device (VISUS) that were used to actuate rotating plug or handling arm rotation: apparently, a characteristic echo of one or several assemblies raised above the level of the others was detected. Several control operations were carried out before it became obvious that the sodium level had to be lowered before proceeding with a periscopic examination. This operation enabled fine-tuning of the imaging system for clearing the handling operations.

Fuel loading took place in a continuous manner and we were able to go through the first critical-ity on August 31, 1973, at 08:15 AM with a core of 91 fuel assemblies. After making all the correc-tions or fine-tuning, the reactor attained its first criticality with 87 subassemblies when the all the control rods are positioned at the top most positions. Loading operations were monitored well by the two neutron flux detectors in the reactor vessel. However, the analysis of the curve indicating the neutron counting rate versus the fuel assembly number became an extremely delicate task because the neutron auxiliary source and neutronic detectors came apparently nearer to the core due to the increasing size of the fuel volume loaded. After a first rod worth measurement, the core configura-tion for power rising was completed (58 Pu assemblies, 48 U235 assemblies, 9 diluents). New mea-surements were undertaken after this loading. Therefore, just 3 days after the first criticality, these showed an available reactivity of 3300 pcm and an available antireactivity of 5300 pcm (1\$ = 440 pcm). This is a particularly satisfactory result: this core is exactly the reference core defined some 2 years ago when a selection had to be made for fuel manufacturing enrichment core, over which exhaustive neutronic, thermal, and dynamic computations had been made. Low-power neutronic tests continued for about 1 month: control rods worth, fuel assemblies and diluents reactivity worth, and power distribution measurements. Generally speaking, measurements proved to be perfectly in agreement with neutronic computations and confirmed the validity of the reference core. After secondary loop sodium refilling, these tests ended by a temperature rise to over 400°C, followed by cooling at 150°C. Such development was anticipated for the fulfillment of a measurement of the temperature coefficient over the 150°C–400°C range. The value obtained (about 3 pcm/C) is quite satisfactory for full power operation.

25.2.2.3 Steam Generators Operation

The final operation required before the plant power buildup was starting the steam generators. A steam sweep of the three steam generator units was started with sodium circuits at 150°C and water flow in the economizers/evaporators. Hydrogen detection signals were more particularly mon-itored during these operations and they did not show any significant variations. Rupture disc break up was simulated on each of the steam generators to test all automatic actuations that would be trig-gered in case of sodium/water reaction. Upon completing all the planned checks, the nuclear plant was ready for its power rise on October 29, 1973.

25.2.3 Power Buildup Plan

The power rise follows the nuclear boiler sodium tests that were specifically characterized by an isotherm operation at 450°C, with fuel loading and neutronic tests. They have enabled to monitor the performance of equipment under high temperature and output conditions including confirmation of overall heat balance with flow and temperature measurements in the primary and secondary circuits. Through several lower power operations, there were many challenges in establishing satisfactory operations of systems and equipment especially in steam–water circuit. Subsequently, the power rise was conducted in order to fulfill two goals: (1) to go ahead step-by-step while constantly monitoring all set requirements for plant nuclear safety and equipment safeguarding and (2) to reach as quickly as possible a significant length of operation under conditions nearing those of rated operation to ensure proper overall plant performance.

The first steam production test was carried out with the steam generator in order to ensure the removal of residual power, to control the safety valves, and to rotate the turbo-alternator with its coupling to the national grid. The plant operating on the turbine bypass circuits then underwent a series of tests to attain rated temperatures and to check overall nuclear boiler operation specifically: core thermal monitoring, fuel failure detection system, steam generator regulation, turbine bypass circuits regulation, and ON/OUT procedures. The first plant power rise was then undertaken in successive steps with in-between stops (scram and quick shutdown) to test the safety system and check equipment performance in corresponding transients. An overall review of the whole plant was made at each given step before proceeding further with the power buildup.

This first power rise was followed by an operation of the plant for a significant duration at a rated operation nearing that of rated regime: during this time, tests are conducted for a detailed surveillance of main equipment performances. Finally, a series of additional power tests are carried out to get an accurate determination of plant rated operation and also to check plant behavior under operational incidents with part loop configurations (two primary pumps out of three and two secondary loops out of three).

25.2.4 IMPORTANT COMMISSIONING TEST RESULTS AND OBSERVATIONS

Natural convection tests were carried out in several successive steps. Each time, the initial condition was as follows: low-power reactor, 100 rpm speed for primary and secondary pumps, sodium temperature range of 150°C–300°C. During one preliminary test, pumps had to be stopped when power reactor was 1 MWth. Although control rods' position was not changed, power drop was observed. During a second test, reactor power was held at 1 MWth after the pumps were stopped, then it was progressively brought to 2, then 4 MWth due to control rods action. The last test consisted in stopping the pumps when reactor power was 4 MWth and in holding this power by control rods action. The observed temperature shifts were close to the calculations. During the last two tests, a stable level was held after a few minutes and this proved the presence of natural convection in the primary circuit. Thereafter, convection in secondary loops was witnessed, cooling taking place through natural thermal losses with limited temperature difference between hot sodium and cold sodium. These tests are not entirely representative of actual conditions, particularly regarding initial distribution of temperatures. However, it is established that natural convection flows can be counted on, and the results obtained showed good comparison with numerical predictions.

It was possible to carry out the progressive tests to control the performance of various equipments during reactor scram transients. During the temperature rise while operating with the turbine bypass circuits, two voluntary scram tests were made at 470°C at 150 MWth. These enabled checking proper performance of the steam generator control in this configuration and the pertinent start of ON/OUT circuit during installation cooling. A test was carried out with the turbo-alternator operating at the attained regime, and two other tests were carried out at 80% and 100% of rated power. With these tests, it was possible to check all the automatic controls triggered by this incident: (1) steam generator depressurizing down to 60 bar compared to outer atmosphere and (2) absence of flow instabilities in the steam generators.

The fuel failure detection and location test by delayed neutrons that were carried out with a special fuel assembly with a bare uranium surface gave us a chance to study counting rate variations in the function of temperature and power, upto the 470°C–120 MW level. The pollution signal has been fulfilled previously under the same conditions. Charts defining pollution signal and sensitivity in the function of output, temperature, and power were outlined. The pollution signal was properly checked up to the presently attained regime. At the end of the test, time delay was measured during a voluntary scram. The measured core outlet sodium temperatures indicated a better power distribution compared to the predictions carried out on the reference core. This had raised the confidence for the next higher power operations.

Phénix "once through"-type steam generators, operating at very high heat flux, with austenite steel superheater, require particularly good quality of water: this is why the total amount of condensed water goes through a treatment installation consisting of cellulose filters and ion exchanger resins. In spite of such precautions, difficulties arose in power rise, during temperature rise in feed water circuits, upon bypass and turbine start-up, or during tests leading to quick output transients. Subsequently, after improving the water quality, there has been a progressive abatement of such difficulties, and at operating levels, an excellent quality of steam generator feed water has been obtained.

From the start of power buildup, the following 10 unplanned shutdowns took place, which were all explained satisfactorily:

1. Unwarranted quick shutdown (the six control rods are driven downward for 90 s) due to turbine trigger data, while the turbine was not yet in operation
2. Normal shutdown (voluntary power reduction to zero)
3. Two quick shutdowns decided because of steam generators feed water bad quality
4. Two scrams triggered by the core temperatures processing system
5. Two quick shutdowns generated by unwarranted triggering of a secondary sodium pump due to a fault of the SHERBIUS electronic control, which was corrected
6. Two voluntary quick shutdowns decided on unsatisfactory operations of electrical power installation auxiliaries

25.2.5　CONCLUDING REMARKS

The most important fact to note is that during all the tests in this project, it was possible to confirm all of the main choices made. Nominal power was reached after all tests completed in a step by step manner: sodium filling began only after operation of auxiliary circuits proved to be good; fuel loading began only after a long period of sodium tests and hot dynamic run; after good plant operation experience on turbine by-pass circuits, the power buildup started with turbine. All the tests were carried out very systematically. Main components characteristics and behavior during transients and their physics and thermal measurements were in good agreement with the predictions.

25.3　BN-600 REACTOR COMMISSIONING EXPERIENCE

The major steps before actual physical commissioning of the reactor are transportation and purification of sodium for filling the reactor and secondary loops, preheating the reactor and filling it with sodium, testing the turbine generators, heating the secondary loops and filling with sodium, and water and acid washing of the pipelines. These activities are described in detail next.

25.3.1　VARIOUS ACTIVITIES PRIOR TO REACTOR START-UP

25.3.1.1　Sodium Transportation and Purification

About 1800 tons of sodium was required to fill the reactor and the secondary loops. Before the sodium arrived, significant activities were completed on electrical installation and commissioning of the electrical heating system for the sodium loops, the reception system, and the systems for storing and purifying the sodium. Before the loops were filled, the air in them was replaced with inert gas by double evacuation and filling with nitrogen. The sodium, after being heated in a tanker, was transferred by electromagnetic pumps in the reception system to storage tanks (three tanks for the first and four tanks for the second loop) of 150 m^3 capacity each together with pipelines. During the accumulation, the tanks were connected to cold filter traps in the second loop to purify the sodium.

The temperature of the sodium in the tanks was maintained at a set level (240°C–250°C) by electrical heating. The equipment in the system for receiving, accumulating, and purifying the sodium form a special set of buildings.

25.3.1.2 Reactor Preheating and Filling with Sodium

The installation of the reactor was completed in June 1979 by the installation of the rotating plugs and the central rotating column. In August 1979, the drive mechanisms for the control and reloading systems were installed in the reactor, whereupon comprehensive tests began. At the same time, the electric drives for the circulation pumps in the first loop were tested along with the speed-control system. All the pumps and mechanisms on the reactor were tested at the respective industries, and hence, the commissioning was not prolonged. After the tin–bismuth eutectic alloy was run into the hydraulic seals on the rotating plugs and the sealing of the reactor checked, one circulation pump was extracted from the reactor and in its place a gas heating pipeline was installed. Gas heating of the reactor was necessary to prevent large thermal load in the metal constructions upon filling with sodium at 250°C and to prevent freezing of the sodium in the reactor tank. Gas heating of the reactor was performed at a rate of 10°C–15°C/day, and after ~15 days the temperature of the devices within the pressure vessel had attained 180°C–230°C. The heating was adjusted with electric heaters in the pipelines feeding gas to the reactor. Then these pipelines were disconnected from the reactor and the preheated circulation pump replaced them.

Before the reactor was filled with sodium, the content of oxygen in the atmosphere was about 0.5% by volume, while the water content was 5 g/m³. The reactor was filled with sodium in December 1979. Immediately afterward, the main circulation pumps were switched on and flushing of the devices within the pressure vessel free from installation contamination was initialized, while the sodium was purified in the cold traps. A few hours after the start of circulation, the working temperature reached 135°C–155°C. The maximum working temperature (about 200°C) was attained after the first rapid flushing of the reactor with sodium at 380°C. The cleaning was completed at a temperature of 115°C. The sodium was purified in a short period largely because the incoming sodium was of high purity.

The state of strain and vibration of the pressure vessel and the device within the reactor was examined during the heating and filling. After filling, the oxygen content in the reactor atmosphere was reduced to 0.001% by volume, while the water content was reduced to 0.02 g/m³. During sodium purification, checks were made on the control and reloading mechanisms with the reactor in hot condition.

25.3.1.3 Turbine Generator Evaluation

Turbine generators were installed between February and December 1979, for which they were evaluated with the steam from a start-up boiler. The oil-supply systems common to the block had first been commissioned, together with the water circulation and technical water supply, as well as the steam supply for equipment needs. Before the turbine generators were commissioned, the lubrication, sealing, and cooling systems for the stator were flushed and commissioned, while the steam-ejector apparatus was also commissioned. Along with the generators, the control systems were also commissioned, and tests were carried out on the safety controls, after which the turbines were connected into the line and were operated under load for 3 days.

25.3.1.4 Preheating of Secondary Loops and Filling with Sodium

The steam generators were installed over 8 months. The modules in these were supplied ready for installation and were filled with nitrogen at 0.3–0.5 kgf/cm². Before the loop was heated and filled with sodium, the pipelines to the third loop (steam–water system) and the modules were subjected to hydraulic tests, after which the modules were drained and filled with nitrogen. The installation and testing of the electric heating systems for the steam generators were very lengthy, extending upto 5 months. The loops were heated in steps of 20°C–250°C. The circulation pumps in the secondary loops were switched on as soon as the loop had been filled with sodium and the filtration traps were operated to purify the metal. The filling of the loops with sodium was done in turn during the period from December 1979 to February 1980 as sodium was accumulated and purified and the loops were prepared for heating. After the loops had been filled with sodium, the circulation pumps were run up and the hydrodynamic characteristics of the second loop were examined. The temperature of the sodium in the loops was maintained at 240°C–250°C by pumps and the electric heating. During the heating and the filling with sodium, measurements were made on the strain state in the steam generators by means of strain gauges and thermometers.

25.3.1.5   Water and Acid Washing of the Steam–Water Systems

In accordance with the technical specifications for removing mechanical pollutants, the flow part of the turbines in the loop was washed with demineralized water by means of standard feed pumps. The high flow rate provided the necessary kinetic energy in the flow. The performance of the washing was monitored based on the material that accumulated on the metal grids at the input to the pumps and also on the transparency of the circulating water. The water washing removed about 200 kg mechanical contaminants, mainly welding products. This was followed by acid washing. The loop was heated to 140°C–150°C with steam from the start-up boiler before the reagents were introduced, which were then fed directly to the deaerators in the loop. The washing was monitored based on the iron content in the solution and also on the content of the complexing agent and the pH. The washed surfaces were passivated.

25.3.2   Physical Start-up of Reactor

Loading the fuel-pin assemblies began on December 28, 1979, when the simulators were replaced by working assemblies. On February 26, 1980, the BN-600 attained criticality after loading 215 assemblies in the low-enrichment zone and 44 in the high-enrichment zone. During the physical commissioning, the performance and temperature of the control devices were determined and the barometric, power, and hydrodynamic effects on the reactivity were examined. There was good agreement with the calculated characteristics. When the core had been fully loaded, the flow rate of sodium through the fuel assemblies was measured and the hydrodynamic characteristics of the primary loop were also examined. The protection system and the interlocks were fully checked before running the system up to working power. A check was made on the operation of the unit when the power supply failed, including power supplies to the pumps in the water-circulation station. Additional measures were implemented to improve the safety of the unit during the physical commissioning: a hydraulic device was developed and installed for protecting the pressure vessel from elevated pressure, and tests were carried out on the equipment and pipelines in the second loop for clear flow conditions, while the protective devices against elevated pressures in the second loop employing forced rupture were replaced by self-rupturing ones. The system for monitoring for leaks of water into the sodium was also evaluated by injecting water into the second loop, and improvements were made to the fire-extinguishing system, while systems for protecting the environment from contamination were also introduced and evaluated. Finally, additional calculations were made on the strength of the equipment and experiments were performed on the primary loop.

25.3.3   Power Rise

Feed water was first supplied to the steam generators on April 2, 1980, and the reactor power was raised to 0.5% of the nominal value. After the quality of the feed water had been brought up to standard on April 6, the reactor power was raised to 5% and the steam generators were checked for steam production. On April 8, the reactor power was raised to 30%, and the turbine generators were connected to the line, while the temperature reached 430°C. After a thorough all-round check, power run-up was performed in the following stages: (1) operation at 30% nominal power for the performance of commissioning and research operations, including commissioning of the steam–water system, and adjustments to the unit regulators; (2) operation at 40%–70% power to continue the commissioning and research operations, refinement of the conditions in the core and in the sodium loops, as well as in the main equipment; and (3) operation at 80% power with the steam–water turbines and commissioning of the water system, with checks on the operation of the equipment with nominal steam parameters and refinement of the physical characteristics of the core and steam generators.

25.3.4 TESTS DURING POWER OPERATIONS

During the first month of operation, checks were made on the planned basic commissioning modes as well as on the shutdown and power regulation. Emergency situations were simulated, particularly as caused by the action of protective systems on the reactor. We recorded the changes in the parameters of the media as well as the stresses and vibrations in the components of the equipment. During operation at 30% power and the associated tests, checks were made on the main design points and on the characteristics of the equipment and recommendations were formulated on improving the schemes and the equipment design. During this period, adjustments were made to the automatic regulators for the various processes, and studies were conducted on the radiation environment, including production buildings and the neighborhood.

From May 14 to June 15, 1980, the BN-600 was shut down for the examination of the equipment and the elimination of minor defects in the auxiliary systems. During this period, the energy consumption in the fuel-pin assemblies was measured and checks were made on the operation of the fuel reloading mechanisms. The flow rates of sodium through the fuel assemblies were measured for comparison with those before commissioning. The covers on the water and steam cavities in the modules of one generator were removed, and the internal components were examined. On June 15, 1980, the unit was again commissioned for further power rises. At the end of June, the power had been raised to 50% with the sodium temperature at the exit from the core being 470°C. In June–August, several steam-generator modules were switched out on account of an elevated hydrogen content in the sodium recorded by systems for monitoring the sealing between loops in the steam generators. About 60% nominal power was reached in mid-August, and 70% at the end of the month. Checks were made on the state of the main equipment, including tests similar to those made at 30% power. The automatic regulators for the processes were adjusted to the higher power levels. In the middle of September, the power was raised to 80%, while the temperature of the sodium at the exit from the core attained 525°C and the test program continued under conditions of long-term operation. At all stages in the power run-up, checks were made on the radiation environment in the production buildings and the neighborhood. The characteristics of the equipment corresponded to the designs, while the control was good and the modes of operation were stable.

25.3.5 CONCLUDING REMARKS

The high skill of the start-up and operating staff and well-prepared start-up documents enabled the commissioning of the BN-600 within a short period. The start-up to power operations in stages and the correct design parameters provided safety in the start-up, testing, and research at all stages. The experience in running up the power demonstrated reliable operation of the equipment and systems in all modes of operation, which will enable one to improve the technology and design of fast reactor systems in the future.

REFERENCES

25.1 Carlisle, C.S., Noordhoff, B.H. (1977). Fast flux test facility startup plan, HEDL-SA-1216 Rev. *ANS Reactor Operations Division Eighth Conference on Reactor Operating Experience*, Chattanooga, TN.

25.2 Noordhoff, B.H., Moore, C.E. (1980). FFTF startup—status and results, HEDL-SA-2133. *International Conference of Liquid Metal Technology for Energy Systems*, Richland, WA, USA, p. 2.

25.3 Redekopp, R.D., Umek, A.M. (1981). Startup of FFTF sodium cooled reactor, slide presentation, HEDL-SA-2371. *Proceedings of the 16th Intersociety Energy Conversion Engineering Conference*, Atlanta, GA.

25.4 Hurd, E.N., Bliss, R.J., Olson, O.L. (1978). Status of FFTF startup program and future FFTF utilization, Energy Contract No. EY-76-G14-2170. Japan Section of the American Nuclear Society, Tokyo, Japan.

25.5 Carie, R., Megy, J., Guillemard, B., Robert, E., Le Marechal, L. (1974). Phénix startup. In: *Conference on Fast Reactor Safety*, Beverly Hills, CA. American Nuclear Society, Los Angeles, CA, pp. 1009–1020, April 2, 1974.

25.6 Guillemard, B., Le Marechal, T. (1974). PHÉNIX survey of commissioning and startup operations. *Proceedings of the International Conference Organized by BNES*, Institution of Civil Engineers, London, U.K., March 11–14, 1974.

25.7 Nevskli, V.P., Malyshev, V.M., Kupnyi, V.I. (1981). Experience with the design, construction and commissioning of BN-600 reactor unit at the Beloyarsk Nuclear Power Station. *Sov. Atom. Energy*, 51, 691–696.

Section V

International SFR Experiences

26 SFR Program in Countries

26.1 INTRODUCTION

There is a widespread perception that existing water-, gas-, and liquid-metal-cooled reactor systems can meet present safety standards quite adequately, and are expected to do so for the foreseeable future. Nevertheless, it is essential to continue the search for greater safety because (1) no opportunity to improve technology should be ignored, (2) safety standards may be raised by the regulatory authorities, (3) nuclear power plants (NPPs) suitable for countries or regions with a less developed infrastructure will be needed, and (4) more cost-effective ways of meeting existing safety standards may be found.

The development of fast reactors has been delayed in countries characterized by advanced market economics, relatively slow growth in primary energy consumption and availability of fossil fuel resources. In due course, when it is economically vital, fast reactors will almost certainly make the major contribution to the world's energy supplies. There will undoubtedly be further improvement of the technology to achieve the highest standard of safety, nonproliferation, environmental protection, and economics. Now, because of delay of fast reactors' commercial introduction, there is an opportunity to investigate alternative technical versions. The objective is to get a complete knowledge of their characteristics to, when the time comes, choose the best reactor design for a concrete realization. Reactor designs using gas, steam, lead, and lead–bismuth alloy coolants are considered possible alternatives.

Considerable experience has been gained in Russia in the course of the development and operation of reactors cooled with lead–bismuth eutectic, in particular, propulsion reactors. Studies on lead-cooled fast reactors are also under way. Preliminary studies on lead–bismuth and lead-cooled reactors and accelerator-driven systems (ADS) have been initiated in France, Japan, the United States, Italy, and other countries. The appropriate choice of the primary coolant is of great significance for achieving high liquid-metal-cooled fast reactor (LMFR) performances. This also determines the main design approaches of LMFR and, to a great extent, the technical and economical characteristics of NPP.

Among all liquid metal coolants, it is sodium that has gained the widest acceptance. The use of sodium as a coolant poses the risk of fire in the case of a leakage and interaction with air or water. Operating experience testifies to the possible coping with this problem, but the quest for excellence calls for future improvement in heavy metal technology.

In the earlier phases of breeder reactor development, especially in the 1950s and 1960s, high-pressure gases, such as helium, CO_2, or superheated steam, were studied. Between 1960 and 1970, H_2O-steam-cooled and D_2O-steam-cooled fast reactor concepts were studied in the United States. Helium-cooled fast reactor concepts have been pursued as an alternative coolant concept in Europe and the United States. Fuel development for a CO_2-cooled fast breeder reactor (FBR) continues on a small scale in the United Kingdom. Lead–bismuth alloy as a coolant was studied in the former USSR for propulsion and land-based reactors. However, the choice of liquid sodium as a coolant and the principal design features of fast reactors were mainly determined in the 1960s along with the requirement of a high power density in the reactor core (about 500 W(t)/cc for metal oxide (MOX) fuel) and use of a weakly moderating material with good heat transfer properties in order to attain a short Pu doubling time and high breeding ratio. The important fact is that sodium is practically noncorrosive to stainless steel.

Availability of excess plutonium produced by thermal reactors and that released from now-redundant nuclear weapons in the United States, the Russian Federation, the European Union, and in some other countries has not focused on the reactors with short doubling time and high breeding ratio. The changing of the strategic scenario beyond the 1980s delayed fast reactors' commercial introduction. Further, some experiences with sodium leaks, particularly in case of sodium-cooled fast reactor (SFR), have slowed down the fast reactor program. Some experts believe that an exploratory research on the different options, including *revisited* ones, should be done again, as in the early stages of fast reactor development.

Not only new innovative ideas like, for example, lead-cooled or lead–bismuth-cooled fast reactors are being studied in some countries now, but almost all of the old reactors discussed can also be part of this innovation: this is the case of gas-cooled high temperature reactors and superheated supercritical steam-cooled fast reactors. This is because of an increase in gas turbine's thermal efficiency from 35% to 50% within 10 years and also due to works on supercritical water/steam high-performance light water reactor (LWR) with thermal efficiency of 44% and high power density. Considerable experience has been gained in the Russian Federation on lead–bismuth eutectic alloy application as reactor coolant. Lead-based alloys are currently being considered for hybrid systems (accelerator-driven fast reactors) in which the coolant could double as the spallation source for driving the core. Techniques to counter a heavy metal coolant's disadvantages are being developed, but inspite of this work and the apparent disadvantages of sodium, the consensus in favor of sodium remains strong. This is demonstrated by the fact that even Russia, which possesses a comprehensive experience in heavy metal coolant technology, has announced that "before lead cooled fast reactor BREST-300 is built, MINATOM will first build a sodium-cooled LMFR BN-800" (E. Adamov, NW, September 23, 1999) [26.1]; in the last few years, sodium has been chosen in both China and the Republic of Korea. This is significant endorsement of sodium as a fast reactor coolant. It is, therefore, essential to clarify as far as possible the scientific issues related to the different innovative options and maintain the technology of fast reactors for a long term to secure nuclear fuel resources and incinerate the long-lived radioactive waste.

LMFRs have been under development for more than 50 years. Twenty LMFRs have been constructed and operated. Five prototype and demonstration LMFRs (BN-350/Kazakstan, Phenix/France, Prototype Fast Reactor/United Kingdom, BN-600/Russian Federation, and Superphenix/France) with an electrical output ranging from 250 to 1200 MWe and a large-scale (400 MWt) experimental fast flux test reactor (FFTF/United States) have gained nearly 110 reactor-years. In total, LMFRs have gained nearly 400 reactor-years of operation. In many cases, the overall experience with fast reactors has been extremely good, the reactors themselves, and more frequently particular components, showing good performance well in excess of their design expectations. They have also been shown to have very attractive safety characteristics, mainly because they are low-pressure systems with a large thermal inertia and negative power and temperature reactivity coefficients. Significant technology development programs for LMFRs are in progress in France, India, Japan, and the Russian Federation. Similar activities are being carried out in a number of other countries at a smaller scale.

SFR programs in Russia, France, the United States, China, India, Japan, Germany, and Korea are described in this chapter. It is intended to present details of the current programs in brief rather than dwelling on historical perspectives. The data presented in this chapter have been extracted from Refs. 26.2–26.4.

26.2 CHINA

The China Experimental Fast Reactor (CEFR) achieved its first criticality in 2010 and was connected to the grid in 2011. The main tasks of CEFR are irradiation of fuel and structural material, accumulation of operation data and experience, development of new technologies to enhance safety and reliability, and validation of the fuel cycle technology at the laboratory scale. According to the

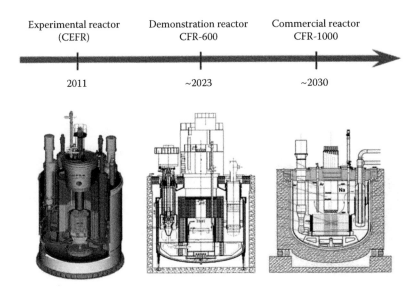

FIGURE 26.1 Fast reactor development strategy in China. (From IAEA-TECDOC-1691, Status of fast reactor research and technology development, IAEA, Vienna, Austria, 2012.)

fuel developing roadmap, MOX and metal fuel will be developed using CEFR. In order to realize the commercialization of fast reactors, "experimental, demonstration, commercial" is adopted as its developing strategy. The demonstration reactor CFR-600 is in the design phase now. CFR-600 will adhere to the rules of GEN IV to meet the safety requirements after the Fukushima nuclear disaster. The shutdown system, residual heat removal system, and confinement will be improved on the basic CEFR. This reactor is expected to accomplish the construction before 2023. After that, CFR-1000 will be constructed.

A 1000 MWe Chinese prototype fast reactor (CDFR) based on CEFR is envisaged with construction starting in 2017 and commissioning as the next step in the Chinese Institute of Atomic Energy (CIAE)'s program. This will be a three-loop, 2500 MWt, pool-type reactor using MOX fuel with an average 66 GWd/ton burnup and run at 544°C with a breeding ratio 1.2, 316 core fuel and 255 blanket assemblies, and a 40-year life. This is CIAE's "project one" CDFR. It will have active and passive shutdown systems and passive decay heat removal. This may be developed into a CCFR of about the same size by 2030, using MOX + actinide or metal + actinide fuel. MOX is seen only as an interim fuel, the target arrangement is metal fuel in a closed cycle. The CIAE's CDFR 1000 is to be followed by a 1200 MWe CDFBR by about 2028, conforming to GEN IV criteria. This will have U–Pu–Zr fuel with 120 GWd/ton burnup and a breeding ratio of 1.5, or less with minor actinide and long-lived fission product recycle. CIAE projections will show fast reactors progressively increasing from 2020 to at least 200 GWe by 2050 and 1400 GWe by 2100. The plan for fast reactor deployment in China is shown in Figure 26.1.

26.3 FRANCE

The French liquid metal reactor technology has demonstrated a number of positive examples of designs, project realizations, and experience in LMFRs construction and operation: experimental reactor Rapsodie (40 MW(t) power, 1967–1983), prototype reactor Phenix (255 MW(e) power, commissioned in December 1973), the large-size LMFR Superphenix (1986–1998). The Phenix reactor has been in operation for ~100,000 h, with the reactor hot components having a temperature of 833 K (560°C) and a thermal efficiency of 45.3% (gross), that is, the highest value in the nuclear power practice; the average burnup was increased from 50,000 to 100,000 MWd/ton, with maximum burnup exceeding 150,000 MWd/ton. A breeding ratio of 1.16 was experimentally confirmed

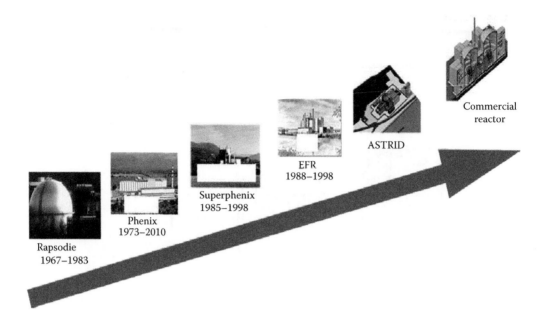

FIGURE 26.2 Fast reactor evolution in France. (From IAEA-TECDOC-1691, Status of fast reactor research and technology development, IAEA, Vienna, Austria, 2012.)

in the Phenix reactor, and hence, the plutonium produced in the reactor was used as the fuel for its core. These levels were reached with eight cores of fuel, amounting to 166,000 fuel pins. After the premature shutdown of Superphenix in 1998, the role of Phenix gained importance as an irradiation facility particularly for long-lived radioactive waste management.

The present French policy in the field of fast reactors development is focused on the design and development of the sodium-cooled ASTRID (Advanced Sodium Test Reactor for Industrial Demonstration) reactor, which is a medium-size (600 MWe) SFR and has been proposed and endorsed at the European level as the reference concept for Generation IV fast reactors.

Deriving from the feedback of experience, high levels of design goals have been set for the ASTRID reactor that would demand innovations to further enhance safety, reduce capital cost, and improve efficiency, reliability, and operability. The reactor will also provide some irradiation capacities especially in order to validate the expected properties for advanced fuel that may be needed for commercial deployment and the ability to burn minor actinides at a larger scale. The high safety requirements demand that severe accidents be postulated and mitigation measures be included accordingly into the design. More independence is expected between all levels of defense-in-depth. The absence of cliff edge effects beyond the design basis would be thoroughly verified. Until a few years back, R&D program was carried out with respect to choice of design options such as (1) loop and pool designs, (2) the choice of energy conversion systems, (3) the choice of 9Cr steel for the secondary systems, (4) the possibility of oxide dispersion–strengthened steel (ODS) as cladding tube material, and (5) the development of concepts for robust in-service inspection and repair (ISI&R) covering the sensors' inspectability, reparability, and robotics. The French evolution of fast reactors is shown in Figure 26.2.

26.4 GERMANY

In Germany, the major fast reactor development started with the Kompakte Natriumgekuhlte Kernreaktoranlage-II (KNK-II) reactor which is an experimental sodium-cooled fast power reactor with a 20 MWe electric output. The KNK reactor was originally designed as a thermal reactor (KNK-I) and was successfully operated from 1971 to 1974. Between 1975 and 1977, the plant was modified into a fast reactor (KNK-II) and provided with an unmoderated core. KNK-II was put into operation in 1977 and was operating successfully till 1991, when the plant was finally shut down. Another reactor, SNR-300, was considered an international project from the very beginning, that is, about 1966. The final arrangement consisted of a three-country cooperation comprising Germany (70%), Belgium (15%), and the Netherlands (15%), involving the manufacturers, the utilities, and the R&D organizations. The first nuclear license of SNR-300, which was necessary to begin construction, was granted in December 1972. The adverse political influence started in Germany as early as the late 1970s, when the usefulness and the safety standards of nuclear power in general and SNR-300 in particular were questioned, was a reason for the announcement by the German Federal Minister of Research and Development in March 1991 about unconditional abandonment of the project after a thorough evaluation of the overall situation. This finally led to the fact that the development of the fast reactor technology was suspended in Germany.

However, later nuclear research concentrated on safety research on nuclear waste disposal (including transmutation) and safety research with a focus on severe accidents of LWRs. R&D activities are being carried out in liquid metal loop facilities with sodium, lead, and lead–bismuth. Modeling activities related to fast reactor safety are done mainly with SASSFR and SIMMER codes, which are used for simulating different phases of hypothetical accidents, including severe accidents in fast reactors.

26.5 INDIA

The Indian nuclear power program is being implemented in three stages taking into consideration limited uranium and vast thorium resources in the country. The first stage consists of investing natural uranium in pressurized heavy water reactors (PHWR). This stage has a potential output of 10 GWe. The second stage involves the large-scale deployment of FBRs with co-located fuel cycle facilities (FCFs) to utilize the plutonium and depleted uranium extracted from the PHWR spent fuel. This stage has a potential output of about 300 GWe. In the third stage, effective utilization of the vast thorium resources is planned. The Indira Gandhi Centre for Atomic Research (IGCAR), established in 1971 at Kalpakkam, is involved in the mission of developing the technology of FBR. A host of multidisciplinary laboratories are established in the center around the central facility of the 40 MWt fast breeder test reactor (FBTR). The FBTR, which went critical in 1985, has given valuable experience in the operation of sodium systems including steam generators, and has served as a test bed for various experiments and fuel irradiation programs. Presently, the construction of indigenously designed MOX-fueled 500 MWe PFBR that started in 2003 is in advanced stage and commissioning activities are under way. The design of PFBR incorporates several state-of-the-art features and is foreseen to be an industrial-scale technoeconomic viability demonstrator for the FBR program. Beyond PFBR, the proposal is to build one twin unit having two reactors with MOX fuel, each with an improved design compared to PFBR, to be commissioned by 2022–2023. Subsequently, toward rapid realization of nuclear power, the department is planning a series of metal-fueled FBRs starting with a metal-fueled demonstration fast breeder reactor (MDFR) to be followed by industrial-scale 1000 MWe metal-fueled reactors. As for the reactor size for MDFR, 500 MWe is proposed in order to capitalize on the experience gained in terms of design, manufacture, and erection of major reactor systems and components of PFBR.

In line with this, India has undertaken the task of designing MDFR-500, and during the conceptualization of the reactor system, proven design concepts as adopted for MOX-fueled FBRs were retained and important variants incorporated with due consideration to experience gained during safety review, manufacturing, and construction of PFBR with a view to improve the overall economics and enhance plant safety. In parallel, on the fuel and material development front, in order to study the irradiation behavior of binary (U–Pu) and ternary (U–Pu–Zr) metallic fuels and related structural materials, an elaborate testing program with FBTR as the test bed followed by a detailed postirradiation examination (PIE) program is drawn. Toward optimizing the MDFR-500 core for

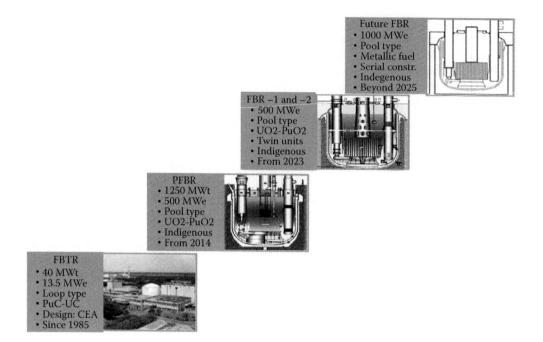

FIGURE 26.3 Fast reactor program in India. (From IAEA-TECDOC-1691, Status of fast reactor research and technology development, IAEA, Vienna, Austria, 2012.)

maximizing the breeding potential (BR ~ 1.3–1.4), several fuel options and design variants have been studied both for sodium and for mechanical bonded fuel pin options. For example, a reference design with U-(19w%)Pu-(10w%)Zr ternary sodium–bonded fuel with a pin diameter of 6.6 mm (T91 clad) and a smeared density of 75% was optimized for a peak linear power of 450 W/cm for an initial target burnup of 100 GWd/ton. Further, thermal hydraulic design was carried out to assess the margins to design safety limits. Detailed studies with apportioned cumulative damage fraction (CDF) limit and conservative assumption of 100% fission gas release have indicated that target burnup could be reached satisfactorily. Parallely, mechanically bonded fuel pin option is being pursued with due consideration to infrastructural constraints for arriving at a judicious choice for fuel composition and type of bonding with the pin. The heat transport (sodium) systems and the balance of plant (BOP) are conceptualized for a compact plant layout. The layout is also drawn with flexibility to accommodate an optional ex-vessel sodium storage tank to facilitate the storage of sodium-bonded fuel subsequent to fuel handling operations. The safety instrumentation and controls are detailed with adequate redundancy and diversity requirements. The plant design is being evolved to be compatible for both coastal and inland sites. The design evolution of fast reactors in India is shown in Figure 26.3.

26.6 JAPAN

Since the issuance of the Framework for Nuclear Energy Policy in October 2005 by the Atomic Energy Commission (AEC) of Japan, the significance of the development of the fast reactor cycle technology has been recognized once again in the national fundamental nuclear energy policy. In March 2006, the Council for Science and Technology Policy of the cabinet office selected the fast reactor cycle technology as one of the key technologies of national importance in the third-term "Science & Technology Basic Plan." After this, the Ministry of Education, Culture, Sports, Science and Technology (MEXT) and the Ministry of Economy, Trade and Industry (METI) investigated the action plans for nuclear technologies and published their reports. In response to the action plans and the review by the MEXT of the result of the Feasibility Study (FS) Phase-II, the AEC decided the "Basic Policy on R&D of FBR Cycle Technologies over the Next Decade" on December 26, 2006.

The experimental fast reactor JOYO (SFR, thermal power: 140 MWt) has been dedicated to developing SFRs. Various irradiation tests were successfully conducted together with the results from the PIE. During the 15th periodical inspection of JOYO, it was found that the irradiation test subassembly *MARICO-2* had bent onto the in-vessel storage rack. This incident necessitates replacing the upper core structure (UCS) and retrieving MARICO-2 for JOYO restart. Design works have almost been completed and 2-year period was estimated for the replacement of the UCS and retrieving MARICO-2. Other major activities included the development of an innovative instrumentation technique of failed fuel detection and location system by means of laser resonance ionization mass spectrometry (RIMS).

Monju is a 714 MWt (280 MWe) loop-type prototype FBR, fueled with Pu–U mixed oxide and cooled by liquid sodium. Monju was restarted on May 6, 2010, and reached its criticality on May 8, 2010. Subsequently, the core confirmation test was conducted till July 2010. Additional core physics data such as control rod reactivity worth were evaluated from the measured data. However in August 2010, fuel handling incident happened where the in-vessel transfer machine (IVTM) dropped down when hanged up by the gripper of the auxiliary handling machine succeeding to the refueling. The Japan Atomic Energy Agency (JAEA) decided to withdraw the IVTM together with the fuel throat sleeve and investigated the withdrawal procedure, and completed the withdrawal and restoration in November 2011. Furthermore, a huge tsunami rushed to the Fukushima Dai-ichi Nuclear Power Station and the reactors were melted down. In order to ensure further safety of the cooling systems, JAEA implemented multiple countermeasures, emergency response procedures, and operational training for earthquakes and tsunamis. The cooling functions of core, ex-vessel fuel storage tank, and spent fuel pool were confirmed by plant dynamics simulations even in these severe conditions.

Based on the Japanese policy, the "Fast Reactor Cycle Technology Development" (FaCT) project was launched in 2006 toward the commercialization of the fast reactor cycle technology. In the FaCT project, both the conceptual design study for the advanced loop-type fast reactor named *Japan sodium-cooled fast reactor* (JSFR) and the development of related innovative technologies have been conducted. In 2010, technical evaluation for the applicability of innovative technologies to JSFR was made, in which all of the technologies were shown to be applicable. After the accidents in Fukushima Dai-ichi Nuclear Power Station (Fukushima NPS) in March 2011, the Energy and Environment Council published a report, "Innovative Strategy for Energy and the Environment," which states that the government will mobilize all possible policy resources to such a level as to even enable zero operation of nuclear power plants in the 2030s. The report also states the nuclear fuel cycle policy including a role of Monju and the promotion of R&D for the reduction of the amount and toxicity level of radioactive wastes. Under these circumstances, further improvement on the safety of next-generation SFRs has been activated based on the lessons learned from those accidents. The SFR development program in Japan is as follows:

*   *FS phase-I*: Conceptual design study on a wide range of various coolant/fuel and selection of four systems (sodium, helium gas, lead–bismuth, and water).
*   *FS phase-II*: Detail comparison on selected concepts and selection of JSFR concept (sodium cooled + MOX fuel).
*   *FaCT phase-I*: Evaluation of key technologies for commercial JSFR.
*   *FaCT phase-II (suspended due to the TEPCO Fukushima Dai-ichi [1F] Accident)*: Demonstration of key technologies and the conceptual design of demonstration JSFR.
*   *After 1F accident*: Design study on safety enhancement was initiated.
*   *Safety Design Criteria (SDC)*: SDC was initiated in October 2010 in GIF to establish global safety requirements for SFRs. The necessity for SDC was reconfirmed by 1F accident and the efforts on SDC were enhanced. Safety design guide (SDG) also should be established secondary to SDC completion.

26.7   KOREA

In order to provide a consistent direction to long-term R&D activities, the R&D action plan for the advanced SFR and the pyro-process was authorized by the Korea Atomic Energy Commission (KAEC) in December 22, 2008, and this long-term plan will be implemented through nuclear R&D programs of the National Research Foundation, with funds from the MEST. A detailed implementation plan is now being developed. The long-term advanced SFR R&D plan will include the construction of an advanced SFR demonstration plant by 2028 as follows: (1) first phase (2007–2011—development of an advanced SFR design concept, (2) second phase (2012–2017)—standard design of an advanced SFR demonstration plant, and (3) third phase (2018–2028)—construction of the advanced SFR demonstration plant.

It has been recognized nationwide that a fast reactor system is one of the most promising nuclear options for electricity generation with an efficient utilization of uranium resources and a reduction of the radioactive wastes from nuclear power plants. The Republic of Korea's LMFR program started with the development of basic design technologies and conceptual design of KALIMER-150. In December 2008, the KAEC authorized an R&D action plan for an advanced SFR and a pyro-process to provide a consistent direction to long-term R&D activities, which was revised later on in 2011. The milestones include design of a prototype SFR by 2017, approval by 2020, and construction of a prototype GEN IV SFR 150 MWe (PGSFR) by 2028. The prototype SFR program includes the overall system engineering for SFR system (NSSS and BOP) design and optimization, development of codes, verification and validation of technologies, major components development, and also development of metal fuel technologies. The status of fast reactor development in Korea is shown in Figure 26.4.

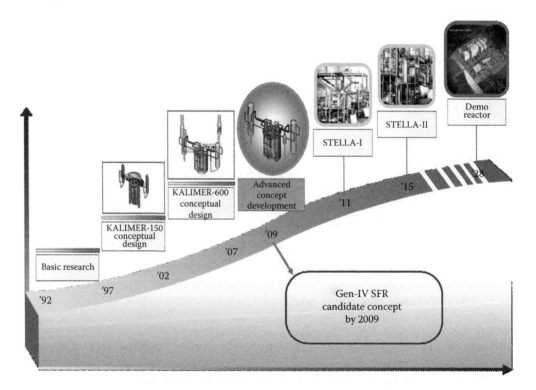

FIGURE 26.4   Status of fast reactor development in the Republic of Korea. (From IAEA-TECDOC-1691, Status of fast reactor research and technology development, IAEA, Vienna, Austria, 2012.)

26.8 RUSSIA

Russia began SFR program in a phased manner starting from test, prototype, and semicommercial reactors (BR-5/10, BOR-60, BN-350, and BN-600) and they were successful in operation. The first fast reactor with sodium coolant and plutonium fuel (BR-5/10) was created in 1958 and was in operation for over 44 years and it was finally shut down on December 6, 2002, and now it is at the preparatory stage of its decommissioning. The BOR-60 test reactor has been in operation since more than 43 years and its license to operate has been extended until 2014. This reactor was used for isotope production and for testing materials and various fast reactor equipments, along with heat and electricity generation. BN-600, a commercial reactor, has been in operation since more than 33 years with an electricity generation of 91 billion kW h at an average load factor of 74.4% (1982–2012). On April 7, 2010, the Beloyarsk nuclear power plant received the Rostechnadzor's license for lifetime extension of the BN-600 power unit up to March 31, 2020. The lifetime of the BN-600 reactor has been extended until March 31, 2020. Current efforts with regard to the fast reactor development in the Russian Federation are directed toward increasing safety margins and improving economics. The detailed design of commercial-size fast reactor BN-800 was completed, and its construction and power unit commissioning are scheduled for completion in 2014. BN-800 is planned to be used for closing fuel cycle and for recycling stocks of weapon-grade plutonium. Future plans include the development of a design of large-scale sodium fast reactor BN-1200 and also to design and construct research sodium-cooled fast reactor *multifunctional research fast reactor* (MBIR), which is scheduled in 2019 at Research Institute of Atomic Reactors (RIAR), Dmitrovgrad. The most important new conceptual design options considered for the BN-1200 design are as follows:

1. Pool-type arrangement of the primary circuit with all sodium systems, including cold traps and chemical-engineering control systems, located within the reactor vessel to eliminate radioactive sodium release outside of the reactor vessel and hence its fire
2. Simplification of a refueling system by elimination of intermediate storage drums of fresh and spent FSAs and provision of high-capacity in-reactor vessel storage (IVS), providing a direct unloading of SFSAs (after their exposure in the IVS) from the IVS into washing cells and further into an exposure pool
3. Transition from a sectional-modular SG scheme to an integral one based on the application of straight-tube, large-capacity modules
4. Enhancement of inherent safety features based on passive principles for reactor protection, passive decay heat removal system through independent loops connected to a reactor vessel.

Other prospective activities on fast reactors with heavy liquid metal coolant are (1) development of the design of the SVBR-100 reactor facility with a lead–bismuth coolant (LBC) and (2) implementation of the R&D on justification of the design of the NPP with the lead-cooled reactor BREST-OD-300 with appropriate revision to the design—construction of a pilot plant with the BREST reactor. The evolution of fast reactor concepts in Russia is shown in Figure 26.5.

26.9 UNITED STATES

EBR-II was the most successful of the U.S. fast reactors. It was a 62.5 MWt, 20 MWe, sodium-cooled, *pool-type* reactor. It achieved criticality with a sodium coolant in November 1963. EBR-II demonstrated the feasibility of a sodium-cooled FBR operating as a power plant. EBR-II had an adjoining FCF that permitted continuous reprocessing and recycling of fuel. In 1967, EBR-II was reoriented from a demonstration plant to an irradiation facility. After

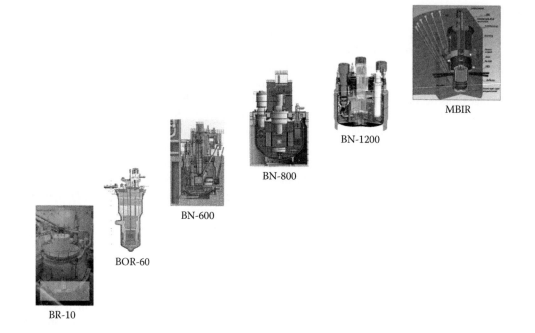

FIGURE 26.5   Fast reactor evolution in Russia. (From IAEA-TECDOC-1691, Status of fast reactor research and technology development, IAEA, Vienna, Austria, 2012.)

cancellation of the Clinch River Breeder Reactor (CRBR) in 1983, the EBR-II reactor and the FCF became the research and demonstration facilities for the integral fast reactor (IFR) concept. The IFR program was terminated and EBR-II began shutdown operations in September 1994, after 30 years of operation. In the United States, the vision for fast reactors now is toward flexible actinide management for fuel cycle missions, that is, to demonstrate the transmutation of transuranics recovered from the LWR spent fuel, and hence to benefit from the fuel cycle closure for nuclear waste management purposes. The advanced burner test reactor (ABTR) is a fast reactor concept that could be the first step in demonstrating the transmutation technologies. It directly supports the development of a prototype full-scale ABR, which would be followed by the commercial deployment of ABRs.

Fast reactor deployment awaits fuel cycle strategic decisions. Hence, efforts are focused on the development of required base technical capabilities and a small set of promising innovative technology options. With no current fast reactor demonstration plan, the focus is on innovative R&D that could allow significant performance improvements. With respect to fuel, the R&D is on the development of transmutation fuels with associated closed fuel cycle, improved performance with extended burnup, and minimum losses and waste. To this end, fabrication, fuel characterization, and irradiation testing are being conducted on new metal alloy fuels (containing minor actinides). Another area of focus is the simulation of diverse, coupled physics including neutronics, thermal fluid dynamics, and structural phenomena. A fully coupled multiphysics, multiscale simulation of an entire reactor core is intended to be demonstrated in a short time by coupling high-fidelity neutronics, thermal, and structural dynamics code using a flexible computational framework. Other work is in the areas of compact fuel handling, vented fuel, twisted tube annular heat exchangers, electromagnetic pumps, alternate heat transport and containment configurations, and supercritical CO_2 Brayton cycle energy conversion systems. Recent U.S. work focused on small (~100 MWe) reactors with unique features, such as long-lived cores. The evolving vision for fast reactors in the United States is shown in Figure 26.6.

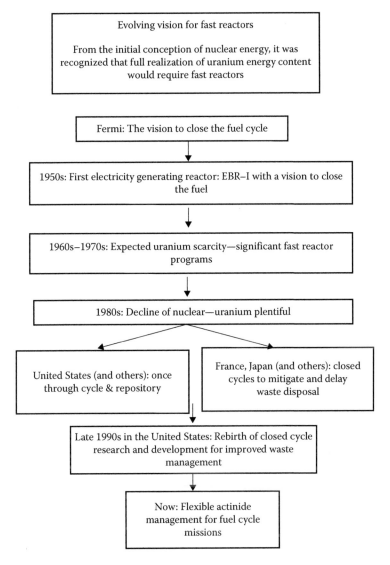

FIGURE 26.6 Fast reactor evolution and vision for future in the United States. (From IAEA-TECDOC-1691, Status of fast reactor research and technology development, IAEA, Vienna, Austria, 2012.)

REFERENCES

26.1 IAEA-TEDOC-1289. (2002). Comparative assessment of thermophysical and thermohydraulic characteristics of lead, lead-bismuth and sodium coolants for fast reactors. IAEA, Vienna, Austria.

26.2 IAEA-TECDOC-1691. (2012). Status of fast reactor research and technology development. IAEA, Vienna, Austria.

26.3 IAEA. (2013). *46th Meeting of the Technical Working Group on Fast Reactors (TWG-FR), Working Material*, Vienna, Austria. http://www.iaea.org/NuclearPower/Meetings/2013/2013-05-21-05-24-TWG-NPTD.html

26.4 IAEA-TECDOC-1569. (2007). Liquid metal cooled reactors: Experience in design and operation. IAEA, Vienna, Austria.

27 Feedback from Operating Experiences

27.1 INTRODUCTION

Beginning in the 1970s, fast reactors have been launched in various countries for R&D and experiment purposes. After generating the essential data, getting industrial and operation experience from sodium components, prototype sodium-cooled fast reactors (SFRs) were deployed in a few countries. Especially in France, a large-size fast reactor of 1200 MWe, Superphénix, was commissioned. All of the world's fast reactors combined have accumulated more than 400 years of reactor operating experience, yielding rich data on the behavior of fuel, fuel safety, sodium components, steam generators, fuel handling equipments, etc. Small experimental reactors, for example, EBR-II, Rapsodie, BOR-60, JoYo, and fast breeder test reactor (FBTR) have provided valuable experience on sodium technology, fuel element design involving fuel choice, cladding and wrapper material, burnup limit demonstration, and material irradiation data. In particular, EBR-II was extensively utilized in the development of sodium-bonded metal fuel. The objective of the U.S. fast reactor program with U–19Pu–10Zr ternary sodium-bonded metal fuel has been successfully demonstrated in EBR-II and the Fast Flux Test Facility (FFTF). With respect to reactor assembly and other components, however, these small reactors are limited in demonstrating structural integrity requirements of commercial fast reactors because the design loading, in particular thermal loading, increases with the size rating of the components. When initiating the design of a new reactor, every design organization considers it prudent to utilize feedback from operating experience. This was done when Superphénix in France and the prototype fast breeder reactor (PFBR) in India were designed, and a similar approach was taken with the design of the European fast reactor. The International Atomic Energy Agency (IAEA) has also contributed to feedback exchange through annual meetings of the International Working Group on Fast Reactors, technical meetings on specific topics, and publications on fast reactor technology. The IAEA has also undertaken work related to knowledge management of fast reactor experiences. SFRs are favored candidates for sustainability of nuclear energy by answering to limited resources of uranium and high-level-waste management by incineration of minor actinides and long-lived fission products. The summary of operation of fast reactors worldwide is given in Table 27.1 for small reactors and in Table 27.2 for medium-sized reactors [27.1].

27.2 DESIGN CONCEPTS

With respect to design concepts and options, major feedback is on the choice of fuel and its performance; choice of core and permanent reactor components materials; fuel handling equipment performance, especially sodium equipments; and steam generator experience. The choice of loop and pool concept is still a question of debate even though, by and large, prototype and industrial-scale reactors have opted for pool design mainly from safety considerations. Of late,

TABLE 27.1
Small-Sized Fast Reactors in the World

Name	MWt	Start	Country	Status
Clementine	0.02	1946	USA	Stopped 1952
EBR-I	1.4	1951	USA	Stopped 1963
BR 2	0.2	1956	Russia	Stopped 1957
BR 5–BR 10	5/10	1958/1973	Russia	Stopped 2002
LAMPRE	1	1961	USA	Stopped 1965
DFR	75	1959	UK	Stopped 1977
EBR-II	60	1963	USA	Stopped 1993
EFFBR	200	1963	USA	Stopped 1972
RAPSODIE	24/40	1967/1970	France	Stopped 1983
SEFOR	20	1969	USA	Stopped 1972
KNK1/KNK2	60	1972/1977	Germany	Stopped 1991
FFTF	400	1980	USA	Stopped 1992
PEC	120		Italy	Given up
FBTR	40	1985	India	Running
JOYO	50	1977	Japan	Running
BOR 60	60	1968	Russia	Running
CEFR	60	2009	China	Running

TABLE 27.2
Medium-Sized Fast Reactors in the World

Name	MWe	Start	Country	Status
EFFBR	100	1963	USA	Stopped 1972
BN 350	150	1972	Kazakhstan	Stopped 1993
Phénix	250	1973	France	Stopped 2009
PFR	250	1974	UK	Stopped 1994
CRBR	350	—	USA	Given up
SNR 300	300	—	Germany	Given up
BN-600	600	1980	Russia	Running
MONJU	280	1992	Japan	To restart
PFBR	500	2014	India	To start

the loop design is being revisited, which remains a choice of the French ASTRID reactor. The major design options and considerations thereof are as follows [27.2]:

- Sodium coolant
- Steam generators
- Intermediate heat exchanger (IHX)
- Sodium pumps
- Fuel handling systems and schemes

27.2.1 Sodium Coolant

Use of a large amount of sodium in the primary coolant system is advantageous from the standpoint of safety, because of the high heat capacity and natural circulation flow of sodium. In spite of these

TABLE 27.3
Summary of the Number of Sodium Leaks

Phénix	SPX	BN-600	BN-350	PFR	DFR	FFTF	MONJU	KNK II	FBTR
31	7	27	15	20	7	1	1	21	6

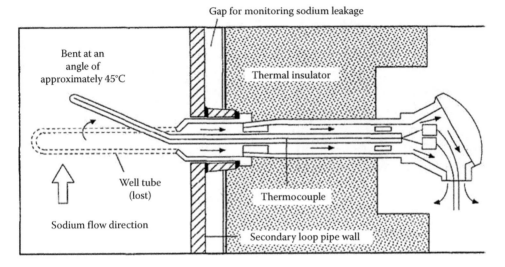

FIGURE 27.1   Monju thermowell sodium leak. (From Guidez, J. et al., *Nuclear Technol.*, 164, 207, 2008.)

positive features, sodium leaks cannot be completely ruled out. A summary of sodium leaks is given in Table 27.3. The worst sodium leak that had a major impact on reactor restart-up occurred in Monju in December 1995 due to a thermowell failure in the secondary sodium circuit piping due to flow-induced vibration leading to a sodium leak of 640 kg (Figure 27.1). Detailed investigations have revealed that the thermowell failed because of high cycle fatigue of the thermowell tube tip due to flow-induced vibration. The geometry also had a steep change in diameter, leading to stress concentration. A survey of sodium leak events in reactors indicates leaks ranging from a few grams to 1825 kg. Figure 27.2 shows an example leak of a few liters on a Phénix reactor secondary purification circuit valve. The main reasons for sodium leak incidents are as follows [27.1]:

- Valve manufacturing defects—an FBTR leak event after 17 years of operation in the primary sodium purification circuit is attributed to a drilled hole almost through the valve body thickness (Figure 27.3).
- Flange joint construction in piping and valve body bonnet joint.
- Defective welds.
- Poor material choices like the use of Type 321 stainless steel.
- Design deficiency due to inadequate piping flexibility or inadequate check on design against flow-induced vibration.
- Thermal striping—different sodium temperature streams mixing at the tee junctions leading to thermal fatigue failure.
- Operator error—pipeline cutting before sodium is frozen.
- Improper melting technique for solid sodium in pipe work.

Apart from sodium leak consequences, safe disposition of huge amounts of radioactive sodium during the decommissioning stage would call for high investment and operational costs and complex

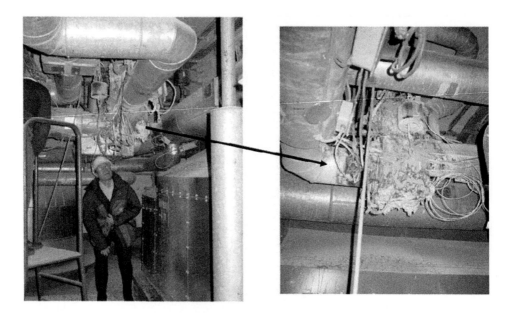

FIGURE 27.2 Examination of the circuit in Phénix following a leak. (From Guidez, J. et al., *Nuclear Technol.*, 164, 207, 2008.)

FIGURE 27.3 Sodium leak from a valve in the purification cabin.

technologies. Further, the opaqueness of sodium poses challenging technological problems for in-service inspection and repair. Also, there are a few specific structural mechanics problems related to sodium—thermal striping and thermal fluctuations—which severely affect the structural integrity of adjoining structures. Despite these facts, the problems have been overcome through suitable design measures: the quantity of sodium used in the reactor needs to be optimized, in particular, possibly minimum size of reactor and heat removal systems should be assured by the application of innovative design concepts.

27.2.1.1 Feedback from the Experiences of Sodium Leaks [27.1]

By close scrutiny and taking into account the advantages of sodium cooling, SFRs do not have any significant disadvantage. Several design features are incorporated to prevent sodium leaks and to detect and mitigate leak effects. The choice of materials with high ductility and efficient and diverse leak detection systems (wire, spark plug, and mutual inductance types) facilitate to demonstrate leak before break justification with comfortable margins. With the safety vessel, provided around the main vessel, the sodium level is maintained to facilitate decay heat removal even after main vessel leaks. All sodium pipes inside the reactor containment building are provided with guard pipe/nitrogen-filled cells to rule out the possibility of sodium fire. The main vessel welds are periodically inspected through robust in-service inspection device, which can travel in the space between main vessel and safety vessel. Apart from these, the provision of leak collection trays, the use of sodium-resistant concrete to safeguard against sodium-concrete interaction, and fire extinguishers and fire fighting facilities are incorporated following the tradition of international SFRs. Advanced computational simulations are employed in the design of various systems relevant to sodium fire studies. Provisions are made in the design for minimizing the consequences of sodium leaks early, detecting sodium leaks, fast dumping of sodium safely, and fighting sodium fire. All the reactors have specific design features for sodium leaking which ensures the leak from the primary sodium system into inserted guard vessel piping cabin and not to air which will result in fire. It is worth mentioning that there have been no leaks in Russian reactors for the last 14 years.

Comprehensive review and modifications consequent to sodium leak events have been carried out in Monju. The modifications include replacing the thermowells with an improved design of shorter length, taper well geometry instead of steep diameter change and new design guide against flow-induced vibration. In addition, improvements have been made to the sodium leak detection system to detect a leak at an early stage and to the sodium drain circuit to reduce the drainage time, the secondary circuit rooms were partitioned to limit the quantity of air availability, thermal insulation was applied for the protection of the concrete floor and walls, and a nitrogen injection system was installed to extinguish sodium fire quickly. A leak before break concept has been well demonstrated in sodium leak events. This can form the design basis for the leak collection tray for piping and provisions of powder for active means of fighting sodium fire. To minimize the risk of a sodium leak from piping tee junctions due to thermal striping, detailed thermohydraulic analysis should be carried out for the steady state and the transient state. Leakage across the seat of a normally closed valve should also be considered. The need and application of thermal mixers should also be carefully assessed. Researchers are also studying some alternative coolants, such as gases (nitrogen, helium, and CO_2), and liquid metals (lead and lead–bismuth).

27.2.1.2 Sodium Aerosols [27.1]

Convection currents in the reactor cover gas containing sodium aerosols have the potential to cause accumulation of sodium in the narrow gap in the rotating plugs, control rod drive mechanisms, and IHX sleeve valve. In 1997, in the BN-600 reactor it was noted that significantly greater efforts were required to move the rotating plug (Figure 27.4). The disassembly operations revealed that the bearing surface was covered with sodium deposits. Sodium residues were removed from the bearing races and narrow gaps of the rotating plug, and new balls and cages were installed. The operation of the EBR-II rotating plug freeze seal had not been trouble free. The seal alloy for the freeze seal was tin–bismuth alloy (42% tin and 58% bismuth). Sticking of the large rotating plug was attributed to the accumulation of intermetallic compounds of sodium and tin and sodium and bismuth in the annulus between the plug wall and the seal through support structure. Periodic cleaning of the annulus was necessary to maintain rotation of the large plug. In prototype fast reactor (PFR), a number of spurious trips occurred because of drops of the absorber rods held by electromagnets because of buildup of sodium deposits on the electromagnet, reducing its efficiency to hold the absorber rod. Lessons learned from sodium aerosol deposits resulted in a basic design change involving selecting a hot roof to minimize sodium deposits and exercising the control rods during operation.

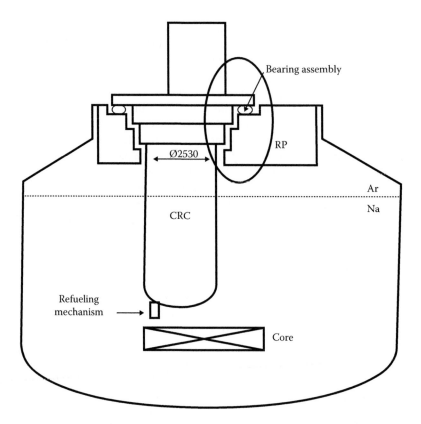

FIGURE 27.4   Aerosol deposition in the BN-600 rotating plug. (From Guidez, J. et al., *Nuclear Technol.*, 164, 207, 2008.)

27.2.2 STEAM GENERATOR PERFORMANCE [27.1]

Except for Superphénix and FBTR, all single-wall steam generators experienced tube leaks during operation. Table 27.4 summarizes steam generator tube leaks in various fast reactors. The principal reasons for leak events can be summarized as follows:

1. *Manufacturing defects*: One leak incident in KNK-II, several tube leak events during sodium filling or the initial years of operation in BN-350 and BN-600, and initial tube leaks in the PFR evaporator are attributed to manufacturing and quality assurance defects. The absence of postweld heat treatment of Cr–Mo tube-to-tubesheet weld in the evaporator caused the poor performance of the PFR evaporator. The poor quality of the bottom-end cap material of the bayonet steam generator design led to tube leaks in BN-350. The tube-to-tubesheet weld of a rolled and welded design has also been responsible for tube leaks.
2. *Fatigue crack*: A combination of design flaws or an inappropriate operating procedure could lead to thermal shocks leading to tube leak. The first four incidents of tube leaks in Phénix were attributed to malfunctioning of the turbine steam bypass circuit during plant start-up, causing a water ingress in the reheater.
3. *Flow-induced vibration fretting*: A large steam–sodium reaction in the PFR super heater involving a rupture of multiple tubes was caused by fatigue failure due to tube fretting against the central flow baffle.

Unlike other heat exchangers, a tube leak in a sodium-heated steam generator could lead to damage to neighboring tubes by impingement wastage—thinning of affected tube overheating due to

TABLE 27.4
SG Performance (Number of Tube Leaks Leading to Sodium–Water Reactions)

FERMI-I	EBR-II	KNK-II	BOR-60	PFR	Phénix	BN-350	BN-600
2	Nil	1	1	40	5	Several (initial) 3 (after 1980)	12

exothermic reaction and corrosive sodium hydroxide formation resulting from the sodium–water reaction. In the early years of SFRs, in the first tube leak event in any given country, significant damage occurred because of the relatively long time taken for the tube leak detection, safety action on water–steam sodium valves, and nitrogen blanketing of the tube side. Typical examples of tube leak events leading to sodium–water reaction are as follows:

- Fermi steam generators faced considerable difficulties during operation. The seamless tubes of the single-wall steam generator had intricate serpentine involutes (Figure 27.5). A major sodium–water reaction occurred just after 2 weeks of operation during the first flow tests in 1962 due to flow-induced vibration directly in front of the sodium inlet nozzles. During the reaction, five additional tubes failed due to a wastage mechanism and overheating, and extensive tube thinning was observed.
- In 1982, the first tube leak incident in the Phénix reheater unit led to 30 kg of steam leaking into sodium. A subsequent tube leak incident led to only 1–4 kg of steam leaking into sodium because of improvements made in the signal processing design and duplication of the nitrogen injection system into the tubes. All the leaks occurred at the butt welds of the reheaters (Figure 27.6).

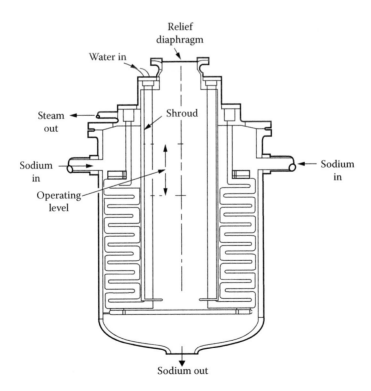

FIGURE 27.5   FERMI reactor steam generator. (From Guidez, J. et al., *Nuclear Technol.*, 164, 207, 2008.)

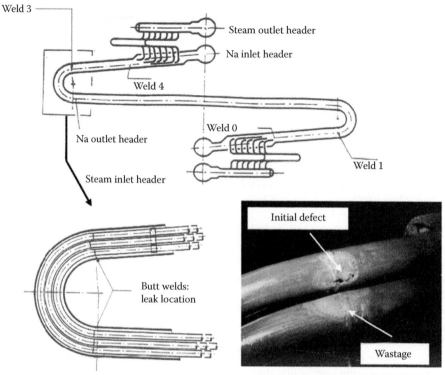

Weld 3

Steam outlet header

Na inlet header

Weld 4

Weld 0

Weld 1

Na outlet header

Steam inlet header

Butt welds:
leak location

Initial defect

Wastage

Steam generator tubes after
sodium–water reaction

FIGURE 27.6 Phénix reheater tube leak. (From Guidez, J. et al., *Nuclear Technol.*, 164, 207, 2008.)

- The worst tube leak event to cause concern for the integrity of the IHX in the SFRs occurred in 1987 in the PFR superheater: 39 tubes failed within 10 s because of overheating with the damage initiated by a tube leak failure due to fretting against the central flow baffle (Figure 27.7). The automatic protection system operated as designed, with actuation of rupture disks, dumping of steam and sodium, and isolation of the steam generator. The maximum IHX pressure during the incident was estimated to be 10.5 bars, which is less than the IHX design pressure, and thus, the integrity of the IHX was not affected because of the safety protection system.
- In BN-600, tube leak events occurred in which the maximum amount of water–steam injected into sodium was 40 kg.

The lessons learned from these incidents that would help in future reactor designing are the following:

- *Material selection*: Severe damage to the austenitic stainless steel due to caustic stress corrosion cracking resulting from a tube leak led to the rejection of austenitic steel material for the superheater, and reheater austenitic stainless steel is not used for the evaporator because of the risk of chloride stress corrosion cracking.
- *Manufacturing*: A tube-to-tubesheet joint is more vulnerable to a leak than the rest of the tube bundle. A better option is to adopt a butt weld for the tube-to-tubesheet joint with a raised spigot from a weld inspection point of view and low operating stress compared to the conventional rolled and welded joint.

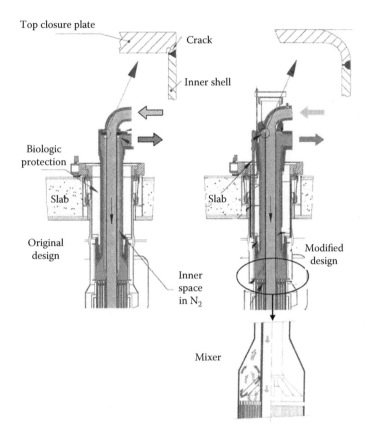

FIGURE 27.7 Design modification on the Phénix IHX. (From Guidez, J. et al., *Nuclear Technol.*, 164, 207, 2013.)

- *Tube bonding*: In current designs, tube-to-tubesheet welds in Cr–Mo material are preheated and heat treated postweld to provide low residual stress and enhanced reliability against stress corrosion. The number of tube-to-tubesheet weld joints is minimized by having long seamless tubes of length as governed by manufacturing and transportation aspects. Metallurgical bonded tubes are preferred to mechanical bonded tubes in the duplex tubing steam generator design. One of the EBR-II superheaters with a mechanically bonded tube design experienced a sudden decrease in superheater outlet steam due to increased thermal resistance caused by a reduction in residual interface contact pressure as a consequence of long-term thermal aging.
- *Protection against sodium–water reaction*: In the current operation of SFRs, the tube leak detection system is not bypassed. The PFR superheater tube leak accident mentioned earlier could have been much milder if hydrogen had been detected in the sodium and the suitable action had been taken of putting the steam generator in a safe configuration of draining and isolation of the steam generator on the water–steam side. The design of tube bundle should permit in-service inspection by reliable periodic nondestructive testing examination, identifying the leaking tube, and deciding how to plug the neighboring tubes. Remote field eddy current testing with an advanced signal analysis method has since been developed to meet the exact inspection regime. Protection against overheating needs to be provided. Incorporation of a rupture disk at the steam generator sodium inlet and outlet are preferred options. The reliability of water–steam dump and nitrogen injection into the tube side can be improved through better designs and/or by duplicating the system.

- *Modular power steam generator*: Twelve leak events in BN-600 led to insignificant loss of power generation because of design provisions of the operating steam generator with the remaining modules (there are eight evaporators, eight superheaters, and eight reheaters for each of three secondary circuits), that is, seven modules of the affected circuit. A modular steam generator design choice is worth considering at the design stage, the number of steam generators being kept as one per circuit with an increased number of secondary circuits or multiple steam generators for each secondary circuit with a relatively fewer number of circuits. The number of steam generators for the reactor is governed by the outage cost in case of a tube leak, the provision or absence of a spare steam generator, tube leak probability, and the impact of the number of steam generators on the overall construction schedule.
- *Tube leak*: The probability of a tube leak is likely to decrease after the first few years of operation as manufacturing defects are revealed in the initial period. BN-600 suffered 12 tube leaks in first 10 years of operation with no incident subsequently since 1991.

27.2.3 SODIUM-TO-SODIUM HEAT EXCHANGER [27.1]

- IHX operational experience, except for the Phénix reactor and a minor incident of drain pipe failure in EBR-II, has not been a concern in terms of loss of plant availability. Sodium leaks from the Phénix IHX took place at the secondary sodium outlet header because thermal loading caused by a difference in the temperatures of the inner and outer shells was underestimated at the design stage. All IHXs were repaired. Design modifications were made to the sodium outlet header including incorporation of a thermal mixer (Figure 27.7).
- Removal, washing, and decontamination of the radioactive primary sodium component have provided valuable maintenance experience. For future SFRs, detailed thermohydraulics of the secondary sodium outlet header needs to be carried out to minimize thermal stresses. For primary sodium in the shell and secondary sodium in the tube, the design option of variable secondary sodium flow in the tubes with higher flow on a number of the outer rows of tubes to have nearly the same secondary sodium outlet temperature of the inner rows of tubes is worth incorporating.

27.2.4 SODIUM PUMPS [27.1]

The performance of the mechanical sodium pumps in the reactors has been good, and the load factor outage due to the pumps is very marginal. Minor incidents occurred in EBR-II, Rapsodie, KNK-II, BOR-60, FFTF, PFR, BN-350, Phénix, and BN-600 sodium pumps with most of the incidents in the early stages of operation. In BN-600, the shaft of the primary pump and the gear couplings of the motor shaft suffered damage due to matching of shaft vibration mode frequencies with the torsional vibration frequencies. Subsequent modifications relating to the design of the shaft and the mode of operation have led to trouble-free operation. In Phénix, vibrations were detected in primary pumps due to a design defect, which led to the hydrostatic bearing bush to be separated from the shaft as a result of hot thermal shock. A similar incident occurred in a secondary pump. Modifications were carried out, and further operation of the pumps was trouble free. In PFR, in 1984, one secondary sodium pump failure occurred because of hydrostatic bearing seizure caused by detachment of the spray fused stellite coating on the shaft sleeve.

Past and current power reactors have had a minimum of three primary and three secondary pumps. Reliable operation of these pumps has led the designers to choose two primary pumps and two secondary pumps for the 500 MWe PFBR, now under construction in India, and two primary and two secondary pumps for the design of the 1500 MWe Japan Atomic Energy Agency SFR (JSFR). This approach is based on robust performance results in reducing SFR capital cost.

27.2.5　Fuel Handling Systems

In FBTR, fuel handling is carried out in the reactor shutdown state with sodium at 453–523 K. Except for a major incident in 1987 in FBTR, fuel-handling operations in all fast reactors have not been a subject of concern. The FBTR fuel-handling incident is described in this section.

During an in-reactor transfer operation, the foot of the subassembly being transferred had been projecting slightly below the heads of the other subassemblies. This resulted in bending of the foot of the subassembly. The heads of the subassemblies in its transfer path also bent. Because of the bent foot, the subassembly could not be loaded in any position. Lifting it up after the trial of lowering its bent foot ejected out a steel reflector subassembly. During further plug rotation, the guide tube interacted with the head of the ejected subassembly and bent. The bent guide tube had to be cut into two pieces for its removal in situ with a special machine developed for this purpose (Figure 27.8). What initiated this incident was the inadvertent bypassing of the plug rotation logic. Subsequent to the incident, all fuel-handling operations are being carried out with a rigorous checklist. It took 2 years to restore the normal operation state from this incident. Since 1989, fuel-handling operations have been smooth and trouble free [27.1].

In FFTF, one of the three fuel transfer ports experienced difficulties in the removal of the port plug during refueling because of tilt of the transfer port nozzle. The tilt was due

FIGURE 27.8　Special cutting tool for guide tube cutting in FBTR.

to oxidizing and swelling of a uranium shield ring around the fuel transfer port. The uranium shield was stored in sealed steel containers, but rupture of the steel casing caused the uranium to swell, forcing the shield against the nozzle flange and causing tilting and stretching of the nozzle. Subsequent power operation was resumed by in situ removal of the uranium shield. Relaxation of the nozzle occurred, and there was no further movement of the nozzle.

Fuel-handling problems have also occurred in EBR-II [27.3]. The second major incident was damage to a fuel assembly during fuel handling in April 1978, bending it so that it could not be removed from the fuel storage basket. Fuel handling in sodium is done without visual reference as all operations are done remotely. When an attempt was made to engage the assembly upper adaptor with the fuel-handling arm as part of the procedure to remove it from the storage basket, it was found that the upper adaptor was out of position and could not be engaged. A technique was developed for profiling the assembly by mechanical means, using the fuel-handling equipment to characterize its position and configuration. Following this work, a mock-up of the storage basket, the deformed assembly, and the fuel transfer system was constructed to develop the tools and procedures for the removal of the assembly. Removal was accomplished in May 1979 using a specially designed shaft and gripper that penetrated one of the nozzles in the cover of the primary tank. Reactor operation was not impacted and fuel handling from the storage tank proceeded normally for other assemblies located within it. The techniques developed and experience gained have proven valuable for fuel-handling system design and proved to be beneficial for the second incident associated with fuel handling at EBR-II. It was found that the damage occurred because the assembly had not been fully seated in the storage basket and when the storage basket was raised, the assembly contacted the lower shield plug of the primary tank cover. On November 29, 1982, a fuel assembly was dropped over the EBR-II core as it was being transferred from the fuel storage tank. The problem was discovered when no assembly was present for the exchange between the transfer arm and the core fuel assembly gripper. Extensive checks were made to verify that the assembly was not located in the storage basket or the transfer arm and then a search began to determine its location. The assembly had been dropped somewhere between the storage basket and the intended location in the core. Care had been taken in design to provide extensive interlocks to ensure that movement of the fuel handling equipment did not begin until assemblies were securely gripped, and manual operation of the transfer operation was such that checks could be made manually. However, in this instance the assembly had become disengaged from the transfer arm and fallen. It was found that the transfer arm and storage basket were misaligned, preventing the assembly's upper adapter to be fully seated and locked before transfer. Mechanical probes were used to locate the assembly and precisely identify its position. As before, a full-scale mock-up was constructed and tools and procedures were developed to retrieve the assembly. The major retrieval tool was a stainless steel cable extending as a loop beyond a stainless steel tube that penetrated the top cover. (A number of spare nozzles had been provided through the cover in the original design, a decision which proved to be very valuable.) The loop was maneuvered into position manually and the noose pulled tight, snagging the assembly's upper adaptor so that it could be retrieved. (This process was aided not only by the ability of the operator to feel resistance but also by acoustic monitors installed in the tank that detected the sound from contact with the equipment.) The assembly was then moved to a position where it hung from the noose and could be engaged by the transfer arm; it was handled normally from that point. The total operation took less than a month and in this case required the reactor to be shut down. However, advantage was taken of the downtime to conduct preventive maintenance normally scheduled for the spring shutdown, so the overall impact on reactor operation was minimized. Over 40,000 fuel assembly transfers were made without incident in the 30 years of operation of EBR-II, so these incidents were certainly rare. However, mishaps during fuel handling can have a significant impact on reactor operation and every reasonable precaution needs to be taken to prevent them. Besides robust fuel-handling systems and extensive interlocks, the EBR-II experience demonstrated the importance of operator tactile feel and acoustic monitoring for the operation of the equipment. Much of the success for the EBR-II fuel handling experience, for example, was due to the fact that motion of the rotating transfer arm was manual, allowing the operator to verify through a *wiggle* test that the arm had successfully engaged the assembly before it was released by the core gripper. Under-sodium viewing technology is now available as another guard against fuel-handling errors.

Another lesson learned from EBR-II operation was the importance of anticipating problems and providing design features to accommodate them. For example, in anticipation of an assembly falling from the transfer arm after it had cleared the core, a catch basket was provided that would funnel the assembly to a position where it could be easily retrieved. Spare nozzles had also been provided to support special operations in the primary tank. Of note, each of the primary pumps was removed for maintenance twice during the course of EBR-II operation, facilitated by designs and equipment that anticipated the need.

In August 1974 in BOR 60, a large number of fuel subassemblies were bent during fuel-handling operations due to inadvertent upward movement of an absorber element above the core SA. There was also an oil leak from one of the secondary sodium pumps. Recently in Joyo [27.4], the top of irradiation test subassembly *MARICO-2* (material testing rig with temperature control) was bent onto the IVS (in-vessel storage rack) and had damaged the upper core structure (Figure 27.9). This incident required replacement of upper core structure and retrieval of MARICO-2 subassembly for Joyo restart.

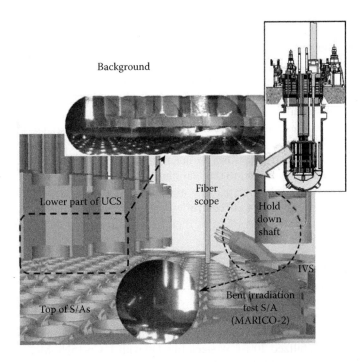

FIGURE 27.9 MARICO assembly bending in Joyo. (From IAEA-CN-199/103, Restoration work for obstacle and upper core structure in reactor vessel of experimental fast reactor "Joyo", Japan Atomic Energy Agency, Paris, France.)

27.3 MATERIAL BEHAVIOR

27.3.1 FUEL

During more than 40 years of intensive multinational development, significant experience has been accumulated on fast reactor MOX fuel pins as follows [27.5]:

- In Europe, more than 7000 pins have reached burnup values of 15 at%. In addition, some experimental pins (with solid or annular pellets) have attained burnup levels of 23.5 at% in PFR and 17 at% in Phénix.
- In the United States, more than 63,500 pins with solid pellets have been irradiated in FFTF under prototypical conditions with more than 3,000 pins at 15 at% burnup and with a maximum burnup level of around 24.5 at%.
- In Japan, 64,000 pins with solid pellets have been irradiated in JoYo and foreign fast reactors, with a maximum burnup level of around 15 at% in JoYo and 15 at% in FFTF.
- In the Russian Federation, a great experience was gained with vibro-compacted MOX fuel. A record high burnup level of about 35 at% was reached with an experimental subassembly in BOR-60 and about 260 standard fuel pins have attained burnup levels of 25–30 at%. More than 4000 fuel pins with pellet MOX fuel were irradiated in BN-350 and BN-600; maximum burnup in BN-600 was 11.8 at%.

27.3.2 CORE STRUCTURAL MATERIALS

The choice for the core structural materials, since the beginning, was largely around the 300 series austenitic stainless steel family for the oxide fuel elements. For the metallic fuel that has been extensively irradiated in the United States, the choice was ferritic steel. Hence, often the material is chosen along with the type of fuel, whether ceramic or metallic. In general, the choice for the clad and wrapper is normally the same. Between the clad and the wrapper, the clad limits the burnup as it reaches the high dpa levels due to excessive clad strain as a result of swelling and in-pile creep rupture. However, the wrapper also would some time place the limit ahead of the clad depending on the subassembly design and the operating conditions. Initial reactors could sustain moderate burnup with the austenitic steel clad. The limitation to the austenitic steel comes from degradation due to void swelling and by embrittlement of ferritic and ferritic–martensitic steels. When used as a wrapper, the limit is often dictated by dimensional limits and the ductility limit rather than by creep rupture. The maintenance of desirable properties is directly linked to the maintenance of a stable microstructure against the action of neutron-induced displacements. The most important requirements for such stability are well-defined specifications and well-controlled production methods. Various reactors in the world have chosen the material according to their design requirements. Three major class of materials used for clad are

- Austenitic steel 316 and its variants in stabilized and cold worked condition
- Martensitic and ferritic–martensitic alloys
- High nickel alloy like PE-16, due to its low swelling properties

With respect to austenitic steel, several varieties have been developed over the years, among which the major one is the Ti-stabilized 20% CW steel, like D9 with 15Cr–15Ni–Mo. Among the ferritic steels, early designs have used HT-9, though of late plain 9Cr–1Mo is being used for wrapper and modified 9Cr–1Mo is proposed for metal fuel pins. PE16 was widely used in the United Kingdom. The United States has used 20% CW D9 in oxide-fueled FFTF and HT-9 in metal-fueled EBR-II. For Russia, France, Germany, Japan, and India, used austenitic steel has been the major choice. If very high burnups, above 20 at%, are to be achieved, the choice becomes complex as austenitic clad and swelling-resistant ferritic would restrict the limit due to bundle–duct interaction and flow reduction through the subassemblies.

Titanium-stabilized (0.4%–0.5%) cold-worked (15%–20%) steels such as French 15–15 Ti alloy and the German DIN 1.4970 alloy (10Cr–Ni–Mo–Ti), around 10,000 pins in Europe, have reached dose values of 100 dpa (NRT) up to 923 K. Around 1000 pins could reach 125 dpa and one experimental assembly containing 217 pins has been taken up to 148 dpa [27.6]. These large numbers of irradiated pins have emphasized the beneficial effects of an adjustment of cold work level and the addition of some minor elements (Ti, C, Si, P), on the swelling and irradiation creep behavior where the increase in incubation dose before swelling occurs is the most important factor. The mechanical properties of these materials have also been accurately investigated through tensile tests performed in both the longitudinal and transverse directions using specimens machined from defueled Phénix fuel elements. The results showed that the mechanical behavior depends not only on the test and irradiation conditions but also on swelling resistance. The effect of irradiation varies with temperature: hardening at low temperatures and softening at high temperatures was observed, associated with dislocation rearrangements, but no recrystallization occurred. It appears that at least up to about 120 dpa, these alloys have sufficiently high temperature strength and adequate ductility in the temperature range of fuel pin cladding.

Based on the results and subsequent development work, advanced reference cladding material was chosen as AIM1 (austenitic improved material number one) in the European context designed to reach 170 dpa in European fast reactors. In parallel, an R&D program has been conducted to develop advanced austenitic cladding materials (10.15 Cr/15.15 Ni type—Ti Nb stabilization—high P content). Ti-stabilized material containing 12 Cr and 25 Ni has been irradiated in Phénix with promising results: its swelling resistance is clearly higher than the best alloy 15.15 Ti, even Si modified. In the United States, 20% CW D9 was demonstrated up to 140 dpa and pin level irradiation was taken up to a record 200 dpa level without any pin failure. However, D9 has been found to suffer from brittle failure when swelling reaches 10% diametral strain, which was attributed to a channel type of fracture. Japan has experienced around 185 dpa with the PNC 316 alloy. Russian clad demonstrated its capability of up to 100 dpa. The Indian choice of 20% CW D9 has been irradiated in its test reactor around 65 dpa with burnup around 112 GWd/t.

In summary, worldwide R&D work on various cladding materials has reached a very high level of understanding of the basic phenomena involved as well as of the operational requirements to be met by a well-designed fuel element. Austenitic alloys (15.15 Ti, 1.4970, PNC 316, D9, PNC 1520) have proved their ability to reach exposure rates as high as 150 dpa and their advanced versions (AIM1, CEA 12.25) are very promising candidates for the current target doses (170 dpa) of commercial fast breeder reactors. Ferritic–martensitic alloys (HT 9, EM 12, PNC-FMS, EP-450) are also able to fulfill these objectives if a limitation on peak cladding temperature is acceptable. For the future, oxide dispersion–strengthened (ODS) steels are under development, which would combine the advantages of both austenitic with high temperature strength and ferritic with its insignificant swelling. ODS steels are being developed to reach the ambitious target of 200 dpa.

27.3.3 OUT-OF-CORE MATERIALS [27.1]

The performance of austenitic stainless steels with the exception of Type 321 stainless steel has been satisfactory in fast reactors. Grades that perform well include Types 304, 304LN, 316, 316L, and 316LN stainless steels. There have been a number of cracks and sodium leaks associated with Type 321 stainless steel welds in the Phénix secondary sodium piping and steam generators and in the PFR superheater and reheater vessel shells. The cracks are attributed to delayed reheat cracking. As a result, Type 321 stainless steel was gradually replaced by Type 316LN stainless steel in Phénix. Because of this experience, stabilized grades Types 321 and 347 stainless steels will not be considered for future fast reactors. The performance of C–0.3 Mo steel (15 Mo 3) in the Superphénix fuel storage drum and sodium tanks constructed for use in SNR 300 was not satisfactory and led to rejection of this grade of steel for SFRs. In some countries, R&D for structural materials is directed toward an improved-grade Type 316 stainless steel with nitrogen added. Cr–Mo steel is being chosen over austenitic stainless steel for secondary sodium piping to improve economics and also to reduce thermal fatigue damage due to Cr–Mo steel's relatively low thermal expansion coefficient and high thermal conductivity.

27.4 SAFETY EXPERIENCE

27.4.1 Reactivity Events

Four negative reactivity trips occurred in Phénix in 1989–1990 on August 6 and August 24, 1989 [27.7]. The first explanation was a possible interference on the measurement channels (modified during the 10-yearly statutory inspection in 1989). The incident on September 14, 1989, was attributed to power variation due to possible gas volume crossing the core. After detailed analysis of this scenario and its consequences, and taking preventive measures, the reactor was authorized to be started up again in December 1989. The fourth event occured on September 9, 1990, which invalidated the gas scenario due to the fact that the larger amplitude of neutron signal was not compatible with gas entrainment effect. The details of the reactivity variations are shown in Figure 27.10. Even though several investigations were carried out, the cause could not be conclusively ascertained. Recently, it has been hypothesized that the event could be due to the movement of the core subassembly caused by sudden vapor bubble impact by the experimental assemblies in which moderator material was put due to inadequate thermal hydraulics requirements.

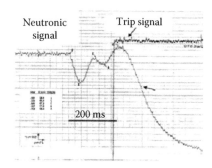

Event	First minimum	Second minimum	Positive reactivity before CR drop
Time	50 ms	150 ms	200 ms
Trip n°4	−0.99$	−0.28$	+0.11$
Trip n°3	−0.43$	−0.11$	+0.07$

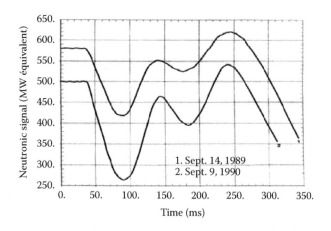

FIGURE 27.10 Phénix negative reactivity events. (From Prulhière, G. and Dumaz, P., Phenix negative reactivity trips, Nuclear Energy Division, CEA, France, 2013.)

FBTR had three positive reactivity transients: one in 1994, one in 1995, and one in 1998 with the reactor operating at 8–12 MW thermal power with relatively low primary sodium flow. In 1994 and 1995, 19 postulations were studied experimentally for the transients, but reasons could not be established. Experimental investigations revealed that the transients in 1998 were reproducible. Reactor operation subsequently has been free of reactivity incidents with primary sodium flow above the threshold value. The most probable reason is thermally induced geometric changes in the core at low flow rates [27.1].

27.4.2 Fuel Pin Cladding Failures

Pin failure is categorized into (1) leaker phase failure and (2) wet rupture. In leaker phase failure, there is a pin hole–type opening in the clad that allows bond gas He, fission gases, and volatile fission products like Cs, I, Br, etc., to escape from the fuel pin. It takes hours to days for these gases to escape. Failure is detected by cover gas monitoring systems. Prolonged stay of failed fuel pin in the reactor results in wet rupture wherein sodium enters the fuel pins resulting in the formation of sodium urano plutonate, Na_3 (U, Pu) O_4, compounds. Wet ruptures are detected by the delayed neutron detector (DND) detectors kept in the IHX mouth. During operation of PFR, Phénix, and KNK-II, there have been 60 fuel pin failures, 41 with exposed fuel and 19 with gas leakers [27.8,27.9]. Table 27.5 shows the details of pin failure due to leaker and wet rupture types. Fuel pin failure rates in different reactors are given in Table 27.6.

TABLE 27.5
Type and Number of Pin Failures

Reactor	No. of Failures with Exposed Fuel	No. of Failures with Gas Leakers
PFR	18	4
Phénix	15	14
KNK II	8	1

TABLE 27.6
Fuel Pin Failure Rates

Country	Reactor	Irradiated Fuel	Failure Rate (%)	Total
USSR	BR-5	–2,490		–61,600
	BR-10	–1,520		
	BOR-60	11,400	<0.5	
	BN-350	–46,200		
France	DFR	41	10	44,650
	Rapsodie-Core 1	4,305		
	Rapsodie-Fortissimo	–17,300	<0.2	
	Phénix	23,002 (>40,000)	<0.01	
USA	SEFOR	648		–2,450
	EBR-II	–1,800		
UK	DFR	–1,000	10	–1,000
DEBENELUX	Rapsodie	73		181
	DFR	108	10	
				–150
Other				–110,000

It can be seen that the fuel failure rate in Phénix is <1 in 10^4 fuel pins irradiated. Based on the international experience, it can be said that a typical fuel failure evolved in three steps:

- Release of fission gas to the primary circuit, which is detected by cover gas monitors
- Penetration of sodium through the cladding defect depending on the defect size, location, burnup, and continued operation, which is detected by DND monitors
- Reaction of sodium with the fuel resulting in secondary failure of clad to accommodate increased volume of reaction products

Pin failures could be due to various reasons, ranging from (1) poor fabrication, (2) fuel–clad chemical interaction (FCCI), to (3) high stress on the clad due to fuel–clad mechanical interaction (FCMI). The experience of failed fuel pins from PFR, Phénix, and KNK-II shows a statistical figure of 1–2 fuel pin failures per cumulated reactor operating year, and the fuel pin had undergone a peak burnup of 190 GWd/t and a dose of about 135 dpa NRT. It was observed that all failures have been reliably detected and located and there was no evidence of failure propagation. Besides, there was no primary sodium contamination. Also, clad failure at low burnup was attributed to manufacturing defects. An analysis of the fuel pin failure in Phénix indicated that failures occurred at random burnup ranging from 0.6% to 12%; some failures were due to FCMI, some were due to rise in power level to full power from 2/3 power resulting in FCMI with a low ductile Inconel clad material. Clad with annealed 316 showed local failures in contrast to CW 316 Ti cladding. From PFR experience, it is noted that pin failure has occurred at a burnup of 0.3% and at a high burnup of 21% and that the timescale of development failure from *gas leaker* to *exposed fuel* was from a few hours to tens of days. The largest single defect had an actual failed area of 5.49 cm^2, while many pins were allowed to continue operation with several hundred square centimeter recoil areas. Analysis of the failures in KNK II indicated that failures occurring at low burnup were due to pin/grid space wire interaction and that occurred at high burnup may be due to end-of-life consideration. Various failures, 60 nos., that occurred in Phénix, PFR, and KNK II have been classified into various categories (Table 27.7). It can be seen from the table that 20% failures occur due to fabrication deficiencies, 22% due to pin–duct interaction, 5% due to FCMI, and 3% due to FCCI. There are about 50% failures where reasons could not be attributed. In BN-600, fuel cladding failures and shutdowns due to loss of integrity of the standard fuel have not occurred since 1999.

TABLE 27.7

Categories of Fuel Pin Failures in Reactors Worldwide

	Class A Manufacturing Deficiencies	Class B Mechanical Interaction, e.g., Pin/Spacer/ Duct Interaction	Class C Fuel–Clad Mechanical Interaction (FCMI)	Class D Fuel–Clad Chemical Interaction (FCCI)	Unclassified
Phénix	6	7	2		14
15 exposed fuel					
14 gas leakers					
PFR	4		1	2	15
18 exposed fuel					
4 gas leakers					
KNK-2 8 exposed fuel	2	6			1
1 gas leaker					
% of 60 failure events recorded	20%	22%	5%	3%	50%

27.4.3 Fuel Melting

Core meltdown accidents have occurred in EBR-I and Enrico Fermi Fast Breeder Reactor. However, the most severe accident, the core disruptive accident (CDA), has not occurred in any FBR. The core meltdown accidents in EBR-I and Fermi reactor are severe accidents, though there was no CDA subsequent to core meltdown [27.10,27.11]. The Enrico Fermi reactor suffered a meltdown of more than two assemblies in 1966. This was directly the result of a blockage of the inlets to those assemblies, and the accident was therefore a flow reduction. Fortunately, the blockage occurred at assemblies in a low rated region of the core while the reactor was being brought up to power. Thus, the power rating at the time of the accident was not high. Here is the sequence of events that followed:

1. Presumably core voiding took place although this was not detected.
2. Fuel cladding failed.
3. Fuel movement within the failed assemblies gave a reactivity decrease. Later, this was interpreted as being due to radial motion induced by the fact that the two assemblies melted close to each other, presumably because they were being cooled by the unaffected subassemblies on their other side.
4. This caused asymmetric fuel movement toward that side of the fuel subassembly.
5. No fuel foaming was observed as had been observed in EBR-I, despite the fact that Fermi also contained metal fuel.

In EBR-I, tests were run involving reactivity ramps at low flows. As the power neared the safety limit, the order to shut the system down was misinterpreted and a slow shutdown was initiated instead of a scram system. This occurred in November 1955. Here is the sequence of events that followed:

- Temperatures at the center of the core exceeded the NaK coolant boiling point.
- The boiling forced molten fuel material outward both above and below the core (the fuel was metal uranium).
- The channels were then blocked by freezing of the core material that then formed a cup into which further fuel material fell, including the tops of the pins.
- Fission products from the 0.1% burnup fuel were released as gas bubbles, forming a froth or foam and thus a porous fuel mass when the fuel froze.
- Forty to fifty percent of the fuel melted before shutdown occurred.

The EBR-I core was very small and its 8 in. diameter core was hardly much bigger than today's assemblies and the fuel was highly enriched metallic uranium (EBR core is shown in Figure 27.11). The consequences of the failure are, therefore, hardly representative of possible events in a large distributed core. They are nevertheless representative of the speed of failure since all this took only a few seconds. Further, EBR-I's initial original core was the only core that exhibited a substantial instability. This proved to be the result of a secondary fuel rod bowing effect, by which a primary prompt positive reactivity change, produced by an original bowing of the fuel rods, was counteracted by a delayed and negative effect. When the system was brought to power, the fuel rod first bowed inward while supported by the first shield plate and the lower grid plate, but sometime later the rod began to be affected by the expansion of the third shield plate, which then effectively moved the fuel element away from the center of the core giving a slow negative reactivity coefficient for large power-to-flow ratios (Figure 27.2). The diagnosis of this effect was very difficult but the problem was solved eventually by using a restrained core to prohibit all bowing, including the original prompt positive effect. The absence of any instability confirmed the analysis.

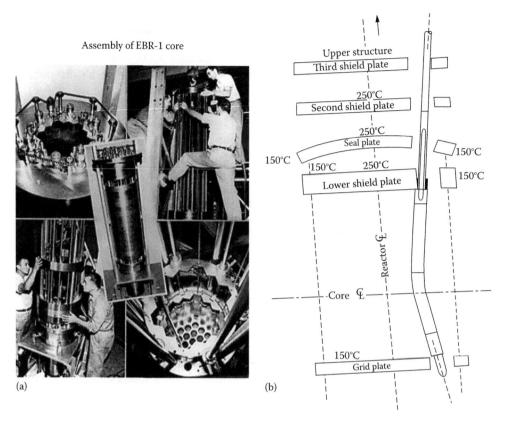

FIGURE 27.11 (a) EBR core and (b) bowed subassembly in EBR core. (From Wigeland, R., Review of safety-related SFR experimental and operational experience in the United States, Idaho National Laboratory, Idaho Falls, ID.)

27.5 OPERATIONAL EXPERIENCE

Major learning from operating and safety analysis of fast reactors can be summarized to be in the areas of [27.2] (1) sodium aerosols blockages, (2) reactivity transients, (3) gas or fluids in primary circuit, (4) materials qualification, (5) core support inspection, (6) monitoring fuel handling in sodium, (7) complete unloading of the core, (8) sodium leak detection, and (9) sodium–water reaction detection. The nature and number of incidents were similar to LWRs, and there was no accident except the partial core meltdown in Fermi 1 reactor.

The learning from the operation feedback can be broadly divided into three aspects:

1. Availability, which is more than 90%—toward this attention is to be paid on long operating cycles and short outages and better components, such as reliable pumps, IHX, SGs, and better materials and circuits that allow the least sodium leaks.
2. Toward safety at the level of the third generation and after the Fukushima nuclear accident—focus should be on primary sodium under extreme conditions, natural convection and margins before boiling, core meltdown knowledge and shutdown, and decay heat removal and confinement.
3. With respect to cost and financial aspects as compared to other electricity generation means, the parameters to be focused are construction and operating costs (reactor + fuel cycle and transport) and in-service inspection, repair or replacement, reliability, and life duration.

Another compilation on the extensive operational experience indicating the positives and the challenges are given next [27.12, 27.13].

27.5.1 POSITIVES

- Fast reactor fuel is reliable and safe, whether metal or oxide.
- Cladding failure does not lead to progressive fuel failure during normal or off-normal reactor operation.
- High burnup of fast reactor fuel is achievable, whether metal or oxide. Acceptable conversion ratios (either as breeders or as burners) are also achievable with either fuel type.
- Sodium is not corrosive to stainless steel or components immersed within it.
- Leakage in steam-generating systems with resultant sodium–water reactors does not lead to serious safety problems. Such reactions are not catastrophic, as previously believed, and can be detected, contained, and isolated.
- Leakage of high-temperature sodium coolant, leading to sodium fire, is not catastrophic and can be contained, suppressed, and extinguished. There have been no injuries from sodium leakage and fire (operation at near atmospheric pressure is an advantage to safety).
- Fast reactors can be self-protecting against anticipated transients without scram when fueled with metal fuel. Load following is also straightforward.
- Passive transition to natural convective core cooling and passive rejection of decay heat has been demonstrated.
- Reliable control and safety-system response has been demonstrated.
- Effective systems for purity control of sodium and cleanup have been demonstrated.
- Efficient reprocessing of metal fuel, including remote fabrication, has been demonstrated.
- Low radiation exposures are the norm for operating and plant maintenance personnel, less than 10% of that typical for LWRs.
- Emissions are quite low, in part because sodium reacts chemically with many fission products if fuel cladding is breached.
- Maintenance and repair techniques are well developed and straightforward.
- Electromagnetic pumps operate reliably.

27.5.2 CHALLENGES

- Steam generators have not been reliable and are expensive to design and fabricate.
- Sodium heat-transport systems have experienced a significant number of leaks because of poor-quality control and difficulty with welds. Also, because of sodium's high thermal conductivity, many designs did not adequately anticipate the potential for high thermal stress on transients.
- Many problems with handling fuel in sodium systems have occurred, primarily because of the inability to visually monitor operations.
- Failure of in-sodium components without adequate means for removal and repair has resulted in costly and time-consuming recovery.
- SFRs have been more expensive than water-cooled reactor systems.
- Reactivity anomalies have occurred in a number of fast reactors, requiring careful attention to core restraint systems and potential for gas entrainment in sodium flowing through the core.
- Operational problems have been encountered at the sodium–cover gas interface, resulting from the formation of sodium oxide that can lead to the binding of rotating machinery, control rod drives, and contamination of the sodium coolant.

REFERENCES

27.1 Guidez, J., Martin, L., Chetal, S.C., Chellapandi, P., Raj, B. (2008). Lessons learned from sodium cooled fast reactor operation and their ramifications for future reactors with respect to enhanced safety and reliability. *Nucl. Technol.*, 164, 207–220, November 2008.

27.2 IAEA. (2013). Prototypes & industrial sodium fast reactors yesterday, today, tomorrow, Jean-François SAUVAGE, Generation IV Projects, IAEA FR 13, Paris, France, March 2013.

27.3 Sackett, J.I. (2008). EBR-II test and operating experience prepared for the US Nuclear Regulatory Commission, INL, Bozeman, MT, December 2008.

27.4 IAEA-CN-199/103. Restoration work for obstacle and upper core structure in reactor vessel of experimental fast reactor "Joyo". Japan Atomic Energy Agency, Paris, France, March 4–7, 2013.

27.5 IAEA. (2007). TECDOC-1569: Liquid metal cooled reactors: Experience in design and operation. IAEA, Vienna, Austria.

27.6 IAEA-TECDOC-1691. *Status of Fast Reactor Research and Technology Development*. IAEA, Vienna, Austria, 2013.

27.7 Prulhière, G., Dumaz, P. (2013). Phenix negative reactivity trips. Nuclear Energy Division, CEA, France.

27.8 Plitz, H., Crittenden, G.C., Languille, A. (1993). Experience with failed LMR oxide fuel element performance in European fast reactors. *J. Nucl. Mater.*, 204, 238–243.

27.9 Warinner, D.K. (1993). LMFBR operational and experimental in-core local-fault experience, primarily with oxide fuel elements, *J. Eng. Gas Turbines Power*, 105(3), 669–678, July 1, 1993.

27.10 Graham, J. (1971). *Fast Reactor Safety*. New York: Academic Press, pp. 149, 280, 287.

27.11 Wigeland, R. Review of safety-related SFR experimental and operational experience in the United States. Idaho National Laboratory, Idaho Falls, ID.

27.12 Sackett, J.I., Grandy, C. (2013). International experience with fast reactor operation and testing. *International Conference on Fast Reactors and Related Fuel Cycles*, May 2013, Paris, France.

27.13 Potapov, O.A. (2013). Operating experience from the BN-600 sodium fast reactor. *International Conference on Fast Reactors and Related Fuel Cycles: Safe Technologies and Sustainable Scenarios (FR13)*. Paris, France, March 4–7, 2013. http://www.iaea.org/NuclearPower/Downloadable/Meetings/2013/2013-03-04-03-07-CF-NPTD/T9.1/T9.1.potapov.pdf

28 Innovative Reactor Concepts for Future SFRs

28.1 MOTIVATION, STRATEGIES, AND APPROACHES

Innovative ideas and concepts are being thought of for future sodium-cooled fast reactors (SFRs), toward achieving challenging objectives, by various countries through a few initiatives. The motivations for developing innovative concepts are development and demonstration of safe nuclear technologies, sustainability, economy, proliferation resistance (PR), amenability for closed fuel cycle program in conjunction with the pressurized water reactor (PWR) fuels, reducing the amount of high-level wastes, reduction in the heat load to the geological disposal, minor actinides (MA) burning to reduce toxicity reduction, and efficient nuclear resource utilization.

Fast reactors have been under development for many decades: primarily conceived as breeders, in recent years their development has also focused on burning high-level waste. The closed fuel cycle that has been demonstrated and experimentally confirmed an effective breeding ratio for these reactors encourages future deployment. Fast reactors promise many advantages: first, and in addition to safe and economical electricity production, they can utilize uranium resources to the greatest extent possible due to the breeding ratio achievable in a fast spectrum reactor, which allows exploiting the energy potential of the fertile nuclides contained in natural uranium resources. Furthermore, radioactive waste is significantly reduced, also thanks to the fact that MA, which represent the long-lived components of high-level nuclear waste, are produced at a lower rate and are even consumed through fission. Finally, fast reactors deal successfully with the issue of U233 production with low U232 production in a uranium–thorium fuel cycle. These facts, in a long-term perspective, support the acceptability and worth of nuclear energy in terms of providing sustainable energy security as well as clean environment. The fast reactor program was started in various countries since 1950, and research and design activities have been carried out for more than 50 years. A significant amount (>50 billion US dollars) of money has been spent on R&D. In spite of these, fast reactor development activities started declining since the mid-1980s and even countries like the United States, who were considered as technology holders, stopped the program. The reason behind this decision has been analyzed to be political rather than scientific. At the same time, a few countries, particularly France, Japan, Russia, India, and China, enhanced their efforts in fast reactor development activities, mainly to realize the faster growth in nuclear contribution and also effective management of high-level waste. However, in recent years, fast reactor programs are being renewed under mega international collaborations, like the Generation IV forum, for example. France has remained committed to the growth of SFRs in particular. It is important to note that out of six reactor systems selected as the most promising systems, to be exploited at the earliest (by 2020–2025) by the Generation IV forum, four are fast reactors. The prospects of various kinds of reactors conceived under Generation IV are covered in Chapter 6. For more details, readers are requested to read Ref. 28.1 and also Chapter 26.

With a global perspective, to motivate the development of fast reactor program in an intensive way that would provide sustainable energy in the twenty-first century and also to create a forum (which does not exist so far) for detailed discussions and establishing collaborations among interested states, a joint study on closed nuclear fuel cycle with fast reactors (CNFC-FRs) was conceived in the technical meeting held at Obninsk, Russia, under IAEA-INPRO in October 2004 by the Russian Federation. China, France, India, the Republic of Korea, and the Russian Federation started the joint study in December 2004 at Vienna. Subsequently, Ukraine joined in 2005 and Canada and Japan became participants in 2006. In fact, these member states cover more than half the world population and these countries are among the predicted large users of energy in the coming decades. Subsequently, the International Core Group (ICG) was formed by the IAEA under the INPRO project. At the national level, teams were formed to take care of various indicators on these specific topics: economy, safety, fuel cycle, infrastructure, and collaborative projects. Toward executing these aspects, dedicated working groups were set up. Apart from this, the scientific and technical committee included national representations from Canada, China, France, India, Japan, the Republic of Korea, Russia, and Ukraine. The IAEA put their best efforts to involve its own experts, task managers, and cost-free experts from a few member states. It organized several meetings for critical review and discussions of the activities at various levels during the period. The technical outcome of the discussions is detailed in this chapter. Also, certain concepts specific to some countries are presented.

28.2   INPRO: CLOSED FUEL CYCLE WITH FAST REACTOR (CNFC-FR)

Essentially, the INPRO is meant to provide the methodologies for the assessment of a nuclear energy system (NES) rather than providing a technical design and design solutions as it is considered too complex given the wide variations among the various countries with respect to their national objectives and the level of technological maturity and time frame for the implementation. An NES assessment uses the INPRO methodology to evaluate a given NES in a holistic manner in order to confirm its long-term sustainability or to identify issues or gaps that need to be addressed (e.g., by defining actions to be taken to move the NES toward a sustainable option of energy supply). The INPRO methodology defines six assessment areas [28.2]:

1. Economics
2. Safety of nuclear reactors and fuel cycle facilities
3. Waste management
4. Proliferation resistance
5. Physical protection (security)
6. Environment (impact of stressors, availability of resources)

These assessment areas define requirements mainly for designers of nuclear facilities. Additionally, an assessment area of the INPRO methodology was defined, covering aspects related to institutional measures or the infrastructure (such as legal framework) required for a nuclear power program that must be established independently by a country, for example, through its government, the operators of nuclear facilities, and/or the national nuclear industry. For example, one such assessment study of fast reactor systems performed within INPRO is described in this section [28.3]. The joint assessment study was part of Phase-1 of the INPRO. Initiated by the Russian Federation, this joint study was performed (2005–2007) by Canada, China, France, India, Japan, the Republic of Korea, the Russian Federation, and Ukraine. The objectives were to determine milestones for the deployment a CNFC-FR system, assess the potential of the system to satisfy the criteria of sustainability as defined in the INPRO methodology, and establish frameworks and identify areas of collaborative R&D work.

28.2.1   SALIENT FEATURES IN THE CNFC-FR DEVELOPMENT

During the first stage of the study, experts from participating countries analyzed relevant data at the country/region/world level, discussed national and global scenarios for the introduction of a CNFC-FR, identified technologies suitable for such a system, and arrived at a broad definition of a reference CNFC-FR to be used for joint assessment. Natural, social, and economic conditions in the countries developing the technology differ to a great extent. Nevertheless, good agreement was found on the inevitability of CNFC-FR and some aspects of technology development. SFR was ascertained as the most mature fast reactor option. The demonstration of a serial commercial SFR with a matching fuel cycle is the first milestone of national programs. A commercial CNFC-FR, deployable in the 15–30 years based on proven technologies, such as sodium coolant, mixed oxide (MOX) pellet fuel, and advanced aqueous reprocessing technology, was defined as a reference system for joint assessment. Variants were found in terms of priorities on the introduction of SFRs, reactor concepts (pool/loop), plant size, fuel cycle options, assessment of costs, and overall perspectives on collaborative research, which are of course inevitable and indeed desirable at the development stage of the technology.

Contrary to rather consistent approaches regarding the near future, the vision of the remote prospect by participants varies to a considerable degree. Innovative concepts based on new coolants are being explored, particularly in Russia (heavy metals) and France (gas). Loop-type commercial fast reactor is being designed in Japan (a deviation from a generic pool-type arrangement) and modular, medium, and small fast reactors are being developed in both Russia and Japan. There is no common viewpoint on the selection of innovative fuels either. Nitride fuel of equilibrium composition is being considered in Russia as an appropriate choice for the lead coolant fast reactor with enhanced inherent safety characteristics. High-dense metallic fuel as an option for providing high breeding ratio is selected by China, India, and the Republic of Korea to assure the needs in nuclear fuel. France is examining carbide fuel for gas-cooled fast reactors.

Remarkable physics of fast reactors and variety of options for closure of the nuclear fuel cycle make it possible to adapt the CNFC-FR to specific national conditions and realize diverse aspirations. At the same time, participants of the project found it prudent to intensify discussions between technology holders to get more agreement on the vision of innovative concepts of the commercial fast reactor of the next generation and consider possibilities for sharing unique, high-cost, and complex facilities to derive R&D results leading to enhanced understanding of fast spectrum reactors.

28.2.2 Highlights of the Results of the Assessment Using the INPRO Methodology

In the second stage of the study, characteristics of the reference CNFC-FR system, and national systems in case some of their parameters deviated from the reference, were identified and assessed for compliance with the criteria of sustainability developed in the INPRO methodology. Safety characteristics of the CNFC-FR systems under operation met the current safety standards. A comparison of a CNFC-FR with a thermal reactor (TR) system showed that disadvantages of the fast neutron system were compensated by its inherent safety features and additional engineered safety measures. Examples of inherent safety features are negative reactivity feedback in cases of power and temperature disturbances, stability of neutron pattern, no poisoning effects, excellent heat transfer characteristics, and high boiling point of sodium that permit to design the reactor coolant system with a very low pressure, resulting in low stored energy of the coolant fluid. Further, the application of probabilistic methods and the introduction of passive safety features in the shutdown and decay removal systems are emphasized. A probabilistic analysis performed in Russia for upcoming SFR designs confirmed that its innovative design features lead to a significant reduced risk of severe accidents, thus relieving the need for relocation or evacuation measures outside the plant site.

In the INPRO environment, two aspects are covered: (1) inputs to an NES that may lead to a depletion of natural resources, such as uranium and zirconium, and (2) outputs from an NES, which represent environmental stressors. A breakthrough potential of CNFC-FR for the most INPRO environmental indicators was identified. Recycling of plutonium and uranium leads to practically inexhaustible resources of fissile material (and fertile material), that is, such a system might de facto be considered as a renewable energy source. This feature of the CNFC-FR was found especially important for countries with expected high growth of nuclear energy demand (e.g., China, India). For countries in this group, fuel assurance through achieving very high breeding is a driving force for developing fast reactors. However, globally, there is sufficient spent fuel available for reprocessing plutonium to be used as fuel. Some countries even plan to burn it. The inefficiency of the development based on noncollinearity of national programs was clearly indicated, and the need for multinational arrangements was emphasized.

CNFC-FR systems avoiding mining/enrichment steps in their fuel cycle show a significantly reduced environmental impact caused by a much lower release of nonradioactive elements compared to current licensed TR systems. The radiation dose on the public is demonstrated to be far below regulatory limits. Thus, it can be considered as a system suitable for large-scale national and global deployment with excellent environmental and health-preserving features, meeting the highest requirements of sustainable development in the area.

The CNFC-FR system meets all INPRO requirements of effective and efficient nuclear waste management. The utilization of plutonium from the spent fuel of test reactors is an important incentive for developing fast reactor technology in countries with a considerable nuclear share in electricity generation and a low or moderate expected growth of nuclear capacities (e.g., France, Japan). By recycling of specific (heat-producing and long-lived) nuclear fission products and minor actinides in addition to plutonium, the CNFC-FR system has the potential to significantly reduce the heat load, mass/volume, and radiotoxicity of high-level waste to be deposited. The reduction of heat load enables to store more waste per volume of rock, and the removal of actinides and specific fission products from the waste decreases the time required to manage nuclear high-level waste from a geological timescale (several 100,000 years) to a civilization timescale (several 100 years).

CNFC-FR has intrinsic features that provide high PR potential. PR can be enhanced by excluding plutonium separation in advanced reprocessing technologies, by producing fresh fuel with a high radiation barrier, by reducing fuel transportation via collocation of fast reactor, and fuel cycle facilities. The system provides eliminating uranium enrichment and avoiding accumulation of Pu in spent fuel (*plutonium mines*) in a once-through fuel cycle. The inherent PR potential of CNFC-FR is especially important when both test reactor and fast reactor systems are considered as components of the holistic NES. However, only a combination of intrinsic features and extrinsic measures can provide effective and cost-efficient PR of the holistic NES based on test reactors and fast reactors.

The industrial infrastructure and human resources to design, manufacture, construct, and operate a CNFC-FR are available in most of the countries participating in the joint study. However, regional or international approaches might require new international legal infrastructure. The CNFC-FR is well suited for and might require such new regional or international arrangements to offer the opportunity for expanding fuel cycle front-end and back-end services on a multinational basis for mutual benefits for both the technology holder and technology user countries.

The designs of currently operating fast reactors are not completely economically competitive against test reactors or fossil power systems, primarily because of the high capital costs of both the reactor and the fuel cycle facilities. The necessary improvements of the design via R&D are integrated into the development programs of the five countries of the joint study, that is, France, India, Japan, the Republic of Korea, and the Russian Federation. Upon implementation of such improvements, in conjunction with low fuel costs for a fast reactor system, national CNFC-FR systems are expected to be competitive in spite of the different economic conditions (different overnight capital costs, discount rates, etc.), within the next 10–20 years in the countries mastering this technology.

28.2.3 RESEARCH AND DEVELOPMENT

To achieve competitiveness in the area of economics, primarily the capital costs of fast reactor systems are to be reduced. The possible measures are, for example, design simplification, reduction of steel consumption by reducing the number of loops and the thickness of main components, elimination or reduction of the size of reactor systems, using more efficient and cheap radiation shielding and a more compact plant layout, and serial construction with the reduction of the time of

TABLE 28.1

Assessment of Electricity Cost of CNFC-FR and Alternative Energy Sources

Country	CNFC-FR (mills $/kWh)	Alternate Source (mills $/kWh)
France	35.41	44.07
India	41.00	45.00
Japan	15.10	26.59
Korea	31.15	34.00
Russia	17.74	24.50

Source: Status of innovative fast reactor designs and concepts, A supplement to the IAEA Advanced Reactors Information System (ARIS), http://aris.iaea.org, Nuclear Power Technology Development Section, Division of Nuclear Power, Department of Nuclear Energy, IAEA, Vienna, Austria, October 2013.

construction. A significant effect of enhanced R&D on capital cost reduction was identified in all countries realizing SFR programs so that the ratio of capital cost per capacity unit of SFRs versus test reactors has been assessed to approach very close to a unit in the new SFR designs (BN-1800, Russia; JSFR, Japan) or even in the SFR under construction (PFBR, India).

Some specific R&D programs to achieve enhanced safety, for example, preventive surveillance and inspection aspects; development of sensors and repair of welds in sodium; development and validation of sodium fire models, sodium pumps with high inertia, automatic negative reactivity insertion devices, and engineering measures to prevent recriticality in molten core configurations were identified.

The joint study concluded that it is possible to identify generic areas for international collaboration such as development and testing of materials, in-service inspection technologies, modeling and validation of codes, and probabilistic methods for safety analysis of fuel cycle facilities. High emphasis has been given to sharing unique and expensive facilities among member states. The economic assessment of electricity from CNFC-FR vis-a-vis alternative energy source is given in Table 28.1.

28.3 CONCEPTS SPECIFIC TO NATIONS

In the framework on innovative fast reactor design and concepts, several countries are developing concepts specific to their design based on their own experience and assessments. The underlying objective is to develop them further to match essentially the Generation IV requirements. Some of the concepts being considered and developed by counties are described briefly in the following sections. The information presented in this section is mainly extracted from the recent publication by the IAEA [28.4].

28.3.1 FRANCE

ASTRID reactor would incorporate certain innovative design options toward safety (Figure 28.1). The core concept involves heterogeneous axial (U-Pu)O_2 fuel with a thick fertile zone in the inner core. The core is asymmetrical and crucible shaped with a sodium plenum above the fissile area. These features are incorporated mainly to get an overall negative sodium void effect when the sodium boiling condition is reached in the event of total loss of primary coolant, which is sought to be achieved by the negative reactivity due to the expansion of the sodium plenum above the core and also due to the heterogeneous fertile zone provided within the active core. A conical inner vessel (redan) is opted to enable extended ISIR access in the case a loop-type design is chosen at the preconceptual design phase of primary and secondary sodium circuit systems.

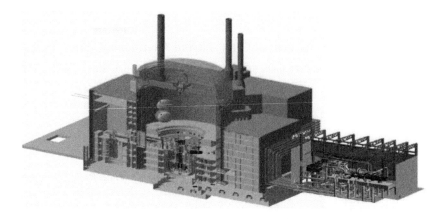

FIGURE 28.1   ASTRID (French). (From Status of innovative fast reactor designs and concepts, A supplement to the IAEA Advanced Reactors Information System (ARIS), http://aris.iaea.org, Nuclear Power Technology Development Section, Division of Nuclear Power, Department of Nuclear Energy, IAEA, Vienna, Austria, October 2013.)

Three primary pumps and four intermediate heat exchangers (IHXs) are incorporated with respect to the reactor assembly. Each secondary sodium loop beginning with the intermediate heat exchanger is associated with modular steam generators. Options are kept open w.r.t. the choice of tertiary fluid between water and gas. This circuit includes chemical volume control system. Core catcher is provided, as a defense in-depth measure in the event of whole core melting, which would cover the entire core and subcriticality would be maintained and ensuring, at the same time, a long-term postaccident phase cooling. The in-service inspectability of the core catcher, like any other safety-related component, is the added feature aimed in the design. The choice of the core catcher location, whether inside or outside the vessel, is kept open and studies are being undertaken by the French. The containment will be designed to resist the release of the mechanical energy caused by a hypothetical core accident or large sodium fires, to make sure that no countermeasures are necessary outside the site (radiation emergency off-site) in the event of an accident.

28.3.2 INDIA

Subsequent to PFBR, the prototype reactor, India is embarking upon an improved design toward economy. Apart from the design and technological challenges that are addressed toward commercial exploitation of FBRs, the economy needs to be improved significantly. In view of the fact that fast reactors are capital-cost-intensive projects, the capital cost needs to be brought down significantly. Particularly, in the Indian context, the interest during construction (IDC) is also very high in the unit energy cost. To achieve this, the construction time should be brought down. For PFBR, being a first-of-its-kind project, capital cost has been high and construction time longer (>9 years). These two values are not acceptable for future plants to have commercialization features. In this respect, India has put considerable efforts to reduce the capital cost as well as construction time of FBR-1 and -2. A few salient features are brought out here.

With the objective of demonstrating the potential of FBRs without linking to breeding, the matured fuel cycle technology with MOX fuel that was adopted for PFBR is retained. Further, for PFBR, the two-loop concept with two primary pumps, two intermediate heat exchangers per loop, and two secondary sodium pumps was selected, after considering comprehensively the associated parameters such as economics, plant availability, the size and number of components, operating experiences, the capacity and capability of Indian industries, and safety aspects. In view of these, the cost of the secondary sodium circuit has been kept as low as possible, and hence, there is a little scope for further reduction in the cost for the secondary sodium circuit except steam generators.

However, use of the twin unit concept for sharing many services, use of advanced shielding materials such as ferro-boron, SS 304 LN in place of SS 316 LN for cold pool components and ferritic steels for sodium piping, three steam generator modules per loop with an increased tube length of 30 m (PFBR has 4 modules per loop with 23 m length), and enhancing the design life from 40 to 60 years have indicated sizeable capital cost reduction.

Apart from these measures, construction experience of reactor assembly components indicates a need for major improvements in the design of the grid plate (large number of sleeves, posing difficulty in assembly, hard facing of large-diameter plates, and heavy flange construction), roof slab (large box-type structure with many penetrations, complicated manufacturing process, time-consuming and difficult-to-handle laminar tearing problems), inclined fuel transfer machine (long manufacturing time and extensive qualification tests), there is no need of large gap between safety vessel and reactor vault and there exists a long time delay in importing large-diameter bearings. Take into account this feedback for future design, the following improvements have been made in reactor assembly: (1) reduction of main vessel diameter; (2) dome-shaped roof slab with conical support skirt under compression; (3) thick plate concept for rotatable plugs; (4) welded grid plate with reduced number of sleeves, reduced diameter of intermediate shell, and reduced height; (5) increased number of primary pipes; (6) inner vessel with single radius torus welded with the grid plate; (7) integrated liner and safety vessel with thermal insulation arrangement; and (8) optimization of main vessel, inner vessel, and safety vessel thickness. These improvements have brought significant benefits with enhanced safety. A schematic of FBR-1 and -2 is shown in Figure 28.2. Additionally, the inclined fuel transfer machine has been eliminated with an addition of two straight pull machines (Figure 28.3).

The safety features incorporated in FBR-1 and -2 are as follows:

Provision of four primary pipes per sodium pump offers higher safety margin for Category 4 primary pipe rupture design basis events.

In-vessel primary sodium purification is adopted to avoid the radioactive sodium being taken outside the primary vessel.

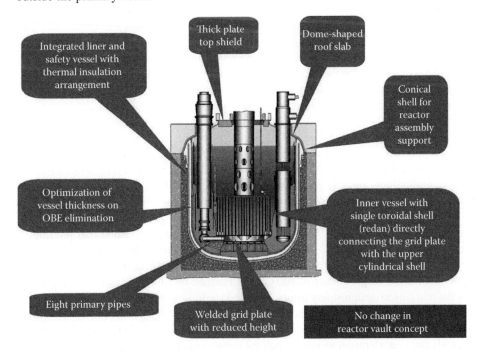

FIGURE 28.2   (See color insert.) Improvements introduced in FBR-1 and -2 reactor assembly (India).

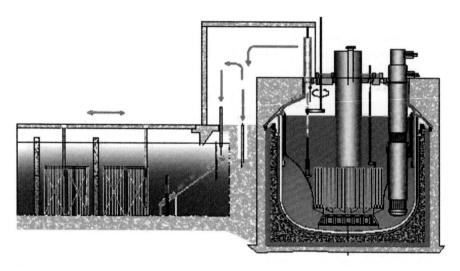

FIGURE 28.3   **(See color insert.)** Fuel handling scheme conceived in FBR-1 and -2 (India).

Stroke limiting device is added to the control and safety rod, the primary shutdown system, to
    guard against inadvertent control rod withdrawal events.
A temperature-sensitive electromagnet is added to the diverse safety rod, the secondary shut-
    down system, leading to enhanced safety.
As an ultimate shutdown system, either a liquid Li–based or B4C granule–based system is
    added to practically eliminate core disruptive accidents.
Three trains each of two diverse concepts (passive, active–passive) for decay heat removal are
    incorporated. Active–passive systems are designed for two-third of the capacity by natural
    circulation.
An innovative multilayer core catcher is installed to handle whole core melt debris.

Subsequently, toward rapid realization of nuclear power, India is planning a series of metal-fueled
FBRs starting with a metal fuel demonstration fast breeder reactor (MDFR) to be followed by
industrial-scale 1000 MWe metal-fueled reactors. As for the reactor size for MDFR, 500 MWe is
chosen in order to capitalize on the experience gained in terms of design, manufacture, and erection
of major reactor systems and components of PFBR. In line with this, India has undertaken the task
of designing MDFR-500 and during the conceptualization of the reactor system, proven design concepts as adopted for MOX-fueled FBRs were retained and important variants incorporated with due
consideration to experience gained during safety review, manufacturing, and construction of PFBR
with a view to improve the overall economics and enhance plant safety. In parallel, on the fuel and
material development front, in order to study the irradiation behavior of binary (U–Pu) and ternary
(U–Pu–Zr) metallic fuels and related structural materials, an elaborate testing program with FBTR
as test bed followed by a detailed postirradiation examination program is drawn. Toward optimizing
the MDFR-500 core for maximizing the breeding potential (BR ~ 1.3–1.4), several fuel options and
design variants have been studied for both sodium- and mechanical-bonded fuel pin options. For
example, a reference design with U-(19 w%)Pu-(10 w%)Zr ternary sodium–bonded fuel with a pin
diameter of 6.6 mm (T91 clad) and a smeared density of 75% is optimized for a peak linear power of
450 W/cm for an initial target burnup of 100 GWd/t. Further, thermal hydraulic design is carried out
to assess the margins to design safety limits. Detailed studies, with apportioned cumulative damage
fraction (CDF) limit and conservative assumption of 100% fission gas release, indicate that target
burnup could be reached satisfactorily. In parallel, mechanically bonded fuel pin option is being
pursued with due consideration to infrastructural constraints for arriving at a judicious choice for
fuel composition and type of bonding with the pin. The heat transport sodium systems and balance

of plant are conceptualized for a compact plant layout. The layout is also drawn with flexibility to accommodate an optional ex-vessel sodium storage tank to facilitate the storage of sodium-bonded fuel subsequent to fuel-handling operations. Safety instrumentation and controls are detailed with adequate redundancy and diversity requirements. The plant design is being evolved to be compatible for both coastal and inland sites.

28.3.3 JAPAN

With respect to the 4S reactor concept (Figure 28.4), the active systems and feedback control systems from the reactor side are totally eliminated, aiming toward higher reliability of components. The same strategy is also adopted for components with rotating parts. No refueling is envisaged during the lifetime of the reactor, and hence, the radioactivity confinement area becomes limited. The design features and concepts envisage prevention of core damage during accidents, confinement of radioactive materials, prevention of sodium leakage, and mitigation of associated impacts in the case of a leak. The concern on proliferation is addressed by the use of uranium-based fresh fuel with U235 enrichment limited to less than 20% by weight and the plutonium content in the spent fuel is designed to be less than 5% by weight. As a strategy, during the reprocessing of the metal fuel, whether in a binary or ternary form, plutonium is always recovered along with minor actinides and other nuclides with high radioactivity and toxicity. Two decay heat removal systems, comprising reactor vessel auxiliary heat removal system and intermediate auxiliary heat removal system, are provided. These characteristics and the systems together with the other passive and inherent safety characteristics associated with the metallic fuel ensure negative sodium void reactivity effect leading to elimination of core damage. Reactor building is supported by seismic isolator for earthquake and the building is reinforced, which can be protected from massive water invasion by maintaining water tightness.

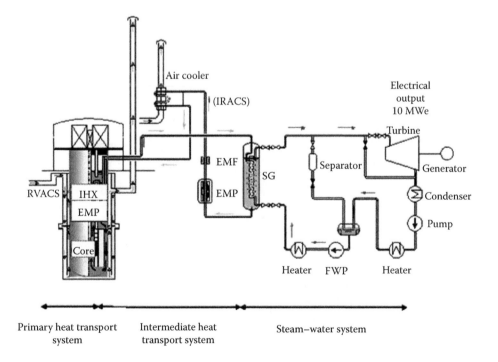

FIGURE 28.4 4S Reactor (Japan). (From Status of innovative fast reactor designs and concepts, A supplement to the IAEA Advanced Reactors Information System (ARIS), http://aris.iaea.org, Nuclear Power Technology Development Section, Division of Nuclear Power, Department of Nuclear Energy, IAEA, Vienna, Austria, October 2013.)

The Japan SFR (JSFR) is expected to achieve the development targets of the FaCT (fast reactor cycle technology development) project and the Generation IV reactor goals by adopting the following advanced key technologies:

1. High burnup core with oxide dispersion–strengthened steel cladding material
2. Safety enhancement with self-actuated shutdown system (SASS) and recriticality free core
3. Compact reactor system adopting a hot vessel and in-vessel fuel handling with a combination of an upper inner structure (slit UIS) with a slit and advanced fuel-handling machine (FHM)
4. Two-loop cooling system with a large-diameter piping made of Mod. 9Cr-1Mo steel
5. Integrated IHX pump component
6. Reliable steam generator (SG) with double-walled straight tube
7. Natural circulation decay heat removal system (DHRS)
8. Simplified fuel-handling system
9. Steel plate reinforced concrete containment vessel (SCCV)
10. Advanced seismic isolation system

These technologies have been evaluated to be suitable for installation in the demonstration JSFR (Figure 28.5).

For safety design, the JSFR concept adopts the defense-in-depth principle to the same level and extent as in LWRs. Two independent reactor shutdown systems with independent/diversified

FIGURE 28.5   JSFR (Japan). (From Status of innovative fast reactor designs and concepts, A supplement to the IAEA Advanced Reactors Information System (ARIS), http://aris.iaea.org, Nuclear Power Technology Development Section, Division of Nuclear Power, Department of Nuclear Energy, IAEA, Vienna, Austria, October 2013.)

signals are installed for the reactor shutdown systems. For the fourth level of defense in depth, SASS is installed providing a passive shutdown capability. The recriticality free core concept plays an important role in ensuring in-vessel retention scenario against whole core disruptive accidents. The initiating phase energetics due to exceeding the prompt criticality has to be prevented by limiting the sodium void worth and the core height. The possibility of molten fuel compaction has to be eliminated by enhancing fuel discharge from the core. The effectiveness of fuel assembly with inner duct structure (FAIDUS) has been confirmed by both in-pile and out-of-pile experiments. The JSFR decay heat removal system consists of a combination of one loop of direct reactor auxiliary cooling system (DRACS) and two loops of primary reactor auxiliary cooling system (PRACS) adopting full natural convection system.

28.3.4 KOREA

PGSFR has a plant capacity of 150 MW(e) and features a proliferation-resistant core without blankets, a metallic-fueled core, pool-type PHTS, and two intermediate heat transport system (IHTS) loops (Figure 28.6). The core adopts a homogeneous configuration in the radial direction that incorporates annular rings of inner and outer driver fuel assemblies. All blankets are completely removed from the core so as to exclude production of high-quality plutonium. The active core has a height of ~90 cm and a radial equivalent diameter of ~1.6 m. The metallic fuel U–Zr (or U–TRUZr) is used as the driver fuel. Each fuel assembly includes 217 fuel pins. All the fuels have a single enrichment of U (or TRU) nuclide. The reactivity control and shutdown system consists of nine control rods assemblies that are used for power control, burnup compensation, and reactor shutdown in response to demands from the plant protection control and systems. The heat transport system of PGSFR consists of PHTS and IHTS, steam generation system, and decay heat removal system (DHRS). PHTS mainly delivers the core heat to IHTS and IHTS as the intermediate system between PHTS, where nuclear heat is generated, and the SGS where the heat is converted to steam. PHTS is pool type in which all the primary components and primary sodium are located within the reactor vessel. Two mechanical PHTS pumps and four IHXs are immersed in the primary sodium pool. IHTS has two loops, and each loop has two IHXs connected to one steam generator and one IHTS pump. Each steam generator has a thermal capacity of ~200 MW(th). The IHTS sodium flows downward

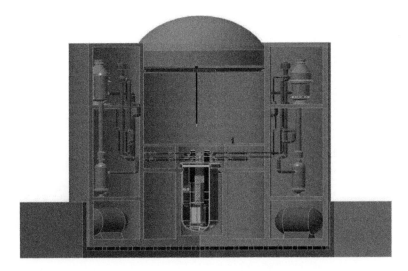

FIGURE 28.6 PGSFR (Korea). (From Status of innovative fast reactor designs and concepts, A supplement to the IAEA Advanced Reactors Information System (ARIS), http://aris.iaea.org, Nuclear Power Technology Development Section, Division of Nuclear Power, Department of Nuclear Energy, IAEA, Vienna, Austria, October 2013.)

through the shell side while the water/steam goes up through the tube side. The IHTS adopts an electromagnetic pump to simplify installation and reduce moving parts. The IHX design conditions were established to prevent water/steam and sodium water reaction products from being discharged into the reactor vessel. The DHRS, one of the safety design features, is composed of two passive decay heat removal systems (PDHRS) and two active decay heat removal systems (ADHRS). DHRS is designed to remove the decay heat of the reactor core after reactor shutdown when the normal heat transport path is unavailable.

PGSFR is designed to be safe against severe accidents caused by earthquakes and tsunamis. DHRS, a combination of passive and active decay heat removal systems, has a sufficient capacity to remove the decay heat in all design basis events without operator involvement by incorporating the principles of redundancy and independency. Double reactor vessels and double piping in IHTS are designed for the prevention of sodium leakage. PGSFR also has a passive reactor shutdown system.

28.3.5 RUSSIA

BN-1200 is a new SFR for serial construction (Figure 28.7). The succession of fundamental engineering solutions proved to be positive for BN-600 and BN-800 and has been preserved in the process of BN-1200 reactor development. The monoblock principle has been used in the BN-1200 power design, that is, one reactor and one turbine. The reactor core comprises assemblies of different types: fuel subassemblies (FSA), boron shield assemblies, and absorber rods. The central part of the core consists of FSA with fuel of similar enrichment and cells with absorber rods. The adopted tank designed for in-reactor storage reduces residual heat release to the safe level for refueling activities and FSA washing and it was selected taking into account the possible elimination of the spent FSA cask. Rows of assemblies with natural boron carbide are arranged behind in-reactor storage to form additional side shielding of in-reactor equipment. The structure of the core FSA represents a hexagonal wrapper tube with a top nozzle attached to one end and a bottom nozzle to the other one. Inside the hexagonal wrapper tube, there are bundles of absorber elements and fuel pins arranged one beneath the other forming a *sodium* cavity between each other producing the

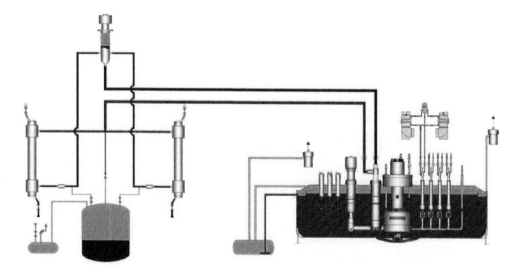

FIGURE 28.7 BN-1200 (Russia). (From Status of innovative fast reactor designs and concepts, A supplement to the IAEA Advanced Reactors Information System (ARIS), http://aris.iaea.org, Nuclear Power Technology Development Section, Division of Nuclear Power, Department of Nuclear Energy, IAEA, Vienna, Austria, October 2013.)

sodium void reactivity effect. MCP-1 is an impeller submersible pump including a check valve and variable frequency electric drive with a stepless control of speed. The reactor plant has three circuits. The primary and secondary circuits are sodium cooled and the third circuit coolant is water/steam. Each of the three circuits is divided into four parallel flows (loops) equally participating in heat transfer from the reactor to the turbine plant of the power unit. Each loop of the secondary circuit consists of an MCP-2 and an SG connected by pipelines. Temperature transfers at the secondary circuit pipelines are compensated via bellows compensators. The MCP-2 is a single-stage vertical centrifugal pump with a free level of sodium in it. The SG is a block-type once-through heat exchanger consisting of two modules with straight heat exchanging pipes. The SG is fitted with automatic protection system in case of intercircuit leaks.

The BN-1200 reactor plant design utilizes a number of new engineering solutions compared to BN-600 and BN-800:

- Primary circuit sodium systems and equipment are completely integrated in the reactor tank, which eliminates radioactive sodium leaks.
- Emergency heat removal system (EHRS) is applied providing natural circulation through all EHRS circuits, including circulation immediately through FSA, thereby increasing the level of power output under allowable temperature conditions in the reactor core.
- Passive shutdown system comprising hydraulically suspended absorber rods and a system of rods responding to sodium temperature variations in the core is used.
- A special containment is provided in the reactor compartment to confine accidental releases from the reactor under beyond design-basis accidents.

Through these solutions better safety performances are expected: (a) probability of severe damage (10^{-6}) to the reactor core is an order of magnitude less than the one required by regulatory documentation, (b) target criterion has been specified, that is, the boundary of protective action zone has to coincide with the boundary of production site for severe beyond design-basis accidents with their probability not exceeding 10^{-7} over reactor-year.

28.3.6 UNITED STATES

The PRISM-311 MWe (Figure 28.8) design uses a modular, pool-type, liquid sodium–cooled reactor. Metallic fuel of ternary alloy is used as the fuel. The reactor employs passive shutdown and decay heat removal features. The reactor is designed to use a heterogeneous metal alloy core. The primary heat transport system is contained entirely within the reactor vessel. The flow path goes from the hot sodium pool above the reactor core through the IHXs, where heat is transferred to the IHTS; the sodium exits the IHX at its base and enters the cold pool. Four electromagnetic pumps take suction from the cold pool and discharge into the high pressure core inlet plenum. The sodium is then heated as it flows upward through the reactor core and back into the hot pool. Heat from the IHTS is transferred to a steam generator where superheated steam is produced. This high-pressure, high-temperature steam drives the turbine generator to produce electricity.

The passive shutdown characteristics of the reactor core provide diverse and independent means of shutdown in addition to the control rod scram. The passive features comprise several reactivity feedback properties including the Doppler effect, sodium density and void, axial fuel expansion, radial expansion, bowing, control rod drive line expansion, and reactor vessel expansion. The negative feedbacks maintain the reactor in a safe, neutronically stable condition. The passive reactor vessel auxiliary cooling system (RVACS) provides primary cooling during all design basis accident conditions and anticipated transients without scram (ATWS). This passive system operates effectively without electricity or operator intervention for an unlimited amount of time. Heat is transferred from the reactor vessel to the containment vessel by thermal radiation and then to the surrounding atmospheric air by natural convection. Redundant decay heat

FIGURE 28.8 PRISM (United States). (From Status of innovative fast reactor designs and concepts, A supplement to the IAEA Advanced Reactors Information System (ARIS), http://aris.iaea.org, Nuclear Power Technology Development Section, Division of Nuclear Power, Department of Nuclear Energy, IAEA, Vienna, Austria, October 2013.)

removal is provided by the ACS, which consists of natural circulation of air past the shell side of the steam generator. The combination of systems allows for reduced plant outages for inspections and maintenance.

28.3.7 OTHER POTENTIAL OPTIONS

In addition to SFRs, several designs of heavy metal_cooled (lead) reactors are under development. Some of them are (1) MYRRHA (SCK•CEN, Belgium), (2) CLEAR-I (INEST, China), (3) ALFRED an ELFR (Ansaldo Nucleare, Europe/Italy), PEACER (Seoul National University, Republic of Korea), BREST-OD-300 (RDIPE, Russian Federation), ELECTRA (KTH, Sweden), G4M (Gen4 Energy Inc., United States). Designs on gas-cooled fast reactors are (1) ALLEGRO (European Atomic Energy Community, Europe), (2) EM2 (General Atomics, United States).

REFERENCES

28.1 U.S. Department of Energy. (2013). DOE EIA 2003 New Reactor Designs. GIF Annual Report 2008; GIF 2014, Technology roadmap update for Gen IV nuclear energy systems, ELSY Project, 2012. http://www.world-nuclear.org/info/Nuclear-Fuel-Cycle/Power-Reactors/Generation-IV-Nuclear-Reactors/.
28.2 IAEA. (2010). Introduction to the use of the INPRO methodology in a nuclear energy system assessment, IAEA Nuclear Energy Series No. Np-T-1.12. IAEA, Vienna, Austria.
28.3 Raj, B., Vasile, A., Kagramanian, V., Xu, M., Nakai, R., Kim, Y.-I., Usanov, V., Stanculescu, A. (2011). Multi-lateral assessment of the fast reactor system as a component of the future sustainable nuclear energy and paths for the system deployment. *J. Nucl. Sci. Technol.*, 48(4), 591–596.
28.4 IAEA. (2013). Status of innovative fast reactor designs and concepts. A supplement to the IAEA Advanced Reactors Information System (ARIS), Nuclear Power Technology Development Section, Division of Nuclear Power, Department of Nuclear Energy, IAEA, Vienna, Austria, October 2013. http://www.iaea.org/NuclearPower/Downloadable/FR/booklet-fr-2013.pdf

Section VI

Fuel Cycle for SFRs

29 Fuel Cycle for SFRs

29.1 INTRODUCTION

As the name suggests, the term *fuel cycle* refers to the life cycle of nuclear fuel—from its first processing to form the fuel element to its processing after irradiation in the reactor, there being variants in the processing steps depending upon the nature of the fuel material, its source, its subsequent use, and the policy of its reuse. These are described in some detail in subsequent sections. To start with, we study the generic variants in the fuel cycle, *open* and *closed*.

29.2 OPEN AND CLOSED FUEL CYCLE

One can envisage broadly two approaches to the concept of nuclear fuel cycle:

1. Once-through or open fuel cycle
2. Closed fuel cycle

Figure 29.1 illustrates open and closed fuel cycles. Essentially, in the *open fuel cycle* the fuel is used in the reactors in a once-through mode. After irradiation in the reactor to the desired burnup level, the fuel is discharged and subsequently stored without any treatment, for ultimate disposal as a waste. In closed fuel cycle, the fertile and fissile materials used in the reactor are processed after their irradiation to recover and reuse them for further generation of energy or production of fissile material. The steps leading to and including the fabrication of the nuclear fuel constitute the *front end* of the nuclear fuel cycle, while reprocessing and waste management form the *back end*.

The schematic diagram of a typical closed fuel cycle, which integrates the fuel cycles of the thermal and fast reactors, is shown in Figure 29.2. In fact, the figure also illustrates the linkage between the thermal reactor and the fast reactor fuel cycle. The figure is self-explanatory. A detailed discussion on the fuels for sodium-cooled fast reactors (SFRs) and the status of the development of the back-end fuel cycle of fast reactors can be found in recent IAEA publications [29.1,29.2].

In the past, many countries had decided to adopt once-through fuel cycle mainly because of concern over proliferation of nuclear materials and also due to economic reasons. Since uranium (U) irradiated in nuclear reactors contains plutonium (Pu) produced through neutron absorption followed by beta decay (refer to Figure 1.5), reprocessing of such fuel is associated with concerns about the recovery of plutonium and its use in weapons production. In many countries, reprocessing is considered to be an uneconomical proposition as long as uranium is available at a reasonable price.

Only some countries—such as France, India, Japan, Russian Federation, and the United Kingdom—have continued their programs in the back-end fuel cycle on a commercial scale. In France, Pu generated from U in thermal reactors is recovered through reprocessing and recycled through refabrication as U–Pu mixed oxide fuel for its utilization in other pressurized water reactors (PWRs) [29.3], and recycling of Pu contributes to nearly 10% of the country's total nuclear energy production [29.4]. In India, the three-stage nuclear program has closed fuel cycle as its focus. The Pu and depleted U recovered from irradiated fuels of thermal reactors (first stage) are used to fabricate fuels for fast reactors (second stage). Recently, the U–Pu mixed carbide fuel used

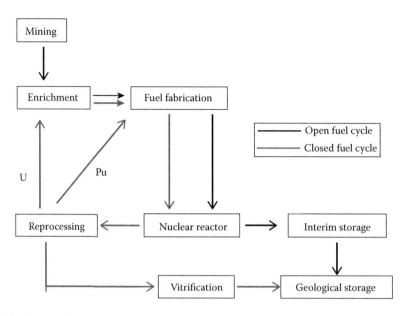

FIGURE 29.1   Open and closed fuel cycles.

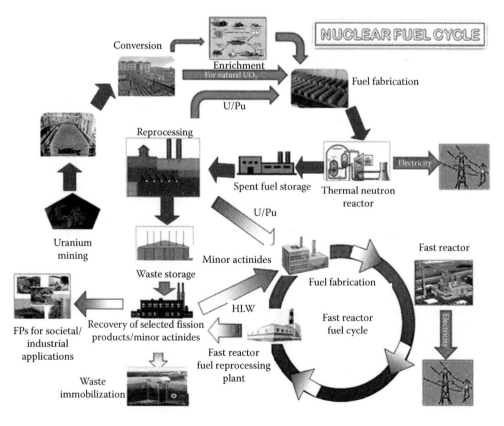

FIGURE 29.2   Thermal and fast reactor fuel cycles and their interrelationship.

in the fast breeder test reactor (FBTR) was reprocessed and the material recovered used for the fabrication of fresh mixed carbide fuel; the refabricated fuel was subsequently introduced in the FBTR for irradiation.

Intermediate between the once-through and the closed fuel cycles, there are a variety of other options. One such option is a *twice-through* cycle [29.5,29.6], wherein the U and/or Pu are recycled once and used in the thermal reactor before disposal.

As mentioned earlier, based on economic and proliferation resistance considerations, once-through fuel cycle is preferred in several countries that do not have a fast reactor program. However, in the case of fast reactors, closed fuel cycle is considered as an inevitable option. During once-through cycle of operation in nuclear reactors, only a small portion of the nuclear fuel is actually burnt. In a typical PWR, approximately 6% heavy nuclides can be burnt in one cycle to produce energy, whereas in a fast reactor up to 20 atom% burnup was achieved [29.7]. In any case, it is clear that more than 80% of the fuel material remains unused in one cycle of irradiation.

29.3 CLOSED FUEL CYCLE FOR FAST REACTORS

The considerations that support the adoption of closed fuel cycle for fast reactors are as follows:

1. The fast reactor fuels invariably contain plutonium as a major component. Plutonium is obtained by reprocessing irradiated uranium-based fuels and is thus a limited and valuable resource. Thus, the use of open fuel cycle, which burns only a part of the plutonium, leads to poor utilization of plutonium.
2. There are several safety and security issues with respect to the direct disposal of plutonium containing fuels after irradiation.
3. The radiotoxicity of irradiated fuels, which contains plutonium and many other *minor* actinides, is very high. With regard to public acceptance, one of the concerns about nuclear waste disposal in a repository is the long time it would take for the reduction in the radiotoxicity of the disposed irradiated fuel to a reference level (this equals the radio-toxicity of raw material used to fabricate 1 ton of enriched uranium, uranium isotopes, and their products). The radiotoxicity of the spent fuel arises mainly from Pu, americium (Am), and curium (Cm) and some long-lived fission products such as Iodine-129 (half-life 1.57×10^7 years) and Technicium-99 (half-life 2.1×10^4 years). When the fuel is disposed directly without the recovery of Pu and other actinides, it might take more than 100,000 years for the radiotoxicity of the spent fuel to be reduced to the reference level [29.8,29.9] because actinides, and particularly plutonium isotopes, have long half-life (see Figure 29.3).
4. The space available for waste disposal is a valuable and limited resource. A geological repository will require a large amount of space for the disposal of irradiated fast reactor fuel due to the high content of radiotoxic actinides and the high heat load generated from the radioactive decay [29.10]. Recovery and burning of the valuable U–Pu as well as some of the minor actinides and long-lived fission products have demonstrated the potential to reduce their radiotoxicity to the reference levels within a period of several hundred years [29.9,29.11] and also to reduce the footprint requirement of the repository by an order of magnitude [29.11]. Thus, the recovery of actinides from irradiated fast reactor fuel is an important step in achieving a sustainable nuclear energy system based on fast reactors.

In the following sections, the discussion is limited to U and U–Pu-based fuels that are used as driver fuels for fast reactors.

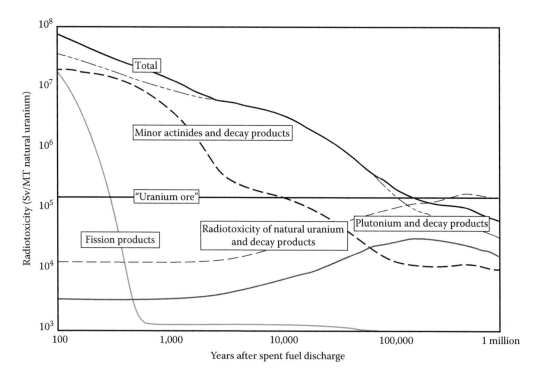

FIGURE 29.3 Ingestion radio toxicity of spent nuclear fuel. (From NEA, Physics and safety of transmutation systems: A status report, NEA Report No. 6090, OECD, Paris, France, 2006.)

29.4 FUEL TYPES

The designs and types of nuclear fuels used in fast reactors are dealt with in Sections 7.1 and 8.1. It can be seen from Table 29.1 [29.12] that a wide variety of nuclear fuel forms are being used in fast reactors. These include metal alloys, oxide, carbide, and nitride. The U–Pu mixed oxide fuel has been the most widely used driver fuel. France, Germany, and Japan use mixed oxide fuel (MOX) in their fast reactors. The United Kingdom and the United States use both metallic and mixed oxide fuel as driver fuel. The Russian Federation has significant industrial-scale manufacturing experience with highly enriched uranium oxide fuel (>20% U-235); it was used as the fuel for BOR-60, BN-350, and BN-600 reactors. Uranium- and plutonium-mixed oxide has also been chosen as the reference fuel for BN-800. In France and the United Kingdom, all the steps for fast reactor fuel cycle including fabrication, irradiation, reprocessing, and refabrication have been demonstrated on an industrial scale with the mixed oxide fuel. Japan also has experience in manufacturing mixed oxide fuel on an industrial scale. In India, the prototype fast breeder reactor (500 MWe), which is under commissioning, will also use mixed oxide as the driver fuel. Industrial-scale manufacturing of the fuel is now under way.

The oxide fuel continues to be a reference fuel for fast reactors because of its advantages in terms of properties, performance, economics, and large-scale experience. Compared to the mixed oxide fuel, experience on carbide and nitride fuels is limited. R&D on these fuels used in fast reactors has been pursued in many countries (e.g., the United States, France, Germany, the United Kingdom, Russian Federation, India, and Japan) and both He-bonded and Na-bonded carbide and nitride fuels have been developed. The BR-10 fast reactor in Russia was operated with uranium nitride and uranium carbide as driver fuels [29.13]. There is, however, very little experience on the reprocessing of these fuels. In India, the FBTR at Kalpakkam has been operating with U–Pu mixed carbide as its driver fuel, and the reprocessing of carbide fuel and its refabrication has been demonstrated.

TABLE 29.1

Fuel Types Used in Fast Reactors

Fuel Type	Reactors (Illustrative List)
Metal and Metal Alloys	
Pu	Clementine, USA
U	EBR-I, USA
U–Cr	DFR, UK
U–Mo	Fermi, DFR
U–Zr	EBR-I, EBR-II, USA
Pu–Al	EBR-I, USA
Pu–Fe (molten)	LAMPRE I, USA
U-Fissium[a]	EBR-II, USA
U–Pu–Zr	EBR-II, USA
Ceramic Fuels	
(Enriched) UO_2	BOR-60, BN-600, Russia; Rapsodie, Phenix, France
$(U, Pu)O_2$	Phenix, France; Joyo, Monju, Japan; PFR, UK; BN-600, CIS; FFTF, USA; KNK-II, Germany
$(U, Pu)C$	FBTR, India
UC, UN	BOR-60, Russia

Source: Kittel, J.H. et al., *J. Nucl. Mater.*, 204, 1, 1993.

[a] U-Fissium: an alloy of noble metal fission products of Mo, Ru, Rh, Pd, Zr, and Nb.

Metal alloys are an attractive option as fast reactor fuels from the point of view of breeding. In fact, the first nuclear reactor to produce electricity was the fast reactor EBR-I, and it used enriched uranium metal as fuel. Metallic fuel was the first fuel to be used in the experimental fast reactors in the United States and the United Kingdom. Subsequently, a large number of metal alloys have been tested for use in fast reactors. The alloying elements included Cr, Zr, and Mo. The Dounreay fast reactor (DFR) in the United Kingdom used U–Mo alloy fuel and also tested U–Cr fuel. EBR-II fuel initially operated with U 5% fissium alloy, where fissium is a combination of fission product elements Nb 0.01%, Zr 0.1%, Pd 0.2%, Rh 0.3%, Ru 1.9%, and Mo 2.4%. Subsequently, U–Zr alloy was used as the driver fuel. It subsequently operated with enriched U–Zr alloy as the driver fuel. Uranium-based metal alloy fuels have been rigorously tested in the United States and have reached high levels of burnup (over 15 at%). However, experience with Pu containing metal alloy fuels is limited. While a number of U–Pu–Zr fuel pins were irradiated in EBR-II and the FFTF in the United States, the complete fuel cycle experience is mainly limited to U–Zr alloy. The integral fast reactor concept proposed in the United States was based on actinide recycle program (ARP), and its fuel cycle was proposed to be closed by reprocessing the spent U–Pu–Zr alloy fuel using pyroelectrometallurgical process and refabrication in hot cell.

29.5 PERFORMANCE REQUIREMENTS OF FAST REACTOR FUELS

As mentioned in Section 29.4, fast reactors are operated with a variety of fuel types such as oxide, carbide, nitride, and metallic alloys. For economic reasons, fast reactor fuels have to be operated to a high level of burnup at relatively aggressive conditions of high linear power and high temperatures. Especially in the case of mixed oxide fuels, because of its poor thermal conductivity and high linear power, the centerline temperature could reach as high as 2000°C and a high gradient of temperature exists across the radius (typically 2500°C/cm) resulting in several thermochemical processes that make the irradiated fuel a complex chemical system.

Fast reactors have to use fuel with higher fissile material enrichment as compared to fuels for thermal reactors because of the lower fission cross sections at high neutron energies. Thus, the fuels either are based on highly enriched uranium or contain significant concentration of Pu, especially in the case of research reactors. For example, the uranium oxide fuel used in BOR-60 had uranium enriched in U^{235} (45%–90%). The U–Pu mixed oxide fuel used in fast reactors has typically 20%–30% Pu content. In the case of FBTR in India, the fuel used is a U–Pu mixed carbide with Pu content as high as 70% (Pu/U + Pu = 0.7).

Pu-containing fast reactor fuels have to be fabricated in glove boxes. The fabrication of metallic, carbide, and nitride fuels has an additional challenge arising from the reactivity of the fuel with oxygen and moisture as a result of which these have to be fabricated in inert-atmosphere glove boxes.

The high temperature of operation in the fast reactor fuels implies that the fuel pins have to be slender in order to restrict the centerline temperature to well below the melting point. A typical fast reactor fuel pin (MOX) contains pellets of fuel with diameter in the range of 6–8 mm as compared to thermal reactor fuel (uranium dioxide [UOX]) where the diameter is in the range of 12–13 mm. Fuel design is covered in detail in Section 8.1.

Due to the lower values of cross sections of the absorption for neutrons, the fast reactor fuels can tolerate higher concentrations of impurities.

29.6 FUEL FABRICATION PROCESSES

A summary of the processes used in the industrial-scale fabrication of MOX fuels in various countries is provided in Figure 29.4 [29.1].

The conventional route for the fabrication of uranium–plutonium mixed oxide fuels for fast reactors involves blending of uranium oxide and plutonium oxide powders, comilling, pelletization, and sintering. The actual process adopted by different countries has minor variations in these steps. The MOX fuel for Rapsodie, PHENIX, and Super PHENIX reactors was produced by *Co*broyage *Ca*darache (COCA) process, which involved optimized ball milling of UO_2 and PuO_2 powders

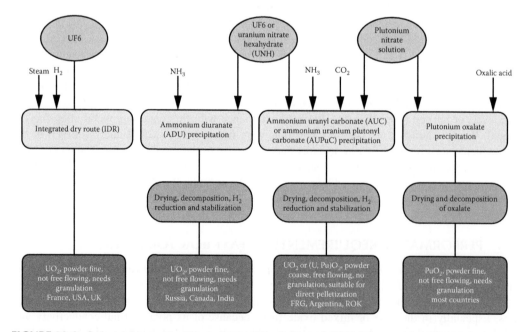

FIGURE 29.4 Industrial processes for producing UO_2, PuO_2, and $(U, Pu)O_2$ powders. (From IAEA, Status and trends of nuclear fuel technology for sodium cooled fast reactors, IAEA Publication STI/PUB/1489, Nuclear Energy Series NF-T-4.1, IAEA, Vienna, Austria, 2011.)

followed by extrusion of lubricated micronized powder through a sieve leading to free-flowing granules that were directly pelletized and sintered [29.1]. The oxide comilling (OCOM) process used by Germany for fabricating the fuel for SNR-300 reactor and the process flow sheet adopted by India for the fabrication of mixed oxide fuel for PFBR have similar features. In the AUPuC (ammonium uranium plutonium carbonate) process developed in Germany, U and Pu were coprecipitated from U and Pu nitrate solutions and micronized, and the AuPuC was subsequently calcined to get $(U, Pu)O_2$ power suitable for direct utilization. The GRANAT process was developed by All Russia Institute for Inorganic Materials (VNIINM). This process involves coprecipitation of uranium, plutonium hydroxides using flocculent followed by conversion to oxides and granulation before pelletization.

These processes are used in the fabrication of MOX fuel in the form of pellets. The Russian fast reactor BOR-60 uses oxide fuel in the form of granules (produced by pyroprocessing). In this case, the fuel pin is fabricated by direct vibrocompaction of oxide granules. Since the product of pyroprocessing is highly radioactive because of the relatively low decontamination factor, the fuel fabrication has to be carried out in hot cells. This process is also amenable to fuel containing americium.

29.6.1 Sol-Gel Process

The sol-gel routes are designed to use the output of reprocessing plant in the form of solutions and convert them into consolidated gel particles, thus eliminating the powder handling and associated hazards. Vibrocompaction of multiple sizes of high-density microspheres in fuel pin cladding has been used to fabricate *sphere-pac* type fuel pins.

The term *sol-gel process* refers to chemical routes that involve gelation of a droplet of sol (or solution) of the desired fuel material into a gel microsphere. These microspheres are washed, dried, and heat treated to make them high-density microspheres. This process is one of the well-studied processes for the preparation of gel microspheres of actinide oxides. The process uses the solutions of the nitrates of uranium, thorium, and plutonium or their desired mixtures. In the internal gelation process, the cooled (~273 K) metal nitrate solutions are mixed with urea and hexamethylene–tetramine (HMTA) solutions in cooled condition (~273 K). The droplets of this mixture come into contact with hot oil (silicone oil, ~363 K) to make gel microspheres. The gel microspheres are washed first with CCl_4 to remove the silicone oil and then with ammonia solution to remove the excess gelation agents. Then they are dried and calcined and then reduced in $N_2 + H_2$ mixture to obtain UO_2 microspheres, which are sintered to get >99% TD microspheres. The external gelation process, also called the SNAM process or gel-supported precipitation method, is being pursued by several countries for the development of coated particle fuels. In this method, a water-soluble polymer is added to the heavy metal solution (or sol), droplets of which are exposed to ammonia vapor to form the gel. Several variants of this process have been developed. For details, see Ref. 29.14.

The sol-gel process for the production of microspheres of ceramics has special advantages with respect to the fast reactor fuels and especially those containing Pu and/or minor actinides. These considerations arise out of the necessity to simplify the fabrication route and to carry out operations at lower temperature in order to reduce radiation exposure to the operators caused by the fabrication activity. In contrast with the conventional powder pellet route, the sol-gel route does not involve handling of powders but involves handling of microspheres that are free flowing. The microspheres can be fabricated to have a degree of softness that could enable them to be pelletized. This process is known as sol-gel microsphere pelletization (SGMP) and has been reported to result in pellets whose performance in the reactor could be similar to the pellets produced through the conventional route [29.14]. The hard microspheres fabricated by the sol-gel process, on the other hand, can be directly packed into the fuel pins by vibrocompaction to densities close to the smear density of approximately 85% that is usually adopted for the MOX fuel for fast reactors. The advantages of SGMP process are as follows:

1. Radiotoxic dust hazard is avoided since fine powder particles of fuel are absent.
2. Dust-free and free-flowing microspheres (diameter 0.2–1.0 mm) facilitate remote processing.

3. Excellent microhomogeneity is ensured in fuel pellets.
4. Fabrication steps and pyrophoricity hazards are reduced for carbide and nitride.
5. Fabrication of both low- (≤85% T.D.) and high-density (≥96% T.D.) fuel pellets of controlled *open* and *closed* porosity is possible.

The sol-gel process has not been implemented on a commercial scale in any country. However, considering its amenability for automated remote operation and the possibilities of reduction in radiation exposure, the sol-gel process is an attractive option for the production of MOX fuel, especially for fuels containing minor actinides that would require remote fabrication due to the high gamma radioactivity of americium. The sol-gel process also has the nitrate solution of the constituents as its feed, which results in microhomogeneity of the constituent elements in the fabricated fuel. Additionally, the product of reprocessing (nitric acid solution of the actinides), without its conversion to oxides, can be used as the feed for sol-gel process, after adding a few steps, which includes the reduction of acid concentration by a suitable chemical process and increase in the concentration of actinides to the desired level. It must be noted that these steps are yet to be demonstrated on a significant scale and this would be a key process in the adoption of the sol-gel process for MOX fuels for the fast reactors. Another important concern is the generation of complex waste solution, albeit in small quantities, from which the actinides have to be recovered/removed prior to disposal.

A typical flow sheet for the fabrication of MOX fuel through the sol-gel route is given in Figure 29.5. The MOX fuel with Pu content as required for the fast reactor can be produced by this route. Most recently, MOX microspheres containing as high as 50% Pu have been produced at Bhabha Atomic Research Centre (BARC) through the sol-gel route [29.15].

29.6.2 Fabrication of Carbide and Nitride Fuels

The carbide fuel can be produced by carbothermic reduction of oxide or by the reaction of metal with a suitable hydrocarbon. In the latter route, the metal is converted to a fine powder through hydriding—dehydriding cycles before its reaction with hydrocarbon. In this method, the product is a mixture of various carbides and the composition cannot be controlled precisely. Also, the product is in the form of a finely divided powder, which is very reactive to oxygen and moisture. Carbothermic reduction route is the preferred route for commercial-scale production of carbides even though the product obtained is relatively impure as compared to the carburization of metals. The carbothermic reduction can be performed through a two-step route—higher carbide is produced in the first step, by reaction with excess amount of carbon, which is followed by the reduction to the lower carbide using hydrogen. In this method, the oxygen impurities can be maintained at a lower level, since the oxygen is effectively removed by the excess carbon used. Carbide can also be produced by a single-step reduction, wherein a calculated amount of carbon is allowed to react with oxide. In either case, an intimate mixture of oxygen and carbon (usually in the form of graphite) is pressed into pellets and the pellets are heated at temperatures in the range of 1873 K in flowing argon or argon–hydrogen stream to remove the carbon monoxide formed and drive the reaction to completion.

$$MO_2 + 3C \rightarrow MC + 2CO \uparrow$$

The oxygen is not, however, fully removed in this step, and the product, therefore, is usually a mixture of $MC_{1-x}O_x$ and M_2C_3 (the intermediate product). Depending upon the Pu content of the oxide, the temperature of the reaction has to be optimized to obtain the maximum removal of oxygen with a minimum loss of plutonium.

The synthesis of nitrides can be carried out, as in the case of carbide, starting from metal or by carbothermic reduction in the presence of nitrogen. The metal can be nitrided by reaction with nitrogen. In the case of uranium, reaction with nitrogen at relatively low temperatures leads

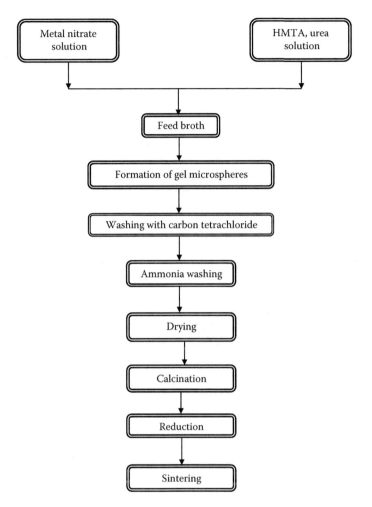

FIGURE 29.5 Typical flow sheet for the fabrication of MOX fuel microspheres by internal gelation process. (From Ashok Kumar, S.L. et al., *J. Nucl. Mater.*, 434, 162, 2013.)

to the formation of sesquinitride, which can be decomposed by heating to high temperature (greater than 1673 K). In the case of plutonium, however, only mononitride is formed. In the case of carbide, the material produced by the direct reaction of the element is in the form of highly reactive powder.

The carbothermic reduction of oxide by flowing nitrogen produces nitride as per the following equation:

$$MO_2 + 2C \xrightarrow{N_2} MN + 2CO \uparrow$$

The nitride product has to be cooled in argon atmosphere since it can react with nitrogen at lower temperatures to form higher nitrides.

29.6.3 FABRICATION OF METALLIC FUELS

Metal alloy fuels are fabricated by injection-casting route and are usually sodium bonded to the clad material. This is because the metal fuels are prone to swelling, and to accommodate the swelling and allow the interconnected porosity to develop and permit gas release, the smear density of the fuel is kept low (around 75% as against 85%–90% used for ceramic fuels). The fabrication technology for U- and Pu-based alloys has also been developed in Japan and Korea, apart from the United States.

29.7 FUEL REPROCESSING

Recovery of U and Pu from the irradiated fuel constitutes a vital process in the fast reactor fuel cycle as the closure of the fuel cycle is important as emphasized in Section 29.1. The reprocessing of fast reactor fuels is carried out by aqueous as well as nonaqueous processes. While the aqueous reprocessing schemes for fast reactor fuels are similar to the schemes employed for thermal reactor fuels, the nonaqueous processes can be considered to be more relevant for fast rector fuel cycle as compared to thermal reactor fuel cycle.

29.8 AQUEOUS REPROCESSING

Worldwide, PUREX (Plutonium URanium EXtraction process), with some variations, has been the preferred aqueous scheme in all the countries that have pursued fast reactor fuel reprocessing. In the United States, France, the United Kingdom, Germany, Russia, Japan, and India, aqueous processes have been used to process the irradiated fast reactor fuels. A typical flow sheet for the PUREX process for fast reactor fuel is given in Figure 29.6.

This generic flow sheet, with some changes in the head-end step, can be applicable to various types of fuels including metal, oxide, carbide, and nitride fuels. After dissolution of the fuel in

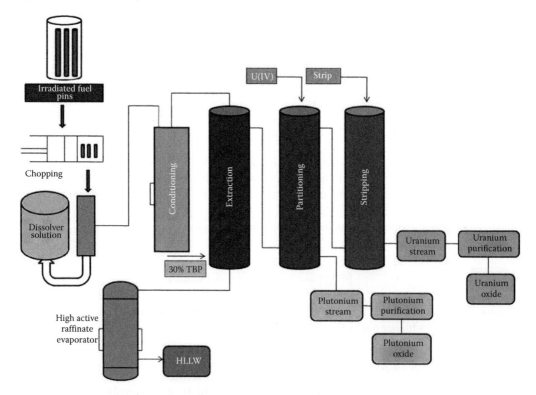

FIGURE 29.6 Typical flow sheet for fuel reprocessing by purex process.

nitric acid, uranyl nitrate and Pu (IV) nitrate are extracted selectively into a solution of tri-n-butyl phosphate (TBP) in normal paraffin diluent and separated from fission products. Through multiple cycles of extraction and stripping, U and Pu can be purified from the fission products to an extent that would permit subsequent handling of the products by conventional methods.

29.9   SPECIAL FEATURES OF FAST REACTOR FUEL REPROCESSING

The main differences between the reprocessing operations for the thermal reactor and fast reactor fuels arise from the following factors:-

1. The high radiation levels of the irradiated fast reactor fuels resulting from high burnup and short cooling times required for reducing the doubling time
2. High content of Pu in the fuel
3. Pyrophoricity and ease of oxidation of some of the fuels (carbide, nitride, and metal)
4. Possible presence of sodium contamination on the fuel elements in case of SFRs

The high Pu content of fast reactor fuel demands that the sizing and design of process equipment take into account the criticality considerations. MOX fuels with high plutonium content are also known to be difficult to dissolve in nitric acid [29.16]. The irradiation of MOX fuel is generally reported to improve the dissolution [29.17]. The distribution of fission product nuclides formed during the fission of ^{239}Pu is slightly different from those formed during the fission of U^{235}. The fission of ^{239}Pu results in the formation of higher proportion of noble metals such as Ru, Rh, and Pd. These metals form intermetallic alloys among themselves and with other constituents, including U and Pu, and these alloys are difficult to dissolve in nitric acid. Vigorous dissolution conditions have to be employed to dissolve these alloys for the recovery of actinides. The increased concentration of platinum group metals also causes phase segregation during the vitrification of the high-level waste. The high plutonium content of the fuels also accentuates the need for comprehensive and precise monitoring systems or analytical techniques that can track the strategic nuclear materials and enable its accounting.

From the chemistry point of view, the phenomenon of third-phase formation assumes high importance in the case of fast reactor fuel reprocessing involving plutonium-based fuel. The extraction of high concentration of plutonium in TBP in the solvent extraction process leads to the splitting of the organic phase beyond a certain limit called limiting organic concentration. The second (heavier) organic phase formed is highly concentrated in plutonium while the lighter phase contains a little or no plutonium. The accumulation of the third phase, thus, has a potential to cause criticality, in addition to process malfunction. This phenomenon is not present in the extraction of U(VI) by TBP and due to the relatively low concentrations of plutonium encountered in thermal reactor fuel reprocessing, the concern of third-phase formation does not arise.

The high radiation levels associated with fast reactor fuels demand a careful optimization of process parameters to ensure a high level of decontamination from fission products. The high radiation levels also cause degradation of extractants and diluents, which leads to the formation of a number of products that are deleterious to the extraction/stripping process. The buildup of degradation products can lead to the retention of plutonium in the organic phase, implying incomplete stripping. Some of the fission products such as Zr also have enhanced retention in the degraded organic phase. To minimize radiation degradation, the solvent extraction process with short cooled/high burnup fast reactor fuels can be carried out using centrifugal extractors that reduce contact time between the aqueous and organic phases thereby reducing the degradation of the organic phase. The treatment of the solvent to remove degradation products is an important step in aqueous reprocessing of fast reactor fuels to reduce waste volumes.

29.10 INTERNATIONAL EXPERIENCE ON FAST REACTOR FUEL REPROCESSING

A key characteristic of fast reactor technology is that used fuel may be reprocessed to recover fissile and fertile materials in order to provide fresh fuel for existing and future nuclear power plants. Several countries had developed fast reactor fuel reprocessing technology based on the LWR fuel cycle experience and demonstrated in their reactors. However government policies in many countries have not yet addressed the various aspects of reprocessing technology towards the closed fuel cycle. Experiences of various countries are summarized below:

- *France*: In Fontenay-aux-Roses, France, a laboratory-scale reprocessing of oxide fuel from the fast reactors RAPSODIE and PHENIX was carried out. Pilot-scale reprocessing was carried out at Marcoule (APM facility) and La Hague (AT1 facility). Fast reactor fuel was also reprocessed at the UP2 La Hague plant by diluting with natural uranium fuel for graphite-moderated gas-cooled reactors.
- *Germany*: In Germany, the studies on the reprocessing of SFR fuels were first carried out at the Karlsruhe Nuclear Research Centre. Test irradiation samples from DFR were processed in the MILLI facility at Karlsruhe at the burnup levels of up to 60 GWd/ton.
- *India*: In India, the uranium- and plutonium-mixed carbide fuel irradiated in FBTR to a burnup of up to 155 GWd/ton has been reprocessed on pilot scale at *CORAL* facility in Kalpakkam [29.18]. It is important to note that the fuel reprocessed had Pu content as high as 70%, and the reprocessing was carried out after a cooling period as short as 2 years. The fuel was reprocessed by PUREX process, and the process flow sheet followed was essentially same as that for oxide fuel. The challenges in reprocessing carbide fuels are discussed in Section 29.11.
- *Japan*: Japan Atomic Energy Agency (JAEA) has been pursuing fast reactor fuel reprocessing technology development since 1975. Laboratory-scale hot experiments on reprocessing MOX fuel from JOYO were carried out in the Chemical Processing Facility (CPF). The PUREX solvent extraction flow sheet was adopted. Approximately 10 kg of fast reactor–spent fuels with a burnup of approximately 40–50 GWd/ton in early times and approximately 100 GWd/ton at peak times have been treated in CPF [29.2].
- *Russia*: At the RT-1 facility of the Mayak reprocessing plant at Chelyabinsk, Russia has demonstrated reprocessing of spent fuel from the fast reactors BN-600 and BN-350.
- *United Kingdom*: The development of fast reactor fuel reprocessing was carried out in the United Kingdom initially at Harwell as well as Dounreay laboratories. A dedicated plant to process the fuel from DFR was set up at Dounreay where it reprocessed 10 tons of DFR fuel. The plant was originally designed to process enriched 235 U (75%) alloy fuel by dissolution in nitric acid medium followed by PUREX process with 20% TBP/odorless kerosene. Later, the plant also reprocessed the oxide fuel of the PFR.
- *United States*: Due to the US decision of not pursuing the reprocessing option, development work for the fast reactor fuel cycle was performed only in laboratories. Oak Ridge National Laboratory (ORNL) together with Argonne National Laboratory (ANL) pioneered many aspects of fast reactor fuel reprocessing including the development of centrifugal contactors, remote handling systems, and hardware prototypes for most unit operations in reprocessing. Limited experience was also obtained with the reprocessing tests of the Fast Flux Test Facility (FFTF) MOX fuel. ORNL performed laboratory-scale dissolution and solvent extraction studies on MOX fuel irradiated up to 220 GW/ton HM burnup.

29.11 PYROCHEMICAL REPROCESSING

Pyrochemical and Pyroelectrochemical processes are particularly appropriate for fast reactor fuels as compared to thermal reactor fuels. Among the fast reactor fuels, pyroprocessing routes have more advantages with respect to metallic fuels than for other fuels, even though these routes have indeed been applied for oxide fuels also. These processes use inorganic salts as process media and do not employ aqueous or organic reagents. Pyroprocesses can be classified into three categories:

1. Methods based on difference in volatility of actinide and fission product compounds—fluoride volatility process and nitrofluor process fall in this category.
2. Pyrometallurgical processes: In these processes, the fuel is maintained in molten state throughout the process. The melt refining process used for reprocessing of EBR-II fuel belongs to this category.
3. Pyroelectrochemical or pyrochemical process: Based on electrochemical/redox behavior of actinides and fission products—salt transport process, salt cycle process, and molten salt electrorefining fall in this category.

The distinctive advantages of pyroprocess for fast reactor fuels are as follows:

1. Fast reactor fuels reach high levels of burnup as compared to thermal reactor fuels and have to be processed after relatively short cooling times due to economic reasons and considerations of reducing doubling time. The nonaqueous processes are better suited for processing the high burnup fuel after short cooling times, because the inorganic salts are resistant to radiation and do not suffer degradation, whereas the organic reagents deployed in aqueous reprocessing are susceptible to chemical and radiological degradation.
2. Since the aqueous and organic media are highly moderating, the concentrations of Pu and other fissile nuclides have to be limited in aqueous extraction systems to avoid the possibility of criticality. In the case of nonaqueous reprocessing schemes, however, much higher quantities and concentrations of fissile materials can be handled because of the absence of moderating media. As a result, the pyroprocess plant is expected to be compact as compared to the reprocessing plant based on aqueous process.
3. The pyroprocess schemes do not produce high-level liquid waste and produce mainly solid waste. It can, therefore, be expected that the cost of waste management could be less in the case of pyroprocessing as compared to aqueous reprocessing. It must be noted that the management of high-level liquid waste (HLLW) is among the most important technological issues with respect to aqueous reprocessing schemes and contributes significantly to the cost of the fuel cycle.

The pyroprocesses also involve relatively a less number of steps, and the operations can be carried out in compact equipment, reducing the size of the plant and thereby contributing to the economy.

29.11.1 Pyroprocessing of Oxide Fuels

Russian Institute for Atomic Reactors (RIAR), Dimitrograd, has developed a pyroprocess, Dimitrovgrad dry process (DDP), which is also known as oxide electrowinning process for reprocessing oxide fuels from fast reactors. It exploits the differences in the thermodynamic stabilities of the oxychlorides of uranium, plutonium, and the fission products and the fact that the oxides of uranium and plutonium are conducting at temperatures above 4000°C.

The process steps of oxide electrowinning process are shown in Figure 29.7. In this process, after the mechanical decladding of pins, the oxide fuels are subjected to chlorination in the presence of molten salts NaCl–KCl or NaCl–2CsCl, wherein uranium oxychloride (UO_2Cl_2) and plutonium tetrachloride ($PuCl_4$) that are formed get dissolved in the salt. In the next step, electrolysis is carried out under inert atmosphere to deposit uranium oxide along with the fission products zirconium, niobium, and the noble metals, Ru, Rh, and Pd, on a graphite cathode. The oxygen potential of the purge gas is then enhanced by increasing the oxygen to chlorine ratio in the gas mixture and under these conditions PuO_2 gets precipitated. The plutonium oxide is of 99.5%–99.9% purity and free from fission products. Further electrolysis results in the deposition of UO_2 along with fission products.

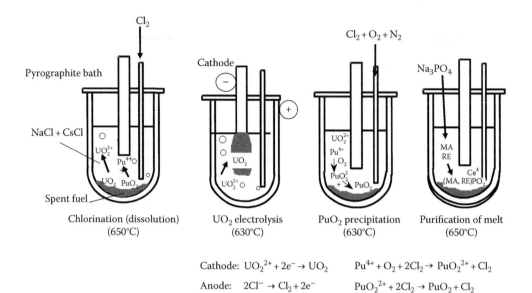

FIGURE 29.7 Process steps of the oxide electro-winning process.

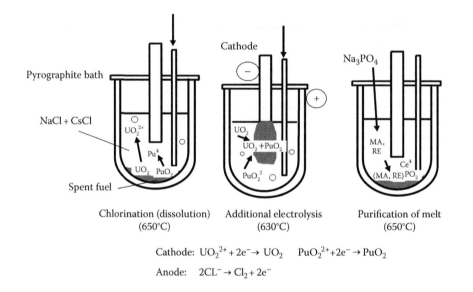

FIGURE 29.8 Schematic of MOX co-deposition by oxide electro-winning process.

The particles in the deposit have a high density of >10.7 g/cm³ and are of less than 1 mm size. The salt occluding the deposit is washed with water and then the salt recovered by the evaporation of water for recycling. The decontamination factors are less than 100, which are acceptable for fast reactors.

Another version of the flow sheet, shown in Figure 29.8, is also used for the fabrication of mixed fuel for fast reactors [29.19,29.20]. In this flow sheet, after the first step of dissolution, the oxygen potential of the gas mixture is adjusted in such a way to stabilize Pu in the salt in the form of its oxychlorides and electrolysis is carried out to codeposit uranium and plutonium oxides on the cathode. The deposit from the process is used for vibrocompaction after the process steps of crushing and sieving. It has also been reported that NaCl–KCl is more suitable for the flow sheet where uranium and plutonium oxides are partitioned and NaCl–2CsCl is more suited for the code-position option.

The process has been used for reprocessing about 12 kg of spent MOX fuel from BOR-60 and about 3.5 kg of spent MOX fuel from BN-350 apart from several kilograms of spent UO_2 fuel from these reactors. Fuel with 24 at% burnup has been reprocessed using this process, and the fuel produced by this process has been taken to a burnup of 32 at%.

The pyrochemical processes also pose several challenges:

1. The decontamination factor (DF) achievable in the pyrochemical processes is lower by orders of magnitude as compared to the aqueous processes. The DF achievable with pyro-processes can range from as low as 100 to values of the order of 10^5 for a few elements. This is to be compared with the values ranging from a minimum of 1000 to as high as 10^6 obtained for most of the elements by aqueous processing. As a result of this low decontamination, the final product obtained by pyroprocessing is highly radioactive. The refabrication of the fuel, therefore, has to be carried out in shielded enclosures. The process steps are also carried out at high temperatures and in inert atmosphere to avoid the oxidation of the metal.
2. The use of corrosive salt mixtures as the media for processing demands attention to corrosion issues and the qualification of components for long life.
3. The commercial-scale experience with aqueous reprocessing is rather significant while experience with pyroprocess is quite limited.

29.11.2 PYROPROCESSING OF METAL FUELS

The advantages offered by pyrochemical processes as compared with aqueous processes are accentuated by the fact that metals can be processed to obtain directly metal as the product, without additional intermediate steps. In view of these advantages, France, Japan, Korea, Russia, and the United States have pursued the development of pyroprocesses such as pyrochemical process for reprocessing of metallic alloy fuels. The pyroelectrochemical process is more specifically dealt with in the following sections. A detailed description of various pyroprocesses can be found in Ref. 29.2.

A typical flow sheet for the pyrochemical processing of metallic fuels is given in Figure 29.9. In this process, the metal fuel pin is chopped into pieces and placed in a basket in a molten salt bath and used as anode. An iron rod or a crucible containing liquid cadmium is used as cathode. The actinides are oxidized to their cations and transported across the salt bath and deposited on the cathode. For the deposition of uranium, iron cathode is used. After scrapping the deposit, it is further processed to remove the adhering salt and then consolidated to obtain the metal ingot. For plutonium deposition, a cadmium cathode is used, and the cathode material with deposit is subsequently heated to distill off the cadmium. The residue contains plutonium, minor actinides, and any residual uranium that follows plutonium. The residue is consolidated and combined with the required amount of uranium for constituting the fuel alloy for refabrication.

The molten salt electrorefining process is based on the relative electrochemical stabilities of various metal/cation systems. Highly stable metals such as zirconium and palladium remains as metals and are not converted to their cations. Metals with highly stable chlorides such as alkali metals get oxidized at the anode and remain in the salt. Those metals whose chlorides have an intermediate stability, including actinides and lanthanides, get transported across the cell and get deposited on the cathode. Due to similar electrochemical potentials, minor actinides such as americium also get transported along with U and Pu and get deposited.

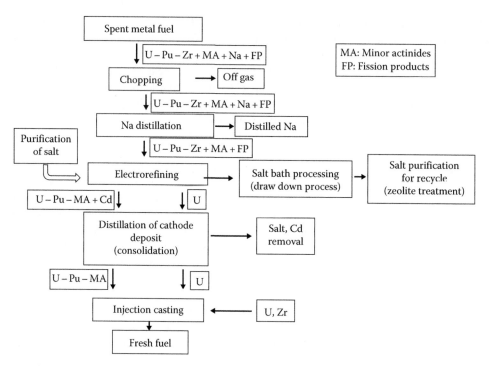

FIGURE 29.9 Typical flow sheet for pyroprocessing of metal fuel by molten salt electrorefining.

29.12 REPROCESSING OF CARBIDE AND NITRIDE FUELS

For reprocessing the carbide fuel, it can be dissolved in nitric acid and the solution processed by PUREX process. The dissolution of carbide fuel with nitric acid produces a number of organic compounds. Nearly 50% of the carbon in the carbide fuel gets converted to carbon dioxide during the dissolution step, while the balance carbon is converted to organic compounds soluble in nitric acid. Only a few of these compounds have been identified and these include oxalic acid and mellitic acid. The organic compounds can complex the actinides and interfere in the extraction of actinides by TBP. Some compounds also get extracted in the TBP and these can interfere in the stripping of Pu from TBP in the nitric acid.

Extended boiling of the dissolver solution helps in destroying the majority of the organic compounds and makes solvent extraction process feasible. Other methods to destroy the organic compounds include chemical destruction using oxidation such as potassium dichromate, electrochemical oxidation using electrogenerated Ce(IV) or Ag(II) or photochemical destruction. To avoid the formation of organic compounds, the carbide can be converted to oxide by heating in air, oxygen, or other oxidizing atmosphere. Conversion of oxide can also be achieved by pyrohydrolysis, which involves heating of the carbide in steam. Pyrohydrolysis results in hydrogen as one of the products. The oxidation step has to be carried out in a controlled manner since the oxide product can be difficult to dissolve in nitric acid if it gets sintered during the conversion step.

At Indira Gandhi Centre for Atomic Research (IGCAR), India, the carbide fuel discharged by various burnup levels up to 150 GWd/ton has been reprocessed by employing PUREX process, after the dissolution of carbide in nitric acid followed by the destruction of organic compounds by heating the solution [29.18]. No interference from the organic compounds was observed in the process steps.

The nitrides are easily dissolvable in nitric acid and the product solution is expected to be compatible with the PUREX process. One of the issues with the nitrides is the production of ^{14}C isotope by (n,p) reaction involving ^{14}N. The dissolution step thus leads to the production of $^{14}CO_2$ for which additional steps of treatment would be required. Barring these differences, the processing of carbide or nitride fuels by aqueous processing is expected to be quite similar to the processing of irradiated MOX fuel with similar Pu content.

The molten-salt electrorefining method of pyroreprocessing used for metallic fuels can also be used for the reprocessing of carbide and nitride fuels taking advantage of their high electrical conductivities. The uranium and plutonium metals deposited on cathode can be converted to carbides and nitrides by suitable methods. In this regard, significant amount of work has been carried out in Japan on the application of the process for nitride fuels. Uranium and plutonium metals deposited on to liquid cadmium cathode have been shown to be convertible to their nitrides by nitridation in presence of liquid cadmium, which is advantageous as it avoids several process steps of consolidation.

29.13 PARTITIONING OF MINOR ACTINIDES

The importance of the recovery of all transuranic elements from the irradiated fast reactor fuels was described in Section 29.2. The term *minor actinide* refers to the transuranium elements Np, Am, Cm, etc., which are formed in small quantities during the nuclear reactions of isotopes of uranium and plutonium with neutrons. Typical routes for the formation of ^{237}Np and ^{241}Am are as follows:

$$^{238}U(n,2n) \, ^{237}U \rightarrow \, ^{237}Np$$

$$^{235}U(n,\gamma)^{236}U(n, \gamma)^{237}U \rightarrow \, ^{237}Np$$

$$^{239}Pu(n, \gamma)^{240}Pu(n, \gamma)^{241}Pu \rightarrow \, ^{241}Am$$

Many of the isotopes of minor actinides have relatively long half-lives. For example, ^{237}Np has a long half-life (2.14×10^5 years) and ^{241}Am has a half-life of 533 years. As a result of the long half-life, the presence or growth of these isotopes of minor actinides in the waste is a matter of concern, since it will take a very long period before the radiotoxicity of the waste, contributed by the minor actinides, is reduced to low levels. This implies that the waste has to be monitored for impractically long periods if minor actinides are present in the waste.

As explained in the earlier sections, the aqueous reprocessing of nuclear fuels involves the dissolution of fuel in nitric acid and the recovery of U and Pu from the dissolver solution by a solvent extraction process. TBP extracts actinide ions in tetravalent and hexavalent oxidation states. The transplutonium elements are present in aqueous solutions in trivalent oxidation state. Therefore, they are not extracted by TBP. Neptunium can be present in tetravalent, pentavalent, or hexavalent oxidation state. Depending upon the oxidation state of Np present in the fuel solution, a part of it or all of it is rejected to the high-level aqueous waste stream. The solution left after the extraction of U and Pu (termed as *raffinate*) thus contains all the fission products and minor actinides. Due to its high level of radioactivity, this solution is called *high level liquid waste* (HLLW).

Several countries have been actively involved in the development of extraction systems for minor actinides from HLLW. The extractants studied include octylphenyl N,N-diisobutyl carbamoyl methyl phosphinoxide (CMPO), diglycolamides, and di-2-ethylhexyl phosphoric acid. While the extraction behavior of minor actinides with these extractants has been fairly well understood, demonstration of the separation process with actual waste solutions has been limited. A systematic, multigroup study was undertaken in Europe as a part of the Europart. More recently, the Actinide reCycling by SEParation and Transmutation (ACSEPT) project funded by European Commission and involving several countries has aimed to develop and demonstrate the actinides separation schemes. A typical scheme for minor actinide partitioning is shown in Figure 29.10. The first step of partitioning provides a product that is a mixture of actinides

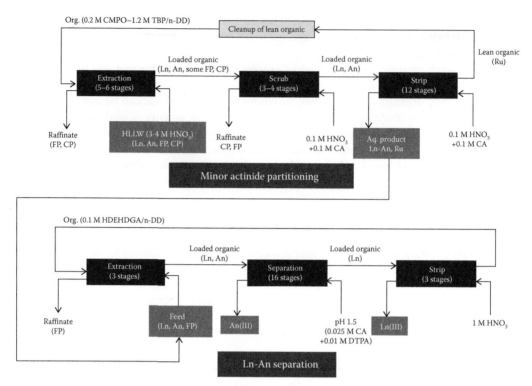

(CMPO: Octyl phenyl N,N-dissobutyl carbamoyl methyl phosphinoxide; CA: Citric acid; DTPA: Diethylenetriamine pentaacetic acid: HDEHDGA:Bis(2-ethylhexyl) diglycolamic acid)

FIGURE 29.10 Typical scheme for minor actinide partitioning.

and lanthanides, and the next step provides a scheme for their separation from each other. For the separation of minor actinides along with lanthanides, as well as minor actinides from lanthanides, several extraction systems have been proposed, but the studies have been confined to laboratory scale.

In the case of fast reactors with metallic fuels, pyroprocessing is used to recover and recycle the fissile material. In pyroprocessing based on molten salt electrorefining, the minor actinides follow Pu in the deposition step, and, therefore, there is no need for a separate recovery process for minor actinides as in the case of oxide fuel, which is processed by PUREX process.

29.14 WASTE MANAGEMENT FOR FAST REACTOR FUEL CYCLE

In addition to HLLW, reprocessing also produces waste solutions of intermediate and low level of radioactivity. However, among the waste solutions the HLLW is of minimum volume, but contains over 95% of the radioactivity. The high-level waste solution is converted into a solid matrix for long-term storage/disposal. The current practice is to immobilize the radioactive waste, with vitrified borosilicate glass as the matrix [29.21]. The solution is mixed with the glass ingredients and evaporated, the residue melted and canned for intermediate disposal. A typical flow sheet for the immobilization of waste is shown in Figure 29.11. Extensive experience has been generated in the immobilization of HLLW in glass matrix. In France, for example, over 7000 tons of HLW glass have been produced until 2012. However, till date, no radioactive waste has been permanently disposed into a repository.

The immobilization of radioactive waste arising out of the reprocessing of fast reactor fuels poses additional challenges because of the high levels of radioactivity associated with the waste

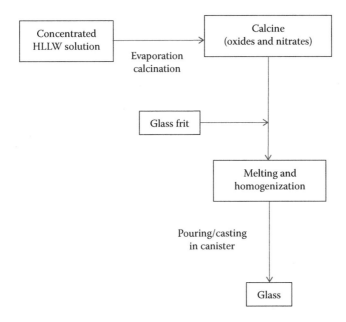

FIGURE 29.11 Flow sheet for immobilization of HLLW in glass matrix.

(arising from the high burnup and relatively short cooling time) and also the presence of higher concentrations of certain fission products that are formed at a higher yield in the fission of Pu-239. The cumulative fission yield of the noble metal palladium in fission of Pu-239, for instance, is nearly 10 times than that in the fission of U-235 (16.6% vs. 2.2%). The noble metals do not have adequate solubility in the glass matrix and, hence, tend to segregate when immobilized in conventional borosilicate glass matrix. Also, the high level of radioactivity associated with HLLW from fast reactor fuel reprocessing implies that the waste solution needs to be diluted in order to limit the degradation of the glass due to radiation damage. As a result, alternate matrices for immobilizing the HLW from fast reactors are under development. The candidates include iron phosphate–based glasses as well as ceramics such as monazite and synthetic rock or *synroc* [29.22]. Monazite is rare earth orthophosphate, and synroc is a multiphase titanate matrix.

The approach here is to treat waste streams with calculated quantities of chemically active additives, so that upon heat treatment and consolidation, a dense polyphase ceramic assemblage (or a single phase, as in monazite) is formed in which the waste elements are incorporated into the lattice positions of known crystalline phases [29.23]. Synroc is an assemblage of four mutually compatible minerals: zirconolite ($CaZrTi_2O_7$), perovskite ($CaTiO_3$), hollandite ($BaAl_2Ti_6O_{16}$), and rutile (TiO_2). Many natural minerals resembling these, containing radioactive elements, have survived in nature under different geochemical conditions for periods of up to two billion years. The synroc minerals are endowed with complex crystal structures consisting of coordination polyhedra of different sizes and shapes into which the HLW elements can get incorporated depending on their ionic size and charge. For instance, hollandite can host the elements Cs, Rb, and Ba; zirconolite can incorporate U, Zr, Np, Pu, and rare earths; and perovskite ($CaTiO_3$) can take up Sr, Np, and Pu. Rutile provides microencapsulation of the minor alloy phases formed by the noble metal fission products; it also increases the mechanical strength of the ceramic monolith. The relative amounts of the minerals in the phase assemblage are determined by the amount of additives. High-temperature and high-pressure operations are required to prepare the ceramic waste forms.

Two types of waste forms are envisaged in the pyrometallurgical reprocessing [29.24]: (1) metallic waste form for the cladding hulls, Zr, and noble metal fission products from the anode basket of the electrorefiner, and (2) ceramic waste form for the alkali, alkaline earth, and rare earth fission products along with a fraction of the eutectic salt bath.

The contents of the anode basket are made chloride free by distilling. Extra Zr is then added and the mixture melted in a casting furnace to form a durable stainless steel containing up to 15 wt% of Zr. This constitutes the metal waste form [29.25]. The molten salt bath containing the fission products is treated with a large excess of zeolite at 773 K. A part of the cations is ion exchanged into the zeolite matrix, and the rest of the salt gets occluded in zeolite cavities. The salt-loaded zeolite is then treated with a glass frit and subjected to hot isostatic pressing or pressureless sintering. This treatment converts the powder mix into glass-bonded sodalite, a glass–ceramic material. This is called the ceramic waste form.

29.15 FAST REACTORS AND MINOR ACTINIDES BURNING

Fast reactors are better suited to burn minor actinides as compared to thermal reactor systems because of higher neutron flux and higher average neutron energy. In a nuclear reactor, the minor actinide isotopes can undergo two types of nuclear reactions: it can absorb a neutron and undergo fission, which would result in the destruction of minor actinide and the formation of fission products. On the other hand, the isotope can also absorb neutron and transform into a higher minor actinide isotope as explained in Figure 29.12.

In order that the minor actinide is effectively *burnt,* that is, fissioned into fragments, the fission cross section has to be significantly higher than the capture cross section. The cross sections for some minor actinide isotopes in thermal and fast neutron spectrum are indicated in Figure 4.2. From this table, it is obvious that the fast spectrum reactors can be more effective for the incineration of minor actinides. Metal-fueled reactors offer a harder spectrum that is more amenable for the incineration of minor actinides since the $\sigma_{\text{fission}}/\sigma_{\text{capture}}$ ratio is higher.

29.15.1 MATRICES FOR MINOR ACTINIDE BURNING

Several approaches are available for burning minor actinides in fast reactors. In one approach (homogeneous), the minor actinides are mixed with the driver fuel of the reactor. In this case, the concentration of the minor actinide is around 5% or less. In another approach (heterogeneous), the minor actinide is contained in specific matrices at high concentration (up to 20%) in special target subassemblies. The matrix can be either an actinide oxide or an inert matrix such as magnesium oxide. In either case, the processing and recycling of the irradiated target/fuel are important to ensure the complete burning of the minor actinides, and processes for such recycling are under development. The development of minor actinide containing fuels is discussed in detail in Ref. 29.26.

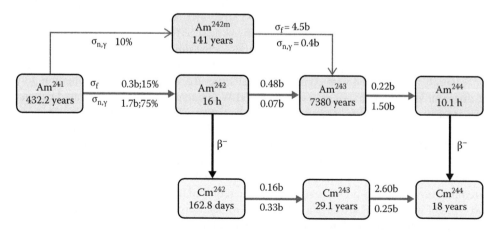

FIGURE 29.12 Transmutation and incineration reactions of Am & Cm isotopes in fast neutron spectrum.

29.16 CONCLUSION

The fast reactor fuel cycle is still in a stage of evolution and the international experience is limited as compared to fuel cycle for thermal reactors. Accordingly, the fast reactor fuel cycle also offers several opportunities for research and development. With the increasing emphasis being placed on safety, economics, minimizing waste generation, nonproliferation, etc., there is ample scope for challenging research programs that could make a significant contribution to the development of fast reactor fuel cycle.

REFERENCES

29.1 IAEA. (2011). Status and trends of nuclear fuel technology for sodium cooled fast reactors. IAEA Publication STI/PUB/1489. Nuclear Energy Series NF-T-4.1. IAEA, Vienna, Austria.

29.2 IAEA. (2011). Status of developments in the back end of the fast reactor fuel cycle. IAEA Publication STI/PUB/1493. Nuclear Energy Series NF-T-4.2. IAEA, Vienna, Austria.

29.3 Gros, J.-P., Drain, F., Louvet, T. (2011). The french recycling industry. *Proceedings of Global 2011 International Conference.* Makuhari, Japan, P.No. 360445.

29.4 Boullis, B., Warin, D. (2010). Future nuclear systems: Fuel cycle options and guidelines for research. *11th Information Exchange Meeting on Actinide and Fission Product Partitioning and Transmutation,* San Francisco, CA.

29.5 Kazimi, M., Moniz, E. J., Forsberg, C. W. (co-chairs) (2011). The future of the nuclear fuel cycle: An interdisciplinary MIT study. Massachusetts Institute of Technology, Cambridge, MA.

29.6 Poinssot, C., Rostaing, C., Greandjean, S., Boullis, B. (2012). Recycling the actinides, the cornerstone of any sustainable nuclear fuel cycles. *Procedia Chem.,* 7, 349.

29.7 IAEATECDOC 1083. (1999). Status of liquid metal cooled fast reactor technology. IAEA-TECDOC-1083, International Atomic Energy Agency, Vienna, Austria.

29.8 Gras, J.-M., Quang, R.D., Masson, H., Lieven, T., Ferry, C., Poinssot, C., Debes, M., Delbecq, J.-M. (2007). Perspectives on the closed fuel cycle – implications for high level waste matrices. *J. Nucl. Mater.,* 362, 383–394.

29.9 NEA. (2006). Physics and safety of transmutation systems: A status report, NEA Report No. 6090. OECD, Paris, France.

29.10 Wigeland, R.A., Bauer, T.H., Morris, E.E. (2007). Status report on fast reactor recycle and impact on geologic disposal, ANL-AFCI-184. Office of Scientific & Technical information, US DOE.

29.11 Bernard, H. (2013). Challenges for Pu recycling in the future: French vision, Pu futures—The science. Cambridge. Los Alamos National Laboratory, USA, 2(2), August 2013.

29.12 Kittel, J.H., Frost, B.R.T., Mustelier, J.P., Bagley, K.Q., Crittenden, G.C., Van Dievoet, J. (1993). History of fast reactor fuel development. *J. Nucl. Mater.,* 204, 1.

29.13 Poplavsky, V.M., Zabudko, L.M., Shkabura, A., Skupov, M.V., Bychkov, A.V., Kisly, V.A., Kryukov, F.N., Vasiliev, B.A. (2012). *Proceedings of International Conference Fast Reactors and Related Fuel Cycles: Challenges and Opportunities FR09,* IAEA, Vienna, Austria, p. 261.

29.14 Nagarajan, K., Vaidya, V.N., Aparicio, M., Jitianu, A., Klein, L.C. (Eds.) (2012). Sol-gel processes for nuclear fuel fabrication. *Sol-Gel Processing for Conventional and Alternative Energy.* Springer Ltd., Chapter 16, 341–373.

29.15 Ashok Kumar, S.L., Radhakrishna, J., Rajesh Kumar, N., Pai, V., Dehadrai, J.V., Deb, A.C., Mukerjee, S.K. (2013). Studies on preparation of (U0.47, Pu00..53) O_2 microspheres by internal gelation process. *J. Nucl. Mater.,* 434, 162.

29.16 Carrott, M.J., Cook, P.M.A., Fox, O.D., Maher, C.J., Schroeder, S.L.M. (2012). The chemistry of (U, Pu) O_2 dissolution in nitric acid. *Procedia Chemistry,* 7, 92.

29.17 Ikeuchi, H., Shibata Asano, Y., Koizumi, T. (2012). Dissolution behavior of irradiated mixed – oxide fuels with different plutonium contents. *Procedia Chemistry,* 7, 77.

29.18 Natarajan, R., Raj, B. (2011). Fast reactor fuel reprocessing technology: successes and challenges. *Energy Procedia,* 7, 414.

29.19 Kofuji, H., Sato, F., Myochin, M., Nakanishi, S., Kormilitsyn, M.V., Ishunin, V.S., Bychkov, A.V. (2007). Mox co-deposition tests at riar for sf reprocessing optimization. *J. Nucl. Sci. Technol.,* 44(3), 349–353.

29.20 Vavilov, S., Kobayashi, T., Myochin, M. (2004). Principle and test experience of the riar's oxide pyro-process. *J. Nucl. Sci. Technol.,* 41(10), 1018–1025.

29.21 Ojovan, M.I., Lee, W.E. (2005). *Introduction to Nuclear Waste Immobilisation*. Elsevier, Amsterdam, the Netherlands.

29.22 Ringwood, A.E., Kesson, S.E., Ware, N.G., Hibeerson, W.O., Major, A. (1979). *Geochem J.*, 13, 141.

29.23 Ringwood, A.E., Kesson, S.E., Reeve, K.D., Levins, D.M., Ramm, E.J. (1988). *Radioactive Waste Forms for the Future*, eds. W. Lutze and R.C. Ewing. North-Holland, Amsterdam, the Netherlands, p. 233.

29.24 Board on Chemical Sciences and Technology, Commission on Physical Sciences, Mathematics, and Applications, National Research Council. (2000). *Electrometallurgical Techniques for DOE Spent Fuel Treatment: Final Report*. The National Academies Press, Washington, DC.

29.25 Janney, D.E. (2003). Host phases for actinides in simulated metallic waste forms. *J. Nucl. Mater.*, 323, 81.

29.26 (2009). Status of minor actinide fuel development. IAEA nuclear energy series document NF-T-4.6. IAEA, Vienna, Austria.

Section VII

Decommissioning Aspects

Section VII

30 Decommissioning Aspects

30.1 INTRODUCTION

Since the early 1960s, several countries have undertaken important fast breeder reactor (FBR) development programs. Test reactors were constructed and successfully operated in a number of countries, which includes Rapsodie (France), KNK-II (Germany), FBTR (India), JOYO (Japan), DFR (the United Kingdom), BR-10, BOR-60 (Russian Federation), and EBR-II, Fermi, and FFTF (the United States). This was followed by commercial-size prototypes (Phénix and Superphénix in France, SNR-300 in Germany, Monju in Japan, PFR in the United Kingdom, BN-350 in Kazakhstan, and BN-600 in Russian Federation). However, from the 1980s onward, mostly for economic and political reasons, fast reactor development in general began to decline. By 1994, in the United States, the Clinch River Breeder Reactor (CRBR) had been cancelled, and the two fast reactor test facilities, FFTF and EBR-II, had been shut down. Thus, in the United States, efforts essentially disappeared for FBR development. Similarly, programs in other nations were terminated or substantially reduced. In France, Superphénix was shut down at the end of 1998; SNR-300 in Germany was completed but not taken into operation, and KNK-II was permanently shut down in 1991 (after 17 years of operation) and is now fully dismantled. Apart from this, PFR was shut down in 1994, BN-350 was shut down in 1998, and Phénix was shut down in 2009. One major consequence of the slowdown in fast reactor technology development programs is that considerable knowledge and experience is currently accumulated in the field of decommissioning of reactors and other sodium facilities.

The term decommissioning generally covers all the technical and administrative activities performed after shutdown of a nuclear installation in order to achieve a predetermined final status. These activities may in particular include equipment disassembly, clean out of premises and soils, demolition of civil engineering structures, processing, packaging, removal and disposal of radioactive and other waste. As many nuclear installations were built between the 1950s and the 1980s, a large number of them were gradually shut down and then decommissioned, particularly over the next 15 years. In 2008, about 30 nuclear installations of all types (electricity generating or research reactors, laboratories, fuel reprocessing plants, waste treatment facilities, etc.) were shut down or were undergoing decommissioning in France. The safety and radiation protection of the decommissioning of these installations, therefore, gradually became major issues. With the specific aspects of decommissioning activities (changing nature of the risks, rapid changes in the installation status, duration of the operations, etc.) ruling out implementation of all the regulatory principles that were relevant during the installation operating period, the nuclear installation decommissioning regulations have evolved gradually since the 1990s. The decommissioning of fast reactors in particular had been so far accomplished in very safe manner without any major events. This is mainly due to judicious adaptation of processes and procedures implemented during the reactor operation phase and the development of safe sodium waste treatment processes. However, on the path to achieving total dismantling, challenges remain with regard to the decommissioning of components after sodium draining.

In this chapter, major differences between the decommissioning aspects of sodium-cooled fast reactor (SFR) and pressurized water reactor (PWR), major activities and principles involved in decommissioning of SFR, decommissioning strategy and issues related to sodium and reactor designs and review and analysis of the international experience gained from the decommissioning of both active sodium loops and SFRs are brought out.

30.2 MAJOR DIFFERENCE BETWEEN DECOMMISSIONING ASPECTS OF SFR AND PWR

The difference of coolant between PWR and SFR is one of the major parameters that bring the differences between these two concepts. In PWR, water is not considered a hazardous chemical product. It can be drained, transferred, stored, and purified with no major chemical hazards, except its own contamination. On the other hand, sodium is considered in itself a hazardous chemical product and will have to be handled with caution as long as it exists in its metallic form. Therefore, the technical decommissioning operations are more complex as far as there exists still metallic sodium inside the plant. Moreover, sodium requires a specific technology and know-how to handle it. Technological aspects that need to be additionally addressed for its safe management are concepts as compared to water coolant: nitrogen or argon cover gas, preheating piping and vessels, smoke and fire detectors, sodium leak detectors, specific in-service inspection of component, etc. The safe conversion of metallic sodium to a stable chemical product for final depository will induce several specific processes (involving several treatment facilities). Advanced sodium treatment processes are today under industrial qualification and development. The feedback from these developments will be useful for future SFR decommissioning programs and only a minor R&D adjustment may be necessary. Intuitively, it can be thought that decommissioning an SFR will be more expensive than a PWR because of the additional steps of sodium waste treatment and handling. Moreover, almost 50% of the nuclear electricity produced in the world is coming from PWRs. Therefore, the reproduction of several matured PWRs for the commercialization will allow a significant gain in experimental feedback, improvement, common decommissioning strategy, and well-trained specialized industry. Currently, for SFRs there are not even two similar design reactors in the world. In Europe, the first SFR standard design should have been the European fast reactor (EFR) project, but it had been achieved and abandoned without realization in 1998.

30.3 MAJOR ACTIVITIES AND CHALLENGES INVOLVED IN DECOMMISSIONING OF SFR

The step by step activities presented in this section are mainly adopted from the methodologies that were proposed for Rapsodie and Superphénix reactors in France. The major steps are shown in Figure 30.1 [30.1]. The first activity is unloading of steel assemblies followed by handling and dismantling of all removable components. This operation includes washing to eliminate the residual sodium and removal of measuring instruments and control rod mechanism from the core and package in dedicated containers. Next operation is draining of the primary sodium into a separate tank and purification to remove Cs-137. The reactor assembly design elements of SFRs that facilitate ease of their decommissioning includes reactor vessel, plugs, fuel handling mechanisms, and other components within the main vessel. The purification campaign consists of passing the liquid sodium through cesium traps. With the removal of sodium and other replaceable structures supported on the top shield, the permanent components of reactor assembly are isolated. The remaining primary sodium is then removed from the primary circuits by in situ treatment with ethylcarbitol. The circuits are decontaminated in three steps: the labile cesium is first removed by alkaline washing, followed by acid decontamination with Ce(IV) (removing 10% of the fixed contamination), and a final phosphatation step. The estimated contamination level was reduced from 5500 Bq/cm^2 to <10 Bq/cm^2. The primary sodium is destructed through the injection of small quantities of liquid sodium into a strong flow of aqueous sodium hydroxide. The dismantling activities can proceed without constraints and the occupational dose could be limited to an acceptable level. In spite of the best effort, experience indicates that there could be some residual sodium in the vessel releasing significant activity. Hence, the reactor assembly is inerted with nitrogen and completely sealed by a cover.

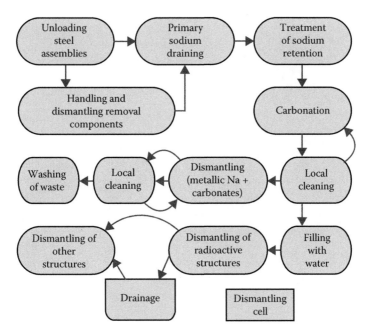

FIGURE 30.1 Major steps in decommissioning of SFR. (From Joulia, E., Superphenix—Creys Malville strategy for dismantling the reactor block: Decommissioning of fast reactors after sodium draining, IAEA-TECDOC-1633, IAEA, Vienna, Austria, 2009, pp. 143–148.)

Radiological inventory of components and coolant must be assessed in advance. Figure 30.2a and b shows the assessment of specific activity distribution in the core of Rapsodie and at various locations for Superphénix, respectively [30.2,30.3]. The general trend would be to define metallic material with low activation under neutron flux and small contamination. Special attention must also be paid on coatings and fuel design. Simplification of means and maintenance facilities must be carefully prepared in advance. Considerable saving is possible in this aspect. Related to the treatment of large-size removable primary components (sodium pumps and IHX), cleaning pits must be designed for plant lifetime including decommissioning estimated time (about 60 years: 40 years operation + 20 years decommissioning). Cleaning pits must be fitted to allow hard decontamination process. Withdrawing components is considered today the reference strategy. But the overall treatment of primary components during the vessel treatment must be technically and economically evaluated. For removable elements of small primary components, similar guidelines could be applicable. Due to low activity, this operation for the treatment of removable secondary components could be done in air, if a specific area is defined in the early design of SFR. The use of existing cleaning pits could be a risk of contamination as it increases activity from very low level wastes to low level wastes.

Design must be done in such a way as to allow the final unloading of nuclear material (fuel and breeder elements) without dummy core (waste minimization). A high defueling rate is required. In normal situation, an unloading factor reaching three assemblies per 24 h should be achieved. The route for failed assemblies must be defined in advance. Cleaning process should minimize the liquid waste effluents and gas consumption and a standardized and efficient process that could treat all assemblies would be a great benefit. Design must be done to allow fully drained control rods with B_4C pellets inside. It would be useful, if the same process as for fuel and breeder elements should be applied. Regarding unloading of remaining assemblies (steel and reflectors), design should be such that they are handled with existing handling devices used during normal operation. If not, designers must device appropriate remote operations to withdraw these assemblies. Such solution must be economically and technically compared with continuing the sodium storage in the reactor vessel.

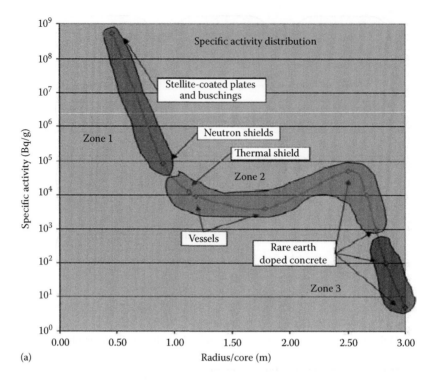

(a)

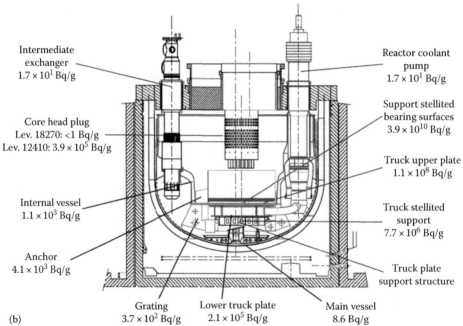

(b)

FIGURE 30.2 (a) Specific activity distribution in reactor core. (From Roger, J. et al., Transformation of sodium from the Rapsodie fast breeder reactor into sodium hydroxide, *Proceedings of the International Conference on European Community on Decommissioning of Nuclear Installations*, Luxembourg, 1994.) (b) Specific activity at various locations in SPX reactor block. (From Marmonier, P. and Del Negro, R., Information about the accident occurred near Rapsodie (March 31st, 1994), *Proceedings of the Technical Meeting on Sodium Removal and Disposal from LMFRs in Normal Operation in the Framework of Decommissioning*, Aix-en-Provence, France, 1997, pp. 136–137.)

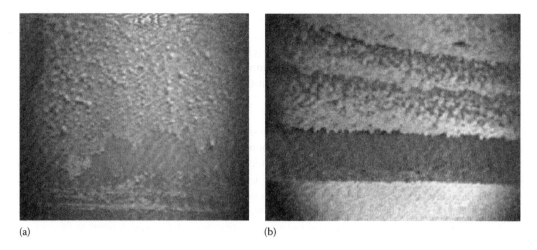

(a)   (b)

FIGURE 30.3 Sodium film/aerosol on metal surfaces: (a) sodium film on vessel wall and (b) aerosols under rotating plug. (From Goubot, J.M. et al., *Decommissioning of the Rapsodie Fast Reactor: Developing a Strategy*, IAEA-TECDOC-1633, IAEA, Vienna, Austria, 2009, pp. 149–160.)

Careful check must be done in early design allowing efficient draining of all pipes and components. Treatment of residual primary sodium remained inside the vessel must be done as thorough as possible. Three methods are generally used for the sodium residues treatment: The first one is the WVN process (without draining of the caustic soda) developed in 1994 for the PFR and DFR; second one is adopted for the FFTF, where the reactor vessel was cleaned in 7 h with super-heated steam; and third one is sodium residues treatment by CO_2–passivation method, adopted for Phénix. These processes have to industrially demonstrate their ability to safely treat sodium in specific cases such as narrow gaps. Improvements in primary vessel design must be done to reduce traces/ films of sodium (Figure 30.3a,b) [30.4], allow better draining, allow better preliminary sodium treatment, minimize wastes, and facilitate future decommissioning operations. To facilitate smooth handling of residual sodium in secondary loops and their decommissioning, secondary loops must be designed with a good final drainage. In that case, the preliminary treatment before decommissioning might be reconsidered. For the steam generator treatment and decommissioning, the strategy followed for the treatment of residual sodium of secondary loops and their decommissioning can be adopted.

Primary vessel dismantling is the most complicated operation. Experience indicated that the underwater decommissioning is generating large volume of liquid effluents, however, unavoidable to respect dose-rate limitation. Therefore, solutions to dispose the wasted water must be found. It should be emphasized that the planning must be optimized. Treatment of exotic sodium components is one of the major concerns still under development for many SFRs. Cold traps should be relooked to allow their easy decommissioning without minimizing their efficiency in operation.

30.4 TECHNOLOGICAL STRATEGIES

IAEA has defined three strategies for decommissioning nuclear installations following their final shutdown [30.4]:

- *Deferred decommissioning*: The parts of the installation containing radioactive materials are maintained or placed in a safe state for several decades before actual decommissioning operations begin (the *conventional* parts of the installation can be decommissioned as soon as the installation is shut down).

- *Safe containment*: The parts of the installation containing radioactive materials are placed in a reinforced containment structure for a period that is long enough to reach a radiological activity level sufficiently low to allow the release of the site.
- *Immediate decommissioning*: Decommissioning is started as soon as the installation is shut down, with no waiting period, although these decommissioning operations can be spread out over a long period of time.

The decision to opt for one or other of the decommissioning strategies is influenced by a large number of factors: national regulations, social and economic factors, financing of the operations, availability of waste disposal routes, decommissioning techniques and qualified personnel, exposure of the personnel and the public to ionizing radiations as a result of decommissioning operations, etc. International practices, therefore, differ from one country to another.

Specific chemical properties of sodium coolant will impose the major guidelines of decommissioning operations. They are related to sodium-specific treatment and can be defined as follows:

- Radiological inventory of components and coolant.
- Functional simplification of means and maintenance facilities.
- Maintaining sodium in liquid state with means other than primary pumps in operation.
- Secondary loops draining in isolation of primary circuits.
- Nuclear material unloading (fuel and breeder elements).
- Unloading of control rods.
- Unloading of other assemblies (steel and reflectors).
- Treatment of removable primary big components (primary pumps and intermediate heat exchanger).
- Treatment of removable primary small components.
- Treatment of removable secondary component.
- Primary sodium treatment.
- Secondary sodium treatment.
- Complementary primary sodium draining.
- Treatment of residual primary sodium remaining inside the vessel.
- Treatment of residual sodium of secondary loops and their decommissioning.
- Steam generator treatment and decommissioning.
- Primary vessel decommissioning.
- Treatment of other Na components: cold traps, cesium traps, bubblers, etc.

The strategy of maintaining the sodium liquid with the means other than primary pumps needs to be economically compared with the option of transferring the sodium to tanks. Leaving the sodium in the primary vessel itself can save money, if tanks already exist. But the primary sodium treatment remains on the critical path. By draining sodium, new wastes are generated (the tanks themselves); however, gains could be realized by optimizing and planning (e.g., sodium treatment and primary vessel decommissioning done in parallel). However, leaving the sodium inside the primary vessel is the preferred option for pool-type reactors. If it is the reference option, the facility to keep sodium in molten state during defueling must be defined at early design stage. For achieving the secondary loops draining, secondary loops and components must be designed to allow good and complete draining. Hot gas sweeping can be a way. Definitive secondary component draining systems (may be by destructive operations) can be thought at the early design stage (pumps, valves, electromagnetic pumps, etc.).

Further, addressing the large quantity of liquid effluents that are generated during the treatment of primary sodium is a major concern because the corresponding wastes are of high volume, consequently, high cost. An innovative strategy for volume reduction and waste minimization would be very valuable. For the secondary sodium treatment, similar treatment to that followed for primary sodium could be adopted or an economic alternative could be thought off. Draining of residual primary sodium in the vessel remains a costly process. It is mainly due to a lack of decommissioning integration early in design. Major progress has to be done to suppress this operation or reduce it to its minimum.

30.5 EXPERIENCE AND FEEDBACK FROM REACTOR DECOMMISSIONING

In comparison with some other GEN IV reactor concepts, SFR concept has the main advantage that prototypes of this system have been built and successfully operated in many countries. Some of these prototypes have been shut down and are presently being decommissioned. The importance of decommissioning this type of reactor cannot be underestimated, in support of the SFR as a viable reactor concept for the future, for the following reasons:

- To provide guidelines on how future reactors should be designed to optimize the performances in terms of operational behavior, inspection, maintenance, and repair.
- To provide information on how future reactors should be designed to reduce decommissioning costs.
- To inform about the performance of construction's materials that have been exposed to sodium, radiations, high temperatures, and temperatures cycling.
- To develop new technologies that will facilitate the implementation of improved operational characteristics of future SFRs.

Thus, even if decommissioning has to be considered as a long-term study, significant cost savings in the decommissioning of future systems can be identified from this actual experimental feedback. Innovations, conceived from the decommissioning experiences in the fields of decontamination processes, robotics, and waste management, etc., will support the successful operation, maintenance, and inspections of future SFRs.

An overview of all the SFRs under decommissioning shows several promising strategies as well as technical issues. The solutions are strongly linked to conditions inherent to the reactor design and to national context (legislation and long-term strategy of respective nuclear industry or countries). Therefore, there was no real harmonization between the different national SFR designs. This lack of harmonization implies that several technical decommissioning solutions are specific to one reactor. As examples, Dounreay fast reactor is the only SFR operating with eutectic NaK coolant. This specific coolant is modifying the decommissioning strategy. Rapsodie, KNK-II, and Superphénix had all decided to unload their assemblies without dummy core. PFR defueling was done with a dummy core. After defueling, primary sodium was drained in one big vessel storage for Rapsodie and in multiples of 200 L drums for KNK-II. Several innovative gadgetries with specific application for the decommissioning of PFR and KNK-II were developed. Figures 30.4 through 30.6 show the sodium tank dismantling machine [30.5], main waste holding module, and chips vacuum system [30.6], respectively, deployed during the machining operations as part of overall decommissioning process. For PFR and SPX, the strategy was to keep primary sodium molten in the reactor vessel until its final treatment. The issue of the primary sodium treatment is made with the NOAH process for Rapsodie, PFR, and SPX (planned for Phénix). For EBR-II, Fermi, FFTF, and BN-350 another process has been chosen. Nevertheless, it is possible to define a general trend on decommissioning SFRs strategy. This trend is moreover, more precise, if the reactors PFR, SPX, and Phénix future project are compared. They have some similarities in their respective design, and a common approach for sodium treatment had been chosen with the NOAH process [30.7].

The experience and feedback from the decommissioning of four SFRs (Fermi, EBR-II, KNK-II, and JOYO) were presented in Ref. 30.8. Sodium residue cleanup for Fermi was accomplished with the help of WVN method for processing NaK. Overall, the steam-processing experience has been positive. However, lessons learned during sodium processing (e.g., difficult access to work areas, confined work spaces, etc.) will prove valuable for future SFR designs and decommissioning plans. EBR-II

FIGURE 30.4 Sodium tank being dismantled after sodium-residue removal. (From Crippe, M., Selection of process parameters for sodium removal via the water vapour nitrogen process, HEDL TME 77-62 UC-79, 1977.)

Pneumatic connecting plate

Pneumatic cylinder

Electric connecting plug

Electromagnets

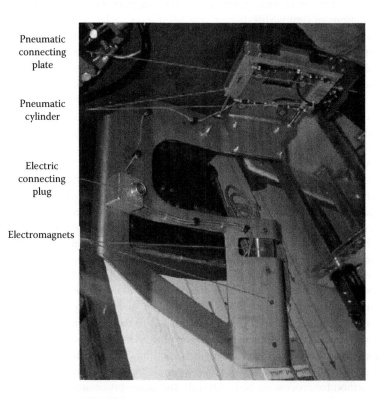

FIGURE 30.5 Main waste-holding module. (From Crippe, M., Selection of process parameters for sodium removal via the water vapour nitrogen process, HEDL TME 77-62 UC-79, 1977.)

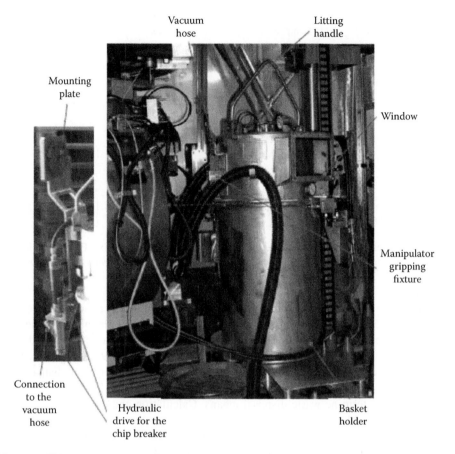

FIGURE 30.6 Chip vacuum system. (From Crippe, M., Selection of process parameters for sodium removal via the water vapour nitrogen process, HEDL TME 77-62 UC-79, 1977.)

was shut down in 1994. After processing of the bulk of the sodium drained from primary and secondary loops, residual sodium was transformed into sodium bicarbonate with the help of moist carbon dioxide (a technique tested at ANL). Remaining sodium treatment with moist carbon dioxide followed by steam and water flush is governed by the Resource Conservation and Recovery Act (RCRA) in the United States. The RCRA mandate issued in 2002 requests removal of all hazardous materials within 10 years. However, the moist carbon dioxide followed by steam and water flush methods is becoming less and less effective, and detailed studies for alternative treatment processes are underway. The KNK-II dismantling works for major reactor components (vessel and rotating plugs), as well as sodium residual cleaning processes at each step of the KNK-II decommissioning, are well documented [30.6]. There were two major lessons learned during these works: first, the recognition of the impact of the underestimation of material embrittlement on the primary vessel dismantling process, and, second, the importance of residual sodium on the outer surfaces of the dismantled pieces and in the various gaps. As part of the Mk-III upgrade program of JOYO experimental fast reactor (implemented to enhance its irradiation capabilities), a 40% power increase was necessary. This entailed the need to replace the intermediate as well as the dump heat exchangers. During replacement works, fuel and liquid sodium were kept in the reactor vessel. Accordingly, impurity ingress had to be prevented and workers' radiation exposure limited. Many important lessons were learned from the replacement work of large components in sodium under difficult conditions like cramped work spaces, closeness to primary sodium boundary and to fuel assemblies with consequent high radiation dose rates, and radioactive sodium treatment. Successful completion of these works and the lessons learned provide valuable insights for future fast reactor designs, construction, operation, and decommissioning.

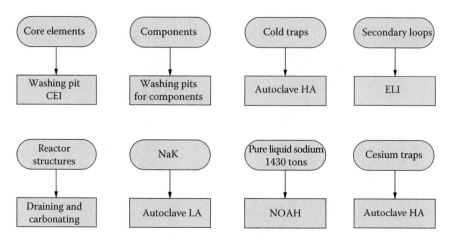

FIGURE 30.7 Process of the treatment of sodium and sodium waste in Phénix. (From Soldaini, M. et al., Phenix plant decommissioning project, *Proceedings of the ANS Topical Meeting on Decommissioning, Decontamination and Reutilization*, Chattanooga, TN, 2007.)

A comprehensive overview of sodium waste characterization and treatment in decommissioning works for BN-350 and Superphénix as well as decommissioning planning for Phénix is presented in [30.8]. The overview indicates the process of treatment of all sodium and sodium wastes generated from Phénix and is schematically shown in Figure 30.7 [30.9]. The overview includes radiological characterization, sodium removal technologies, and reactor block dismantling strategy. Radiological characterization of activation and contamination is needed for all decommissioning projects to determine waste disposal means. Calculation and measurements are needed to identify all isotopes. Minor incidents linked to temporary sodium storage have occurred in the past; therefore, needs for the temporary storage of sodium must be planned ahead of time. Based on the results of experiments on the behavior of sodium residues in air, it can be concluded that their storage in air can be performed safely for a limited time period, provided dry air is being used. Safe storage has been demonstrated for components containing sodium residues, that is, safe storage rules are defined for such components, and the use of vinyl/plastic minimized. For BN-300, safety considerations lead to bicarbonization as the method chosen for the treatment of sodium residues. For Superphénix cold traps, based on testing as well as on safety and efficiency considerations, the hot WVN method was selected as the preferred option for the treatment of sodium residues. Superphénix decommissioning plans foresaw the removal of all removable components from the reactor first. The most radioactive components will be removed underwater to significantly reduce dose rates. In the next step, carbonation [30.10,30.11] will be performed, followed by water fill. Some sodium is expected to remain after carbonation and react during the water fill. Different evaluations lead to different conclusions on the ability of the carbonation process to completely remove sodium residues. The most important factors determining the location of the sodium bulk draining point are linked to the particular design features of the reactor and its circuits. In turn, the residual sodium treatment process to be put in place is mainly determined by how judiciously the right bulk sodium drainage method was chosen and how successfully it was implemented. In Rapsodie, the amount of residual sodium in the primary vessel is estimated to be 100 kg of sodium aerosols and oxides in the gas blanket zone and 80 kg of metallic sodium, including approximately 20 kg on surfaces of structures and 60 kg in retentions. This was later confirmed by borescope examination of the primary vessel (Figure 30.8) [30.12]. The dose rate was measured in October 1989 and the maximum values (approximately 2×10^3 Gy/h) were found near the diagrids. The maximum activation (up to 10^9 Bq/g) is found on the satellite-coated bushings located just beneath the core. These measurements played an important role in proceeding further with the decommissioning plan with suitable modifications.

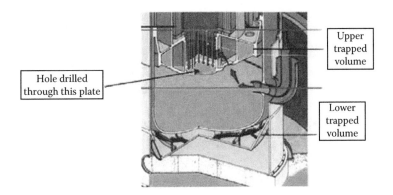

FIGURE 30.8   Trapped (nondrainable) sodium in Rapsodie. (From Farabee, O.A. and Church, W.R., Closure of the fast flux test facility: History, status, future plans, Technical Report DOE-0322-FP-Rev 0, 2006.)

A strong requirement to future fast reactor designs is the incorporation of design measures facilitating sodium drainage and cleaning operations. Following are the major feedbacks derived from the experience:

- Keep primary sodium inside the reactor vessel until its final online treatment. This choice is justified because primary vessel answers to all safety requirements to store active sodium. And no other tank is required to store this large amount of sodium. The design and construction of additional storage tanks would have implied costs and more metallic low-level radioactive wastes.
- Keep sodium in liquid state with additional heating devices adapted on the primary vessel.
- First operation consists of unloading the fuel and fertile elements plus control rods to remove radioactive material risk. In case of Phénix and SPX, steel assemblies are also withdrawn. In parallel, secondary loops are drained and isolated from primary circuits. They can be decommissioned early.
- Draining provisions or access to the vessel bottom for the installation of final bulk sodium draining with better access from bottom and external sides.
- Fuel-handling machines with better access to remote areas of the tank.
- Leak-tight safety vessel.
- Materials selection to minimize potential for high-level activation products (avoiding nickel-based materials) and high corrosion resistance.
- Build shielding with removable blocks for ease of removal and do not include hard-to-dispose-of materials in these blocks.
- Access for cleaning and drainage in baffle-plate area or other structures at the top of the reactor and simpler geometry to prevent pooling of sodium. If possible, it is preferable to eliminate undrainable areas.
- Improve defueling capability and speed.
- Improve ability to remove high-activity components.
- Ability to remotely disassemble and remove components.
- Modular rotating plug to ease removal and disassembly.
- Improve maintainability of reactor block internals, which will also improve the ability to disassembly (Figure 30.9) [30.4].
- Ability to heat the reactor block to keep sodium in liquid state during fill, draining, and subsequent decommissioning activities.
- Provision of instrumentation and inspection channels for in-service inspection and for use during ultimate decommissioning.
- Accessible coupons for use as activation samples for radiological characterization as well as for machine-tool qualification placed throughout reactor block.

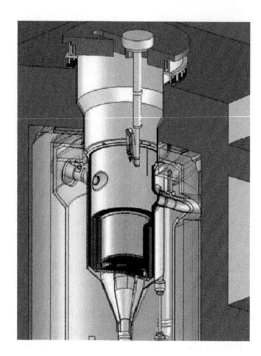

FIGURE 30.9 Improved dismantling procedure for reactor block. (From Goubot, J.M. et al., *Decommissioning of the Rapsodie Fast Reactor: Developing a Strategy*, IAEA-TECDOC-1633, IAEA, Vienna, Austria, 2009, pp. 149–160.)

- Maintain primary sodium purification to minimize plugging temperature during decommissioning.
- Design considerations to ease graphite removal or replacement of graphite as moderator.
- High priority to design considerations for ease of removal of highly activated components.
- Capture on video the whole construction.
- Keep samples of all construction materials.
- Produce and keep detailed as-built drawings with indication of materials, weights, dimensions, surface areas, etc.
- Design simple drainage methods for all components (e.g., channels large enough for flow and no captive volumes); if not possible, leave access for postoperation penetration of components.
- Define general accessibility requirements.
- Recognize the need for compromise between design goals and decommissioning goals.
- Innovative core designs that eliminate intermediate heat exchangers (direct heat transfer from primary sodium to steam generator).
- Innovative valve designs without bellows.
- Innovative sealing systems that eliminate cleaning and residues removal problems.
- R&D to develop the diaphragm valve (temperature- and material-related problems).
- Innovative core designs that eliminate pumps in the primary and/or secondary loop (natural convection core designs).
- Innovative cold-trap designs with removable, replaceable, and washable cartridge.
- Avoid NaK systems (additional hazards outweigh the benefits).
- Accessible location, minimum bends, no low-point traps, and minimize imbeds.
- Consider once-through forced flow designs based on the experience.

Typical comparison on the estimated decommissioning costs of several SFRs is shown in Figure 30.10 [30.13]. In France, the French electricity supplier, EDF, is providing funds for the future decommissioning

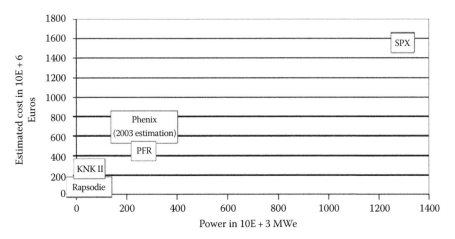

Legend: KNK-I: German experimental sodium fast reactor.
Rapsodie: French experimental sodium fast reactor, Phenix: French prototype sodium fast reactor.
PF: British prototype sodium fast reactor.
SPX: French demonstration sodium fast reactor.

Note: For KNK-II and Rapsodie it has been assumed a potential electrical yield of 0.4 because these reactors were producing only thermal power.

FIGURE 30.10 Comparison of the estimated decommissioning cost of several SFRs. KNK-II, German experimental sodium fast reactor; Rapsodie, French experimental sodium fast reactor; Phénix, French prototype sodium fast reactor; PFR, British prototype sodium fast reactor; SPX, French demonstration sodium fast reactor. *Note*: For KNK-II and Rapsodie it has been assumed a potential electric yield of 0.4 because these reactors were producing only thermal power. (From Sauvage, J.F., Preparing for the dismantling of the Phénix plant, *Proceedings of the Seventh WANO FBR Meeting*, Beloyarsk NPP, Russian Federation, 2004.)

of the current reactors as well as for future wastes and used-fuel management disposal. This fund is defined as a tax included in the electricity price. Decommissioning price is estimated up to 10%–15% the reactor capital cost. Fuel management from mining to waste disposal, reprocessing, and transport is estimated to around 20% of the total capital cost. These values are based on estimations and are mainly settled for commercialized PWRs. Therefore, the objective of GEN IV reactor is to be economically competitive to PWR even in the decommissioning costs. Indeed, among the technology goals GEN IV reactors have to fulfill, the last two have direct or indirect links with decommissioning tasks.

30.6 SUMMARY

The decommissioning of fast reactors to reach safe enclosure presented no major difficulties, and this had been accomplished mainly through judicious adaptation of processes and procedures implemented during the reactor operation phase and the development of safe sodium waste treatment processes. However, on the path to achieving total dismantling, challenges remain with regard to the decommissioning of components after sodium draining. Hence, there is a need for in-depth scientific and technical exchanges on this topic among international specialists.

REFERENCES

30.1 Joulia, E. (2009). Superphenix—Creys Malville strategy for dismantling the reactor block: Decommissioning of fast reactors after sodium draining. IAEA-TECDOC-1633. IAEA, Vienna, Austria, pp. 143–148.
30.2 Roger, J., Latge, C., Rodriguez, G. (1994). Transformation of sodium from the Rapsodie fast breeder reactor into sodium hydroxide. *Proceedings of the International Conference on European Community on Decommissioning of Nuclear Installations*, Luxembourg.

30.3   Marmonier, P., Del Negro, R. (1997). Information about the accident occurred near Rapsodie (March 31st, 1994). *Proceedings of the Technical Meeting on Sodium Removal and Disposal from LMFRs in Normal Operation in the Framework of Decommissioning*, Aix-en-Provence, France, pp. 136–137.

30.4   Goubot, J.M., Fontaine, J., Berson, X., Soucille, M. (2009). Decommissioning of the Rapsodie fast reactor: Developing a strategy. IAEA-TECDOC-1633. IAEA, Vienna, Austria, pp. 149–160.

30.5   Crippe, M. (1977). Selection of process parameters for sodium removal via the water vapour nitrogen process. HEDL TME 77-62 UC-79.

30.6   Hillebrand, I., Brockmann, K., Minges, J. (2005). Decommissioning KNK—First steps in dismantling of the high activated reactor vessel. *Proceedings of the Eighth WANO-Meeting*, Karlsruhe, Germany.

30.7   Mason, L., Rodriguez, G. (2002). The disposal of bulk and residual alkali metal. *Proceedings of the Spectrum 2002*, Reno, NV.

30.8   IAEA-TECDOC-1633. (2009). Decommissioning of fast reactors, after sodium draining. IAEA, Vienna, Austria.

30.9   Soldaini, M., Deluge, M., Rodriguez, G. (2007). Phenix plant decommissioning project. *Proceedings of the ANS Topical Meeting on Decommissioning, Decontamination and Reutilization*, Chattanooga, TN.

30.10 Sherman, S.R. (2005). Technical information on the carbonation of the EBR-II reactor. Technical Report INL/EXT-05-00280 Rev 0.

30.11 Rodriguez, G., Gastaldi, O. (2001). Sodium carbonation process development in a view of treatment of the primary circuit of Liquid Metal Fast Reactor (LMFR) in decommissioning phases. *Proceedings of the Eighth International Conference on Radioactive Waste Management and Environmental Remediation*, Bruges (Brugge), Belgium.

30.12 Farabee, O.A., Church, W.R. (2006). Closure of the fast flux test facility: History, status, future plans. Technical Report DOE-0322-FP-Rev 0.

30.13 Sauvage, J.F. (2004). Preparing for the dismantling of the Phénix plant, *Proceedings of the Seventh WANO FBR Meeting*, Beloyarsk NPP, Russian Federation.

BIBLIOGRAPHY

Alphonse, P. (2005). Superphenix decommissioning: Technical present status. *Proceedings of the Eighth WANO FBR Group Meeting*, Karlsruhe, Germany.

Autorités de Sûreté Nucléaire Française (ASN). (2004). Pertinence de la stratégie de démantèlement des réacteurs de première génération d'EDF, Letter Sent by French Safety Authority to EDF, http://www.asn. gouv.fr/domaines/demantel/UTO003.pdf.

Bertel, E., Lazo, T. (2003). Decommissioning policies, strategies and costs: An international review, facts and opinions. *NEA News Autumn*, No. 21.2.

Demoisy, Y., Thomine, G., Rodriguez, G. (2001). ATENA—Project of storage and disposal plant for radioactive sodium wastes. *Proceedings of the Eighth International Conference on Radioactive Waste Management and Environmental Remediation (ICEM'01)*, Bruges (Brugge), Belgium.

NEA Report No. 4373 (2003). Electricité nucléaire, Quels sont les coûts externs? (Nuclear Electricity, what are the external costs?), Développement de l'énergie nucléaire, agence pour l'énergie nucléaire organisation de coopération et de développement economiques, Paris, France.

OECD Report. (2003). Decommissioning nuclear power plants: Policies, strategies and costs, Développement de l'énergie nucléaire, agence pour l'énergie nucléaire organisation de coopération et de développement economiques, Paris, France.

Rodriguez, G., Gastaldi, O. (2001). Sodium carbonation process development in a view of treatment of the primary circuit of Liquid Metal Fast Reactor (LMFR) in decommissioning phases. *Proceedings of the Eighth International Conference on Radioactive Waste Management and Environmental Remediation (ICEM'01)*, Bruges (Brugge), Belgium.

Rodriguez, G., Gastaldi, O., Baque, F., Recent sodium technology development for the decommissioning of the Rapsodie and Superphénix reactors and the management of sodium wastes. *Nucl. Technol.*, 150(1), 100–110.

Shimakawa, Y., Kasai, S., Konomura, M., Toda, M. (2002). An innovative concept of a sodium cooled reactor to pursue high economic competitiveness. *Nucl. Technol.*, 140, 1–17.

Section VIII

Domains of High Relevance
to SFR: Typical Examples

31 Material Science and Metallurgy

31.1 INTRODUCTION

The core of a sodium-cooled fast reactor (SFR) comprises ceramic oxide/carbide/metal fuel pellets stacked in thin-walled tubes and loaded into a long hexagonal duct to form the fuel bundle that is immersed in a pool of liquid sodium (Figure 31.1). The heat transport system consists of a primary sodium circuit, secondary sodium circuit, and steam–water system (see Chapter 8 for details).

In an SFR, the core structural materials are at a temperature of 670–820 K and are exposed to a fast neutron fluence of 1021–1022 n/cm². The efforts at the science-based development of core structural materials for SFRs that include the advanced austenitic stainless steels (SS), ferritic steels, and oxide dispersion–strengthened (ODS) steels are focussed to address the issues of the materials under irradiation and high temperature. The materials issues that are crucial to SFRs pertain to void swelling of structural materials under fast neutron irradiation and the degradation of mechanical properties at elevated temperatures (873 K) and under irradiation (discussed in detail in Section 6.2).

This chapter presents the materials science details of core structural material and their compatibility with the coolants, namely, liquid sodium and the fuel pellets. The materials science and metallurgical aspects, including welding, of permanent reactor structural materials and steam generator materials and the issues related to the hardfacing of reactor structural materials are also presented in this chapter.

31.2 CORE STRUCTURAL MATERIALS

Structural materials for fast reactor core components have evolved continuously over the years resulting in substantial improvement in the performance. The first-generation materials belonged to austenitic stainless steel types 304 and 316. These steels quickly reached their limits because of unacceptable swelling at doses higher than about 50 dpa. The term dpa is conventionally used for quantifying the effects of irradiation on structural materials in terms of the average number of displacements that each atom undergoes. Many years of research and development of austenitic SS have managed to extend the low-swelling transient regime. Based on the basic understanding of void swelling phenomena, the strategies for mitigating void swelling in austenitic steels are focused on tailoring the microstructure through the incorporation of dislocations by cold working and incorporation of nanodispersoids in the steel matrix. These act as trapping sites for point defects altering their effective diffusivities and increasing the recombination rate and delay the steady-state void growth. The other strategy is to use ferritic/martensitic (including ODS) steels with improved mechanical properties that inherently exhibit low void swelling (Figure 31.2). Various advanced materials have been developed in different countries for use as clad and wrapper materials for their oxide fuel–based SFR programs (see Tables 7.4 and 7.5 in Chapter 7 for details). High chromium (9–12 wt%) ferritic/martensitic steels are considered the long-term solution for SFR core structural materials because of their inherent void swelling resistance and lower shift in ductile-to-brittle transition temperature (DBTT) on neutron irradiation. While several alloys in this class have excellent swelling resistance to doses even up to 200 dpa, their creep resistance decreases drastically above 823 K, making them unsuitable for clad tube applications. Another concern is the increase in DBTT due to irradiation.

Materials issues in FBR

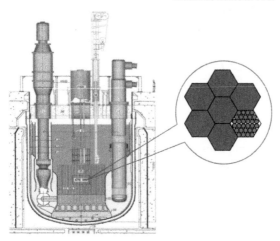

Clad and wrapper

- High dose ~ 10^{22} n/cm
- High temperature : 670–820 K

❖ Void swelling

❖ Mechanical properties
- High-temperature creep
- Ductile-to-brittle transition temperature

❖ Compatibility with Na

❖ Compatibility with fuel
- Fuel–clad mechanical interaction
- Fuel–clad chemical interaction

Materials
- Modified austenitic steels
- Ferritic steels
- Oxide dispersion–strengthened steels

FIGURE 31.1 Schematic of the cross section of prototype fast breeder reactor at Kalpakkam showing the fuel bundle immersed in liquid sodium. The various materials issues that are pertinent to the clad and wrapper subjected to high neutron dose at high temperature.

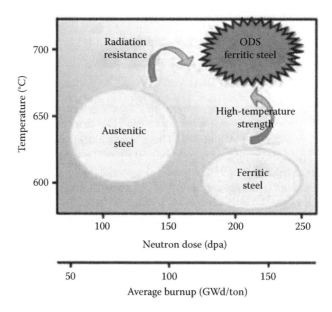

FIGURE 31.2 Usage regimes of high-temperature radiation-resistant steels.

Consequently, extensive studies involving the modification of composition and initial heat treatments have been carried out to improve the fracture toughness of these ferritic–martensitic steels. The presently used 9% Cr ferritic–martensitic steels with low sulfur and phosphorus have the lowest increase in DBTT on irradiation. Efforts have been made to further optimize this composition especially with respect to silicon content. Target burnup levels of up to 200 GWd/ton set for the next generation of SFRs with higher operating temperatures would require the development of ferritic/martensitic ODS steel with adequate creep strength for clad tube application. Synthesis of this alloy, using prealloyed powders of the steel and nanosized yttria particles, necessitates a complex powder metallurgy route followed by hot and cold mechanical processes steps to produce the clad tubes.

31.3 RADIATION-RESISTANT STEELS

From the earlier discussion on the basics of radiation damage, it is seen that while the neutron-induced damage is initiated at the atomic level, the consequent microstructural changes such as the growth of voids, dislocation loops, and precipitates, which eventually influence the physical properties of materials, occur on a range of scales, from clusters of a few atoms to features comparable to the grain size. This inherent multiscale nature of the problem makes it a challenge to uniquely predict the evolution of materials behavior under the nonequilibrium condition of irradiation field. The development of radiation-resistant materials usually takes several iterations of irradiation, testing, and modification before material with optimum properties can be qualified for use in a reactor. The worldwide effort on the development of radiation-resistant materials has followed the approach that encompasses (1) controlled production of defects using ion beams from accelerators, (2) detailed investigation of defects and microstructure following irradiation using an array of experimental techniques such as transmission electron microscopy (TEM) and positron annihilation spectroscopy, and (3) computer simulations.

31.4 ION BEAM SIMULATION

The charged particle irradiation produces damage at an enhanced rate (~10^3 times faster than neutrons from available sources) and hence has proved effective in the screening of materials. In addition to an accelerated damage, ion beam irradiation offers other advantages: (1) control of experimental parameters (temperature, flux, energy, and environment) is far better from those that can be achieved in reactor conditions and (2) ion-irradiated specimens are generally not radioactive, unlike specimens irradiated in reactors that may be highly radioactive and require examination in hot cells. Using heavy-ion irradiation, displacement damage ~100 dpa can be obtained over a short period of few hours. The effect of helium produced by (n, α) reactions can be simulated in accelerator experiments by preimplanting appropriate concentration of helium in the sample or by performing dual beam ion irradiation experiments in which two accelerators are used, one for injecting helium and another for irradiation by heavy-ion beam, which produces displacement damage at very high rate.

As an illustration of ion beam studies, Figure 31.3 shows the results of swelling studies in titanium-modified 14Cr–15Ni steel (alloy D9) that have been developed for use as fuel cladding and wrapper materials in SFRs [31.1]. In alloy D9, TiC forms as fine precipitates through appropriate thermomechanical treatments, which act as sites for the increased recombination of vacancies and interstitials resulting in a suppression of void swelling. It is seen from Figure 31.3 that the magnitude of swelling and the peak swelling temperatures are significantly different for the alloys with a marginal difference in Ti concentration. In order to understand the drastically different behavior of the two alloys with respect to void swelling, detailed investigations on the nucleation and growth of TiC precipitates have been carried out using positron lifetime measurements. These studies along with the combination of macroscopic swelling measurements with experimental techniques examining the defects at atomic scale have proved valuable in screening the alloys for radiation resistance.

31.5 COMPUTER SIMULATION

While rate theory models have been the workhorse for simulating radiation damage, in the last decade, multiscale modeling strategies are being implemented internationally [31.2] to understand materials behavior under irradiation. In such an approach, the simulation begins at the atomistic level with ab initio and molecular dynamics (MD) techniques, moves through the mesoscale using lattice kinetic Monte Carlo (KMC) and dislocation dynamics (DD), and ends with the macroscale using finite element methods and continuum models.

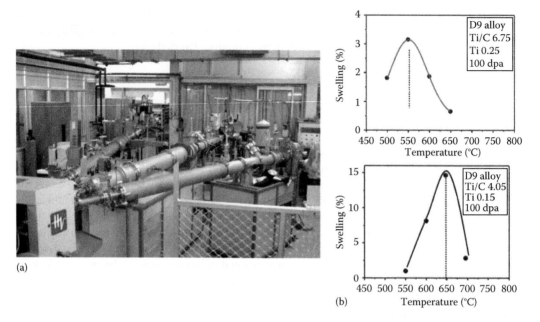

FIGURE 31.3 (a) Photograph of the accelerator facility at IGCAR, Kalpakkam and (b) the results of swelling studies on two different compositions of Ti-modified steels.

The modeling paradigm is illustrated with an example on the investigation of nanoyttria precipitates in ODS alloys that are emerging as promising structural material for advanced nuclear reactors. The ODS alloys derive their strength from the dispersion of oxygen-rich Y–Ti–O nanoclusters in the ferritic matrix [31.3]. The size distribution of a fine dispersion of nanoparticles, which improves the mechanical properties, depends on the alloy composition and processing methodologies. There is a considerable research interest on the structure and composition of nanooxide phases and their thermal and radiation stability.

In these studies, ab initio density functional theory (DFT) calculations are used to obtain information on the energetic of point defects in Fe matrix, such as the formation migration and binding energies of point defects and small defect clusters [31.4]. The cascade structure in Fe formed by the primary knock-on atom is investigated using MD simulations and is used to address the effect of nanoparticles and grain boundaries on the cascade generation. In these MD simulations, the proper interatomic potentials for such calculations are derived from DFT calculations. At the microscale, these cascades lead to a primary state of damage with the production of defects, whose evolution in turn affect the material's macroscopic properties. This microscopic evolution, the interaction, annihilation, or clustering of defects are studied using KMC simulations [31.5]. In the results shown in Figure 31.4, the lattice KMC simulations have been used to study the clustering of Y–Ti–O clusters in Fe matrix, in particular to address the crucial role of Ti in refining the nanoparticle size distribution and comparison with TEM observations. The presence of nanoparticles influences the mechanical properties such as the yield strength and these are investigated using discrete dislocation dynamics simulations. The parameters from the DD simulations form inputs for finite element calculations.

The given examples illustrate as to how the science-based approach comprising controlled ion beam irradiation, in-depth investigation of defects, and computer simulations is being used toward the development of radiation-resistant steels. Enormous progress in the development of radiation-resistant steels has accrued from the tailoring of microstructure, in particular the incorporation of nanodispersoids in steels that can provide good stability to high-dose neutron irradiation and also provide strength at high temperature.

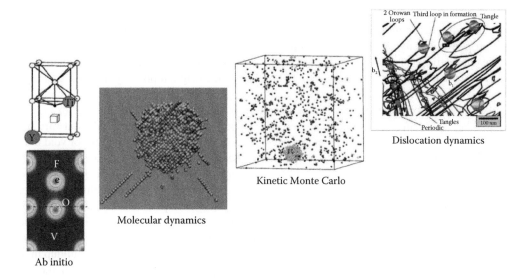

Ab initio

Molecular dynamics

Kinetic Monte Carlo

Dislocation dynamics

FIGURE 31.4 Multiscale modeling of ODS steels encompassing ab initio calculations of point defect energetics, molecular dynamics simulations of cascade structure in Fe, kinetic Monte Carlo simulations of the formation of Y–Ti–O clusters in Fe, and discrete dislocation dynamics simulations of the interaction of dislocations with nanoclusters in Fe.

31.6 COMPATIBILITY OF CLAD MATERIAL WITH COOLANT AND FUEL

When ferritic steels are used in a system with austenitic SS with sodium as a coolant, transport of various elemental species to and from the ferritic steel can occur, depending on the relative activity of the given elements in both matrices. Extensive decarburization up to 70% was observed in EM12 steel in flowing sodium at 923 K. The economy of SFRs requires reprocessing of spent fuel elements to extract the plutonium for recycling. While the rate of dissolution of austenitic SS in nitric acid is insignificant, the ferritic steels dissolve more rapidly in acid, with complete dissolution of EM12 steel specimens occurring in less than 10 h in boiling nitric acid. It has also been found that the rate of dissolution of EM12 steel is more than 300 times higher than that for 316 SS. This is extremely undesirable as it introduces other metallic ions in the solution along with uranium and plutonium compounds, thereby contributing to an increase in the quantity of radioactive solid waste.

In SFRs, the heat generated by nuclear fission in the fuel is conducted through the clad material and is extracted by liquid sodium coolant. Therefore, the compatibility of the clad material with sodium especially the corrosion behavior of austenitic/ferritic steel clad material in liquid sodium is an important consideration. It is known that sodium promotes corrosion in two ways: (1) corrosion produced by dissolution of alloy elements into sodium, and (2) corrosion produced through a chemical reaction with the impurities in sodium, especially the dissolved oxygen. In a system with a thermal gradient, dissolution of alloy elements into sodium can occur continuously as a function of temperature, temperature gradient, and the dissolution and deposition rates of alloy constituents in the sodium circuit. The chemical reaction with the impurities in sodium can be controlled through impurity control techniques. The important factors associated with the sodium environmental effect on the clad material are immersion time, temperature, dissolved oxygen, sodium velocity, and alloy composition. The corrosion behavior of 12% Cr steel is the same as that of conventional austenitic/ferritic steels, and there is no effect of sodium on the mechanical strength. In ODS steel under the high sodium velocity condition and at temperatures above 950 K, weight gain is caused largely by nickel activity gradient mass transfer from the sodium circuit structural materials via the sodium. However, there was no tensile strength reduction due to nickel diffusion, and the strength of the surface-degraded layer produced by nickel diffusion is maintained by the fine yttria particles in the ODS steel [31.6].

Most SFRs utilize mixed uranium–plutonium oxide as fuel. Due to the low thermal conductivity of the oxide fuel, removal of fission heat results in high central temperatures of over 2273 K. The high power density of the fuel, in conjunction with the small radius of the fuel rod and the efficient convective heat removal by the flowing coolant, produces large temperature gradients. The temperature gradient in the fuel causes rapid migration of the original porosity in the as-fabricated fuel radially inward. The mechanism of pore motion is by vapor transport, with the solid on the hot face of a pore in a temperature gradient exerting higher vapor pressure than the solid of the cold face, and the vapors diffuses through the inert gas trapped in the pore. This radially outward flow of matter causes the pore to move in the opposite direction, resulting in a central hole or void in the fuel. Although migration porosity occurs in a short time, the microstructure is not static after the initial restructuring as the mixed oxide is plastic at temperatures above ~1673 K. Hence, the hot central portion of the fuel deforms because of compressive stresses due to fuel–cladding mechanical contact. The outer cooler annulus of the fuel, which is brittle, develops extensive network of radial cracks and is subject to irradiation creep that significantly alters the stress distribution in the fuel, and thereby affects the intensity of mechanical interaction between fuel and cladding. Also, as some of the fission products are gaseous; they either escape from the fuel to cause pressure buildup in the clad or precipitate into bubbles that swell the fuel and bring it into mechanical contact with the clad.

Along with the chemical evolution of the mixed oxide fuel during irradiation, the temperature gradient causes most of the chemical species in the fuel to migrate radially: (1) plutonium moves toward the center of the fuel, thereby degrading heat transport by placing the fissile heat source further from the heat sink (the liquid sodium coolant) than in the as-fabricated fuel and (2) oxygen is transported to the cladding, thereby enhancing corrosion. Some fission products move up the thermal gradient, while others such as caesium, tellurium, and iodine move down the gradient to the cladding where they accelerate corrosion. This attack can produce uniform but shallow corrosion or deep intergranular corrosion.

In metallic fuels, both mechanical and chemical interaction between the clad material and the fuel may result in clad failure. Fuel–clad mechanical interaction (FCMI) arises from applied stress when the element design restrains fuel swelling and may result in the plastic deformation of the clad. The major source of stress in metallic fuel elements is the buildup of fission gas in bubbles and/or existing tears. Internal fission gas pressure builds up rapidly with burnup and metallic fuel is rather plastic during fissioning. To allow sufficient free swelling of the metallic fuel, an as-built smear density of ~75% allows free fuel swelling of ~30% at which point porosity becomes largely interconnected and open to the outside of the fuel. This results in the release of a large fraction of the fission gas to a suitably large plenum at the top of the element. Fuel elements containing U–Pu–Fissium (Fs), U–Pu–Ti, and U–Pu–Zr tested with a variety of clad materials and a range of smear densities and plenum volumes showed that high burnup operation in excess of 10 at% without clad failure is achievable with smear densities of 75% or lower and plenum-to-fuel volume ratio of 0.6 or higher.

The clad strain of low-smear-density small-plenum fuel element is primarily creep strain due to FCMI, while that of high-smear-density small-plenum fuel element is mainly due to irradiation-induced clad swelling. As the presence of open (interconnected) fission gas porosity is the key to reducing FCMI, at high burnup there is an effect of accumulation of low-density solid fission products on this porosity. The accumulation of nongaseous fission products contributes to the net volume change in three ways: (1) volume increase due to nonsoluble fission products, (2) volume decrease due to fission of U and Pu, and (3) volume increase due to increase in Zr, Mo, and Nb, which are soluble in fuel matrix. As the total volume increase due to nonsoluble fission products is ~1.2% per percent of burnup, an as-built smear density of 75% would change to 81% at 10 at% burnup and to 90% at 20 at% burnup. In this high burnup range, the initially open porosity will become increasingly closed off and fission gas pressure in the fuel will build up more rapidly, since less generated gas will vent into the plenum. In effect, this represents the onset of possibly significant FCMI. However, in irradiations with 75% smear-density fuel, FCMI has not been observed up to 18% burnup because the deformation in austenitic stainless steel clad at high burnup is so large that it compensates for any solid fuel volume increase. The deformation in austenitic SS clad is caused mainly by swelling, and any irradiation creep strain can be accounted for by plenum pressure alone; hence, no evidence of FCMI. In nonswelling 12Cr–1Mo–0.3V–0.5W (HT9) clad metallic fuel elements, the clad deformation at 16 at% burnup is only ~1% with very little FCMI as the clad creep strain could be accounted for by plenum pressure only. Only when solid fission products increase the fuel smear density to well above 85%, FCMI becomes clearly evident. Thus, even with nonswelling clad, significant FCMI can be avoided at high burnup, if the as-built fuel smear density is kept at 75%.

Fuel–clad chemical interaction (FCCI) is one of the most crucial considerations for the choice of material for the clad of a metallic fuel element. FCCI is a complex multicomponent diffusion problem, and the characterization of fuel–clad interdiffusion is exceedingly difficult because of the number of alloy components involved. The U and Pu in the fuel and the rare-earth fission products generated during irradiation have a propensity of interacting metallurgically with the clad material at high temperature. During steady-state irradiation, solid-state interdiffusion may occur across the fuel/clad interface. The high temperatures generated due to fission, adjacent to the fuel–clad interface, can cause the melting of interdiffusion layers on the fuel and internal clad surfaces and can lead to additional dissolution of the fuel thereby causing liquid-phase penetration into the clad material. FCCI is affected by composition, temperature, and contact time at the fuel–clad interface and results in reducing the effective thickness of the clad by an amount commonly called *wastage*. The fuel constituents and fission products can penetrate the clad up to 100 μm. FCCI also causes lowering of the solidus temperature. The fuel smear density and the type of clad material have a bearing on FCCI, with fuels of high smear density having lower clad wastage. FCCI is also very strongly influenced by the concentration of lanthanide fission products at the clad interface. The lanthanide pickup in the clad material can be as high as 20 wt%. It has been identified that the reaction is between Fe in the clad and the heavy metals with the clad constituents, especially Fe and Ni, penetrating the fuel matrix significantly leading to depletion of these clad constituents in the zone adjacent to the fuel/clad interface. The U–Ni and U–Fe eutectic compounds formed have fairly low melting points (e.g., 1008 K for UFe_3) and can cause problems in transient over-power incidents. Higher clad wastage occurs with alloy D9 owing to preferential migration of more Ni compared to Fe or Cr with Ni enhancing the diffusion of U and Pu into the clad. The interaction between U and austenitic SS begins at 673 K and at 873 K the growth rate of the interaction layer is about 3 μm/day [31.7]. In the temperature range 1033–1073 K, almost complete dissolution of the austenitic SS in U occurs within 24 h. Use of sodium layer between fuel rod and clad decreases interaction significantly; however, this does not solve the problem.

Out-of-pile (diffusion couple and dilatometry) and in-pile studies have shown that U reacts with Fe in the clad forming low melting eutectics, while addition of Pu to the fuel results in (1) the increase in the rate of attack and (2) the decrease in the temperature at which melting occurred in the diffusion zone. Addition of at least 10% Zr to the fuel results in (1) the increase in the rate of diffusion, that is, attack and (2) the increase in the temperature at which melting occurs at the interface. Melting temperatures with fuels of varying composition and different clad materials are generalized and used for identifying temperature zones for melting (Figure 31.5). Melting occurs at lower temperatures for high-Pu fuel, with the melting temperature being lowest for low-Cr steel.

Addition of Zr to metallic fuel improves chemical compatibility between fuel and austenitic SS clad by suppressing interdiffusion of fuel and clad constituents. A Zr-rich layer has been observed near the surface of U–Pu–Zr fuels, which retards molten phase formation at the fuel/clad interface [31.8]. Formation of the Zr-rich layer is related to the availability of N, the primary source of which appears to be the steel clad in which N is present as impurity element [31.8]. In view of this, 316N SS with 600 ppm N has been tailored for enhanced compatibility than alloy D9 and HT9 with 40–50 ppm N. The formation of reaction layers is influenced by the interdiffusion of the alloy constituents and N through both the reaction layers and the matrix alloy. The relative solubility of N in Zr is higher than that in U with two distinct solid solution phases being formed: U-rich (U, Zr)N

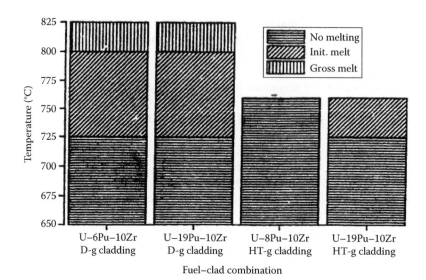

FIGURE 31.5   Schematic of fuel–clad compatibility test results.

and Zr-rich (Zr, U)N. The Zr–N layer acts as an effective barrier to the metal atoms and prevents molten phase formation by reaction of actinides with Fe and Ni at high temperatures. For the development of metallic fuel for BN-350, 30–50 μm thick protective barrier-layer coatings of Mo and Zr with upper admissible temperatures of 1173 and 1023 K, respectively, were chosen [31.9]. Ni plays an important role in fuel–clad interdiffusion, and the presence of Ni increases the diffusion layer thickness. In titanium-modified austenitic stainless steel, Ti stabilizes O and N. Therefore, N is not available for layer formation and is, therefore, more susceptible for FCCI attack. On the other hand, the HT9 steel that contains very little Ni is expected to behave like type 440 SS.

31.7   REACTOR STRUCTURAL MATERIALS

Austenitic stainless steels of type 316 and its modified grades including the closely related variant 316LN are the preferred materials for out-of-core high-temperature nuclear steam supply system components of SFRs (except for the steam generators) due to their adequate high-temperature low-cycle fatigue and long-term creep properties, compatibility with the liquid sodium coolant, ease of fabrication, good weldability, and commercial availability. This austenitic stainless steel used as reactor structural material differs from the conventional grade 316 stainless steel with respect to close control on composition to avoid scatter in mechanical properties. Understanding the microstructural changes, dislocation evolution, and damage mechanisms during long-term creep and creep–fatigue deformation in this material [31.10,31.11] has enabled the development of robust creep life prediction models that can predict lives under service conditions not covered by laboratory testing [31.12].

   In general, austenitic stainless steels have relatively poor resistance to intergranular stress–corrosion cracking (IGSCC) in chloride and caustic environments. The 316 stainless steel welds exposed to marine environments can fail by IGSCC in the heat-affected zone due to the combined influence of sensitization and the presence of residual stresses introduced during welding. A nitrogen-alloyed low-carbon (0.03% maximum) version of this steel, 316LN stainless steel, is now considered as the most preferred material for the high-temperature reactor structural components of SFRs. In 316LN stainless steel, 0.06–0.08 wt% nitrogen is added to compensate for the loss in solid solution strengthening due to the reduced carbon content. This has the effect of substantially increasing the creep rupture life (Figure 31.6). The beneficial effects of nitrogen arise due to the higher solubility of nitrogen in the matrix than the carbon, reduction in stacking fault energy,

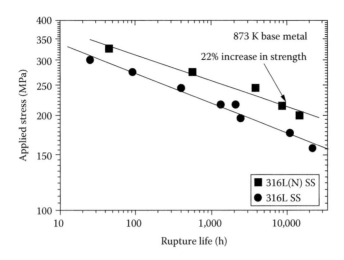

FIGURE 31.6 Influence of nitrogen on creep properties of 316LN stainless steel.

and the introduction of strong elastic distortions into the crystal lattice giving rise to strong solid solution hardening [31.13]. Nitrogen also affects the diffusivity of chromium in austenitic stainless steels leading to retardation in coarsening of M23C6-type carbides, thereby retaining the beneficial effects of carbide precipitation at longer durations [31.14,31.15].

With a view to further increasing the design life of structural components of future SFRs, nitrogen-enhanced 316LN stainless steel has also been developed with superior tensile, creep, and low-cycle fatigue properties compared to 316LN stainless steel containing 0.07% nitrogen [31.16]. Increase in nitrogen content in the range 0.07–0.22 wt% substantially increases the creep rupture strength of 316LN stainless steel (Figure 31.7), and this has been ascribed to the decreasing tendency for subgrain formation leading to a uniform distribution of dislocations in the steel with 0.22% nitrogen. While tensile and creep strength increases with increasing nitrogen content, the low-cycle fatigue life peaks at 0.14% nitrogen content. Further, increasing nitrogen content in 316LN stainless steel has beneficial effects on the resistance to pitting corrosion, sensitization, stress corrosion

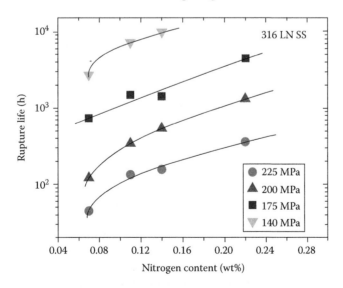

FIGURE 31.7 Influence of nitrogen content on creep properties of 316LN stainless steel at 923 K. (From Mathew, M.D. et al., *Mater. Sci. Eng.*, A535, 76, 2012.)

cracking and corrosion fatigue. However, in aged condition, the 0.22% nitrogen-containing steel has inferior resistance to stress–corrosion cracking and corrosion fatigue compared to the steels with 0.14 and 0.07 wt% nitrogen, respectively.

Studying the hot cracking behavior of the 316LN stainless steels is an important consideration during welding as these austenitic stainless steels, especially the nitrogen-enhanced versions, solidify without any residual delta-ferrite in the prior-austenitic mode. It is now generally accepted that rather than the residual delta-ferrite content, it is the formation of delta-ferrite as the primary phase during solidification that reduces hot cracking susceptibility. As nitrogen is an austenite stabilizer, it decreases the occurrence of primary delta-ferrite solidification, and hence, the susceptibility to hot cracking increases. Further, nitrogen addition enhances the segregation of phosphates and sulfides at the grain boundaries that promote cracking. Varestraint and hot-ductility tests are used to evaluate the hot cracking susceptibility of the alloys [31.17]. While the brittleness temperature range (BTR) can be evaluated by Varestraint tests, hot ductility tests conducted on Gleeble thermomechanical simulator can determine the nil-strength temperature (NST), nil-ductility temperature (NDT), and ductility-recovery temperature (DRT) of materials through simulations comprising heating to different temperatures below the melting point and applying tensile load to fracture the specimens. For example, these weldability studies have shown that the difference between NST and DRT of 50 K for the 0.22% nitrogen steel is higher than that of the steels containing 0.07% nitrogen (40 K) and 0.14% nitrogen (30 K), indicating the high susceptibility of 0.22% nitrogen steel to hot cracking. Varestraint tests also confirmed that the 0.22% nitrogen steel has the highest hot cracking susceptibility. In view of the high hot cracking susceptibility of these 316LN stainless steels, special welding procedures and welding consumables have to be developed and qualified to enable welded fabrication of SFR components made of these materials.

31.8   STEAM GENERATOR MATERIALS

Niobium-stabilized 2.25Cr–1Mo steel is used in the steam generators of earlier SFRs. Modified 9Cr–1Mo (grade 91) steel is now being used in constructing steam generators of present-day SFRs in view of their improved creep strength coupled with high thermal conductivity, low thermal expansion coefficient, and immunity to stress corrosion cracking in chloride and aqueous mediums compared to austenitic stainless steels. The grade 91 steel is generally used in the normalized and tempered condition that gives rise to tempered martensite structure. In this grade 91 steel, the additions of vanadium, niobium, and nitrogen ensure intragranular precipitation of highly stable vanadium and niobium MX-type carbonitride particles on tempering and during creep exposure [31.18] contribute to high creep strength.

Creep strength of the fusion welded joint of these chromium–molybdenum ferritic steels is considered to be a life-limiting factor. In actual structures fabricated by welding, a high percentage of failures occur in the heat-affected zone [31.19,31.20]. The detailed microstructure in the heat-affected zone of ferritic steels is extremely complex and is controlled by the interaction of thermal fields produced by the heat input from the welding process and the phase transformation and grain-growth characteristics of the materials being welded [31.21]. Further modifications in microstructure can occur as a result of tempering either during the later stages of welding and post-weld heat treatment or during service. These microstructures that generally vary from those of the wrought base material through transformed heat-affected zones to cast weld metal can have greatly different mechanical properties. As a consequence, premature cracking occurs in the intercritical region of the heat-affected zone leading to reduction in creep rupture life, commonly termed as type IV failure. The weld joints of these ferritic steels possess lower creep rupture life than the base steels. The chemical composition of the grade 91 steel can be altered by microalloying with about 100 ppm of boron and control nitrogen content to less than 100 ppm. The boron-added grade 91 steel exhibits better resistance to type IV cracking with less reduction in creep rupture strength of the weld joint compared to that of the base metal (Figure 31.8) [31.22].

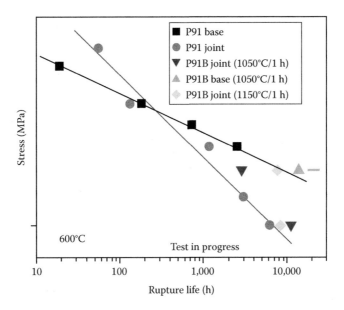

FIGURE 31.8 Effect of boron on the creep rupture life of modified 9Cr–1Mo steel and its weld joint.

31.9 HARDFACING

In SFRs, austenitic stainless steels are used as the reactor structural material, and the liquid sodium coolant acts as a reducing agent and removes the protective oxide film present on the stainless steel surface of the in-sodium components. Many of these components are in contact with each other or have relative motion during operation and their exposure at high operating temperatures (typically 823 K) coupled with high contact stresses can result in self-welding of the clean metallic mating surfaces. In addition, the relative movement of mating surfaces could lead to galling, a form of high-temperature wear, in which material transfer occurs from one mating surface to another due to repeated self-welding and breaking at contact points of mating surfaces. Further, susceptibility to self-welding increases with temperature. Hardfacing of the mating surfaces is widely used in the components of SFRs to avoid self-welding and galling.

Cobalt-base hardfacing alloys (e.g., stellite) have been traditionally used very extensively for high-temperature application in many critical hardfacing applications due to their excellent wear-resistance properties. However, when cobalt-base alloys were used in a nuclear reactor environment, the cobalt-60 isotope formed due to irradiation enhances the radiation dose rate to operating personnel during handling, maintenance, or decommissioning of the hardfaced components. Hence, there is an emerging trend of avoiding the use of cobalt-base alloys for the hardfacing of nuclear power plant components. Nickel-base hardfacing alloys (e.g., colmonoy) were developed mainly to replace the cobalt-base alloys for avoiding induced radioactivity problems in SFR applications. Accordingly, selection of suitable hardfacing materials for various components is preceded by detailed induced radioactivity, dose rate, and shielding computations to ensure that induced radioactivity from hardfaced components is kept to the minimum for maintenance and decommissioning purposes and also to reduce the shielding thickness required for the component-handling flask, which in turn would reduce the flask weight, size of handling crane, and loads on civil structures.

Based on radiation dose rate and shielding considerations during maintenance, handling, and decommissioning, nickel-base colmonoy hardfacing alloy is chosen to replace the traditionally used cobalt-base stellite alloys [31.23]. Studies on the effect of long-term aging of this nickel-base hardface deposit on austenitic stainless steel substrate demonstrated that colmonoy deposits after exposure at service temperatures up to 823 K would retain adequate hardness well above R_C 40 at the end of the components' designed service life of up to 40 years. Further, based on detailed metallurgical studies, including residual stress measurements after thermal cycling, the more versatile plasma-transferred arc welding process is chosen for the deposition of the colmonoy hardfacing alloy, so that the width of the dilution zone can be controlled by optimizing the deposition parameters.

31.10 SUMMARY

Science-based developments of materials for SFRs have been made possible by controlled experiments coupled with simulations. Enormous progress in the development of radiation-resistant steels has accrued from the tailoring of microstructure, particularly the incorporation of nanodispersoids in steels facilitating good stability to high-dose neutron irradiation and improved strength at high temperatures. The controlled additions of Ti and P to austenitic Fe–Cr–Ni alloys have been demonstrated to produce fine dispersions of TiC precipitates that provide dramatic improvement in void swelling resistance during high-dose irradiation compared to standard Fe–Cr–Ni alloys. Similarly, the nanosize Y–Ti–O-rich particles in ODS alloys appear to exhibit good stability under irradiation and provide significant strength advantages over ferritic–martensitic alloys at high temperatures. An added benefit of nanometer-sized oxides is their important role in healing radiation damage. Further, creep behavior at elevated temperatures and under irradiation for austenitic, ferritic, and ODS steels, the DBTT of ferritic/martensitic steels and their compatibility with the liquid sodium coolant as well as the fuel pellet are crucial issues that are addressed for the choice of appropriate material for the clad and wrapper applications in SFRs.

The preferred reactor structural materials for elevated temperature applications in SFRs are variants of type 316 austenitic stainless steels, in which the carbon content is reduced to less than 0.03 wt% to avoid sensitization, especially in the heat-affected zone in welded components and controlled nitrogen is added to compensate for the loss in solution strengthening due to reduction is carbon content. Increasing nitrogen content up to about 0.14 wt% in these steels improves tensile, low-cycle fatigue, creep, and corrosion resistance properties. The hot cracking susceptibility during welding of these enhanced-nitrogen steels needs careful consideration through the use of appropriate welding procedures and special-purpose welding consumables.

Chromium–molybdenum ferritic steels such as the modified 9Cr–1Mo steel are used as the materials of construction for the steam generators of SFRs. The welded joints of these ferritic steels are prone to type IV cracking due to accelerated creep deformation in the intercritical region of the heat-affected zone. Controlled addition of boron and nitrogen is suggested as a possible way to alleviating this problem.

Various components of SFRs encounter wear of adhesive or abrasive nature and sometimes erosion. Hardfacing by weld deposition has to be used to improve the resistance to high-temperature wear, especially galling, of mating surfaces in sodium. Based on radiation dose rate and shielding considerations during maintenance, handling, and decommissioning, nickel-base colmonoy hardfacing alloy is now being used to replace the traditionally used cobalt-base stellite alloys.

REFERENCES

31.1 Mannan, S.L., Chetal, S.C., Baldev, R., Bhoje, S.B. (2003). Selection of materials for prototype fast breeder reactor. *Trans. Indian Inst. Met.*, 56, 155.

31.2 Samaras, M., Victoria, M., Hoffelner, W. (2009). Advanced materials modelling—E.U. perspectives. *J. Nucl. Mater.*, 392, 286.

31.3 Odette, G.R., Alinger, M.J., Wirth, B.D. (2008). Recent developments in irradiation-resistant steels. *Annu. Rev. Mater. Res.*, 38, 471.

31.4 Murali, D., Panigrahi, B.K., Valsakumar, M.C., Sharat, C., Sundar, C.S., Raj, B. (2010). The role of minor alloying elements on the stability and dispersion of yttria nanoclusters in nanostructured ferritic alloys: An ab initio study. *J. Nucl. Mater.*, 403, 113–116.

31.5 Jegadeesan, P., Murali, D., Panigrahi, B.K., Valsakumar, M.C, Sundar, C.S. (2011). Lattice kinetic Monte Carlo simulation of Y-Ti-O nanocluster formation in BCC Fe. *Int. J. Nanosci.*, 10, 973.

31.6 Furukawa, T., Kato, S., Yoshida, E. (2009). Compatibility of FBR materials with sodium. *J. Nucl. Mater.*, 392, 249–254.

31.7 Hofman, G.L., Walters, L.C. (1994). Metallic fast reactor fuels. *Mater. Sci. Technol.*, 10A, 1–43.

31.8 Kaufman, A. (ed.) (1962). *Nuclear Reactor Fuel Elements: Metallurgy and Fabrication.* Interscience Publishers, London, U.K.

31.9 Akabori, M. et al. (1994). Reactions between U-Zr alloys and nitrogen. *J. Alloys Compd.*, 213/214, 366–368.

31.10 Mathew, M.D. et al. (1988). Creep properties of three heats of type 316 stainless steel for elevated temperature nuclear applications. *Nucl. Technol.*, 81, 114.

31.11 Mathew, M.D. et al. (1997). Dislocation substructure and precipitation in type 316 stainless steel deformed in creep. *Trans. JIM*, 38, 37.

31.12 Wolf, H. et al. (1992). *Mater. Sci. Eng. A*, 159, 199.

31.13 Shastry, G. et al. (2005). *Trans. IIM*, 58, 275.

31.14 Sasikala, G. et al. (2000). *Trans. IIM*, 53, 223.

31.15 Sasikala, G. et al. (1999). Creep deformation and fracture behaviour of a nitrogen-bearing type 316 stainless steel weld metal. *J. Nucl. Mater.*, 273, 257.

31.16 Mathew, M.D. et al. (2012). *Mater. Sci. Eng. A*, 535, 76.

31.17 Srinivasan, G., Divya, M., Albert, S.K., Bhaduri, A.K., Klenk, A., Achar, D.R.G. (2010). *Weld. World*, 54, R322–R332.

31.18 Vitek, J.M., Klueh, R.H. (1983). *Metall. Trans. A*, 14A, 1047.

31.19 Laha, K. et al. (2007). *Metall. Mater. Trans. A*, 34A, 58.

31.20 Brett, S.J. (2001). Identification of weak thick section modified 9 chrome forgings in Service: 3rd EPRI conference on advances in materials technology for fossil power plants. In: Viswanathan, R. et al. (ed.), *Advances in Materials Technology for Fossil Power Plants.* University of Wales, Swansea, Wales, pp. 343–351.

31.21 Alberry, P.J., Jones, W.K.C. (1977). *Met. Technol.*, 4, 360.

31.22 Das, C.R. et al. (2011). *Metall. Mater. Trans. A*, 42A, 3849.

31.23 Bhaduri, A.K., Albert, S.K. (2007). Development of hardfacing for fast breeder reactors. In: Glick, H.P. (ed.), *Materials Science Research Horizons.* Nova Science Publishers, New York, pp. 149–168, Chapter 5.

32 Chemical Sensors for Sodium Coolant Circuits

32.1 INTRODUCTION

Liquid sodium, when pure, is chemically compatible with the structural steels of heat transfer circuits. But the presence of dissolved impurities like oxygen and carbon in it, even at a parts per million (ppm) level, can lead to corrosion and mass transfer in these circuits. When oxygen concentration in sodium is high, transport of radioactive nuclides such as ^{54}Mn from the reactor core to other areas would be enhanced [32.1]. Similarly, high carbon activity levels in sodium can lead to carburization of the steels [32.2]. Hence, it is necessary to monitor these impurities in the coolant continuously using reliable sensors. In the steam generator section of the reactor, high-pressure steam (~150 bars) and liquid sodium at near-ambient pressure are separated by a ferritic steel tube with a wall thickness of ~4 mm. Although these steam generator components are subjected to very strict quality assurance examinations before their installation in circuit, development of a defect during their service that ultimately results in a steam leak into sodium is a possibility. Sodium–water reaction is highly exothermic and produces gaseous hydrogen and corrosive molten NaOH. Molten sodium hydroxide would cause corrosion and erosion of steels and this would accelerate the spread of the defect. It would also initiate attack on nearby healthy ferritic steel tubes resulting in a cascade of failures [32.3,32.4]. Therefore, it is essential to detect a steam leak at its inception itself for initiating remedial actions. When the temperature of the sodium coolant is above 400°C, which would be the case when the fast reactor operates at full power, NaOH and hydrogen gas formed during sodium–water reaction dissolve into sodium. Since this causes increase in dissolved hydrogen concentration in sodium, continuous monitoring of hydrogen levels in the sodium is important for the detection of this type of leak. However, the temperature of the sodium coolant would be low when the reactor is under start-up conditions or under low-power operations. The reaction/dissolution of NaOH and hydrogen gas into sodium is kinetically hindered at low temperatures. Monitoring of dissolved hydrogen in liquid sodium would not be the reliable method to detect a steam leak under these conditions. At low temperatures, the gaseous hydrogen formed is transported as bubbles in the flowing sodium and gets collected in the argon cover gas of the coolant circuit. Since the volume of the argon cover gas plenum is low, the rate of increase in the hydrogen partial pressure in it would be significantly high and the event of a leak can be detected reliably by continuous monitoring of hydrogen in the cover gas.

32.2 SENSORS FOR MONITORING DISSOLVED HYDROGEN IN LIQUID SODIUM

Hydrogen reacts with sodium metal to form NaH(s), which is not thermodynamically very stable. The equilibrium pressure of hydrogen when sodium is saturated with NaH(s) at ~440°C is 1 bar. The solubility of hydrogen in sodium is also low, and the temperature dependence of this solubility is given by the following expression [32.5]:

$$\log\left(\frac{C_H^s}{ppm}\right) = 6.467 - \frac{3023}{T} \tag{32.1}$$

where C_H^s and T are the solubility of hydrogen in sodium (in ppm) and T is the temperature in K. Sodium in heat transfer circuits is purified by cold trapping and the general dissolved hydrogen concentrations in sodium would be in the range of 50–100 ppb. The equilibrium hydrogen pressure, p_{H_2} when NaH concentration in sodium is below its solubility, is given by Sievert's law:

$$C_H = k_s \sqrt{p_{H_2}} \tag{32.2}$$

where
   C_H is the dissolved hydrogen concentration in sodium
   k_s is the Sievert's constant

Sievert's constant for hydrogen in sodium is found to be temperature independent but weakly dependent on oxygen concentration [32.6]. When C_H and p_{H_2} are expressed in ppm and Pa, respectively, it has a value of $120.4 + 0.737\,C_O$ where C_O is the dissolved oxygen concentration in sodium expressed in ppm. Sievert's relationship shows that the equilibrium hydrogen pressures in purified liquid sodium in the heat transfer circuits would be in the range of 0.01–0.1 Pa. An event of steam leak into sodium increases the dissolved hydrogen level above the prevailing background level and hence increases the equilibrium hydrogen pressure. Detection of steam leak into sodium involves the measurement of the sudden increase in hydrogen partial pressure. Hydrogen measurement systems can be classified as diffusion-based hydrogen monitors and electrochemical sensors.

32.2.1 DIFFUSION-BASED HYDROGEN MONITORS

The equilibrium hydrogen pressures in purified sodium themselves are amenable for measurement by physical methods. In all these monitors, a thin-walled metal membrane separates the liquid sodium from the gas measurement system and hydrogen from sodium diffuses through the metal membrane into this gas measurement system [32.7–32.11].* The gas measurement system operates either under high vacuum [32.7–32.9] or under a flow of inert gas such as argon [32.10,32.11]. In the former method, invariably a pumping module incorporating a sputter ion pump is used to maintain the high vacuum in the gas measurement system. Nickel or stainless steel is used as the material for the membrane which is in the form of either tubes or bellows. Under conditions of high vacuum, a steady flux of hydrogen gas into the system is established and this flux is proportional to dissolved hydrogen concentration in sodium. A quadrupole mass spectrometer (QMS) that is tuned to the mass of 2 (corresponding to hydrogen gas) is employed to measure this flux. The current of the ion pump is also sometimes measured since it is proportional to the steady flux of hydrogen. Schematics of the nickel membrane assembly in the tubular form and the gas measurement system, which has been used in the fast breeder test reactor at IGCAR, India, are shown in Figure 32.1. These types of diffusion-based monitors are capable of measuring as low as 0.01 ppm of hydrogen in sodium, but the output signal from them is low under these conditions, since the output is directly proportional to the hydrogen concentration. These monitors have been used in different fast reactors, but they are characterized by the following operational shortcomings:

1. The entire monitor module is somewhat bulky and complex owing to the size of the vacuum system and the QMS and requires frequent periodical maintenance. The capacity of the ion pump also varies with time, which influences the performance of the monitor.

---

* Hence, the nomenclature diffusion-based hydrogen monitors.

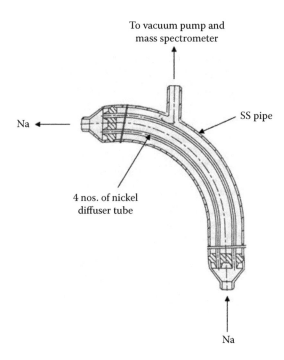

FIGURE 32.1 Schematic of tubular form of nickel membrane in diffusion-based hydrogen monitors.

2. Deposition of the corrosion products on the sodium side of the membrane followed by their diffusion into it would alter the permeability coefficient of hydrogen. This as well as the change in capacity of the vacuum pump demands frequent calibration of the monitor.
3. A failure of the membrane would result in flooding of liquid sodium into the gas measurement chamber with subsequent nonavailability of the monitor for prolonged periods.

In the other type of diffusion-based hydrogen monitor, a low and constant flow of an inert gas is maintained over the metallic membrane to collect the diffusing hydrogen. The membrane is generally in the form of a long coil made of thin-walled nickel tube and dipped into liquid sodium and the inert gas employed is argon [32.10,32.11]. Since the thermal conductivity of hydrogen is about an order higher than that of argon, the difference in the thermal conductivity of the gas stream before its entry into and after its exit from the coil is measured using a thermal conductivity detector (TCD) to obtain the hydrogen partial pressure. These data are related to dissolved hydrogen concentration in sodium after suitable calibration. Although this sensor is simple with respect to operation and maintenance, its lowest detection limit is determined by the characteristics of the TCD and inherent fluctuations of the inert gas flow. Generally, TCDs in this type of configuration can measure reliably approx. 5×10^{-5} bar or 5 Pa hydrogen pressures of the gas phase (corresponds to ~1 ppm of dissolved hydrogen in sodium) and can detect a change in concentration of about 5%–10% only. However, these types of monitors are well suited for the measurement of hydrogen in argon cover gas (see later).

32.2.2 Electrochemical Hydrogen Sensors

Hydrogen dissolved in sodium can be measured by sensors based on electrochemical principles. An electrochemical hydrogen sensor that operates in liquid sodium can be represented as:

$$\underset{Fe}{P_{H_2}^{Na}} \left|\right| Electrolyte \left|\right| \underset{Fe}{P_{H_2}^{ref}}$$

Sodium containing dissolved hydrogen and the reference electrode that maintains a fixed and predictable hydrogen partial pressure is separated from the electrolyte by suitable thin-walled metallic membranes. Generally, iron is the material of choice from the chemical compatibility and hydrogen permeability viewpoints. The iron membranes enable the separation of the electrolyte from the sample and electrodes but establish the equilibrium hydrogen pressures of sample and reference electrodes at the respective Fe–electrolyte interfaces. The electromotive force (EMF) that develops across the electrolyte is given by Nernst relation:

$$E = \frac{RT_M}{2F} \ln\left(\frac{P_{H_2}^{ref}}{P_{H_2}^{Na}}\right) \tag{32.3}$$

where
  the references $P_{H_2}^{ref}$ and $P_{H_2}^{Na}$ are partial pressures of hydrogen in reference and sample electrodes, respectively
  R is the universal gas constant
  F is the Faraday constant
  T_M is the operating temperature of the sensor

Since $P_{H_2}^{ref}$ and T_M are constants and by Sievert's relation $P_{H_2}^{Na}\alpha(C_H)^2$, EMF can be expressed as

$$E = A + B\log C_H \tag{32.4}$$

The change in the output of the sensor for a change in hydrogen concentration in sodium is represented by $\left(\partial E / \partial C_H\right)_T = 1/C_H$. Hence, the sensitivity of electrochemical sensors is high at low hydrogen concentrations. Electrochemical sensors are compact and simple to operate and do not demand frequent calibrations. In principle, the electrolyte can be either a proton or a hydride ion conductor. The electrolyte should possess high conductivity with the transport number of the ion, $t_{ion} > 0.99$ under the hydrogen partial pressures prevailing in sodium systems. It should also be thermochemically stable under those conditions for the cell to follow the Nernst relation. Currently, known proton conductors decompose at high temperatures that prevail in sodium systems and also require moisture for their stability and hence cannot be used in these electrochemical sensors. C.A. Smith of Berkeley Nuclear Laboratories, United Kingdom, initially proposed a solid-state hydride ion-conducting electrolyte based on a $CaCl_2$–CaHCl system for use in sodium systems [32.12]. Later work by Sridharan et al. from IGCAR, India, showed the inadequacies of this electrolyte [32.13]. Based on detailed studies on thermochemical and electrical properties of several electrolyte systems, they also showed that $CaBr_2$–CaHBr is a suitable electrolyte for use in sodium systems [32.14]. The schematics of the electrochemical sensor using this electrolyte is given in Figure 32.2. These sensors employ a mixture of CaO and MgO along with CaH and Mg metals that fixes the hydrogen pressure at the reference electrode. These sensors operate at 450°C and utilize compact instrumentation with high input impedance to measure the cell EMF. They have been demonstrated to operate satisfactorily at low hydrogen concentrations in sodium and respond instantaneously to an increase in hydrogen levels in sodium [32.15].

32.2.3 HYDROGEN MONITORS FOR USE IN ARGON COVER GAS

Monitoring of hydrogen in argon cover gas is the reliable method for the detection of steam leaks, if it occurs under low-power conditions or during start-up of the reactor. Diffusion-based hydrogen

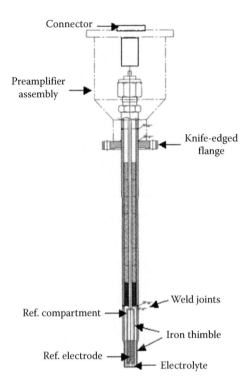

FIGURE 32.2   Schematic of an electrochemical hydrogen sensor.

monitoring system with QMS (or by measuring the ion current of the sputter ion pump) can be employed for continuously monitoring hydrogen levels in the cover gas. Alternatively, a nickel coil membrane assembly with a stream of argon gas passing through it can be used for this purpose [32.16]. The nickel coil assembly, made from 7 m long nickel tube of 2.5 mm diameter with 0.25 mm thickness and operating at 500°C in cover gas plenum of sodium systems, is shown in Figure 32.3. As mentioned in Section 2.1, a TCD measures the hydrogen partial pressure in the argon gas that exits the coil and the detection limit of this system is ~30 vppm of hydrogen and can reliably measure at least up to 0.1% of hydrogen.

In order to extend the detection limit of this system down to a few vppm of hydrogen, semi-conducting oxide-based sensors can be suitably modified and used. Tin oxide is a n-type oxide semiconductor and the semiconducting behavior arises due to its nonstoichiometry. Exposure of a thin film of this n-type tin oxide to oxygen-containing gas ambient, for example, air, leads to chemisorption of oxygen molecules from the ambient as charged ions such as O_2^-, O^-, and so on, by depleting the excess electrons from the film. The film therefore possesses high resistivity. When reducing

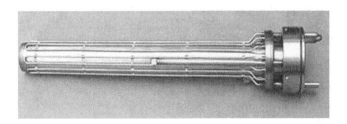

FIGURE 32.3   Coil assembly made from long and thin nickel tubes for use in monitoring hydrogen in cover gas.

gases such as hydrogen are introduced into the gas ambient, the chemisorbed oxygen ions react with them and release the electrons back into the thin film causing the conductivity of the film to increase. At low hydrogen levels, this increase in conductivity is linearly dependent on the hydrogen concentration and this property can be exploited for measuring trace levels of hydrogen in argon gas exiting the TCD. Since these sensors operate in a gas ambient containing oxygen, a controlled amount of oxygen is introduced to the gas stream coming out from TCD before it is analyzed by the sensor. A typical sensor comprises a thin film of tin oxide deposited on one side of an alumina substrate, which is provided with a platinum heater on the rear side. The film is maintained at 623 ± 2 K and is housed in a low-volume chamber (~8 mL), which has provisions for gas inlet and outlet. This sensor shows a linear response to hydrogen from 2 to 80 vppm, and thus, the detection limit of cover gas hydrogen monitoring system could be brought down to 2 vppm of hydrogen [32.17].

32.3 SENSORS FOR MONITORING CARBON ACTIVITY IN LIQUID SODIUM

The primary sources of carbon impurity in sodium metal are as follows: (1) the carbon released from the graphite anodes of the Downs cell where it gets produced and (2) the trace level carbonaceous impurities such as greases present over the structural components of the coolant circuits. Carbon does not form stable carbide with sodium metal, although thermodynamically unstable sodium acetylide, Na_2C_2, can be formed as a metastable compound in sodium systems. The chemistry of carbon in sodium is complex depending upon the source of carbon introduced into it. In liquid sodium, carbon exists in active (dissolved) and inactive (suspended fine particulates) forms and the active form determines the chemical activity of carbon in sodium. Solubility of carbon in sodium is very low [32.18]. It is to be pointed out that the total carbon content in a sodium sample can be very high compared to the solubility limits but the sodium may not be saturated with the active carbon. Because of its presence in active and inactive forms, carbon impurity in sodium cannot be purified by cold trapping. The carbon levels in sodium coolant of a heat transfer circuit can increase in the event of a leak of the hydrocarbon oil coolant through the seals of the centrifugal pump shafts. This oil leak would result in increase in the carbon activity levels in sodium that would result in carburization of the structural steels, which is detrimental to their mechanical properties. By continuously monitoring carbon activity levels in sodium, such events could be detected.

32.3.1 DIFFUSION-BASED CARBON MONITORS

Two types of diffusion-based monitors have been reported in the past. Both use thin-walled iron membranes that are exposed to sodium at high temperatures. Carbon in sodium diffuses through the iron membrane. Carbon activity at the other side of the membrane is maintained near zero by a suitable chemical reaction that consumes the carbon diffusing out. Under these conditions, the carbon activity difference across the iron membrane is essentially equal to the carbon activity in sodium. The flux of carbon that diffuses through the iron membrane would therefore be proportional to the carbon activity in sodium. The carbon-bearing gaseous product of the reaction is continuously flushed out of the membrane and analyzed suitably. Concentration of the gaseous product can be related to carbon activity in sodium. In the first carbon monitor reported by the United Nuclear Corporation, USA [32.19], the iron membrane was in the form of a small and thin-walled cup with an active area of a few square centimeters. This cup welded to a long stainless steel pipe was exposed to sodium at 973 K so as to enhance the flux of carbon across the membrane. A steady flow of argon containing moist hydrogen was maintained over the membrane for the following decarburizing reaction to occur:

$$[C]_{Fe} + H_2O(g) \rightarrow CO(g) + H_2(g) \tag{32.5}$$

Carbon monoxide formed by this reaction was swept away by the gas stream, catalytically converted to methane and analyzed by flame ionization detectors (FID). The low active area of the monitor limited the total amount of carbon that diffused. Hence, the measureable lowest carbon activity in sodium was in the range of 10^{-2}. Further, operation of the membrane at very high temperatures and use of moist gases for the decarburizing reaction are the major shortcomings of this type of carbon monitor. In the diffusion-based carbon monitor developed at the Atomic Energy Research Establishment, Harwell, UK [32.20], the iron membrane was in the form of a coil made from thin-walled iron tube with a high surface area ($\sim$600 cm^2). The inside of the iron membrane was preoxidized to form a thin layer of FeO on it. The coil was kept immersed into liquid sodium at a fixed temperature between 500°C and 550°C and a steady flow of high purity argon gas was maintained through it. Carbon that diffused through iron membrane underwent the following reaction at the other side and maintained the carbon activity level close to zero:

$$FeO(s) + [C]_{Fe} \rightarrow CO(g) + Fe(s) \tag{32.6}$$

In this monitor also, the carbon monoxide that formed was swept away, catalytically converted to methane, and analyzed by FID. The use of large area membrane and sensitive FIDs configured in differential mode enabled the monitor to measure carbon activity levels in the order of 10^{-4} also. The need to constantly maintain the entire inner surface area of iron membrane coated with FeO (which otherwise would lead to reduction in the formation of CO and to erroneous results) and the possibility of reduction of this oxide by hydrogen, which also diffuses from liquid sodium through the iron membrane are major shortcomings of this type of carbon monitor. Both the diffusion-based monitors are bulky because of the gas manifolds and carbon monoxide measurement systems.

32.3.2 ELECTROCHEMICAL CARBON SENSORS

Molten carbonate-based electrochemical cell for measuring carbon activity in sodium was first reported by Salzano and coworkers from Berkeley Nuclear Laboratories, UK [32.21]. A molten eutectic mixture of Li_2CO_3 and Na_2CO_3 taken in a thin-walled iron membrane serves as the electrolyte, and graphite is used as the reference electrode. The half-cell reaction in the sensor is represented in the following:

$$C + 3O^{2-} \rightarrow CO_3^{2-} + 4e^- \tag{32.7}$$

The EMF of the cell is related to carbon activity in sodium a_C^{Na} by Nernst equation:

$$E = -\frac{RT}{4F} \ln a_C^{Na} \tag{32.8}$$

However, attempts by other workers to measure carbon potential in liquid sodium by the same method were not satisfactory. Problems such as mixed potential development due to parasitic reactions resulting in whisker growth and carbon deposition, drift in potential due to instability of reference electrode, need for a long period of equilibration, disagreement with carbon activity as against the expected one, and poor life had been reported. S. Rajan Babu and coworkers of IGCAR, India, carried out a detailed thermochemical analysis of the molten carbonate salts that can support several ionic species other than those represented by reaction (32.7). Based on these evaluations and other electrochemical investigations, they standardized the method to assemble the electrochemical carbon sensors that could be used to measure carbon activity levels down to 10^{-3} in sodium systems [32.22]. A schematic of the electrochemical carbon sensor is shown in Figure 32.4. These sensors operate at 625°C and require simple instrumentation.

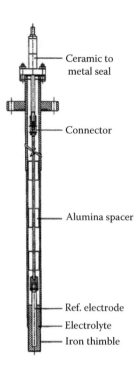

Ceramic to metal seal

Connector

Alumina spacer

Ref. electrode

Electrolyte

Iron thimble

FIGURE 32.4   Schematic of electrochemical carbon sensor for use in sodium systems.

32.4   SENSORS FOR MONITORING OXYGEN IN LIQUID SODIUM SYSTEMS

The sources for oxygen impurity in liquid sodium are trace moisture and oxygen gases present in the input argon used for cover gas circuits and impurities that are adsorbed on fresh steel surfaces on their introduction into sodium. In case of the secondary circuit, in addition to these sources, significant levels of oxygen could be introduced into sodium if a steam leak occurs at the steam generator. Sodium forms a stable oxide (Na_2O) with oxygen and the solubility of this oxide in sodium is also low [32.23]. The oxygen partial pressures of the following equilibrium with dissolved sodium oxide in sodium are extremely low:

$$2\,Na(l) + \tfrac{1}{2}O_2(g) \Leftrightarrow \left[Na_2O\right]_{Na} \tag{32.9}$$

Unlike in the case of dissolved hydrogen in sodium, these low oxygen pressures cannot be measured by conventional pressure measurement techniques. Solid oxide electrolyte-based electrochemical sensors are used for this measurement. Among the two common solid oxide electrolytes—stabilized zirconia and doped thoria—only the latter is suited for measuring low oxygen partial pressures that prevail in liquid sodium. The electrochemical cell can be represented as follows:

$$P_{O_2}^{Na} \left| Electrolyte \right| P_{O_2}^{ref} \tag{32.10}$$

The electrolyte is generally yttria-doped thoria (YDT) and is in the form of a tube closed at one end with a reference electrode consisting of a suitable metal and its oxide for fixing $P_{O_2}^{ref}$. Metallic tin mixed with tin oxide is a common choice for this since tin exists as liquid metal at operating temperatures of the oxygen sensor (~400°C). However, thoria-based ceramic electrolytes pose a problem with respect to sintering of the compacted powders. When the powders are prepared by conventional techniques, the required temperature for sintering the compacts is very high (~2000°C). Significant grain growth that takes place during sintering at these temperatures and the attendant poor thermal shock resistance of this ceramic led to early failures of the sensors made from long tubes of YDT. Work on sensors with electrolytes in the form of short thimbles that were either brazed or glass soldered to metallic parts so that the entire ceramic section of the sensor could be immersed into sodium has been carried out in the past [32.24–32.27]. This configuration avoids temperature gradients along the ceramic section and is anticipated to increase the life of the sensors in sodium systems. Development work on oxygen sensors for use in sodium is still continuing in several laboratories [32.28–32.31]. Recently, nanocrystalline yttria–doped thoria powders prepared by a novel combustion methods were used for shaping the thimbles that could be sintered at low temperatures (~1700°C) with zinc oxide as a sintering aid [32.32]. This could reduce the grain growth in the final products and the thimbles made of these powders had a helium leak rate less than 10^{-9} std. L/s indicating their impervious characteristics.

REFERENCES

32.1 Borgstedt, H.U., Mathews, C.K. (1987). *Applied Chemistry of the Alkali Metals*. Plenum Press, New York.
32.2 Natesan, K. (1975). *Met. Trans. A*, 6, 1143.
32.3 Hans, R., Dumm, K. (1977). *Atom. Energy Rev.*, 15, 611.
32.4 Hori, M. (1980). *Atom. Energy Rev.*, 18, 707.
32.5 Whittingham, A.C. (1976). *J. Nucl. Mater.*, 60, 119.
32.6 Gnanasekaran, T. (1999). *J. Nucl. Mater.*, 274, 252.
32.7 Vissers, D.R., Holmes, J.T., Barthlome, L.G., Nelson, P.A. (1974). *Nucl. Technol.*, 21, 235.
32.8 Lecocq, P., Lannou, L., Masson, J.C. (1973). *Proceedings of Conference on Nuclear Power Plant Control and Instrumentation*, Prague, Czech Republic. IAEA, Vienna, Austria, p. 613.
32.9 Lions, N., Cambillard, E., Pages, J.P., Chantot, M., Buis, H., Baron, J., Langlois, G., Lannou, L., Viala, M. (1974). Special instrumentation for PHENIX. *Proceedings of International Conference on Fast Reactor Power Stations*, BNES, London, U.K., Paper A. 33, pp. 522–535.
32.10 Davies, R.A., Drummond, J.L., Adaway, D.W. (1971). Detection of sodium water leaks in PFR secondary heat exchangers. *Nucl. Eng. Int.*, London, 16(181), 493–495. XP002266399.
32.11 Davies, R.A., Drummond, J.L., Adaway, D.W. (1973). *Proceedings of Conference on Liquid Alkali Metals*, BNES, p. 93.
32.12 Smith, C.A. (1972). An electrochemical hydrogen concentration cell – with application to sodium systems. CEGB Report, RD/B/N/2331, Berkeley Nuclear Laboratories, Central Electricity Generating Board, U.K.
32.13 Sridharan, R., Mahendran, K.H., Gnanasekaran, T., Periaswami, G., Varadha Raju, U.V., Mathews, C.K. (1995). *J. Nucl. Mater.*, 223, 72.
32.14 Joseph, K., Sujatha, K., Nagaraj, S., Mahendran, K.H., Sridharan, R., Periaswami, G., Gnanasekaran, T. (2005). *J. Nucl. Mater.*, 344, 285.
32.15 Jeannot, J.Ph., Gnanasekaran, T., Sridharan, R., Ganesan, R., Augem, J.M., Latge, C., Gobillot, G., Paumel, K., Courouau, J.L. (2009). *Proceedings In-Sodium Hydrogen Detection in the Steam Generator of Phenix Fast Reactor: A Comparison between Two Detection Methods. ANIMMA International Conference*, Marseille, France.
32.16 Mahendran, K.H., Sridharan, R., Gnanasekaran, T., Periaswami, G. (1998). *Ind. Eng. Chem. Res.*, 37, 1398.
32.17 Prabhu, E., Jayaraman, V., Gnanasekar, K.I., Gnanasekaran, T., Periaswami, G. (2005). *Asian J. Phys.*, 14, 33.
32.18 Longson, B., Thoreley, A.W. (1970). *J. Appl. Chem.*, 20, 372.
32.19 Caplinger, W. (1969). Carbon Meter for Sodium. USAEC report, UNC-5226, March 1969.
32.20 Asher, R.C., Harper, D.C., Kirstein, T.B.A. (1980). *Proceedings of the Second International Conference on Liquid Metal Technology in Energy Production*, Richland, WA, pp. 15.46–15.53.
32.21 Salzano, F.J., Newman, L., Hobdell, M.R. (1971). *Nucl. Technol.*, 10, 335.
32.22 Rajan Babu, S., Reshmi, P.R., Gnanasekaran, T. (2012). *Electrochimica Acta*, 59, 522.
32.23 Noden, J.D. (1972). General equation for the solubility of oxygen in liquid sodium, Central Electricity Generating Board. Berkeley, England, UK, Report RD/B/N 2500, p. 17.

32.24 Reetz, T., Ulmann, H. (1974). *Kernenergie*, 17, 57.
32.25 Roy, P., Bugbee, B.E. (1978). *Nucl. Technol.*, 39, 216.
32.26 Jung, J. (1975). *Nucl. J. Mater.*, 56, 213.
32.27 Jakes, D., Kral, J., Burda, J., Fresl, M. (1984). *Solid State Ionics*, 13, 165.
32.28 Nollet, B.K., Hvasta, M.G., Anderson, M.H. (2013). *Proceedings of International Conference on Fast Reactors and Related Fuel Cycles*, FR-13, p. 317.
32.29 Shin, S.H., Kim, J.J., Jeong, J.A., Choon, K.J., Choi, S.I., Kim, J.H. (2012). *Transactions of the Korean Nuclear Society Meeting*.
32.30 Shin, S.H., Kim, J.J., Jung, J.A., Choi, K.J., Choi, S.I., Kim, J.H. (2013). *Proceedings of International Conference on Fast Reactors and Related Fuel Cycles*, FR-13, p. 329.
32.31 Gabard, *M.*, Tormos, B., Brissonneau, L., Steil, M., Fouletier, J. (2013). *Proceedings of International Conference on Fast Reactors and Related Fuel Cycles*, FR-13, p. 228.
32.32 Jayaraman, V., Krishnamurthy, D., Ganesan, R., Thiruvengadasami, A., Sudha, R., Prasad, M.V.R., Gnanasekaran, T. (2007). Development of yttria-doped thoria solid electrolyte for use in liquid sodium systems. *Ionics*, 13, 299.

33 Robotics, Automation, and Sensors

33.1 INTRODUCTION

Automation and robotic technology have traditionally been quite popular in the manufacturing industry mostly in automotive sector and electronic industries. At a significant chronological juncture, this technology was first welcomed for applications in hostile environments like nuclear and space applications for obvious reasons that necessitated further advancements in this exciting domain. Automation and remote handling with robotic technology play a crucial role in almost all facets of nuclear fuel cycle: fuel fabrication, postirradiation examination (PIE), in-service inspection (ISI) of plants and components, fuel reprocessing and waste management, and, last but not the least, decommissioning of nuclear plants. In nuclear reactors, the synergistic effects of radiation, temperature, and lack of space make the use of automated, robotic, and remote handling devices imperative for any unforeseen intervention for repair or maintenance and for periodic inspection to assess the integrity of the components or the subsystems. The typical demands and challenges to the robotics and automation community in the nuclear energy sector are depicted in Figure 33.1.

Robots and automated inspection systems make great use of sensors for position location, distance assessment, object recognition, and guidance. Apart from this, sensors enable nondestructive examination of components. These sensors, also called the end effectors, exploit the recent advances in optics, infrared, ultrasonic, magnetic, electromagnetic, and other related domains to enable development and deployment of automated robotics systems for inspection of inaccessible components, especially for detection and imaging of corrosion and weld defects.

The nuclear fuel cycle opens up a plethora of opportunities for deploying robots, often as the only feasible means to achieve an end. Development and implementation of various robotic and automated devices could be justified by the safe and efficient operation of nuclear reactors and associated plants. The impact of remote handling, automation, and robotics in the various facets of the fast reactor fuel cycle is shown in Figure 33.2.

Throughout the world, automated devices and robotic systems have been developed in active collaboration with industry for carrying out ISI. The use of robotics, automation, and sensors assumes great importance to ensure plant safety, increase plant availability, and in the context of aging management. The developments in robotics and automation in this fascinating area have been discussed in the following sections, with a specific reference to the automation, robotics, and sensors for the fast reactor fuel cycle.

33.2 IN-SERVICE INSPECTION OF COMPONENTS OF FAST BREEDER REACTOR

In fast breeder reactor (FBR), primary containment structure, primary circuit equipment, and reactor internals require remote ISI devices and techniques. The primary containment structures comprise the main reactor vessel (MV), the safety vessel (SV), and the reactor roof structure (Figure 33.3). One of the main in-service surveillance requirements for the fast reactors is monitoring the external boundary of the MV using in situ spark plug detectors and this has to be performed on a continuous basis when the reactor is at power. The MV is guarded against any sodium leakage by the outer SV that is placed concentric to it. The continuous surveillance of the integrity of the vessel

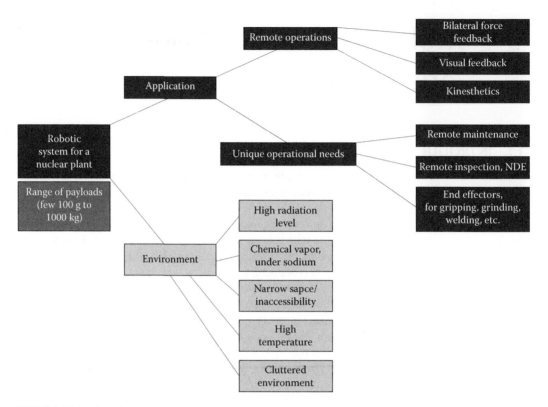

FIGURE 33.1   Requirement of robotics and automation in the nuclear fuel cycle.

FIGURE 33.2   Impact of remote handling and robotics in the fast reactor fuel cycle.

is supplemented by the periodic examination of the MV, SV, and core support structure, performed when the reactor is shut down using remote-controlled robotic vehicle.

Automated devices have been developed and used in fast reactors all over the world for the remote inspection of MV and SV. A remote-controlled four-wheeled drive device known as MIR [33.1] has been developed and used in France for the in-service inspection of MV and SV of SUPERPHENIX 1 fast reactor (Figure 33.4). A remote-controlled wheeled vehicle called MOLE [33.2,33.3] has

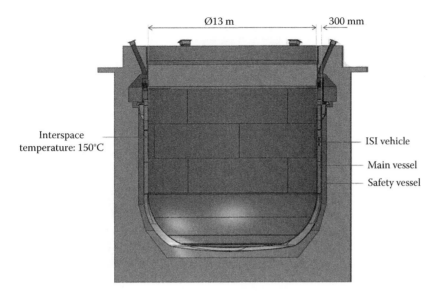

FIGURE 33.3 Front sectional view of a typical pool-type FBR showing the primary containment structures and the interspace gap.

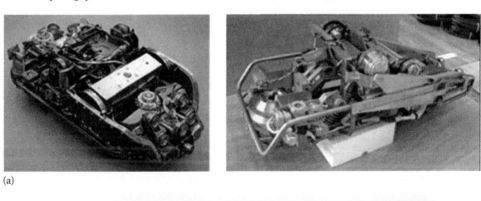

(a)

(b)

FIGURE 33.4 (a) MIR device designed to inspect welding in the Superphenix MV. (From Asty, M. et al., Super Phenix 1: In-service inspection of main and safety tank weldament. *Specialists Meeting on In-service Inspection and Monitoring of LMFBRs*, Bensberg, Federal Republic of Germany, 1980, pp. 20–22.) (b) The two versions of the MOLE developed in Japan. (From Matsubara, T. et al., Development of remotely controlled in-service inspection equipment for fast breeder reactor vessel, *Proceedings of an Internal Symposium on Fast Breeder Reactors: Experience and Trends 2*, Lyons, France, 1985, pp. 501–508; Tagawa, A. et al., *J. Power Energy Syst.*, 1(1), 3, 2007.)

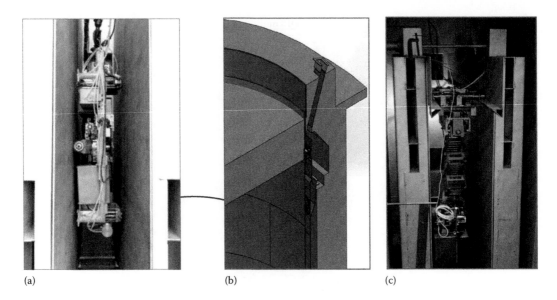

(a)   (b)   (c)

FIGURE 33.5 **(See color insert.)** (a) Main ISI device during mock-up trials. (b) Depiction of the main ISI device in the interspace. (c) One of the inspection devices in the test chamber.

been developed for FBR vessels in Japan (Figure 33.5). In Germany, an automated device on guide rails on SV was conceptualized for the SNR 300 fast reactor, for carrying out visual and volumetric examinations.

In the United Kingdom, development of special remote equipment and techniques was conceptualized to inspect the reactor vessel and internals of commercial demonstration fast reactor (CDFR) [33.4]. The developments include a *links manipulator under sodium viewing* system to view the reactor internals submerged in sodium, an *automated-guided vehicle* (AGV) to be used to survey the externals of the reactor vessel, and the snake manipulator used to gain access to restricted areas such as the vessel support and roof structures.

A comprehensive ISI system has been planned for the inspection of MV and SV of 500 MWe FBR, which is under construction at Kalpakkam, India. The inspection involves volumetric inspection of welds on the MV and SV using ultrasonic technique and visual inspection of the surfaces of MV and SV during fuel handling operations.

TABLE 33.1

Vehicle Design Concepts Using Magnetic Principles

Characteristic Designs of Magnetic Vehicles		
Magnetic-Wheeled Vehicle	**Magnetic-Tracked Vehicle**	**Magnetically Coupled**
Magnets are placed in wheels	Magnetic strips are fixed to the belt	Permanent magnets are fixed to the base of vehicle and, hence, the gap is maintained
Steering is easy	Steering is by skidding	Maintaining constant gap is an issue
Less area of contact	Large area of contact; so less powered magnets can be used	As gap varies, attraction force and, hence, traction force varies
Powerful magnets required		Suitable for plain surfaces

Choice of magnets

Temperature limit for neodymium–iron–boron (NdFeB) is 160°C

Temperature limit for samarium–cobalt (SmCo) is 300°C

The comprehensive ISI system consists of independent devices that can move inside the interspace between main and SVs carrying inspection modules for eddy current, ultrasonic, and visual inspections. Each device is provided with navigation camera modules for guiding the devices in the interspace during ISI and essential sensors such as resolvers, temperature sensors, inclinometers, linear variable differential transducer (LVDT), load cells, and so on, for monitoring and controlling the robotic vehicles during ISI.

As a natural evolution, improved design for future reactors would present tighter constraints for ISI. A reduced MV–SV interspace gap and using ferritic steel for SV are likely to be the major design changes in future FBRs. Various methods used in industrial magnetic vehicles are summarized in Table 33.1. Magnetic field coupling along with ferritic SV could be effectively used to design a compact and novel vehicle to traverse the annular interspace.

33.3 REMOTE HANDING TOOLS AND ROBOTIC DEVICES FOR NUCLEAR FUEL CYCLE FACILITIES

Handling and manipulation of workpiece and gadgets behind the wall at a distance are inevitable in nuclear facilities. Immense attempts have been made by many countries and agencies to develop a remote handling device with high degree of dexterity and reliability for use inside the radioactive containment cell known as hot cells where human entries are restricted due to high levels of radioactivity. Figure 33.6a shows a view of a hot cell facility. Remote tongs, master-slave manipulators (MSMs), electrical MSM, in-cell crane, power manipulator, and servo manipulator are the major remote handling devices being used for nuclear applications.

Among the remote handling devices, MSMs are the prime workhorses in nuclear fuel fabrication, spent fuel examination cells, and reprocessing plants where large quantities of highly radioactive materials are handled. The MSM, as the name implies, has a master arm and a slave arm, which are mechanically coupled through linkages, wires, pulleys, and chains. By virtue of their simple/ergonomic design and bilateral force reflection coupling, their performance is highly reliable and robust. Figure 33.6b shows an operator using an MSM in a hot cell looking through the viewing window. The design features of these devices have undergone continuous evolution over the years. Manipulators with better force reflection, higher degrees of freedom, higher payload capacity, and servo manipulators with electromechanical coupling mounted on overhead trolleys and fitted with vision systems to extend their reach and work volumes are few of the examples.

In the nuclear fuel fabrication facilities, remotization and automation result in enhanced safeguards and product quality, which give improved protection for radiation workers; accountability of fuel material; and increased productivity. Improved fuel performance, representative fabrication rates, and reduced fuel costs can be achieved only as a result of improved product quality and

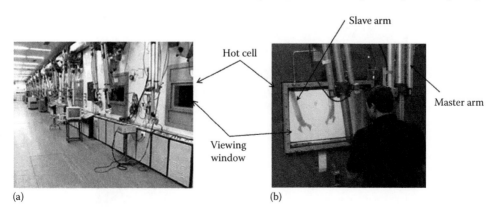

(a)   (b)

FIGURE 33.6   (a) View of a hot cell facility with MSMs. (b) Operator using a mechanical MSM in a hot cell.

productivity. Implementations of modern concepts like intelligent processing, where feedback from inspection is given to the process for online correction, can result in nearly zero defect products. Automation also facilitates higher levels of documentation of all important parameters of fuel pellets, end plugs, fuel pins, etc., and accountability of fuel material. Large-scale fuel fabrication using U^{233} needed in the thorium fuel cycle is possible only in a fully automated fabrication facility as radiation level due to daughter products of U^{232} can be very high.

Plutonium recycling in the form of mixed oxide (MOX) fuel is now a mature industry in Europe with successful operational experience in large-scale MOX fuel fabrication plants and with MOX fuel loading in reactors in both France and Germany [33.5]. The MELOX plant in France is one of the large-scale fuel fabrication plants in the world, with an efficient and modern MOX fuel fabrication facility. Its new automated design enables the plant to produce assemblies at a rate well above one 500 kg assembly per day. TVEL Corporation of Russia is one of the leading manufacturers of nuclear fuel. Fuel rods are manufactured in automated production lines equipped with state-of-the art instrumentation and test equipment. In 1980, secure automated fabrication (SAF) project was established in the United States with the goal to design, build, and operate a remote process for manufacturing breeder reactor fuel pins with mixed uranium/plutonium oxide as the fuel for the Fast Flux Test Facility (FFTF). The Power Reactor and Nuclear Fuel Development Corporation (PNC) in Japan, which includes plutonium fuel development facility (PFDF) and plutonium fuel fabrication facility (PFFF), was constructed to convert the developed fuel production technology on a large scale. Based on the experience of MOX fuel fabrication technology developed in PPFF, the plutonium fuel production facility (PFPF) was constructed and commissioned to demonstrate the mass production technology of MOX fuel. It employed remote and automated technology, and the fabrication of the initial fuel for the FBR *Monju* was completed in this facility [33.6,33.7].

Nuclear fuel fabrication is one of the areas where robot-assisted automation is essential in achieving close tolerances and large throughputs while reducing the radiation exposures to the operators. For instance, a selective compliance assembly robotic arm (SCARA) (Figure 33.7) can be used for handling green fuel pellets and inspection and handling of sintered pellets and their sorting, stacking, and loading into fuel pins during the remote fuel fabrication. These automated/robotic devices are used with tactile sensors and machine vision systems.

Reprocessing/refabrication of fuel for the metallic fuel cycle involves pyrochemical or pyrometallurgical methods and is possible only with remote/automated operations not only because of the

FIGURE 33.7   SCARA.

high radiation level from the high burnup, short-cooled fuel but also due to poor decontamination from fission products in pyroreprocessing. Metallic fuel cycling requires hot cell technology with inert atmosphere (with control of oxygen and moisture in ppm level) for reprocessing as well as refabrication. All process operations and associated handling have to be necessarily carried out using remote techniques. Power manipulator and in-cell cranes have been of great use inside radio-active containment boxes and hot cells for carrying out specialized remote handling operations for various process steps in pyrochemical reprocessing facilities.

Reprocessing of spent nuclear fuel to recover valuable fissile material is a complex chemical process that requires continuous monitoring of process parameters at various stages. Reprocessing of the spent fuels of nuclear plants involves handling of highly radioactive materials in aggressive chemical environments through remote techniques. The handling of spent fuels in a disordered environment with high levels of radioactivity often inevitably necessitates unplanned remote inter-vention for inspection, decontamination, or repairs. Dissolver, process tanks, and pipe lines and waste storage tanks are the critical equipment and components in the fuel reprocessing plants that require periodic inspection to assess their integrity. Various devices have been developed and used for remote inspection/repair of dissolver vessel, drip trays, process tanks, and waste tanks in repro-cessing plants. There has been a need for the development of devices for inspection of the waste-vault tanks and drip trays to assess the healthiness and integrity of these tanks. It is also necessary to have remote collection and analysis of the samples of the process solution.

Dissolver is one of the important equipment in a nuclear fuel reprocessing plant. After chopping of the irradiated/spent fuel pins in a chopper, the bulk material will be fed to the dissolver vessel where dissolution is carried out in boiling concentrated nitric acid. The final dissolver solution contains several fission products and ions. Failure of the dissolver will cause leakage of radio-active material–contained liquid to the operating area, which is not acceptable. Hence, periodic inspection of the dissolver vessel is essential to assess the effects of corrosion-induced damage. Remote repair and maintenance in the back end of the fuel cycle, especially in reprocessing plants, is attempted in few countries in view of the nature of the environment and the consequences of a probable incident that can result in the release of radioactivity outside the cell. In 1980s, an improve-ment program was undertaken at the Tokai reprocessing plant in Japan, using a floor-rambling remote inspection vehicle [33.8] for the ISI of the surface of drip tray. A well-known example of remote technique applications in the head-end part of reprocessing plants is the remote inspection and maintenance operation conducted to repair the dissolver at Tokai-mura reprocessing plant in Japan [33.9]. Leakage occurred in two of the dissolvers of Tokai reprocessing plant at Tokai works, Japan, and the leak was examined and repaired using a remote weld device shown in Figure 33.8.

Another example is the repair work performed at the La Hague plant in France, making use of a specially designed and built remote-controlled system. In this context, it was a similar work previously conducted at Tokai reprocessing plant. At the British Nuclear Fuels Limited (BNFL), Sellafield reprocessing plant, United Kingdom, the dissolver vessel branch pipes (Figure 33.9a) have been inspected for corrosion-induced damages and repaired using a dextrous multijointed manipu-lator REPMAN (Figure 33.9b) [33.10,33.11].

For corrosion loss detection in pipes, an internal rotary inspection system (IRIS) has been developed in which the pulses from an ultrasonic transducer are reflected by an inclined mirror [33.12]. A single actuation–based ISI device for pipes has also been developed for the inspection of long pipes in repro-cessing plants [33.13]. The lack of space and the high radiation in an unstructured environment neces-sitate the development of field specific devices for campaign rather than the use of standard products.

Remote inspection device using a CCD-based camera was used for the visual inspection of the dissolver vessel made of titanium [33.14] in a small-scale fast reactor fuel reprocessing plant in India (Figure 33.10).

Advanced remote inspection devices have been used for the remote inspection and maintenance of dissolver vessel in fast reactor fuel reprocessing plants. These devices are designed to conduct ISI using immersion ultrasonic technique for detecting the wall thinning as an evolution from the

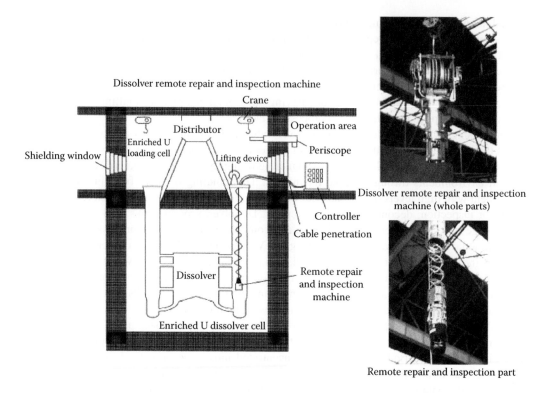

FIGURE 33.8 The dissolver and the remote inspection and repair at Tokai reprocessing plant. (From Yamamuraa, O. et al., *Prog. Nucl. Energy*, 50(2–6), 666, 2008.)

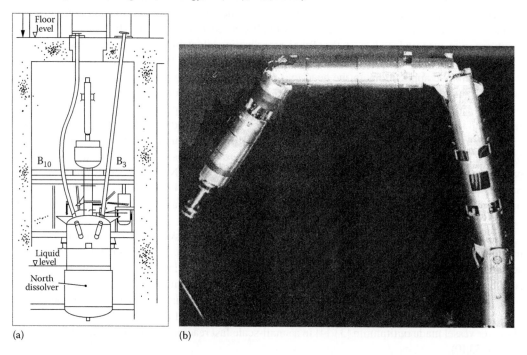

FIGURE 33.9 (a) Layout showing the constraint in accessing one of the dissolver vessels at BNFL, UK. (b) Repair manipulator REPMAN for UT and visual inspection and remote welding of branch pipe connections on the dissolver vessel (shown in (a)—B_3 B_{10}).

FIGURE 33.10 Remote inspection device for the small-scale reprocessing plant dissolver.

previous one. In order to conduct the examination using remote techniques, the device is bestowed with three major degrees of freedom—along three axes for scanning the inner surface of the vessel. The system also provides a means to conduct visual examination using CCD camera [33.13] (Figure 33.11).

Inspection of the tanks and vessels containing highly active process solution located in inaccessible area/cell is a challenging task. Manipulators such as snake/elephant's trunk manipulator and multilink manipulator have been developed and used worldwide for deploying fiberscopes and miniature CCTV cameras to examine areas of interest in regions of restricted access. A monoarticulated chain with a minimum of two controlled motions for the end effector was used to perform the inspection of the tanks and drip trays using a CCD camera [33.15] (Figure 33.12).

33.3.1 Inspection Robots

Automation and robotics are essential in nuclear waste management, especially large-scale waste management of highly active waste streams and it is always better done using automated concepts. The most important among these is the remote inspection of waste vaults as well as remote decontamination.

Camera-mounted mobile robots are useful for inspection and surveillance of waste vault and its contents, where the prevalent high radiation precludes human entry. A four-wheeled robotic device with a camera having pan/tilt was used in the waste vault for inspection and surveillance purpose. The ISI was carried out in the waste vault associated with the pilot plant for FBR fuel reprocessing in India using the mobile robot and images of the waste farm tanks were acquired using the on-board camera (Figure 33.13).

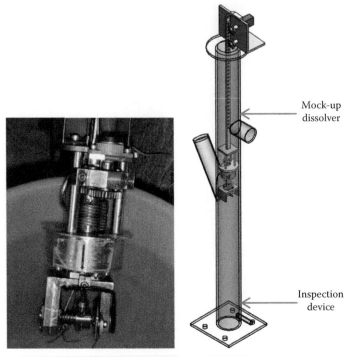

Inspection head of the 3-axis scanner

FIGURE 33.11   3-Axis scanner and the mock-up dissolver.

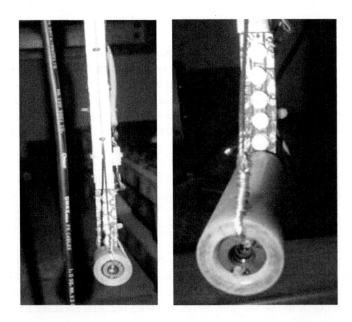

FIGURE 33.12   Monoarticulated chain for the remote inspection of space below containment box.

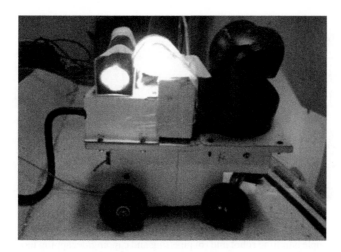

FIGURE 33.13   Mobile robot for the ISI of pilot-plant waste vault.

FIGURE 33.14   Mobile robotic system with manipulator for ISI of large-scale waste vault.

A mobile robotic system (Figure 33.14) has been developed into an autonomous programmed robot and implemented with complete wireless controls and operation [33.16] for the ISI of a large-scale FBR fuel reprocessing facility in India. The ISI is to be done initially using a camera that can relay the visual data wirelessly to the base station. The mobile robot has been designed to move on a guideway inside the waste vault surrounding all waste farm tanks. The robot has a serving manipulator, with five degrees of freedom and can reach various parts of the process tanks and, hence, closer inspection of the tanks is possible.

Automated remote sampling and analysis play an important role in the continuous monitoring of process parameter in a reprocessing plant because of the highly radioactive nature of the process solution. Automation makes the remote analysis faster and reliable and facilitates speedy control of process parameters. Chemical analysis of nuclear fuel reprocessing solutions requires greater accuracy for better accounting of fissile material.

Automated sampling system using a self-propelled vehicle that traverses the periphery of the process cell on a track system and collects sample bottles from track-mounted sample stations has been developed for fuel reprocessing applications at the Oak Ridge National Laboratory (ORNL) [33.17]. For accurate chemical analysis of process solutions, automated, remotely maintainable pipetter using microprocessor control was developed and implemented at the Idaho Chemical Processing Plant (ICPP),

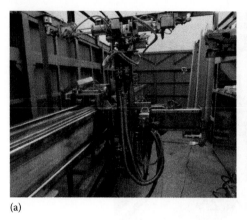

(a)   (b)

FIGURE 33.15   (a) Sample handling robot and (b) pipetting and capping/decapping modules of an automated remote sampling and analytical system.

United States [33.18]. An automated process fluid sampling system with a robotic sampling vehicle has been demonstrated in the remote operation and maintenance demonstration facility at ORNL [33.19]. The design of this equipment incorporates modular construction and special remote-handling features to facilitate remote maintenance. An automated remote sampling and analytical system developed for a fast reactor fuel reprocessing facility in India is shown in Figure 33.15a and b [33.20].

33.4   SENSORS FOR ROBOTICS AND AUTOMATION

Sensors are the feedback devices that enable device automation. Moreover, sensors also relay information about the current state of sites thereby enabling users to take corrective necessary action. The need of the hour is to have intelligent data processing from sensors and world-wide developments are geared toward this trend.

33.4.1   SENSORS FOR NONDESTRUCTIVE EXAMINATION

Sensor is the main link between the component being inspected and the nondestructive examination (NDE) instrument. The effectiveness of an automated inspection system comprising a robotic arm or a manipulator depends on the sensors or the end effectors used. For a nondestructive inspection of critical welds in inaccessible regions and for the detection of leaks and corrosion damage, optical, fiber optic, ultrasonic, electromagnetic, and infrared sensors are very effective. Working in this line, several novel sensors and sensing methodologies have been developed for fast reactors and associated fuel cycle applications [33.21–33.24]. These include electromagnetic acoustic transducers (EMATs), high-temperature PZT-lead zirconate titanate (Pb[Zr(x)Ti(1-x)]O3) (PZTs), and eddy current, giant magnetoresistive (GMR), and fiber optic sensors.

For couplant-free high-temperature ultrasonic examination of reactor components, EMATs are very effective and a very good alternative to piezoelectric transducers. Such transducers can generate and detect sound waves in electrically conducting materials through electromagnetic mechanisms by the use of a set of magnets and wires, as shown in Figure 33.16a. Figure 33.16b shows the spiral coil EMAT transmitter designed and developed at the Indira Gandhi Centre for Atomic Research (IGCAR), India, to generate bulk shear and longitudinal waves. This EMAT development is aimed toward the ultrasonic examination of main vessel welds at 200°C without the need of couplant.

Under-sodium viewing at high temperatures is an important requirement in sodium-cooled fast reactors for navigation during structural inspection. Under-sodium viewing is also needed to locate and identify loose parts under sodium, when required. In this regard, using a high-temperature

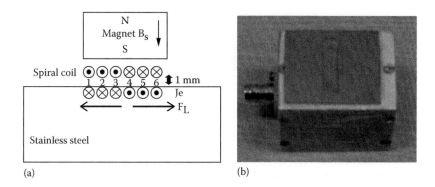

(a)   (b)

FIGURE 33.16 (a) Schematic of a spiral coil EMAT and (b) spiral coil EMAT developed at IGCAR.

PZT transducer and with a specially designed θ-Z manipulator, the automated ultrasonic imaging has been successfully performed at 550°C, in collaboration with CEA, France. Figure 33.17 shows the photograph of a steel plier and its ultrasonic image generated using θ-Z mode scanning. Figure 33.18 shows the photograph and under-sodium ultrasonic 3-D image of a set of elbows, generated by the ensemble of various slices of C-scan images. Recently, lead titanate–based ferroelectric glass ceramic PZTs have been developed for possible use in high-temperature ultrasonic examination up to 250°C. The scanning procedure and the 3-D image algorithm will be integrated with robotic devices for the under-sodium viewing of reactor internals of sodium-cooled fast reactors.

For the location of weld centerline in main vessel welds by ultrasonic examination, a novel high-temperature eddy current probe that can withstand 300°C and a liftoff of 10 mm has been developed. Eddy current–based position sensor has been developed to monitor the free-fall time of diverse safety rods in diverse safety rod drive mechanism (DSRDM) in the control plug of FBRs. It is not possible for the sensor to have any wired connectivity with the instrumentation. To circumvent this problem, an indirect excitation through inductive coupling has been incorporated in the

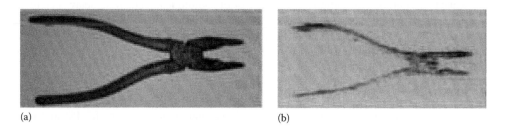

(a)   (b)

FIGURE 33.17 Photograph of pliers and its ultrasonic image obtained by θ-Z scanning. (a) Actual image and (b) ultrasonic image.

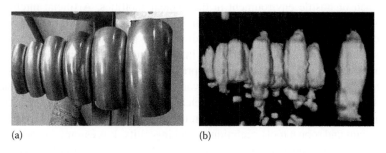

(a)   (b)

FIGURE 33.18 Photograph a set of elbows and their under-sodium ultrasonic image at 550°C. (a) Actual image and (b) under sodium ultrasonic image.

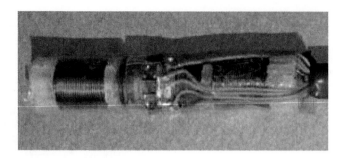

FIGURE 33.19   GMR-array sensor for magnetic flux NDE of tubes.

design of the sensor. The results have established the feasibility of using inductively coupled eddy current position sensor for DSR in FBR.

For the detection of intergranular corrosion in waste vault reprocessing tanks, an eddy current– GMR sensor has been developed. High-sensitive remote field eddy current probes as well as GMR array sensors have been developed for ISI of installed ferromagnetic steam generator tubes of FBRs. These sensors would be deployed in the robotic devices for automated ISI of steam generator. Array sensors offer high sensitivity and possibility for the nondestructive imaging of defects and working in this direction, GMR and eddy current array sensors have been developed. Figure 33.19 shows the GMR array sensor probe developed for magnetic-flux-leakage-based inspection of steam generator tubes.

Leak detection in the coolant loops of nuclear reactors is critical for the safety and performance of the reactors. Presently, techniques such as wire type–, spark plug–, and mutual inductance–based leak detectors are used and they cannot detect the location of the leaks. On the other hand, fiber optic sensor provides sensing over the entire length of the fiber. Raman distributed temperature sensor system that uses an optical fiber encased in stainless steel has been developed. Experiments were successfully conducted in the test section of LEENA (Leak Experiment in Natrium) facility in IGCAR, India, simulating a leak in a horizontal stainless steel pipe. Any temperature rise seen by the fiber sensor on contact with sodium is an indicator of sodium leak. The time delay seen by each sensor along with its spatial location across the cross section of the insulation is useful to reconstruct the temporal sequence of the leaking and the percolating sodium. Studies reveal that Raman distributed temperature sensor system can be used to dynamically detect the sodium fire in real time.

33.5   SUMMARY

An overview of the current developments of robotic and automation systems for ISI and remote operations in FBRs and associated facilities illustrates the range and diversity of requirements for the devices and the remote techniques. Successful ISI campaigns around the world for FBR and its plants have generated enough confidence to undertake and carry out complex inspections in a difficult-to-reach or hostile environment in the nuclear facilities. Artificial intelligence, image processing, and machine vision will further enhance the capabilities of the robotic and automated systems, especially concerning the obstacle finding, path planning, and accurate estimation of defects and discontinuities. In certain situations, information from several sensors may have to be combined incorporating the sensor/data fusion approaches for obtaining a comprehensive picture of the inspected regions. Any special development work has its own price tag and time factor that can be justified by the high price of downtime in a nuclear plant. The trend is toward more sophisticated and autonomous devices, though the majority of remote work will continue to be carried out using conventional elegant and robust tools, techniques, and ingenuity. It is essential to adopt a concurrent approach for extending the concept of remote inspection devices to repair tasks to carry out unforeseen repairs of critical equipment in nuclear plants.

REFERENCES

33.1 Asty, M., Vertet, J., Argus, J.P. (1980). Super Phenix 1: In-service inspection of main and safety tank weldament. *Specialists Meeting on In-service Inspection and Monitoring of LMFBRs*, Bensberg, Federal Republic of Germany, pp. 20–22.

33.2 Matsubara, T., Yoshioka, K., Tsuzuki, S., Matsuo, T., Nagaoka, E. (1985). Development of remotely controlled in-service inspection equipment for fast breeder reactor vessel. *Proceedings of an Internal Symposium on Fast Breeder Reactors: Experience and Trends 2*, Lyons, France, pp. 501–508.

33.3 Tagawa, A., Ueda, M., Yamashita, T. (2007). Development of the ISI device for fast breeder reactor MONJU reactor vessel. *J. Power Energy Syst.*, 1(1), 3–12.

33.4 Fenemore, P. (1987). Developing remote techniques for liquid metal reactors. *Nucl. Eng. Int*, 32(397), 55–56, 58.

33.5 Singh, A.P., Rajagopalan, C., Rakesh, V., Rajendran, S., Venugopal, S., Kasiviswanathan, K.V., Jayakumar, T. (2011). Evolution in the design and development of the in-service inspection device for the Indian 500 MWe fast breeder reactor. *Nucl. Eng. Des.*, 241, 3719–3728.

33.6 IAEA TECDOC 1433. (2005). *Remote Technology Applications in Spent Fuel Management*. IAEA, Vienna, Austria.

33.7 Debes, M. (2002). MOX fuel development in Japan. *International Seminar on MOX Utilization*, Tokyo, Japan, 2002.

33.8 Tsuji, N., Yamanouchi, T., Akahashi, K., Furukawa, H. (1987). Development and improvement of reprocessing technology at the Tokai reprocessing plant. *International Symposium on the Back End of the Cycle Product: Strategies and Options*, IAEA, Vienna, Austria, pp. 433–444.

33.9 Yamamuraa, O., Yamamoto, R., Nomurab, S., Fujiic, Y. (2008). Development of safeguards and maintenance technology in Tokai reprocessing plant. *Prog. Nucl. Energy*, 50(2–6), 666–673.

33.10 Jones, E.L. (1988). Remote handling developments for inspection and repair of highly active reprocessing plant. *Remote Techniques for Inspection and Refurbishment of Nuclear Plants*, BNES, London, U.K., pp. 43–48.

33.11 Jones, E.L. (1990). Remote handling and robotics in the BNFL Sellafield reprocessing plants. *Proceedings 38th Conference on Remote Systems Technology*, San Francisco, CA, Vol. 2, pp. 31–36.

33.12 Subramanian, C.V., Joseph, A., Ramesh, A.A., Raj, B. (1998). Wall thickness measurements of tubes by internal rotary inspection system (IRIS). *Proceedings of the Seventh European Conference on Non-Destructive Testing*, Copenhagen, Denmark.

33.13 Dhanapal, K., Singh, A.P., Rakesh, V., Rajagopalan, C., Rao, B.P.C., Venugopal, S., Jayakumar, T. (2013). Remote devices for inspection of process vessel and conduits. *Procedia Eng.*, 64, 1329–1336.

33.14 Dhanapal, K., Sakthivel, S., Balakrishnan, V.L., George, S.J., Rajagopalan, C., Venugopal, S., Kasiviswanathan, K.V. (2010). Poster: Experience and significance of the remote inspection of dissolver vessel in the pilot fast reactor fuel reprocessing plant. *National Seminar on Recent Advances in Post Irradiation Examination & Remote Technologies for Nuclear Fuel Cycle (RAPT-2010)*, Kalpakkam, India, September 23–24, 2010.

33.15 Rajagopalan, C., Rakesh, V.R., George, S.J., Chellapandian, R., Venugopal, S., Kasiviswanathan, K.V. (2009). Robotic devices for in-service inspection of PFBR and reprocessing plants. *International Conference on Peaceful Uses of Atomic Energy*, New Delhi, India.

33.16 Venugopal, S., Rajagopalan, C., Rakesh, V., Shome, S.N., Roy, R., Banerji, D. (2010). Design of remotely operated mobile robot for inspection of hazardous environment. *National Seminar on Recent Advances in Post Irradiation Examination & Remote Technologies for Nuclear Fuel Cycle (RAPT-2010)*, Kalpakkam, India, September 23–24, 2010.

33.17 Evans, J.H., Schrok, S.L., Mouring, R.W. (1989). Remote automated sampler development. *American Nuclear Society, Third Tropical Meeting on Robotics and Remote Systems*, Charleston, SC.

33.18 Dykes, F.W., Shurtliff, R.M., Hencheid, J.P., Baldwin, J.M. (1979). Rugged, remotely maintainable pipetter using microprocessor control. *27th Conference on Remote System Technology*, San Francisco, California, November 1979.

33.19 Burgess, T.W. (1986). The remote operation and maintenance demonstration facility at the oak ridge national laboratory, Spectrum '86, American Nuclear Society, *International Topical Meeting on Waste Management and Decontamination and Decommissioning*, CONF-860905–6, Niagara Falls, New York, September 14–18, 1986.

33.20 Chellapandian, R., Rajagopalan, C., Venugopal, S., Kasiviswanathan, K.V. (2009). Robotic sampling techniques for the analysis of process fluids in nuclear reprocessing plants. *National Conference on Robotics and Intelligent Manufacturing Process*, Centre for Intelligent Machines and Robotics, Corporate R & D Division, Bharat Heavy Electricals Limited, Hyderabad, India.

33.21 Sharmal, B.L., Mohanbabu, M., Gopalakrishna, M., Bandyopadhyay, M., Ramu, A. (2001). Experience on core shroud inspection TAPS reactors. *National Seminar & Exhibition on Role of NDE in Residual Life Assessment & Plant Life Extension*, Lonavala, India, December 7–9, 2001, p. 60.

33.22 Raj, B., Jayakumar, T. (1997). Recent developments in the use of non-destructive testing techniques for monitoring industrial corrosion. *Proceedings of the International Conference on Corrosion CONCORN*, Mumbai, India, Vol. 97, pp. 117–127, December 1997.

33.23 Ramesh, A.S., Subramanian, C.V. et al. (1998). Internal rotary inspection system (IRIS)—An useful NDE tool for tubes of heat exchangers. *J. Non-Destructive Eval.*, 19(3–4), 22–26.

33.24 Jansen, H.J.M., Festen, M.M. (1995). Intelligent pigging development for metal loss and crack detection. *Insight*, 37, 421–425.

34 Operator Training Simulators for Fast Breeder Reactors

34.1 INTRODUCTION

Operational safety is of utmost importance for any nuclear power plant. A well-trained operator is an asset for any nuclear plant. Hence, operator training plays an important role for the safe operation of the reactor in both normal and abnormal conditions. It is achieved by providing comprehensive training to the operators in all states of the plant operation: reactor in shutdown condition, start-up of reactor, reactor in operation, start-up of fuel handling, and reactor in fuel-handling state. Though class room training is important, a comprehensive hands-on training can be provided only through full-scope, replica-type operator training simulators (OPTS). This training is given not only to new operators, but also on a periodic basis to the qualified operators as a refresher course to maintain their skills at peak levels. The high operating safety and high efficiency record of the nuclear power stations in several countries are largely due to high-quality training imparted to operating staff through computer-based OPTS.

These computer-based OPTSs play a major role in imparting the best possible training to the operators, in view of the flexibilities and functionalities they provide. They incorporate all the features that allow the operator to be trained for normal and abnormal plant conditions covering the full spectrum of reactor operation, including plant transient conditions and design basis events under various categories. They play a key role in developing the required skill sets of the operator, utilizing the knowledge acquired through classroom training and field training. The training platform can be used for providing newer insights and enhancing the decision-making capability of the plant operators. It provides an opportunity for the control room operators to practice, in advance, the system operating procedures and other related plant operations by responding to alarms/indications/controls on the control panel and control consoles in the simulator control room. One of the significant developments in nuclear power plant training methodology is the addition of full-scope replica simulators in the operator training program. Some of the extra features and benefits of computer-based operator training as against in-plant training are listed as follows:

1. It is possible to impart full-fledged operation training and qualify the operator even before the actual plant is commissioned.
2. OPTS can be initialized to any one of the operational states, or it can be switched from one operating state to another without any delay. This is a very useful feature for operation training and significantly contributes toward reducing the training time.
3. Most of the malfunctions and emergency conditions that are mandatory for training purposes cannot be allowed to occur in the reactor because of safety reasons. This will also cause hindrance to normal operation. OPTS offers this flexibility with ease and elegance.
4. If the trainee operator makes a mistake on a real plant, the instructor has to step in and correct the error for reasons of safety. It would be more desirable to let the trainee see the consequences of his error to make the most lasting impression of the proper operating procedures. Features like context freezing, backtracking, etc., facilitate these functions.

Computer simulation involves the application of mathematical models to real systems to derive its characteristics and behavior without actually constructing or operating the system concerned.

To build OPTS, the subsystems and components that contribute toward the plant dynamics and those associated with the operation and control functions of the reactor are identified to start with. Their individual behavior and the interlink among them are mathematically modeled. This mathematical model is then translated into a computer code—the simulation software—that will eventually run in the simulator computer.

This simulation software will take inputs from operator-activated control panels, consoles, etc., and process it in accordance with the plant model and give the computer-generated output on the control panel or console by actuating the appropriate devices or their equivalents. This software also supports an instructor station from which a training instructor can interact with the computer software to give appropriate training lessons to the operator, insert different plant scenarios, introduce malfunctions, monitor the operator response, and evaluate his performance.

34.2 TYPES OF SIMULATORS

The training simulators are broadly classified based on two parameters: the extent of plant to be covered in simulation and fidelity in replication of plant control room. Based on the extent of plant to be covered, the simulators are classified as full-scope and part-task simulators, and based on the fidelity in replication of plant control room, the simulators are classified as replica and nonreplica simulator (Figure 34.1).

34.2.1 FULL-SCOPE SIMULATOR

The full-scope OPTS provides horizontal coverage of all the main systems of the reference plant. It incorporates detailed modeling of all the systems of the reference plant with which the operator interfaces in the actual control room environment. The operator will get comprehensive training on all subsystems.

34.2.2 PART-TASK SIMULATORS

The part-task simulator is focused only on specific plant systems. These systems are represented with features of a full-scope simulator; in this way, detailed mathematical modeling of the referenced plant systems is included, and just a part of the actual control room is replicated with all

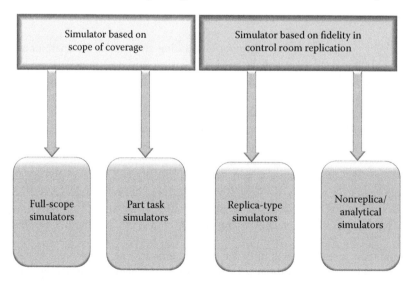

FIGURE 34.1 Types of simulators.

key instrumentation, controls, and alarm signals. The systems that are not included in the human–machine interface (HMI) are simulated with a reduced scope or not simulated and considered as always *on service*, just to satisfy the interactions of the main systems.

34.2.3 REPLICA-TYPE SIMULATORS

In replica-type simulators, the main control room is replicated (one to one) in all aspects with respect to reference plant control room extending to hardware panels, consoles, operator desktops, chairs, lights, layout, and the ambience. The main advantage of the replica-type simulator is its ability to do strict procedural training with respect to location and function of each instrument and control on the panels.

34.2.4 NONREPLICA-TYPE SIMULATORS

In nonreplica simulators, all important indicators and controls are emulated by computers with CRT displays called virtual panels, and no control panels or consoles will be present. This will not create control room atmosphere. However, operator will get some amount of training, and the system will be less costly.

34.3 OPERATOR TRAINING SIMULATOR

The role of simulators in imparting plant-related knowledge to the operators is well recognized, especially with respect to main systems, safety-critical and complex systems of the plant. Here, the full-scope replica simulator and part-task simulators are found to play a major role. The plant OPTS is basically a training tool designed to replicate the steady-state and dynamic behavior of the plant in response to operator actions (Figure 34.2). It is a combination of mathematical

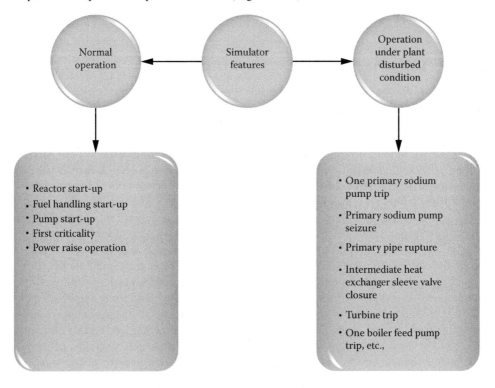

FIGURE 34.2 Features of a training simulator.

models representing the plant components, control system emulation, and HMI. The operators are trained to carry out the plant operations on the simulator platform as per the scenario depicted by the instructor, observe the response as they would occur in the real plant, and understand the plant dynamics.

34.3.1 Resource Identification for Building a Training Simulator

The first and foremost factor to be considered is identification of resources. Here, resources refer to man, machine, and money. The main challenge here is the identification of human resource for developing the process models, availability of domain experts in the related field, acquiring necessary software tools for building the models, and capability to build in-house models wherever necessary. Assuming that enough budget allocation is available, setting up of a hardware platform is the next requirement, which can be installed with a high-end machine as a simulation server and number of work stations connected in a local area network. Procurement of software tool and other accessories would be a parallel activity, which needs to be loaded on the simulator server to complete the setup.

34.3.2 Model Development Platform

The model development platform will be used for developing the process models and testing the models extensively before they can be subjected to verification and validation process. The model development platform consists of simulation computer and work stations connected in a local area network. The server is loaded with UNIX operating system and simulation tool. The work stations are loaded with X-Windows software for accessing the simulation server. The tool facilitates integration of external code developed for various applications. The process models can be developed and tested individually from the work stations. On completion of model development, the integrated run of the simulator can be carried out by accessing the simulation server from any of the work stations.

34.3.3 Deployment Platform for Conducting Operator Training

The models once developed need to be deployed for conducting training. Essentially, a training center has to be built for accommodating a full-scope replica OPTS (Figure 34.3). The simulator control room needs to be replicated with respect to reference control room in terms of selection and installation of control panels and operator consoles including panel color, display and alarm indications, layout, lighting arrangement, seating arrangement, and the ambience. The specification for training simulator has to be carefully drawn based on the latest approved design documents and drawings of plant main control room.

The simulator control room layout design also needs to be in tune with the reference plant layout. This can comprise simulator control room, instructor room, backup control room, handling control room, engineers' room, local control center room, etc. (Figure 34.4). The following example shows one such training simulator replicating the main control room of a fast breeder reactor (FBR).

The simulator hardware architecture consists of simulation computers, input/output (I/O) systems, control panels, operator information consoles, instructor station, simulation network, power supply and distribution system, backup control panel, and handling control panel. The simulation computer/server facilitates computation and execution of various mathematical models of the reactor components in real time. The server is interfaced with the hardware panels through I/O systems, and the signal communication is established using local area simulator network. The operator console handles overall monitoring of the most important and frequently used control signals. The instructor station facilitates control and monitoring of simulator operations/operator actions and conducting training sessions. All plant scenarios are loaded from here for carrying out

FIGURE 34.3   Full-scope replica OPTS of FBR.

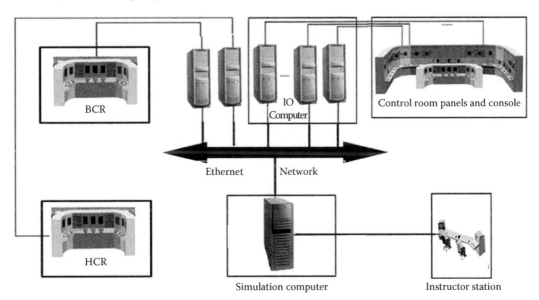

FIGURE 34.4   Hardware architecture of FBR simulator.

training program for the operators. Backup control panel is a part of the training simulator and serves to handle emergency situations like the inaccessibility of main control room in case of fire. Similarly, handling control panel forms a part of hardware architecture and facilitates initial fuel loading and subsequent fuel handling operations.

34.3.4   SOFTWARE ARCHITECTURE OF SIMULATOR

Generally, the software tool for simulator should have the capability to build the process models representing the process components, execute, and provide the input and output signals in a format compliance with the plant data. In addition, the software tool should facilitate generating

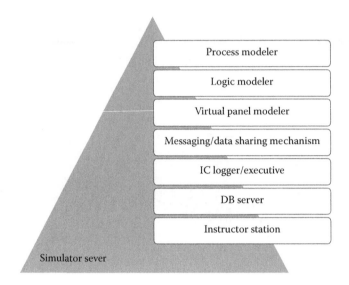

FIGURE 34.5   Simulation software architecture.

control and logic models representing the process control logics for real representation of the plant. Apart from these, the tool should have graphical user interface and capability to generate system flow sheets, display the process signal profiles, dynamic display of simulated signals on the flow sheets, development of virtual control panels consisting of alarms, process signal displays, and controls for plant monitoring and control. Additional capability includes establishing communication between various processes, control and synchronization of operations of the various simulator components, mechanism for data sharing, save-and-restore functions with the use of a database server, display/messaging of information about the state of the simulator, and provision for user interface to simulator environment. The software should cater to conducting training sessions through instructor station by loading all plant scenarios and monitoring simulator operations/operator actions.

Figure 34.5 provides a typical software architecture of a training simulator of an FBR.

The software architecture consists of three major components: process modeler, logic modeler, and virtual panel modeler. The process modeler is used for developing process models; the control logic modeler is for developing control logics for process components, and the virtual panel modeler is for developing virtual control and console panels for monitoring the process parameters. All other modules are supporting modules like messaging and data sharing mechanism for establishing communication between process, logic, virtual panel, and database; the IC logger to save and restore information about the state of the simulator; executive to control and synchronize the operation of various components of the simulator; instructor station to provide user interface to the simulator environment; and DB server to store and retrieve all the data pertaining to simulated models.

34.3.5   Identification of Systems/Plant States to Be Modeled

Development of a full-scope replica OPTS calls for concrete ideas about the training requirements with respect to the plant. The reactor systems to be modeled including various plant states need to be identified by having elaborate discussions with the available experts in the relevant field. This should take into account the operational experience gained on FBR and the event analysis report of the respective plant under consideration and major events that call for attention at international forum on nuclear power.

Figure 34.6 indicates the systems and various plant conditions identified for a FBR training simulator.

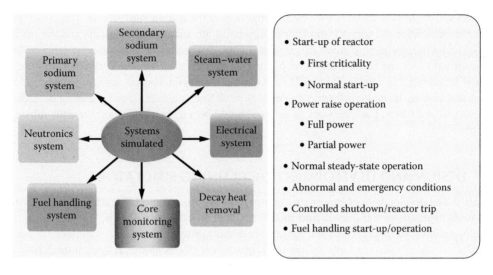

FIGURE 34.6 Simulated systems and plant conditions of FBR.

34.3.6 Generation of Scope Document

A scope document has to be generated elaborating the scope of simulation, systems considered for simulation, and systems that are not considered for simulation. The bench mark transients to be simulated along with the system/component/device and related malfunctions also should form the essential elements of the training simulator. The procedures that are necessary for the operator to be trained like reactor start-up procedure and emergency operating procedure should be covered in the scope document. The hardware setup and software tool for development as well as deployment need to be outlined in the document for completion.

34.3.7 Building the Competency Level for Model Development

All engineers, identified for design and development of simulated process models, need to be trained and made competent enough to build the simulator. The skill development should include general introduction to hardware architecture, simulation server installation, operating system installation, usage of simulator software tools, their basic components, functions of each component, and configuration management. This can be achieved by arranging series of lectures on plant design details and operation details toward good understanding about the plant dynamics under steady-state and under transient conditions.

34.3.8 Formation of Internal Verification and Validation Team

Formation of independent verification and validation (IV&V) team in the initial stage will prove to be useful in the long run. An IV&V team consisting of domain specialists has to be constituted to streamline the model development activities. The job of IV&V team includes component data verification, validation of process models, logic models, and performance checking under steady-state and transient conditions. Free flow of information between the teams is highly essential for appropriate knowledge transfer and provides means for clarifications regarding the design and development of process models.

34.4 BASIC SIMULATOR MODEL

Any training simulator will have three main components, that is, process model, logical model, and virtual panel model. The process model representing actual functioning of the system can be modeled using process modeler. The logic model can be developed using logical modeler, representing interlocks and controls associated with each component of the reactor subsystems. The virtual panel model can be developed using VP modeler representing the actual control panels and the operator consoles including alarm indications, display, and controls. The interface between the process, logic, and the virtual panel can be established by passing the input/output/feedback signals across them.

In the initial stage, a conceptual model with the main components will suffice the model development and testing. It can be followed by detailed modeling at a later stage by adding devices like pipes, valves, and filters. Each network generated should be a representative of a reactor subsystem of the plant.

34.5 DESIGN AND DEVELOPMENT OF TRAINING SIMULATOR

The design and development of training simulator involves process model development using simulation tools available in the market and model development that needs to be built in-house with special attention with respect to uniqueness of the nuclear power plant.

In general, the simulation tools satisfy the requirement of conventional system modeling, that is, process model development pertaining to steam–water system and electrical systems. Nuclear core being unique to each power plant depends upon the type of reactor, that is, thermal reactor or fast reactor. The core configuration and the cooling systems are quite different for the type and choice of nuclear reactor. The available simulation tools may not have the capability to build such models, and they have to be developed in-house with the technical expertise available in the center.

34.5.1 STEPS INVOLVED IN THE DEVELOPMENT OF TRAINING SIMULATOR

Development of training simulator requires good system knowledge, understanding of various connected processes and associated equipments in each system, and overall knowledge about the plant. The acquisition of knowledge could be through work experience, attending technical lectures, discussions with operation and design experts, and carrying out a detailed study on the design and operation aspects of the plant. Figure 34.7 provides various steps involved in the design and development of process model.

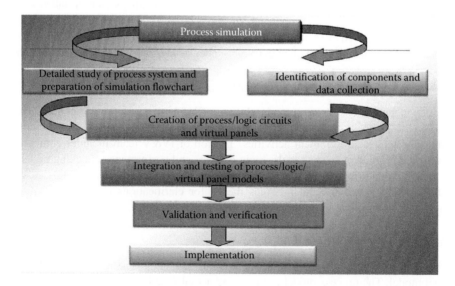

FIGURE 34.7 Steps to build a training simulator.

34.5.2 Development of Process Simulator

Reference is drawn from the scope document for carrying out the reactor systems to be modeled. The development of process simulator involves modeling of various process networks consisting of components and devices. Each subsystem will have a number of components and devices connected together to represent a process function. The major components of each system that are required to be modeled need to be identified in the initial stage.

A sample list of components identified for an FBR training simulator is as follows:

Neutronics core, control safety rod, diverse safety rod, primary/secondary sodium pumps, heat exchangers, cold/hot sodium pool, surge tank, steam generator, turbine generator, main condenser, condensate extraction pumps, condenser cooling water pumps, low-pressure heaters, deaerator, boiler feed pumps, high-pressure heaters, feed water control station, valves, filters, connecting pipes, and associated logics and controls.

Simulator flow sheets need to be prepared based on the main system flow sheets indicating the components to be modeled. This is followed by the collection of component specifications and process data by referring to design and operation documents and related drawings. Then, the process networks are developed by modeling the individual components using the data collected and connecting them appropriately for testing the performance. One has to ensure satisfactory functioning at this stage before further development can be carried out. While modeling, proper naming convention has to be followed for easy identification of system components and devices. A set of rules can be formulated for naming the models in order to fall in line with the naming of actual system components.

The logic model and virtual panel models are needed to be developed simultaneously for all identified systems based on the inputs taken from the operation notes and process and instrumentation diagrams (P&I diagram) and approved panel drawings. It should represent the actual system interlocks and the control panel alarm and displays, respectively. In general, the input process signals are taken from the process model, processed by the logic model as per the set points/thresholds and interlocks. The logic model generates output signal and sends it through I/O unit for displaying on the virtual panel/control panel and automatic control of the process components and devices. Figure 34.8 indicates a typical logic model developed for feed water system control and virtual control panel for monitoring the feed water system parameters.

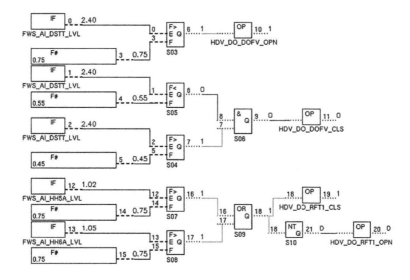

FIGURE 34.8 Logic model developed for feed water system.

34.6 INTEGRATION AND PERFORMANCE TESTING

On completion of the development of process/logic and virtual panel models, all the models need to be integrated and tested thoroughly. Integration and performance testing is the most crucial phase in the process model development and implementation cycle. It is the most challenging part of the training simulator development process. Utmost care should be taken while integrating the process models with the other subsystem models, which can be carried out by checking the linking parameters. Another important aspect of integration and testing is to check and verify the communication between various processes, cycle time of each process, logical conditions of various components, display of associated alarms and system parameters, correct functioning of process, logics and controls, etc.

Integration and testing can be carried out in multiple stages depending upon the requirement. Mainly, the testing can be carried out by integration of process model with the logic and virtual panel models and integration of process model with other reactor subsystem models.

The performance testing is carried out to verify whether the models developed are as per the design philosophy and the accuracy of closeness of the model with respect to system design. The performance testing also provides a platform for checking the degree of completeness and accuracy in meeting the training requirements of the simulator. Simulator performance testing comprises operability and scenario-based testing. Simulator performance testing is conducted in a fully integrated mode to check the correct functioning of the process model under steady-state and transient conditions of the plant.

34.6.1 INSTRUCTOR STATION FACILITY FOR LOADING PLANT SCENARIOS

An instructor station facilitates loading of plant scenarios for performance study and system analysis. It also provides a platform for conducting training sessions for the operators by monitoring the simulator operations, operator actions, and response. The other important features include loading of various reactor states, capturing of relevant process signals, and forcing a specific condition of a process. The important commands that are operable from the instructor station include RUN, STEP, BACKTRACK, FREEZE, REPLAY, and SNAPSHOT. The RUN command is used to execute the models in an integrated mode, which is a continuous and synchronized operation. The STEP command allows execution of models for a period of one simulator cycle (200 ms) and is mainly used for analysis purpose. The FREEZE command suspends all calculations and executions with respect to models temporarily. The REPLAY command recomputes the output variables in response to the stored input signals, and the SNAPSHOT command facilitates capturing of the current state of the simulator at a particular instant. Simulator instructor is provided with an option of BACKTRACK and REPLAY for checking the previous plant operating state. The BACKTRACK command enables the snapshot function to store the status of the simulator for a selected time interval of 2/5 min. The instructor can start the simulator from a predefined state, create various incidents/malfunctions for training, freeze a condition for explaining, backtrack a condition to explain the trainee's behavior, etc. Operator's response to each malfunction/event will be recorded and evaluated to check the correctness of operation. The backtrack feature allows such an analysis for improving the performance of the operators. Operators can also be trained on various plant operating procedures.

34.6.2 PERFORMANCE TESTING UNDER STEADY STATE

Simulator performance testing should be conducted in a fully integrated mode to check the correct functioning of the process model under steady-state and transient conditions of the plant. This is mainly to check the degree of completeness and the accuracy of the models built. As per the standards, steady-state performance testing needs to be carried out at three different power levels

(100%, 50%, 25%) spanning at 50% of the operating range. At each level, the process simulator has to be run in the integrated mode for few hours before going into further analysis. The profiles can be captured through continuous operation over a power range by recording the input and output parameters of the simulated components, which may include variations of flow and temperature, level changes, and pressure variations. The recorded simulated parameters should be compared with the reference plant data.

The mass balance and thermal balance checking also need to be carried out carefully. Any discrepancy noticed should be addressed and solved before any further development. After achieving satisfactory results, one can move on to transient testing.

In the absence of plant data, design data are taken as basis for performance evaluation. A comparison table is prepared as shown later in text for the evaluation of the performance testing. The percentage error indicates the closeness of process simulation with the reference unit. Lower the error, higher the closeness indicating successful integration of the models.

34.6.3 PERFORMANCE TESTING UNDER TRANSIENT STATE

Transient simulation and analysis is the most important phase in the development and implementation process of a full-scope replica OPTS. The OPTSs are qualified based on the steady state and transient performance of the process models in an integrated approach.

The transient simulation is carried out mainly to ensure that the dynamic behavior of the plant under various abnormal conditions is always toward safe operation of the plant. The transient simulation testing can be carried out by loading various plant scenarios from the instructor station. In any nuclear power plant, the event analysis report is prepared by the design experts after extensive studies and analysis of dynamic behavior of the plant under various abnormal conditions to ensure that the plant parameters do not cross the stipulated design safety limits. This ensures safe operation of the plant even under the worst possible condition. Hence, event analysis report forms the basis for evaluating the process simulator for transient testing.

As per the ANSI 3.5 standard meant for nuclear power plant simulator, the model behavior under system transient conditions should be as close as possible to the plant behavior with the error percentage limited to 15–20. As the plant data will not be available during the initial phase of the plant commissioning, the event analysis/plant safety analysis report will be taken as a reference document for the comparison of simulated test results.

34.6.4 BENCHMARK TRANSIENTS

The bench mark transients represent a list of important transients identified for simulation, based on the plant safety analysis report. Generally, they are used for evaluating the process models and qualifying the training simulator for the purpose for which it has been built. The bench mark transients for an FBR should cover major events like uncontrolled withdrawal of one control rod, one primary sodium pump trip, one secondary sodium pump trip, one primary/secondary sodium pump seizure, IHX sleeve valve closure, primary pipe rupture, one CEP trip, both CEPs trip, one BFP trip and not taken over by standby, both BFPs trip, one or both CWP trip, turbine trip, reactor power set back, loss of feed heating due to failure of heaters, generator trip, class IV power supply failure, and station blackout.

34.7 VERIFICATION AND VALIDATION OF TRAINING SIMULATOR

This is the most important phase of the simulator that qualifies the simulator for implementation to do the intended function. Verification testing is performed by comparing the simulated component to the original requirement to ensure that each step in the model development process completely incorporates all the design requirements.

Validation testing is performed by comparing the simulated process parameters to the actual system parameters of the plant either in stand-alone mode or integrated mode. Normally, verification and validation testing is done along with the system experts, who are basically system designers.

All the simulated models should be subjected to verification and validation in order to qualify the process simulator for training purpose. An IV&V team consisting of number of domain experts specialized in process design and instrumentation and control design can be specially constituted to carry out the validation process. The team can be entrusted with the responsibility of verification and validation of the process models before implementation. Wherever deviations are noticed beyond the limits specified by the standard, modification and model tuning have to be carried out and as per the suggestion by the IV&V team.

34.8   IMPLEMENTATION

Once the process models are approved by the IV&V team, porting and implementation is the next step to be followed. The process models can be ported to the operator training platform by taking up backup using an appropriate media. Suitable porting procedure needs to be established for smooth installation and commissioning of a training simulator. It is appropriate to have a checklist of activities to be performed prior to porting of simulated models. The important activities with respect to any training simulator will include commissioning of control room panels, simulation server, uninterrupted power supply system, networking of simulation server, panel I/O systems, and display stations. Establishing a procedure for checking the signal communication between the server and the panels/display stations is equally important.

On successful porting and testing, the simulator will be subjected to verification and validation process once again. The approval needs to be obtained for the simulator training platform using which the plant operators will be trained. It is advisable to train a group of people identified for maintaining the training simulator regarding starting and running of a simulator, loading various plant scenarios from instructor station, and the use of instructor station commands like RUN, STEP, BACKTRACK, FREEZE, REPLAY, and SNAPSHOT.

34.9   CONFIGURATION MANAGEMENT OF TRAINING SIMULATOR

The configuration management of training simulator is essential after successful implementation and commissioning of the training simulator platform. The main purpose of establishing a configuration management practice is to maintain and operate the system more effectively and efficiently during its life cycle. The moment a system is made operational and put into full-fledged usage, the configuration management practice needs to be followed meticulously to maintain the system. Configuration management practice allows the vertical growth of the system while ensuring the system performance and compliance to the design requirements. Following are some of the salient features of the effective configuration management practice:

- Keeping pace with the technological changes with respect to hardware platform (server and development nodes)
- Periodic checking and revision of scope document and system requirement specification with respect to plant
- Keeping track of the modifications in the running plant
- Methodology for implementing the changes in a periodic manner
- Version control of simulator models and related documents
- Methodology for verification and validation as a part of version control measure

34.10   REFERENCE STANDARDS

One should follow the available reference standards for developing an OPTS for a nuclear power plant. The standards include *ANSI-3.5-1998* for operator training and examination, *IAEA–TECDOC-995*, and *IAEA–TECDOC–1411*. These standards establish criteria for the degree of simulation, performance, and functional capability of the instrumentation and controls of the simulated control room.

Section IX

Economics of SFRs with
a Closed Fuel Cycle

35 Economics of SFRs with a Closed Fuel Cycle

35.1 INTRODUCTION

Advanced fast reactors have attained a high level of reliability and environmental cleanness. With fuel breeding, the fast reactors are in a position to supply fuel themselves for the foreseeable future at almost any rates of development of power generation. Experts consider that at the present stage of development, one of the main limitations to the introduction of fast reactors into power generation could be their insufficient economic competitiveness in electricity generation in comparison with existing power sources (thermal reactors, organic-fuelled power stations, etc.). In particular, sodium-cooled fast reactors (SFRs) are costlier than the current pressurized water reactors (PWRs). In order to make SFRs economically competitive, there is a need to analyze the reasons for the high capital cost and means to reduce the cost in stages to make them competitive to well-established PWRs and, subsequently, fossil power plants. The main objective of this chapter is to make the readers familiarize the associated economic aspects of SFR with reference to technological issues. Only specific parameters that are important to be considered are addressed. There are many references on this topic, for example, "Appendix D: Economics Calculational Approach in the Fast Spectrum Reactors" by Alan E. Walter et al. [35.1] have brought out this topic more comprehensively with mathematical approach. The readers are encouraged to refer the same for more details.

35.2 OVERALL PERCEPTION ON THE ECONOMY OF SFRs

The economic assessment of various power-generating options indicate that, in case of higher-calorie fuels, the capital costs of machines and plants for the fuel burning and its conversion into usable energy forms are deciding factors, which need certainly to be reduced to achieve cost competitiveness. Nuclear fuel has higher specific energy capacity by many orders compared to conventional fuels. Enormous thermal energy is released per core unit volume. For example, power densities of the liquid-metal-cooled fast reactor (LMFR) core reach 500 kW/L and more. No respective analogues are possible in stationary fossil-fueled steam supply systems (FSSSs). In spite of this, the nuclear-fueled steam supply systems (NSSSs) are comparatively bulky and expensive. The reasons for this can be explained as follows:

In conventional power plants, a fossil-fuelled steam generator (FSG) represents the FSSS. The FSG entirely involves the whole complex of processes: fuel burning, heating up and evaporation of water, steam superheating and reheating, etc. The fuel and water are supplied to the FSG, and the generated steam is delivered to the turbines. In the comparable NSSS, these processes are separated: in PWR and LMFR, for example, heat is generated in the reactor, and then it is transported to the steam generator (SG) for water heating, evaporation, and steam superheating and even reheating (in some SFRs). So in the NSSS, in contrast to the FSSS, a new sophisticated rather critical and expensive system is used for reactor cooling and heat transport from the reactor to the SG. Second difference is that the high linear power ratings and thermal inertia (decay heat release): considerable heat accumulated in fuel pins (the temperature at the pin centerline being about 2000°C or even more) calls for the highly reliable reactor cooling system including relevant coolant, circulating, and heat-removing equipment.

Third one is the high requirement of safety and reliability, which calls for several diverse and redundancy systems. As a rule, the reactor design includes a number of parallel heat transfer loops. The realization of the heat release and steam generation processes in different equipment units, the presence of a ramified system for the heat transport from the reactor to the SG including pumps, heat exchangers, pipelines, and other components manufactured of high-quality materials and technologies according to *nuclear* specifications are some of the causes (in the LMFR, apparently the main one) of NSSS's high capital costs. Other causes of higher costs of NSSS compared to those of FSSS are (1) radioactivity accompanying physical processes in the nuclear reactor calls for using expensive materials for the neutron and biological shield and (2) need for the systems of prevention and minimization of detrimental failure effects calls for both materials consumption and building architecture of the nuclear power plant (NPP) in large proportion.

Although comparatively low fuel costs can make up for the high capital costs to some extent and, in turn, make the electricity price of the existing NPP competitive with fossil-fuelled power plants, the negative factor of NSSS's high cost still remains. In the 1990s, an important stage of conceptual design studies of large monolithic fast reactor power plants of 1300–1500 MW(e)—Superphenix-2 (SPX-2) in France, commercial demonstration fast reactor (CDFR) in the United Kingdom, and Schneller Natrium Gekuhlter Reaktor-2 (SNR-2) in Germany—was completed. Then, European countries concentrated their efforts on the development of the European fast reactor (EFR) project of 1500 MW(e) and the USSR and, afterward, Russia concentrated on the development of BN-800 and BN-1200 fast reactor designs. The Russian developers possess a well-proven fast reactor technology owing to the construction and successful operation of four LMFRs. Estimates show that the total expenditure, to this date, on the fast sodium-cooled reactor technology in Russian Federation is approximately US$12 billion [35.2].

The economic competitiveness of fast reactors depends on the uranium price. It is well known that the main aim of the extensive introduction of fast breeder reactors into nuclear power generation is for the considerable expansion of the fuel and energy base due to the effective utilization of the natural uranium (and not only the fissile U-235 isotope, as in thermal reactors), and also possibly thorium. As the calculations shown in Figure 35.1 [35.3], the economic advantages of fast reactors manifests if there is an limitation for reserves of cheap natural uranium. At the present

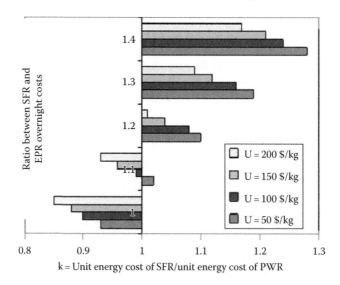

FIGURE 35.1   SFR to PWR energy price for various uranium prices. (From Loaec, C. and Linet, F.L., The French scenario, *IAEA-INPRO Meeting on Joint Study on an Innovative Nuclear Energy System Economics*, Cadarache, France, February 6–9, 2007.)

time, there is no other such assimilated source available regarding the production of fuel and electric power as fast breeder reactors.

It is a necessity for the SFR designers to develop advanced/innovative reactor designs to make the reactor competitive to other energy options.

35.3 ECONOMIC ASSESSMENT OF INTERNATIONAL SFRs

35.3.1 COMMON FACTORS IN THE CONSIDERATION OF PLANT ECONOMICS

The first and the most obvious of the common factors is size (the economy of scale). If the design is only marginally different, the overnight capital cost of a larger unit is significantly cheaper than for a smaller version. The smaller size on one hand allows them to be more accessible to modularization, that is, construction and deployment of a larger number of standardized units. Modularization reduces the requirements for more expensive and time consuming on-site construction and also allows more factory fabrication. Still, modularization is considered a common factor, because it is also employed in the most recent large plant designs and thus has to be comparatively evaluated. Similar considerations apply for the deployment of multiple units at a single site. The obvious advantages are the sharing of infrastructure and better utilization of site material and human resources. Both small and large plants can be deployed in multiples at a single site, and, in fact, several multiunit sites with thousands of installed MWe do exist.

Another factor that needs parallel evaluation is learning. It is well known that an nth-of-a-kind (NOAK) plant costs less than a first of a kind (FOAK) because of the lessons learned in the construction and deployment of earlier units. The learning curve generally flattens out after five to seven units. Comparing a 350 MWe and a 1400 MWe plant, the NOAK is reached after approximately 2100 MWe for the small modular reactors (SMRs) and 8400 MWe for the large plant. Thus, 18 more units of the SMR can take advantage of the learning factor before the large plant is able to *catch up*. Learning is definitely an advantage for the SMRs in the early stages of the market before it eventually equalizes as the market for both designs mature.

In addition to the learning *worldwide* (it does not matter where the units to reach the nth are built), there is also learning *on-site*, obtained from the construction of successive units. Specific characteristics of smaller reactors are smaller size, simpler design, increased modularization, higher degree of factory fabrication, and serial fabrication of components leading to a shorter construction time. The unit cost of smaller reactors is of course a fraction of the cost of a larger plant. This reduced requirement of the front-end investment can be *the* critical factor for a utility or country with limited resources. Finally, the combination of the reduced front-end investment and the shorter construction time makes it possible to minimize the cash flow through progressive construction/ operation of multiple modules deployed in succession. Assuming that the construction of a module starts when the preceding one initiates operation, the power-producing module will finance construction of the following one.

The case for the fast reactor has always been strategic in nature. By utilizing uranium in fast reactors rather than in thermal reactors, the energy potential of uranium is increased approximately 60 times, and it can be ranked as a major world energy resource. This technology, therefore, bestows a degree of energy independence on countries that have invested in it. Broadly for SFR, the focus of the optimization process could be associated both at the conceptual level and at the level of design parameters/options.

35.3.2 ECONOMICS OF NUCLEAR POWER PLANTS: GENERAL APPROACH

For a country that possesses all the requisite scientific knowledge and technological base, the need for venturing into large-scale deployment of nuclear power is mainly guided by economics. The economic competitiveness of nuclear power generation, as compared to other modes, is however largely influenced by the prevailing rates of fissile material in the domestic market or being made available to a particular country through international markets. The scenario is very complex and, hence, difficult to generalize as it would vary from country to country. It is therefore prudent to appreciate various factors that influence the economics of nuclear power, and the important ones, construction time and output rating, are discussed briefly in the following.

Before starting the discussions, it must be highlighted that the severe challenge posed by the necessity to reduce the emissions of greenhouse gases, especially in the electricity generation sector, has led to renewed interest in the construction of new NPPs. These would initially replace the aging stock of existing reactors, then meet electricity demand growth, and may eventually replace some of the fossil-fired electricity-generating plants. In the longer term, the promise is that a new generation of NPPs could be used to manufacture hydrogen, which would eventually replace the use of hydrocarbons.

35.3.2.1 Construction Time

An extension of the construction time beyond the forecast does not directly increase costs, but indirectly it increases the interest during construction and it is also seen often as a symptom of problems in the construction phase such as design issues, site management problems, or procurement difficulties which will reflect in higher construction costs. In a competitive electricity system, long-forecast construction times would be a disadvantage because of the increased risk related to changes in circumstances, making the investment uneconomic during its construction stage and because of the higher cost of capital in a competitive environment.

35.3.2.2 Output Rating

The maximum output rating of the plant will determine how many kilowatt-hours of saleable power the plant can produce. Particularly, if problems of corrosion and poor design have meant that most of the plants cannot sustain operation at their full-design rating, then it is to be thoroughly investigated. For the more widely used designs worldwide, plant *derating* has not been an important issue in recent years, and most plants have been able to operate at their design level. Indeed, in some cases, changes to the plant after it has entered service—for example, use of a more efficient turbine or increase in the operating temperature—have meant that some plants are able to operate at above-design rating. For future projects, there is still a small risk for unproven designs that the plant will not be able to operate at as high a rating as planned, but this risk is probably quite small compared to other risks incurred.

It is widely observed and accepted in engineering that the cost of a piece of equipment is not directly proportional to its capacity; rather the proportionality is through the power law:

$$K = a + bY^n$$

where
  K is the cost
  Y is the capacity
  a and b are constants
  n is the scale exponent

The relative value of the constant a with respect to K can be very important in the ratio of the costs for two power plants of different sizes. In addition, economy of scale is valid only over a limited range; it is questionable that an empirical law with fixed constants can be valid over a wide range as, for example, from 100 to 1300 MWe.

Engineers have traditionally looked at the forward, or prospective, or *bottom up*, structure of the costs in estimating the scaling exponent n, while economists have customarily looked at the overall costs retrospectively. As the overall costs include several social, regulatory, and economic elements and have become an increasingly larger portion of total costs during the 1970s and the early 1980s, cost scaling has become less evident. The following correlation is based on the data collected from the 46 nuclear units of PWRs completed between 1971 and 1978 in the United States:

$$\frac{\$}{\text{KWe}} = 6.41 f_1 f_2^{-0.105} f_3 f_4 f_5 \text{MW}^{-0.2} f_6^{0.577}.$$

f_1 = Based on the location of plant (e.g., 1.28 if location is northeast in the United States, 1 elsewhere)

f_2 = The number (1 or more) of reactors the architect/engineer has been involved in

f_3 = 0.903 if multiple unit, 1 if single unit

f_4 = 1.34 if *dangling*, that is, if the unit was still under construction at the time of analysis but some cost numbers had already been available

f_5 = 1.20 if the unit has cooling towers, 1 if once-through cooling

f_6 = Cumulative nuclear capacity in the nation

Considerable data are required to generate an empirical correlation of this nature. Considering that relatively less number of fast reactor units have been built either in a given country or worldwide, it is, however, difficult to arrive at such specific correlation as given. Nevertheless, the same could be used as an indicative measure.

For example, an assessment of the generating cost of CDFR was carried out by a joint UK nuclear industry team in the period 1987–1988. The assessment showed that the demonstration plant would have a generating cost about 20% more than a PWR of the type then planned for a small series under UK conditions, but a follow-on fast reactor without the FOAK costs would be competitive with a PWR. It was also shown that no one element dominated the generating cost of the follow-on plant so that any attempt to reduce the cost significantly would have to address all elements.

35.3.3 EXPERIENCES FROM INTERNATIONAL SFR WITH REFERENCE TO ECONOMY

Based on the experience of fast reactor design and construction, the capital cost distributions between the main NSSS systems and components are observed as seen in Figure 35.2 [35.4]. The comparison with the 1400 MW(e) PWRs plants, built at the same period, showed that the cost per unit power installed and electricity production of Superphenix-1 (SPX-1) is much

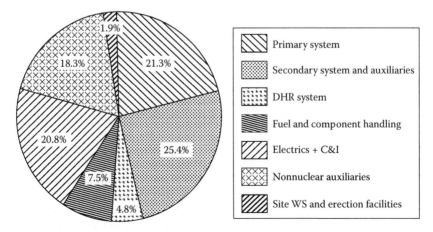

FIGURE 35.2 Capital costs of SFR NSSS hardware breakdown. (From Rapin, M., Fast breeder reactor economics, presented at *the Royal Society Meeting on the Fast-Neutron-Breeder Fission Reactor*, London, England, May 24–25, 1989.)

TABLE 35.1

Power Generation Cost by SPX-1 and PWR-P′4, Centimes/kWh

Cost Category	PWR-P′4 (1400 MW(e))	SPX-1 (1200 MW(e))
Capital investment	8.5	22.6
Fuel	5.3	10.0
Operation and maintenance (O&M)	3.2	5.0
Power generation costs	17.0	37.6

Source: Rapin, M., Fast breeder reactor economics, presented at *the Royal Society Meeting on the Fast-Neutron-Breeder Fission Reactor*, London, England, May 24–25, 1989.

higher than a PWR. The annual capital investment charge is the largest component of the unit power generation cost. As shown in Table 35.1, the cost of electricity produced by SPX-1 was, substantially, ~2.2 times higher than the cost of electricity produced by a PWR (1400 MWe). SPX-1 had more than 2.6 times higher capital investment costs than a PWR-P′4 (Table 35.1). That is why when designing the next French LMFR Superphenix-2 (SPX-2), the issue of choosing an optimum type of arrangement, reactor, and equipment design was raised again. The primary goal was clearly to cut the costs of SPX-2 on the basis of the design and construction of SPX-1. The important step in this direction was increasing the power generation capacity from 1240 to 1540 MW(e). An experience in construction and installation work at the SPX-1 plant has shown that such a design is rather complicated, expensive, and its realization entails a number of technical problems, which increases the time of construction (the plant originally was to have been commissioned in 1982), and a capital investment. It was understood that construction costs should be decreased by avoiding complex structural configuration, such as circular walls and domed roofs, reducing the number of components and their weights as well as building volumes, using simple and compact design, and limiting the number of safety-grade systems. The fast reactor design optimization was carried out based on the reactor operation and construction experience. In the USSR, such optimization studies were carried out for BN-800 from the experience of BN-600 design and operation, in France, SPX-2 from SPX-1, and in Germany, SNR-2 from SPX-1 and SNR-300.

An economic assessment of BN-600 and WWER-1000 reactors reported in Ref. 35.5 is also presented here. BN-600 has been operating solely for commercial electricity production at comparatively high reliability for more than 30 years, which is a milestone in the implementation of LMFR technology. Therefore, it seems that the most suitable reactors in use, when comparing technical and economic characteristics of thermal and fast reactors in the Russian Federation, were undoubtedly the BN-600 constructed as the No. 3 unit at the Beloyarsk site and the WWER-1000/V-320 reactor plant, constituting the No. 5 unit at the Novo-Voronezh site. Both these units were constructed on sites that already had the needed infrastructure. It is worth mentioning that the degree of development and the solution of the safety matters of these reactors are significantly different: the first one is semicommercial pool-type SFR and the second one is WWER-1000, the pilot power unit representing typical water cooled/water-moderated thermal reactor. In view of the absence of more comparable analogs, this comparison was made of the specific natural and cost indices of fast and thermal reactors. It should be mentioned that the absolute values of the cost characteristics of comparable power units, due to their development and different design organizations, are nonindicative, and therefore, no emphasis can be placed on them. Besides the analysis of the technical and economic characteristics of BN-600 and WWER-1000, some specific indices for another power station are also given later in text for comparison. Bringing other types of power unit into analysis should facilitate an understanding of the specific nature of the units under comparison.

TABLE 35.2

Major Natural and Economical Indices of WWER-1000 and BN-600

Parameters	WWER	BN-600
Power, MW(e)/net thermal efficiency, %	940/31.3	600/40.6
Weight of equipment, shielding/construction metal, ton/MWe	38.0	57.2
Volume of reinforced concrete structures, m³/MWe	180	170
Work required to carry out construction and assembly work on commercial structures, man-day/kWe	3.0	3.5
Specific capital investment on construction work (in comparison with WWER-1000), %	100	96
Total specific capital investment	1.00	1.5

Source: Rineiski, A.A., *Atomnaya Enegia*, 53, 807, 1982.

35.3.3.1 Comparison of Major Natural Economical Indices

Because of the difference in net electrical output between BN-600 and WWER-1000, a suitable indicator for an economic comparison of both plants can be taken as the *specific* weights (ton/MWe), volumes (m³/MWe), and construction men-days (man-day/kWe). Table 35.2 gives the most important natural indices and certain relative cost indices of power stations with thermal and fast reactors. It can be seen from Table 35.3 that the difference in the specific natural (metal content) and cost indices of thermal and fast reactors is not so significant if we consider that the fast reactor being compared has a considerably lower capacity. Analysis of the cost of power stations of different power ratings shows that when the power of a unit is doubled, up to 12% of specific capital investments are saved. Therefore, if BN-600 is to have an identical electrical power rating to that of the thermal reactor, the specific capital investments in the fast reactor could be reduced by 8%–9% [35.6]. Based on the analysis conducted, the following conclusion can be drawn about the cost of BN-600 and WWER-1000. Under comparable conditions (identical electrical capacity, region and period of construction, degree of serial production, number of units on the site), the difference in capital investment for fast/thermal reactors would be 20%–30%. To achieve competitiveness, this difference of electric energy cost in principle could be compensated by the efficient fuel breeding by fast reactors. The experience in building BN-600 indicates that the processes of constructing nuclear power stations with fast and thermal reactors do not vary substantially. Somewhat increased specific capital investments for fast reactors in comparison with WWERs are explained by a higher total cost of equipment. It is important to note that the specific construction costs are almost equal. The reason for the higher cost of equipment for fast reactors, as can be seen from Table 35.3, is their wider range and higher per-unit metal content. A reduction of the metal content of the plant, and thus bringing down the use of expensive safety-grade steels, is one of the most important challenges facing design and planning of fast reactors.

35.3.3.2 Material Consumption and Cost Indices

In order to clarify the reasons for increasing the material consumption and cost of fast reactors in comparison with thermal reactors, all the equipment of the plants were divided into functional groups, with an apportionment for each of the natural and cost indices (Table 35.3). In the constitution of the BN-600 NPP, the greatest material content and the most expensive is the pool-type reactor itself: its specific material content is a factor of three times greater than for the WWER-1000 reactor. Further, for BN-600, components are arranged according to cost in this order: equipment of the secondary circuit (SG, sodium pumps and pipelines, and auxiliary systems), electrical equipment, equipment of auxiliary buildings and structures, control and measuring equipment, and refueling equipment and technological systems. The greatest material content and the most expensive

TABLE 35.3

Specific Indices of Material Consumption and Cost for BN and WWER

Items	Proportion of Total Cost, %		Specific Metal Content, ton/MWe	
	BN-600	W-1000	BN-600	W-1000
Reactor block:	25.2	11.1	7.7	2.5
Vessel and in-vessel components: primary pumps, IHXS (for BN-600), primary circuit pipes, neutron and thermal shields, refueling and control rod mechanisms; sodium and gas auxiliary systems				
Process equipment:	26.1	31.7	6.2	5.7
Steam generators (SGs), secondary circuit pumps (for BN-600), pipes from reactor to SGS, secondary sodium, gas and water (for WWER-1000) auxiliary systems				
Fuel recharging and handling systems (FRHS)	5.0	3.1	2.3	2.5
Figures on NSSS + FRHS	56.3	42.9	16.2	10.7
Turbine with main and auxiliary equipment, station pipe work with valves and other elements	5.8	21.5	16.0	14.5
Station auxiliary building equipment and structures	10.0	10.8	7.7	5.0
Station electrical equipment	10.3	7.5	—	—
Station control and instrumentation	12.2	15.8		
Metal structures: steel liners on the inner surface of all containment cells including smothering bottom catch pans (for BN-600), shield slabs, compartment doors, covers of manholes	5.4	1.5	17.3	9.6
Total figures			57.2	39.8

Source: Rineiski, A.A., *Atomnaya Enegia*, 53, 807, 1982.

in the constitution of the WWER-1000 NSSS is the process equipment (SGs, coolant circulation circuit, and its auxiliary systems). It should be mentioned that the specific cost of the equipment of the turbine hall of BN-600 is substantially lower than the corresponding cost of WWER-1000, despite the use at BN-600 of turbines of smaller (by 2.5 times) unit power. The latter is explained by the use of three high thermal efficiency standard turbines each of 200 MW(e) power rating at BN-600, whereas for WWER-1000, two special turbines each of 500 MW(e) rating with reduced steam parameters and thermal efficiency. The BN-600 has been operating solely for commercial electricity production at comparatively high reliability for more than 30 years, which are milestones in the proven and implementation of LMFR technology.

35.4   FUTURE DIRECTIONS: TECHNOLOGICAL CHALLENGES

SFRs have a number of positive engineering characteristics, such as high thermal efficiency (higher by one-third than for existing thermal reactors), compact core, effective heat transfer owing to liq-uid-sodium-coolant-resulted compactness of the SG and intermediate heat exchanger (IHX) and low pressure in the reactor vessel and heat exchange equipment (0.15–0.2 MPa), and delivery pipelines (0.6–0.8 MPa). On the basis of the most general considerations, it might have supposed that the dif-ference in thermal efficiency alone should basically compensate the difference in cost per kilowatt caused by the introduction of an intermediate circuit, although practice has shown that a reduction in the number of circuits does not necessarily reduce cost. As an example of this is the single-loop (direct cycle) boiling water reactors, which are not cheaper than the two-loop PWR. With a reduc-tion in the number of heat transfer loops, the requirements on leak tightness, choice of materials, and maintenance and repair of the relatively massive branched steam/power generating part of the nuclear power unit are made more stringent, which significantly reduce the gain from the exclusion of compact intermediate circuits. Hence, optimization of the number of systems, components, and equipment should be carried out comprehensively, considering several aspects such as economical, technological, and safety-related aspects.

The low pressure in the reactor vessel equipment, the compact core, and the small heat exchange surface area make it possible to locate the core and heat exchange equipment inside thin-walled vessels, which, in right design, could lead to improvement in the safety characteristics. However, this decision should be taken by investigating the maintenance, in-service inspection, manufacturability, transportation and erection aspects, etc.

A cost reduction and reliability improvement studies should be conducted and analyzed in a comprehensive manner. For the complete realization of potentially feasible SFRs from the point of view of cost, metal content, and reliability, a new SG and other equipment design solution will be required, different from those being used at the present time. For example, the EFR single-vessel SG of a 600 MWt has been verified favorable based on the successful operating feedback of SPX SG. The advanced SG for EFR is a once-through straight tube type, without welds in the long tubes other than one at each end required to attach the tube to the tubesheet. Such developments call for technological demonstrations by manufacturing industries.

In the pool reactor, there is a relatively large expenditure of metal on the so-called shielding structures: neutron, thermal and biological shields, and supporting structures (support ring, strong back, main circulation pump, and IHX mountings).

There is no other way to look at this than to treat it as the price paid for a high neutron flux and the use of sodium as a coolant. Since these structures are disposed inside the reactor vessel in the coolant, it is essential to use high-quality manufacturing technology and materials (at the present time, this is stainless austenitic steel) with high demands on the purity of the surface treatment. As a result, their cost comes close to the cost of the main equipment.

Therefore, the most pressing task for the fast reactor designers is to reduce the outlay on metal through a reduction in its weight and the use of low-grade steels or their substitutes and also through effective segregation, for example, by providing sufficient distance between the core and equipment, which is essential for maintenance and repair. In some reactor designs, sodium that lies between the core and the IHX plays a major role in shielding against activation through an increase in the distance between them. Although this measure leads to an increase in the diameter or height of the vessel, it enables the thickness and weight of the in-vessel steel shielding to be reduced. This gives a diminution in the total weight of the reactor. Such approach of reduction of the cutting of the in-vessel shielding has been achieved through the development of integral layouts with a horizontal vessel. In these versions, the distance between core and IHX may be set, without increasing the vessel diameter, such that the sodium layer wholly performs the functions of a neutron shield.

The use of loop grouping solutions allows a number of in-vessel structures to be done away with. It is reckoned that if a positive solution to certain essential technical problems is found, the specific metal content and the mass of sodium of the primary circuit may be reduced in the loop layout. It is reckoned also that the high self-power SG presents considerable potential as regards the savings on metal content.

Apart from the aspects mentioned earlier, the following specific directions can be studied toward improving economics of pool-type SFR:

- Reducing the weight of equipment and systems by cutting the number of heat removal loops and components; reducing the number of IHXs and primary and secondary pumps; housing all radioactive systems (cold traps and other) in the reactor vessel; and cutting the number of the secondary circuit sodium and gas auxiliary systems owing to layout optimization and usage of block-type SG.
- Using a maximum degree of design and engineering decisions adopted for previous reactors and proved by the experience of their long operation period.
- Reducing the weight of materials for in-vessel shield and fuel subassembly handling and recharging systems by excluding a drum to store irradiated fuel subassemblies (FSAs).

Further, the detailed analyses also showed that the efficiency of the reactor plant engineering and economics were improved through optimization of all components responsible for the high–electricity generation costs: capital costs by reduction of specific weight of the NSSS, fuel cycle cost by higher fuel burnup, improvement of equipment reliability, and extension of service life to 60 years.

35.5 APPROACH TO ECONOMICS OF SFR IN INDIA: A CASE STUDY

In order to improve the economy or in other words unit energy cost (UEC), it is very important to know the contribution of various parameters toward UEC. This will facilitate in understanding how much can be gained in terms of reduction in UEC by varying different parameters that contribute to UEC. For the prototype fast breeder reactor (PFBR), the pie chart (Figure 35.3) gives the absolute and percentage-wise contribution of various cost components toward the UEC. A close analysis of this chart indicates that the contribution of return on equity is the highest followed by fuel cycle cost and depreciation, which together constitute approximately 80%. The total project cost is the contributor for the return on equity. Among these three, the depreciation cost is based on the government policy of 5% depreciation per year, and hence it is decided that the cost optimization study can focus on the remaining two, that is, return on equity and fuel cycle cost. As capital cost is contributing to the *return on equity*, any substantial reduction in the capital cost will definitely reduce the UEC. A close look at the capital cost breakup details for PFBR (Figure 35.4) indicates that reactor assembly, sodium circuits, steam water system, and electrical power system together constitute ~53% of total project cost. However, as the last two systems, that is, steam water system and electrical power system are almost standardized, further work to reduce cost needs to be done in other two systems, that is,

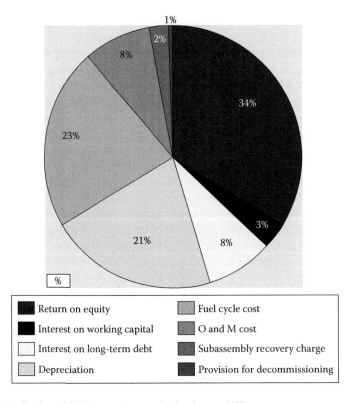

FIGURE 35.3 Distribution of UEC constituents: (i) absolute and (ii) percentage.

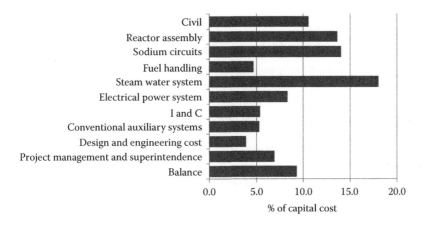

FIGURE 35.4 PFBR capital cost breakup.

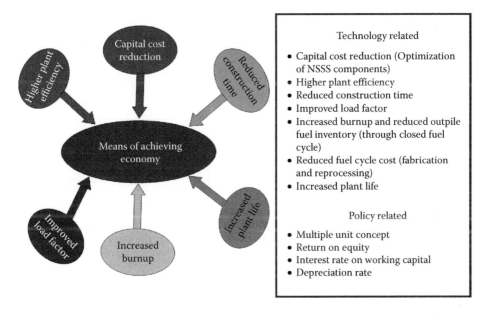

FIGURE 35.5 Means of achieving economy.

reactor assembly and sodium circuits. However, efforts are to be made toward cost optimization in the earlier systems too. As can be noticed, project management alone constitutes about 6.9%, which offer lot of scope for reduction by adopting modern management practices.

The capital cost would be reduced by adopting the measures such as increase in burnup, increased thermal efficiency, increase in capacity factor, reduced construction time, multiple unit construction, and policy measures on financial parameters such as depreciation rate, debt equity ratio, and interest rate (Figure 35.5). With the optimization of all these listed parameters, it is possible to achieve considerable reduction in UEC at a constant money value basis (Table 35.4). The challenge is to reduce that portion of UEC that is due to capital cost by adopting innovative means considering that PFBR design has already been largely optimized. Various measures have been deliberated that require further study ranging from short-term study to long-term R&D depending on the nature of their influence on the design toward economy (Figure 35.6).

In this chapter, innovative design features that could bring down the capital cost are focused.

TABLE 35.4

Preliminary Estimation of UEC with Various Parameters for SFRs in India

Current PFBR Unit Energy Cost Is Taken as the Reference
(Thermal Efficiency—40%, Capacity Factor—62.8%, Burnup—100 GWd/ton, Construction Time—7 Years, and Plant Design Life—40 Years)

S. No.	Effect of Change in Parameters	Expected Reduction (%)
1	Thermal efficiency from 40% to 40.5%	1.0
2	Burnup from 100 GWd/ton to 200 GWd/ton	5.0
3	Capacity factor from 62.8% to 80%	18.6
4	Thermal efficiency from 40% to 40.5% and capacity factor 62.8% to 80%	19.3
5	Thermal efficiency from 40% to 40.5%, capacity factor from 62.8% to 80%, and burnup from 100 to 200 GWd/ton	23.3
6	Twin units of 500 MWe, thermal efficiency 40.5%, capacity factor 80%, and burnup 200 GWd/ton (capital cost reduction 10% and O&M cost reduction of 5% for twin-unit construction)	32.3
7	Twin units of 500 MWe, thermal efficiency 40.5%, capacity factor 80%, and burnup 200 GWd/ton, reduced construction time from 7 to 5 years (capital cost reduction of 10% and O&M cost reduction of 5% for twin-unit construction)	34.8
8	Twin units of 500 MWe, thermal efficiency 40.5%, capacity factor 80%, and burnup 200 GWd/ton, construction time from 7 to 5 years and plant life from 40 to 60 years (capital cost reduction of 10% and O&M cost reduction of 5% for twin-unit construction)	39.4

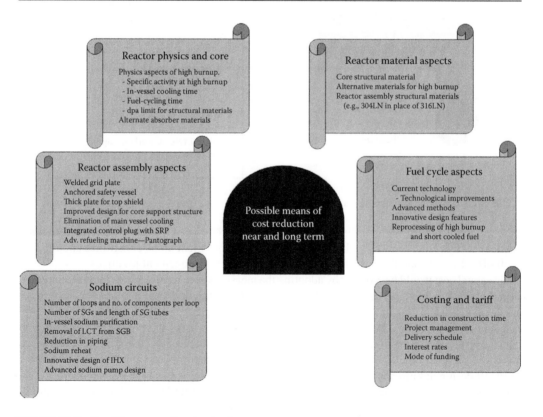

FIGURE 35.6 Possible means of cost reduction.

35.5.1 Design Improvements Considered for Reactor Assembly Components

For the reactor assembly, several design options were analyzed, and only those design features that are feasible to be incorporated in the next series of reactors were considered. The options include the following:

- Machining of penetrations in roof slab and rotatable plugs to reduce the annular gaps to achieve reduction in overall roof slab/main vessel diameter. There is an added advantage of machining of penetrations also, that is, the amount of complementary shielding required will reduce significantly.
- Small rotatable plug, integrated with control plug.
- Adopting welded grid plate, reduction in grid plate diameter, and removal of grid plate sleeves for permanent shielding subassemblies.
- Increased number of pipes supplying cold primary sodium to grid plate.
- Optimization of main vessel diameter with diameter checks at core level, radial and circumferential checks at top shield level, etc.
- Toroidal-shaped inner vessel with provision for in-vessel transfer post.
- Dome-shaped roof slab with separate arrangement for shielding in top axial direction.
- Integrated liner and safety vessel with thermal insulation arrangement.

Besides these, several features contributing to enhanced safety and overall design simplifications were incorporated. Preliminary economic study indicates that a reduction of about 25% is possible in the specific steel consumption of future FBR with PFBR as the reference, which is quite significant considering the already optimized design of PFBR. Also, the reduced size of vessel leads to a reduction in primary sodium inventory and reduced concrete volume. The primary sodium inventory would come down to around 1000 tons as against the present 1100 tons for PFBR. With higher reactor power, further reduction in specific sodium inventory is possible. Figure 35.7 shows the schematic sketch of the improved features of the reactor assembly of future FBR.

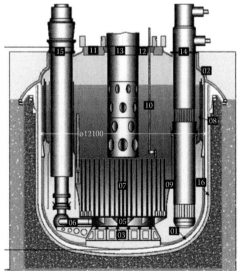

01-Main vessel	02-Roof slab	03-Core sup. structure
04-Core catcher	05-Grid plate	06-Primary pipe
07-Core	08-Thermal baffle	09-Inner vessel
10-Transfer arm	11-LRP + support	12-SRP + support
13-Control plug	14-IHX	15-Primary pump
16-Safety vessel and thermal insulation		

FIGURE 35.7 Improved reactor assembly concept for future SFR in India.

35.5.2 Design Optimization of Secondary Sodium System

Optimization studies on the number of heat transport systems and its components have been carried out. It is seen that less number of components (except for the SG) and loops in the primary and secondary sodium circuits can greatly reduce the capital cost and construction period and improve the capacity factor. For PFBR, two primary sodium pumps, four IHXs, and two secondary loops are selected. Each secondary loop consists of one secondary sodium pump, two IHXs, and four SGs. This was arrived at based on a detailed review. The same configuration is envisaged for the future commercial breeder reactors also. The basic idea of the present configuration is to achieve economy by having lesser number of components, thereby increasing the safety aspect, reducing the capital cost, and increasing plant capacity by deploying spare critical components such as pumps, IHX, and SGs in order to reduce the downtime of the plant.

It is proposed to change the material of construction of all the sodium circuits except primary and sodium/argon circuits, which may see radioactivity in the plant's lifetime. For piping and components working under 400°C, the material of construction shall be 2.25Cr–1Mo, and for the piping and components operating at and above 400°C, it can be modified 9Cr–1Mo. The SG and sodium-to-air heat exchanger of safety-grade decay heat removal system are already made of modified 9Cr–1Mo. The total material cost savings work out to be significant. Lower thermal expansion coefficient of the proposed material will make the piping layout more compact leading to lesser number of spring and seismic supports. The loads on the component nozzles will also be lesser due to low thermal expansion coefficient of these materials. Higher thermal conductivity will lower the thermal shocks seen by the piping during plant transients. Sodium valves, particularly lower-sized valves, are bellow sealed having stainless steel bellows. Sodium is hermitically sealed with this bellow in the valve. These valves will call for dissimilar welding between low-alloy steel and stainless steel. This issue is being addressed.

35.5.3 Summary of the Indian Program

Fast breeder reactor with closed fuel cycle is an inevitable technology option for providing energy security for India. PFBR is a technoeconomic demonstrator and also a forerunner in the series of SFRs planned in India. Beyond PFBR, economic competitiveness is important for rapid commercial deployment of SFRs. Roadmap with comprehensive R&D for the large-scale deployment of SFRs with improved economy and enhanced safety is being pursued in a systematic manner at Indira Gandhi Centre for Atomic Research (IGCAR), India.

REFERENCES

35.1 Walter, A.E., Todd, D.R., Tsvetkov, P.V. (2012). *Fast Spectrum Reactors*. Springer Publishers, New York.
35.2 Kuriswa, K. et al. (1977). An optimization study of a demonstration fast breeder reactor plant, appeared in optimization of sodium-cooled fast reactors. *Proceedings of the International Conference*, BNES, London, U.K., November 28–December 1, 1977.
35.3 Loaec, C., Linet, F.L. (2007). The French scenario. *IAEA-INPRO Meeting on Joint Study on an Innovative Nuclear Energy System Economics*, Cadarache, France, February 6–9, 2007.
35.4 Rapin, M. (1989). Fast breeder reactor economics. Presented at *the Royal Society Meeting on the Fast-Neutron-Breeder Fission Reactor*, London, England, May 24–25, 1989.
35.5 Mourobov, V.M. (1998). Liquid-metal-cooled-fast reactor (LMFR) development and IAEA activities. *Energy*, 23(7/8), 637–648.
35.6 Rineiski, A.A. (1982). Comparison of technical and economic characteristics of modern nuclear power stations with fast and thermal reactors. *Atomnaya Enegia*, 53, 807–815, December 1982.

Index

For Product Safety Concerns and Information please contact our EU
representative GPSR@taylorandfrancis.com Taylor & Francis Verlag GmbH,
Kaufingerstraße 24, 80331 München, Germany

Printed and bound by CPI Group (UK) Ltd, Croydon, CR0 4YY
06/05/2025
01861285-0001